CALCULUS
Mathematics and Modeling

UPDATED PRELIMINARY EDITION

CALCULUS
Mathematics and Modeling

Wade Ellis, Jr.
West Valley College

William C. Bauldry
Appalachian State University

Joseph R. Fiedler
California State University, Bakersfield

Frank R. Giordano

Phoebe T. Judson
Trinity University

Ed Lodi
West Valley College

Richard Vitray
Rollins College

Richard D. West
United States Military Academy

 **ADDISON-WESLEY**

An imprint of Addison Wesley Longman, Inc.

Reading, Massachusetts • Menlo Park, California • New York • Harlow, England
Don Mills, Ontario • Sydney • Mexico City • Madrid • Amsterdam

Senior Acquisitions Editor: Laurie Rosatone
Project Manager: Bess Deck
Assistant Editor: Ellen Keohane
Senior Production Supervisor: Peggy McMahon
Text Designer: Rebecca Lemna
Copyediting: Anne Scanlan-Rohrer, Two Ravens Editorial Services
Cover Photo:©1996 PhotoDisc, Inc.
Print Buyer: Evelyn Beaton

Page 64, Galileo: photo courtesy of The Bettmann Archives, New York, NY.
Page 83, Golf Ball Bounce, © The Harold E. Edgerton 1992 Trust, Courtesy of Palm Press, Inc.
Page 405, Newton; page 138, Leibniz; page 439, Curie: photos courtesy of
The Bettmann Archives, New York, NY.

Library of Congress Catalog Card Number 96-80401
ISBN 0-201-33860-2

1 2 3 4 5 6 7 8 9 10 — DOC — 02010099

PREFACE

Philosophy

In the seventeenth century, Newton and Leibniz created the tools of analysis, later known as Calculus and Differential Equations, in studies of optics, astronomy, motion, and area. The concepts of derivative, integral, and limit arose naturally in their studies of how measurable quantities change. In the 300 years following their work, the explosion of ideas and techniques which they inspired has resulted in an enormous body of mathematical knowledge. The sheer volume of this knowledge has resulted in a separation of the originally unified concepts of Calculus and Differential Equations into distinct topics studied in a variety of courses. As a result of this separation, many students of mathematics never obtain a global understanding of the material. Such understanding is necessary for creative and effective application of these concepts when the student is challenged with new situations in mathematics and modeling. We attempt in this text to reunite these mathematical tools in a single study of problems in a contemporary context using modern computational devices.

Role of Computational Technology

Implementation of our philosophy requires a computational environment in which the student can move seamlessly between symbolic, numeric, graphic, and textual contexts. Such environments have existed for a decade or more in the form of computer algebra systems. With the introduction of the TI-92 (and now the TI-89), the required environment is now available and affordable for a wider audience. The preliminary edition of the text is therefore written to make consistent and essential use of these devices.

Student Outcomes

Students will come away form this course with a conceptual and practical knowledge of the derivative and the integral. They will have experience in applying these calculus concepts in the context of science, engineering, economics, and mathematics. Moreover, they will have available to them the twin tools of difference equations and differential equations to analyze situations from these disciplines. Students will have a unified and robust view of mathematics through their early and repeated experience with modeling and with a variety of interpretations of the fundamental ideas of the course. The students who have used these materials report that the frequent Reflections are a valuable resource in improving their ability to read and write mathematics. In addition, the students develop their ability to communicate the deep and subtle ideas of Calculus and Differential Equations through the frequent writing exercises and group activities (Think and Share). At the completion of the course, students are well prepared to pursue further mathematics studies in Linear Algebra, Differential Equations, and Functions of Several Variables.

Distinctive Features

Reflections.   To encourage active reading by the students, "Reflections" are included in the body of the text. In a Reflection, the students are asked to briefly consider a question about what has just been read or connect one idea with another. Reflections appear frequently at the beginning of the text. As students become more familiar with the types of questions they should ask themselves while reading mathematics, Reflections appear less often. Students, through these Reflections, can improve their ability to read mathematics.

Writing.   The students are encouraged to actively participate in their own learning by writing. The authors have found that requiring students to write daily responses to the Reflections appearing in the assigned reading is an effective pedagogical tool. This allows a frequent exchange of ideas among students and between students and instructor which encourages students to engage the ideas and gives the instructor greater insight into student progress. Additionally, twenty-five per cent of the exercises in the text require writing. Students are frequently asked to interpret the results of computations in practical situations and to describe the meaning of the mathematics in those contexts that are primarily mathematical in nature. In particular, in the early chapters, students are asked to describe the behavior of functions and sequences relating to mathematical models of situations and of abstract functions represented as graphs. In the second and third parts of the text, the student frequently needs to interpret tables, graphs, and symbols involved in investigating the solution to differential equations as mathematical constructs and models of real world phenomena.

Problems in Context.   Throughout the book mathematical ideas and concepts are introduced and developed in the context of significant practical applications from chemistry, biology, medical sciences, decision sciences, and mathematics. This basic approach to the book carries over into the exercises where applied problems appear prominently, but not to the exclusion of purely mathematical problems.

Think and Share.   The authors believe that the communication of mathematics both in writing and verbally sharpens students' understanding of mathematics and provides students with the ability to transfer their use of mathematics from mathematics courses to physics, or chemistry, or economics courses, or to the engineering workbench. Group work is incorporated in the text through the feature called "Think and Share." It is the authors' intent and hope that the instructor finds it appropriate to devote class time in the course to these activities. Students will learn from such in-class activities how to participate effectively in group work outside class. Group activities make (for many students) an enormous difference in how much mathematics they can learn, master, and apply in other courses. There also are group activities in the exercises to take advantage of what students learn about group work in class.

Discovery Activities.   The authors believe that students profit from problems that engage them in a question before a canonical answer is given. A variety of discovery activities are, therefore, an integral part of the text. For example, there is a trial-and-error or guess-and-check aspect to most of the modeling activity based on difference equations in the first several chapters. Similarly, discovery approaches are used in the development of the derivative and also in investigating functions arising from differential equations by conjecturing their critical features from tables, graphs, and symbols. There are no special discovery activities, but rather exercises that frequently require the students to explore results. We do not depend upon the student to create knowledge. Rather we capitalize on the ideas that occur when students try to come up with their own ways of working through a mathematical concept.

Required Exercises.   Throughout the text a simple dagger (†) precedes some exercises. This indicates an exercise that introduces or develops an idea that will be exploited later in the text. The authors believe that these exercises must be assigned and discussed as an integral part of the course.

Projects.   The authors strongly believe that extended projects allow students to see mathematics in action and to reflect on how the mathematics works. Such extended activities are an important part of an individual's mathematical development. Our text includes a selection of such extended activities at the end of every chapter that involve many of the ideas that the student has encountered.

Technology.   The power of the TI computer symbolic algebra graphing calculators (TI-89 and TI-92) is exploited throughout the book. The authors believe that computer symbolic algebra technology in conjunction with the technology of the graphing calculator allows the first course in college or university mathematics to change significantly to the benefit of the mathematical understanding of students. We hope and intend that the instructor model appropriate use of technology in class. To that end, we have included copious marginal notes in the first two chapters in support of effective use of the TI computer algebra system calculators. There is also a tutorial appendix for the novice calculator user. Throughout the text, shaded Technology Notes are used to introduce new features or more sophisticated applications of the calculator. The instructor, of course, sets the appropriate degree of use of technology in the course. The text assumes unrestricted student access to a TI computer algebra system calculator in the reading and in working through Reflections, Exercises, Think and Shares, and Projects.

Summary Remarks.   Just as the authors believe in the importance of lectures by the instructor, so too, they believe in the usefulness of summary remarks in the text. These remarks are not intended as a rehash of what happened in each section, but supply a statement of what was learned and why. They place each chapter in the context of the course as a whole.

Supplementary Exercises.   Supplementary exercises are included at the end of each chapter. In addition to some routine exercises that allow the student to validate his or her mastery of techniques from the chapter, there are some exercises that help the student integrate material from several sections with material from prior chapters.

Interludes.   Interludes include material that is not central to the development of the text, but can be used by students to further expand their understanding.

Advice to the Authors.   The authors are eager to receive comments, suggestions, and complaints on this preliminary edition so as to improve the text. We can be reached at

    E-mail: honolulu@cs.cs.appstate.edu

Mail:

        c/o Laurie Rosatone
        Addison Wesley Longman
        One Jacob Way
        Reading, MA 01867

Acknowledgments

The authors would like to thank the following reviewers:

Charlene Beckmann
Grand Valley State University

Patrick Boyle
Santa Rosa Junior College

Chris Burditt
Napa Valley Community College

William Davis
The Ohio State University

Susan Ganter
Worcester Polytechnic Institute

Paul Latiolais
Portland State University

Robert Lopez
Rose-Hulman Institute of Technology

Paula Maida
American University

Kirby Smith
Texas A&M University

Rebecca Stoudt
Indiana University of PA

The authors would also like to thank Alethea Vitray of Orlando, Florida, and Amy Miller of Green Bean Graphics for the art work they prepared for the book. In addition, we thank Jane C. Ellis for the administrative work she did while the authors labored in Honolulu and Mountain View. We thank Lenore Parens and Paul Lorczak for their developmental editing of the original manuscript. And thanks to the 1998-99 calculus students of Bakersfield South High School and to calculus students at Appalachian State University, Rollins College, and Trinity University for their invaluable feedback. We thank Peggy McMahon of Addison-Wesley for her unstinting assistance in the production of the book and Art Ogawa of TeX Consultants for developing the typesetting macros and for his help in the final typesetting process. We thank Rebecca Rosser for her help in entering text and typesetting commands. We would also like to thank Anne Scanlon-Rohrer for her timely copyediting of the text. We thank Jennifer Wall and Joe Vetere for their cheerful energy and unfailing assistance. Finally, we are grateful to Laurie Rosatone for her ongoing belief in, and intellectual support of, this project from its inception and her diplomacy in working with so many authors.

To the Student

The authors have written this book with you in mind. We believe that you will learn most effectively if you actively participate in the learning process. In an effort to engage you in the mathematics we have used a variety of approaches to the material. We have placed questions called Reflections throughout the book to help you focus on the concepts and ideas in the text as you read. In addition, we have provided extended exercises, called Think and Share in the text, that we feel are important for you to discuss with other students in the class.

Good writing *is* good thinking. The authors believe that you can improve your understanding of mathematical ideas by writing carefully about them. Thus, many of the exercises you will be assigned require you to write your ideas and thoughts about mathematical concepts and problems. These writing problems are sometimes Group Exercises that require you to encapsulate and interpret the consensus of your working group in a written report.

The purpose of computation is never numbers, but insight.[†] Throughout the text, we encourage you to use the TI-89 or TI-92 to make computations— numerical, graphical, and symbolic—that give insight into the concepts of the calculus and to allow you to investigate situations that otherwise would be computationally burdensome. In some cases, hand or mental computations are more fruitful than machine-based computations. We trust that this course will develop your ability to choose appropriate computational strategies to solve problems you encounter.

The authors are eager to receive comments, suggests, and corrections on this preliminary edition so as to improve the text for other students. We can be reached at

E-mail: honolulu@cs.cs.appstate.edu

Mail:

c/o Laurie Rosatone
Addison Wesley Longman
One Jacob Way
Reading, MA 01867

[†] A paraphrase of a quote from R.W. Hamming.

CONTENTS

CALCULUS: MATHEMATICS AND MODELING   1

P A R T   I   CALCULUS — THE MATHEMATICS OF CHANGE   3

C H A P T E R   1   **Modeling Change   5**

1.1   Drugs in the Body   5
1.2   Patterns of Accumulation   15
1.3   A Model for Natural Growth   25
1.4   Investigating Limits   36
1.5   Continuous and Piecewise Continuous Models   46
Summary   59

C H A P T E R   2   **Measuring Change   67**

2.1   Analyzing a Discrete Function   67
2.2   Contructing Models from Patterns   78
2.3   Divided Differences and Average Rate of Change   89
2.4   Rate of Change   98
2.5   Rate of Change as a Function   111
Summary   118

C H A P T E R   3   **The Derivative: A Tool for Measuring Change   127**

3.1   The Derivative   127
3.2   Rules, Rules, and More Rules   141
3.3   The Chain Rule   152
3.4   Finding Features of a Continuous Function   159
3.5   Properties of Continuous Functions   176
3.6   Optimization: Finding Global Extrema   181
3.7   Implicit Differentiation and Its Applications   193
3.8   Modeling Motion with Parametric Equations   203

3.9 Partial Derivatives 213
Summary 219

CHAPTER 4 **The Definite Integral: Accumulating Change 227**

4.1 Rate and Distance 227
4.2 Sums of Products 237
4.3 Error Bounds for the Left and Right Endpoint Methods 246
4.4 The Definite Integral 254
4.5 Other Methods and Their Error Bounds 262
4.6 Applications 270
4.7 The Fundamental Theorem of Calculus, Part I 287
Summary 296

A THEORETICAL INTERLUDE The Truth About Limits 303

CHAPTER 5 **The Integral: Theory, Applications, and Techniques 311**

5.1 Rate, Accumulation, and the Fundamental Theorem
   of Calculus, Part II 311
5.2 The Antiderivative Concept 319
5.3 Finding Antiderivatives Using Properties and Formulas 328
5.4 Applications 334
5.5 Using the Chain Rule in Finding Antiderivatives 348
5.6 Techniques of Integration 356
Summary 362

A TECHNICAL INTERLUDE Further Techniques of Integration 371

PART II **MODELING WITH CALCULUS 391**

CHAPTER 6 **Modeling with the Derivative 393**

6.1 One Day in the Life of a Modeler 393
6.2 Warming and Cooling 400
6.3 Population Modeling 406
6.4 Euler's Method 413
6.5 Slope Fields 420
6.6 Errors in the Model Construction 427
Terminology 435
Summary 436

A THEORETICAL INTERLUDE Existence and Uniqueness Theorems 443
Uniqueness 443
Existence 446
The General Existence and Uniqueness Theorem 450

CHAPTER 7   **Solving Differential Equations   453**

7.1   Integration and Separation of Variables   454
7.2   Linear Differential Equations   461
7.3   Errors in Euler's Method   469
7.4   Improving Euler's Method   474
7.5   Advanced Numeric Techniques   481
Summary   487

CHAPTER 8   **Modeling With Systems   495**

8.1   Spirals of Change: You Are What You Eat   496
8.2   Modeling   506
8.3   Numerical Solutions: Iteration and Euler's Method   515
8.4   Symbolic Solutions of Systems of Differential Equations   528
8.5   Bungee Jumping   536
Summary   542

CHAPTER 9   **Power Series: Approximating Functions with Functions   555**

9.1   Polynomial Approximation of Functions   555
9.2   Using Polynomial Approximations   560
9.3   How Good Is a Good Polynomial Approximation?   565
9.4   Convergence of Series   573
9.5   Power Series Solutions of Differential Equations   582
Summary   587

CHAPTER 10   **Optimization of Functions of Two Variables   593**

10.1   Optimization with Two Variables   593
10.2   Vectors, Lines and Planes   598
10.3   Tangent Vectors and Tangent Lines in Three Dimensions   607
10.4   Tangent Planes   614
10.5   The Gradient Search   622
Summary   628

APPENDIX A   **TI-89 Computer Algebra System Tutorial   631**

A.1   The TI Computer Algebra System Tutorial   631
A.2   Troubleshooting: Things that Go Bump in the Night   649

APPENDIX B   **TI-92 Computer Algebra System Tutorial   653**

B.1   The TI Computer Algebra System Tutorial   653
B.2   Troubleshooting: Things that Go Bump in the Night   672

CONTENTS

APPENDIX C   **Solving Equations with the TI Calculator   675**

APPENDIX D   **The SlopeFld Program   685**

Index   689

CALCULUS: MATHEMATICS AND MODELING

This book is about change. In it we develop some of the sophisticated techniques of calculus, the branch of mathematics that measures and analyzes how quantities change over time.

We begin, in Part I, by investigating functions, especially functions that are descriptive models of how things change in the real world. We then develop and apply the techniques of differentiation and integration, the two tools of calculus that allow us to analyze functions in detail.

We continue our study of change in Part II by using calculus to create and explore a new type of equation, a differential equation. A differential equation can model change based on the laws of nature that effect change and can give insight into why things change. The solutions to such equations are the functions we use in Chapter 1 to describe and model change. Thus, in this book we will study functions as descriptive models of change, and we will study equations that show why these functions occur in modeling change. Through an understanding of calculus, we can look beyond mere descriptions to uncover some of the mysteries of the real world.

I CALCULUS — THE MATHEMATICS OF CHANGE

Change plays an important role in nearly every aspect of our lives. Over time, we change our minds, our moods, our likes and dislikes, our weight, our age, and our income. Similarly, the world around us changes temperature, pressure, size, and speed. Mathematics has little to say about changes in mood or disposition because such changes are difficult or impossible to measure. On the other hand, changes in measurable quantities, like temperature, income, or heart rate are expressly suited to the application of mathematics in general and calculus in particular.

Measurable quantities typically fall into one of two broad categories; discrete or continuous. Discrete quantities have natural gaps between all possible values. Thus, the number of patients with a particular disease or the number of memory locations in a computer are both discrete quantities since their values must be nonnegative integers and all integers are separated by gaps of size 1. Prices are also discrete since different prices are separated by gaps of size $0.01.

We think of continuous quantities as having no gaps. For example, we view the distance traveled during a trip as continuous since we can't move from one spot to another without following an unbroken path between the two locations. We don't suddenly disappear from one spot and pop up somewhere else (even the Star Trek transporters require a beam). Nevertheless, when we measure continuous processes the limitations of our measuring instruments force us to generate discrete data. A stopwatch accurate to a thousandth of a second will be unable to determine the winner between two racers whose times are within a millionth of a second. A film of the race may look continuous when played but it actually consists of thousands of photographs each one recording a discrete instant in time. Consequently, even when investigating a continuous process we often begin with a discrete model. A **model** is a mathematical description of a real world phenomena. This description will involve some or all of the following: definitions, assumptions, data, relationships or equations.

Calculus provides a mathematical bridge between the discrete and the continuous through the concept of limits. With limits we can convert from discrete representations of phenomena to continuous ones and back. Discrete models better fit discrete measurements and are easier to understand than continuous ones. We can study changes in discrete models by simply computing differences (subtraction) and sums (addition). On the other hand, continuous models can provide insights which are difficult or impossible to obtain any other way.

We begin to study the modeling process and investigate the bridge between discrete and continuous models in Chapter 1. In Chapter 2, we exploit continuous models through the introduction of the concept of average and instantaneous rate of change. In Chapter 3, we use the derivative, the generalized concept of rate of change, to study the behavior of functions. In Chapter 4 and 5, we investigate sums, the definite integral, and antidifferentiation.

Think and Share

Break into groups of two or three. Think up two original real world quantities that you would consider to be discrete, and two that you would consider to be continuous. Be prepared to present your quantities to the class and to defend your choices.

C H A P T E R

1

MODELING
CHANGE

1.1
Drugs in the Body

1.2
Patterns of Accumulation

1.3
A Model for Natural Growth

1.4
Investigating Limits

1.5
Continuous and Piecewise Continuous
Models

Summary

In this chapter, we begin our study of change by investigating examples in medicine, finance, demography, and mechanics. We will develop discrete models to describe these situations and in so doing illuminate the model building process. Analysis of these models will lead us naturally to the concept of a limit. Finally, we will use the limit concept to clarify what we mean by the word "continuous".

1.1

Drugs in the Body

Doctors rely on drugs to change the body's response to infection and pain (both physical and psychological). The amount of a drug in the patient's body determines the effects of the drug. Too little will not help; too much may hurt or even kill. Hence, it is important to accurately model the patient's drug level. In this section, we will look at how the concentration of a drug in the blood after a single dose changes over time.

Single Doses and Difference Equations

Our first example involves the rate at which human kidneys filter molecules from blood, a crucial factor in determining drug dosages. Consider the amount of penicillin in the blood of a patient as a function of time. The kidneys filter blood largely by passive filtration. In this process, small, water-soluble molecules (for example, penicillin, nicotine, caffeine, novocaine) migrate across the membranes in the kidneys by osmotic pressure. The rate at which these molecules leave the blood is proportional to the concentration of molecules in the blood.[†] As a practical result the kidney removes a fixed proportion of molecules from the blood during any fixed time period.

E X A M P L E   1   **Painless Dentistry—The Novocaine Example**

Novocaine (Procainamide Hydrochloride), injected as an anesthetic for minor surgical and dental procedures, is eliminated from the body primarily by the kidneys. In one hour, a person with normal kidney function removes approximately 20% of the Novocaine in the blood. We can use this fact to predict, at one-hour intervals, the amount, or level, of Novocaine remaining in the body after a single injection of 500 mg. We use u_n to represent the level of Novocaine corresponding to the nth observation. We model the initial level of Novocaine immediately after the 500 mg injection with the equation

$$u_1 = 500$$

Each hour following the injection, the patient's body eliminates 20% of the drug present at the beginning of that hour; hence, we can compute u_2, the level of Novocaine after one hour

$$u_2 = u_1 - 0.2u_1 = (1 - 0.2) \cdot u_1 = 0.8 \cdot u_1 = 0.8 \cdot 500 = 400.0 \qquad \textbf{(1.1)}$$

Each subsequent amount can be similarly computed from the term just before it. Study Equations (1.2) to verify that they are a symbolic representation of this idea over several hours.

$$
\begin{aligned}
u_2 &= u_1 - 0.2u_1 = 500(0.8) = 400.0 \qquad \textbf{(1.2)} \\
u_3 &= u_2 - 0.2u_2 = 400(1 - 0.2) = 320.0 \\
u_4 &= u_3 - 0.2u_3 = 3200.8 = 256.0 \\
&\ \ \vdots \\
u_n &= u_{n-1} - 0.2u_{n-1} = 0.8u_{n-1}
\end{aligned}
$$

To generate your own sequence of values using your TI calculator, go to the MODE *dialogue box and set* Graph *... to* SEQUENCE. *Then go to the* Y = *editor and enter* u1(n) = u1(n-1) - 0.2*u1(n-1) ui1 = 500. *On the* Home *screen, enter* u1(1), u1(2), U1(3), u1(4) *to display the values. You may also use the* Table *feature to display these values.*

In the margin you will see directions on how to create your own sequence of Novocaine levels using your TI-89 or TI-92 calculator. Look for similar notes throughout the text.

The last equation in the list is called the *recursion equation* or *recursion formula* because it provides the iterative process needed to obtain the future term, u_n, from present term, u_{n-1}. The equation specifying the value of u_1 is called the *initial condition*. The recursion formula and the initial condition taken together are referred to as a *difference equation*.

[†]Actually the rate of removal is proportional to the difference in concentrations on both sides of the membrane and is directed towards the region of lower concentration. But other biological processes keep the concentration on the kidney side of the membrane low.

Notice that the recursion formula follows a general pattern expressed by the equation:

$$future = present + change.$$

Often, this simple pattern plays a fundamental role in the model building process. In our example the negative term for change indicates the decreasing level of Novocaine.

We can create a graph of the Novocaine values with respect to time by first entering values in the WINDOW editor as shown in the margin and in Figure 1.1.

You produce the graph in Figure 1.3
with the following WINDOW *editor*
settings:

```
nmin=1,nmax=25,
plotstart=1,plotstep=1,
xmin=0,xmax=25,xscl=5,
ymin=0,ymax=500, and yscl=100.
```

You display the plot by going to the
GRAPH *screen.*

FIGURE 1.1   WINDOW Settings

```
F1- F2-
Tools Zoom
nmin=1.
nmax=25.
plotStrt=1.
plotStep=1.
xmin=0.
xmax=25.
xscl=5.
ymin=0.
ymax▾500.
  ̶ ̶ ̶ ̶ =100
```

Figure 1.3 displays a graph of the patient's Novocaine level at one-hour intervals for 24 hours following the initial injection. Directions for creating this table and graph appear in the margin.

The ordered set of numbers $\{u_n\}$, arising from this difference equation is called a *sequence*. A clear algebraic pattern emerges if we rework our computations in terms of the initial dosage u_1. Since 20 % of the Novocaine has been eliminated after one hour, 80% remains. Therefore $u_2 = 0.8u_1$. Thus, we can represent all of the predicted levels in terms of u_1.

$$u_1 = 500$$

The table of values shown in Table 1.2
is available using TABLE.

FIGURE 1.2   Times (n hrs) and
Novocaine (u1 mg)

n	u1		
1.	500.		
2.	400.		
3.	320.		
4.	256.		
5.	204.8		

u1(n)=u1(n-1)-.2*u1(n-1)

FIGURE 1.3   Novocaine (mg) in the Body vs.
Time (hrs)

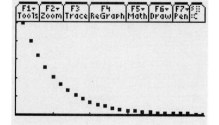

$$u_2 = 0.8u_1$$
$$u_3 = 0.8u_2 = 0.8(0.8u_1) = 0.8^2 u_1$$
$$u_4 = 0.8u_3 = 0.8^3 u_1$$
$$\vdots$$
$$u_n = 0.8u_{n-1} = 0.8^{n-1} u_1$$

▶ **FIRST REFLECTION**

Our computations show that $u_3 = 320.0$. How many hours have passed since the initial injection of Novocaine at that point? How many hours have passed since the initial injection of Novocaine when the level is given by u_n?   ◄

The last equation,

$$u_n = 0.8^{n-1} \cdot 500, \tag{1.3}$$

allows us to compute u_n directly from u_1 without computing any other terms in the sequence $\{u_n\}$, and is called a **closed form** for the sequence. Unfortunately, most sequences arising from difference equations do not have closed forms.

Closed forms can serve another purpose beyond simplifying computations. Even though we have chosen to observe the level of Novocaine at discrete one hour time intervals; the actual process of Novocaine elimination by the kidneys is continuous. The level of Novocaine does not stay at 500 mg for the first hour and then suddenly drop to 400 mg. Instead, the level gradually decreases over the entire hour. Our discrete sequence does not provide any information about the level between the hourly observations. However, by a leap of faith, we can use the closed form for the sequence to conjecture (guess) a model for the level of Novocaine at any time. On the right of Equation (1.3), we simply replace the index $n - 1$, which is the number of hours since the initial injection, with the variable t to represent any time since that injection. In this way, we obtain the function

Go to the MODE *dialogue box and set*
Graph ... *to* FUNCTION.
Then go to the Y = *editor and enter*

y1(x) = 500*(0.8)^x

$$N(t) = 0.8^t \cdot 500, \tag{1.4}$$

where $N(t)$ is the level of Novocaine at time t.

▶ **SECOND REFLECTION**

What value of n in the discrete model corresponds to $t = 2$ in the continuous model? What value of t in the continuous model corresponds to $n = 1$ in the discrete model?   ◄

Using the continuous model, we can conjecture that the level of Novocaine after one and a half hours rounded to the nearest mg will be $N(1.5) = 0.8^{1.5} \cdot 500 = 358$ mg. Graphing the function N we obtain the smooth curve shown in Figure 1.5.

► **THIRD REFLECTION**

When will less than 10% of the initial dose remain in the patient's system? After a full day, what percent of the original injection remains? Explain why this is desirable. ◄

The graph in Figure 1.5 is produced with the same WINDOW *editor settings as the graph in (Figure 1.3).*

FIGURE 1.4 Y= Editor for the Function $N(t) = 500 * (0.8)^t$

FIGURE 1.5 Novocaine (mg) in the Body as a Continuous Function

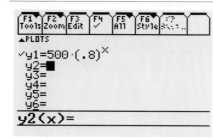

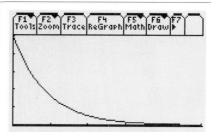

With the GRAPH *screen open, select the* Trace *feature and cursor to the left or right until the y-coordinate,* yc: *reads roughly 50.*

Be careful. People are very good at recognizing patterns and consequently are highly tempted to play "connect the dots". Even though the smooth curve seems to match the scatter plot of Figure 1.3 quite well, we have not truely justified our "leap of faith". The need for such justification may become more apparent to you after you've considered the next example.

Repeated Drug Doses and More Difference Equations

The administration of antibiotics to combat bacterial infection poses a different problem. The antibiotic needs to be maintained at an effective level long enough to kill the invading bacteria but the level must never be large enough to poison the patient. One such drug, amoxicillin, is a broad spectrum antibiotic that is eliminated rapidly from the body. A patient with normal kidney function will eliminate half (0.5) of the body's amoxicillin in one hour. To maintain an effective level, large doses must be administered frequently.

E X A M P L E 2 **Antibiotics — The Amoxicillin Example**

Let's track the level of amoxicillin in a patient's system under the regimen of ingesting 500 mg every four hours. We assume a fanatically punctual patient who arises in the middle of the night to take the prescribed medication.

Upon taking the first pill the patient has 500 mg of amoxicillin in their system.

$$u_1 = 500$$

Four hours later, most of the original dose has been eliminated. At the point of taking the second pill, the patient has 500 mg of new medication plus a small fraction of the amoxicillin left from the first pill. The fraction left from the first pill can be computed as in the previous example.

$$u_2 = 500 + (0.5)^4 u_1 = 531.25 \qquad \textbf{(1.5)}$$

▶ **FOURTH REFLECTION**

Why is 0.5 raised to the power four in the formula for u_2 in Equation (1.5)? What number would be raised to the fourth power if slightly less than half (0.48) were eliminated each hour?   ◀

The accuracy of the displayed computations (10 decimal places) would seem to exceed the accuracy of the model since the 500 mg dose is probably measured to the nearest milligram. We occasionally record the 10 decimal places for you to check your work on the calculator.

With each subsequent dose, the amount of amoxicillin in the patient's system is calculated as 500 mg of new medication plus a fixed proportion, $(0.5)^4$, of the amoxicillin that was in the patient's system immediately after taking the previous dose.

$$u_3 = 500 + (0.5)^4 u_2 = 533.203125$$
$$u_4 = 500 + (0.5)^4 u_3 = 533.3251953125$$
$$\vdots$$
$$u_n = 500 + (0.5)^4 u_{n-1}$$

To generate your own sequence of values, go to the MODE dialogue box and set Graph . . . to SEQUENCE. Then go to the Y = editor and enter
`u1(n) = 500 + (0.0625)*u1(n-1)`
`ui1 = 500.`

Notice that u_n represents a momentary maximum level of amoxicillan since as soon as the drug is ingested the liver starts removing it from the blood system. Consequently, we will refer to the various values of u_n as local maximums. A graph of the sequence $\{u_n\}$ for the first four days of the drug regimen is shown in Figure 1.6. Recall that medication is taken at intervals of four hours, so each day is represented by six sequence elements. Apparently the local maximum levels stabilize rather rapidly.

You produce 1.6 with the following WINDOW editor settings:
`nmin=1, nmax=25,`
`plotstart=1, plotstep=1,`
`xmin=0, xmax=25, xscl=2,`
`ymin=0, ymax=800, and yscl=100`

FIGURE 1.6   Local Maximum Levels of Amoxicillin at Four-hour Intervals — Asymptotic Stability

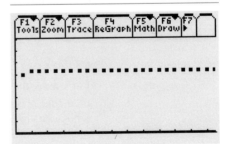

▶ **FIFTH REFLECTION**

According to the graph and the above computations are the local maximums increasing or decreasing? Do the later values appear to be close to some fixed value? Estimate what that value might be.   ◀

It is vital that the amount of a drug in the blood does not grow without bound. Any substance is toxic at high enough concentrations, and fatal drug overdoses are a frequent topic of news reports.

► **SIXTH REFLECTION**

If you were the patient, would you be satisfied with the information contained in the graph alone as evidence that the drug amount would remain safe?   ◄

The numeric and graphic representations support the notion that the later values of u_n are getting close to a fixed value. We call such a value when it exists a *limit value* or *limit point*. For our example, the level of amoxicillin will not exceed this limit value. How can the limit value be determined?

The numeric calculations of u_n and the graph indicate that we can estimate the limit value by calculating the value of u_n for large n. For our example n needn't be particularly large to obtain a very good estimate. Unfortunately, other sequences are not so well behaved. How large n must be to provide a good estimate of the limit value varies. Even in our example, how can we be sure the levels of amoxicillin don't start growing more rapidly on the fifth or sixth day? To reach a higher level of comfort with our result we seek to determine the limit value via an analytic method.

Using L to stand for the unknown limit value, we symbolically represent the idea that u_n approximates L with $u_n \approx L$ for large n. The same reasoning implies that $u_{n-1} \approx L$ for large n. On the other hand, according to the recursion equation

$$u_n = 500 + (0.5)^4 u_{n-1}.$$

By replacing both u_n and u_{n-1} with L we obtain

$$L = 500 + (0.5)^4 L.$$

We have cheated a bit here since strictly speaking we should have replaced the $=$ with $\approx$. We justify using "$=$" by noting that if L is a limit point and n is large enough then the approximation will be very nearly exact. Notice that by solving the equation, we can find L without specifying how large n must be to guarantee a good approximation.

► **SEVENTH REFLECTION**

Solve this equation for L to find the limit value of the level of amoxicillin for our example. How does this result compare to your previous estimate in the Fifth Reflection?   ◄

A patient whose drug level reaches the limit value after a particular dose will continue to reach that value after each successive dose. In other words, if u_{n-1} equals the limit value then so does $u_n, u_{n+1}, \ldots$. Values of this type are called *rest points*.

REST POINTS

The **rest points** of the recursion formula $u_n = F(u_{n-1})$ are the solutions to the equation $u_n = F(u_n)$ or $x = F(x)$.

In the model of repeated doses of amoxicillin, the limit value is also a rest point which explains why we were able to find the limit value by solving an equation. Unfortunately, not all limit values are rest points. We will explore these two concepts in the exercises and in subsequent sections.

Form groups of three and discuss the following series of questions. Once your group has reached consensus, report and defend your answers to the class as a group.

Think and Share

- Suppose that each dose of amoxicillin were 1000 mg instead of 500 mg. Would there still be an equilibrium level that is stable (that is, a rest point)?
- Will there always be a rest point no matter what dosage is used?
- Suppose the doses are administered every hour. Will there still be an upper limit on the level of drug?
- Is there any time between doses that would cause the drug level to grow without bound?

EXERCISE SET 1.1

In Exercises 1 and 2, classify each quantity as discrete or continuous and justify your decision.

1. **a.** pills in a bottle
   b. amount of drug in your body
   c. time to complete a test
   d. amount of rainfall on a given day
   e. days in an academic quarter (or semester)
2. **a.** chips in a chocolate chip cookie
   b. time to complete a mathematics project
   c. crimes committed on campus in a year
   d. the percentage of mothers who work outside the home
   e. distance between classes

In Exercises 3 and 4, identify each expression as either an initial condition, a recursion formula, or a difference equation.

3. **a.** $u_1 = 4$
   b. $u_n = u_{n-1} + 2$ and $u_1 = 0$
   c. $u_n = \dfrac{\cos(u_{n-1})}{n}$
   d. $u_0 = 3$
   e. $u_n = \dfrac{u_{n-1}}{2}$

4. **a.** $u_n = 0.3u_{n-1}$
   b. $u_1 = 1$
   c. $u_{n-1} = u_{n-2} - 1$
   d. $u_n = (0.5)^4 u_{n-1}$ and $u_1 = 300$
5. The sequence $3, 9, 27, 81, 243, \ldots$, has $u_n = 3u_{n-1}$ as a recursion formula and $u_n = 3^n$ as a closed form formula for the nth term. For each of the sequences indicated below, find both a recursion formula and a closed form formula for the nth term, u_n.
   a. $2, 4, 8, 16, 32, 64, \ldots$
   b. $1, \frac{1}{2}, \frac{1}{4}, \frac{1}{8}, \frac{1}{16}, \frac{1}{32}, \frac{1}{64}, \ldots$
   c. $1, 3, 7, 15, 31, 63, \ldots$
   d. $1, -1, 1, -1, 1, -1, \ldots$
   e. $7, 10, 13, 16, 19, 22, \ldots$
   f. $1, 4, 9, 16, 25, 36, \ldots$

In Exercises 6–11, record the first ten elements of the sequence generated by the difference equation. If possible, find a closed form for the sequence.

6. $u_n = 2u_{n-1}$
       $u_1 = 1$

7.
$$u_n = \frac{1}{2}u_{n-1}$$
$$u_1 = 1$$

8.
$$u_n = 2u_{n-1}$$
$$u_1 = 0$$

9.
$$u_n = -2u_{n-1}$$
$$u_1 = 1$$

10.
$$u_n = \frac{-1}{2}u_{n-1}$$
$$u_1 = 1$$

11.
$$u_n = 2 + u_{n-1}$$
$$u_1 = -2$$

In Exercises 12–15, record the first 10 elements of the sequence generated by the difference equation. Find all rest points for the recursion formula.

12.
$$u_n = 10 + 0.5u_{n-1}$$
$$u_1 = 5$$

13.
$$u_n = u_{n-1}(2 - u_{n-1})$$
$$u_1 = 0.5$$

14.
$$u_n = u_{n-1}(3 - u_{n-1})$$
$$u_1 = 10$$

15.
$$u_n = 8 + 0.4u_{n-1}$$
$$u_1 = 10$$

16. Suppose a dental patient is injected with 800 mg of Novocaine and that this patient does not have normal kidney function. That is, only 15% of the Novocaine in the blood is removed in one hour.
  a. Find a recursion formula with which to model the level of Novocaine.
  b. Find a difference equation that models the level of Novocaine.
  c. Find a closed form for the equation which expresses u_n in terms of u_1.
  d. How many hours must pass before less than 10% of the initial dose remains? Justify.

  e. Find a function that models the amount of Novocaine in the system at any time. Explain.
  f. Classify the function you obtained in the preceding part as continuous or discrete and justify.

17. Suppose 600 mg of a certain antibiotic is administered every three hours, and that the body eliminates 40% of the antibiotic in one hour.
  a. Model the maximum level of the antibiotic with a difference equation. Be careful to think about how much antibiotic is left, not how much has been eliminated.
  b. Does the maximum level stabilize? If so, what is this "stable" value?
  c. Produce a graph to depict the amount of the antibiotic in the body during the first 12 hours.

†**18.** †Show that the sequence determined by any difference equation of the form
$$u_n = a \cdot u_{n-1}$$
$$u_1 = b$$
has the closed form $u_n = a^{n-1} \cdot b$. If $b > 0$ and $0 < a < 1$, the sequence is said to exhibit *exponential decay*. If $b > 0$ and $1 < a$, the sequence is said to exhibit *exponential growth*. Discuss the reasonableness of these terms.

†**19.** The difference equation
$$u_n = 500 + (0.5)^4 u_{n-1}$$
$$u_1 = 500$$
models the amount of amoxicillin in a patient's body on a regimen of 500 mg of amoxicillin at the beginning of every four-hour period. This is the maximum amount in blood during that period. Establish a difference equation that models the *minimum* amount of amoxicillin in the patient's body during each four hour period.

20. The interest earned by a standard savings account during one compounding period is a fixed proportion of the amount of money in the account. If the annual interest rate is 6% and the compounding period is one month, then the interest earned in one month is $\frac{0.06}{12}$ times the amount in the account at the beginning of the month.
  a. Write a difference equation to represent the amount in the account at the end of n months as a function of the amount in the account at the end of $n - 1$ months. (Let A_n be the amount in the account after n months.)

---

†The †ed exercises are required since they are referred to later in the book.

b. Suppose the account is started with $500. Use your difference equation to compute the amount of money in the account at the end of each month for the first year.

21. Most patients continue to feel numbness as long as the amount of Novocaine remaining in their body exceeds 10 mg. For about how long after the 500 mg amount of Novocaine is administered should a patient expect to continue feeling numbness?

†22. The graph of Figure 1.6 displays the momentary maximum levels of amoxicillin in the body. Complete it with a rough hand-drawn sketch to show the level of amoxicillin in the patient's body as a function of time t (measured in hours) for

$$0 < t < 24.$$

†23. Novocaine is given to heart patients to correct arrhythmias. A normal drug regimen for a 180-pound patient might be one gram taken at six-hour intervals.

  a. Write a difference equation that models the maximum amount in this drug regimen.

  b. Write a difference equation that models the minimum amount in this drug regimen.

  c. Produce a graph that shows the amount in the patient's body over the first three days of this drug regimen.

  d. Determine the rest points for each difference equation you produced in the first two parts.

24. Here is an old conundrum from Victorian England: In far away Brazil there is a species of water lily that doubles the size of its pad each day and grows to fantastic proportions. If such a water lily is planted at the center of a one-acre pond, it will grow to cover the pond in exactly 30 days. Counting the day the lily is planted as day one, on what day will the water lily cover half of the surface of the pond?

25. One theory of athletic training is that at any one period, an athlete has a stable theoretical optimal level of performance and that a month of ordinary training will decrease the difference between the athlete's current actual level of performance and that athlete's theoretical level by a fixed percentage. Ten-year-old Niki Peck can currently swim the 100 meter breaststroke in 148 seconds. Her father assures us that her current theoretical best time is 130 seconds. If a week of practice allows her to gain 10% of the difference between her current time and her theoretical best, describe her expected times over the next 12 months.

26. *Group Exercise* Prozac™ (Fluoxetine Hydrochloride) is approved for use to treat depression and obsessive-compulsive disorder. A half-life of roughly nine days (half the substance is destroyed every nine days) makes it an unusually long-lived drug in the body. Typical regimens for treatment are daily (morning) oral doses of between 20 and 80 mgs.

  a. Use the fact that half of the Prozac in a patient's blood is eliminated in nine days to find a closed form for the amount of Prozac remaining in a patient's system after a single dose.

  b. Convert this closed form to a function that gives the percentage of the dose as a function of time measured in hours.

  c. Use your function from the second part above to determine the amount of a single dose that remains in a patient's system 24 hours after ingestion.

  d. Write a difference equation that models the maximum amount (in mgs) of Prozac in a patient's blood on daily regimens of 20 mg, 50 mg, and 80 mg.

  e. Determine the rest points for each of your models and produce graphs that show the increase and stabilization of the dosage.

  f. For each of your models, determine the time it takes from initiation of the drug regimen for the maximum dose to reach 90% of the stable level.

  g. As with all effective medications, Prozac can have marked side effects such as chills, angina, arrhythmias, anemia, and bronchitis. These usually disappear after the drug is eliminated from the body. For a patient who is maintained at the stable level in each of your drug regimens, how long after stopping the drug regimen does it take for the levels of Prozac in that patient's body to fall below 5 mg?

1.2

Patterns of Accumulation

In the last section, we studied the accumulation of drugs in the body as the result of repeated doses. We understood the change in drug level by viewing the accumulation as a sum of incremental changes. Studying the notion of accumulation (summation) eventually led to the development of calculus. Not surprisingly, calculus cannot be fully understood without understanding summations.

By building on the repeated dosage problem, we can learn a bit more about summations and develop a language that facilitates their use. We can then apply this knowledge to analyze new situations involving accumulation.

How Change Accumulates

When modeling the Novocaine injected into a patient, we were able to obtain a closed form formula for u_n the amount of Novocaine in the patient's system at the nth observation. Can we develop such a formula for the model of repeated doses of amoxicillin?

E X A M P L E 1 **The Amoxicillin Example Revisited**

Recall that we modeled the maximum level of amoxicillin in a patient after n doses with the recursion formula

$$u_n = 500 + (0.5)^4 u_{n-1}.$$

To obtain a closed form formula, we look at the terms of the sequence u_n as an unevaluated sum in order to find a pattern.

$$u_1 = 500$$
$$u_2 = 500 + (0.5)^4 u_1 = 500 + 0.0625 u_1$$
$$u_3 = 500 + 0.0625 u_2$$
$$= 500 + 0.0625(500 + 0.0625 u_1)$$
$$= 500 + 0.0625 \times 500 + 0.0625^2 \times 500$$
$$u_4 = 500 + 0.0625 u_3$$
$$= 500 + 0.0625(500 + 0.0625 \times 500 + 0.0625^2 \times 500)$$
$$= 500 + 0.0625 \times 500 + 0.0625^2 \times 500 + 0.0625^3 \times 500$$
$$u_5 = 500 + 0.0625 u_4$$
$$\vdots$$

We have deliberately refrained from evaluating the sums since that would have obscured the developing pattern.

▶ **FIRST REFLECTION**

Write u_5 and u_6 as unevaluated sums (see the final expression for u_4). These sums should *not* contain any terms with u's in them. Such sums occur frequently in mathematical modeling.   ◀

Right off the bat we have a notational problem. Even though the simplest way to represent a sum is to write out all of the terms as we have done above, such representations are terribly inconvenient. Consider the complexity of the sum you obtained for u_6 in the reflection. Suppose you had been asked to write the sum for u_{100}? Such a sum would have 100 terms!

Fortunately, when the sum exhibits a pattern there are a number of devices to shorten the notation. Perhaps the most intuitive device is an *ellipsis*. An ellipsis is a series of three dots ($\ldots$) used to indicate the omission of letters or words. Using this convention we would write u_{100} as

$$u_{100} = 500 + 0.0625 \times 500 + 0.0625^2 \times 500 + \cdots + 0.0625^{99} \times 500.$$

▶ **SECOND REFLECTION**

Use ellipsis notation to complete the right side of the following equation:

$$u_n = \underline{\hspace{5cm}},$$

a closed form formula for the level of amoxicillin.   ◀

Finding a Pattern.   An ellipsis should only be used in a mathematical expression when it is reasonable to assume that the reader can fill in the missing symbols. This requires that the expression exhibit a pattern. Moreover, the terms that are included in the shortened expression must be carefully chosen to allow the reader to recognize that pattern. Thus, although ellipsis notation simplifies communicating summations, its interpretation requires the reader to figure out a pattern that has been merely suggested. Happily, human beings are very good at discovering patterns, so ellipsis notation is both popular and powerful. Unhappily, human beings are very inventive and discover patterns that the writer may not have intended or even considered. The expression $3 + 5 + 7 + \cdots + 19$ could be interpreted as the sum of odd integers from 3 to 19, which incidentally is 99. On the other hand, some people might decide that

$$3 + 5 + 7 + \cdots + 19 = 75 \tag{1.6}$$

by interpreting the sum to include only the primes between 3 and 19.

▶ **THIRD REFLECTION**

Give two possible values for the sum $2 + 4 + \cdots + 32$. Describe the interpretation you used to obtain each value.   ◀

Specifying the Pattern.   We can eliminate false patterns by supplying an explicit description of the terms to be summed. This places the burden of cleverness on the writer who must create an explicit description, usually an algebraic formula, of the terms in order to avoid any ambiguity.

To represent the sum of the odd integers from 3 to 19, we require an algebraic formula for odd numbers. We can generate such a formula by noticing that every odd number is one more or one less than an even number and every even number is two times some integer. Consequently, if n is any integer then $2n$ is even, and either $2n - 1$ or $2n + 1$ can be used to represent an odd number. By convention, the letters i, j, k, l, m, and n are used to indicate integers; so we might use any of $2i - 1$, $2i + 1$, $2j - 1$, $2j + 1$, $2k - 1$, $2k + 1$, $2m - 1$, $2m + 1$, $2n - 1$, $2n + 1$ to indicate an odd integer. Picking $2k + 1$ as our formula for odd numbers we write

$$3 + 5 + \cdots + (2k + 1) + \cdots + 19. \tag{1.7}$$

Expression (1.7) is unambiguous, shorter than writing out the entire summation, and nearly as easy to read as the ellipsis in Expression (1.6).

▶ FOURTH REFLECTION

Write a summation formula for u_n in the Second Reflection which includes a "kth" term.   ◀

The three features that make Expression (1.7) unambiguous are the formula for the terms to be added and the explicit starting and ending numbers (3 and 19).

Summation (Sigma) Notation.   Since the eighteenth century, mathematicians have used an extremely compact notation for summations. This notation has become especially important as the way we communicate sums to machines, which have no facility at all for guessing patterns. There are four parts to this notation.

1. The notation starts with the symbol (Σ), the Greek letter that corresponds to 'S' for "summation."
2. An unambiguous algebraic formula for the terms to be summed is given directly following the Σ. The variable used in this formula is called the *index of summation*.
3. An equation specifying the first value of the index of summation appears immediately *below* the Σ. This value is called the *lower limit of the summation*.
4. The last value of the index of summation appears immediately *above* the Σ and is called the *upper limit of the summation*.

Using this *sigma notation*, mathematicians write

$$3 + 5 + \cdots + (2k + 1) + \cdots + 19 = \sum_{k=1}^{9} (2k + 1).$$

The TI calculator notation is
$$\Sigma(2*k+1,k,1,9).$$
The Σ symbol is found in the Calc *(F3) menu.*

FIGURE 1.7   Sum Notation

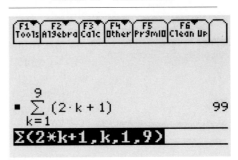

In this case, we found the lower and upper limits by solving the equations $2k + 1 = 3$ and $2k + 1 = 19$, respectively.

▶ **FIFTH REFLECTION**

Does the sum $\displaystyle\sum_{i=1}^{9}(2i - 1)$ represent the same sum as $\displaystyle\sum_{k=1}^{9}(2k + 1)$? Justify your answer.   ◀

The Closed Form Formula.   After factoring out the 500, we have three closed form expressions for u_n, the level of amoxicillin after the nth dose:

$$u_n = 500 \times (1 + 0.0625 + 0.0625^2 + \cdots + 0.0625^{n-1})$$
$$= 500 \times (1 + 0.0625 + 0.0625^2 + \cdots + 0.0625^k + \cdots + 0.0625^{n-1})$$
$$= 500 \times \sum_{k=0}^{n-1} 0.0625^k \qquad\qquad\qquad\qquad\qquad\qquad \textbf{(1.8)}$$

The sum in these expressions falls into a widely studied class of sums which are called *finite geometric series*.

GEOMETRIC SERIES

A sum of the form

$$\sum_{k=0}^{n-1} a^k = 1 + a + a^2 + \cdots + a^k + \cdots + a^{n-1}$$

is called a **finite geometric series** of n terms.

The TI calculator uses this form to evaluate geometric sums. Be sure that a, k, and n have not been defined (use F6 *to* Clear a-z *if they have been) and look at the result of the commands* $\Sigma(a^k,k,0,n-1)$ *followed by* comDenom(ans(1)).

▶ **SIXTH REFLECTION**

What value should be assigned to a in the finite geometric series formula to obtain the sum in our expression for u_n?   ◀

One remarkable feature of finite geometric series is that they equal a simple rational expression. Indeed, whenever $a \neq 1$, it can be shown that

$$\sum_{k=0}^{n-1} a^k = \frac{1-a^n}{1-a}. \tag{1.9}$$

▶ **SEVENTH REFLECTION**

To expand the finite geometric series $\sum_{k=0}^{n} a^k$ we write it as a sum displaying all the terms. Hence, $\sum_{k=0}^{3} a^k$ expands to $1 + a + a^2 + a^3$. Expanding the left side of Equation (1.9) for $n = 4$ produces the equation

$$1 + a + a^2 + a^3 = \frac{1-a^4}{1-a}.$$

Multiply both sides by $1 - a$ and simplify to show that this equation is correct. Repeat the process for $n = 5$. ◀

By applying Equation (1.9) to the summation in $u_n = 500 \sum_{k=0}^{n-1} (0.0625)^k$ with $a = 0.0625$, we obtain

For the same sum the TI calculator returns

$533.333 \cdot (0.0625)^n ((16.)^n - 1.)$

$$u_n = 500 \left(\frac{1 - (0.0625)^n}{1 - 0.0625} \right).$$

▶ **EIGHTH REFLECTION**

Evaluate u_{10} by using the "with" operator just below the = key (2nd K on the TI-92) as follows:

```
500*(1-(0.0625)^n)/(1-0.0625)|n=10
```

◀

You should not be surprised by the value for u_{10} found in the reflection. We have already argued that as n increases, the sequence $\{u_n\}$ approaches the rest point for the recursion formula. In that argument, however, we were forced to "cheat" by replacing $\approx$ with $=$.

We can use our new knowledge of geometric series to clarify the situation. First, we rewrite the above expression as the difference of two fractions;

$$500 \left(\frac{1 - (0.0625)^n}{1 - 0.0625} \right) = 500 \left[\frac{1}{1 - 0.0625} \right] - 500 \left[\frac{(0.0625)^n}{1 - 0.0625} \right].$$

Notice that 0.0625^2 is closer to 0 than 0.0625. Furthermore, 0.0625^3 is closer to 0 than 0.0625^2. The higher the value chosen for n the closer 0.0625^n is to 0. Thus for large values of n the second fraction on the right is close to zero and the whole expression is nearly equal to

$$\left(\frac{500}{1 - 0.0625} \right),$$

our limit value. As an added bonus, the expression $500 \times \dfrac{(0.0625)^n}{1 - 0.0625}$ represents the exact difference between u_n and the limit value.

We can apply the same reasoning to the general finite geometric series $\sum_{k=0}^{n-1} a^k$. Once again we start with Equation (1.9) and split the right hand side into the difference of two fractions,

$$\sum_{k=0}^{n-1} a^k = \frac{1 - a^n}{1 - a} = \frac{1}{1 - a} - \frac{a^n}{1 - a}. \qquad \textbf{(1.10)}$$

We need to be a bit careful with the next step. The second fraction on the right will be close to 0 for large values of n as long as $-1 < a < 1$. Thus, for large values of n and $-1 < a < 1$, $\sum_{k=0}^{n-1} a^k$ should approximately equal $\dfrac{1}{1 - a}$ since $\dfrac{a^n}{1 - a}$ will be very close to zero.

▶ **NINTH REFLECTION**

Calculate the values of $\dfrac{(0.3)^n}{1 - 0.3}$ and $\dfrac{(0.7)^n}{1 - 0.7}$ for $n = 10, 50, 100$. What happens to the value of $a^n/(1 - a)$ as n increases for $a = 0.3$ and $a = 0.7$?   ◀

On the TI calculator, enter
`limit(a^n,n,∞) | 0 < a and a < 1.`

To describe the behavior of $\dfrac{a^n}{1 - a}$, exhibited in the Ninth Reflection, we use the language of limits and say "when $0 < a < 1$, the limit of $\dfrac{a^n}{1 - a}$ as n grows to infinity is 0."

▶ **TENTH REFLECTION**

Use Equation (1.9) to find a closed form expression for

$$\sum_{k=0}^{n-1} \left(\frac{1}{2} \right)^k.$$

Compute the value of this closed form expression when $n = 5$, when $n = 15$, and when $n = 25$.   ◀

Think and Share

Break into groups of three. While one member of each team uses the formula $\dfrac{1}{1 - a}$ to determine the value of the "infinite sums" $\sum_{k=0}^{\infty} a^k$ for $a = \pm\frac{1}{2}, \pm 1, \pm\frac{3}{2}$, other members of the team should determine approximate values for $\sum_{k=0}^{100} a^k$ and $\sum_{k=0}^{101} a^k$ for $a = \pm\frac{1}{2}, \pm 1, \pm\frac{3}{2}$. Notice that some values of a are negative. Describe when, in terms of the value of a, it makes sense to say the summation $\sum_{k=0}^{\infty} a^k$ is equal to $\dfrac{1}{1 - a}$.

The results of your Think and Share are represented symbolically as

$$\sum_{k=0}^{\infty} a^k = \frac{1}{1-a}, \quad \text{when } 0 < |a| < 1. \tag{1.11}$$

Putting It All Together

In Section 1.1, we examined two models of the amount of drugs in a patient's body as time passed. Simple physical and biological considerations gave rise to difference equations of two types:

$$u_n = a \cdot u_{n-1}$$
$$u_1 = b \qquad\qquad\qquad\qquad \text{Type I}$$

and

$$u_n = b + a \cdot u_{n-1}$$
$$u_1 = b \qquad\qquad\qquad\qquad \text{Type II.}$$

FIGURE 1.8 $0 < a < 1$ **FIGURE 1.9** $a > 1$

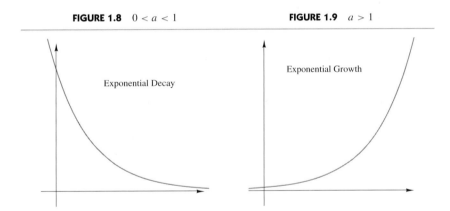

Exponential Decay   Exponential Growth

Exponential Growth and Decay.   Sequences arising from difference equations of the first type had the easily recognized closed forms $u_n = b \cdot a^{n-1}$ and are said to be *exponential*. When $0 < a < 1$, the sequence u_n decreases (since $u_n < u_{n-1}$) and models *exponential decay* (see Figure 1.8). When $a > 1$, the sequence u_n increases and models *exponential growth* (see Figure 1.9). These models are so well known and widely used that the phrase "exponential growth" is used metaphorically for any rapid and unbounded increase.

Limited Growth.   The second type of difference equations produces sequences that tend toward a stable asymptotic value only if $|a| < 1$. These sequences also have a closed form involving exponential functions, although their discovery required somewhat more elaborate mathematical computations. To find the closed form using exponentials we use the ideas of this section to rewrite the term u_n as a summation and then apply Equation (1.9):

$$u_n = b \cdot \sum_{k=0}^{n-1} a^k = b \cdot \frac{1 - a^n}{1 - a}. \tag{1.12}$$

When $|a| < 1$ the expression $b \cdot \dfrac{1 - a^n}{1 - a}$ very nearly equals $\dfrac{b}{1 - a}$ for large values of n; which implies $\dfrac{b}{1 - a}$ is the limit value for the sequence. On the other hand, solving the equation

$$u_n = a \cdot u_n + b$$

for u_n, we find that $\dfrac{b}{1 - a}$ is also a rest point; which justifies our previous assertion that the limit value for this sequence is also a rest point.

▶ **ELEVENTH REFLECTION**

Use Equation (1.11) to write an equation for u_n, the level of amoxicillin, as a rational expression which does not include a summation. Consider the general difference equation of Type II. Suppose that $|a| > 1$. Will $\dfrac{b}{1 - a}$ still be a limit value? Will $\dfrac{b}{1 - a}$ still be a rest point? Explain your assertions.   ◀

Geometric series appear in a broad range of applications other than drug regimens. A project at the end of this chapter explores the use of geometric series in finance.

We have found closed forms for both of our difference equation models. Such formulas are extremely useful for simplifying computations and for understanding the model. Unfortunately, not all difference equations have a closed form as we will see in the next section.

EXERCISE SET 1.2

1. Express each of the following sums using Σ-notation. Be sure to indicate the index, starting value, and ending value.
   a. $1 + 2 + 3 + \ldots + k + \ldots + 20$
   b. $1 + 4 + 9 + \ldots + k^2 + \ldots + 2500$
   c. $1 + 4 + 7 + \ldots + (3k + 1) + \ldots + 28$
   d. $1 - 2 + 4 - 8 + \ldots + ((-1)^k \cdot 2^k) + \ldots - 2048$

2. Interpret each of the following colloquial descriptions using Σ-notation.
   a. The sum of the first 5 positive cubes (starting with 1^3, 2^3).
   b. The sum of cubes from 1^3 to n^3.
   c. The sum of the first thousand reciprocals (starting with 1/1, 1/2, 1/3).

   d. The sum of the squares of the first two hundred positive integers.
   e. The sum of the first five fractions of the form $\dfrac{1}{n \cdot (n + 1)}$ (starting with 1/2, 1/6, 1/12).
   f. The sum of all fractions of the form $\dfrac{1}{n \cdot (n + 1)}$.

3. Expand each summation using ellipsis notation. Then use the TI calculator to evaluate each sum.
   a. $\displaystyle\sum_{k=1}^{20} k$
   b. $\displaystyle\sum_{k=1}^{20} 5k$

c. $\displaystyle\sum_{k=1}^{20} 1$

d. $\displaystyle\sum_{k=1}^{20} (5k+1)$

e. $\displaystyle\sum_{k=0}^{11} (-2)^k$

f. $\displaystyle\sum_{k=1}^{n} k^3$

g. $\displaystyle\sum_{n=1}^{\infty} \frac{1}{n\cdot(n+1)}$

4. Determine if each equation is true or false. Justify your decision.

a. $\displaystyle\sum_{k=1}^{5} 3 = 15; \quad \sum_{k=1}^{n} 3 = 3n$

b. $\displaystyle\sum_{k=1}^{10} 3k = 3\sum_{k=1}^{10} k$

c. $\displaystyle\sum_{k=1}^{14} (5k+2) = 5\sum_{k=1}^{14} k + \sum_{k=1}^{14} 2$

d. $\displaystyle\sum_{k=1}^{100} k^2 = \sum_{k=1}^{10} k^2 + \sum_{k=10}^{100} k^2$

e. $\displaystyle\sum_{k=0}^{\infty} 5\left(\frac{4}{9}\right)^k = 5\sum_{k=0}^{\infty}\left(\frac{4}{9}\right)^k$

f. $\displaystyle\sum_{k=1}^{\infty}\left(\frac{1}{4}\right)^k = \sum_{k=0}^{\infty}\left(\frac{1}{4}\right)^k - 1$

In Exercises 5 and 6, evaluate the given summations using the closed form given in Equation (1.9) or Equation (1.11) where appropriate. Then use the TI calculator to check your results.

5. a. $\displaystyle\sum_{k=0}^{10} 0.3^k$

b. $\displaystyle\sum_{k=0}^{10} 2^k$

c. $\displaystyle\sum_{k=0}^{10} 1^k$

d. $\displaystyle\sum_{k=0}^{\infty}\left(\frac{1}{3}\right)^k$

e. $\displaystyle\sum_{k=0}^{15} (-3)^k$

f. $\displaystyle\sum_{n=0}^{10}\left(\frac{3}{5}\right)^n$

g. $\displaystyle\sum_{n=0}^{\infty}\left(\frac{3}{8}\right)^n$

6. a. $\displaystyle\sum_{k=0}^{\infty} 5\left(\frac{2}{3}\right)^k$

b. $\displaystyle\sum_{k=0}^{\infty}\left(\frac{5}{7}\right)^k$

c. $\displaystyle\sum_{k=0}^{6} 5\left(\frac{5}{4}\right)^k$

d. $\displaystyle\sum_{k=0}^{\infty} 3\left(\frac{1}{5}\right)^k$

e. $\displaystyle\sum_{k=0}^{\infty}\left(\frac{2}{5}\right)^k$

f. $\displaystyle\sum_{k=0}^{8} 5\left(\frac{3}{8}\right)^k$

† **7.** Determine if each equation is true for all sequences a_k. If there are examples for which the equation is false, give one of these "counterexamples."

a. $\displaystyle\sum_{k=0}^{n} 1 = n$

b. $\displaystyle\sum_{k=0}^{n} a \times a_k = a \times \sum_{k=0}^{n} a_k$

c. $\displaystyle\sum_{k=0}^{n} (a_k + b_k) = \left(\sum_{k=0}^{n} a_k\right) + \left(\sum_{k=0}^{n} b_k\right)$

d. $\displaystyle\sum_{k=0}^{n} (a_k \times b_k) = \left(\sum_{k=0}^{n} a_k\right) \times \left(\sum_{k=0}^{n} b_k\right)$

e. $\displaystyle\sum_{k=0}^{n} (a_k - b_k) = \left(\sum_{k=0}^{n} a_k\right) - \left(\sum_{k=0}^{n} b_k\right)$

f. $\displaystyle\sum_{k=0}^{n}\left(\frac{a_k}{b_k}\right) = \frac{\left(\displaystyle\sum_{k=0}^{n} a_k\right)}{\left(\displaystyle\sum_{k=0}^{n} b_k\right)}$

† **8.** When treating a patient with a bacterial infection, a physician may inject a large dose of amoxicillin prior to starting the oral administration. This initial injection assures a rapid attack on the invading bacteria and reassures the patient.

   a. Establish a difference equation that models the maximum amount of the drug in the body based on a treatment regimen beginning with an initial injection of 1000 mg followed by 500 mg of amoxicillin at four-hour intervals.

   b. Record and graph the first six terms in the sequence determined by your difference equation.

   c. Does the initial dose affect the long-term behavior of the drug regimen?

9. Suppose 600 mg of a certain antibiotic is administered every three hours and that the body eliminates 40% of the antibiotic in one hour. (Refer to Exercise 17 in Section 1.1.)

a. Establish a recursion formula to model the maximum level of the antibiotic after n doses.

b. Write a closed form expression for u_n as an unevaluated sum that does not contain any terms with u's in them. (See First Reflection in this section.)

c. Write a closed form expression for u_n using summation notation. Is your sum a finite geometric series? Explain.

d. Evaluate u_{10}.

e. Is there a limit value for your geometric series? Explain.

f. Does the maximum amount of antibiotic in the body reach a limit value? Explain.

10. Let the sequence $\{u_n\}$ be determined by the difference equation

$$u_n = a \cdot u_{n-1} + b$$
$$u_1 = c.$$

a. Use summation notation to express u_n in term of a, b, and c.

b. Give a compact closed form for u_n in terms of a, b, and c.

†11. The TI calculator command `when(` is used to create piecewise defined functions. The syntax for `when(` as displayed from the CATALOG as shown in Figure 1.10 is `CONDITION, TRUE, [,FALSE] [,UNDEF]`.

FIGURE 1.10   The Command `when(` as Found in the CATALOG

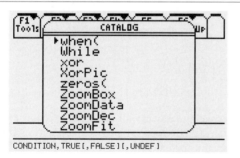

The intended meaning of this brief syntax statement can be paraphrased as, "When this condition is true, use the first formula, otherwise use the second formula or leave the function undefined."

a. Enter the function `when(x < -3, 8, x^2 - 1)` as $y1(x)$ in the Y= editor and examine several plots of the function in different windows. Produce a hand-drawn graph that shows the behavior of this function.

b. From the HOME screen remembering that definitions are entered with `q`, enter

`Define f(q) = when(q < -3, 8, q^2 -1)`. Record the values of $f(-5)$, $f(-4)$, $f(-3)$, $f(-2)$, $f(2)$, $f(3)$, and $f(5)$.

c. Interpret, in complete sentences, what the function `when(x < -3, 8, x^2- 1)` does.

d. Use the `when(` command to graph the function

$$g(x) = \begin{cases} 3 & \text{if } x < 2 \\ 3 - (x - 2) & \text{if } 2 \le x \end{cases}$$

†12. *Group Exercise* Novocaine is given to heart patients to correct arrhythmias. A normal drug regimen for a 180-pound patient might be one gram taken at six-hour intervals. Recall that the body eliminates 20% of the Novocaine in the body each hour.

a. Express the maximum amount (just after taking the nth pill) of Novocaine in the patient's body as a sum.

b. Express the minimum amount (just before taking the nth pill) of Novocaine in the patient's body as a sum.

c. Exhibit a closed form for the maximum and minimum amounts in the regimen described.

d. Modeling the amount of drug in a patient's body during the period between pills presents a special challenge. For example, when our patient takes the third pill (12 hours after starting the therapy) it initially adds one gram of Novocaine and decays exponentially from there. It contributes zero grams to our patient's system until it is swallowed. To describe this effect as a function of x, we need a function that is zero until $x = 12$ and then decays as a power of 0.8 from one gram. We can make such a function with the command `when(x < 12, 0, (0.8)^(x - 12))`. Give formulas for functions that describe the effect of the fourth and fifth pills on the patient's system.

e. Give a formula for the amount of Novocaine in the patient's body caused by the nth pill.

f. Use summation notation to give a function that represents the amount of Novocaine in the patient's body during an three-day course of therapy under the indicated regimen.

g. Produce a hand-drawn graph showing the amount of Novocaine in the patient's body during an eight-day course of therapy under the indicated regimen. Include the curves that indicate the maximum and minimum dosages you derived above.

13. *Group Exercise* An American cigarette contains 11 mg of nicotine. Of this, 0.9 mg are extracted by the average smoker. Half of the nicotine present in a healthy human is removed from the body in 90 minutes. Develop a difference equation model for the amount of nicotine in the body of a smoker who consumes a pack (20 doses) a day. From your model determine the long-term dosage of nicotine in the body of

such a smoker. Sketch a graph showing the estimated level of nicotine in the smoker's body during a day of smoking. This will require that you develop an explicit scenario as to when your hypothetical smoker lights up.

14. Sums of polynomial functions also have closed forms but they are often more cumbersome than the closed form of the geometric series. There is, however, a way of writing polynomials using what are called *rising factorials*, which displays a nice pattern. Rising factorials are defined by the following sequence.

$$x^{[1]} = x$$
$$x^{[2]} = x(x + 1)$$

$$x^{[3]} = x(x + 1)(x + 2)$$
$$x^{[4]} = x(x + 1)(x + 2)(x + 3)$$
$$\vdots$$
$$x^{[m+1]} = x^{[m]}(x + m)$$

Examine the summations $\sum_{k=1}^{n} k^{[m]}$ for particular values of m to discover a pattern. Give an algebraic argument to support your pattern. What does this pattern tell us about the summation of any polynomial?

1.3

A Model for Natural Growth

In Section 1.1, we initiated our study of change with a simple example, a single drug administered to a single patient. We ignored a whole host of possibly relevant factors including variations in the patient's metabolic rate, the possible effects of other drugs, and the stress associated with a visit to the dentist. The examples in this section (spread of a rumor, growth of yeast in fermentation, and natural increase of populations) require consideration of the interaction between individual elements. We will develop techniques to represent and measure such interactions. Including these interactions leads to difference equations which have no closed form.

Creating the Model

In the modeling process, we study the situation or problem to determine those quantities that are changing, how the changes are measured, and how variables might interact.

EXAMPLE 1   **Interaction of Individuals — The Rumor Example**

Consider the way a rumor spreads in a small community (like a college campus), or a global village (like the World Wide Web). A rumor, true or false, usually starts with a single person or a small group of people. The rumor spreads by verbal interactions: someone relates the rumor to someone else who hasn't yet heard it. The process continues until everyone knows the news or it ceases to be of interest. In order for the rumor to spread, there must be an interaction between someone who knows the rumor and someone who does not know the rumor. Who knows whom, who is a known gossip, or when key people learn the rumor, are all factors that affect the rumor's actual spread. Including such details would create an impossibly complicated model. Instead we consider the bigger picture. Every person who knows the rumor has the potential to interact with every person who doesn't know the rumor. Consequently, we make the conceptually unsophisticated, but computationally easy, estimate that, *the number of people who learn the rumor in any one time interval is proportional to the product of the number who know the rumor and the number who don't know the rumor.* We use "proportional" rather than "equals" because only a portion of the possible interactions take place in a given time interval and not every interaction results in transmission of the rumor.

▶ **FIRST REFLECTION**

What quantity is changing in this example? How should the speed at which the quantity changes when only a few people know the rumor compare to the speed at which it changes when about half of the people know the rumor? If b individuals know the rumor in a total population of 300, how many do not know the rumor and how might you use our simple estimate to represent the number of potential interactions between those who know and those who don't know?   ◄

Building the Model.   In building a difference equation model, we will need an algebraic representation of time, the number of people who know the rumor, and the number of people who *don't* know the rumor. As usual we let n represent one more than the number of time intervals that have passed. We take u_n to represent the number of people who know the rumor after $n-1$ time intervals. We let Pop represent the total size of the population of interest; so, $Pop - u_n$ represents the number of people who *don't* know the rumor at time n. Finally, we use b to represent the number of people who start the rumor and a to represent the constant of proportionality that governs the rate at which the rumor spreads. Our simple estimate leads to the difference equation model

Pop represents the total population capable of learning the rumor. Babies, for example, are poor rumormongers and so are not part of the population of interest.

$$u_n = u_{n-1} + a \cdot u_{n-1} \cdot (Pop - u_{n-1}) \qquad \textbf{(1.13)}$$
$$u_1 = b$$

Notice that the interpretations of a and b require that $0 < a < 1$ and $0 < b \leq Pop$.

▶ **SECOND REFLECTION**

Explain why $0 < a < 1$. How would you expect the value of a for an extremely juicy piece of gossip to compare to its value for a more mundane rumor?   ◄

Variables and Parameters.   Our model uses the variables n, u_n, Pop, b, and a. That's a lot of variables! However, each of the variables falls into one of two categories. When modeling the spread of a particular rumor in a particular population, the values of Pop, b, and a remain constant. Quantities which remain constant in a given context but vary from context to context are typically referred to as *parameters*. Normally, we reserve the word *variable* for the quantity (or quantities) which may change within a given context. Thus, in our example both u_n and n are considered to be variables.

▶ THIRD REFLECTION

Equation (1.13) can also be used to model the spread of an infectious disease. How would each of the parameters and variables be interpreted in such a model?   ◀

Exploring the Model

The difference equation we have proposed in Equation (1.13) contains three parameters, a, b, and Pop that depend on the details of a particular situation. In practice, these can be difficult to estimate. To understand the roles played by parameters, we look at some examples.

Provided the population is reasonably large, its actual numerical size is unlikely to affect the *way* a rumor spreads. So we treat u_n and the population itself as proportions and assign Pop the value 1.00. A rumor always seems to start in a small group, say 1% to 10% of the population. We will examine a quartet of scenarios in which the initial value b is taken to be 1%, 4%, 7%, and 10% of the population.

The parameter a depends on how much the population interacts and the likelihood that the rumor will be passed on during any given interaction. In the following simulations, we take a to be 0.05 which assumes that, during any given time period, the number of people who learn the rumor equals roughly 5% of all the *conceivable* interactions between people who know the rumor and those who don't. In Figures 1.11, 1.12, 1.13, and 1.14, we track the spread of the rumor over the first 200 time periods by graphing the number of people who know the rumor (u_n) as a function of time (n). To do this, we use the difference equation

$$u_n = u_{n-1} + 0.05u_{n-1}(1 - u_{n-1}), \qquad (1.14)$$

where n takes on integer values and u_n is a positive number between 0 and 1. The times are integer values and the population that knows the rumor is a positive real number between 0 and 1.

All four graphs start out fairly flat which indicates that early on the rumor spreads slowly. Then the graphs become steeper and steeper until they reach a point with second coordinate about 0.5 which indicates that the rumor spreads faster and faster until about half the population has heard it. After this point the graphs become less and less steep until they are nearly flat. Thus, after half the population has heard the rumor it spreads less and less rapidly until eventually it is hardly spreading at all (presumably because nearly everyone has already heard the story). Formally, we describe the graph as increasing throughout with a horizontal asymptote for large values of time n. The point at which the graph is steepest is called an *inflection point* or *Point of diminishing returns*. To the left of the inflection point of this graph we describe the graph as *concave up*; to the right, the graph is *concave down*.

In the MODE *dialogue box set* Graph . . . *to* SEQUENCE

In the Y= *editor enter* u1(n)=u1(n-1)+ 0.05*u1(n-1)*(1-u1(n-1)) *and set* u1 *to 0.01, 0.04, 0.07, and 0.10, respectively*

WINDOW *settings include* nmin=1, nmax=200, plotStrt=1, plotStep=1, xmin=0, xmax=200, ymin=0, ymax=1.1

You can turn the axes Labels . . . ON *after selecting* Format . . . *from the* Tools *menu in either the* Y= *or* Window *editors or while in the* GRAPH *screen itself.*

▶ FOURTH REFLECTION

Use the `Trace` feature to estimate the x- and y-coordinates of the inflection point in each of the graphs displayed. How do the y-coordinates compare? How do the x-coordinates change as b increases?   ◀

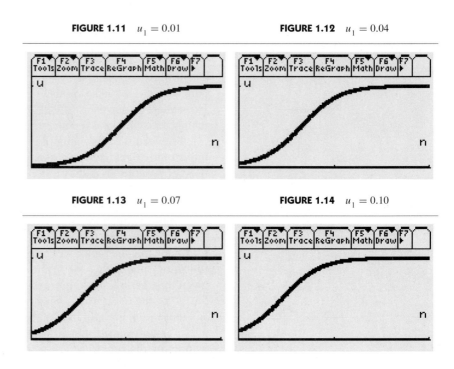

FIGURE 1.11 $u_1 = 0.01$ **FIGURE 1.12** $u_1 = 0.04$

FIGURE 1.13 $u_1 = 0.07$ **FIGURE 1.14** $u_1 = 0.10$

In recognition of their resemblance to the letter "S", these graphs are called *sigmoid* graphs. The resemblance to the letter "S" is rather weak, but the name is widely used.

By solving the equation $x = x + 0.05x(1.00 - x)$ for x we find two rest points, 0 and 1.00.

▶ FIFTH REFLECTION

Explain how the spread of the rumor to the left of the inflection point is different from its spread to the right of the inflection point. Explain how the values, 0 and 1, make sense, as rest points, in terms of the problem being modeled.   ◀

Think and Share

Let's look at how altering the parameters changes the graph of the function. Break into groups of three. In each group, have one member modify the model by changing the parameter a from 0.05 to 0.09. The other team members should change their values for a from 0.05 to 0.13, and 0.01, respectively. Each team member should obtain a representative graph of their modified model and estimate the coordinates for the point of inflection and for the time at which 90% of the population knows the rumor. According to your results, can a community effectively suppress the spread of a rumor (or disease) by slowing the transmission alone?

E X A M P L E   2   **The Logistics of Bread and Wine — the Yeast Example**

Yeast colonies are central to the production of bread, wine, and beer. Not surprisingly, the growth of yeast under carefully controlled conditions has received a great deal of study. The data in Table 1.1 from a classic study by the German biologist R. Carlson shows the weight of a colony of yeast as a function of time.

TABLE 1.1   Time vs. Weight of Yeast

Time (min)	10	20	30	40	50	60	70	80	90	100
Weight (grams)	9.6	29.0	71.1	175	351	513	594	640	656	662

Entering and Displaying the Data

You can use your calculator to display the yeast data by first entering it in a data table and then plotting it as described in the *Data/Matrix Editor* subsection of Appendix A. The plot in Figure 1.16 looks suspiciously familiar. It's sigmoid!

Figures 1.15 and 1.16 display a data table and a plot of the yeast data of the yeast data, respectively.

FIGURE 1.15   The Data Table ᵞᵉᵃˢᵗ   **FIGURE 1.16**   A Scatter Plot of the Data Table
                                                        ᵞᵉᵃˢᵗ on [0, 100] × [0, 662]

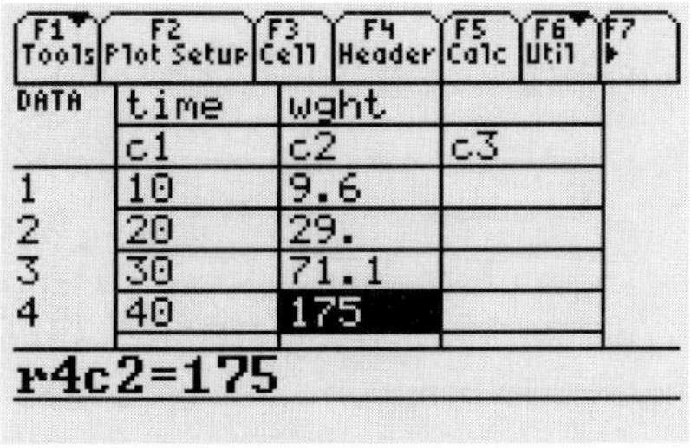

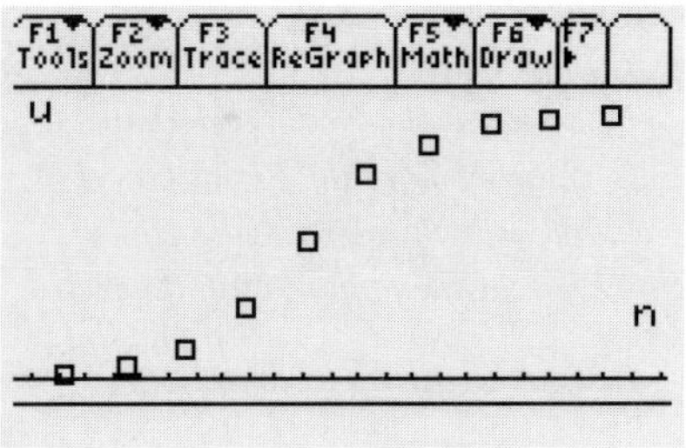

Why should the growth of a yeast colony resemble the spread of a rumor?

Building the Model.   We begin with our basic pattern:

$$future = present + change.$$

Observing that new yeast cells only come from existing yeast cells, we make the assumption that the number of new cells, or the change, is proportional to the current number of cells. Using u_{n-1} for the present weight of the colony and u_n for the future weight our pattern produces the equation

$$u_n = u_{n-1} + r \cdot u_{n-1},$$

where r is the constant of proportionality. This model should remind you of the earlier model of Novocaine levels and predicts an exponential pattern of growth with $u_n = u_1 \cdot (1 + r)^{n-1}$.

► SIXTH REFLECTION

Explain why the difference equation $u_n = u_{n-1} + r \cdot u_{n-1}$ predicts that

$$u_n = u_1 \cdot (1 + r)^{n-1}.$$

◄

We cannot accept this model, however, because it fails to predict the limiting behavior exhibited by the data. An adequate model must predict this limit. Two principal factors halt the growth of a yeast colony, even under favorable conditions. First, yeast needs sugar for growth, and only limited amounts are available to the colony during a given period of time. Second, each yeast cell excretes alcohol that poisons other organisms, including other yeast cells, competing for the available sugar. (This poisonous effect of alcohol prevents the alcohol level in wine from exceeding 18% no matter what the initial amount of sugar.) Both factors inhibit growth through competition within the colony.

Competition is a form of interaction between individuals. The details of this interaction may be hidden from us, but we should be able to estimate the *number* of interactions. In the rumor model, we needed the same sort of estimate. The spread of a rumor requires an interaction between two individuals. We reasoned that the number of times this happened was proportional to the *product* of u_{n-1} (the number of people available to share the rumor) and $Pop - u_{n-1}$ (the number available to receive the rumor). Likewise, a competitive interaction requires two individuals, where one can be thought of as an aggressor and the other can be thought of as a victim. In the case of yeast, each cell is both an aggressor and a victim. Hence, we estimate the effect of aggression as proportional to the product of the number of potential aggressors u_{n-1}, and the number of potential victims, again u_{n-1}. Taking "p" as the constant of proportionality leads us to the difference equation

$$u_n = u_{n-1} + r \cdot u_{n-1} - p \cdot u_{n-1}^2 \qquad\qquad \textbf{(1.15)}$$
$$u_1 = B.$$

Designating one yeast cell as aggressor and another as the victim is perhaps a little anthropomorphic. Besides, every cell plays both roles simultaneously. However, the metaphor is fruitful even if fanciful.

To use this difference equation as a model of the spread of rumor, the parameters B, p, and r must be positive.

► SEVENTH REFLECTION

Why is there a minus sign in front of the p in the model? We used the recursion equation $u_n = u_{n-1} + a \cdot u_{n-1}(Pop - u_{n-1})$ to model the spread of a rumor. Explain why this equation has the same structure as the recursion equation for the growth of yeast.

◄

Estimating the Parameters.

The parameter B represents the initial weight of the colony and so can be set to 9.6. We are left to determine values for p and r. A bit of algebra tells us that our recursion equation has rest values 0 and r/p. Just as in the model for the spread of a rumor, we assume these rest points correspond to limit values which can be reasonably estimated using the data in Table 1.1. From the data, we take 690 as our estimate of the limit value for the colony and conclude that $690 = r/p$ or $p = r/690$.

Solve $x = x + r * x - p * x^2$ for x.

We have reduced our problem of determining values for the parameters to the problem of determining a value for r. Under the admittedly arbitrary assumption that 10% of all the possible interactions within the yeast colony are deadly we assign the value 0.1 to r. Figure 1.17 shows the graph generated under this assumption plotted with the original data. Other choices for r may give a better fit.

Plotting the Model Against the Data — How Well Did We Do?

You can visually judge how well this model fits the data.

Break into groups of three. First, graph the data plot for yeast as described earlier. Examine the plot and the data and determine your estimate for the (asymptotically) maximum value for the weight of the yeast colony. Store your estimate as a variable *maxval*. Set the mode for Graph … to SEQUENCE. In the Y= editor, enter

Think and Share

$$\text{u2(n) = u2(n-1)+r*u2(n-1) - (r/maxval)*u2(n - 1)^2}$$

and

$$\text{ui2 = 9.6.}$$

Back in the HOME screen store 0.1 to r. In the GRAPH window, you should get an image similar to that in Figure 1.17. If your model stops too soon, go to the WINDOW menu and set the value of nmax to 100. By resetting r and ui2 to different values, find a difference equation that gives an excellent (or merely acceptable) fit to the yeast data. Reconcile these values with your partners and report one set to the class as a group.

Type your estimate on the Entry line of the HOME screen. Then hit the STO *key, type* maxval, *and hit the ENTER key.*

FIGURE 1.17   Fitting a Difference Equation to the Data of yeast

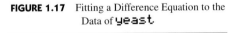

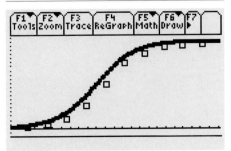

After you are done, return to the HOME *screen and clear the variables* r *and* maxval *by entering* DelVar r, maxval. *Clear the difference equation for* u2(n) *from the* Y = *editor.*

The Discrete Logistic Model.   We have generated the difference equation

$$u_n = u_{n-1} + r \cdot u_{n-1} - p \cdot u_{n-1}^2 \tag{1.16}$$
$$u_1 = B$$

with the parameters r, p, and B all positive numbers as a simple model for biological growth. Throughout the text we will refer to this difference equation as the *discrete logistic equation*. The recursion equation of the model fits the general pattern

$$future = present + change,$$

with change represented by the expression $r \cdot u_{n-1} - p \cdot u_{n-1}^2$. The first term in this expression for change models growth using the parameter r to represent the *natural rate of increase*. The second term models factors which inhibit growth. The equation has two rest values, 0 and r/p and the rest point r/p is called the *carrying capacity* of the environment.

▶ **EIGHTH REFLECTION**

What is the *carrying capacity* of a bridge? What is being carried in the expression *carrying capacity of an environment*?   ◀

E X A M P L E   3   **A Chaotic Population**
_____

Insects, amphibians, and annual plants have relatively large reproductive rates which can result in chaotic, population outbreaks. Such outbreaks are actually predicted by the logistic model. Below we examine the sequence of values predicted by increasing the parameter r to 3.

The Style Line *has been selected for these plots while in the* Y= *editor.*

$$u_n = 4 \cdot u_{n-1} - 0.005 u_{n-1}^2$$
$$u_1 = 100$$

TABLE 1.2   A Chaotic Population Generated by u_n

1	2	3	4	5	6	7	8	9	10	11	12	...
100	350	788	49	185	568	658	466	778	86	307	757	...

Figure 1.18 shows a graph of the first 50 values of the sequence.

The WINDOW *settings include*
nmin=1, nmax=50,
plotstrt=1, plotstep=1,
xmin=0, xmax=50,
ymin=0, ymax=1000.
Also, the Style Thick *has been selected for this plot while in the* Y= *editor.*

FIGURE 1.18   Graph of a Chaotic Population

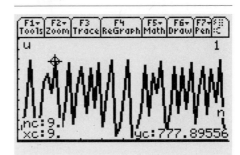

Values in the table have been rounded to represent whole individuals.

The graph does not reveal a discernable pattern and looking over a longer time scale does not help. Figure 1.19 shows a graph of the first hundred values.

The WINDOW *settings include*
nmin=1, nmax=100,
plotstrt=1, plotstep=1, xmin=0,
xmax=100, ymin=0, ymax=1000

FIGURE 1.19   Graph of a More Chaotic Population

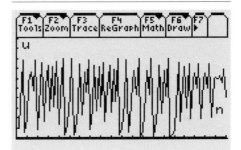

Putting It All Together

In the first three sections we created discrete models of a number of biological and sociological processes with the simple ideas of difference equations and summations. Difference equations generate sequences of values that can provide valuable information about the process being modeled. We examined three types of difference equations in detail, and observed that different behaviors can arise from changes in the parameters in even the simplest of these equations. Sequences generated by difference equations of the form

$$u_n = a \cdot u_{n-1}$$
$$u_1 = b$$

may be increasing, decreasing, constant, or alternating. They may be bounded or grow to infinitely large values depending on the specific values of a and b. By solving the equation $x = a \cdot x$ for x, we find the unique rest point 0. By examining the closed form $u_n = b \cdot a^{n-1}$, we see that the rest point is also a limit value when $b = 0$ or $|a| < 1$.

Sequences generated by

$$u_n = a \cdot u_{n-1} + b$$
$$u_1 = b$$

model limited growth and have $\frac{b}{1-a}$ as their unique rest point. Once again, by examining the closed form

$$u_n = \frac{b}{1-a} - a^n \cdot \frac{b}{1-a} \qquad \textbf{(1.17)}$$

we see that the rest point is a limit value when $b = 0$ or $|a| < 1$.

Finally, we examined the discrete logistic equation

$$u_n = u_{n-1} + r \cdot u_{n-1} - p \cdot u_{n-1}^2$$
$$u_1 = b$$

In this case we found two rest points: 0 and r/p. For some choices of the parameters, one of the two rest points is a limit value. For other choices no limit value exists. We cannot determine which sequences have limit values from a closed form because this difference equation has no closed form.

Limit values play a crucial role in understanding discrete models. When they exist, limit values predict long range behavior. Even more importantly, limit values of sequences and the associated concept of limits of functions provide a crucial link between the continuous and the discrete. In the next section, we will look more closely at limit values to begin to understand that link.

▶ **NINTH REFLECTION**

All of the recursion equations for the three models we've studied are of the form $u_n = F(u_{n-1})$. Other than the names of parameters, how is the function F for the discrete logistic equation different from the functions for the other two? ◀

EXERCISE SET 1.3

1. Explain why the rest values $x = 0$ and $x = 1.00$ make sense for our model of the spread of a rumor.

2. A small town high in the mountains of North Carolina has a population of 50,000 people. One day an itinerant mathematician arrives and announces to a colloquium of 50 people that she can prove mathematically that the earth will cease to exist within five years. Suppose that in any one daytime period roughly 5% of all conceivable interactions between people in the town result in the mathematician's startling announcement being passed on. Approximately how many days will pass before half the town has heard about the predicted end of the world?

3. The model for the spread of a rumor through a population could be interpreted as describing the spread of some diseases. What are the characteristics of a disease that might make this model applicable? Give two examples of such diseases.

4. A calculus student came to class one day with a very bad cold. The following data table contains the number of days since the introduction of the cold and the number of students ill on each day. Eventually, they all caught the cold.

TABLE 1.3   Day vs. Number Ill

Day	1	2	3	4	5	6	7	8	9
Number Ill	1	2	4	6	9	12	15	15	18

a. Obtain and describe the shape of a data plot for this scenario. Explain why you might have expected this shape.

b. Find a difference equation that gives a reasonable fit for the data. If this is not possible, explain why it is impossible.

5. In this exercise, you will be asked to compare growth models.

   a. Explain why an exponential growth model of the form used in the Novocaine example would be inappropriate for data collected from the spread of the rumor or disease.

   b. On the other hand, would an exponential growth model be appropriate for the spread of a highly contagious and deadly disease like the Black Plague? Explain.

6. Show (and explain the relevant computations in a form that another student might be willing to read) that $\frac{r}{p}$ is a rest value for

$$u_n = u_{n-1} + r \cdot u_{n-1} - p \cdot u_{n-1}^2$$
$$u_1 = B.$$

7. A population (measured annually) of white-tailed deer in a confined habitat is modeled by

$$u_n = 1.15u_{n-1} - 0.0005u_{n-1}^2$$
$$u_1 = C.$$

   a. Verify that the carrying capacity implied by the model is 300 individuals.

   b. Describe how the different initial population sizes affect the long-term size of the herd by observing the graphs of the size of the herd over the first 50 years if the initial population is 500, 400, 300, 200, 100, or 50.

8. Suppose that hunting is allowed in the habitat of the deer herd in the preceding exercise, and that 10 % of the deer are taken yearly in the hunt.

   a. Give a difference equation that models this new situation.

   b. Determine the rest values of your difference equation.

   c. Describe how different initial population sizes affect the long-term size of the herd by observing the graphs of the size of the herd over the first 50 years if the initial population is 500, 400, 300, 200, 100, or 50.

9. Suppose that hunting is allowed in the habitat of the deer herd in the preceding exercise, and that 10 deer (instead of 10% of the deer) are taken yearly in the hunt.

   a. Give a difference equation that models this new situation.

   b. Determine the stable values of your difference equation.

   c. Describe how the different initial population sizes affect the long-term size of the herd by observing the graphs of the size of the herd over the first 50 years if the initial polulation is 500, 400, 300, 200, 100, or 50.

10. In setting the allowable take of deer, the State Department of Natural Resources may try to set the harvest at a number known as the *maximum sustainable harvest*. That is, the largest harvest size that does not eventually eliminate the herd is chosen. Determine the maximum sustainable harvest for a deer herd whose population is modeled by the following difference equation:

$$u_n = 1.15u_{n-1} - 0.0005u_{n-1}^2$$
$$u_1 = 500.$$

11. Observe the first 25 terms of the sequence generated by

$$u_n = 3.5u_{n-1} - b \cdot u_{n-1}^2$$
$$u_1 = 100$$

    for b equal to 0.001, 0.008, and 0.01. How do these sequences compare to the sequence of Example 3?

12. *Group Exercise*

    a. Show (and explain the computations in a way that another student might be willing to read) that the carrying capacities for the recursion formulas

$$u_n = 1.5u_{n-1} - 0.005u_{n-1}^2$$
$$u_n = 3.5u_{n-1} - 0.005u_{n-1}^2$$
$$u_n = 4u_{n-1} - 0.005u_{n-1}^2$$

    are 100, 500, and 600, respectively.

    b. On one set of axes, graph both of the difference equations with recursion formula

$$u_n = 1.5u_{n-1} - 0.005u_{n-1}^2$$

    but with the initial conditions $u_1 = 100$ and $u_1 = 101$.

    c. On one set of axes, graph both of the difference equations with recursion formula

$$u_n = 3.5u_{n-1} - 0.005u_{n-1}^2$$

    but with the initial conditions $u_1 = 500$ and $u_1 = 501$.

    d. On one set of axes, graph both of the difference equations with recursion formula

$$u_n = 4u_{n-1} - 0.005u_{n-1}^2$$

but with the initial conditions $u_1 = 600$ and $u_1 = 601$.

 e. If you were to use models based on each of the three recursion formulas in (a) to determine future populations, how accurately must you determine the initial population in each case?

13. Apply the recursion formula

$$u_n = 3.5u_{n-1} - 0.005u_{n-1}^2,$$

twice to obtain a formula for u_{n+1} in terms of u_{n-1}. Solve the resulting equation for those values for which $u_{n-1} = u_{n+1}$.

14. Repeat the analysis of the preceding exercise with the recursion formula

$$u_n = 1.5u_{n-1} - 0.005u_{n-1}^2.$$

To find all real and complex solutions use the command `csolve(`.

15. *Group Exercise* Repeat the analysis of the preceding exercise with the recursion formula

$$u_n = r \cdot u_{n-1} - 0.005u_{n-1}^2.$$

For what range of values of r might you expect an alternating sequence to emerge? Test your hypothesis by tracking the sequences generated by fivedifferent choices for r. From your examples, discuss how quickly, if at all, the alternating pattern emerges.

1.4

Investigating Limits

In the puppet play[†] that unfolds below, Zeno misconstrues some of the statements spoken by the Pupil. For example, Zeno hears "line in the street" instead of "lion in the street."

Dramatis personae: Zeno, Pupil, Lion

Scene: The school of Zeno at Elea.

Pupil. Master! There is a lion in the streets!

Zeno. Very good. You have learned your lesson in geography well. The fifteenth meridian, as measured from Greenwich, coincides with the high road from the Temple of Poseidon to the Agora – but you must not forget that it is an imaginary line.

Pupil. Oh no, Master! I must humbly disagree. It is a *real* lion, a *menagerie* lion, and it is coming toward the school!

Zeno. My boy, in spite of your proficiency at geography, which is commendable in its way – albeit essentially the art of the surveyor and hence separated by the hair of the theodolite from the craft of a slave – you are deficient in philosophy. That which is real cannot be imaginary, and that which is imaginary cannot be real. Being is, and non-being is not, as my revered teacher Parmenides demonstrated first, last, and continually, and as I have attempted to convey to you.

Pupil. Forgive me, Master. In my haste and excitement, themselves expressions of passion unworthy of you and of our school, I have spoken obscurely. Into the gulf between the thought and the word, which, as you have taught us,

[†]Adapted with apologies from *Zeno's Paradoxes*. Edited by Wesley C. Salmon, pp. 1-3, Bobbs-Merrill Company, Inc.

is the trap set by non-being, I have again fallen. What I meant to say is that a lion has escaped from the zoo, and with deliberate speed it is rushing in the direction of the school and soon will be here!

Zeno. O my boy, my boy! It pains me to contemplate the impenetrability of the human intellect and its incommensurability with the truth. Furthermore, I now recognize that a thirty-year novitiate is too brief – *sub specie aeternitatis* – and must be extended to forty years, before the apprenticeship proper can begin. A real lion, perhaps; but really arriving here, absurd!

Pupil. Master...

Zeno. In order to run from the zoological garden to the Eleatic school, the lion would first have to traverse half the distance.

The lion traverses half the distance.

Zeno. To complete his journey, the lion would then have to traverse half of the remaining distance. Having done so, once again the lion would find himself faced with the task of completing half of the remaining distance. And so no matter how long the lion runs always he will find himself with half of the previous distance yet to be traversed.

The lion bursts into the schoolyard.

Pupil. O Master, run, run! He is upon us!

Zeno. And thus, we have proved that the lion could never complete the journey from the zoological garden to here, the mere fantasy of which has so unworthily filled you with panic.

The pupil climbs an Ionic column, while the lion devours Zeno.

Pupil. My mind is in a daze. Could there be a flaw in the Master's argument?

Do not be misled by the apparent simplicity of Zeno's reasoning. His argument (one of a number of similar arguments known collectively as *Zeno's Paradoxes*) points to profound difficulties in our understanding of time and reality. We cannot attempt to address those difficulties here (philosophy is hard); but, we can examine the underlying mathematical construct.

Guessing Limits of Sequences

Did you recognize the sequences in Zeno's description of the lion's trip? Several sequences can be constructed. For example, consider the sequence whose nth term represents the portion of the trip still to be completed by the lion at the nth stage. For this sequence, $u_1 = 1$ since the lion starts out with the entire trip still to be completed. Also, if u_{n-1} is the portion left for the lion at the $(n-1)$st stage then half of that proportion will be left at the next stage. Hence, u_n will be half of u_{n-1}. In the language of difference equations, we represent the sequence as

$$u_n = \frac{1}{2} u_{n-1}$$
$$u_1 = 1$$

where u_n represents the portion of the total trip still to be traversed by the lion at the nth stage. (Notice the similarity between this sequence and the one we used for metabolizing Novocaine).

To examine this sequence numerically we construct the table of values shown in Table 1.4

TABLE 1.4 n vs. u_n

n	1	2	3	4	5	6	7	8	9	10
u_n	1	0.5	0.25	0.125	0.0625	0.03125	0.015625	0.007812	0.003906	0.001953

The table suggests that we can make the value of the sequence as close to 0 as we like simply be choosing a large enough value for n. The graph of the data in Table 1.20 supports this notion.

FIGURE 1.20   Graph of $u_n = \frac{1}{2}u_{n-1}$

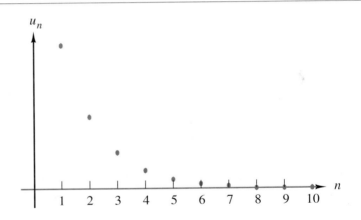

Moving to the right on the graph we see that the y-coordinates are closer to the horizontal axis which corresponds to the values of the sequence getting closer to 0. Based on this evidence we guess that 0 is the limit value of the sequence. Written in mathematical notation our guess becomes $\lim_{n\to\infty} u_n = 0$ which is read "the limit of u_n as n goes to infinity equals 0".

Be careful! Tables and graphs provide evidence for a limit value but they are not conclusive. Adding 0.000001 to every value of the sequence would change the limit value without changing the graph above.

▶ FIRST REFLECTION

How would the limit value change if 0.000001 is added to every term of the sequence? How would the appearance of the graph change?   ◀

We can construct a second sequence to represent the part of the trip the lion has already completed. Letting d_n represent the proportion of the trip completed by the lion at the nth stage, we represent the sequence with the difference equation

$$d_n = d_{n-1} + \frac{1}{2}(1 - d_{n-1})$$
$$= \frac{1}{2}d_{n-1} + \frac{1}{2}$$
$$d_1 = 0$$

This difference equation resembles the limited growth model we used for repeated drug dosages.

▶ **SECOND REFLECTION**

Explain why $d_1 = 0$. Interpret $1 - d_{n-1}$ in terms of the lion's trip. Create a table of values and a graph for this sequence. Guess the limit value for this sequence. ◀

Tables and graphs provide useful information regarding limit values but their accuracy is inherently imprecise. The entries in a table are only accurate to the finite number of decimal places displayed and graphs generated by technology cannot be more precise than the resolution of the screen on which they appear. Furthermore, we can only tabulate or graph a finite number of the terms in the sequence. Even if we graph the first million terms how can we be sure that the next million terms follow the same pattern? We can improve our understanding of sequences by examining closed forms.

Our first sequence has closed form $u_n = \frac{1}{2}^{n-1}$, (determined just as for the Novocaine example). From this form we see that u_n must be positive for any n. Furthermore, the value of u_n will be as close as we like to 0 as long as n is sufficiently large. For example, suppose we would like the value of u_n to be as close as 0.01 to 0. We need only insist that n be bigger than 8 because $\frac{1}{2}^{8-1}$ is less than 0.01 and any larger value of n will result in a u_n value even closer to 0.

The TI calculator computes limit values from closed forms. For this example, enter

`limit(1/2^(n-1),n,∞).`

▶ **THIRD REFLECTION**

How large would n have to be if we would like the value of u_n to be as close as 0.001 to 0. (Note that because the terms are positive we only require that $u_n < 0.001$). ◀

We can use our analysis of the closed form for our sequence as the basis for a definition of limit value.

LIMIT OF A SEQUENCE

L is called the limit value of the sequence u_n if the values of the sequence are as close as we like to L for all n sufficiently large. In mathematical notation, we write $\lim_{n \to \infty} u_n = L$ which is read "the limit of u_n as n goes to infinity equals L."

The definition still contains some ambiguity. The phrase "as close as we like" lacks precision as does the phrase "sufficiently large". Note the use of "for all" in the definition. This condition must be included. Consider the following sequence:

$$1, \frac{1}{2}, 1, \frac{1}{3}, 1, \frac{1}{4}, 1, \frac{1}{5}, \dots \tag{1.18}$$

We would not want to claim 0 as a limit for this sequence since, no matter how large a value we choose for n, there will be terms with higher indices that have value 1. Thus, we cannot force *all* the terms with sufficiently large n to be within $\frac{1}{2}$, say, of 0.

▶ **FOURTH REFLECTION**
Verify that $d_n = 1 - \frac{1}{2}^{n-1}$ is a closed form for our second sequence d_n by computing some of the initial terms. Does the closed form appear to justify your previous guess for the limit value? Explain in terms of the definition.   ◀

Rest Points and Limit Values

Recall that at rest points the sequence maintains the same value for successive terms. Stated algebraically: $u_n = u_{n-1}$. Thus, to find the rest point(s), we replace u_{n-1} with u_n in the recursion equation and solve for u_n. For our first example the recursion equation $u_n = \frac{1}{2}u_{n-1}$ becomes $u_n = \frac{1}{2}u_n$ which has $u_n = 0$ as its only solution. For the second example, we solve $d_n = \frac{1}{2}d_n + \frac{1}{2}$ to obtain the unique solution $d_n = 1$. In both cases, the rest points are also limit values. Although the two concepts of rest points and limit values are related, they are not the same.

To find rest points for the general difference equation

$$u_n = F(u_{n-1})$$
$$u_1 = b.$$

we simply solve the equation $u_n = F(u_n)$. Thus, the value of b has no effect at all on the value of the rest points. In other words, rest points depend on the recursion equation but are independent of the initial condition. Limit values on the other hand depend on both the recursion equation and the initial condition. To understand this distinction, we consider a particular example of the discrete logistic equation.

E X A M P L E   1   **Rest Points Are Not Necessarily Limit Values**

Consider the difference equation

$$u_n = 2u_{n-1} - u_{n-1}^2$$
$$u_1 = b$$

Solving the equation $u_n = 2u_n - u_n^2$ for u_n we find two rest points: 0 and 1. Notice that these values are determined without specifying a value for b. Whether or not these rest points are also limit values depends on the value of b. We consider the question: "For what values of b will a given rest point also be a limit point? ".

We can answer the question easily if the sequence starts out at a rest point. For our example, if $b = 0$, then $u_1 = 0, u_2 = 0, ..., u_n = 0, ...$ Every term in the sequence equals 0, so 0 is a limit value of the difference equation for this initial condition. Similarly, if $b = 1$, then 1 is a limit value for the sequence. More generally, any rest point will be the limit value of a sequence if the rest point is taken as the initial value of the sequence.

What happens if we choose other values for b? The logistic equation does not have a closed form to help us answer this question. Instead we employ trial and error. To begin with, we consider initial values of b between our two rest points. Table 1.5 shows the first eight terms of the sequences with initial values 0.1, 0.5, 0.8, 0.9, and 0.99.

TABLE 1.5 n vs. u_n for Initial Values Less Than One

n \ Initial Value	0.1	0.5	0.8	0.9	0.99
1	0.1	0.5	0.8	0.9	0.99
2	0.19	0.75	0.96	0.99	0.9999
3	0.3439	0.9375	0.9984	0.9999	1.00000
4	0.56953	0.99609	1.00000	1.00000	1.00000
5	0.81470	0.99998	1.00000	1.00000	1.00000
6	0.96556	1.00000	1.00000	1.00000	1.00000
7	0.99882	1.00000	1.00000	1.00000	1.00000
8	1.00000	1.00000	1.00000	1.00000	1.00000

All of these sequences appear to have limit value 1. On the basis of this information, we guess that the sequence generated by our difference equation has limit value 1 when $0 < b < 1$. This implies 0 is not the limit value for the sequence even if b is a very small (as small as you like) positive number.

Note that this implies that 0 is not a limit value no matter how small a positive value we choose for b.

► **FIFTH REFLECTION**

Generate the sequence with initial value $b = .001$. How many terms must you take before your technology shows the sequence values to be 1?   ◄

What if we choose b to be bigger than 1. Table 1.6 shows the first eight terms of the sequences with initial values 1.1, 1.5, 1.9, 2, and 2.5.

TABLE 1.6 n vs. u_n for Initial Values Greater Than One

n \ initial value	1.1	1.5	1.9	2	2.5
1	1.1	1.5	1.9	2	2.5
2	0.99	0.75	0.19	0	-1.25
3	0.9999	0.9375	0.3439	0	-4.0625
4	1.00000	0.99609	0.56953	0	-24.6289
5	1.00000	0.99999	0.81498	0	-655.841
6	1.00000	1.00000	0.96566	0	$-431439.$
7	1.00000	1.00000	0.99882	0	-1.86140×10^{11}
8	1.00000	1.00000	0.99999	0	-3.46482×10^{22}

The first three sequences behave nicely. They start above 1 but immediately jump to a value between 0 and 1 and then approach 1 as in the previous case. What about the other two sequences? When $b = 2$, the second term of the sequence jumps to 0, the other rest point and stays there (of course); so, 0 is the limit value. When $b = 2.5$ the sequence jumps clear past 0 into negative numbers and then just explodes into larger and larger negative numbers. This last sequence does not have a limit value.

Based on this information we make the following conjecture (guess) regarding limit values. The value 0 is a limit value when $b = 0$ or $b = 2$; the value 1 is a limit value when $0 < b < 2$; and there is no limit value when $b < 0$ or $b > 2$.

► **SIXTH REFLECTION**

Generate the sequences with initial values $b = -0.5$ and $b = 2.1$. Do these sequences support the conjecture? Explain which part of the conjecture applies to these sequences.   ◄

We have not yet developed the tools to justify our conjecture. That justification will have to wait until after we investigate first differences in Chapter Two.

We have seen that rest points are not always limit values. Are limit values always rest points? So far, every limit value we have discovered has also been a rest point. For a large class of important difference equations this will continue to be the case. Nevertheless, in general limit values do not have to be rest points

EXAMPLE 2 **Proving a Limit**

Consider the following difference equation which might be used to model an oscillating electrical phenomena which is dampening over time

$$u_n = \frac{\cos(u_{n-1})}{n^2}$$
$$u_1 = 1$$

Taking our standard approach to finding the limit value, we compute initial terms of the sequence which are shown in Table 1.7.

TABLE 1.7 n vs. u_n

n	1	2	3	4	5	6	7	8	9	10
u_n	1	0.1351	0.1101	0.0621	0.0399	0.0278	0.0204	0.015628	0.0123	0.0100

From the table we might conjecture (guess) that the sequence has limit value 0. Given the meager amount of data, however, we are not particularly confident about our guess. Fortunately, we can make an argument to justify our choice of 0 for the limit value.

We know that values of the cosine function are always between -1 and 1. Using this fact, we see from the recursion equation that u_n must lie between $\frac{-1}{n^2}$ and $\frac{1}{n^2}$. These numbers can be made as close to 0 as we like simply by choosing a large enough value for n. For example, if we want to guarantee that u_n is as close as 0.01 we need only insist that n be bigger than 10. We conclude that 0 is indeed the limit value for this sequence. Is 0 a rest point?

Suppose that $u_{n-1} = 0$. Computing u_n from the recursion equation we obtain: $u_n = \frac{\cos(0)}{n^2} = \frac{1}{n^2}$. No matter what value n may have $\frac{1}{n^2}$ does not equal 0; so, $u_n \neq u_{n-1}$ and 0 is not a rest point.

▶ SEVENTH REFLECTION

Use the sharpened argument of Example 2 to prove $\lim\limits_{n \to \infty} \dfrac{8}{n^3} = 0$. ◀

Properties of Limits of Sequences

We can use the definition of the limit of a sequence to show that the following properties hold.

Suppose that c is a real number and $\lim\limits_{n\to\infty} a_n$ and $\lim\limits_{n\to\infty} b_n$ both exist.

(1) $\lim\limits_{n\to\infty} (a_n + b_n) = \lim\limits_{n\to\infty} a_n + \lim\limits_{n\to\infty} b_n$

(2) $\lim\limits_{n\to\infty} (c \cdot a_n) = c \cdot \lim\limits_{n\to\infty} a_n$

(3) $\lim\limits_{n\to\infty} (a_n \cdot b_n) = \lim\limits_{n\to\infty} a_n \cdot \lim\limits_{n\to\infty} b_n$

(4) $\lim\limits_{n\to\infty} \dfrac{a_n}{b_n} = \dfrac{\lim\limits_{n\to\infty} a_n}{\lim\limits_{n\to\infty} b_n}$ if $\lim\limits_{n\to\infty} b_n \neq 0$

These properties allow us to determine limits of new sequences from already familiar sequences as in the next example.

E X A M P L E   3   **A Proof Using the Properties**

Determine $\lim\limits_{n\to\infty} \dfrac{1}{n^2}$. By property (3)

$$\lim_{n\to\infty} \frac{1}{n^2} = \lim_{n\to\infty} \frac{1}{n} \cdot \lim_{n\to\infty} \frac{1}{n} = 0 \cdot 0 = 0.$$

Sharpening the Argument [Optional].   Our argument above that 0 is the limit of $u_n = \dfrac{\cos(u_{n-1})}{n^2}$ in Example 2 lacks mathematical precision. We have demonstrated that we can get the values of the sequence as close as 0.01 to 0 but we have not really shown that we can duplicate the feat for even smaller values. Strictly speaking, we have to show we can make the terms as *small as we like* no matter how *small* that may be. Consequently, instead of using a specific number, we use a variable to represent the required smallness. Traditionally we use the Greek letter ϵ for this purpose.

So, suppose we would like to guarantee that the terms of the sequence all are as close as ϵ to 0. We will assume that ϵ is positive since negating a number does not make it any closer to 0. As before u_n must lie between $\dfrac{-1}{n^2}$ and $\dfrac{1}{n^2}$ and we need only guarantee that these numbers are as close as ϵ to 0. In other words, we want to guarantee that $\dfrac{1}{n^2} < \epsilon$ for large enough values of n. Equivalently, we want $\sqrt{\frac{1}{\epsilon}} < n$ for large enough values of n. The last inequality follows from the unboundedness of the integers. No matter how large $\sqrt{\frac{1}{\epsilon}}$ might be, we can always find an integer N that is bigger. All the terms in the sequence with subscript n greater than that integer N will be closer than ϵ to 0. We have strengthened the argument by not relying on a particular value for ϵ. We formulate the following revised definition for the limit of a sequence to incorporate our use of ϵ and N.

LIMIT OF A SEQUENCE - REVISED

L is called the limit value of the sequence u_n if given an ϵ greater than 0 there exists an integer N such that $|u_n - L| < \epsilon$ for all $n > N$.

As above, the variable ϵ represents how close we would like the sequence to be to the limit value. The inequality $|u_n - L| < \epsilon$ provides a mathematical expression of the idea that u_n is as close as ϵ to L.

Summary

The concept of limit plays a pivotal role in calculus. Many key ideas find their clearest expression in the language of limits. Our discussion has focused on limits of sequences but the limit concept can be used in a variety of contexts beyond sequences. In the next section, we will use limits to introduce the concept of continuity.

EXERCISE SET 1.4

In Exercises 1–8
   a. Examine each sequence numerically by constructing a table of values. (Use the Table feature of your calculator to do this.) Guess a limit value for each sequence or explain why you think the sequence has no limit value.
   b. Find the rest points of each sequence to determine if the limit value is also a rest point.

1.
$$u_n = \frac{1}{3}u_{n-1}$$
$$u_1 = 1$$

2.
$$u_n = \frac{1}{3}u_{n-1} + 12$$
$$u_1 = 5$$

3.
$$u_n = 1.5u_{n-1}$$
$$u_1 = 2$$

4.
$$u_n = 7 - 0.4u_{n-1}$$
$$u_1 = 0$$

5.
$$u_n = 7 - 0.4u_{n-1}$$
$$u_1 = 10$$

6.
$$u_n = 3.5u_{n-1} - u_{n-1}^2$$
$$u_1 = 1$$

7.
$$u_n = 3.5u_{n-1} - u_{n-1}^2$$
$$u_1 = 3$$

8.
$$u_n = 3.5u_{n-1} - 0.5u_{n-1}^2$$
$$u_1 = 1$$

In Exercises 9–12, conjecture values of b so that a given rest point will also be a limit value for the sequence.

9.
$$u_n = 3.5u_{n-1} - 0.5u_{n-1}^2$$
$$u_1 = b$$

10.
$$u_n = 5u_{n-1} - u_{n-1}^2$$
$$u_1 = b$$

11.
$$u_n = 7u_{n-1} - u_{n-1}^2$$
$$u_1 = b$$

12.
$$u_n = 1.5u_{n-1} - 2u_{n-1}^2$$
$$u_1 = b$$

13. Explain the meaning of rest point and limit value. Can a rest point be a limit value? Why or why not?

14. Does a rest point depend on the value of u_1 in any way? Explain.

15. Do we need to know a value for u_1 in order to determine a limit value for a sequence? Explain.

16. Consider the recursion formula

$$u_n = u_{n-1}(4 - u_{n-1}) - 2.$$

The rest points for this formula are 1 and 2. In this exercise you will be investigating sequences which begin with values close to one of these two rest points. Observe the first seven (or as many as necessary) terms of the sequence generated by this recursion formula for the following initial conditions (You need not record these values). For each initial condition use your observations to guess a limit value.

a. $u_1 = 0$
b. $u_1 = 0.5$
c. $u_1 = 0.9$
d. $u_1 = 0.99$
e. $u_1 = 1$
f. $u_1 = 1.1$
g. $u_1 = 1.5$
h. $u_1 = 1.9$
i. $u_1 = 2$
j. $u_1 = 2.1$

Based on your observations, which initial conditions result in a sequence whose limit value is 1? Which initial conditions result in a sequence whose limit value is 2? Which initial conditions result in a sequence with no limit value?

In Exercises 17–20, let $a_n = 1 + \frac{1}{3}^n$, $b_n = 2 - \frac{1}{5}^n$, and $c_n = -\frac{1}{n^2}$.

k. Find $\lim_{n\to\infty} (a_n + b_n)$.
l. Find $\lim_{n\to\infty} (a_n \cdot b_n)$.
m. Find $\lim_{n\to\infty} (b_n - c_n)$.
n. Find $\lim_{n\to\infty} (5 \cdot c_n)$.

In Exercises 21 and 22, use an ϵ and N argument to verify the given limit.

o. $\lim_{n\to\infty} \frac{2}{n^2} = 0$
p. $\lim_{n\to\infty} 3 - \frac{1}{n} = 3$

17. Consider the difference equation

$$u_n = \frac{\cos(u_{n-1})}{n^3}$$

$$u_1 = 0.5$$

Observe the first seven terms of the sequence. Based on your observations, guess a limit value for the sequence. In the recursion equation replace u_{n-1} with your limit value and compute u_n. Does the value you compute for u_n equal the limit value exactly? Explain why your limit value is or is not a rest point.

1.5

Continuous and Piecewise Continuous Models

Mathematical models enable us to investigate reality by analyzing mathematical representations with simple numerical tools. Discrete models fit naturally with discrete data and can be analyzed via the operations of addition and subtraction. Unfortunately, such models may miss important features of the process under study. As a result, critical questions may be impossible to answer with the discrete model alone. In such situations, we often try to improve the fit by replacing the discrete model with a continuous one.

A Continuous Model: Painless Dentistry.   Recall the discrete model for a dental patient's Novocaine level given by the closed form formula:

$$u_n = 0.8^{n-1} \cdot 500.$$

This model predicts the level of Novocaine at hourly intervals following the injection but tells us nothing about the level between the hourly observations. In Section 1.1, we made a leap of faith to create a function which predicts the level of Novocaine at all times by replacing $n - 1$ with the variable t to obtain,

$$N(t) = 0.8^t \cdot 500,$$

Be careful. $N(t)$ is not the only continuous function which satisfies the stated conditions. We could obtain a different continuous model by linear extrapolation, simply drawing line segments between each of the points of the discrete model.

where $N(t)$ is the level of Novocaine t hours after the injection. This model agrees with the discrete model for integer values of t. It also supplies Novocaine levels at any time, not just at hourly intervals, and predicts a gradual decrease in the level of Novocaine without any sudden drops. We can look at the graph of $N(t)$ in Figure 1.21 for evidence of the last condition since a sudden drop in the Novocaine level would correspond to a break in the graph.

Intuitively, a *continuous* function is one whose graph has no breaks or holes (we will make this idea more precise shortly). From our graph, $N(t)$ appears to be a continuous function. On the other hand, the calculator graph actually consists of a finite number of pixels which cannot reveal breaks or holes below the resolution of the screen.

FIGURE 1.21 Graph of $N(t)$ on $[0, 24] \times [0, 500]$

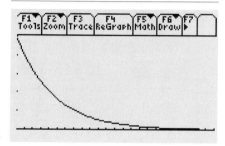

► **FIRST REFLECTION**

Create a graph of $N(t)$ similar to the one shown. About how many pixels are used in your graph? What is the approximate time gap between each pixel? (You should be able to answer both questions by tracing the graph: the first question by counting the number of times you press the right cursor key to trace from one end of the graph to the other; the second by observing the change in xc as you trace.) Answers will vary depending on WINDOW settings. ◄

No matter how small the resolution of our screen, the graph of $N(t)$ will consist of a finite number of points. In other words, calculators and computers always generate discrete representations of the function being graphed. The continuity of $N(t)$ must follow from its formula, not from the graph. Unfortunately, our notion of continuity refers to the graph, not the formula. For functions defined algebraically by a formula, we require an algebraic definition of continuity. Before proposing such a definition, let's take a closer look at the idea of continuity.

One Equation – Two Models. Starting with the closed form for a discrete model we can always create a continuous model by simply replacing the index variable n with the real variable t.

EXAMPLE 1 **The Lion Model**

Recall the closed form $d_n = 1 - \left(\frac{1}{2}\right)^{n-1}$ for the proportion of a trip completed by a lion at stage n. This discrete model provides the lion's progress at various stages but fails to tell of the lion's location in between those stages. The lion does not move by disappearing from one spot and then suddenly appearing at another, but travels continuously, so we should be able to represent his trip with a continuous model. Replacing n with the variable x we create the continuous model $D(x) = 1 - \left(\frac{1}{2}\right)^x$.

► **SECOND REFLECTION**

Compare a graph of the sequence d_n and with a graph of the function D. How are they related? Does the function D appear to be continuous? Does D seem to be a reasonable model for the lion's trip? Discuss whether or not the function D supports or refutes Zeno's argument.   ◄

As in the previous example, the function D agrees with the sequence (d_n) for integer values of x (algebraically, $D(n) = d_{n+1}$ if n is an integer), predicts the proportion of the trip traveled by the lion in between the stages, and exhibits a gradual increase in the proportion of the trip the lion has completed. One difficulty with this model involves the interpretation of the variable x. If x represents time, then according to D each stage of the trip requires 1 time unit. On the other hand, the later stages of the trip, which are much shorter than the earlier stages, should take much less time to be completed (unless the lion runs at slower and slower speeds).

Notice that d_n falls into the class of limited growth equations of the form

$$u_n = c - c \cdot a^n \tag{1.19}$$

► **THIRD REFLECTION**

Equation (1.12) in *Putting it All Together* in Section 1.2 on limited growth can be expanded as follows:

$$u_n = b \cdot \left(\frac{1 - a^n}{1 - a}\right) = b \cdot \left(\frac{1}{1 - a} - \frac{a^n}{1 - a}\right).$$

Find the value of c in Equation (1.19) in terms of a and b.   ◄

E X A M P L E   2   **Amoxicillin Revisited**

The closed form of the model for repeated doses of amoxicillin

$$u_n = 500 \left(\frac{1 - (0.0625)^n}{1 - 0.0625}\right) \tag{1.20}$$

also falls into the class of limited growth equations.

If we replace the index n with the variable t, we obtain the function

$$A(t) = 500 \left(\frac{1 - (0.0625)^t}{1 - 0.0625}\right). \tag{1.21}$$

FIGURE 1.22   Graph of $A(t)$ on $[0, 24] \times [0, 600]$

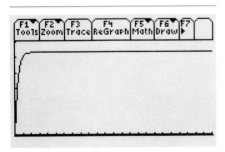

From the graph in Figure 1.22, A does appear to be a continuous function.

Does A provide a reasonable model of the level of amoxicillin?

Recall that u_n from the discrete model represents the level of amoxicillin in the patient's system immediately *after* taking the nth dose. Administering this dose results in a jump in the level of the drug. After the jump, however, the action of the kidney gradually decreases this level. Although continuous, A fails miserably to properly represent the level of amoxicillin. Notice that A is always increasing while the level of amoxicillin should be decreasing at all times except when a new dose is taken.

▶ FOURTH REFLECTION

For what values of t does $A(t)$ accurately predict the level of amoxicillin? Draw a rough sketch to describe the actual level of amoxicillin as explained in the previous paragraph.　　　　　　　　　　　　　　◀

Each dose causes a jump in the level of amoxicillin. Such jumps are called *jump discontinuities* and cannot be represented by a continous function. We can investigate the first jump discontinuity by finding a function to model the level of amoxicillin for the first 8 hours.

Recall that half of the amoxicillin is eliminated every hour. For the first four hours, before the patient takes the second dose, we can model the level of amoxicillin in the same way as we modeled the level of Novocaine. In function notation, we write

$$h(t) = 500(0.5)^t \quad \text{when} \quad 0 \le t < 4.$$

To define h on the TI calculator use the when *command:*
Define h(t)=
when(t<4,500(0.5)^t,
531.25(0.5)^(t-4))

At the end of the first four hours, the patient takes the second dose and the amoxicillin level reaches 531.25 mg. (as we saw in section 1). Once again the liver kicks in to eliminate the drug and we can model the next four hours with the equation

$$h(t) = 531.25(0.5)^{t-4} \quad \text{when} \quad 4 \le t < 8.$$

Notice that we must subtract 4 from t in the exponent (Why?). We can write a formula for the level of amoxicillin during the first 8 hours by putting our two equations together.

TABLE 1.8　Table of Values for $h(t)$ centered at 4

t	$h(t)$
3.	62.500
3.9	33.493
3.99	31.467
3.999	31.272
3.9999	31.252
4.0	**531.25**
4.0001	531.21
4.001	530.88
4.01	527.58
4.1	495.67
5.0	265.63

$$h(t) = \begin{cases} 500(0.5)^t & \text{when} \quad 0 \le t < 4 \\ 531.25(0.5)^{t-4} & \text{when} \quad 4 \le t < 8 \end{cases}$$

Figure 1.23 shows the graph of h.

We can easily recognize the jump discontinuity from the graph. Notice that the TI calculator has drawn a nearly vertical line connecting the two pieces of the graph even though no such line should exist! To investigate the discontinuity numerically, we generate tables of function values for input values close to 4 in Table 1.8.

The values of h for inputs less than 4 are very different from the ones for inputs greater than 4. When t is less than 4, the sequence of function values appear to have a limit value close to 31.25.

We express this by saying "the limit of $h(t)$, as t approaches 4 from the left, is 31.25." We write this as $\lim_{t \to 4^-} h(t) = 31.25$. On the other hand, when t is greater than 4 the sequence of function values appear to have limit value

FIGURE 1.23   Graph of $h(t)$ on $[0, 8] \times [0, 600]$

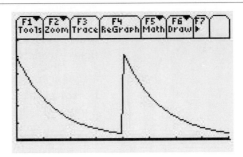

close to 531.25. We say this time that the limit of $h(t)$, as t approaches 4 from the right, is 531.25. We write $\lim_{t \to 4^+} h(t) = 531.25$.

▶ **FIFTH REFLECTION**

Graph $f(x) = \sqrt{x^2 - 25}$ and $g(x) = \sqrt{\sin(x)}$ on your TI calculator. Each function is defined by a single expression but has several discontinuities. Is each function discontinuous on intervals or just at single points?   ◀

Limits of Functions

We now state a formal definition of left- and right-sided limits of functions.

LEFT- AND RIGHT-SIDED LIMITS OF FUNCTIONS

For a function f, we say L is the limit of f as x approaches a from the left and we write $\lim_{x \to a^-} f(x) = L$ if $f(x)$ can be made arbitrarily close to L as x approaches a through values less than a. More formally, $\lim_{x \to a^-} f(x) = L$ if, for **any** sequence $\{x_n\}$ of numbers less than a with $\lim_{n \to \infty} x_n = a$, $\lim_{n \to \infty} f(x_n) = L$.

We say L is the limit of f as x approaches a from the right and we write $\lim_{x \to a^+} f(x) = L)$ if $f(x)$ can be made arbitrarily close to L as x approaches a through values greater than a. More formally, $\lim_{x \to a^+} f(x) = L$ if, for **any** sequence $\{x_n\}$ of numbers greater than a with $\lim_{n \to \infty} x_n = a$, $\lim_{n \to \infty} f(x_n) = L$.

The TI calculator will compute right and left limits using the `limit(` *command found in the* `Calc` *menu. To compute the lefthand limit of the function h enter* `limit(h(x),x,4,⁻1)`. *Note that the −1 indicates the limit is to be taken from the left. To compute the limit from the right simply replace the ⁻1 with a 1.*

Look carefully at the notations for the left and right hand limits. They only differ in the $+$ or $-$ sign after the a below the lim. Be alert to this difference! It's easy to miss. The $-$ sign indicates that the inputs must be below a (or on the minus side of a). This does **not** mean that the terms of the inputs must be negative.

Jump discontinuities occur whenever the left and right limits exist but are not equal. Many important functions have no jump discontinuities. In particular, continuous functions have no jump discontinuities. Many other functions have only a few jump discontinuities. Our eight hour model of the level of amoxicillin had only one. For any value of a in the domain of h other than 4 we will find that $\lim_{x \to a^-} h(x) = \lim_{x \to a^+} h(x)$.

▶ **SIXTH REFLECTION**

Compute $\lim_{x \to 2^-} h(x)$ and $\lim_{x \to 2^+} h(x)$. How do the values compare?   ◀

In order to say that the $\lim_{x \to a} f(x) = L$, both the left- and right-sided limits must exist and be equal. We make the following definition.

LIMIT OF A FUNCTION

For a function f, we say L is the limit of f as x approaches a if $\lim_{x \to a^-} f(x) = L = \lim_{x \to a^+} f(x)$. We write $\lim_{x \to a} f(x) = L$.

Once again be particularly careful about the notation. The only difference between the notation for the limit of f and the limit from the left or the right is the absence of a $+$ or $-$ as a superscript of the number a under the lim.

The next examples shows that we must be careful in how we let x approach a.

EXAMPLE 3 **A Sine-Based Sequence**

Consider the function, $f(x) = \sin\left(\frac{2\pi}{x}\right)$. To compute the $\lim_{x \to 0^+} f(x)$, we consider the sequence of inputs $1, \frac{1}{2}, \frac{1}{3}, \ldots, \frac{1}{n}, \ldots$. Notice that all the terms in this sequence are greater than 0 and the limit of the sequence equals 0. These inputs generate the outputs, $\sin(2\pi), \sin(4\pi), \sin(6\pi), \ldots$. The sine of an integer multiple of π is always 0 so all of the terms in the sequence of function values are 0. Based on this information we would predict $\lim_{x \to 0^+} f(x) = 0$. Unfortunately, our prediction would be dead wrong. Consider the input sequence $\frac{1}{1 + \frac{1}{4}}, \frac{1}{2 + \frac{1}{4}}, \ldots, \frac{1}{n + \frac{1}{4}}, \ldots$. Once again, all of the terms in the sequence are greater than 0 and the limit of the sequence equals 0. This sequence generates the outputs $\sin(2\pi + \frac{\pi}{2}), \sin(4\pi + \frac{\pi}{2}), \sin(6\pi + \frac{\pi}{2}), \ldots$. All of these outputs equal 1, not 0. By changing our input sequence, we get a different sequnce of output values. For this function, we say the limit as x approaches 0 from the right does not exist.

The TI calculator uses the `limit(` *command to compute the limit of a function. Simply omit the plus or minus 1. To compute the limit at 4 of h enter* `limit(h(x),x,4)`. *What result does the TI calculator return?*

Holes (or Removable Discontinuities)

For the model of amoxicillin, $\lim_{x \to 4^-} h(x) = 31.25$ and $\lim_{x \to 4^+} h(x) = 531.25$; hence, $\lim_{x \to 4} h(x)$ does not exist because even though the left and right limits exist they are not equal.

A function f cannot have a jump discontinuity when $\lim\limits_{x \to a^-} f(x) = \lim\limits_{x \to a^+} f(x)$. Does that mean that if the left and right limits are equal then the function must be continuous?

E X A M P L E   4   **A Discontinuous Function**

Consider the function $f(x) = \dfrac{x^3 - 8}{x - 2}$. Does this function have a limit at 2? In other words, do the left and right limits exist and are they equal?

To investigate this question numerically fill in the missing entries in the following table.

You can find the missing entries by defining $f(x) = \dfrac{x^3 - 8}{x - 2}$ on the entry line of your TI calculator, and entering $f(1.9)$, $f(1.99)$, etc.

TABLE 1.9

x	1.9	1.99	1.999	1.9999	2	2.0001	2.001	2.01	2.1
$f(x)$	11.41								12.61

The table suggests that the $\lim\limits_{x \to 2} f(x) = 12$ because the outputs for $f(x)$ approach twelve as x approaches two from the left and from the right. Thus, we conjecture that $\lim\limits_{x \to 2} f(x) = 12$. Notice that when arriving at this conjecture we consider sequences that approach 2 from the left or the right but our sequences never contain the number 2. For this particular example, we cannot evaluate $f(2)$ because we cannot divide by zero. When we say $\lim\limits_{x \to 2} f(x) = 12$, we are saying we can make the value of f as close as we like to twelve by making the value of x close to two. Be aware that we are saying nothing at all about what the function does when $x = 2$.

The TI calculator will compute the limit if you enter `limit((x^3-8)/(x-2),x,2)` *(or* `limit(f(x),x,2)` *if you've defined f as in the previous margin note).*

The limit exists at 2 and the function does not have a jump discontinuity at 2. Nevertheless, the function cannot proceed smoothly from the left of two to the right of two because $f(2)$ is undefined. There must be a break. When the limit exists but does not equal the value of the function, we say that function has a *hole* or a *removable discontinuity*.

FIGURE 1.24 Graph of $f(x)$ on $[-3, 3] \times [-5, 20]$

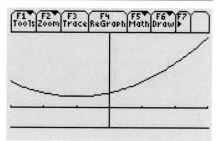

Removable discontinuities are much more difficult to recognize graphically than jump discontinuities. The graph of $f(x) = \dfrac{x^3 - 8}{x - 2}$ in Figure 1.24 gives no clue of the hole at 2.

▶ **SEVENTH REFLECTION**

What are the closest pixels to 2 on the graph of the function in the window $[-3, 3] \times [-5, 20]$? Is a pixel computed for $x = 2$? How can you change the `Xmin` and `Xmax` values so that such a pixel computation is attempted by the calculator? Can you cause the `xc` value in Trace mode to be 2 with the x-intervals set to $[0, 7.9]$ or $[0, 3.95]$ on the TI-89 (or $[0, 23.8]$ and $[0, 11.9]$ on the TI-92)? Is such a hole exhibited when `xc` is 2?   ◀

FIGURE 1.25   A Graph to Illustrate Limits

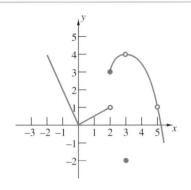

When first encountering removable discontinuities one might well dismiss them as being exotic and therefore unimportant. Nothing could be further from the truth! Removable discontinuities play a central and extremely important role in differential calculus.

Finding Limits from Graphs.   Often, limits of functions can be easily estimated from graphs. Graphs can provide a very good "feel" for the behavior of a function. Jump discontinuities in particular tend to be quite obvious.

EXAMPLE 5   **Finding Limits from a Graph**

In Figure 1.25, the open circle means that the point is not included on the graph while the black dot indicates that the point is included. We know that $f(2) = 3$ because the black dot corresponds to $x = 2$ and $y = 3$. On the other hand, $f(5)$ is undefined because when $x = 5$, there is no value given for $f(x)$. There is no black dot on the vertical line $x = 5$. The value of f at 3 is -2, that is, $f(3) = -2$.

We see that $\lim\limits_{x \to 2^-} f(x) = 1$ because we can make the value of $f(x)$ (the y-coordinate on the graph) as close as we like to one by making x a number less than but close to two. (Did you remember that the $\lim\limits_{x \to 2^-}$ means to approach two from the left?) One way to see this limit is to place your pencil tip on the straight line segment which joins the point $(0,0)$ to the open circle at $(2,1)$, and then trace the line to the open circle reading the coordinates of the points you traverse as you go. You should see x values approaching 2 from the left, and $f(x)$ or y values approaching 1 from below.

On the other hand,

$$\lim_{x \to 2^+} f(x) = 3.$$

This time, place your pencil tip on the parabola-like curve to the right of $x = 2$ and trace the curve down toward the black dot. As x approaches 2 from the right, $f(x)$ approaches 3 from above.

The function f has a jump discontinuity at 2 because the left-sided limit is not equal to the right-sided limit. It follows that

$$\lim_{x \to 2} f(x)$$

does not exist. Look at the graph carefully and realize that $f(2) = 3$, but $\lim\limits_{x \to 2} f(x)$ does not exist.

Think and Share

Work in pairs to find each of the following limits from the graph in Figure 1.25 and then compare your answers with another pair. Identify any discontinuities as a jump discontinuity or a removable discontinuity.

Think and Share

1. $\lim_{x \to 1^+} f(x)$
2. $\lim_{x \to 3} f(x)$
3. $\lim_{x \to 5} f(x)$
4. $\lim_{x \to 0^-} f(x)$
5. $\lim_{x \to 0^+} f(x)$

Defining Continuous Functions. For centuries mathematicians used the word *continuity* to describe a function whose graph consisted of a single piece. The idea seemed so obvious that no one bothered to propose a formal definition. In the early 1800s, however, physicists and mathematicians found themselves working with classes of functions defined by infinite sums. For many of these functions, even a property as simple as continuity could not be determined based on intuitive notions. A clear and unambiguous definition of continuity became essential. The emerging concept of limit provided a tool for creating such a definition.

We have seen that if the left and right limits are not equal at a point a then the function has a jump discontinuity. Furthermore, if the two limits are equal to each other but do not equal the value of the function then the function has a hole. Taking these difficulties into account leads to a precise definition of continuity.

CONTINUITY AT A POINT

Let a be a number in the domain of a function f. The function f is considered to be **continuous** at a if

$$\lim_{x \to a} f(x) = f(a).$$

In the preceding example, $f(x)$ is continuous everywhere except at 2, 3, and 5. We say $f(x)$ is **discontinuous** at these points.

Although we have defined what we mean by continuity at a point, we have not yet defined what we mean by a continuous function. Generally, we say **a function is continuous** if it is continuous at every point in its domain. Thus, $N(t)$, $D(t)$ and $A(t)$ are all continuous functions. The function $h(t)$ which models the level of amoxicillin for an eight hour period is not continuous because of the jump discontinuity at 4. Nevertheless, there is only one discontinuity and such a function may be referred to as being piecewise continuous. Functions which only have a finite number of discontinuities are referred to as **piecewise continuous**.

Continuous functions have useful properties that discontinuous functions do not have. In particular, if we wish to compute $\lim_{x \to a} f(x)$ at a and f is continuous at a, then according to the definition we can simply compute $f(a)$.

E X A M P L E 6   **Finding A Limit Symbolically**

Recall the function $f(x) = \dfrac{x^3 - 8}{x - 2}$. We have claimed that $\lim\limits_{x \to 2} f(x) = 12$ based on very scant evidence (only 8 points). If the actual limit were 12.000001, our data points might look exactly the same. We can never be entirely sure of the value of a limit simply by looking at a finite number of points. Graphs also lack precision. Obtaining a precise value for the limit requires an algebraic (or symbolic) approach.

By factoring the numerator, we see that

$$f(x) = \frac{(x - 2)(x^2 + 2x + 4)}{x - 2}$$

which seems to be equal to the expression

$$x^2 + 2x + 4.$$

Notice, however, that when $x = 2$ the expression

$$\frac{(x - 2)(x^2 + 2x + 4)}{x - 2}$$

is undefined while $x^2 + 2x + 4 = 12$. So the two expressions are not the same! Nevertheless, for all values of x other than 2, the two expressions are equal. To compute the limit of f at 2 we consider sequences of inputs that approach 2 from below (left) and sequences that approach 2 from above (right). None of the terms in the sequences are ever taken to be equal to 2. In other words, for all the values of x that we consider when finding the limit at 2 the two expressions are equal. Thus,

$$\lim_{x \to 2} \frac{(x - 2)\left(x^2 + 2x + 4\right)}{x - 2} = \lim_{x \to 2} x^2 + 2x + 4.$$

This is an important observation because the function $g(x) = x^2 + 2x + 4$ is continuous at 2. We justify our assertion that 12 equals the limit of f as x approaches 2 by writing:

$$\lim_{x \to 2} f(x) = \lim_{x \to 2} \frac{x^3 - 8}{x - 2} \tag{1.22}$$

$$= \lim_{x \to 2} \frac{(x - 2)\left(x^2 + 2x + 4\right)}{x - 2} \tag{1.23}$$

$$= \lim_{x \to 2} x^2 + 2x + 4 \tag{1.24}$$

$$= (2)^2 + 2(2) + 4 = 12 \tag{1.25}$$

Notice that we used the continuity of $g(x) = x^2 + 2x + 4$ to compute $\lim\limits_{x \to 2} x^2 + 2x + 4$ by simply replacing x with 2. We've also cheated a bit by assuming that $g(x)$ is continuous. Actually, all polynomial functions are continuous but we will not prove that assertion here.

► **EIGHTH REFLECTION**

What happens if you try to replace x with 2 in the expression $\dfrac{x^3 - 8}{x - 2}$?   ◄

Finding limits algebraically provides an accurate and reliable result. Unfortunately, the algebraic manipulations may be difficult or even impossible to carry out. In such cases, we often predict limits by generating tables, by using technology or by examining graphs.

A Continuous Model of a Discrete Phenomenon.   Even for a discrete phenomenon, a continuous model can reveal general trends which might otherwise go unnoticed. Recall the discrete logistic model for the growth of yeast colonies from Section 1.3.

Continuous models are less reasonable for animal or plant populations that breed and reproduce once a year as do herds of caribou or fields of dandelions.

$$u_n = u_{n-1} + r \cdot u_{n-1} - p \cdot u_{n-1}^2 \qquad (1.26)$$
$$u_1 = B.$$

The number of individuals in a population must be an integer and so is inherently a discrete phenomenon. However, yeast colonies contain a great many individuals which reproduce independently and rapidly, so we may well wish to represent their growth with a continuous model.

Unfortunately, this difference equation does not have a closed form. Consequently, we cannot apply the trick we used for the Novocaine, the lion's trip, and the repeated doses of amoxicillin. When no closed form exists (and in practice this happens more frequently than not), we cast about in our inventory of known functions for one which seems to fit the characteristics of the observed data.

What happens if you delete the restrictions $0 < c < 1$?

There are several families of continuous functions whose graphs exhibit the characteristic sigmoid shape of the yeast data. One such family of functions is that of the form $f(x) = \dfrac{a}{1 + b \cdot c^x}$ where a and b are positive and $0 < c < 1$. We can use this function to develop a continuous model for the yeast data.

Think and Share

Break into groups. Graph the data plot for yeast as described earlier. Examine the plot and the data and determine your estimate for the (asymptotically) maximum value for the weight of the yeast colony. Store your estimate of the maximum weight to the parameter maxval. Make a guess for the weight of the yeast colony at time zero and store your guess to the parameter initval. Set your Graph ... mode to FUNCTION. In the Y= editor enter

$$y1(x) = a/(1 + b * c^x).$$

Return to the HOME screen and enter the equations

$$\texttt{maxval = limit(y1(x),x,}\infty\texttt{) | 0 < c and c <1} \qquad \textbf{(1.27)}$$

to produce an equation which determines the value of a. Store your value to a. Again, on the home screen, enter the equation

$$\texttt{initval = y1(0)} \qquad \textbf{(1.28)}$$

Think and Share

to produce an equation in b. Solve the equation for b and store your solution to b. Store 0.5 to c. Go to the GRAPH window and compare your function and the data plot. By adjusting the value of c (and possibly a and b), find a function that gives what you feel is an excellent fit to the $\texttt{yeast}$ data. Reconcile these values with your partners and record the function your group adopts below.

Report your group's function to the class as a team.

Putting It All Together

To build a model of an event, we begin by collecting data. The discrete nature of the data then naturally leads us to formulate discrete models. In many cases, such models provide adequate information for our purposes. Often, however, a deeper understanding of the event requires a continuous model.

If the discrete model has a closed form representation, we can acquire a continuous model simply by viewing that representation as a continuous function. Sometimes, as with the Novocaine model and to a lesser extent the model of the lion's trip, the continuous function provides a reasonably good model. In other cases, such as with the repeated doses of amoxicillin, the continuous function provides an utterly useless model. In particular, if we are studying an event that has discontinuities, then we cannot expect a continuous funtion to provide a useful model. Finally, most discrete models do not have a closed form; and, for these models, we must fall back on our collection of known continuous functions.

Examining discontinuities led us to define the limit of a function at a point. using our notion of the limit of a sequence. We have learned how to find limits graphically (by observing a graph), numerically (by substituting numbers close to the point), and symbolically (by using algebra). In addition, we have used limits to give a precise definition of continuity both at a point and on a domain.

Limits play a critical role in mathematics. They bridge the gap between the finite and the infinite. We will see that limits are at the heart of the two primary operations of calculus: differentiation and integration.

EXERCISE SET 1.5

1. Show that you understand the meaning of discrete and continuous by explaining why a continuous function is appropriate to model the amount of a drug in the body when a single dose is taken and yet is utterly useless when repeated doses are taken. Illustrate your explanation.

2. We have correctly asserted in the text that if we have a closed form for a discrete model, we can always create a continuous model by simply replacing the index variable $n - 1$ with the real variable t. Please explain why, although the assertion is true, it is not always appropriate to do so.

3. Draw the graph of a function (you do not need an equation for your function) that has a removable discontinuity at $x = 2$. Explain, in terms of limits, why the discontinuity is removable.

4. Draw the graph of a function (you do not need an equation for your function) that has a jump discontinuity at $x = 2$. Explain in terms of limits, why the discontinuity is considered to be a jump discontinuity.

5. Explain, in terms of the graph of a function, what it means for a function to be continuous.

6. Draw the graph of a function (you do not need an equation for your function) that has a limit value at 3 when x approaches 2 from the left and a limit value of -1 when x approaches 2 from the right.

7. Draw the graph of a function (you do not need an equation for your function) that is undefined at $x = 2$ and has a limit value of 3 when x approaches 2.

8. Draw the graph of a function (you do not need an equation for your function) that is undefined at $x = 3$ and has a limit value of 2 when x approaches 3.

Refer to the graph of $j(x)$ in Figure 1.26 to find each limit in Exercises 9–22. Explain each answer in a sentence or two. If a limit does not exist, explain why. Answer in the spirit of the explanations given in the text in Example 5.

9. $\lim\limits_{x \to 1^-} j(x)$

10. $\lim\limits_{x \to 1^+} j(x)$

11. $\lim\limits_{x \to 1} j(x)$

12. $\lim\limits_{x \to -1^-} j(x)$

13. $\lim\limits_{x \to -1^+} j(x)$

14. $\lim\limits_{x \to -1} j(x)$

15. $\lim\limits_{x \to 2^-} j(x)$

16. $\lim\limits_{x \to 2^+} j(x)$

17. $\lim\limits_{x \to 0} j(x)$

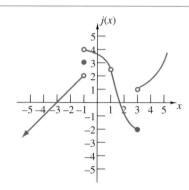

FIGURE 1.26 A Discontinuous Function

18. $\lim\limits_{x \to -\infty} j(x)$

19. $\lim\limits_{x \to -2} j(x)$

20. $\lim\limits_{x \to 3^+} j(x)$

21. $\lim\limits_{x \to 3^-} j(x)$

22. $\lim\limits_{x \to 3} j(x)$

Refer again to the graph of $j(x)$ in Figure 1.26 to find the functional values in Exercises 23–26. Explain why you believe your answers to be correct.

23. $j(-2)$

24. $j(-1)$

25. $j(1)$

26. $j(3)$

Refer to the graph in Figure 1.26 to answer the following questions.

27. Find all values of x for which j is discontinuous. Explain your answer using clear and complete sentences.

28. Find all values of x for which j has a jump discontinuity. Explain your answer using clear and complete sentences.

29. Find all values of x for which j has a removable discontinuity. Explain your answer using clear and complete sentences.

30. Is it possible for a jump discontinuity to be removable? Explain your answer using clear and complete sentences.

In Exercise 31–36, conjecture the limit (or guess the limit) by substituting numbers and by graphing. Then prove your conjecture to be true by finding the limit algebraically.

31. $\lim\limits_{x \to 1} \dfrac{x^2 + x - 2}{x - 1}$

32.
$$\lim_{x \to -2} \frac{x^2 + x - 2}{x + 2}$$

33.
$$\lim_{x \to 5} \frac{x^2 - 25}{x - 5}$$

34.
$$\lim_{x \to -1} \frac{x^3 + 1}{x + 1}$$

35.
$$\lim_{x \to 1} \frac{2x^2 + 3x - 2}{x + 2}$$

36.
$$\lim_{x \to 1} \frac{3x^2 - 2x - 1}{x^2 - 1}$$

37. Is $f(x) = \dfrac{\sin(x)}{x}$ continuous at $x = 0$? Justify.

†**38.** *Group Exercise* In addition to the visual evidence of comparing a data plot to the graph of a continuous model in the GRAPH screen, we can quantify how well an individual model fits a data set. Open the Y= editor and enter the function you recorded as your continuous model for the yeast as y1(x). Now go to the Data/Matrix editor and open your file yeast. Select the cell containing c3 and enter c3 = y1(c1) in the editing line. The values of your continuous model will appear in the third column for comparison to the actual data in the second column. If you have found a good fit, the values in columns c2 and c3 should be very close and have no discernible pattern. The closeness of the values is an indication of how close or "good" the fit is; the lack of a pattern (randomness of differences between columns) "suggests" that the type of curve (linear, quadratic, sigmoid, exponential) may be correct. You judge for yourself how good the fit is. To quantify these ideas, select the cell containing c4 and enter c4 = c3−c2 in the editing line. This will produce the differences mentioned above, called *residuals* of your model, in column 4. The following numbers are possible measures of the accuracy of your model (TI calculator commands for the fifth through eighth columns of the Data/Matrix editor are given).

- The sum of the residuals. c5=sum(c4)
- The largest absolute value of your residuals. c6=max(abs(c4))
- The sum of the absolute values of your residuals. c7=sum(abs(c4))
- The sum of the squares of your residuals. c8=sum(c4^2)

Choose the number you think best measures how well your model fits the data.

39. Functions of the form $\dfrac{A \cdot x^2}{\theta^2 + bx^2}$ are known as Hill's functions and for positive A, b, x, and θ give sigmoid graphs. With a starting value $A = 1000$, experiment with values of A, b, and θ to fit the yeast data.

40. The arctan functions of the form $A \tan^{-1}(Bx + C) + D$ are familiar from trigonometry and for positive A give sigmoid graphs. With starting values $A = 235$ and $D = 350$, experiment with values of A, B, and C to fit the yeast data.

41. Hyperbolic tangent functions (tanh(can be found in the CATALOG) of the form $A \tanh(B \cdot x + C) + D$ give sigmoid graphs. With starting values $A = 300$ and $D = 350$, experiment with values of A, B, and C to fit the yeast data.

42. Explain, in your own words, the meaning of each of the following terms. Write your explanations in clear and complete sentences. Include annotated diagrams whenever appropriate.
   a. $\lim_{x \to 3} g(x)$
   b. continuous
   c. discontinuous
   d. continuity at a point

Summary

We have developed and analyzed a number of models in this chapter. A mathematical model is, necessarily, a simplification of a complex situation. Mathematical models are interesting and fun because of their power to predict the future.

We express the fundamental approach we have used to construct our models in the equation

$$future = present + change.$$

Our principle is simple, but there is considerable technique involved in the implementation. Even making an effective visual display of information from our models is a problem of considerable subtlety. The algebraic tools we have brought to the modeling process involve differences. Difference equations are discrete models; functions are continuous models. Most real-world situations can be approached from either the perspective of a discrete or a continuous model. Indeed, information obtained from one type of model may give us insight into another type of model.

Finally, data from the real world *must* play a key role in the modeling process. Models can be beautiful constructs on their own, but they are successful only when they give us reasonable and verifiable predictions. As you make and explore your own models, you need to cultivate a healthy respect for data as a guide and standard in modeling. Even though a mathematical model is an abstraction and, in principle, can never be a full description of the world, it can nonetheless be a powerful tool for understanding our world. Welcome to the adventure.

SUPPLEMENTARY EXERCISES

1. Find the rest points of the difference equation

$$u_n = u_{n-1} - \frac{1}{4} u_{n-1}^2$$
$$u_1 = 1.$$

   What is the long-term behavior of u_n?

2. Write the sequence generated by the geometric series

$$x_n = \sum_{k=0}^{n} \frac{1}{3^k}$$

   as a difference equation.

3. To demonstrate exponential decay to a group of 60 Young Scholars, Jim Bishop drops a hard rubber ball to a hard, level floor from a height of five and a half feet. The ball rebounds to 90% of the distance it fell on this and every subsequent bounce. For example, on the first bounce, the ball rebounds to $5.5 \times 0.9 = 4.95$ ft. He estimates that it took the ball 0.6 seconds to drop the original five and a half feet. Use infinite summations to predict how long the ball bounces before coming to rest.

4. In North Carolina, the following formula is used to compute state income tax on adjusted gross income (*AGI*) over $25,000.

   If $\$25,000 \le AGI < \$50,000$,

   then $Tax = \$7,500$ plus

   15% of the amount over $25,000.

   If $\$50,000 \le AGI$,

   then $Tax = \$7,500$ plus

   10% of the amount in excess of $50,000.

   Write the tax function as a piecewise function. Use the when to produce a graph.

5. Argue for or against using a continuous function to model each scenario. You do not need to find any equations.
   a. The cost of mailing a first class letter.
   b. The amount of water remaining in a pan as water boils.
   c. The amount of gasoline in the tank of your car between fill-ups.
   d. The amount of gasoline in the tank of your car during a drive from Maine to Texas.

6. Explain why $\sum_{k=0}^{\infty} a^k = \frac{1}{1-a}$ when $0 < |a| < 1$. Provide an original example.

7. Explain why $\sum_{k=1}^{\infty} a^k = \left(\sum_{k=0}^{\infty} a^k \right) - 1$. Provide an original example.

8. Suppose 500 mg of a certain antibiotic is administered every 6 hours, and that the body eliminates 25% of the antibiotic in one hour.
   a. Find a difference equation to model the local maximum levels of antibiotic in the body. Is there asymptotic stability? Explain.
   b. Find a difference equation to model the local minimum levels of antibiotic in the body.

9. Explain why the graphs in the rumor and yeast examples have similar shapes. What is the significance of the inflection point in each graph?

10. Explain how to find the rest points and limit values of a difference equation.

11. Explain briefly (in your own words) and illustrate each concept:
    a. Removable discontinuity
    b. Jump discontinuity
    c. Continuity at a point

12. Explain why graphical evidence of a limit value cannot always be considered to be conclusive.

13. Draw the graph of a function f that has an inflection point at $x = 1$.

14. Use the when(command to graph the piecewise defined function $g(x) = \begin{cases} x^2 - 1, & \text{if } x < 1 \\ 2 - x^2, & \text{if } 1 \leq x \end{cases}$, with graphing window $[-4, 4] \times [-16, 16]$.
    a. Decide, based on your graph, whether or not g is a continuous function. Explain.
    b. Determine algebraically whether or not g is a continuous function.
    c. Do your answers to part (a) and (b) agree? If not, which answer is correct. Justify.

In 1904, Helga von Koch invented a curve called the *Koch Snowflake*. *Koch's Island* is the area enclosed in her curve. To construct the curve, do the following: start with an equilateral triangle having two-inch edges. Remove the middle third of each edge and replace it with a smaller equilateral triangle. Continue this process forever. When you're finished, you have Koch's Snowflake. (See Figure 1.27.) Exercises 10–13 refer to Koch's Snowflake.

FIGURE 1.27 Constructing Koch's Snowflake

15. Calculate the area of the first island.

16. Determine the following:
    a. The area of one of the small triangles added to the first island.
    b. The number of small triangles added to the first island.
    c. The area of the second island.

17. Determine the following:
    a. The area of one of the small triangles added to the second island .
    b. The number of small triangles added to the second island.
    c. The area of the third island.

18. Write a difference equation for the area of the nth Koch Island.

PROJECTS

Project One: Geometric Series and Financial Calculations

Geometric sums are used daily to make financial calculations. All of us have experienced what is called the *time value of money*. Twenty dollars now is more valuable than twenty dollars next year. This time value is what loans are all about.

Present and Future Value. By a legal tradition dating back to the time of Hammurabi,[†] the *Future Value*, FV, of a loan with *Present Value*, PV, com-

---

[†] A Babylonian king of the 18th century B.C.E. after whom a code of laws is named.

pounded for *n periods* at *interest rate, r*, is computed by the *Equation of Value*:

$$FV = PV(1+r)^n \qquad (1.29)$$

For example, a deposit of $3000 (the present value) in a savings account which earns interest at a rate of 2% each period has a future value after nine periods of

$$FV = \$3000(1 + 0.02)^9 = \$3585.28.$$

Sometimes the future value is known and the present value is to be computed. Thus, we calculate the present value of a future payment of $240 at the end of 48 payment periods at an interest rate of 2% by solving

$$240 = PV(1 + 0.02)^{48}$$

for PV. The solution tells us that the present value of a payment deferred for so long is only $92.77. The loss in value due to time is referred to as the *discount*.

Use the TI calculator command

`solve(240= pv*(1+0.02)^48,pv)`

Installment Loans and Payment Size.   Often debts are repaid with installments throughout the life of the loan rather than with a lump sum payment at the end. Car loans, for example, are typically repaid in monthly installments which may well stretch over four to five years (that is, from 48 to 60 months). We use summation and the equation of value to calculate the size of payments necessary to repay an auto loan of $12,000 at a 10.8% Annual Percentage Rate (APR) for five years. The contract calls for the purchaser to make 60 payments of a fixed size (one each month for five years) to the financing institution. We give the variable name, *pay*, to the amount of one such payment. The contract also fixes the computation of the present value of the kth payment as

$$pay = PV_k * \left(1 + \frac{0.108}{12}\right)^k.$$

Notice that the annual interest rate has been divided by 12 to convert it to a monthly interest rate. Solving for PV_k gives us the present value of the kth payment as

$$PV_k = pay * \left(1 + \frac{0.108}{12}\right)^{-k}.$$

To calculate the present value of *all* the payments we simply add them all up:

$$\sum_{k=1}^{60} pay * \left(1 + \frac{0.108}{12}\right)^{-k}.$$

Since the present value of the loan also is known to be $12,000, we determine the payment size by solving the equation

$$12000 = \sum_{k=1}^{60} pay * \left(1 + \frac{0.108}{12}\right)^{-k}$$

for *pay*.

1. Verify that $pay = \$259.71$ by using the closed form for finite geometric sums and by using the `solve(` and `Σ(` commands on your TI calculator.
2. Suppose that the buyer makes a $1000 down payment for the car described in the example so that the present value of the loan is only $11,000. What is the new payment size?

Project Report. Write a report on your investigations. Address your report to your classmates, assuming the material is new to them. Describe your reasoning and include all necessary background information. A minimal project report must include:

1. Written responses to both numbered exercises.
2. Suggestions for possible extensions of your observations and further questions for exploration.

Project Two: Variety in Behavior

Populations newly introduced into an environment are frequently modeled with the discrete logistic equation. The introduction of yeast into a bread dough or malt, disease organisms into an immunologically naive population, and rabbits into the Australian Outback are all reasonably modeled by this difference equation. For some values of r and p, as we have seen, the discrete logistic equation generates a sequence exhibiting a sigmoid shape. This is by no means the only type of behavior modeled by the discrete logistic equation, as we know from Section 1.3, Example 3.

Let's see what the logistic model predicts for the growth of a species that breeds annually and has a large litter. Such species include rabbits, dandelions, salamanders, and butterflies living in a temperate climate. For our example, we postulate a species that produces, on average, 2.5 young that survive to adulthood the next year. That is, given unlimited resources, every adult produces 2.5 additional adults in the next generation, so $r = 2.5$. We also assume a modest level of competition, keeping b relatively small, say $b = 0.005$. With a first generation of 100, Equation (1.16) becomes

$$u_n = u_{n-1} + 2.5u_{n-1} - 0.005u_{n-1}^2$$
$$u_1 = 100.$$

Using this difference equation we generate the table of values in Table 1.10 and the graph in Figure 1.28.

TABLE 1.10   Discrete Logistic Table

1	2	3	4	5	6	7	8	9	...
100	300	600	300	600	300	600	300	600	...

The WINDOW *settings include*
nmin=1, nmax=25,
plotstrt=1, plotstep=1,
xmin=0, xmax=25,
ymin=0, ymax=700.

FIGURE 1.28   Graph of Discrete Logistic Table

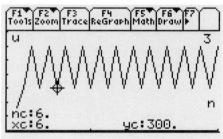

1. Compute the nonzero rest point for this model.
2. Interpreting this value as the carrying capacity of the environment, describe why the alternating population levels makes sense.
3. Would you characterize the carrying capacity as a limit value for the given sequence? Explain why or why not.
4. If we reset u_1 to the carrying capacity, what is the resulting sequence of values?
5. Would you characterize the carrying capacity as a limit value for this new sequence?

Project Report.   Write a report on your investigations. Address your report to your classmates, assuming the material is new to them. Describe your reasoning and include all necessary background information. A minimal project report must include:

1. Written responses to the five numbered exercises.
2. Suggestions for possible extensions of your observations and further questions for exploration.

Project Three: An Odd Sum and Mathematical Induction

The geometric series is not the only summation that has a closed form. In fact every summation whose general term is a polynomial has a closed form that is also a polynomial. Of these summations, there are a number that enjoy considerable notoriety. The sum of the first n odd numbers is one used by Galileo in his analysis of projectiles. Look at the sum of the first several odd

FIGURE 1.29   Galileo 1564 – 1642

integers systematically:

$$1 + 3 = 4$$
$$1 + 3 + 5 = 9$$
$$1 + 3 + 5 + 7 = 16$$
$$1 + 3 + 5 + 7 + 9 = 25$$
$$1 + 3 + 5 + 7 + 9 + 11 = 36$$

We can see a pattern emerge. Consulting a calculator, we find that

$$\sum_{k=1}^{n} 2k - 1 = n^2. \qquad \textbf{(1.30)}$$

The following illustration even suggests an algebraic explanation. The expla-

FIGURE 1.30   The Sum of Odd Integers

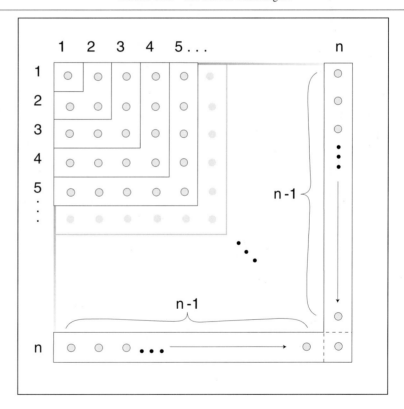

nation — such as it is — is that consecutive squares differ by an odd integer. Once we notice that the pattern is initiated, we can see that it continues, since:

$$\sum_{k=1}^{n+1}(2k-1) = \sum_{k=1}^{n}(2k-1) + (2(n+1)-1) \qquad (1.31)$$
$$= n^2 + (2n+1)$$
$$= (n+1)^2$$

The reader should be aware that we just proved our conjecture by a process called *mathematical induction*. We showed the conjecture true for $n = 2, 3, 4, \ldots, 6$. We then assumed the conjecture true for the first n odd numbers when we recognized the pattern and wrote the sum in the form $\sum_{k=1}^{n}(2k-1) = n^2$ in Equation (1.30). When we computed the sum in Equation (1.31), we showed the conjecture to be true for the next odd integer (by adding the next odd integer $(2(n+1)-1)$ to both sides of Equation (1.31).

1. Explain why proving the conjecture true for the next odd integer proves the conjecture true for *ALL* odd integers.

2. Consult your calculator to obtain a closed form formula for the sum $\sum_{k=1}^{n} k$ and use the method of mathematical induction to verify your formula.

3. Consult your calculator to obtain a closed form formula for the sum $\sum_{k=1}^{n} k^2$ and use the method of mathematical induction to verify your formula.

4. Consult your calculator to obtain a closed form formula for the sum $\sum_{k=1}^{n} k^3$ and use the method of mathematical induction to verify your formula.

5. Examine the expanded forms of your formulas and describe any patterns or regularities that you find.

Project Report.   Write a report on your investigations. Address your report to your classmates, assuming the material is new to them. Describe your reasoning and include all necessary background information. A minimal project report must include:

1. Written responses to the five numbered exercises.
2. Suggestions for possible extensions of your observations and further questions for exploration.

C H A P T E R

2

MEASURING CHANGE

2.1

Analyzing a Discrete Function

2.2

Contructing Models from Patterns

2.3

Divided Differences and Average Rate of Change

2.4

Rate of Change

2.5

Rate of Change as a Function

Summary

In this chapter, we use differences to analyze discrete functions and to motivate the use of a new concept called the derivative in analyzing continuous functions. We learn how to use differences to determine interesting properties of a mathematical model: Where is the *change* most rapid? Where does the graph of the function *peak*? What are the *largest* and *smallest* values assumed anywhere in the domain of the function?

In many situations, we are asked to *maximize* profits, *minimize* cost, or find the *best* solution. The processes of modeling and optimization (finding the maximum or minimum) allow us to solve such problems.

Although we introduce the notion of derivative as rate of change using tables and graphs, a formal definition for the derivative will be delayed until Chapter 3.

2.1

Analyzing a Discrete Function

To the naive stock market observer, the price of a stock often seems to go up and down at random. More sophisticated observers attempt to analyze the future behavior of a stock from its past behavior.

E X A M P L E 1 **Richer or Broker**

Just before going off to college, Tony sells his aging car for $1000 and deposits the money in a savings account. He plans to buy another "clunker" next summer with this money and the interest earned on it.

When his roommate gets wind of this plan, he can't believe it.

"You've got to invest it," he says. "I have a wonderful broker named Heidi. Here's her number, give her a call." Tony is a bit nervous about risking his car fund, but he decides it can't hurt to at least talk to Heidi.

"You've called at just the right time," Heidi says. "I just got a tip on a wonderful stock, California Multi Media (CMM), that's about to take off. You should invest right away." Still feeling nervous about the whole thing, Tony eventually decides to go ahead and invest his $1000. (It won't kill me to walk for a summer, he thinks.) He buys 17 shares of the stock on Monday for $58.33 per share. Everyday he checks the afternoon newspaper to see how his stock is doing and records its price in a table (see Table 2.1.

TABLE 2.1   CMM Stock Prices

Day	0	1	2	3	4
Stock Price	58.33	59.04	61.10	62.27	61.62

He's pleased with the way the stock is performing, as the share price climbs early in the week, and isn't too worried about the minor drop off on Friday. Monday is a different story. The price on CMM drops to $59.26, and Tony immediately calls Heidi.

"You're still ahead of where you started," says Heidi. "You shouldn't panic. Wait one more day before you do anything." Tuesday the stock drops to $58.31. Tony gets Heidi on the phone determined to unload his stock before his "car" goes up in smoke. Her attitude stops him cold.

"Isn't it wonderful how much better your stock is doing," Heidi gushes. "I'll bet you're glad you held on to it now." Anxiously, Tony takes a stab at dampening Heidi's enthusiasm.

"Didn't the value of the stock drop?" he whimpers.

"Yes, of course," says Heidi. "But the drop today was much less than the drop yesterday! This stock is going to turn around in no time."

Tony gives up. He decides to kiss goodbye to all thoughts of having a car the following summer. When the stock drops to $57.92 on Wednesday and then $57.86 on Thursday, he doesn't even consider contacting Heidi. On Friday the stock price rises to $58.00. Tony begins to wonder if maybe Heidi was right. All weekend he is on pins and needles thinking about what might happen with his stock on Monday. He even goes back and fills in the blank values on his table (he had stopped looking at his table since the stock began to fall).

On Monday the stock rises again to $59.11. Tony continues to keep track of his stock as the price rises over the next eight business days (see Table 2.2). On Wednesday, with the stock at its highest value of $72.15 after rising every day for eight straight business days, Heidi calls.

TABLE 2.2 The Total Picture

Day	0 M	1 T	2 W	3 Th	4 F	5 M	6 T	7 W	8 Th
Stock Price	58.33	59.04	61.10	62.27	61.62	59.26	58.31	57.92	57.86
Day	9 F	10 M	11 T	12 W	13 Th	14 F	15 M	16 T	17 W
Stock Price	58.00	59.11	62.33	67.84	68.55	69.96	71.37	71.96	72.15

To create your own plot set up a data table called stock. *Enter Tony's table in* stock *with Day as* c1 *titled* n *and Stock Price as* c2 *titled* a(n). *Use the* Define *option in the* Plot Setup *window to designate* x... *as* c1 *and* y... *as* c2. *Use* ZoomData *from the* Zoom *menu in the* Window *editor to obtain the scatter plot in Figure 2.1.*

"Time to get rid of that junk CMM stock," she says. "How about buying some nice shares of Texas Microchip?"

Let's see if we can figure out why Heidi thought she knew what the stock was going to do. As a first step, we have plotted the data from Tony's table in Figure 2.1.

FIGURE 2.1 CMM Stock

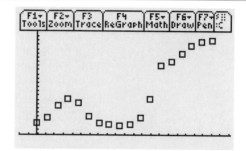

Tony became seriously concerned when the value of his stock dropped to $58.31 which was less than he paid for it. The two values just before $58.31 are $61.62 and $59.26. Thus, from Friday to Monday the stock dropped $2.36 but from Monday to Tuesday it only dropped $0.95. These values which are called *first differences* or *forward differences* provide important clues about the behavior of the data.

You can define a function to compute all the first differences at once. On the home screen enter

Define fd(q)=shift(q,1)-q.

With q as the name of the sequence, this statement creates a sequence of first (or forward) differences. The function shift(q,1) *moves the list* q *up one so that* shift(q,1)[2] *is* q[3]. *To use* fd *to compute the first differences for Tony's stock values, type* fd(c2) *on the Entry line next to* c3= *and press* ENTER. *Column 3 will fill with the first differences of column 2. Figure 2.2 shows what part of the table should look like. We will refer to this sort of table as a **difference table**. To obtain the Δ symbol, press* 2nd CHAR, *select* Greek *and then the character.*

We can use a sequence to represent the values of Tony's stock by letting $a_0 = 58.33$, $a_1 = 59.04$, etc. In general, a_n is the nth stock value in Tony's difference table (or the nth entry in column c2 of your table). Notice that the first differences are themselves a sequence. The symbol Δ (the Greek letter delta) is commonly used to represent differences, so a convenient notation for the sequence of **first differences** is $\Delta a_0, \Delta a_1, \Delta a_2 \ldots$, where $\Delta a_0 = a_1 - a_0$, $\Delta a_1 = a_2 - a_1, \ldots$ In general,

$$\Delta a_n = a_{n+1} - a_n.$$

Since each interval has the same length, $\Delta n = 1$ day, we can consider Δa_n to be the change in the stock price per day.

The subscripts of the Δa_n terms are a bit confusing, because the first differences provide information about the change *between two points* in time

FIGURE 2.2 First Difference Table for Stock Data

F1 Tools	F2 Plot Setup	F3 Cell	F4 Header	F5 Calc	F6 Util	F7 Stat

DATA		a(n)	Δa(n)	
	c1	c2	c3	c4
1	0	58.33	.71	
2	1	59.04	2.06	
3	2	61.1	1.17	
4	3	62.27	-.65	

c3=fd(c2)

rather than *at a particular moment*. The value of Δa_n says just as much about a_{n+1} as it does about a_n. For predictions, what it says about a_{n+1} often carries more weight. As a result, first differences are often referred to as forward differences.

► FIRST REFLECTION

What is the value of Δa_{17}? Why? ◄

Looking at the difference table, we see that $\Delta a_0 = 0.71$. This corresponds to the fact that the value of the stock increased by \$0.71 during the first day. The first differences tell us whether the stock is increasing or decreasing and by how much. Tony's stock increased for the first three days; so, Δa_0, Δa_1, and Δa_2 are all positive with the $\Delta a_1 = 2.06$ corresponding to the greatest increase.

From the first differences, it is difficult to see why Heidi was not concerned after the value of Tony's stock dropped to \$58.31. At that point, the first (forward) differences had been negative for three straight days. Heidi, however, was taking a deeper look at the data. We can get a clue from Heidi's statement that "... the drop today was much less than the drop yesterday..."

In terms of first differences, Heidi was noticing that Δa_5 was "greater" (less negative) than Δa_4. We can find precisely how much greater by computing

$$\Delta a_5 - \Delta a_4 = (-.95) - (-2.36) = 1.41. \qquad \textbf{(2.1)}$$

We've just computed a *second difference*. Second differences form yet another sequence. We use the symbol Δ^2 for second differences, so the previous computation can be written $\Delta^2 a_4 = 1.41$. In general, $\Delta^2 a_n = \Delta a_{n+1} - \Delta a_n$.

The subscripts for second differences are even more confusing than the subscripts for first differences. The second difference, $\Delta^2 a_n$, is computed by subtracting the change in the sequence immediately before a_{n+1} with the change in the sequence immediately after a_{n+1}. Symbolically,

$$\Delta^2 a_n = \Delta a_{n+1} - \Delta a_n = (a_{n+2} - a_{n+1}) - (a_{n+1} - a_n). \qquad \textbf{(2.2)}$$

Thus, the second difference $\Delta^2 a_n$ is most strongly tied to a_{n+1}. It is even more of a forward difference than the first difference.

► **SECOND REFLECTION**

Why are the values of $\Delta^2 a_{16}$ and $\Delta^2 a_{17}$ undefined?   ◄

*To add the second differences for Tony's data to your table simply apply the **fd** function to **c3** by entering **fd(c3)** on the Entry line next to **c4=**. Use the **Format** option in the **Tools** menu to set cell width to 5. Will you remember to reset it?*

Figure 2.4 includes the second differences for Tony's data in column **c4**.

FIGURE 2.3 Data Plot

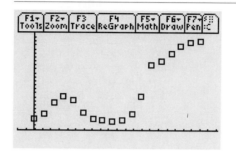

FIGURE 2.4 Complete Difference Table

n	$a(n)$	$\Delta a(n)$	$\Delta^2 a(n)$
0	58.33	0.71	1.35
1	59.04	2.06	−0.89
2	61.1	1.17	−1.82
3	62.27	−0.65	−1.71
4	61.62	−2.36	1.41
5	59.26	−0.95	0.56
6	58.31	−0.39	0.33
7	57.92	-0.06	0.2

Looking at the second differences reveals a striking feature of the data. Of the first six second differences, the largest one is $\Delta^2 a_4 = 1.41$. The value of $\Delta^2 a_4$ provides information about changes in the sequence immediately before and after a_5. Notice on your calculator that the actual value of the stock takes its biggest drop from the fourth to the fifth day (see Figure 2.4). Thus, the changes *in the changes* of the value of the stock looked most promising at precisely the same moment that the actual value of the stock looked bleakest ($58.31). This was the value that first prompted Tony to call Heidi about selling. Notice that the first differences lag one day behind and the second differences lag behind by two days. Thus, the value 1.41 cannot be computed until the seventh-day value, $58.31, is known.

The second differences compute the change in the change. If the second difference is positive and the sequence is decreasing, it will tend to decrease less rapidly. If the second difference is positive and the sequence is already increasing, it will tend to increase even more rapidly. Heidi projected that, although the value of the stock was decreasing, it would decrease less and less until eventually it would begin to increase. The first differences tell us when the shift from decreasing to increasing actually occurs.

► **THIRD REFLECTION**

Assuming that Heidi was looking at second differences, why wasn't she more upbeat the first time Tony called her to sell his stock when its value was $59.26? Hint: What was the latest second difference she knew at that time?   ◄

► **FOURTH REFLECTION**

Find the smallest value of a_n in the table. Look at the values of Δa_{n-1} and Δa_n. What do you observe? Explain your observations.   ◄

In groups, let's use part of the stock data to make sure we understand what the first and second differences tell us. To focus on the behavior of the stock from day 4 to day 10, change the graphing window to $[3.5, 12] \times [57, 65]$ (horizontally from 3.5 to 12 and vertically from 57 to 65).

Think and Share

1. Describe the shape of the data plot from day 4 to day 11. From the graph, discuss the first differences for these days: which are positive, which are negative? Describe how the first differences are related to the shape of the data plot for these days.

2. Are the first differences increasing or decreasing? If the first differences are increasing, will the second differences all be positive or negative? If the first differences are decreasing, will the second differences all be positive or negative? Are the second differences for these days positive or negative? How are these second differences related to the shape of the data?

3. Does the graph from 4 to 11 have a maximum or a minimum?

4. Create column `c5=-c2+100` and then create a data plot of `c1` and the new `c5` using `Plot Setup` and `ZoomData`. Now change the graphing window on your plot as follows: horizontally from 3.5 to 11; vertically from 35 to 45.

For day 4 through 11, repeat parts 1 through 3 again for this new data plot.

So far we have used first and second differences to make numerical conclusions about our data. These quantities can also be viewed graphically.

The Properties of the Graph of a Sequence

First Differences. First differences have a fairly straightforward graphical interpretation. If the first difference is negative, then the graph of the sequence must be falling from left to right (see Figure 2.5). On the other hand, if the first difference is positive, then the graph must be rising (see Figure 2.6).

$$\Delta a_{n-1} = a_n - a_{n-1}$$
$$\Delta a_n = a_{n+1} - a_n$$

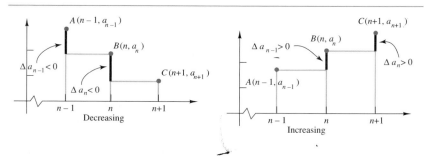

FIGURE 2.5 Decreasing Sequence   **FIGURE 2.6** Increasing Sequence

Looking back, compare the table of data for Tony's stock with the graph. Whenever Δa_n is positive, a_{n+1} is greater than a_n and the graph rises (remember the first difference looks "forward"). Also, when Δa_n is negative, the graph falls.

There are two places in Tony's table where the first differences change sign. The first corresponds to a maximum and the second corresponds to a minimum. In general, the first differences change sign from positive to negative at a local maximum and from negative to positive at a local minimum (see Figure 2.7).

FIGURE 2.7 Local Minimum and Local Maximum

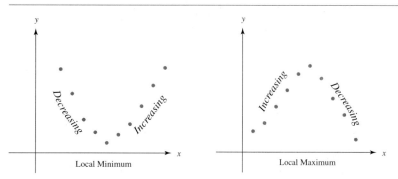

FIGURE 2.8 Concavity and Inflection Point

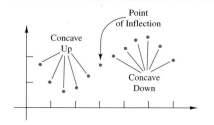

Second Differences. Second differences can help us to determine where the graph of a sequence is "bending up" or "bending down," and even where the bending changes direction. Let's consider the graph of a sequence to visualize these properties. The region where the graph of the sequence is "bent upward" is called **concave up**. Similarly, the region where the graph is "bent downward" is called **concave down**. An **inflection point** is where a graph changes concavity. These properties are labeled in Figure 2.8 and will be defined in terms of the *second differences* of a sequence.

Figures 2.9 and 2.10 illustrate two graphs of points

$$A = (n - 1, a_{n-1}), \quad B = (n, a_n), \quad \text{and} \quad C = (n + 1, a_{n+1}).$$

The three points differ horizontally by one unit, so the first differences $\Delta a_{n-1} = a_n - a_{n-1}$ and $\Delta a_n = a_{n+1} - a_n$ compute the slopes of the line segments AB and BC. The second difference $\Delta^2 a_{n-1} = \Delta a_n - \Delta a_{n-1}$ then represents the change in the slopes of the line segments AB and BC. We can use the sign of the second difference $\Delta^2 a_{n-1}$ to determine whether these slopes are increasing or decreasing. Notice that when $\Delta^2 a_{n-1} > 0$, the slope of the line segment BC is greater than the slope of the line segment AB and the graph at point B is concave up as illustrated in Figures 2.9 and 2.11. Likewise, when $\Delta^2 a_{n-1} < 0$, the slope of the line segment BC is less than the slope of the line segment AB and the graph at point B is concave down as illustrated in Figures 2.10 and 2.12. *Note especially that $\Delta^2 a_{n-1}$ determines the concavity at a_n* (based on $a_{n-1}, a_n,$ and a_{n+1}). An inflection point is a point where the concavity changes. Thus, (n, a_n) is an inflection point if the sign of $\Delta^2 a_n$ differs from the sign of $\Delta^2 a_{n-1}$. Inflection points can be extremely important.

FIGURE 2.9   Decreasing Concave Up   **FIGURE 2.10**   Increasing Concave Down

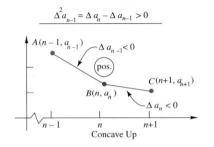

FIGURE 2.11   Increasing Concave Up   **FIGURE 2.12**   Decreasing Concave Down

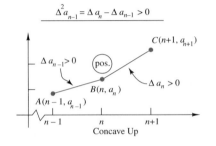
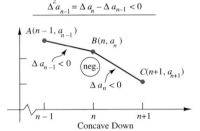

► **FIFTH REFLECTION**

Look back at the graph of Tony's data. Locate the second inflection point. How does the concavity change at that point? How did Heidi interpret this change in concavity?   ◄

We summarize our observations in Table 2.3.

TABLE 2.3   Difference Tests

Property of a Sequence at Index k	Forward Difference Tests
Increasing	$\Delta a_k > 0$
Decreasing	$\Delta a_k < 0$
Relative Maximum	$\Delta a_{k-1} > 0$ and $\Delta a_k < 0$
Relative Minimum	$\Delta a_{k-1} < 0$ and $\Delta a_k > 0$
Concave Up	$\Delta^2 a_{k-1} > 0$
Concave Down	$\Delta^2 a_{k-1} < 0$
Inflection Point	$\Delta^2 a_{k-1}$ and $\Delta^2 a_k$ have different signs.

The following example will help illustrate the ideas in the table.

EXAMPLE 2 **Completing and Analyzing a Difference Table**

We have entered values for n, a_n, Δa_n, and $\Delta^2 a_n$ in Table 2.4. We used information from Table 2.4 to make the other entries. In the $n = 3$ row, we entered *dec* in column 5 because $\Delta a_3 < 0$, *rel max* in column 6 because $\Delta a_2 > 0$ and $\Delta a_3 < 0$, *down* in column 7 because $\Delta^2 a_3 < 0$, and *no* in column 8 because concavity did not change.

▶ SIXTH REFLECTION

Describe the behavior of a sequence $a(n)$ at index k where $\Delta a_k < 0$ and $\Delta^2 a_{k-1} < 0$. Sketch a sequence that displays this behavior at index k. ◀

▶ SEVENTH REFLECTION

Complete Table 2.4 and explain how you obtained the entries in the $n = 4$ row. ◀

TABLE 2.4 Table of Differences

n	a_n	Δa_n	$\Delta^2 a_n$	Increasing or Decreasing	Rel max or Rel min	Concavity
0	−55.5	38	−20	inc		—
1	−17.5	18	−14	inc	no	down
2	0.5	4	−8	inc	no	down
3	4.5	−4	−2	dec	rel max	down
4	0.5	−6	4			down
5	−5.5	−2	10			up
6	−7.5	8	16			up
7	0.5	24	22			up
8	24.5	46	28	inc		up
9	70.5	74	undef.	inc		
10	144.5	undef.	undef.	—	—	—

Formal Definitions

Up to this point we have been relying on an intuitive understanding of a sequence as a list of numbers. We can formalize this notion using the concept of a function.

SEQUENCE

A *sequence* is a function whose domain is the set of either all positive integers or all nonnegative integers and whose range is a subset of the real numbers.

The TI calculator uses function notation when defining sequences in the Y= *window.*

To illustrate, consider the sequence $a_0 = 1, a_1 = 2, a_2 = 4, \ldots, a_n = 2^n, \ldots$. Viewed as a function on the nonnegative integers this sequencs maps 0 to 1, 1 to 2, 2 to 4, etc. We can even use function notation by writing $a(0) = 1, a(1) = 2, a(2) = 4, \ldots, a(n) = 2^n, \ldots$.

FIRST DIFFERENCE

The *first difference* for any sequence

$$A = \{a_1, a_2, a_3, \ldots\}$$

is defined as follows

$$\Delta a_1 = a_2 - a_1$$
$$\Delta a_2 = a_3 - a_2$$
$$\Delta a_3 = a_4 - a_3$$

and, in general

$$\Delta a_n = a_{n+1} - a_n$$

for any value of $n > 0$.

SECOND DIFFERENCE

The *second difference* for any sequence

$$A = \{a_1, a_2, a_3, \ldots\}$$

is defined as follows

$$\Delta^2 a_1 = \Delta a_2 - \Delta a_1$$
$$\Delta^2 a_2 = \Delta a_3 - \Delta a_2$$
$$\Delta^2 a_3 = \Delta a_4 - \Delta a_3$$

and, in general

$$\Delta^2 a_n = \Delta a_{n+1} - \Delta a_n$$

Putting It All Together

First and second differences allow us to analyze discrete data both numerically and graphically. Roughly speaking, first differences relate how quickly the data is increasing or decreasing while second differences provide information regarding concavity (or bending) of the data. We will encounter these ideas again in the context of continuous functions when we study the derivative.

EXERCISE SET 2.1

1. Complete the difference table (Table 2.5) at the end of this exercise set. Identify critical features of the graph of the sequence in columns 5, 6, and 7.

TABLE 2.5   Sequence for Exercise 1

n	a_n	Δa_n	$\Delta^2 a_n$	Increasing or Decreasing	Rel max or Rel min	Concavity
0	78					
1	77					
2	76					
3	65					
4	77					
5	78					
6	79					
7	78					
8	77					
9	75					

2. The data in Table 2.6 represent the speed of an automobile in mph and the braking distance in feet. Graph the data and build a difference table. Use the first and second differences to predict the braking distance for speeds of 85 mph and 90 mph.

TABLE 2.6   Braking Distance*

Speed	20	25	30	35	40	45	50	55	60	65	70	75	80
Dist.	42	56	74	91	116	142	173	209	248	292	343	401	464

* Based on tests conducted by the U.S. Bureau of Public Roads

3. The data in Table 2.7 give the length of a bass in inches and its weight in ounces (for six bass). Graph the data and build a difference table. What do the first and second differences predict about the weight of a 20-inch bass?

TABLE 2.7   Bass Length and Weight

Length (in)	12	13	14	15	16	17	18
Weight (oz)	17	17	18	21	27	36	49

4. Table 2.8 displays the number of days since the beginning of an epidemic in a small community and the number of people ill on each day. Graph the data and build a difference table. What do the first and second differences tell you about how the epidemic is spreading?

TABLE 2.8   Day and Number Sick

Day	1	2	3	4	5	6	7	8	9	10
Number Sick	1	2	4	8	11	13	15	17	17	18

5. Sketch a sequence that has a relative minimum at a_4. What can you say about $\Delta^2 a_3$?

6. Sketch a sequence that satisfies the following conditions:
$\Delta a_k > 0$ for $0 \leq k < 3$
$\Delta a_k < 0$ for $3 \leq k \leq 5$.

7. Explain the significance of an inflection point in terms of how fast a rumor or a disease is spreading.

8. Build a difference table for the sequence $a_n = 2n - 3$. How do the first and second differences vary with n?

9. Build a difference table for the sequence $a_n = n^2 - 5n + 4$. How do the first and second differences vary with n?

10. **a.** Build a difference table for the sequence $a_n = 10n^2 + 4n - 13$, $n = 0, 1, 2, 3, 4, 5, 6$. Comment on the second differences.

b. Consider the sequence $a_n = b_1 n^2 + b_2 n + b_3$, $n = 0, 1,$ 2, 3, 4, 5 where b_1, b_2, and b_3 are arbitrary constants. Show that the second differences are constant.

11. Show that the second differences of the following sequences are constant. What type of function might represent such sequences? Plot the sequence on the TI calculator.
   a. $\{0, 1, 4, 9, 16, 25\}$
   b. $\{4, 5, 2, -5, -16, -31, -50\}$
   c. $\{1, 4.1464, 10.757, 20.831, 34.37, 51.372, 71.839\}$

In Exercises 14–19, use the first differences to determine whether the function is increasing or decreasing on the interval [2, 5].

12. $f(x) = -2x^2 + 5x - 4$

13. $g(x) = 0.12x^2 - 3x - 5$

14. $h(x) = \sin(x)$

15. $f(x) = x^3 - 4x^2 - 3x + 2$

16. $g(x) = \ln(x)$

17. $h(x) = 2^x$

In Exercises 20–22, use second differences to find at least one interval on which the function is increasing at an increasing rate or explain why such an interval does not exist.

18. $g(x) = 0.12x^2 - 3x - 5$

19. $h(x) = \sin(x)$

20. $g(x) = \ln(x)$

21. Explain why a function is increasing if its first differences are positive. Include a diagram.

22. Explain why a function is decreasing if its first differences are negative. Include a diagram.

2.2

Contructing Models from Patterns

Raw data comes to us in many forms: in newspaper articles and charts, in textbook examples, in measurements from laboratory experiments, and in technical reports. In this section, we create models from patterns revealed by graphs and by first and second differences.

Sun Microsystems's Yearly Revenues

Approximating Data with a Linear Function.　After years of studying linear functions, you should be pleased to know that they play an important role in modeling. The general linear function is represented symbolically in various ways: sometimes as $f(x) = m \cdot x + b$, sometimes as $f(x) = a \cdot x + b$, sometimes as $f(x) = a_0 + a_1 \cdot x$, or even $f(x) = a + b \cdot x$ — depending on the context. In the next example, we use linear functions to model the revenues of Sun Microsystems.

The TI calculator uses the form
$$f(x) = a \cdot x + b.$$

E X A M P L E　1　**Sun Microsystems's Revenues**

The revenues for Sun Microsystems for the period 1986 to 1996 are recorded in Table 2.9. We wish to model this data and use our model to predict future revenues in the years 2001 and 2010.

TABLE 2.9

Year	86	87	88	89	90	91	92	93	94	95	96
Revenues (in millions)	210	538	1052	1765	2466	3221	3589	4309	4690	5902	7095

To duplicate the table create a data file titled sunreven *with the* Data/Matrix *editor. Enter the year and revenue data in columns* c1 *and* c2. *To obtain the graph in Figure 2.14, define a* Plot *with column* c1 *for* x . . . *and* c2 *for* y. . . . *Use the* ZoomData *command to graph the plot.*

Figure 2.13 shows the beginning of a table in which the first column represents the years and the second column contains the corresponding Sun Microsystems revenue in millions of dollars. (To help keep the size of numbers manageable, the years column uses year 1 for 1986).

The data points plotted in Figure 2.14 appear to lie nearly in a straight line. In order to make predictions about the future of Sun Microsystems's revenues,

FIGURE 2.13 Data Table for Revenues for Sun Microsystems: 1986 through 1996

FIGURE 2.14 Scatter Plot for Revenues for Sun Microsystems: 1986 through 1996

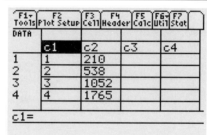
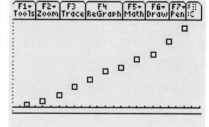

we would like to find the equation of a line that matches our data plot.

▶ **FIRST REFLECTION**

Use two points from the data (or visually estimate a slope and intercept) to get a quick estimate for a linear equation that fits the data. Record your equation as y1(x) and press ReGraph F4 to display both the graph of the data and your line. How well does your line match the data? ◀

The Linear Least Squares Line. Even though the data plot looks reasonably linear, the task of finding a particular linear model can be challenging. Many lines may fit the data well. Which one do we pick? One choice uses the popular curve fitting method of *linear least squares regression* or simply a *linear regression*. This method minimizes the sum of the squares of the differences between the actual data values and those predicted by the line. We have not yet developed the tools to go into the theory behind linear regression (those tools will be developed in chapter 10) but we can still apply the method using technology. Follow the directions in *Fitting Linear Functions to Data* to find the equation of the linear regression function for Sun Microsystems.

The method of linear regression produces the function

$$y2(x) = 663.69 \cdot x - 815.15$$

The TI calculator abbreviates linear regression as LinReg.

FITTING LINEAR FUNCTIONS TO DATA

Bring up the Data/Matrix editor with the sunreven data table. Open the Calc menu. From the Calculation Type... submenu choose LinReg. In the resulting dialogue box, enter c1 for x... and c2 for y.... Select y2(x) as the place to store RegEQ. This last selection will store the regression equation we are about to calculate as y2(x) in the Y= editor for future reference. Press ENTER to calculate and store the regression equation. In addition to the coeficients a and b, a parameter called the correlation coefficient, which is denoted *corr* by TI calculators, indicates how closely the line fits the data. The correlation coefficient always lies between -1 and 1. For very good fits the correlation coefficient will be very close to 1 or -1. The square of the correlation coefficient, denoted R^2, also indicates goodness of fit but avoids negative numbers. Thus, good fits result in values of R^2 which are close to 1.

---

where the coefficients are given to the nearest hundredth. A graph of the data points along with the linear function reveals a very close fit (see Figure 2.15).

FIGURE 2.15   Scatter Plot and LinReg Lineof the Data

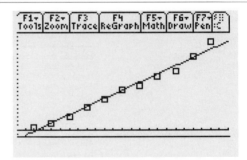

Notice that the line seems to touch all the data points except the two corresponding to 1994 and 1996.

In this example we chose a linear model by recognizing a pattern in a plot of the data. For very large data sets, scatter plots may be difficult to construct or hard to read. Also, computers and calculators do not recognize patterns in graphs as easily as humans. As an alternative, we can use first differences to provide numeric support for a linear model.

First Differences in Linear Functions

In Table 2.10, you can write down first differences generated by odd number inputs to the linear function $f(x) = 3x - 5$.

TABLE 2.10   First Differences

x	$f(x)$	**First Differences**
1	−2	
3	4	
5	10	
7	16	
9	22	
11	28	
13	34	
15	40	

▶ **SECOND REFLECTION**

What are the differences between successive values of x? What are the differences between successive $f(x)$ values? How are these differences related to the slope of the linear function? ◀

In the language of sequences, Table 2.10 contains the first differences of the sequence of values of x and of the corresponding sequence values of $f(x)$.

Figure 2.16 shows the x values, the $f(x)$ values and the first differences for the $f(x)$ values.

The first differences are all the same! To see why, we compute $f(x+2) - f(x)$ which is

$$f(x+2) - f(x) = (3(x+2) - 5) - (3x - 5) = 3x + 6 - 5 - 3x + 5 = 6.$$

Notice that x does not appear in the result. In words, "if we increase the argument or input for f by 2 from x to $x+2$, then the value for f increases by 6 regardless of the value of x. If we take the quotient of the increase in f over the increase in x, 6/2, we obtain the slope, 3.

▶ **THIRD REFLECTION**

Will the first differences for any linear function $g(x) = mx + b$ always be a constant if the x values are always increased by 2? Justify your response by computing $g(x+2) - g(x)$. What if the x-values are always increased by 5? By h? If the first differences for a linear function are always constant, how are they related to the slope, m? ◀

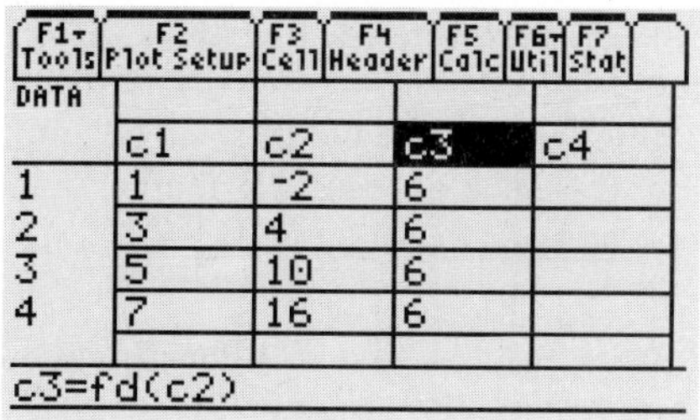

Let's reexamine our original data.

Determining Linearity from First Differences

E X A M P L E 2 **Sun Microsystems's Revenues Revisited**

The function fd applied to the Sun Microsystem data produces the first differences displayed in Figure 2.17. Notice that although these differences are not constant many of them are close to 700. Let's calculate the mean. A mean close to 700 might suggest that the data can be reasonably modeled using a linear function. We obtain a value of 688.5 for the mean of the first differences as shown in Figure 2.18. We use this mean to approximate the slope of a linear model, and estimate the y-intercept by subtracting our slope approximation from the first revenue value. Thus, the y-intercept is -478.5 and we obtain a third linear model for the data:

$$f(x) = 688.5x - 478.5.$$

To compute first differences for the function values in Table 2.10 using the Data/Matrix *editor, create a new data table with the x-coordinates in column* c1 *and the y-coordinates in* c2 *and define* c3 *to be* fd(c2). *(Recall that* fd=shift(q,1)-q *computes first differences of the sequence of numbers in the list* q.)

FIGURE 2.16 First Differences Calculated

F1▾ Tools	F2 Plot Setup	F3 Cell	F4 Header	F5 Calc	F6▾ Util	F7 Stat
DATA						
	c1	c2	c3	c4		
1	1	-2	6			
2	3	4	6			
3	5	10	6			
4	7	16	6			
c3=fd(c2)						

To calculate the mean of these first differences, we erased the formula from c3 *and deleted the undefined entry in the 11th row of the* c3. *We then entered* approx(mean(c3)) *in the first row of the* c4 *column.*

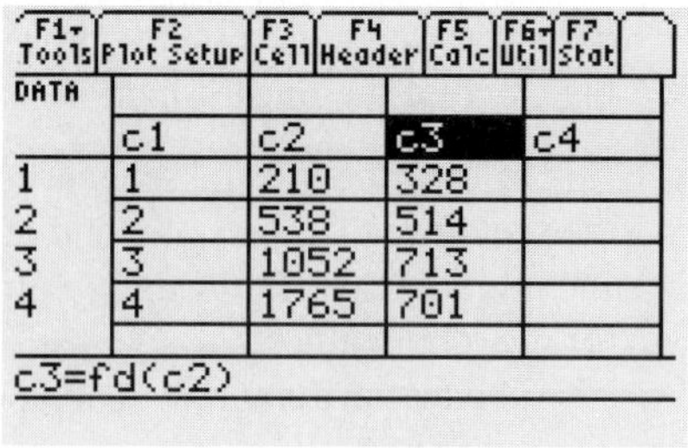

FIGURE 2.17 First Differences 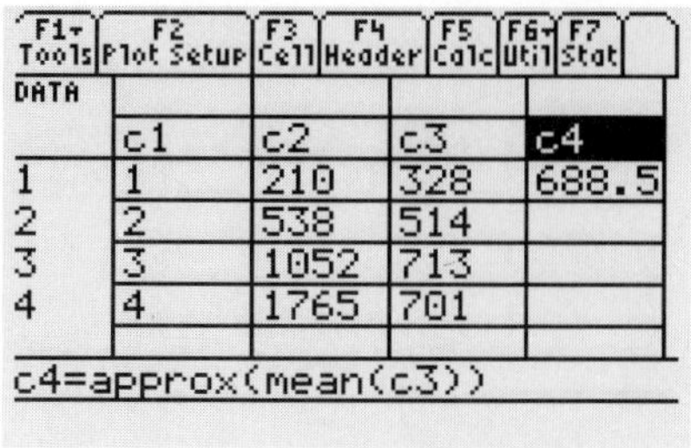   **FIGURE 2.18** The Mean of theFirst Differences

▶ **FOURTH REFLECTION**

Why can we use the average of the first differences for the slope without dividing by the increment from x-coordinate to x-coordinate? Which of the three ways of computing a linear approximation to the Sun Microsystems revenue data gives the best fit and why? Which method do you understand best?   ◀

In the next example we extend our approach to a non-linear situation.

E X A M P L E   3   **A Bouncing Golf Ball: Graphics Approach**

The picture in Figure 2.19 contains photographic images of a golf ball bouncing on a hard surface. The images were recorded on the same photograph using a strobe light flashing 300 times per second. Notice that the path of the golf ball bears a striking resemblence to the graph of a parabola. Scientists frequently focus on such patterns in nature in the hope that they will reveal some fundamental property of the universe.

The flight of the golf ball can be viewed as a combination of horizontal and vertical motion. We restrict our attention to the vertical motion and leave the horizontal motion as an exercise. A careful inspection of the photograph reveals that between the fifth and sixth images from the left of the photograph the golf ball bounced on the hard surface. To avoid the complications of the bounce, we focus on the motion beginning with the sixth image from the left. Using the vertical scale provided in the figure, we measure the distance from the bottom of each image to the hard surface (the bottom of the ball is relatively easy to locate). To include time, we recall that the strobe light flashes every 300th of a second; hence, the golf ball images occur every 0.03 seconds. The time values and our height measurements are recorded in Table 2.11. Notice that to simplify the table, we have used 0.03 seconds as our unit of time. As a further simplification, we assign time 0 to the moment when the strobe flash created the image of the sixth ball.

The photograph is not life-size. Determining actual distances requires a scaling factor relating measurements in the photograph to the real world. A library search reveals that United States Golf Association approved golf balls

FIGURE 2.19 "Golf Ball Bounce" photographed by Harold E. Edgerton

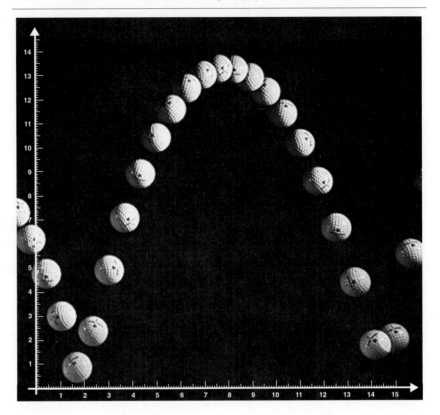

must be at least 1.680 inches in diameter. To obtain a scaling factor, first determine the diameter of one of the balls using the scale in the picture. Assuming the smallest approved diameter for the golf ball, divide your measurement into 1.68 to obtain the scaling factor.

▶ FIFTH REFLECTION

Why would manufacturers avoid oversized golf balls?   ◀

Figure 2.20 displays the time data in column $c1$, the measured height data in column $c2$, and the scaled height data in column $c3$.

Figure 2.21 displays a graph of the scaled heights as a function of time.

TABLE 2.11 Time vs. Vertical Locations

Time in 0.03 Seconds	0	1	2	3	4	5	6	7	8	9	10
Vertical Location	1.7	4.3	6.4	8.2	9.7	10.9	11.7	12.3	12.6	12.5	12.2
Time in 0.03 Seconds	11	12	13	14	15	16	17				
Vertical Location	11.6	10.7	9.5	7.9	5.9	3.6	0.9				

To duplicate Figure 2.20 use
`Data/Matrix` *editor to create a table*
named `golfdata` *with the data from*
Table 2.11 in columns `c1` *and* `c2`*. Use*
your scaling factor to compute column
`c3` *from column* `c2`*.*

FIGURE 2.20 `golfdata`

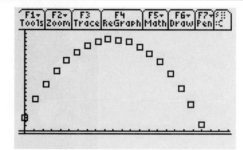

F1▾ Tools	F2 Plot Setup	F3 Cell	F4 Header	F5 Calc	F6▾ Util	F7 Stat
DATA						
	c1	c2	c3	c4		
1	0	1.7	2.197			
2	1	4.3	5.557			
3	2	6.4	8.271			
4	3	8.2	10.6			
c3=1.2923*c2						

FIGURE 2.21   Scaled Height vs. Time Using `ZoomData`

F1▾ Tools	F2▾ Zoom	F3 Trace	F4 ReGraph	F5▾ Math	F6▾ Draw	F7▾ Pen

Just as the photograph appears to show the golf ball following a parabolic arc, the plot of the height versus time also appears to be parabolic.

▶ SIXTH REFLECTION

Is the graph from Figure 2.21 a plot of the path followed by the golf ball in the photograph? How can you tell? Do the vertical and horizontal variables represent the same quantities? (We will explore this point in the exercises). ◀

The Quadratic Least Squares Fit.   As in the Sun Microsystems example, we can use regression to obtain a function which models the vertical location of the golf ball. However, because the data appears to lie on a parabola, we use quadratic rather than linear regression and obtain the function $f(x) = -0.201063x^2 + 3.367398x + 2.271727$. The R^2 value of 0.999428 indicates a very good fit indeed.

To find the quadratic regression model
using the `golfdata` *table follow the*
steps given in Fitting Linear Functions
to Data using `c1` *for* `x...`*,* `c3` *for* `y...`*,*
choosing `QuadReg` *from the*
`Calculation Type...` *submenu*
instead of `LinReg`*, and storing the*
result in `y4(x)`*.*

The Differences Approach.   The first differences represent the vertical distance traveled by the golf ball during each time interval. Figure 2.22 displays these first differences for the first four data entries. Examining these differences (`c4`) in all the rows of your TI calculator table, we see that some differences are positive and some are negative. When the differences are positive, the distance from the golf ball to the surface is increasing. In other words, the ball is moving up. Notice that the positive differences get smaller as time passes. Thus the ball travels a smaller and smaller distance during each equally placed time interval which indicates the ball is slowing down.

Figure 2.23 takes the analysis one step further by computing the second differences of the height data. Notice that almost all of them are about the same, in other words, the second differences are nearly constant.

Add first differences to your table by entering fd(c3) *into the entry line next to* c4=.

FIGURE 2.22 golfdata: First Differences of Heights   **FIGURE 2.23** golfdata: First and Second Differences

	c1	c2	c3	c4
DATA				
1	0	1.7	2.197	3.36
2	1	4.3	5.557	2.714
3	2	6.4	8.271	2.326
4	3	8.2	10.6	1.938

c4=fd(c3)

	c2	c3	c4	c5
DATA				
1	1.7	2.197	3.36	-.646
2	4.3	5.557	2.714	-.388
3	6.4	8.271	2.326	-.388
4	8.2	10.6	1.938	-.388

c5=fd(c4)

▶ **SEVENTH REFLECTION**

Describe the behavior of the first differences when they are negative. Interpret the meaning of these numbers in terms of the flight of the golf ball. Plot the first differences against time. What sort of pattern do you observe? Plot the second differences against time. Again, describe the pattern. How does your plot of the second differences relate to the statement that they are "nearly constant"? Examine the first and second differences for our quadratic regression model by defining c6=y4(c1), c7=fd(c6), and c8=fd(c7). Describe what you observe.   ◀

Recognizing Quadratic Functions from First and Second Differences

We have seen that linear relations produce data with constant (or nearly constant) first differences. We have also found patterns in the first and second differences of the parabolic data obtained from the vertical flight of the golf ball. Are these patterns common to all parabolas?

We can investigate this question by examining tables of first and second differences generated from quadratic functions. Figure 2.24 displays first and second differences for the function $f(x) = 3x^2 - 16x + 5$ applied to the sequence 0, 1, 2, ...

Figure 2.25 shows a plot of the first differences against x, along with a graph of the function f.

▶ **EIGHTH REFLECTION**

How do the positive and negative first differences relate to the behavior of the function f? Can you use the first differences to estimate where the vertex of the parabola will occur? Check your guess by determining the vertex of the parabola either algebraically or by graphing. The second differences are

To create the table in Figure 2.24 define
the function f on the home screen. Use
the Data/Matrix *editor to create a*
new data table called quadrat. *Enter*
seq(i,i,0,10,1) *in the entry line*
c1=, *enter* f(c1) *in the entry line* c2=,
enter fd(c2) *in the entry line* c3=, *and*
enter fd(c3) *in the entry line* c4=. *To*
see all four columns at once, you may
need to adjust the cell width using the
Format... *command in the* Tools
menu .

If you have created the data table
quadrat *you can create a new table*
simply by redefining f .

FIGURE 2.24   Differences for
$f(x) = 3x^2 - 16x + 5$

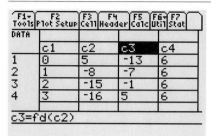

FIGURE 2.25 x vs First Differences and f

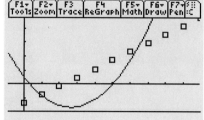

constant. What does this suggest about the path followed by the plot of the first differences? What property of the plot of the first differences can be deduced by observing that the second differences are all positive?   ◄

Break into groups to create tables of first and second differences for the functions.

$$f(q) = 2q^2 - 16q + 5$$
$$f(q) = -2q^2 - 16q + 5$$
$$f(q) = -4q^2 - 16q + 5$$
$$f(q) = 4q^2 - 16q + 5$$

Think and Share

1. Plot the first differences for these functions. What kind of function does the plot indicate?
2. Plot the second differences for these functions. What kind of function does the plot indicate?
3. Generalize your observations by making a conjecture regarding the first and second differences of any quadratic function.
4. Change the interval length to 0.5. (You can do this most easily by putting seq(x,x,0,10,0.5) in the entry line c1=. How does this change affect the first and second differences? Does your conjecture still appear to hold?
5. What kind of expression would you expect for the second differences of a cubic equation? Check your conjecture with a few examples.

First and Second Differences Symbolically

We can use algebra to verify our observations based on tables and graphs. Consider the quadratic function $f(x) = 3x^2 - 16x + 5$. To determine first differences for this function, we used the function $fd(f(x))$. A bit of simplification reveals that $f(x + 1) - f(x) = 6x - 13$. Thus, the first differences are generated by a linear function. Thus, the second differences of f are the first differences of a linear function. Given our earlier discoveries regarding linear functions, we should not be surprised that the second differences of f are constant.

More generally, consider the quadratic function $f(x) = ax^2 + bx + c$. Computing the first difference for an interval of length one, we find that

$$f(x + 1) - f(x) = 2ax + a + b. \qquad (2.3)$$

The parameters a and b will be constant for any particular quadratic function; so this function will always be linear.

▶ **NINTH REFLECTION**

Suppose we vary the length of the interval. Compute $f(x + 2) - f(x)$ for the quadratic function $f(x) = ax^2 + bx + c$. Is the coefficient of x the same as in Equation (2.3)? How can you tell? Is the result still linear? Repeat the process for $f(x + 0.5) - f(x)$. Is it still linear? ◀

Putting It All Together

Patterns in data often reveal important underlying fundamental relations. Humans posess a talent for recognizing patterns in pictures or graphs. (Sometimes that talent leads us to see patterns even when they don't exist!) We can reinforce visual impressions by matching first and second differences with the patterns of such differences for known classes of functions. If the first differences for a data table based on regularly spaced inputs are nearly constant, for example, then the relation is nearly linear. If the second differences are nearly constant, then the relation has a quadratic flavor.

Being numeric, differences are better suited than graphs to analysis by technology such as computers or calculators. Moreover, differences are defined algebraically and hence can be analyzed algebraically. We will see that this analysis of first and second differences forms the basis for the study of change known as differential calculus.

EXERCISE SET 2.2

1. The percentage of funding for public elementary and secondary mathematics education provided by the federal government during the 1980's is given below. The years 81–82 correspond to $t = 1$, 82–83 correspond to $t = 2$, and so on.

TABLE 2.12 Percentage of Funding for Public Elementary and Secondary Mathematics Education

t	1	2	3	4	5	6	7	8	9
Percentage	7.4	7.1	6.8	6.6	6.7	6.4	6.3	6.2	6.1

a. Use the first and the last data points to find an equation of a straight line (linear model) with which to fit the data.

b. Find the `LinReg` linear equation with which to fit the data. Report your equation here, retaining the decimal places.

c. Which line do you think would make your favorite mathematics professor the happiest, the one obtained in part (a) or in part (b)?

2. a. Explain why you think a linear function would be an appropriate model for the data on the Dow Jones Industrial Index in Table 2.13. Base your explanation on the data plot and the first differences.

b. Pick two likely points and use them to find a linear model for the data. Use `LinReg` to obtain a second linear model for the data set. What index is predicted by each model for the year 2000?

TABLE 2.13   Dow Jones Industrial Index

Year	1988	1989	1990	1991
Index	2060.8	2508.9	2678.9	2929.3

Year	1992	1993	1994
Index	3284.3	3524.7	3794.2

3. The data in Table 2.14 below gives National Science Foundation figures for the amount of money in millions spent by the United States federal government on mathematics research in the years 1980 through 1990. This information is reported in the March issue of *The Mathematics Teacher*, the Journal of the National Council of Teachers of Mathematics. This exercise was suggested by Tom Walters of the Los Angeles Unified School District.

TABLE 2.14   NSF Funding

Year	1980	1981	1982	1983	1984	1985
Math Research	91	118	128	134	151	184

Year	1986	1987	1988	1989	1990
Math Research	185	205	212	230	245

a. Plot the data in the table, and use this plot to judge whether the data is linear. Justify.

b. Determine if the data is linear by observing the first differences. Describe your results.

c. Use the `LinReg` linear approximation command of the TI calculator to create a linear function that models the data. Use your model to predict federal government spending on mathematics research in the year 2000 and the year 2010.

d. Compute the average of the first differences. Use this average along with an appropriate *y*-intercept to obtain a second linear function to model the data. Use your model to predict federal government spending on mathematics research in the year 2000 and the year 2010.

e. Which linear model predicts the greatest total research funding in the year 2020? Explain your answer.

f. Do the second differences indicate that a quadratic model could be appropriate for this data? Justify.

4. One measure of productivity is the revenue per employee (total revenue/number of employees). AtSun Microsystems, the revenue per employee is given in Table 2.15. Find a model for this data. Compare the model with the Sun Microsystems revenue model.

TABLE 2.15   Revenue per employee

Year	86	87	88	89	90	91	92	93	94	95	96
Revenues (in thousands)	104	134	164	195	225	256	286	316	347	377	408

5. Measure the horizontal locations of the golf ball images in Figure 2.19 starting with the sixth image. Enter the data in a table and compute the first differences. Explain why the first differences indicate that the horizontal motion is a linear function of time. Find a linear function to model the relation between the horizontal locations and time. Determine the horizontal velocity in feet per second. (Be careful about your units and don't forget the conversion factor).

6. Show that the second differences of the following sequences are constant and find a quadratic function $a_n = f(n)$ that fits the data points.
   a. $\{0, 1, 4, 9, 16, 25\}$
   b. $\{4, 5, 2, -5, -16, -31, -50\}$
   c. $\{1, 4.1464, 10.757, 20.831, 34.37, 51.372, 71.839\}$

7. Explain how to analyze *first* differences to determine the appropriateness of a *linear* model.

8. Explain how to analyze *second* differences to determine the appropriateness of a *quadratic* model.

9. Make a conjecture regarding the first and second differences of any quadratic function. (See the Think and Share following the Eighth Reflection.) Support your conjecture by providing two original examples.

10. What kind of expressions would you expect for the first, second, and third differences of a cubic function? Justify with at least one original example.

11. What is the degree of an expression for first differences, if the function is of degree 3 (cubic)?

12. What is the degree of an expression for first differences if the function is of degree *n*? You do not need to prove your answer, but you should justify it with a few simple examples.

13. We know that the first differences of a linear function are constant. Does this mean that constant first differences (for a fixed subinterval length) imply a linear function? Explain.

14. We know that the first differences of a quadratic function are linear. Does this mean that linear first differences (for a fixed subinterval length) imply a quadratic function? Explain.

15. Do constant second differences imply a quadratic function? Explain.

16. Use the horizontal measurements from Exercise 5 and the vertical measurements given in Table 2.11 to find a quadratic function which models the vertical location as a function of the horizontal location. How does the graph of this function relate to the parabolic arc followed by the golf ball? Using your quadratic function and the linear function from the previous problem, determine a quadrative function which models the vertical location as a function of time. Comment on similarities and differences between this model and the ones given in the text.

17. According to the laws of physics, the vertical motion of the golf ball should be modeled by a function of the form

$y(t) = -(g/2)t^2 + v_y t + y_0$, where y_0 is the initial height, v_0 is the initial vertical velocity, and g is the gravitational constant. Estimate the value of g from one of the models of the golf balls vertical flight (be sure to include units). Convert your estimate for g into units of ft/sec^2. How does the estimate compare to g's commonly known value of $32 ft/sec^2$?

2.3

Divided Differences and Average Rate of Change

For regular data, data generated from regularly spaced input values, patterns revealed by first and second differences often suggest useful models. Unfortunately, not all data is regular. Furthermore, we often wish to go beyond the question of how much the data changes. Having your bank balance increase by 100 dollars in a year differs dramatically from having your bank balance increase by 100 dollars in a week. To handle such situations, we adjust our computation of first differences to obtain a quantity known as a divided difference.

Irregular Data and Divided Differences

E X A M P L E 1 **Ponderosa Pines**

A logging company which plants and grows trees must plan when to harvest the trees in order to optimize the lumber produced. Such planning depends on a number of factors including the rate at which the trees grow, the desirability of the type of wood, and the number of board feet of wood produced by trees of certain sizes. To develop a model of production versus size, the company might well collect the sort of data displayed in Table 2.23 which displays measurements of girth and number of board feet for 10 trees randomly selected from a forest of ponderosa pines.

TABLE 2.16 Ponderosa Pine

Girth (in)	53	60	63	72	78	88	100	119	122	129
Board Feet	19	26	32	48	61	86	128	214	232	275

To duplicate the plot, create a table labeled pines, *enter the girth data in column* c1, *the board feet data in column* c2, *create a plot of* c1 *versus* c2 *and invoke the* ZoomData *command.*

The plot of this data displayed in Figure 2.26 reveals a definite pattern in the relation between the number of board feet and the girth of the tree.

FIGURE 2.26   Girth vs. Board Feet Using `ZoomData`

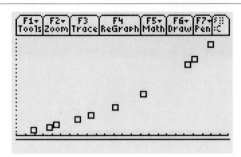

As we might expect, the larger the girth of the tree the greater number of board feet the tree will produce; but, the pattern suggests a stronger relation.

▶ **First Reflection**

Do you expect the first differences for the number of board feet to be positive or negative. Use `LinReg` to fit a line to the data and `QuadReg` to fit a parabola to the data. Which one fits better according to the correlation coefficient? What sort of pattern would you expect for the first differences if the data was regular. Explain your response.   ◀

Following our techniques from the previous section, we examine a graph of the first differences for the number of board feet as displayed in Figure 2.27.

Although positive, the first differences do not follow an easily recognized pattern. In cases like this one, where the original data is very nearly quadratic, we would expect the first differences to be very nearly linear. What went wrong?

Notice that the girths of the trees are not regularly spaced. (One cannot guarantee that the girths of ten randomly selected trees will differ by a uniform amount). Specifically, the girth increases by 7 inches from the first tree to the second tree while it only increases by 3 inches from the second tree to the third tree. In situations where the first differences of the input variable are not constant, we say the data is *irregular*. For the pine trees, the irregular differences between the values of the girth distort the pattern of changes in the number of board feet. To reduce the distortion, we can compute the increase in the number of board feet per unit increase in girth. Simply divide the first differences of the board feet by the first differences of the girths. Thus, from the first tree to the second tree the production increased from 19 to 26 board feet while the girth increased from 53 to 60 inches for an average increase of

$$\frac{26 - 19}{60 - 53} = 1 \text{ board foot per inch.}$$

We refer to these quotients of first differences as *divided differences*. A display of the plot of the divided differences for the pine tree data is shown in Figure 2.28. From the plot we see that the divided differences appear to follow the expected linear pattern.

FIGURE 2.27   Girth vs. First Differences Using `ZoomData`

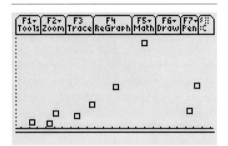

Let c3 = fd(c2), *create a plot of* c1 *versus* c3. *Turn off the plot of* c1 *versus* c2 *and invoke the* `ZoomData` *command.*

FIGURE 2.28   Girth vs. Divided Differences Using `ZoomData`

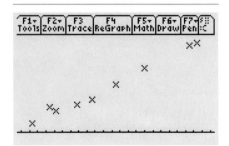

To create the plot of the divided differences, set c4 = fd(c2)/fd(c1), *create a plot of* c1 *versus* c4. *Turn off the plot of* c1 *versus* c3 *and invoke the* `ZoomData` *command*

▶ SECOND REFLECTION

The original pine tree data was ordered. The smallest tree apeared first then the next smallest tree, etc. Why was this ordering needed when considering first differences? Does the same problem occur with divided differences?   ◀

Average Rate of Change

The **average rate of change** of a function over an interval is the change in the value of the function over the interval divided by the length of the interval.

Understanding differential calculus requires an understanding of this concept. Fortunately, average rates of change appears in many contexts, both in mathematics and in our everyday lives. Divided differences compute average rates of change for discrete data. For a distance function, the average rate of change or average velocity during any time interval is the change in the distance function during that time interval divided by the time spent traveling.

E X A M P L E 2 **Daring Turnpike Drive**
_____

Susan and Joey were delighted not to see any patrol cars on the turnpike as they sped home for fall break. Joey kept track of their progress by recording the time they had been traveling at each tool booth. (Table 2.17 below contains Joey's distance versus time data). Even though they frequently exceeded the posted 70 mile per hour speed limit they managed to complete the 150-mile trip in just two hours without getting a speeding ticket. Unfortunately for them, their fare card had been stamped with the time and location of both their entry and exit. Imagine their outrage when they were pulled over and ticketed by the highway patrol as they left their exit's tollbooth.

TABLE 2.17 Distance vs. Time

t in hours	0	0.19	0.35	0.55	0.77	1.1	1.43	1.75	2
Distance in miles	0	13	26	42.5	60	85	110	135	150

▶ THIRD REFLECTION

Should Susan and Joey fight the ticket in court? If you were their lawyer, would you agree to represent them on a contingency basis (you only get paid if you win)? Justify, using mathematical reasoning.   ◀

The data in Joey's table tells the story of their trip. If we only consider the distances between the toll booths, however, we will miss the point of the story. According to the distance data $13 - 0 = 13$ miles separate the first and second toll booths and $26 - 13 = 13$ miles separate the second and third toll booths. These identical distances, however, do not reflect the experience of Susan and

Joey on these two stretches. To see this we must include the time information and compute average speed. The average speed between the first and second toll booths equals

$$\frac{13 - 0}{0.19 - 0} = 68.421 \text{ miles per hour}$$

while the average speed between the second and third toll booths equals

$$\frac{26 - 13}{0.35 - 0.19} = 81.25.$$

On average, Susan and Joey traveled 12.829 miles per hour faster during the second stretch than they did during the first stretch. On the second stretch, their average speed exceeded the speed limit by more than 10 miles per hour.

To compute the average speed over the various stretches of Susan and Joey's trip, we compute the length of the stretch divided by the time it took to complete it. Since the table contains total distance traveled to get the length of any particular stretch we compute the change in the total distance traveled for that stretch. Similarly, to get the time for any particular stretch we compute the change in the time over that stretch. Thus,the average speed equals the average rate of change of the distance traveled per interval of time. To compute the average speed we divide the first differences of the distance data by the first differences of the time data. In short, the average rate of change equals the divided difference.

▶ **FOURTH REFLECTION**

Create a table containing Joey's time and distance data. Compute the average speed over each stretch of the trip. Which two consecutive toll booths are farthest apart? During which stretch were Susan and Joey likely to be most anxious about speed traps? Use your table to justify your response.   ◀

Average Rate of Change for Continuous Functions

We can use the average rate of change to understand continuous functions in a manner which is analogous to our use of first differences and divided differences to understand discrete sequences.

EXAMPLE 3
_____

Consider the function

$$f(x) = 2x \cos(2.5x).$$

To find the average rate of change of this function over the interval [1, 5], we compute

$$\frac{f(5) - f(1)}{5 - 1} \approx 2.9.$$

FIGURE 2.29 Graph of $f(x) = 2x \cos(2.5x)$ on $[-1, 6] \times [-10, 10]$

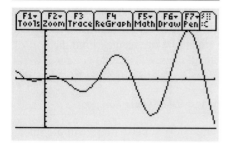

From this, we can say that on the interval $[1, 5]$ f increase *on average* by 2.9 for every increase of 1 in x. Be careful about that phrase "on average"! Consider the graph of f shown in Figure 2.29.

We can see that the value of the function increased on the interval $[1, 5]$ because the point $(5, f(5))$ is higher than the point $(1, f(1))$. Notice that this does *not* imply that the graph is increasing on the entire interval. To the contrary, at $(1, f(1))$ the graph starts out decreasing, then it increases, then it decreases again, and finally it increases to the point $(5, f(5))$.

The average rate of change provides the general trend of a function on an interval but hides the details of that trend. General trends can be quite useful. An executive trying to evaluate the effectiveness of a particular strategy may only have time to consider the general trends ("the bottom line"). On the other hand, average rate of change can be quite misleading. Although the function f shows an average increase on the interval $[1, 5]$, that increase is the result of several large up and down swings and the function is about to take a rapid plunge producing a negative average rate of change on the interval $[1, 6]$.

▶ **FIFTH REFLECTION**

How can you tell from the graph that f has a negative average rate of change on the interval $[1, 6]$? Verify this observation by computing the average change of f on the interval $[1, 6]$. ◀

Before continuing, we introduce a bit of helpful notation.

THE DELTA SYMBOL

The symbol Δf, read "delta f", means **the change in** f. Similarly, Δx refers to **the change in** x. We write $\dfrac{\Delta f}{\Delta x}$ to represent the change in f *with respect to x* or the average rate of change of f over an interval of length Δx. Similarly, the symbol $\dfrac{\Delta y}{\Delta x}$ means *the change in y with respect to x*.

FIGURE 2.30 Table of Divided Differences

	c1	c2	c3
1	1	-1.602	2.7369
2	2	1.1346	.94516
3	3	2.0798	-8.792
4	4	-6.713	16.691

r1c1=1

We can improve our understanding of the changes in f over the interval $[1, 5]$ by computing its average rate of change over smaller intervals. For example, by dividing $[1, 5]$ into intervals of length 1 we obtain the entries in column c1 of the Data table in Figure 2.30. Column c2 contains f of the entries in c1 and column c3 contains the average rate of change over each of the intervals (the values for $x = 5$ are in the next row (not shown) of the Data table.

The the third entry in c3 is negative which reflects the decrease in f from $(3, f(3))$ to $(4, f(4))$. The length 1 intervals are too large, however, to show the decrease in the function which occurs immediately after the point $(1, f(1))$. To see that decrease and get more detailed information on the changes in f, we can take even smaller intervals.

► **SIXTH REFLECTION**

Create a table by dividing the interval $[1, 5]$ into subintervals of length 0.1 and computing the average rates of change on those intervals. Do these average rates of change reflect the decrease at the beginning of the interval? Which subinterval corresponds to the largest average rate of change? Find the part of the graph that correspond to this largest rate of change. What do you observe? ◄

E X A M P L E   4   **Average Rates of Change for a Linear Function**
_____

For linear functions, average rates of change have a particularly simple interpretation. Consider the function $g(x) = 2x + 3$. To find the average rate of change of this function over the interval $[1, 5]$ we compute

$$\frac{\Delta g}{\Delta x} = \frac{g(5) - g(1)}{5 - 1} = \frac{13 - 5}{4} = 2.$$

To find the average rate of change of this function over the interval $[1, 2]$, we compute

$$\frac{\Delta g}{\Delta x} = \frac{g(2) - g(1)}{2 - 1} = \frac{7 - 5}{1} = 2.$$

In both cases, the average rate of change equals the slope of the line. Is this always true? Suppose we compute the average rate of change of g over the interval $[a, b]$. Using the formula for g, we have

$$\frac{\Delta g}{\Delta x} = \frac{g(b) - g(a)}{b - a} = \frac{(2b + 3) - (2a + 3)}{b - a} = \frac{2b - 2a}{b - a} = 2.$$

So, no matter what interval we consider, the average rate of change always equals the slope of the line.

► **SEVENTH REFLECTION**

Consider the general linear function $h(x) = mx + b$. Show that the average rate of change of h on the interval $[a, b]$ equals m. ◄

The result of the reflection leads to the following general statement regarding the average rate of change of a linear function.

AVERAGE RATE OF CHANGE: LINEAR FUNCTION

For the linear function $f(x) = mx + b$ the average rate of change of f on any interval equals m.

Average Rate of Change as the Slope of a Secant Line

We can also interpret the average rate of change graphically. For a function f, mathematicians use the term "secant line" to refer to a line which joins two points on the graph of f (this is *not* the same as the trigonometric definition of "secant"). Consider our earlier function $f(x) = 2x \cos(2.5x)$. The line through the point $(1, f(1))$ and the point $(5, f(5))$ is a secant line. Figure 2.31 displays the graph of f and this secant line.

To obtain the graph of f with the secant line return to the GRAPH *window. Select* line *from the pencil,* Pen, *menu on the* GRAPH *screen. Locate the cursor at the point $(1, f(1))$ and press* ENTER. *Then locate the cursor at the point $(5, f(5))$ and press* ENTER.

FIGURE 2.31   Graph of $f(x)$ and Secant Line

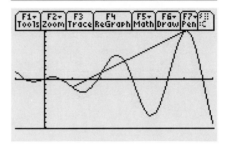

The line rises from left to right because the value of $f(5)$ is greater than $f(1)$. To find the slope of the secant line we use the formula

$$\frac{y_2 - y_1}{x_2 - x_1}$$

and compute

$$\frac{f(5) - f(1)}{5 - 1} \approx 2.895.$$

We made the exact same computation to determine the average rate of change of f on the interval $[1, 5]$.

More generally, *the slope of the secant line joining the point $(a, f(a))$ to the point $(b, f(b))$ equals*

$$\frac{f(b) - f(a)}{b - a},$$

the average rate of change of f over the interval $[a, b]$. This simple observation will prove to be extremely useful.

Back to the Turnpike

Joey and Susan are arguing about Joey's table. Joey claims that the table cannot be used to make the case that they were speeding.

"The table can only be used to compute average speed," says Joey. "Average speed is not the same as the speed on our speedometer. Just because our average speed was over the speed limit that doesn't mean we were speeding."

"Yes it does!!" says Susan. "You can't average 80 miles per hour if you never travel at 80 miles per hour."

Joey is right that average speed is not the same as the speed on a speedometer. The later speed is a special case of an *instantaneous rate of change*. Susan is right too! She understands that average rate of change and the instantaneous rate of change are related by an important theorem with the name *Mean Value Theorem*. We give an informal statement of the theorem here.

MEAN VALUE THEOREM

If the graph of a function f is smooth (no sharp corners) and continuous (no breaks, holes, or jumps) on an interval $[a,b]$, there will always be at least one number between a and b where the (*instantaneous rate of change*) is equal to the (*average rate of change*).

EXERCISE SET 2.3

1. Consider the data in Table 2.18. Plot of the points in the table

TABLE 2.18

x	0	10	16	34	38
y	0	100	256	1156	1444

using the ZoomData feature to obtain a good plot. What sort of pattern do you observe? If the data were regular, what sort of pattern would you expect for the first differences? Plot the first differences and the divided differences. Describe the patterns your observe. Which is a better measure of the relation between the x and y?

2. Consider the data in Table 2.19. Plot of the points in the table

TABLE 2.19

x	0	50	75	87	93
y	3	2503	5628	7572	8652

using the ZoomData feature to obtain a good plot. What sort of pattern do you observe? If the data were regular, what sort of pattern would you expect for the first differences? Plot the first differences and the divided differences. Describe the patterns your observe. Which is a better measure of the relation between the x and y?

3. Consider the data inTable 2.20. Plot of the points in the table

TABLE 2.20

x	0	15	40	54	75	87	93	113
y	−215990	−91115	−7990	−206	3385	19693	35947	148887

using the ZoomData feature to obtain a good plot. What sort of pattern do you observe? If the data were regular, what sort of pattern would you expect for the first differences? Plot the first differences and the divided differences. Describe the patterns your observe. Which is a better measure of the relation between the x and y?

4. Table 2.21 displays the world running records at various distances as of October 1996.

TABLE 2.21   World Records

Distance (meters)	100	200	400	800	1000
Time (seconds	9.84	19.32	43.29	101.73	132.18
Distance (meters)	1500	2000	3000	5000	10000
Time (seconds	207.37	287.88	445.11	764.39	1598.1

a. Create a plot of the Record times versus the distances. Use the ZoomData feature to obtain a good plot. Describe the pattern you observe. If the data were regular, what sort of pattern would you expect for the first differences?

b. Plot the first differences once again using the `ZoomData` feature. Explain why the plot does or does not meet your expectations.

c. Plot the divided differences. Ignoring the first two points, what sort of pattern do you observe? Explain in terms of running why the first two points don't follow the pattern.

5. Table 2.22 displays the average distance from the sun in millions of miles and the time of one revolution in days for each of the nine planets of our Solar System.

TABLE 2.22   Planet Revolutions

Distance (10^6 miles)	35	67	93	142	484
Time (days)	88	225	365	687	4332
Distance (10^6 miles)	887	1765	2791	3654	
Time (days)	10760	30684	60188	90467	

Create a plot of the Record times versus the distances. Use the `ZoomData` feature to obtain a good plot. Does your plot indicate a relation between the time of revolution and the distance from the sun? Plot the first differences and the divided differences. Describe the patterns your observe. Which is a better measure of the relation between the distance from the sun and the time of revolution? Justify your response by referring to the data.

6. Find an expression for the change in the function $f(q) = aq^2 + bq + c$ on the interval $[u, v]$. How does your expression depend on u? Find an expression for the average rate of change of f on the interval $[u, v]$. How does your expression depend on u? Which expression better represents change for a quadratic relation?

For each of the functions in Exercises 7–10: Compute the average rate of change of f from x_1 to x_2. Round your answer to two decimal places when appropriate. Interpret your result graphically.

7. $y(x) = 5x - 4$, $x_1 = 1$ and $x_2 = 7$

8. $y(x) = x^2 - 5x$, $x_1 = -1$ and $x_2 = 3$

9. $y(x) = 5 - 4\left(\dfrac{1}{2}\right)^x$, $x_1 = 0$ and $x_2 = 5$

10. $y(x) = \dfrac{2x^3 + 2x + 100}{x^2 + 1}$, $x_1 = 0$ and $x_2 = 5$

Your TI calculator has a function that will automaticly cahculate an average rate of change, `avgRC(`. The formula `avgRC(f(x),x,h)` computes $(f(x+h) - f(x))/h$, the average rate of change of f over the interval $[x, x + h]$.

For each of the functions in Exercises 11–14, calcuate their average rate of change over the interval $[x, x + h]$.

11. $f(x) = c$

12. $f(x) = mx + b$

13. $f(x) = ax^2 + bx + c$

14. $f(x) = ax^3 + bx^2 + cx + d$

15. What patterns can you find in the preceeding calculations?

For each of functions, f, in Exercises 15–20, graph both f and `avgRC(f(x), x, .1)` on a single set of axes.

16. $f(x) = -3x + 5$

17. $f(x) = x^2 - 5x + 4$

18. $f(x) = \sin(x)$

19. $f(x) = \sin(2x)$

20. $f(x) = 2^x$

21. $f(x) = 3^x$

22. Table 2.23 displays measurements for 10 ponderosa pines: the number of board feet in the tree, and its diameter in inches. Graph the data and build a difference table. What do the first and second differences tell you about the board feet in a tree of diameter of 45 inches? The intervals are of unequal lengths. Does this affect your prediction?

TABLE 2.23   Ponderosa Pine

Diameter (in)	17	19	20	23	25	28	32	38	39	41
Board Feet	19	25	32	57	71	113	123	252	259	294

23. Table 2.24 displays the total weight lifted by contestants in different weight categories during the weightlifting competition in the 1976 Montreal Olympics. Weight measurements are in pounds.

Graph the data and build a difference table. What do the first and second differences predict about the number of pounds a 300-pound person could lift? The intervals are of unequal lengths. Do our differences make sense?

24. Use the information in Table 2.25 to identify the x and y coordinates of the maximum, minimum, and inflection points for a certain mystery function. Justify, of course, but be brief.

TABLE 2.24   Weightlifting

Class	Max body wt	Total wt lifted
Flyweight	114.5	534.6
Bantamweight	123.5	578.7
Featherweight	132.5	628.3
Lightweight	149	677.9
Middleweight	165.5	738.5
Light-heavyweight	182	804.7
Middle-heavyweight	198.5	843.3
Heavyweight	242.5	881.8

TABLE 2.25   Difference Table for a Mystery Function

x	$f(x)$	Δf	$\Delta^2 f$
-4.0	28.0	6.125	-4.25
-3.5	34.25	1.875	-3.5
-3.0	36.0	-1.625	-2.75
-2.5	34.375	-4.375	-2.0
-2.0	30.0	-6.375	-1.25
-1.5	23.625	-7.625	-0.5
-1.0	16.0	-8.125	0.25

25. Table 2.26 gives the time in seconds that an average athlete takes to swim 100 meters freestyle at age x.

TABLE 2.26   Age vs. Time

x (yrs.)	10	12	14	16	18	20	22	24	26	28	30
Time (sec.)	84	70	60	58	54	51	50	49	51	53	57

a. Obtain a scatter plot of the data. Decide, from the shape of your plot, whether a linear or quadratic model is appropriate. Explain your decision and use the F5 Calc menu of the Data/Matrix editor to obtain either a linear or quadratic function with which to model the data.

b. Support your choice of model by analyzing first and second differences. Explain carefully.

c. What is the time predicted by your model for an athlete 20 years old?

26. Do divided differences always give more information about a function's behavior than first differences do? Explain.

27. Agree or disagree with the following statement and justify your opinion.

The way to select a model for a data set is to superimpose graphs of the different regression functions on the scatter plot of the data. The one that passes through the most points is the one to choose.

2.4

Rate of Change

In the previous section, we introduced the concept of divided differences to describe change in discrete data. We now consider divided differences and continuous functions to develop the *derivative* concept. The *derivative* will allow us to measure rates of change for continuous functions.

E X A M P L E   1   **Surfing in Bombay**

Even though you may not live near an ocean, you are probably familiar with the concept of high and low tides. Although the exact measurements of high and low tides are useful, sometimes different information is needed. Because the speed at which the tide comes in and goes out affects the size of waves, an international surfer is concerned about the tides. She wants to know how fast the water is rising or falling at a given time. In particular, our surfer wants to know at what rate the water level is rising at the beginning of a competition in Bombay at 11:00 A.M. The data showing depth of water at a point near the shore in Bombay, India, over a 14-hour period from sunrise to sunset are reproduced in Table 2.27. To determine how fast the tide is changing, we need to model the depth of the water as the tide moves in and out as time passes.

TABLE 2.27   Hour of the Day vs. Depth in Feet (Using a 24 Hour Clock)

Hr. of Day	6	7	8	9	10	11	12	13	14	15	16	17	18	19	20
Depth in Ft	2.5	2.7	4.5	7.5	11.1	14.5	17.0	18.0	17.2	15.0	11.7	8.0	4.9	2.9	2.5

SinReg is available on the TI-89 and the TI-92 Plus, but not on the TI-92.

Enter this data into the `Data/Matrix` editor and obtain a scatter plot using `ZoomData`. Our plot of the data is displayed in Figure 2.32. You can use the `SinReg` feature of the `Calc` submenu in the `Data/Matrix` editor to store a periodic function (the tides are periodic) in `y1`. The stored function is approximately

$$y1(x) = 7.8\sin(0.47x + 1.69) + 10.2.$$

The plot of the data and a graph of $y1$ are displayed in Figure 2.33.

FIGURE 2.32   Plot of the Bombay Tide Data   **FIGURE 2.33**   Graph of the Model

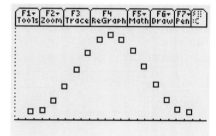
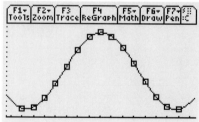

▶ **FIRST REFLECTION**

The period of a sinusoidal function of the form $f(x) = a\sin(bx + c) + d$ is $2\pi/b$. What is the period of $y1$ and what does this number mean in terms of the time between high and low tides? Between low tides?   ◀

We know a good deal about linear functions from the previous section. This function is not linear as we can see from the graph displayed in Figure 2.32. You will be glad to know that all is not lost: we can still apply what we have learned about rates of change of linear functions. But first, let's take a more detailed look at linear functions.

FIGURE 2.34   Leaking Water Tank, Standing Upright

Zero Rate of Change

E X A M P L E   2   **The Leaking Water Tank**

Examples!@The Leaking Water Tank Consider an old, neglected water tank. The tank is filled to a depth of five feet. Water is poured into the tank at the same rate at which it is leaking out. (The bottom of the tank has partially rusted and has holes in it. See Figure 2.34.)

Table 2.28 gives the depth at several times.

TABLE 2.28   A Non-Leaking Tank: Time vs. Depth

t in minutes	1	2	3	4	5	6	7
Depth in feet	5	5	5	5	5	5	5

Although the water is moving, its depth remains at five feet and the rate at which the depth is changing is zero. The numerator of any divided difference will be zero (a constant) since the depth is constant. In this case, the numerator of a divided difference will be (5-5).

Figure 2.35 shows a graph of the depth of the water as a function of time and Figure 2.36 shows a graph of the rate of change of the depth as a function of time.

We indicate that the depth h of the water in the tank depends upon time by identifying water depth as a function of time, using the functional notation $h(t) = 5$ where $h(t)$ represents the depth of water at time t. In a similar manner, we can show that the rate of change of the water depth is zero by writing $dh/dt = 0$, where dh/dt represents the rate of change of water depth h at time t.

FIGURE 2.35   Depth as a Function of Time   **FIGURE 2.36**   Rate of Change of Depth

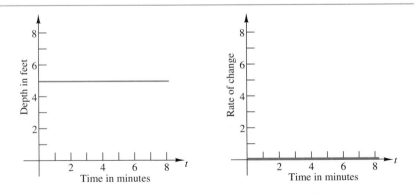

The symbol dh/dt means *derivative of h with respect to t* and it also means *rate of change of h with respect to t*. Another symbol used to indicate a derivative is $h'(t)$. We only introduce the notation at this point. We will have much more to say about the derivative in the next chapter.

FIGURE 2.37 Non-Leaking Water Tank, Standing Upright

If an amount is unchanging, and can therefore be described by a constant function, its rate of change (or derivative) is zero. This fact will be important to us throughout our study of calculus.

A Nonzero Constant Rate of Change

We now consider a situation in which functional values change by a constant nonzero amount. This type of behavior is modeled with a linear function.

EXAMPLE 3 **The Non-Leaking Water Tank**

Consider another water tank in good condition in the shape of a cylinder. The tank is standing vertically on one of its circular ends. (See Figure 2.37.)

This time, as we fill the tank with water at a constant rate, we observe that the depth increases steadily, also at a constant rate. The graph of water depth as a function of time, Figure 2.38, is a straight line which rises from left to right. The rate of change of the depth function is constant and unchanging, so its graph, Figure 2.39, is a horizontal line. In this case, the divided difference $\Delta y/\Delta x$ will always be the slope of the line.

FIGURE 2.38 Water Depth h as a Function of Time **FIGURE 2.39** Rate of Change of Water Depth as a Function of Time

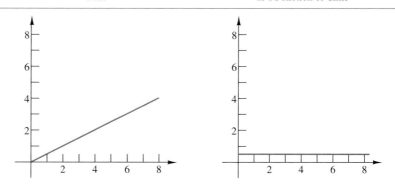

We can also generate tables to express these relationships. (See Table 2.29 and Table 2.30.)

TABLE 2.29   Time vs. Depth

t in minutes	1	2	3	4	5	6	7
Depth in feet	0.5	1	1.5	2	2.5	3	3.5

TABLE 2.30   Time vs. Rate of Change

t in minutes	1	2	3	4	5	6	7
dh/dt	0.5	0.5	0.5	0.5	0.5	0.5	0.5

The notation we use to represent the water depth as a function of time is $h(t) = 0.5t$, where $h(t)$ measures the depth of water at time t. The notation we use to indicate that the water depth h increases by 0.5 feet per minute is $dh/dt = 0.5$, or $h'(t) = 0.5$.

Putting It All Together

We now generalize the concepts of Examples 2 and 3. The graph of a constant function is a horizontal line and is used to describe a relationship between two quantities in which the output remains unchanged for any input (the leaky tank example). We say that the rate of change is zero. We *do not* say that the rate of change is nothing nor do we say that there is no rate of change.

The graph of a nonconstant linear function is a straight line which rises or falls from left to right. Such a function is used to describe a relationship between two quantities in which the output increases or decreases by the same amount for each fixed change in the input. The graph in Figure 2.40 illustrates the case in which the output increases by the same amount for each fixed change in the input (the upright water tank example). The rate of change is a positive constant and the graph of the function rises from left to right.

As you probably expected, relationships like the ones in Examples 2 and 3 can be modeled by linear functions of the form $f(x) = mx + b$ whose graphs are straight lines with rate of change m. We make the important observation that the rate of change for a linear functon equals the slope of the function. Note that when a linear *amount* function is increasing, the associated *rate* function is positive.

On the other hand, the graph in Figure 2.41 shows a relationship between two variables in which the output decreases by the same amount for each fixed change in the input.

▶ SECOND REFLECTION

Estimate the rate of change of the function graphed in Figure 2.41. What property of your estimate indicates that the function is decreasing?   ◀

*If $y = f(x)$, **input** refers to the number substituted for x and **output** refers to the resulting y or $f(x)$ value.*

*The **Amount** function refers to the original function and the **rate** function is derived from the amount function by calculating the rate of change of the amount function.*

FIGURE 2.40 Increasing Amount Function   **FIGURE 2.41** Decreasing Amount Function

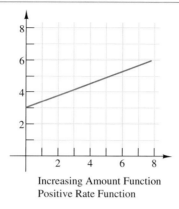

Increasing Amount Function
Positive Rate Function

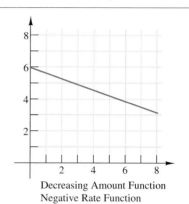

Decreasing Amount Function
Negative Rate Function

Think and Share

Break into groups of four and then divide into teams of two. One team will make up one real-life example that exhibits zero change, and the other team will create one that exhibits constant negative change. For each example, estimate the rate of change. After each two-person team has its examples, the two teams will reunite and share their examples. The four members will then decide which example to share with the class. Each team will produce three ways to illustrate its example: (1) by a table of values; (2) by a graph; and (3) by function notation.

Local Linearity: Slope at a Point

As we observed in Examples 2 and 3, rate of change and slope are the same for linear functions. Is there a way to generalize the concept of slope to nonlinear functions? One way we can approximate the slope of a nonlinear function at a point is by finding the slope of the straight line resulting from successive zooms at that point. We will devote much effort to elaborating this idea in order to find rates of change of nonlinear functions.

Remarkably, the graphs of most functions appear to be linear at a point if we zoom in far enough on a graph at that point. This property of looking like a line is what we will refer to as *local linearity*. Local linearity allows us to analyze a complicated function as though it were linear. Since most of the functions we will deal with in this course are not linear, the concept of local linearity will prove useful.

E X A M P L E   4   **Surfing in Bombay Revisited**

To find the rate of change at 11:00 A.M., we attempt to "straighten" out the wave by zooming in on the point where $x = 11$. To observe the local linearity at $x = 11$, we need to plot the graph over a very small interval around $x = 11$.

We do this by zooming in. Figure 2.42 shows the graph and the point chosen for the first zoom.

FIGURE 2.42   Preparing to Zoom at $x = 11$

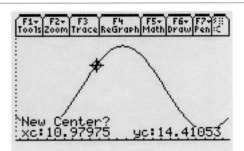

Figure 2.43 shows the first zoom. You should see a similar graph on your screen.

FIGURE 2.43   First Zoom at $x = 11$

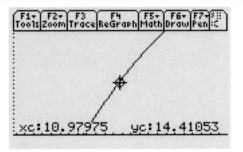

Already, the graph looks nearly straight. Continue zooming in until your graph appears to be a straight line. Even though our second zoom looks straight, we will zoom a third time as shown in Figure 2.44.

Use the Trace feature to pick two points on this "straight" line, and use these two points to find the slope of the line. We choose the points (10.99802, 14.51747) and (11.00134, 14.52758) and find the slope to be

$$\frac{14.52758 - 14.51747}{11.00134 - 10.99802} = 3.045181.$$

(Your calculated slope may not match this exactly, but it should be close.)

▶ THIRD REFLECTION

How does *slope* compare to *divided differences*?   ◀

FIGURE 2.44   Third Zoom at $x = 11$

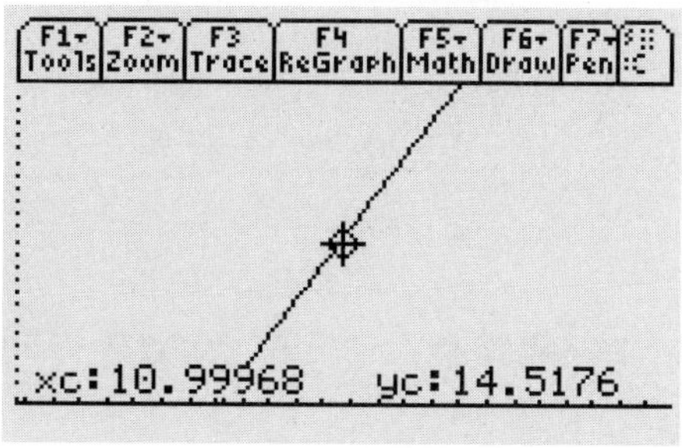

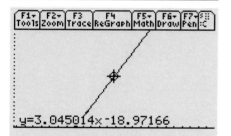

We now graph the line tangent to the curve at the point where $x = 11$ using the instructions in *Drawing a Tangent Line*.

An intuitive definition of a tangent line is: a line is considered to be tangent to a curve at $x = c$ if, in a small interval containing c, the line touches the curve only at $x = c$. Remember the idea of tangent line to a circle. We will be more precise later.

DRAWING A TANGENT LINE

In the GRAPH screen, open the Math menu. Select A:Tangent. Type 11, and press enter. A tangent line is drawn at the selected point and the equation of this tangent line is displayed on the screen.

FIGURE 2.45   Tangent Line at x=11

You should see a line which very nearly coincides with the graph of

$$y1(x) = 7.8\sin(0.47x + 1.69) + 10.2.$$

In fact, it should be nearly impossible to tell which is the tangent line and which is the sine wave. The equation of the tangent line at approximately $(11, 14.52349)$ (see *Drawing a Tangent Line* submenu) is

$$y = 3.045014x - 18.97166.$$

We see from its equation that the tangent line has slope 3.0450, rounded to four decimal places. See Figure 2.45 for the graph we obtained.

We estimated the slope of the tangent line at $x = 11$ to be 3.0450 by finding the slope between two points that were close to each other on the curve. Using this result, we estimate that the water level is changing at a rate of 3.0450 feet per hour on this interval. Because the slope is positive, we know that the water is rising, or that the tide is coming in.

FINDING THE ZOOM INTERVAL

The WINDOW editor always displays the x-range and the y-range for plots. To find the endpoints of the interval over which a function is graphed, go to the WINDOW editor and read the values for xmin and xmax. The zoom interval is $[xmin, xmax]$.

If we were to continue zooming in and making the interval around $x = 11$ smaller and smaller, our estimate of the rate of change would approach the value of the slope of the tangent line. We conclude that the rate of change of a function at a point is equal to the slope of the tangent line to the graph of the function at that point. Since rate of change and slope are equal for linear functions, we make the following definition to extend this notion to nonlinear functions.

RATE OF CHANGE OF A FUNCTION AT A POINT

The rate of change, or slope, of a function f, at a point $(a, f(a))$ is equal to the slope of the tangent line to the graph of f at $(a, f(a))$.

To further clarify this idea about *rate of change at a point*, note the graph in Figure 2.46 where we have sketched the original tide data, the function $y1(x) = 7.8\sin(0.47x + 1.69) + 10.2$, and the tangent line at $x = 11$.

FIGURE 2.46   Tangent Line to the Graph of $y1(x) = 7.8\sin(0.47x + 1.69) + 10.2$ at $x = 11$

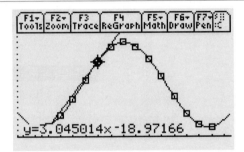

It should be clear that the slope of the tangent line at $x = 11$ does not approximate the slope, or rate of change of the curve except for those values of x that are very, very close to $x = 11$.

LOCAL LINEARITY

The graph of almost any function can be made to appear like a straight line at a point if the function is graphed over a small enough interval containing that point. The rate of change, or slope, of a function at a point is *approximately equal to* the slope of the "straight" line which results from successive zooms.

AN IMPORTANT OBSERVATION

The *exact rate of change*, or slope, of a function at a point is *defined to be* the slope of the tangent line to the curve at the point of tangency.

Think and Share

Break into groups of four and then divide into teams of two. Use the zoom procedure to *estimate* how fast the water is rising (or falling) in Bombay at 3:00 P.M. Then use the tangent line to find the rate of change (slope). Note that we mean the same thing whenever we say *slope* or *rate of change*. Each group of two should describe its procedure and interpret its results in complete sentences for the other group.

*In fact, "smooth" and "local linear"
describe the same phenomenon.*

Zooming in at a point on a smooth function causes the function to look more and more linear. When we zoom in, we look at a smaller and smaller interval of the domain of the function. You obtain the same effect as zooming in by drawing a secant line from the point of tangency to a point on the graph very close to it. Using such nearby points, we compute the average rate of change over a small interval. Figure 2.47 shows the slope of a line joining a point on the graph of the function directly above a to a point on the graph of the function directly above the point $a + h$ where h is a small distance. Such lines between points on the graph are called secant lines and approximate the position of the tangent line if the distance mearsured along the horizontal axis is small. The slopes of such lines are the average rate of change of the function on the small interval and approximate the slope of the tangent line or the *instantaneous rate of change* of the function. Thus, the slope of the tangent line is the instantaneous rate of change of the function and can be approximated by secant lines that are drawn based on small intervals or changes in x.

FIGURE 2.47   Secant Line Between $(a, f(a))$ and a Nearby Point

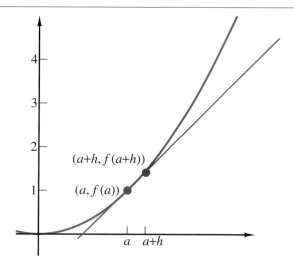

EXERCISE SET 2.4

Each graph in Exercises 1–4 represents an amount function. For each of these amount functions, (a) produce a table that is equivalent to the graph; (b) write a function rule that models the behavior exhibited by each graph; and (c) express the corresponding rate of change function as a graph, as a table, and as a function rule.

1. An amount function.

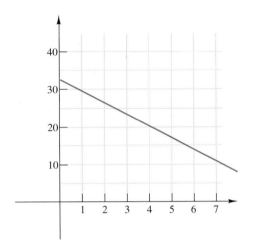

2. An amount function.

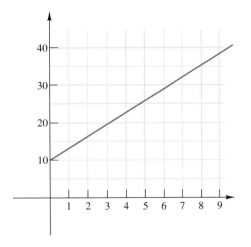

3. An amount function.

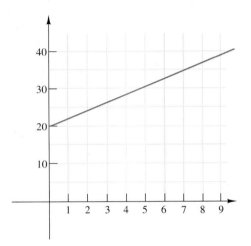

4. An amount function.

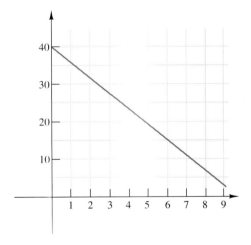

Each table in Exercises 5–8 represents an amount function. For each exercise, (a) draw a graph to represent the table data; (b) write a function rule which could be used to produce the table data; and (c) find the rate of change function as a graph, as a table, and as a function rule.

5.

t in minutes	1	2	3	4	5	6	7
Depth in feet	2	4	6	8	10	12	14

6.

t in hours	.5	1	1.5	2	2.5	3	3.5
Distance s in miles	25	50	75	100	125	150	175

7.

t in seconds	1	2	3	4	5	6	7
Velocity in ft/sec	25	23	21	19	17	15	13

8.

t in minutes	1	2	3	4	5	6	7
Depth in feet	5	5	5	5	5	5	5

Each function rule in Exercises 9–12 represents an amount function. For each amount function, (a) draw a graph of this amount function and label axes and scale clearly; (b) make a table for this amount function; and (c) find the rate of change function as a graph, as a table, and as a function rule.

9. $f(x) = 3x - 2$

10. $h(r) = -5r + 7$

11. $j(m) = 4$

12. $k(t) = 0.2t$

Each graph in Exercises 13–16 represents a rate of change of a quantity over time. For each rate function, (a) provide an example of a real-life quantity which might change in the manner described by the graph, and provide a clearly written description of the behavior of your quantity; and (b) sketch a graph of the corresponding amount function completely labeled.

13. A rate function.

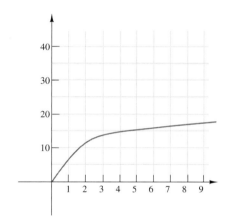

14. A rate function.

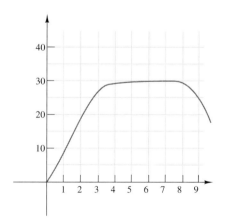

15. A rate function.

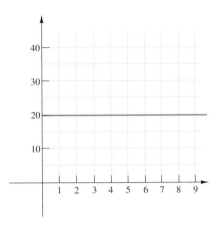

16. A rate function.

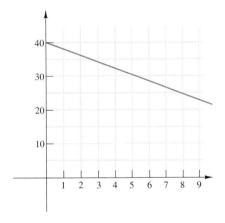

17. Consider a conical water tank, 100 feet tall with its vertex down. The diameter at the top is 80 feet. (a) Draw a diagram to represent this tank. (b) Imagine that you fill the tank with water at a constant rate. Draw a plausible graph of the water depth as a function of time. Be sure to label both axes and specify appropriate units. (c) Describe the rate of change function. Obviously the depth of water is increasing, but how is it increasing?

18. Consider a spherical water tank with diameter of 100 feet. (a) Draw a diagram to represent this tank. (b) Imagine that you fill the tank with water at a constant rate. Draw a plausible graph of the water depth as a function of time.

Be sure to label both axes and specify appropriate units. (c) Describe the rate of change function. Obviously the depth of water is increasing, but how is it increasing?

19. Imitate the procedure explained in Example 4 of this section to estimate the slope of a nonlinear function of your choice at a point of your choice. Do this by successive zooms and choosing two points. Compare your result to that obtained by a tangent line constructed at the point of interest. Explain your work. Provide diagrams.

20. Follow the zooming procedure explained in Example 4 of this section to estimate the slope of the graph of the function whose rule is $x^{2/3}$ at the point $x = 0$. What is the slope of the tangent line at $x = 0$? Is it possible to draw a tangent line to the curve at $x = 0$? Explain and illustrate all your answers.

21. Follow the zooming procedure explained in Example 4 of this section to find the slope of the curve arcsine($\sin(x)$) at $x = \pi/2$. Explain why you can never obtain a straight line, no matter what interval about $x = \pi/2$ you choose. Is it possible to draw a tangent line to the curve at $x = \pi/2$? Explain.

22. Follow the zooming procedure explained in Example 4 of this section to find the slope of the curve $|x - 3|$, at $x = 3$. Explain why you can never obtain a straight line, no matter what interval about $x = 3$ you choose. Is it possible to draw a tangent line to the curve at $x = 3$? Explain.

23. Follow the zooming procedure explained in Example 4 of this section to find the slope of the curve $|x^2 - 5|$, at $x = \sqrt{5}$. Explain why you can never obtain a straight line, no matter what interval about $x = \sqrt{5}$ you choose. Is it possible to draw a tangent line to the curve at $x = \sqrt{5}$? Explain.

24. Follow the zooming procedure explained in Example 4 of this section to find the slope of the graph of $f(x) = (x - 1)^{2/3}$, at $x = 1$. Is it possible to draw a tangent line to the curve at $x = 1$? Explain.

Explain, in your own words, the meaning of each of the following terms. Write your explanations in clear and complete sentences. Include annotated diagrams whenever appropriate.

25. rate of change

26. constant function

27. linear function

28. dy/dz

29. increasing function

30. increasing rate of change

31. local linearity

32. locally linear

2.5

Rate of Change as a Function

In this section we will see that with every *well-behaved* function we can associate a rate (or slope) function called the *derived function* or *derivative*. Later, we will clarify what we mean by a well-behaved function, but for now, we consider a function to be well-behaved if its graph is smooth, with no sharp corners and no breaks, jumps, or holes in its graph.

The function $f(x) = 8\cos\left(\frac{\pi}{6}x\right) + 10$ with which we modeled the Bombay tide in Section 2.1 is a smooth, well-behaved function. When we are able to find its associated rate function, we will be able to determine its instantaneous rate of change at any time t.

We begin by looking at a simplified, but similar, example to develop a sense of the slope (derivative) function. We will find the slope function for $f(x) = \sin(x)$.

Plotting the Slopes: The Derived Function

When we estimate the slope of a function at several points on its graph, and plot these slopes, we obtain a new function. The new function which reveals much about the original function has many names. It is sometimes called the *slope* or *derived* function but most commonly it is simply referred to as the *derivative*.

FIGURE 2.48 $f(x) = \sin(x)$

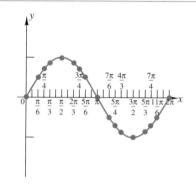

E X A M P L E 1   **The Derived Function of the Sine**

The function $f(x) = \sin(x)$ is graphed in Figure 2.48. We will estimate the slope of the tangent line to the curve at each of the the points marked on the graph, and create a table of slopes corresponding to the x-values:

$$0, \frac{\pi}{6}, \frac{\pi}{4}, \frac{\pi}{3}, \frac{\pi}{2}, \frac{2\pi}{3}, \frac{3\pi}{4}, \frac{5\pi}{6}, \pi, \frac{7\pi}{6}, \frac{5\pi}{4}, \frac{4\pi}{3}, \frac{3\pi}{2}, \frac{5\pi}{3}, \frac{7\pi}{4}, \frac{11\pi}{6} \text{ and } 2\pi.$$

Tables 2.31 and 2.32 contain the slopes estimated at these points by zooming in on the curve until it appears linear, and then finding the slope between two points on the resultant "straightened-out" curves. (Refer to Section 2.4 for the technique for estimating the slope of a function at a point via successive zooms.) The entry corresponding to $x = \frac{\pi}{3}$ has been left blank intentionally.

Estimate the slope at $x = \frac{\pi}{3}$ by successive zooms, fill in the missing value in the table, and check with a friend to confirm your answer. The missing slope entry can also be obtained by observing that a tangent line to the curve has the same slope at $x = \frac{5\pi}{3}$ as it has at $x = \frac{\pi}{3}$.

▶ FIRST REFLECTION

Explain why the slope at $x = \frac{5\pi}{3}$ equals the slope at $x = \frac{\pi}{3}$. Find another pair of x-values where the original function has the same slope.   ◀

TABLE 2.31

x	0	$\frac{\pi}{6}$	$\frac{\pi}{4}$	$\frac{\pi}{3}$	$\frac{\pi}{2}$	$\frac{2\pi}{3}$	$\frac{3\pi}{4}$	$\frac{5\pi}{6}$
Zoom Slope	.9871	.8655	.7014		0.001	−.4801	−.7013	−.8654
Exact Slope	1	$\frac{\sqrt{3}}{2}$	$\frac{\sqrt{2}}{2}$		0	−.5	−$\frac{\sqrt{2}}{2}$	−$\frac{\sqrt{3}}{2}$

TABLE 2.32

x	π	$\frac{7\pi}{6}$	$\frac{5\pi}{4}$	$\frac{4\pi}{3}$	$\frac{3\pi}{2}$	$\frac{5\pi}{3}$	$\frac{7\pi}{4}$	$\frac{11\pi}{6}$	2π
Zoom Slope	−.9805	−.8657	−.7015	−.4932	.001	.4992	.7013	.8652	.9907
Exact Slope	−1	−$\frac{\sqrt{3}}{2}$	−$\frac{\sqrt{2}}{2}$	−.5	0	.5	$\frac{\sqrt{2}}{2}$	$\frac{\sqrt{3}}{2}$	1

Using the slopes in the tables, we plotted the points (x, slope at x) shown in Figure 2.49. In your book, connect the points we have plotted with a smooth curve. Note that the original function, $f(x) = \sin(x)$ is a periodic function. It makes sense that the derived function should also be periodic.

FIGURE 2.49   Slopes Plot

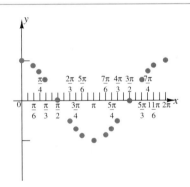

To summarize the results of this example, if
$f(x) = \sin(x), f'(x) = \cos(x).$

▶ **SECOND REFLECTION**

Guess the name of the derivative function from the graph in Figure 2.49.   ◀

NOTATION AND AN INFORMAL DEFINITION

The notation we use for the derived function is $f'(x)$, which is read "f prime of x." **Slope function**, **derived function**, or **derivative** are all names for the function obtained from an original function by calculating slopes. Please be aware that since we are only estimating the slopes, our derivative function is also an estimate. We will learn to find exact derivative functions in Chapter 3. Our estimates in this section are close enough for graphing purposes. We use the terms *zoom slope* in Tables 2.31 and 2.32 for the estimated values and *exact slope* for the more precise values obtained by methods to be presented later.

E X A M P L E 2   **The Derived Function for the Bombay Tides**

Let's return to our Bombay tide example to see if we can provide our surfer with tide rates for the entire day. We can do this by using the techniques of the previous example; that is, by estimating how fast the water level is changing at various times and then using those estimates to generate a graph of the derivative function. Recall that we modeled the Bombay tides with the function $y1(x) = 8\cos\left(\frac{\pi}{6}x\right) + 10$ where x represents the number of hours past midnight (so $x = 0$ is midnight, $x = 1$ is 1:00 a.m., etc.). Enter and graph this function as $y1(x)$ on your calculator. Use the zoom technique to fill in estimates for the first three entries in Table 2.33. Do *not* try to fill in the entire table.

COMPUTING THE SLOPE WITH TWO POINTS

Trace to the first point. To transfer coordinates to the HOME screen, press ◇ (-) on the TI-89 or ◇ H on the TI-92. Trace to the second point and press ◇ (-) (or the equivalent) again. Go to the HOME screen. Enter `ans(1) – ans(2)`. The result is the ordered pair $(\Delta x, \Delta y)$. Use these numbers to compute the slope.

TABLE 2.33

x	0	1	2	3	4	5	6	7	8	9	10	11	12
Zoom Slope													

You should be confident at this point that, given enough time, you could use the zoom technique to complete the table. The process, however, is cumbersome and time-consuming. Fortunately, the TI calculator is capable of making these computations automatically using the `nDeriv` command as shown in *Using the* `nDeriv` *Command*.

USING THE NDERIV COMMAND

Define $y2(x) = \text{nDeriv}(y1(x),x)$ (you should still have $y1(x) = 8\cos(\frac{\pi}{6}x) +$ 10 in the Y= editor). To obtain the nDeriv command, you can simply type the word nDeriv on the command line in the Y= editor or you can find it in the A:Calculus sub-menu of the MATH menu. Use the TblSet dialogue box to start the table at 0 and set the table increments to 1. Then bring up the table.

The second column in the table created in *Using the* nDeriv *Command*, labeled y1, displays the water level predicted by our model at the time corresponding to x. The third column, labeled y2, displays the TI calculator numerical estimate for how fast the water level is changing (that is, the rate of change of y1). Compare the first three entries in the y2 column to the ones you found using the zoom technique. They should be fairly close. If they are not, go back over your work to see if you can find the error. Start by checking that both y1(x) and y2(x) were entered correctly in the Y= editor. Once you have reconciled any differences between your estimates and those of the TI calculator, use the TI calculator table to complete the table in your book (remember to use the values in the y2 column, not the y1 column). Plot the points from your table onto the graph grid in Figure 2.50 and connect the points you plot with a smooth curve.

FIGURE 2.50   Grid for Plot of nDeriv(y1(x),x)

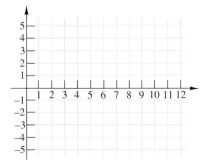

▶ THIRD REFLECTION

Use your graph to estimate the speed at which the water level is changing at 10:30 A.M. At what time in the first 12 hours does it appear that the current is decreasing most rapidly (this corresponds to the time when the tide will be shifting most rapidly) ?   ◀

Use the TI calculator to graph y2(x). Make sure your Window settings fit the graph and be patient. This graph requires some computation. Set xres to 5 to speed things up a bit (remember to reset it when your are done). How does the TI calculator graph of the derivative function compare to yours?

E X A M P L E   3   **The Same Derivative?**

Use your TI calculator and the nDeriv command to sketch the graphs of $f(x) = x^2$ and $g(x) = 3 + x^2$ and their derivatives. (You should be able to simply redefine y1(x) to be first f (and graph) and then g using the same $y2(x) = \text{nDeriv}(y1(x),x)$ that you used previously.) Figure 2.51 gives the table for f and f' while Figure 2.52 shows the graphs of f and f'.

▶ FOURTH REFLECTION

Find a symbolic expression for f', either from the table or the graph.   ◀

FIGURE 2.51   Table for f and f' FIGURE 2.52   Graphs of f and f'

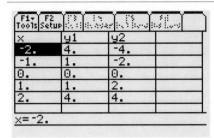
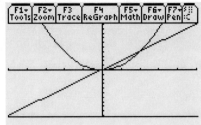

Figure 2.53 and Figure 2.54 are the table and graphs for g and g'. What is a symbolic expression for g'?

FIGURE 2.53   Table for g and g' FIGURE 2.54   Graphs of g and g'

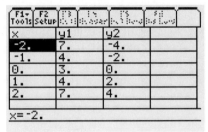
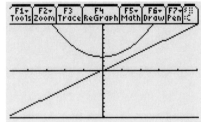

You should have found the same symbolic expression for both $f'(x)$ and $g'(x)$. In general, if two functions have graphs that are vertical translations of each other, so that their equations differ by a constant, then they will have the same derivative. It turns out that the converse is also true; that is, if two functions have the same derivative then they must differ by a constant. Verifying this fact, however, is not easy and will be delayed until Chapter 5.

▶ FIFTH REFLECTION

What does the sign of the derivative tell you about the behavior of the original function? How does this compare with what we have learned about first differences?   ◀

EXAMPLE   4   **Existence of the Derivative**

At the beginning of this section, we remarked that associated with all well-behaved functions was the derived function, or derivative, where well-behaved meant no holes, breaks, or sharp corners. Let's look at what can go wrong with a function that is not well-behaved.

Define y1(x) = abs(5−x^2), *and* y2(x) = nDeriv(y1(x),x). *Use the window with x-range [−5,5] and y-range [−10,10].*

Let $f(x) = |5 - x^2|$. Sketch the graph of f and $f'(x)$ simultaneously over the interval $[-5, 5]$. Your sketch should look like Figure 2.55.

FIGURE 2.55 $f(x) = |5 - x^2|$ and its Derivative Function

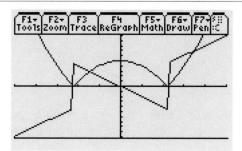

There is something peculiar about the graph of $f'(x)$. Recall that to create a graph, the TI calculator simply plots points and then joins them with a curve. As a result, the calculator may draw a curve between two points that should not be connected. In particular, the graph of the derivative function should not have the two nearly vertical segments in it.

Regraph the absolute value function without the derived function this time, and observe the sharp corners in the graph when $y = 0$. (See Figure 2.56.)

FIGURE 2.56 $f(x) = |5 - x^2|$

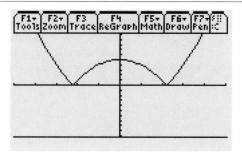

We will now zoom in at the point where $x = \sqrt{5}$ to see why there is a problem. Our first zoom is shown in Figure 2.57. (Your graph should look similar.)

FIGURE 2.57   The Graph of a Function with a Corner

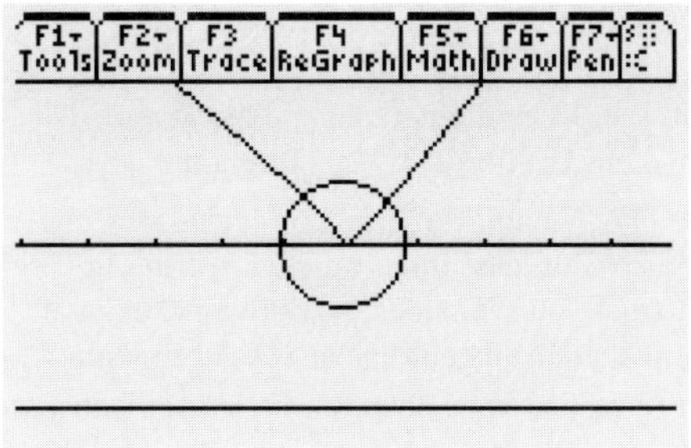

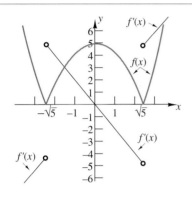

FIGURE 2.58   Graphs of $f(x)$ and $f'(x)$

The Koch snowflake in the Chapter 1 Supplementary Exercises is one such unusually ill-behaved function.

Think and Share

The symbolic form of the derivative is a rule in terms of the variables used in the original function. A numerical derivative, on the other hand, is just the collection of calculated slopes.

Continue to zoom in at this sharp corner. No matter how many zooms you try, you cannot "straighten" out the kink. This is because the curve is not locally linear at the sharp corners. (If necessary, see Section 2.5 to review the concept of local linearity.) The function is not well-behaved. It is continuous (there are no breaks or holes), but it is not smooth (there are sharp corners). As a result, the derived function does not exist at either of the sharp corners on the graph. We say that the function does not have a derivative at $x = \sqrt{5}$ or that the function is not *differentiable* at $\sqrt{5}$. The graph of $f(x)$ and the derived function should look like the graph in Figure 2.58. Notice that the derived function (the straight-line segments in the graph) exists over three separate intervals. Unless a function is unusually ill-behaved, we can find intervals over which the derivative does exist.

Break into groups. Consider the function $f(x) = 1/x^2$ on the closed interval $[-1,1]$. Do you think the derivative, $f'(x)$, exists at 0? Explain your reasoning. Can you tell where the derivative of a function does not exist?

Putting It All Together

We have seen, in this section, that associated with every well-behaved function is another function that we call the derived function, the derivative function, or simply the derivative. The derivative measures rate of change and is used to analyze the behavior of the original function. We plotted a derivative function by finding slopes at points along the graph of the original function and plotting these slopes. We used a built-in command called nDeriv to find approximate values of the derivative function that we used to create its graph. Except in the case of the functions $f(x) = x^2$ and $g(x) = 3 + x^2$, we did not find an equation or rule for our derivative function. The derivative function exists whenever the original function is smooth and continuous, and, when the derivative exists, we can plot its graph even though we have not developed tools for obtaining its rule in symbolic form. We will develop the tools and techniques for finding the derived function symbolically in the next chapter.

EXERCISE SET 2.5

1. Let $f(x) = x^{\frac{1}{3}}$. Graph this function and zoom in on the graph at $x = 0$ at least three times. Describe in a sentence or two what you see. Does the derivative exist at $x = 0$? Explain. Is there a tangent line at $x = 0$? Explain.

2. Let $f(x) = (x - 1)^{\frac{2}{3}}$. Graph this function and zoom in on the graph at $x = 1$ at least three times. Describe in a sentence or two what you see. Does the derivative exist at $x = 1$? Explain. Is there a tangent line at $x = 1$? Explain.

† 3. Use the nDeriv command to graph the numerical derivatives for $\sin(2x)$, $\sin(3x)$, and $\sin(\frac{x}{2})$. What is the relationship between the derivative functions and the derivative of $\sin(x)$? Explain, in a sentence or two, the role of the coefficients of x (the constants 2, 3, and $\frac{1}{2}$) in the height of the derivatives' graphs. What would the graph of the derivative of $\sin(kx)$ look like, compared to the graph of the derivative of $\sin(x)$? Let k be less than zero, between zero and one, and greater than one.

4. Use the nDeriv command to graph the numerical derivatives for $\cos(2x)$, $\cos(3x)$, and $\cos(\frac{x}{3})$. What is the relationship between the derivative functions and the derivative of $\cos(x)$? Explain, in a sentence or two, the role of the coefficients of x (the constants 2, 3, and $\frac{1}{3}$) in the height of the derivatives' graphs. What would the graph of the derivative of $\cos(kx)$ look like, compared to the graph of the derivative of $\cos(x)$? Let k be less than zero, between zero and one, and greater than one.

5. Graph the numerical derivatives for $f(x) = x^3$ and $g(x) = x^3 - 5$. Describe the graphs of the derivatives and describe how the graphs of f and g are related.

6. Graph the numerical derivatives for $f(x) = x^4$ and $g(x) = x^4 + 12$. Describe the derivative graphs and describe how the graphs of f and g are related.

7. Graph the numerical derivatives for $f(x) = x^2$ and $g(x) = (x - 3)^2$. Describe any similarities or differences you observe between the graphs of f' and g'. How are the graphs related? (If a graph is the same as another except

for a horizontal shift, we say the graphs are horizontal translations or we say the graph of one is shifted to the right or left.)

8. Graph the numerical derivatives for $f(x) = |x^2 - 4|$ and $g(x) = |x^2 - 1|$. Describe any similarities or differences you observe between the graphs of f' and g'. Are the graphs related? The dot option in the Style menu of the Y= editor prevents the TI calculator from joining points that should not be connected.

9. Graph the numerical derivatives for $f(x) = x^2$, $g(x) = 3x^2$, and $h(x) = \frac{1}{2}x^2$. Describe the effect of the constants 3 in $g(x)$ and $\frac{1}{2}$ in $h(x)$ on the graphs of $g'(x)$ and $h'(x)$.

10. Describe the graph of $g'(x)$ if $g(x) = |x^2 - 3x + 2|$. Include a sketch of $g'(x)$.

11. Sketch the graphs of two different functions that will have the same derivative. You do not need to have a rule or equation for your functions. Please be original and creative, and show that you understand the concept illustrated by Example 3. Explain how you know that your two functions have the same derivative.

Explain, in your own words, the meaning of each of the following terms. Write your explanations in clear and complete sentences. Include annotated diagrams whenever appropriate.

12. numerical derivative
13. slope function
14. point of inflection
15. concave up
16. horizontal translation
17. vertical translation
18. local linearity
19. differentiable
20. smooth curve
21. symbolic derivative

Summary

In Chapter 2, we developed the ability to analyze and understand discrete and continuous functions. We studied first differences (also called forward differences) so that we could investigate how discrete functions (sequences and two-column data tables) increase and decrease, and how they bend up and down. We learned to gather information about how a function increases and decreases from the sign of its first differences: when are the differences positive, when are they negative? We can obtain further information about how a function bends by careful comparison of the magnitude of its first differences. We can determine which way a function is bending from the sign of its second differences: when are they positive, when are they negative? We looked at divided differences as an extension of first differences for irregular data.

First differences can be used with continuous functions by selecting sets of input values and then proceeding as if we were working with discrete functions. As the input values are selected closer and closer together, the continuous function can be studied in greater detail. This is in contrast to discrete functions where the input values are a fixed (smallest) distance apart. In Chapter 1, we studied inherently discrete yeast data using continuous techniques. In this chapter, we have studied continuous functions using discrete techniques.

We have also studied continuous functions using an analogue to first differences: the first derivative. The derivative indicates when a continuous function that has a derivative is increasing or decreasing. In the continuous case, the sign and magnitude of the first derivative tell us the direction and rate of change of the original function. You will learn about second derivatives in the next chapter. They have much in common with second differences.

SUPPLEMENTARY EXERCISES

1. Consider the graph of a sequence of interest to you, perhaps one that we studied in Chapter 1 or 2. In terms of the behavior modeled by the difference equation, what does a concavity of 25 at some point A mean when compared to a concavity of 2 at point B?

2. What can we conclude about the behavior of the tangent function from the difference table for

$$a_n = \tan(n), n = 1, 2, 3, \ldots?$$

Use radian measure. Explain your answer fully.

3. Enter the following table of data (believed to be quartic in nature) into the Data/Matrix editor.

a. Build a difference table to determine the relative minimum, relative maximum, and inflection points.

b. Use the QuartReg (a least-squares fit to a quartic polynomial) to calculate a quartic model for the data. Store the regression polynomial in $y1(x)$. Use the first derivative (nDeriv command) of $y1(x)$ to determine the relative minimums and relative maximums of the polynomial model.

c. Compare the values obtained from the two methods.

4. Use complete sentences to explain the concept of local linearity. Provide sketches of functions that are locally linear and of functions that are not locally linear.

Exercises 5 and 6 refer to Table 2.35.

TABLE 2.34

x	−1.5	−1.0	−0.5	0.0
$f(x)$	10.702	8.7673	9.8416	9.8537
x	0.5	1.0	1.5	2.0
$f(x)$	8.3776	9.6898	6.8670	6.4045

TABLE 2.35

Day	1	2	3	4	5	6	7	8	9	10
Number sold	10	18	38	80	106	130	145	168	170	180

5. Suppose you are a stockholder in the Bexar BonBon Company. The data in Table 2.35 show the number of items sold (in thousands) over a 10-day period. The current manager fit a linear regression model to the data and recommended the hiring of several more employees. Explain why you should vote to replace the current manager when your next voting proxy comes.

6. What model should be used to fit the data given in Table 2.35? On what day should management have ordered decreased production? Explain.

7. Let $g(x) = |\sin(x)|$. Is $g(x)$ locally linear at $x = \pi$? Explain and include a sketch.

8. Explain the difference between discrete and continuous functions. Draw a graph of a discrete function and of a continuous function. You do not need to have equations for your functions, you just need to show what the difference is.

9. Explain the relationship between first difference and derivative. When does it make sense to talk about first differences and derivatives?

10. Skuzzie Baker, a purely fictitious drug user, ingests 100 mg of some nasty substance every three hours. Forty percent of the substance is eliminated by her overworked liver every hour.

   a. Find a recursion formula to model the maximum amount of the substance in her body — just after taking the nth dose. Does your recursion formula have a stable point? Justify.

   b. Analyze a graph to determine whether the maximum amount of the substance in her body is increasing or decreasing. Is the rate at which the increase or decrease occurs increasing or decreasing? Explain.

   c. An amount of 130 mg or more of this nasty substance results in the death of the drug user. Will Skuzzie Baker live or die? Explain.

11. If f is a continuous function on a closed interval, which of the following statments are true and which are false? You must justify your decisions.

   a. If $f(-5)$ and $f(3)$ are both positive, we know that there are no solutions to $f(x) = 0$ between $x = -5$ and $x = 3$. Name the property involved in this question.

   b. The function f will have a maximum value but it is quite possible for the function not to have a minimum value. Again, name the property.

   c. There will be at least one x-coordinate between -5 and 3 where the instantaneous rate of change is equal to the average rate of change over the interval $[-5, 3]$. Name the property.

   d. If $f(-5)$ and $f(3)$ have opposite signs (one positive, one negative), we know that there is exactly one solution to $f(x) = 0$ between $x = -5$ and $x = 3$. Name the property involved in this question.

12. Sketch the graph of a smooth and continuous function $f(x)$ on the interval $[-4, 4]$ such that:

$$f(-4) = 3, \quad f(0) = -5, \quad \text{and} \quad f(4) = 6.$$

   a. Find the average rate of change of the function your sketched over the interval $[-4, 4]$. Show how you do this.

   b. Indicate all points on your graph where the instantaneous rate of change is equal to the average rate of change over the interval $[-4, 4]$. Estimate the corresponding x-value. What property guarantees that you can find these points?

   c. Estimate the instantaneous rate of change of the function you sketched when $x = -1$. Explain how you do this.

13. If the first and second differences for the data in a mystery data table are all positive, is it appropriate to model the data with a quadratic function? Explain.

PROJECTS

Project One: Exploring Forward Differences

For the quadratic function $a_n = n^2 - 4n + 4$, we saw that the second differences were constant. In the bouncing ball data (Table 2.11), we see that the second differences are nearly constant and that the data is reasonably represented by a quadratic function. The questions that follow will (1) verify that all quadratics have second differences that are constant, and (2) provide evidence that all sequences with constant second differences (with equal domain (n) subinterval lengths of 1) can be represented by a quadratic.

1. Build a difference table for the sequence $a_n = 10n^2 + 4n - 13$, $n = 0, 1, 2, 3, 4, 5, 6$. Are the second differences constant?

2. Consider the sequence $a_n = b_1 n^2 + b_2 n + b_3$, $n = 0, 1, 2, 3, 4, 5$ where b_1, b_2, and b_3 are arbitrary constants. Show that the second differences are constant.

3. Show that the second differences of the following sequences are constant. Find a quadratic function $a_n = f(n)$ that fits the data points.
   i. $\{0, 1, 4, 9, 16, 25\}$
   ii. $\{4, 5, 2, -5, -16, -31, -50\}$
   iii. $\{1, 4.1464, 10.757, 20.831, 34.37, 51.372, 71.839\}$.

4. In this problem we attempt to show that any sequence with constant second differences can be represented by a quadratic function. Let $a_1, a_2, \ldots$ be a sequence with all of the second differences equal to the constant c.

   a. For this sequence, $c = \Delta^2 a_1 = \Delta a_2 - \Delta a_1$. Thus, $\Delta a_2 = \Delta a_1 + c$. Similarly, $\Delta a_3 = c + \Delta a_2 = c + c + \Delta a_1 = 2c + \Delta a_1$. Find Δa_4 in terms of c and Δa_1. Define a function of n which gives Δa_n in terms of c and Δa_1.

   b. The equation $\Delta a_1 = a_2 - a_1$ can be rewritten as $a_2 = a_1 + \Delta a_1$. Similarly, $a_3 = a_2 + \Delta a_2 = a_1 + \Delta a_1 + \Delta a_2$ and $a_4 = a_3 + \Delta a_3 = a_1 + \Delta a_1 + \Delta a_2 + \Delta a_3$. Write a general formula for a_n in terms of $a_1, \Delta a_1, \Delta a_2, \ldots, \Delta a_{n-1}$.

   c. Use the function you found in the first part of this problem to replace the first differences in your general formula to obtain an equation for a_n in terms of a_1, Δa_1, and c. Compute a closed form expression for a_n. View your result as a function of n. What do you observe?

5. This problem extends the investigation of differences to a general cubic function. Let f be a function defined by the formula $f(q) = aq^3 + bq^2 + cq + d$. Compute $f(x + h) - f(x)$. What type of expression did you obtain? This expression provides the first differences for a sequence of values of f generated from inputs that differ by h. What type of expression will you obtain for second differences? Explain how you know.

6. Let g be the function defined by the formula $g(q) = \sin(q)$. Find the sequence of values of g for the inputs 0.0, $\frac{\pi}{10}, \frac{2\pi}{10}, \frac{3\pi}{10}, \ldots, \frac{20\pi}{10}$ (use 0.0 to get decimal approximations). Find the first differences for your sequence of g values. Plot the first differences against the inputs. Describe the pattern. Find a multiple of the cosine function whose graph reasonbly matches the plot of the first differences. How does your multiple compare to the number $\frac{\pi}{10}$?

7. Let h be the function defined by the formula $h(q) = 3^q$. Find the sequence of values of h for the inputs $-5, -4, -3, \ldots, 10$. Find the first differences for your sequence of h values. Plot the first differences against the inputs. Describe the pattern. What do the first differences reveal about the way that h increases and decreases? Compute the second differences. What do the second differences reveal about the shape of the graph of h?

8. Let s be the function defined by the formula $s(q) = \sqrt{q}$. Find the sequence of values of s for the inputs $0, 1, 2, 3, \ldots, 10$. Find the first differences for your sequence of s values. Plot the first differences against

the inputs. Describe the pattern. What do the first differences reveal about the way that s increases and decreases? Compute the second differences. What do the second differences reveal about the shape of the graph of s?

9. Let u be the function defined by the formula $u(q) = \frac{1}{q}$. Find the sequence of values of u for the inputs $1, 2, 3, \ldots, 10$. Find the first differences for your sequence of u values. Plot the first differences against the inputs. Describe the pattern. What do the first differences reveal about the way that u increases and decreases? Compute the second differences. What do the second differences reveal about the shape of the graph of u? Repeat the problem for the inputs $-10, -9, -8, \ldots, -1$.

10. Table 2.36 represents two sets of data, each reflecting the distance s traveled by a ball rolling down a plane inclined at $30°$. (This is a famous experiment by Galileo). In each case, find a lower and upper bound for the average speed by computing the change in distance divided by the change in time? Why are these bounds closer together in the second data set than in the first?

11. Figure 2.59 displays experimental data taken from a ball rolling down a plane inclined at $30°$. The distance represents the cumulative distance traveled down the plane after n seconds.

   a. Is the relationship quadratic?

   b. Duplicate Galileo's experiment by collecting your own data. To approximate the experiment, you should choose materials so that the air resistance is negligible and the motion down the plane frictionless. What materials do you choose?

TABLE 2.36   Rolling Ball

t	0.8	0.9	1.0	1.1	1.2
s	5.12	6.48	8	9.68	11.52
t	0.98	0.99	1.0	1.01	1.02
s	7.68	7.84	8	8.16	8.38

FIGURE 2.59

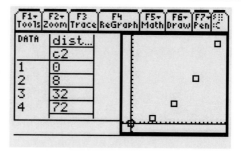

Project Report.   Write a report on your investigations. Address your report to your classmates, assuming the material is new to them. Describe your reasoning and include all necessary background information. A minimal project report must include:

1. Written responses to items 1-11 above.
2. Hand-drawn graphs and displayed calculations to support your conclusions.

3. Suggestions for possible extensions of you observations and further questions for explorations.

Project Two: Surfing the Tangent Lines — Finding the Roots

In this project, you will develop a powerful method for approximating the zeros or roots of an equation $f(x) = 0$. The method is called *Newton's Method*, and exploits the local linearity of polynomials. Begin by graphing the function $f(x) = -0.29x^2 - 0.53x + 0.33$. We want to find the root of the equation $f(x) = 0$ that lies between $x = 0$ and $x = 1$. Do each step as you come to it. Record all your answers in order.

1. Define $y1(x) = -.29x^2 - .53x + .33$ in the Y= editor.
2. Graph $y1(x)$ in a graph window with the settings: xmin: -5; xmax: 5; ymin: -4; ymax: 4.
3. Draw a tangent line at $x = 3$ using the Tangent option in the Math menu, with $x = 3$.
4. Use the 6:Vertical option in the pencil menu Pen to draw a vertical line at the point where this tangent line crosses the x-axis. Use the Cursor button to locate the vertica! line so that it appears to be at the x-intercept of the tangent line, and press ENTER to draw the vertical line. Capture the x-, y-coordinates of your cursor location to the HOME screen by pressing ◇ (-) on the TI-89 or ◇ H on the TI-92 and then HOME. You can then bring the point down to the Entry line, and edit it so that you can store the x-coordinate as x_1.
5. Verify that the value of x_1 is correct by solving for the x-intercept of your tangent line.
6. Draw a tangent line to the curve at $x = x_1$ (where the vertical line intersected the curve). When the calculator prompts for the point of tangency, you can type x1, if you actually stored the value. Your graph should now show two tangent lines, one vertical line, and the original parabola.
7. Zoom in on the graph at the point where the second tangent line crosses the x-axis. Unfortunately, the tangent lines and the vertical line disappear in the zoom process. Never mind. Redraw the second tangent line.
8. Draw a vertical line at the point where this tangent line crosses the x-axis. Store the x-coordinate of this line now, and store it as x_2.
9. What do you notice about the values of x_1 and x_2, in terms of the root of $f(x) = 0$?
10. In principle, you can create a sequence of x-values, $x_1, x_2, x_3 \ldots$, by alternately drawing tangent lines, vertical lines, and zooming in, until your x_n differed from x_{n-1} by a quantity less that 0.00005, or any other predetermined error bound. Continue this process until (x_n differed from x_{n-1} by a quantity less that 0.00005. How many values were necessary?
11. Write an equation for the tangent line to $y = f(x)$ at $x = x_{n-1}$ in terms of $f(x_{n-1})$ and $f'(x_{n-1})$.
12. Use your formula for a tangent line to derive a formula for x_n in terms of x_{n-1}, f, and f'.

Project Report.   Write a report on your investigations. Address your report to your classmates, assuming the material is new to them. Describe your reasoning and include all necessary background information. A minimal project report must include:

1. Written narrative of the process described above.
2. Hand drawn graphs in support of your computations.
3. Written responses to questions 9, 10, 11, and 12.
4. Suggestions for possible extensions of your observations and further questions for exploration.

Project Three: Differences and Polynomials

There are a number of ways of writing polynomials: the expanded and factored forms that you are familiar with are but two of many. One way that is especially convenient for computing differences is *falling factorials*. The n-th falling factorial $x^{\langle n \rangle}$ is defined for nonnegative values of n by the sequence:

$$x^{\langle 0 \rangle} = 1$$
$$x^{\langle 1 \rangle} = x$$
$$x^{\langle 2 \rangle} = x(x - 1)$$
$$x^{\langle 3 \rangle} = x(x - 1)(x - 2)$$
$$\vdots$$
$$x^{\langle m+1 \rangle} = x^{\langle m \rangle}(x - m + 1).$$

The TI calculator notation for $x^{\langle n \rangle}$ is `nPr(x,n)`.

For negative values of n falling factorials are defined by:

$$x^{\langle -1 \rangle} = \frac{1}{x + 1}$$
$$x^{\langle -2 \rangle} = \frac{1}{(x + 1)(x + 2)}$$
$$x^{\langle -3 \rangle} = \frac{1}{(x + 1)(x + 2)(x + 3)}$$
$$\vdots$$
$$x^{\langle -(m+1) \rangle} = x^{\langle -m \rangle} \cdot \frac{1}{x + m + 1}.$$

1. Find A, B, and C so that $2x^2 - 5x + 5 = A \cdot x^{\langle 2 \rangle} + B \cdot x^{\langle 1 \rangle} + C \cdot x^{\langle 0 \rangle}$.
2. Find A, B, C, and D so that $3x^3 - 4x^2 + 2x - 9 = A \cdot x^{\langle 3 \rangle} + B \cdot x^{\langle 2 \rangle} + C \cdot x^{\langle 1 \rangle} + D \cdot x^{\langle 0 \rangle}$.
3. Explain how to express any polynomial as a sum of multiples of falling factorials.
4. Compute $\Delta\left(x^{\langle m \rangle}\right)$ for $m = -3, -2, -1, 0, 1, 2, 3,$ and 4.

5. Examine first differences computed above to discover a pattern. Give an algebraic argument to support your pattern. What does this pattern tell us about the first difference of any polynomial?

Project Report.   Write a report on your investigations. Address your report to your classmates, assuming the material is new to them. Describe your reasoning and include all necessary background information. A minimal project report must include:

1. Responses to items 1-5 above.
2. Justification and explanations for your calculations.
3. Suggestions for possible extensions of your observations and further questions for exploration.

3

THE DERIVATIVE: A TOOL FOR MEASURING CHANGE

3.1
The Derivative

3.2
Rules, Rules, and More Rules

3.3
The Chain Rule

3.4
Finding Features of a Continuous Function

3.5
Properties of Continuous Functions

3.6
Optimization: Finding Global Extrema

3.7
Implicit Differentiation and Its Applications

3.8
Modeling Motion with Parametric Equations

3.9
Partial Derivatives

Summary

3.1

The Derivative

In Section 2.5, we learned how to estimate the value of the derivative at various values of x by zooming in on its graph until the graph appeared linear and then finding the slope of the resulting "straight" line. We also used the TI calculator nDeriv command to compute numerical derivatives. In both cases we used the values we found to draw graphs of derivative functions. The graph of the derivative proved to be extremely helpful in analyzing the behavior of the original function (think of the infectious disease example in which we discovered where the disease could be classified as under control). Unfortunately, graphs suffer from at least two major drawbacks: they are not exact and they can be misleading. In this section, we develop a precise definition of the derivative which will allow us to obtain exact values for the derivative when needed and to verify the validity of its graph. In order to understand and apply the definition, we will also need to learn more about the concept of limit.

FIGURE 3.1   The Graph of $f(x)$ with a Secant Line

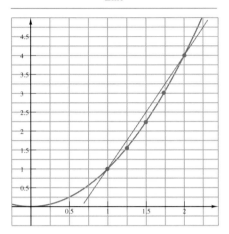

The Derivative at a Point

We begin the development of the precise definition of the derivative with an example.

E X A M P L E   1   **Approximating the Tangent Line Slope using Secants**

Consider the function $f(x) = x^2$. Suppose we want to know the slope of the curve when $x = 1$. To refine our previous numerical techniques, we consider points on the curve near $x = 1$ that get successively closer to 1, for example, $x = 2$, 1.75, 1.5, and 1.25. We construct secant lines from the point on the curve with $x = 1$ using these points. The first of these secant lines from $(1, 1)$ to $(2, 4)$ is displayed in Figure 3.1.

Draw in the rest of the secant lines using the values $x = 1.75$, 1.5 and 1.25 on Figure 3.1. Do these secant lines seem to be better and better approximations for the tangent line? We can compute the slopes of the secant line and, thus, approximations to the slope of the tangent line at $x = 1$ as displayed in Equations (3.1) using Δx and Δy to symbolize the change in x and y.

$$\frac{\Delta y}{\Delta x} = \frac{f(1+1) - f(1)}{(1+1) - 1} = \frac{4-1}{1} = 3, \quad \Delta x = 1 \tag{3.1}$$

$$\frac{\Delta y}{\Delta x} = \frac{f(1+0.75) - f(1)}{(1+0.75) - 1} = \frac{2.0625}{0.75} = 2.75, \quad \Delta x = 0.75$$

$$\frac{\Delta y}{\Delta x} = \frac{f(1+0.5) - f(1)}{(1+0.5) - 1} = \frac{1.25}{0.5} = 2.5, \quad \Delta x = 0.5$$

$$\frac{\Delta y}{\Delta x} = \frac{f(1+0.25) - f(1)}{(1+0.25) - 1} = \frac{1.5625 - 1}{0.25} = 2.25, \quad \Delta x = 0.25$$

▶ **FIRST REFLECTION**

Compute the value of $\dfrac{\Delta y}{\Delta x}$ for $\Delta x = 1/10$. Is your value near 2?   ◀

Notice that each of the computations is of the form:

$$\frac{f(a + \Delta x) - f(a)}{\Delta x} \quad \text{or} \quad \frac{f(a + h) - f(a)}{h} \tag{3.2}$$

where in our computations $\Delta x = 1$, 0.75, 0.5, 0.25 (Δx is often replaced with h in calculator computations). Such quotients of differences are called *difference quotients*. Why is the denominator Δx (or h) a difference?

DEFINING THE DIFFERENCE QUOTIENT

The **difference quotient** can be interpreted as the slope of the secant line through $(a, f(a))$ and $(a + \Delta x, f(a + \Delta x))$ or as the average rate of change of f over the interval $[a, a + \Delta x]$ and is equal to $\dfrac{f(a + \Delta x) - f(a)}{\Delta x}$.

Your TI calculator has a function that computes this difference quotient. The expression `avgRC(` $f(x), x, h$ `)` computes the slope of the secant line for the function f from the point $(x, f(x))$ to the point $(x + h, f(x + h))$. Here the h value is the distance moved along the horizontal or x-axis when we move on the graph of f from $(x, f(x))$ to the point $(x + h, f(x + h))$.

▶ **Second Reflection**

Clear the one character variables by pressing **F6**. Enter the expression

$$\texttt{avgRC(f(x), x, h)}.$$

How does the resulting expression compare with Equation (3.2)?   ◀

The next Think and Share gives you a way to generate many pictures of the process of approximating the tangent line with secant lines as shown in *Drawing Multiple Secant Lines*.

Think and Share

Break into groups. In this Think and Share, you will use the `For...EndFor` statement to animate the secant lines. Recall that we use `h` for Δx in TI calculator calculations. Enter the `Define` and `For...EndFor` statements of *Drawing Multiple Secant Lines*. Select and enter the `ClrDraw` command from the `Draw` menu of the GRAPH window to clear the secant lines. Experiment with another `For... EndFor` statement by letting n go from (a) one to nine in steps of two and (b) one to 101 in steps of 25. Are you convinced that the secant lines approach the tangent line as h approaches zero?

Experiment with the function $f(x) = 4\sin(x)$ at $x = 1$ and at $x = 2$.

DRAWING MULTIPLE SECANT LINES

First, enter $y6(x) = x^2$ in the Y= editor. Then set the graphing WINDOW to: `xmin = -1, xmax = 3, ymin = -1, ymax = 5`. Next, in the HOME screen, enter

$$\texttt{define msec(x,h) = avgRC(y6(x),x,h)}$$

The `avgRC` command computes the average rate of change of $y6(x)$, with respect to x, over an interval of length h.

The collection of statements separated by colons in the `For...EndFor` statement (Equation (3.3)) below draws five secant lines between $(1, y6(1))$ and the five points of the form $(a, y6(a)$ where $a = 1 + 1, 1 + 1/2, 1 + 1/3, 1 + 1/4,$ and $1 + 1/5$. Enter the following statement while (still) in the HOME screen.

$$\texttt{For n,1,5,1: DrawSlp 1,y6(1),msec(1,1/n): EndFor} \qquad (3.3)$$

The `For` statement takes n from one to five in steps of one. The `DrawSlp` statement draws a line through the point $(1, y6(1))$ with slope `msec(1,1/n)` (the slope of the secant line from $(1, y6(1))$ to $(1 + 1/n, y6(1 + 1/n))$). The `For` statement causes five secant lines to be drawn. The `EndFor` statement ends the `For` statement.

Now that we have a graphical understanding of the way secant lines approach the tangent line as h approaches zero, let's look at the numbers.

E X A M P L E 2 **Approximating the Tangent Line Slope Using a Data Table**

Be sure that y6(x) is still x^2 in the Y= editor. Enter the Data/Matrix editor with a new data table named secslp (for secant slopes). Enter the following values for the first three columns of the data table (for the first formula, highlight c1 and press ENTER to access the Data/Matrix editor Entry line):

c1=seq(1+0.1^k,k,1,5)

c2=y6(c1)

c3=(y6(c1)-y6(1))/(c1-1)

Look at the table. Do all five of the values of the secant slopes appear? Do these values get closer and closer to 2?

▶ **THIRD REFLECTION**

After completing Example 2, increase the number of x values available in the table to 15 by changing the seq(statement. Scroll down the table in the c3 column to see if the secant slope values get closer and closer to 2. Are the values ever exactly 2 or are they always 2.? Why does undef appear in the table? Approximate the slope of the curve for different points on the curve. Change the first argument of the seq(statement for c1 to look at points whose first coordinate or x values are $x = 2.6, 3.4, -2$. ◀

L I M I T   N O T A T I O N   A N D   T E R M I N O L O G Y

The mathematical notation to say that the slope of the tangent line to the graph of $f(x) = x^2$ at $x = 1$ gets closer and closer to 2.0 as Δx approaches zero is

$$\lim_{\Delta x \to 0} \frac{f(1 + \Delta x) - f(1)}{\Delta x} = 2.0.$$

This limit is a *two-sided limit*. In order for this limit to exist, the left-sided and right-sided limits must both exist and must be equal (see Section 1.5).

Putting It All Together

Average Rate of Change.   In the language of rates of change, we recall that the average rate of change over the interval $[a, a + \Delta x]$ is computed the same way we compute the slope of the secant line connecting the point $(a, f(a))$ to the point $(a + \Delta x, f(a + \Delta x))$. In other words, it is the change in f over an interval divided by the length of that interval. Thus, when we calculate the slope of the secant line $(f(a + \Delta x) - f(a))/\Delta x$, we are also calculating the average rate of change of f over the interval $[a, a + \Delta x]$. This is how `avgRC(f(x),x,h)` on our calculators gets its name. Again, we use the term *difference quotient* for

$$\frac{f(a + \Delta x) - f(a)}{\Delta x}.$$

▶ **FOURTH REFLECTION**

For the difference quotient $\frac{f(a+\Delta x)-f(a)}{\Delta x}$, what happens geometrically when Δx is zero? ◀

Instantaneous Rate of Change. The instantaneous rate of change at a point, on the other hand, is defined to be the slope of the tangent line to the graph at that point. In the expression $\lim\limits_{\Delta x \to 0} \dfrac{f(a + \Delta x) - f(a)}{\Delta x}$, the difference quotient $\dfrac{f(a + \Delta x) - f(a)}{\Delta x}$ by itself represents the slope of a secant line (denoted *msec*). Thus, when we apply the limit operation to this difference quotient, the resulting expression $\lim\limits_{\Delta x \to 0} \dfrac{f(a + \Delta x) - f(a)}{\Delta x}$ represents both the instantaneous rate of change of f and the slope of the tangent line (which we denote *mtan*). The slope of the tangent line is the limiting value of the slopes of the secant lines. Hence, the tangent line is the limiting position of these secant lines.

THE DERIVATIVE AND DIFFERENCE QUOTIENT

$$difference\ quotient = \begin{cases} msec,\ \text{slope of secant line} \\ \text{average rate of change over an interval of length } \Delta x \\ \dfrac{f(a + \Delta x) - f(a)}{\Delta x} \end{cases}$$

whereas

$$derivative = \begin{cases} mtan,\ \text{slope of tangent line at a point} \\ \text{instantaneous rate of change at a point } (a, f(a)) \\ \lim\limits_{\Delta x \to 0} \dfrac{f(a + \Delta x) - f(a)}{\Delta x} \end{cases}$$

If this limit exists, the result is denoted $f'(a)$, the derivative of f, evaluated at $x = a$.

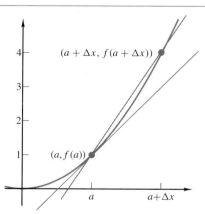

FIGURE 3.2 A Secant Line and the Tangent Line at $x = a$

The picture in Figure 3.2 illustrates the secant line through the points $(a, f(a))$ and $(a + \Delta x, f(a + \Delta x))$ and the tangent line at the point $(a, f(a))$. We are ready to state a precise definition for the derivative at a point.

THE DERIVATIVE AT A POINT

Let f be a function and let a be a number in the domain of f. If the two-sided limit as Δx approaches zero exists, $f'(a) = \lim\limits_{\Delta x \to 0} \dfrac{f(a + \Delta x) - f(a)}{\Delta x}$ is called the derivative of f, evaluated at the point where $x = a$.

E X A M P L E   3   **Finding the Derivative Function Symbolically**

Find the derivative of the function $f(x) = x^2$ at $x = 1$ using the limit definition of the derivative at a point.

We begin with the difference quotient:

$$\frac{f(1 + \Delta x) - f(1)}{\Delta x} = \frac{(1 + \Delta x)^2 - 1^2}{\Delta x}$$
$$= \frac{1 + 2\Delta x + (\Delta x)^2 - 1}{\Delta x}$$
$$= \frac{2\Delta x + (\Delta x)^2}{\Delta x}.$$

Now, using the definition of the derivative, we take the limit as Δx goes to zero, and

$$f'(1) = \lim_{\Delta x \to 0} \left(\frac{2\Delta x + (\Delta x)^2}{\Delta x} \right)$$
$$= \lim_{\Delta x \to 0} (2 + \Delta x) = 2.$$

Of course, this value matches the results we obtained graphically and numerically.

Continuity and the Derivative at a Point

Recall the definition of continuity at a point.

CONTINUITY AT A POINT

Let a be a number in the domain of a function f. The function f is considered to be *continuous* at a if

$$\lim_{x \to a} f(x) = f(a).$$

The expression $2 + \Delta x$ is a continuous (linear, even) function of Δx. Continuous functions have useful properties that discontinuous functions do not have. In particular, if we wish to compute $\lim\limits_{x \to a} f(x)$ and f is continuous at

FIGURE 3.3 The Graph of f

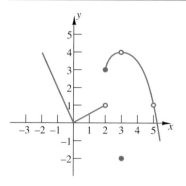

a, then according to the definition, we can simply compute $f(a)$. This is why we were able to simply substitute 0 for Δx when we computed $\lim\limits_{\Delta x \to 0} 2 + \Delta x$.

A derivative cannot exist at a point where a function is discontinuous.

▶ **FIFTH REFLECTION**

Attempt to draw tangent lines to the graph of f in Figure 3.3 at $x = 2$, $x = 3$, and $x = 5$. What difficulties do you encounter at 2, 3, and 5? ◀

Caution: in Figure 3.3, f is continuous at $x = 0$ (there is no break or hole in the graph and $\lim\limits_{x \to 0} f(x) = f(0)$), but f is not **differentiable** (does not have a derivative) at $x = 0$. Do you remember why f does not have a derivative at $x = 0$?

There are also functions that are continuous (throughout their domain), but not differentiable. For example, in Figure 3.4 $f(x) = |x - 4|$ is not differentiable at $x = 4$ because there is a corner there. The limit of the difference quotient from the left is -1, and from the right it is $+1$.

FIGURE 3.4 $f(x) = |x - 4|$

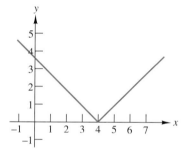

The Derivative as a Function

Now that we have a method for finding the derivative or rate of change of a function at a point on the curve, we will look at finding a general expression for the derivative at any point on a curve. If, for each value of x in the domain of a function, we can find a single value of the slope of the tangent line, then we can create a table. Such a table is a representation of the *derivative function*.

We have looked at the value of the derivative at a point graphically (estimating by zooming), numerically (estimating with tables), and algebraically (exactly with limits). Here we will use hand drawing to approximate the tangent line and numerical computation to find the slope of the tangent and, thus, the slope of the curve at several points.

▶ **SIXTH REFLECTION**

Complete Table 3.1 by drawing the tangent line to the curve $f(x)$ displayed in Figure 3.5 for each value of x in the table and then approximating its slope using the grid provided. Can you guess a rule for f' from your table? ◀

FIGURE 3.5 The Graph of $f(x) = x^2$

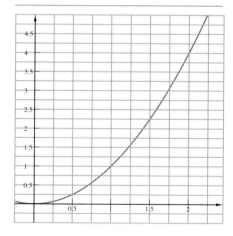

TABLE 3.1 f' for x from 0 to 2

x	0.00	0.25	0.50	0.75	1.00	1.25	1.50	1.75	2.00
$f'(x)$									

▶ **SEVENTH REFLECTION**

Use the difference quotient

$$\frac{f(x + h) - f(x)}{h}$$

with $h = 0.001$ to complete Table 3.2 using

```
avgRC(x^2,x,0.001)|x={0,1/4,1/2,3/4,1,5/4,3/2,7/4,2}.
```

Guess a rule for f'. Does this rule match the rule in Reflection 6?   ◄

TABLE 3.2 f' for x from 0 to 2

x	0.00	0.25	0.50	0.75	1.00	1.25	1.50	1.75	2.00
$f'(x)$									

Definition of the Derivative

We can now give a precise definition of the derivative as a function.

THE DERIVATIVE FUNCTION

Let $f(x)$ be a function of x. The derivative with respect to x of $f(x)$, denoted $f'(x)$, is also a function of x and is the limit of the difference quotient as Δx approaches zero. This limit, if it exists, is named the derivative function, and is denoted by the symbol $f'(x)$. We write

$$f'(x) = \lim_{\Delta x \to 0} \frac{f(x + \Delta x) - f(x)}{\Delta x}.$$

To designate the derivative evaluated at $x = a$, we write $f'(a)$.

► **EIGHTH REFLECTION**

Read the definition above carefully. How does it differ from the definition of derivative at a point given on page 132?   ◄

Look closely at the difference quotient in the definition. For a continous function, if $\Delta x \to 0$, then the numerator $f(x + \Delta x) - f(x)$ approaches 0. As a result, we must simplify the difference quotient algebraically to find the derivative symbolically using the definition.

The process can involve great algebraic prowess even for relatively simple functions. Sometimes students become so bogged down in algebraic manipulations, they forget why they are finding a derivative in the first place. We will avoid this problem by using the power of the computer algebra system built into the TI calculator to do the algebra for us. To this end, we write a *script* which we will use to find derivatives, using the definition of derivative. (If you don't know about scripting, don't worry. You're about to learn.)

A script is a lot like a program, but no programming is needed to write a script. You'll see.

Creating and Working with a TI Calculator Script

A **script** is a set of commands you have saved. You access a script (saved commands) through the Text editor. You create a script by first entering commands on the HOME screen.

WRITE A SCRIPT

Clear the HOME screen. ENTER, on the HOME screen, each of the commands shown below.

    define f(x) = x^2

    f(x+h)

    expand (f(x+h),x)

    f(x+h) - f(x)   Note the automatic simplification.

    (f(x+h)-f(x))/h $h \neq 0$ in this step.

    limit((f(x+h)-f(x))/h,h,0)

SAVE A SCRIPT

Select Save Copy As . . . from the Toolbox menu. Type *diffquo* in the Variable edit box. Press ENTER twice to save the name and return to the HOME screen. (See Figure 3.6.)

Congratulations! You have just created a script.

Open the Script.   Go to the Text editor in the APPS menu. Select open, and press ENTER. Use the cursor button to select the variable *diffquo* in the OPEN dialogue box, and press ENTER twice. You should see your script. Ours is shown in Figure 3.7.

We are now ready to edit and use the script document named *diffquo*. By changing the definition of $f(x)$ in the first line, we can use our script to find the derivative of functions besides $f(x) = x^2$ using the limit definition. Meanwhile, the TI calculator will handle the algebra. Observe the work performed by the TI calculator as you step through the script to see that you understand each algebraic manipulation.

Edit and Use the Script.   The cursor should be blinking just before the D in Define on the first line of your text. Press ENTER and move the cursor back up to the top line. The Clear command option in the Command menu will

FIGURE 3.6   The Save Copy as Dialogue Box   **FIGURE 3.7**   The Script Screen in the Text Editor

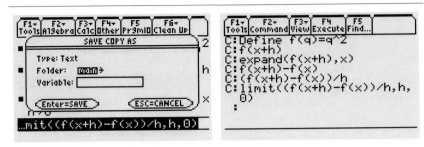

remove the *C* from the top line so that this line will now be a comment and not a command. Title your script; we called ours "derivative by the definition." Use the cursor button to move to the second line. Use the `Command` option in the `Command` menu to change the comment to a command.

Use the `Script View` option in the `View` menu to split the screen. so that you can enter the script commands at the top of the window, and see the results on the HOME screen at the bottom of the window. To toggle back and forth between the script (at the top) and the HOME screen (at the bottom), press 2nd APPS. The active window will be outlined with a thick line.

Activate the HOME screen (right side), clear it, and return to the script screen. Your TI-89 calculator window should now look like ours in Figure 3.8.

With the cursor blinking on the second line of the script, the first command line, press F4 to execute the command. The result will appear on the home screen and the cursor should move to the next command line. Continue to press F4 to execute each command. Figure 3.9 shows our window after we executed the first two commands.

The screen will be split vertically if you are using a TI-92.

FIGURE 3.8   Split Screen with Script Window Active   **FIGURE 3.9**   Running the Script

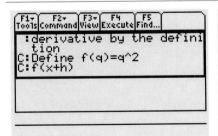
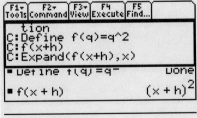

Finish executing all the commands, and then go back up to the top of the script and change the definition of $f(x)$ to x^3. Go through the script again, using only the F4 key to execute each command, and your result this time should be $3x^2$. You have just discovered that if $f(x) = x^3$ then $f'(x) = 3x^2$.

Press 2nd Quit *twice to access the* HOME *screen; select* Current *in the* APPS Text *editor to access the current script;* Script view *in the* View *menu for split screen; and* 2nd APPS *to toggle between script and home screens.*

Think and Share

If you inadvertently press ENTER instead of F4, nothing will be executed. You will be on a new line without a *C:*. You simply use the Backspace key to return to the previous command line and then use the F4 key to execute.

Break into groups of four with two teams of two. Return to the HOME screen (see margin note).Use your *diffquo* script to find the derivative function for each of the following. One team should do the odd-numbered problems and the other team should do the even-numbered ones. See if you can make up a rule for finding the derivative of a power function, $f(x) = x^n$. Use the rule you make up to find the derivative of $f(x) = x^{27}$. Check with your *diffquo* script.

1. $f(x) = x^4$
2. $f(x) = x^5$
3. $g(x) = x^{1/2}$
4. $h(x) = 1/x$
5. $h(x) = 1/x^2$
6. $h(x) = 1/x^3$

Now that We've Got the Derivative, What Can We Do With It?

In Chapter 2, we used the numerical derivative to sketch graphs of functions and their derivatives. We now can use the exact derivative to investigate properties of the original function.

NOTE OF WARNING

Now that we have defined the derivative as a function, we can use functional notation to designate the derivative evaluated at a particular point. For example, if we wanted to know the slope of the function $f(x) = x^2$ at $x = 1$, we would simply evaluate $f'(x) = 2x$ at $x = 1$ and get $f'(1) = 2(1) = 2$. The derivative, evaluated at $x = 2$, is $f'(2) = 2(2) = 4$.

Caution: Sometimes students write

$$f(x) = x^2 = 2x = 2$$

when finding the derivative of x^2 evaluated at $x = 1$. This **must not** be done! Please remember to write your mathematics properly. Never use the equal sign except to designate equal quantities. Write

$$f(x) \quad = \quad x^2$$
$$f'(x) \quad = \quad 2x \qquad \text{the derivative at any point.}$$
$$f'(2) \quad = \quad 4 \qquad \text{the derivative evaluated at x = 2.}$$

FIGURE 3.10   Gottfried Leibniz 1648 – 1716

Alternate Notation for the Derivative

If we designate the change in y as $\Delta y = f(x + \Delta x) - f(x)$ and the change in x as Δx, we will have an alternate name for our difference quotient.

$$\frac{\Delta y}{\Delta x} = \frac{f(x + \Delta x) - f(x)}{\Delta x}$$

Taking the limit as Δx approaches zero, we write

$$\lim_{\Delta x \to 0} \frac{\Delta y}{\Delta x} = \frac{dy}{dx}.$$

The $\dfrac{dy}{dx}$ notation is called Leibniz notation because it was introduced by Gottfried Leibniz, a philosopher and mathematician who is credited with creating calculus. It is just a way to designate *derivative of y with respect to x*. If we want to evaluate dy/dx when $x = a$, we write

$$\frac{dy}{dx}\bigg|_{x=a}.$$

The act or process of finding a derivative is called **differentiation**. If it is possible to find a derivative at every point in the domain of a function f, we say f is everywhere differentiable or that f is differentiable on its domain.

Recall the function $f(x) = |5 - x^2|$ of Section 2.5. We found that $f'(-\sqrt{5})$ and $f'(\sqrt{5})$ did not exist because the tangent line was not unique at either of the input values $-\sqrt{5}$ and $\sqrt{5}$. We say, in this case, that f is not differentiable at $x = -\sqrt{5}$ or $\sqrt{5}$. Alternatively, we could say f is differentiable except at $x = -\sqrt{5}$ and $\sqrt{5}$.

There are many symbols which can be used to designate the derivative. If we start with the notation $f(x) = x^2$, we write $f'(x) = 2x$. If, on the other hand, we differentiate both sides of the equation $y = x^2$, we write $dy/dx = 2x$ or $y' = 2x$. You should read the symbol ds/dt as the derivative of s with respect to t and understand that this derivative represents the instantaneous rate of change of the quantity s with respect to the quantity t. As you continue your study of calculus, you will find different notations are useful in different contexts.

Unusual Expressions for the Derivative

Sometimes the TI calculator gives answers that we may not recognize or understand.

E X A M P L E   4   **The Signum Function**

Change the definition of $f(x)$ to $|x|$ in your *diffquo* script using the TI calculator **abs** function. Run through your script to find the derivative function. The final

FIGURE 3.11 $f(x) = |x|$ and $f'(x) = sign(x)$

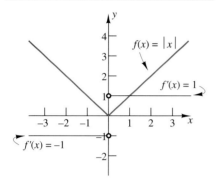

output is ᴤi9n⟨x⟩ which is the abbreviated form for the *signum function*. You may not be familiar with this function. The *signum function* is defined to be:

$$sign(x) = \begin{cases} -1, & x < 0 \\ \text{undefined}, & x = 0 \\ 1, & x > 0. \end{cases}$$

Graph the $|x|$ function and observe that if x is less than zero, the slope is negative one and if x is greater than zero, the slope is positive one. Thus, if $x < 0$, the derivative is negative one, and if $x > 0$, the derivative is one. Recall that the derivative does not exist at $x = 0$.

We have sketched both $f(x)$ and the *signum function* in Figure 3.11.

Think and Share

Break into groups. Use the script you created in this section to find $f'(x)$ for $f(x) = |x^2 - 5|$. Find a symbolic expression for $f'(x)$ without using the word *signum*.

Summary

In this section, we have defined the derivative of a function at a point and the derivative of a function as a function. You should be prepared to state these definitions to anyone at any time in your academic career; they are primary definitions of calculus. We have learned how to use these definitions to find formulas for derivatives of specific functions (with the help of the TI calculator). Before the advent of computer algebra systems, computing derivatives from the definition was so difficult and time-consuming that it was rarely done. Instead, a number of shortcut techniques were developed that enabled the computation of many derivatives without resorting to the definition. Although these techniques are less important for computational purposes than they were in the past, they still shed light on the nature of the derivative by revealing some of its fundamental properties. In the next section, we will generalize our results and you will use the TI calculator to "discover" some of these techniques or rules. We hope you will use them often. Happy rule playing!

E X E R C I S E   S E T   3 . 1

Clear the **HOME** screen and clear all single character variables before beginning these exercises.

1. Let $f(x) = |x^2 - 4|$. Show that $f'(2)$ does not exist by evaluating the difference quotient at 2 for positive and negative values of h. Find $\lim\limits_{h \to 0^-} \dfrac{f(2+h) - f(2)}{h}$ and $\lim\limits_{h \to 0^+} \dfrac{f(2+h) - f(2)}{h}$. Does $\lim\limits_{h \to 0} \dfrac{f(2+h) - f(2)}{h}$ exist? Does $f'(2)$ exist? Justify your conclusions.

2. Let $f(x) = |x^2 - 4|$. Produce two hand-drawn graphs of $f(x)$.
   a. On one of the graphs show a tangent line at $x = 2$ as the limit of secant lines for positive h.
   b. On the other graph show a tangent line at $x = 2$ as the limit of secant lines for negative h.
   c. Is the tangent line at $x = 2$ unique? Does the tangent line exist at $x = 2$? Does $f'(2)$ exist? Justify all answers.

3. Let $f(x) = |x^2 - 4|$. Use the zoom method of Example 4 from Section 2.5 to decide whether or not $f(x)$ is locally linear at $x = \sqrt{2}$. Is there a tangent line at $x = \sqrt{2}$? Does $f'(2)$ exist?

4. Let $g(x) = |x^2 - 2|$. Use the zoom method of Example 4 from Section 2.5 to decide whether or not $g(x)$ is locally linear at $x = 2$. Is there a tangent line at $x = 2$? Does $g'(2)$ exist? Does $g'(\sqrt{2})$ exist?

In Exercises 5–7, find the derivative at the indicated point by evaluating the difference quotient with Δx approaching zero through positive and negative values.

5. Find $f'(2)$ for $f(x) = x^3 - 5x + 1$.

6. Find $g'(3)$ for $g(x) = \dfrac{1}{x}$.

7. Find $h'(-2)$ for $h(x) = \sqrt{x + 3}$.

In Exercises 8–11, find each of the following limits symbolically with pencil and paper using correct limit notation. Confirm your results with the TI calculator `limit(` command. Then confirm your answer numerically by substituting numbers and graphically by drawing an appropriate graph.

† 8. $\displaystyle\lim_{x \to 2} \frac{x^2 - 5x + 6}{x - 2}$.

9. $\displaystyle\lim_{x \to ^-3} \frac{x^2 + x - 6}{x + 3}$.

10. $\displaystyle\lim_{h \to 0} \frac{f(3 + h) - f(3)}{h}$. where $f(x) = x^2$

11. $\displaystyle\lim_{h \to 0} \frac{g(4 + h) - g(4)}{h}$. where $g(x) = x^2$

In Exercises 12–19, find the indicated derivative, using the definition of the derivative at a point.

12. Let $f(x) = 2x^2 - 3x + 4$. Find $f'(1)$. Verify your answer numerically.

13. Let $g(x) = -x^2 + 5x - 2$. Find $g'(2)$. Verify your answer numerically.

14. Let $h(x) = 3x^2 + 2x - 4$. Find $h'(-2)$. Verify your answer numerically.

15. Let $f(x) = 2x^3 - 3x^2 + 3x - 2$. Find $f'(0)$. Verify your answer numerically.

16. Let $g(x) = 2x - 5$. Find $f'(a)$. Verify your answer with a graph.

17. Let $h(x) = 7$. Find $f'(4)$. Verify your answer with a graph.

18. Let $f(x) = \dfrac{2}{x - 2}$. Find $f'(1)$. Verify your answer with a graph.

19. Let $g(x) = x^2 - 4x + 4$. Find $f'(2)$. Verify your answer with a graph.

†20. *Group Exercise.* Animate the secant lines to the graph of $f(x) = 4 + x - 2x^3$ at $x = 0$ using the `For...EndFor` statement and the `DrawSlp` command. Change $1/n$ to

$(1/2)^n$ or $(2/3)^n$ in the `DrawSlp` command to avoid division by zero. Let n vary between 0 and 3 to see the behavior to the right of zero. How should you let n vary to see the secant lines to the left of zero?

In Exercises 21–28, use the script you created in this section to find the derivative of the given function.

21. $f(x) = x^5 + 2x^3$

22. $f(x) = \sin(x)$

23. $g(x) = \cos(x)$

24. $f(x) = \tan(x)$

25. $h(x) = \sqrt{x}$

26. $f(x) = 3e^x$

27. $g(x) = e^{x^2}$

28. $k(x) = \ln(x)$

29. Use the script you created in this section to find $g'(x)$ for $g(x) = |x^2 - 2x - 15|$. Express $g'(x)$ without using the word *signum*. It may help to graph $g(x)$ and $g'(x)$.

30. Use the script you created in this section to find $f'(x)$ for $f(x) = |x^2 + 1|$. Explain why this answer does not make reference to the *signum* function.

31. Use the script you created in this section to find $f'(x)$ for $f(x) = |x^2 + 8x + 15|$. Explain why this answer does not make reference to the *signum* function.

†32. *Group Exercise* Define $y1(x)$ as $9 - x^2$ on the HOME screen. Use your `diffquo` script to find the derivative of $f(x) = 9 - x^2$. Define $dy(x)$ as the derivative of $y1(x)$. ENTER the following three statements as a one-line statement to animate the tangent lines to the graph of $y1$.

```
for n,-5,5,1:
drawslp n, y1(n),dy(n):
endfor
```

  After you have successfully animated the tangent lines to the curve, provide answers or explanations for the following:

a. Explain the meaning of the `For...EndFor` statement.

b. Explain what the `DrawSlp` command does.

c. The graph of $y1(x)$ is increasing at a decreasing rate until $x = 0$. Explain how the slopes of the tangent lines in your animation bear this out.

d. What can be said about the graph after $x = 0$, in terms of increase (decrease) at increasing (decreasing) rates?

33. Does the symbol $\dfrac{\Delta y}{\Delta x}$ mean the same thing as the symbol $\dfrac{dy}{dx}$? Explain carefully. Show that you understand the meaning of both expressions.

34. Explain, in your own words, the meaning of each of the following terms or symbols. Write your explanations in clear and complete sentences. Include annotated diagrams whenever appropriate.

a. secant line
b. tangent line
c. *msec*
d. *mtan*
e. average rate of change
f. instantaneous rate of change
g. derivative at a point
h. difference quotient
i. `For k,1,100,2 EndFor`

j. $\lim\limits_{x \to 3} g(x)$
k. continuous
l. discontinuous
m. continuity at a point
n. derivative at a point
o. derivative as a function
p. *sign(x)*
q. $\dfrac{dx}{dt}$
r. $\dfrac{\Delta y}{\Delta x}$

3.2

Rules, Rules, and More Rules

In the previous section, we learned to calculate the derivative at a point, we defined the derivative as a function in its own right with the formula

$$f'(x) = \lim_{\Delta x \to 0} \frac{f(x + \Delta x) - f(x)}{\Delta x},$$

and we used our definition to find formulas for particular derivative functions (with the help of the TI calculator). In this section, we introduce a new feature of the TI calculator, the differentiation operator, and use it to develop rules and techniques for finding derivatives. As we pointed out at the end of the previous section, rules are really shortcut techniques that provide ways of computing derivatives that are simpler than using the definition. Mastering rules will give you a deeper understanding of the nature of the derivative operation. As an added bonus, you will be developing your *symbol sense* that will give you confidence and help you in your further study of mathematics.

The intent of this section is to allow you to discover rules of differentiation for yourselves. The process of discovering the rules is much more important than the rules themselves. Maintain a spirit of discovery as you read and experiment. We will eventually give you the rules, but we want you to have as much fun learning them as possible.

The Power Rule

The TI calculator has been taught (programmed) the same rules of differentiation that students have been forced to memorize for years. We will tap into the TI calculator's "knowledge" by using the TI calculator differentiation operator (see *The TI Calculator Differentiation Operator* below) to see how many of these differentiation rules we can discover for ourselves.

THE TI CALCULATOR DIFFERENTIATION OPERATOR

Clear the 1-character variables. You use ⏐2nd⏐ and ⏐8⏐ to access the differentiation operator. This is NOT the same as the letter "d" on the QWERTY keyboard. Use the differentiation operator ∂(to enter the command ∂((x^3,x).

On the HOME screen Entry line call up the differentiation operator, ∂(. Finish the command by typing x^3,x) and pressing ENTER.

The differentiation operator allows the TI calculator to calculate derivatives symbolically using the rules it has been taught. After completing the directions in the *The TI Calculator Differentiation Operator*, you should see $\frac{d}{dx}(x^3)$ on the left of the screen, and $3x^2$ on the right. Congratulations! You have just found (once again) that the derivative of $f(x) = x^3$ is $f'(x) = 3x^2$.

The ,x) means differentiate "with respect to x" and tells the TI calculator to compute the instantaneous rate at which x^3 changes as the value of x changes. The TI calculator translates ∂(and the ,x) on the Entry line to $\frac{d}{dx}$ on the HOME screen. When we are differentiating with respect to x, we refer to x as the independent variable. We must always specify the independent variable. For example, if we entered ∂((x^3,z), the output will be zero since, as the value of z changes, the value of x^3 does not change, that is, x^3 acts like a constant when differentiating with respect to z. Try it for yourself!

Think and Share

Break into groups. You will be investigating functions of the form $f(x) = x^n$. Use the differentiation operator, ∂(, to find the derivative of each of the following functions: x^2, x^3, x^4, x^5, and x^6. Watch for a pattern and make up a *personal power rule*. Use your power rule to guess the derivative of x^{28}. Verify with the TI calculator.

Now find the derivative of $x^{4/3}$, $x^{7/5}$, $x^{9/8}$, and $x^{1/2}$. Does your personal power rule still work?

So far we have chosen n to be either a positive integer or a positive fraction. Will your power rule still hold true if n is any real number, positive or negative? Experiment with several different values of n, some negative and some positive. Try some irrational numbers like π or $\sqrt{2}$. Rewrite the TI calculator output so that x only appears in the numerator. What if $n = 0$? (See Figure 3.12 and the margin note for a slick way to experiment.) Write your power rule in complete sentences along with its symbolic representation.

▶ FIRST REFLECTION

Write your personal power rule in complete sentences along with its symbolic representation.   ◀

Symbolically, the Power Rule is
$$\frac{d}{dx}x^n = nx^{n-1}.$$

Use the "with" (|) symbol (2nd K on the TI-92) and then edit the Entry line, changing n each time.

FIGURE 3.12   Power Rule Examples

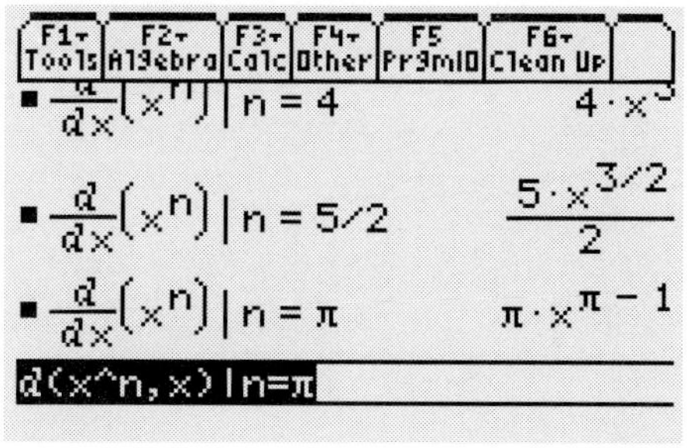

How Well Do You Understand? Can You Undo What You Have Done?

If you really understand the rule, you will be able to reverse the process; that is, given a derivative, you should be able to find a function that has that derivative. The art of finding this function (and believe us, it is an art) is called **antidifferentiation** and the function you find is called an **antiderivative**. Thus, x^3 is an antiderivative of $3x^2$. Antidifferentiation is almost always an exciting pastime and so we give you a few examples to try. Find an antiderivative of $2x$; that is, find $f(x)$ such that $f'(x) = 2x$. Take a wild guess and then find the derivative of your guess. If the derivative is $2x$, you guessed correctly. Try another: find $h(x)$ if $h'(x) = x^3$.

The Constant Multiplier Rule

The Constant Multiplier Rule is
$$\frac{d}{dx}kf(x) = k\frac{d}{dx}f(x).$$

What happens to a derivative if a function is multiplied by a constant? Find the derivatives of $5x^3$, $7x^3$, πx^3, $-7x^3$. Using the fact that the derivative of x^3 is $3x^2$, deduce a *constant multiplier* rule. Use your rule to find the derivative of $12x^5$.

▶ SECOND REFLECTION

Write your constant multiplier rule in complete sentences along with its symbolic representation.   ◀

The Sum Rule

We know that the derivative of x^3 is $3x^2$, by the Power Rule, and that the derivative of $\sin(x)$ is $\cos(x)$ from a previous section. What is the derivative of $x^3 + \sin(x)$? Could it be as simple as $3x^2 + \cos(x)$?

Break into groups. Find the derivative of

Think and Share

$$x^3 + \sin(x),$$
$$x^4 + x^7,$$
$$x^{-5/3} - \sin(2x)$$

Think and Share

with your TI calculator. Make up your own *sum rule*. Find the derivative of $x^3 + \sin(x) + x^4$. What happens when the sum includes more than two terms?

Clearly, the TI calculator has been programmed to output the trigonometric functions first, and then to output power terms in order of descending exponents. Using pencil and paper, we would differentiate each term as we come to it. To see if you have discovered the derivative of a *sum* rule, find $f'(x)$ if

$$f(x) = x^{2/3} - 4\sin(x) - 13x^2$$

without your calculator. Verify your result with the TI calculator.

▶ **THIRD REFLECTION**
Write your sum rule in complete sentences along with its symbolic representation. ◀

Remark for the future: the differentiation operator is called a linear operator because it satisfies the constant multiplier rule and the sum rule.

The Product Rule

What is the derivative of $x^3 \cdot \sin(x)$? Given our experience with the sum rule we might well guess $(3x^2) \cdot \cos(x)$. Unfortunately, we would be wrong. The derivative of a product is **not** equal to the product of the derivatives. Check this by differentiating $x^3 \cdot \sin(x)$ with your TI calculator.

Think and Share

Break into groups. Find the derivative of the following functions with the TI calculator. Find a pattern, and develop your own *personal product rule*: $x^3 \sin(x)$, $x^7 \sin(x)$, and $x^{5/2} \sin(x)$. If you don't see a pattern, make up some more examples and try them. Once you have recognized a pattern, use it to make up a rule. Using your rule and pencil and paper, find the derivative of $x^5 \sin(x)$. Verify your result with the TI calculator. If your own personal rule did not work, differentiate more products until you can find the correct pattern.

▶ **FOURTH REFLECTION**
Write your product rule in complete sentences along with its symbolic representation. ◀

▶ **FIFTH REFLECTION**
Let's try finding another antiderivative. If $h'(x) = 2x\sin(x) + x^2\cos(x)$, what is $h(x)$? ◀

The Quotient Rule

What is the derivative of the quotient $\dfrac{\sin(x)}{x^5}$? The TI calculator can differentiate a quotient, but it has not displayed the quotient rule. The TI calculator has been programmed to turn quotients into products by introducing negative exponents. It can therefore use the product rule and does not seem to know the quotient rule at all. Figures 3.13, 3.14, and 3.15 show how the TI calculator differentiates the quotient $\dfrac{\sin(x)}{x^5}$. We have entered the problem as a quotient and as a product to show that the TI calculator uses the product rule each time. We have also used the common denominator command, comDenom, to display the result as a simple quotient.

FIGURE 3.13 As a Quotient   **FIGURE 3.14** As a Product

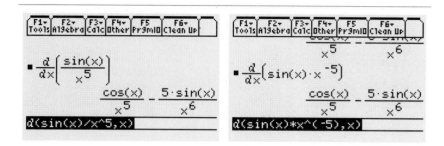

FIGURE 3.15 Display with a Common Denominator

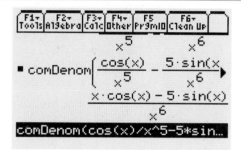

Unfortunately, we cannot guess the quotient rule by observing the output as we have done in previous examples. With a bit of manipulation, however, we can get the TI calculator to produce the symbolic representation of the rule.

▶ SIXTH REFLECTION

Clear the one-character variables on your machine. Use the differentiation operator to compute the derivative of $\dfrac{f(x)}{g(x)}$. Simplify your result using the comDenom command in the Algebra menu. You have obtained a symbolic representation of the quotient rule. Write the rule in complete sentences along with its symbolic representation.   ◀

Break into groups. Here is another antidifferentiation problem for you:

Think and Share

$$f'(x) = \frac{x^5 \cdot \cos(x) - \sin(x) \cdot 5x^4}{x^{10}}$$

Find an expression for $f(x)$.

We give a summary of the rules and then justify three of them.

DIFFERENTIATION RULES: A SUMMARY

The Constant Multiplier Rule
The derivative of a constant times a function equals the constant multiplied by the derivative of the function. If $g(x) = kf(x)$, then

$$g'(x) = (kf(x))' = kf'(x).$$

The Sum Rule
The derivative of the sum of functions equals the sum of the derivatives of the functions. If $h(x) = f(x) + g(x)$, then

$$h'(x) = f'(x) + g'(x).$$

The Product Rule
The derivative of the product of two functions equals the derivative of the first function times the second, plus the first times the derivative of the second function. If $h(x) = f(x) \cdot g(x)$, then

$$h'(x) = f'(x) \cdot g(x) + f(x) \cdot g'(x).$$

The Quotient Rule
The derivative of the quotient of two functions equals the denominator times the derivative of the numerator minus the numerator times the derivative of the denominator, all divided by the denominator squared. If $h(x) = \dfrac{f(x)}{g(x)}$, then

$$h'(x) = \frac{g(x) \cdot f'(x) - f(x) \cdot g'(x)}{[g(x)]^2}.$$

The Power Rule
The derivative of x raised to a power equals the power multiplied by x raised to the power minus one. If $f(x) = x^n$, then

$$f'(x) = nx^{n-1}.$$

Proofs of the Differentiation Rules

We will verify three differentiation rules using three different approaches. We will prove the Sum Rule using the limit definition of derivatives, we will prove the Product Rule using a geometric approach, and the Quotient Rule using symbolic manipulations based on the rules of algebra.

A Proof of the Sum Rule.   The derivative of the sum of functions equals the sum of the derivatives of the functions or, if $h(x) = f(x) + g(x)$,

$$h'(x) = f'(x) + g'(x).$$

We can establish this rule algebraically. If $h(x) = f(x) + g(x)$, then

$$h'(x) = \lim_{\Delta x \to 0} \frac{h(x + \Delta x) - h(x)}{\Delta x}$$

$$= \lim_{\Delta x \to 0} \frac{(f(x + \Delta x) + g(x + \Delta x)) - (f(x) + g(x))}{\Delta x}$$

$$= \lim_{\Delta x \to 0} \frac{f(x + \Delta x) - f(x) + g(x + \Delta x) - g(x)}{\Delta x}$$

$$= \lim_{\Delta x \to 0} \frac{f(x + \Delta x) - f(x)}{\Delta x} + \lim_{\Delta x \to 0} \frac{g(x + \Delta x) - g(x)}{\Delta x}$$

$$= f'(x) + g'(x).$$

A Proof of the Product Rule. The derivative of the product of two functions equals the derivative of the first function times the second, plus the first times the derivative of the second function, or

$$h'(x) = (f(x) \cdot g(x))' = f'(x) \cdot g(x) + f(x) \cdot g'(x).$$

We can establish this rule with the help of the geometric diagram in Figure 3.16 originally developed by Isaac Newton. If $h(x) = f(x) \cdot g(x)$, then

$$h(x + \Delta x) - h(x) = f(x + \Delta x) \cdot g(x + \Delta x) - f(x)g(x)$$

and we can interpret this as the difference between the area of a large rectangle and the area of a smaller rectangle. With this interpretation, we get from Figure 3.17 that

$$f(x + \Delta x) \cdot g(x + \Delta x) - f(x)g(x) = (A_1 + A_2 + A_3 + A_4) - A_1$$
$$= A_2 + A_3 + A_4.$$

FIGURE 3.16 Newton's Area Diagram to Analyze $f(x + \Delta x)g(x + \Delta x) - g(x)f(x)$

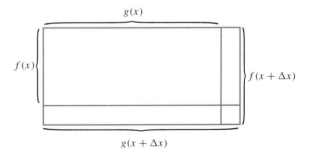

FIGURE 3.17   The Difference of Two Areas

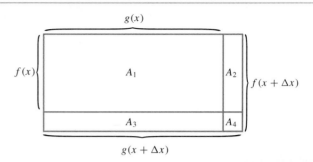

We can consider these areas individually as in Figure 3.18.

FIGURE 3.18   The Areas Considered Algebraically

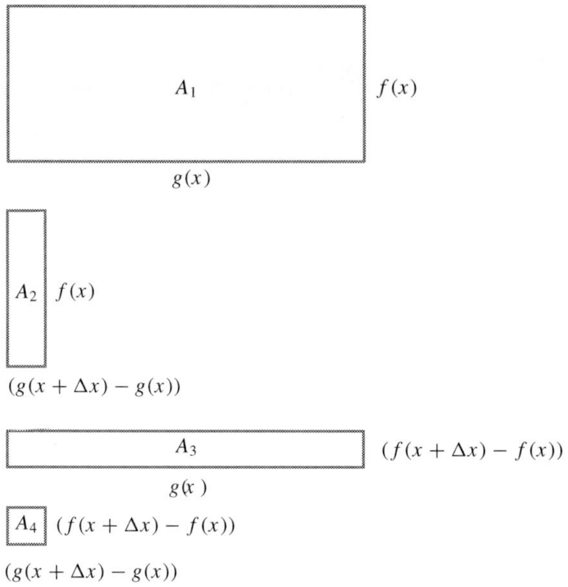

A_1 $f(x)$ $A_1 = f(x) \cdot g(x)$

$g(x)$

A_2 $f(x)$ $A_2 = f(x) \cdot (g(x + \Delta x) - g(x))$

$(g(x + \Delta x) - g(x))$

A_3 $(f(x + \Delta x) - f(x))$ $A_3 = g(x) \cdot (f(x + \Delta x) - f(x))$

$g(x\,)$

A_4 $(f(x + \Delta x) - f(x))$ $A_4 = (f(x + \Delta x) - f(x))$
$\qquad\qquad\qquad\qquad\qquad\qquad\qquad \cdot (g(x + \Delta x) - g(x))$

$(g(x + \Delta x) - g(x))$

So we break

$$\lim_{\Delta x \to 0} \frac{f(x + \Delta x) \cdot g(x + \Delta x) - g(x) \cdot f(x)}{\Delta x}$$

into the sum of three limits:

$$\underbrace{\lim_{\Delta x \to 0} \frac{f(x) \cdot (g(x + \Delta x) - g(x))}{\Delta x}}_{\text{First limit}} + \underbrace{\lim_{\Delta x \to 0} \frac{g(x) \cdot (f(x + \Delta x) - f(x))}{\Delta x}}_{\text{Second limit}}$$

$$+ \lim_{\Delta x \to 0} \left(\frac{f(x + \Delta x) - f(x)}{\Delta x} \cdot (g(x + \Delta x) - g(x)) \right).$$

$$\underbrace{\phantom{+ \lim_{\Delta x \to 0} \left(\frac{f(x + \Delta x) - f(x)}{\Delta x} \cdot (g(x + \Delta x) - g(x)) \right).}}_{\text{Third limit}}$$

The first limit is

$$\lim_{\Delta x \to 0} f(x) \cdot \lim_{\Delta x \to 0} \frac{g(x + \Delta x) - g(x)}{\Delta x} = f(x) \cdot g'(x).$$

The second limit is

$$\lim_{\Delta x \to 0} \frac{f(x + \Delta x) - f(x)}{\Delta x} \cdot \lim_{\Delta x \to 0} g(x) = f'(x) \cdot g(x).$$

The third limit is

$$\lim_{\Delta x \to 0} \frac{f(x + \Delta x) - f(x)}{\Delta x} \cdot \lim_{\Delta x \to 0} [g(x + \Delta x) - g(x)] = f'(x) \cdot 0 = 0.$$

Now, adding the three limits

$$f(x) \cdot g'(x) + f'(x) \cdot g(x) + 0,$$

yields

$$h'(x) = f'(x) \cdot g(x) + f(x) \cdot g'(x).$$

A Proof of the Quotient Rule.   The derivative of the quotient of two functions equals the denominator times the derivative of the numerator, minus the numerator times the derivative of the denominator, all divided by the denominator squared or

$$\left(\frac{f(x)}{g(x)} \right)' = \frac{g(x) \cdot f'(x) - f(x) \cdot g'(x)}{[g(x)]^2}.$$

We can establish this rule using the Product Rule and algebraic manipulations. If $h(x) = \dfrac{f(x)}{g(x)}$, then $h(x) \cdot g(x) = f(x)$.

So

$$h'(x) \cdot g(x) + h(x) \cdot g'(x) = f'(x)$$

$$h'(x) \cdot g(x) + \frac{f(x)}{g(x)} \cdot g'(x) = f'(x)$$

$$h'(x) \cdot g(x) = f'(x) - \frac{f(x)}{g(x)} \cdot g'(x)$$

$$h'(x) \cdot g(x) = \frac{g(x)f'(x) - f(x)g'(x)}{g(x)}$$

$$h'(x) = \frac{g(x)f'(x) - f(x)g'(x)}{[g(x)]^2}$$

Think and Share

We call this game "Beat the Machine" and it is played in pairs. For each of the following functions, one player should compute the derivative using the TI calculator and the other player should compute it by hand using the rules. Whoever finishes first wins, of course, but don't cheat. If you are using the TI calculator, you must enter the function on the machine and use the differentiation operator, even though you may already know the result. (The person working by hand is under no such restrictions and may blurt out the solution at any time.) Take turns using the machine and computing by hand. Which functions are easier by hand and which are easier using the machine?

1. $f(x) = x^2$
2. $f(x) = \sin(x) + x\cos(x)$
3. $f(x) = \dfrac{\cos(x)}{5x^3}$
4. $f(x) = x^2 \sec(x)$ Hint: $\sec(x) = \frac{1}{\cos(x)}$

Summary

Studying rules of differentiation serves several purposes. It improves your symbol sense and gives you the means for solving simple differentiation problems by hand. You have probably already abandoned using the TI calculator to find the derivative of $f(x) = x^5$ or $g(x) = \sin(x)$. Your familiarity with the rules may also lead you to abandon using the TI calculator to compute the derivative of $k(x) = x^5 \sin(x)$ or $h(x) = \sin(x) + x^2$.

The reason for studying rules of differentiation is not to aid in computations. For example, we will not expect you to compute by hand the derivative of $f(x) = \dfrac{x^2 - 7x}{\sin(x)} + \dfrac{\cos(x)}{1 - 3x^2 - 4x^5}$ and simplify the result. We will expect you to be able to solve such problems, but, with the aid of your TI calculator, you should have no trouble meeting that expectation.

The most important reason for studying rules of differentiation is to expand your understanding of the derivative. The derivative is additive (the derivative of a sum is the sum of the derivatives) but it is not multiplicative (the derivative of a product is not the product of the derivatives). Constants factor through the derivative. These facts illuminate the very essence of differentiation.

You are fortunate that you have (or have access to) a powerful computer algebra system that will perform your tedious algebraic manipulations for you, and you can carry it with you everywhere you go. Until very recently, only professional mathematicians with giant computer resources had the luxury of computer algebra systems to do their computations. Using the TI calculator to perform much of the algebraic manipulation involved in the study of calculus will permit you more time to concentrate on concepts. You will learn more, and understand it better. Caution: we are not saying that calculus will be easier. We are saying that your understanding will be enhanced, as will your ability to solve problems and your facility in transferring your knowledge to other disciplines.

EXERCISE SET 3.2

For Exercises 1–6, use the following table. Use the information

x	$f(x)$	$f'(x)$	$g(x)$	$g'(x)$
-1	3	4	2	5
0	4	-3	1	3
1	-1	1	-1	-2
2	0	2	0	-1

about f and g in the table to evaluate $h(x)$ and $h'(x)$ for the indicated x-value. If a value cannot be found, explain why.

1. $h(x) = 3f(x)$, $x = -1$
2. $h(x) = 5g(x)$, $x = 0$
3. $h(x) = f(x) + g(x)$, $x = 1$
4. $h(x) = f(x) \cdot g(x)$, $x = 2$
5. $h(x) = \dfrac{f(x)}{g(x)}$, $x = 1$
6. $h(x) = \dfrac{g(x)}{f(x)}$, $x = -1$

In Exercises 7–10, f and g are the functions defined by the graphs in Figure 3.19. If, for some reason, you are unable to estimate a requested value, explain why.

7. Let $h(x) = f(x) \cdot g(x)$. Estimate the value of $h(0)$ and $h'(0)$.
8. Let $j(x) = f(x) + g(x)$. Estimate the value of $j(-3)$ and $j'(-3)$.
9. Let $k(x) = -2g(x)$. Estimate the value of $k(-1)$ and $k'(-10)$.
10. Let $k(x) = f(x) \cdot g(x)$. Is $k(x)$ increasing or decreasing at $x = 1$?

Exercises 11–16 involve trigonometric functions. You may need to recall some of your favorite trigonometric identities to recognize the TI calculator output.

11. Find the derivative of $\tan(x)$ by expressing $\tan(x)$ in terms of $\sin(x)$ and $\cos(x)$ and using the quotient rule. Apply some well-known trigonometric identities to simplify your answer so that it contains only one term.
12. How will your professor know if you use the TI calculator to differentiate $\tan(x)$? Explain. What famous trigonometric identity does the TI calculator use?

FIGURE 3.19

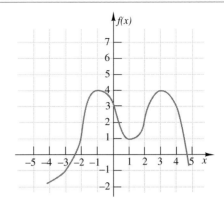

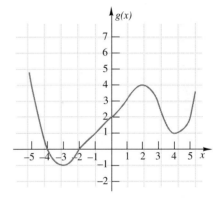

13. Find the derivative of $\cot(x)$ by expressing $\cot(x)$ in terms of $\sin(x)$ and $\cos(x)$ and using the quotient rule. Apply some well-known trigonometric identities to simplify your answer so that it contains only one term.
14. Find the derivative of $\sec(x)$ by expressing $\sec(x)$ as the reciprocal of $\cos(x)$ and using the quotient rule.
15. Find the derivative of $\csc(x)$ by expressing $\csc(x)$ as the reciprocal of $\sin(x)$ and using the quotient rule.
16. Produce a list of the derivatives of the six trigonometric functions. You may obtain these derivatives using the TI calculator, but you should then list them in some sort of order and try to recognize a pattern. Describe this pattern.
17. Restate, using Leibniz notation, each of the rules given in the Differentiation Rules at the end of this section. (Let u and v be differentiable functions of x, and write your rules in terms of these functions.)

3.3

The Chain Rule

The Chain Rule for Linear Functions

Linear Functions and First Differences.   Linear functions have simple derivatives, as we know. If $f(x) = 3x + 2$, then the slope of the tangent line is 3 at every point on the graph of f and the derivative is also always 3. What happens to the output of f when we change the value of the input variable by some small amount say, Δx? In general, starting with an input value of x, we have the following computations.

$$f(x) = 3x + 2$$
$$f(x + \Delta x) = 3(x + \Delta x) + 2$$
$$= 3x + 3\Delta x + 2$$
$$= 3x + 2 + 3\Delta x$$
$$= f(x) + 3\Delta x$$

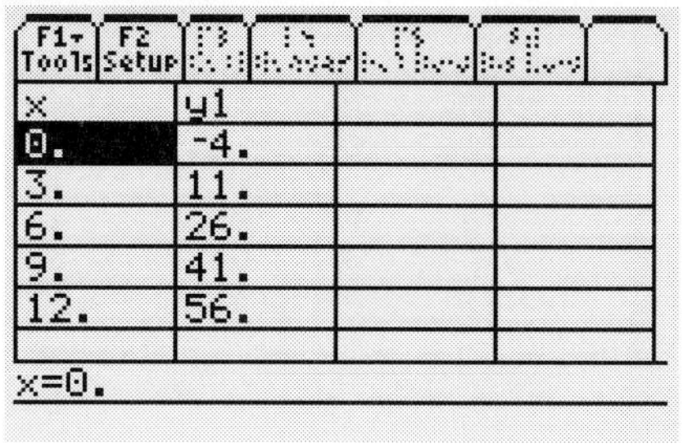

Thus, if the input changes by Δx, then the output changes by $3\Delta x$. Put more simply, the change in the output is 3 times the change in the input.

FIRST REFLECTION

We can look at this numerically using the TABLE feature of the TI calculator. Enter the linear function $f(x) = 5x - 4$ for y1(x). In the TblSet dialogue box, set tblStart to 0 and Δtbl to 3. Now press TABLE to display a table as shown in Figure 3.20. Let's examine the table. When the x-values change by three, by how much do the y1 values change? Are you surprised? Change the Δtbl to 4 in the TblSet dialogue box. Can you guess how much the y1 values change for an x value change of 4? Press TABLE to check your guess. What happens if you use a negative number for the Δtbl value.

SECOND REFLECTION

How is the relationship between an x value change and the associated f (or y1) value change for our function related to what we know about first differences for linear functions?

Composing Linear Functions.   The composition of two functions f and g is again a function, for example, $h(x) = g(f(x))$. We would like to know what the derivative of this composition $h(x)$ is. To find out, we begin with a specific example using numbers to gain some insight into differentiating a composite function.

THIRD REFLECTION

If $f(x) = 5x - 4$ and $g(x) = 2x - 3$, what is $g(f(x))$?

FIGURE 3.20   Table for $f(x) = 5x - 4$

EXAMPLE 1 **Differentiating a Composition of Linear Functions**

Let $f(x) = 5x - 4$ and $g(x) = 2x - 3$. What is the change in the output of the composite function $h(x) = g(f(x))$ when the input changes by 2?

We do this computation in two stages. First, when the input to f changes by 2, its output changes by 10. Second, when the input to g changes by 10, its output changes by 20. Thus, it seems that a change of 2 in the input to the composite function $h(x) = g(f(x))$ changes its output by 20.

Let's find the derivative using our TI calculators. Enter the following functions in the Y= editor.

```
y1(x)=5x-4
y2(x)=2x-3
y3(x)=y2(y1(x))
```

In the TblSet dialogue box, enter 1 for `tblStart` and 2 for Δtbl. Be sure to press **ENTER** twice to save this dialogue box. Display a table by pressing **TABLE** (see Figure 3.21).

When the input to f changes by 2, its output changes by 10. When the input to g changes by 2, its output changes by 4. The **y3** column shows what happens when the input to g (based on the output from f) changes by 10. When the input to the composite function changes by 2, its output changes by 20 as you can see by comparing the **x** and **y3** columns.

FIGURE 3.21 Composite Function Table

x	y1	y2	y3
1.	1.	-1.	-1.
3.	11.	3.	19.
5.	21.	7.	39.
7.	31.	11.	59.
9.	41.	15.	79.

x=1.

More generally, increasing the input to f by Δx increases the output of f by $5\Delta x$. Increasing the input to g by $5\Delta x$ increases the output of g by $10\Delta x$. Thus, increasing the input to h by Δx increases the output of h by $10\Delta x$.

From our formulas for f, g, and h, we see that $f'(x) = 5$, $g'(x) = 2$ and $h'(x) = 10$. The fact that the derivative of h is the product of the derivative of f and the derivative of g is not an accident. We can see this for our linear functions by simply typing **y2(y1(x))** on the Entry line of the **HOME** screen. We see that $h(x) = 10x - 11$.

▶ FOURTH REFLECTION

Make the computation of $h(x)$ by hand to see exactly how the product comes into play. ◀

THE CHAIN RULE FOR LINEAR FUNCTIONS

If h is the composition of two linear functions f and g, then the slope of $h(x) = g(f(x))$ is the product of the slopes of f and g. In other words, the derivative of h is the product of the derivative of f with the derivative of g. Symbolically,

$$h'(x) = f'(x) \cdot g'(x)$$

The Chain Rule in General

E X A M P L E   2   **Differentiating a Composition of Non-linear Functions**

Now suppose that h is the composition of two non-linear (but differentiable) functions. For example,

$$f(x) = x^2 + 1,$$
$$g(x) = x^3, \text{ and}$$
$$h(x) = g(f(x)) = (x^2 + 1)^3.$$

Even though f and g are not linear, they are locally linear. At any point in their domains they are well approximated by a tangent line whose slope can be determined by differentiation.

In particular, suppose we wish to determine $h'(3)$, the instantaneous rate of change of h at 3. Notice that $h'(3)$ is the slope of the line which approximates h at 3. Also, when $x = 3$, f is well approximated by a line with slope $f'(3) = 6$ (Why?).

With linear functions, we did not need to be concerned with where the derivative of the "outside" function g was taken since the derivative was everywhere the same. For non-linear functions, we have to be concerned.

When the input to f is 3, the output from f is 10. Thus, the input to g is 10. So, when the input to h is 3, the input to g is 10 and the function g is well approximated by a line tangent to g at 10 when x is 3. That line has slope $g'(10) = 300$. (Why?)

Applying the chain rule for linear functions we have that

$$h'(3) = f'(3) \cdot g'(f(3)) = f'(3) \cdot g'(10) = 6 \cdot 300 = 1800.$$

▶ FIFTH REFLECTION

Determine $h'(x)$ using your calculator by entering the expression $(x^2 + 1)^3$, expanding it, and then (by hand) differentiating the resulting polynomial. Does the value of this derived polynomial at $x = 3$ equal the value we obtained above for $h'(3)$?.   ◀

T H E   C H A I N   R U L E

If h is the composition of two functions g and f, that is, $g(f(x))$, then the derivative of h at x is the product of the derivative of f at x with the derivative of g at $f(x)$. Symbolically,

$$h'(x) = f'(x) \cdot g'(f(x)) \text{ or}$$
$$= g'(f(x)) \cdot f'(x).$$

We were able to use $g'(x)$ instead of $g'(f(x))$ in the "The Chain Rule for Linear Functions" because the derivative of a linear function is the same at any input value as has been mentioned. For linear functions, $g'(x) = g'(f(x))$. Thus, we could rewrite the chain rule for linear functions to match the more general chain rule.

Finding Derivatives of Composite Functions

The chain rule is used in differentiating composite functions like

$$h(x) = \sin(x^4),$$
$$j(x) = \sin^4(x) = [\sin(x)]^4, \quad \text{or} \tag{3.4}$$
$$k(x) = (\sin(x) + x^5)^3. \tag{3.5}$$

See Figures 3.22 and 3.23 for the TI calculator version of these computations. This rule has both theoretical and practical importance.

FIGURE 3.22 The Chain Rule for $f(x) = \left(x^5 + \sin(x)\right)^4$

FIGURE 3.23 The Chain Rule for $f(x) = \sin(x^4)$ and $g(x) = (\sin(x))^4$

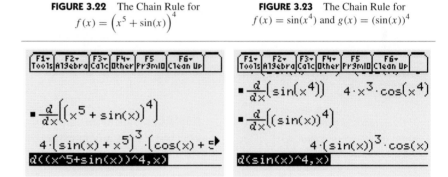

To find the derivative of $h(x) = \sin(x^4)$ by hand, we first analyze h into two other functions; that is, find functions f and g such that $h(x) = g(f(x))$. It may help to think of f as the "inner" function and g as the "outer" function. For our example, we choose $f(x) = x^4$ and $g(u) = \sin(u)$. We let $u = f(x)$ to simplify the use of the chain rule. To obtain h, the input to g equals the output of f, and does not equal x.

Think and Share

Break into groups. Use the TI calculator to compute the derivative of $h(x) = \sin(x^2)$ directly. Decompose $h(x)$ into $g(u) = \sin(u)$ and $f(x) = x^2$. Differentiate the inner function, $f(x)$, with respect to x. Obtain $g'(u)$, the derivative of the outer function, by differentiating $g(u) = \sin(u)$ with respect to u. Replace u with $f(x) = x^2$ in order to obtain the derivative entirely in

Think and Share

terms of x. Use your expression for $g'(u)$ and the formula $f(x) = x^2$ to write down a formula for $g'(f(x))$ in terms of x alone. Compare $h'(x)$ to $f'(x)$ and $g'(f(x))$. Do you see a relation? Repeat the process for $j(x)$ and $k(x)$ in Equations (3.4) and (3.5) (you will have to decompose these functions first). Make up your own rule. Check your rule by differentiating $\sin(7x)$, $(\sin(x))^7$, and $(x^2 - 3)^4$.

▶ **SIXTH REFLECTION**

Once you are confident your chain rule is correct, write it in complete sentences along with a symbolic representation.   ◀

As a challenge, try differentiating $(\sin(5x))^8$. The computations for this function are more involved since it requires two applications of the chain rule.

Using the Leibniz Notation.   Writing the chain rule in Leibniz notation can be helpful. In the Think and Share, we have $y = \sin(u)$ where $u = x^2$. The output variable y is a function of u and u is a function of x, so we compute $\dfrac{dy}{du}$ and $\dfrac{du}{dx}$. If we replace u with x^2 in the equation for y, we see that y can also be viewed as a function of x, so it makes sense to compute $\dfrac{dy}{dx}$. Using these expressions, we write the chain rule in a succinct symbolic form as

$$\frac{dy}{dx} = \frac{dy}{du}\frac{du}{dx}.$$

Actually, Leibniz developed this notation with this "simplification" in mind.

It looks as if the terms on the right are fractions and we are canceling the du. This is *not* what is happening, but it does provide an easy way to remember the chain rule.

▶ **SEVENTH REFLECTION**

Test your understanding of the chain rule by finding antiderivatives for the following functions. Remember, you are just reversing the process of differentiation, and you can always differentiate your answer to see if you are correct.

1. If $f'(x) = 2\cos(2x)$, find an expression for $f(x)$.
2. If $\dfrac{dy}{dx} = 3(\sin(x))^2 \cos(x)$, find an expression for y.   ◀

Think and Share

Break into groups.

1. Differentiate e^x, e^{2x}, $e^{\sin(x)}$, and e^{x^2}. Formulate a rule for differentiating e^u where u is a differentiable function of x.
2. Find antiderivatives for $3e^{3x}$, $10xe^{5x^2}$, and $\cos(x)e^{\sin(x)}$.
3. Differentiate $\ln(x)$, $\ln(x^2 + 1)$, and $\ln(\sin(x))$. Formulate a rule for differentiating $\ln(u)$ where u is a differentiable function of x.
4. Find antiderivatives for $\dfrac{5}{x}$, $\dfrac{3x^2}{x^3 + 1}$, $\dfrac{\cos(x)}{\sin(x)}$.

One further example will help to solidify your understanding of composite functions and the Chain Rule.

E X A M P L E   3   **Practice with the Chain Rule**

Find the derivative of each of the following functions.

1. $h(x) = (3x^2 - 4x + 5)^5$
2. $h(x) = \cos(x^2)$
3. $h(x) = \ln(\sin(x))$
4. $h(x) = e^{x^2 + 3x}$

Solution:

1. We choose $f(u) = u^5$ and $g(x) = 3x^2 - 4x + 5$. Then $h(x) = f(g(x))$ and $h'(x) = 5(3x^2 - 4x + 5)^4 \cdot (6x - 4)$.

2. We choose $f(u) = \cos(u)$ and $g(x) = x^2$. Then $h(x) = f(g(x))$ and $h'(x) = -\sin(x^2) \cdot 2x$.

3. We choose $f(u) = \ln(u)$ and $g(x) = \sin(x)$. Then $h(x) = f(g(x))$ and $h'(x) = \dfrac{1}{\sin(x)} \cdot \cos(x) = \tan(x)$.

4. We choose $f(u) = e^u$ and $g(x) = x^2 + 3x$. Then $h(x) = f(g(x))$ and $h'(x) = e^{x^2 + 3x} \cdot (2x + 3)$.

We conclude this section with a table of rules that use the Chain Rule to extend our differentiation capabilities.

DIFFERENTIATION WITH THE CHAIN RULE

The Chain Rule

The derivative of the composition of two functions equals the derivative of the outer function evaluated at the inner function times the derivative of the inner function. If $h(x) = f(g(x))$, then

$$h'(x) = f'(g(x)) \cdot g'(x).$$

The Extended Power Rule

If $h(x) = (f(x))^n$, then

$$h'(x) = n(f(x))^{n-1} \cdot f'(x).$$

The Extended Cosine Rule

If $h(x) = \cos(f(x))$, then

$$h'(x) = -\sin(f(x)) \cdot f'(x).$$

DIFFERENTIATION WITH THE CHAIN RULE CONT'D

The Extended Sine Rule

If $h(x) = \sin(f(x))$, then

$$h'(x) = \cos(f(x)) \cdot f'(x).$$

The Extended Secant Rule

If $h(x) = \sec(f(x))$, then

$$h'(x) = \sec(f(x)) \tan(f(x)) \cdot f'(x).$$

The Extended Tangent Rule

If $h(x) = \tan(f(x))$, then

$$h'(x) = \sec^2(f(x)) \cdot f'(x).$$

The Extended Exponential Rule

If $h(x) = e^{f(x)}$, then

$$h'(x) = e^{f(x)} \cdot f'(x).$$

The Extended Logarithm Rule

If $h(x) = \ln(f(x))$, then

$$h'(x) = \frac{1}{f(x)} \cdot f'(x).$$

EXERCISE SET 3.3

Find the derivative of each of the functions in Exercises 1–4. You may assume that f is a differentiable function; that is, $f'(x)$ exists for all x, and that n is a constant. Your derivatives will involve both f and f'.

1. $g(x) = [f(x)]^5$; $h(x) = f(5x)$; $j(x) = 5f(x)$

2. $g(x) = [f(x)]^n$; $h(x) = f(nx)$; $j(x) = nf(x)$

3. $g(x) = \sin(f(x))$; $h(x) = f(\sin(x))$; $j(x) = f(x)\sin(x)$

4. $g(x) = e^{f(x)}$; $h(x) = f(e^x)$; $j(x) = \dfrac{e^x}{f(x)}$

In Exercises 5–16, find the derivative of the given function. You may wish to check your answer with a TI calculator.

5. $f(x) = \sin(x^5)$.

6. $g(x) = \ln(\sec(x))$.

7. $h(x) = \ln(2x^4)$.

8. $f(x) = \sqrt{3x^2 - 5x + 2}$.

9. $g(x) = \cos(\ln(x))$.

10. $h(x) = (3x^5 - 5x + 2)^{21}$.

11. $j(x) = \tan(x^2) + 5$.

12. $k(x) = (\ln(x))^7 - 5x^4$.

13. $f(x) = \sec(x^2)$.

14. $g(x) = \sin^2(x) + \cos^2(x)$.

15. $h(x) = \dfrac{x^2}{\ln(x^2)}$.

16. $f(x) = \sin^4(x) \cdot \cos(x^3)$.

In Exercises 17–20, find the formulas for f and g so that $h(x) = f(g(x))$. Confirm your result by defining the two functions on the TI calculator and entering $f(g(x))$.

17. $h(x) = \sqrt{x^2 - x + 1}$.

18. $h(x) = (x + e^x)^4$.

19. $h(x) = \ln(x^3)$.

20. $h(x) = (\ln(x))^3$.

For Exercises 21 and 22, use the following table. Use the

x	$f(x)$	$f'(x)$	$g(x)$	$g'(x)$
-1	3	4	2	5
0	4	-3	1	3
1	-1	1	-1	-2
2	0	2	0	-1

FIGURE 3.24

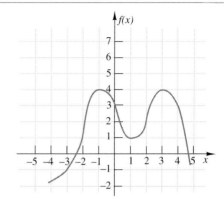

information about f and g in the table to evaluate $h(x)$ and $h'(x)$ for each of the x-values provided in the table. If a value cannot be found, explain why.

21. $h(x) = f(g(x))$
22. $h(x) = g(f(x))$

In Exercises 23–26, f and g are the functions defined by the graphs in Figures 3.24 and 3.25. If, for some reason, you are unable to estimate a requested value, explain why.

23. Let $j(x) = f(g(x))$. Estimate the value of $j(2)$ and $j'(2)$.
24. Let $h(x) = g(f(x))$. Is h increasing or decreasing at $x = 0$? Justify.
25. Let $m(x) = g(x^3)$. Is $m(x)$ increasing or decreasing at $x = 1$?
26. Let $n(x) = (g(x))^3$. Is $n(x)$ increasing or decreasing at $x = -1$? Justify.

27. Find the derivative of $\sec(x)$ by expressing $\sec(x)$ as $(\cos(x))^{-1}$ and using the chain rule. Then express $\sec(x)$ as $1/\cos(x)$ and use the quotient rule. Reconcile your answers if they seem to be different.

28. Find the derivative of $\csc(x)$ by expressing $\csc(x)$ as $(\sin(x))^{-1}$ and using the chain rule. Then express $\csc(x)$ as $1/\sin(x)$ and use the quotient rule. Reconcile youranswers if they seem to be different.

FIGURE 3.25

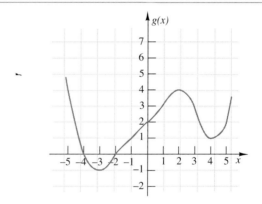

3.4

Finding Features of a Continuous Function

In Chapter 2, we introduced first and second differences in oder to characterize the properties of the graph of a sequence. We saw that we could apply these techniques to the graphs of continuous functions as well, to find the approximate locations of maximum and minimum values, or inflection points. In this section, we extend this analysis of a continuous function by examining the values of the first and second derivatives. We will find that the results for continuous functions using first and second derivatives are analogous to the results for discrete functions using first and second differences.

We begin with a simple example that illustrates the interconnection between a function and its derivative function.

E X A M P L E   1   **Graphing a Function and Its Derivative**

The graphs of $f(x) = x^2$ and its derivative function, $f'(x) = 2x$ are shown in Figures 3.26, 3.27, and 3.28.

FIGURE 3.26 $f(x) = x^2$

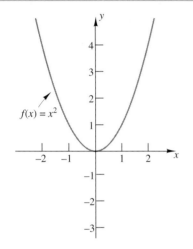

FIGURE 3.27 $f'(x) = 2x$

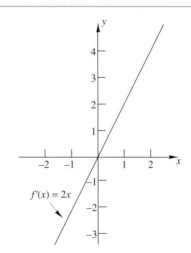

FIGURE 3.28 $f(x) = x^2$ and $f'(x) = 2x$

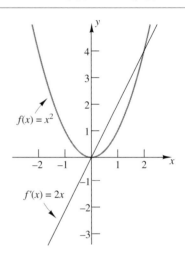

Observe that when f is decreasing (the f graph is falling as x moves from left to right), f' is negative (the graph of f' is below the x-axis.) On the other hand, when f is increasing (the graph of f is rising as x moves from left to right), f' is positive (the graph of f' is above the x-axis). Notice that there are two functions involved here, so we must be very careful with our language to avoid confusion. We must never say something like "*when it is increasing, it is positive,*" when we really mean "*when a function is increasing, its derivative is positive.*" In all our discussions, we need to clearly distinguish between the function and its derivative.

We will see that the sign of the derivative, $f'(x)$, can tell us a lot about the behavior of the original function, $f(x)$, (just as the first difference did for discrete functions). Notice also that when the original function has a minimum value (the function changes from decreasing to increasing), the derivative has a value of zero. We say that the slope of the tangent line is zero or that the graph of the original function has a horizontal tangent line at that point.

Also note that, in this example, f decreases at an increasing rate until f changes direction, and begins to increase at an increasing rate. The derivative, $f'(x)$, is increasing throughout, so its derivative (the derivative of the derivative) must be positive. We call the derivative of the derivative function the **second derivative**, and we designate it with a double prime notation, $f''(x)$. We read this "f double prime of x," "the second derivative of f with respect to x," or simply "the second derivative of f."

Think and Share

Break into groups of four and then into teams of two. One team should draw the graph of a function which increases at a decreasing rate and then increases at an increasing rate. The other team should draw the graph of a derivative function which decreases throughout its domain. Teams should swap their drawings and try to draw the graph of the original function (from the derivative graph) and the derivative function (from the original graph).

Let's look at a more involved example.

EXAMPLE 2 **Heat Shield for a Space Shuttle**
_____

Imagine you are an engineer designing a heat shield for a space shuttle. The shuttle must withstand the high temperatures associated with reentry. After careful testing, you have found a function that models the temperature at any point in the shield exactly 0.01 seconds after reentry. Surprisingly, research showed that the temperature would be the same at every point on the surface of the shield and is higher inside the shield than on its surface. The function is

$$u(x) = \frac{2}{\pi} e^{-0.01\pi^2} \sin(\pi x) + \frac{2}{\pi} e^{-0.04\pi^2} \sin(2\pi x) + \frac{2}{3\pi} e^{-0.09\pi^2} \sin(3\pi x)$$

(3.6)

with $0 \leq x \leq 1$ where $u(x)$ is the temperature (in thousands of degrees) at a depth x (in centimeters) of the shield.

Your materials engineers must determine if materials exist that can withstand the given temperature distribution without burning up or breaking apart, leaving the capsule unshielded. They have asked you several important questions:

• What is the maximum temperature and where does it occur?
• What is the greatest change in temperature per unit depth?
• Where is the temperature changing the fastest?

You begin by graphing the function in the graphing window [0,1] horizontally and [0,1] vertically as shown in Figure 3.29. From the graph, there appears to be a maximum temperature at about $x \approx 0.25$. This is a useful observation but the materials engineers are going to need much more precise information.

The rate of the temperature change is represented by the slope of the graph. From the graph alone, however, it is difficult to determine where the magnitude of the slope is greatest.

FIGURE 3.29   Temperature vs. Depth on $[0, 1] \times [0, 1]$

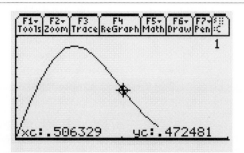

Graphs are good for qualitative analysis; that is, to get an overall feel for what is happening. However, graphs are not always sufficiently precise, especially for sophisticated functions. Further, the above snapshot is but one instant in time ($t = 0.01$). Many more such snapshots will have to be analyzed efficiently and accurately. The other engineers also need to know how sensitive your conclusions are to changes in various parameters of the situation; for example curvature of the shield, smoothness of the shield, and angle of entry. Changing these parameters may create alternative temperature distributions that could allow a greater variety of feasible construction materials to be considered. The amount of number crunching that is required to create graphs of many functions in detail is daunting. To analyze this family of functions at the required level of accuracy quickly and efficiently, you need more powerful tools. Symbolic analysis provides these tools!

We'll get back to this problem, but first we need to investigate how derivatives can help us locate important features on the graph of a continuous function.

Properties of the Graph of a Continuous Function

One of the most fundamental and important properties of a function is whether it is increasing or decreasing. A function is increasing if increasing the input increases the output. Conversely, the function is decreasing if increasing the input decreases the output. (Consider a demand function from economics: increasing the price of a product generally results in decreased sales.) These ideas are straightforward, but we give a precise formal definition to avoid any confusion.

INCREASING AND DECREASING FUNCTIONS

Let f be a function defined on an interval I and let x_1 and x_2 be any two points on I. We say f **is increasing** on I if $x_1 < x_2$ implies that $f(x_1) < f(x_2)$, and f **is decreasing** on I if $x_1 < x_2$ implies that $f(x_1) > f(x_2)$.

We can determine whether a function is increasing or decreasing quite readily from its graph. If the function is increasing, its graph will rise as we move from left to right. On the other hand, if the function is decreasing, its graph will fall as we move from left to right.

FIGURE 3.30   Increasing and Decreasing Functions

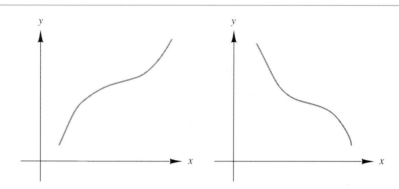

We saw for sequences that the sign of the first difference tells us immediately whether the function is increasing or decreasing. A similar situation exists for continuous functions.

Draw several tangent lines to the graphs of the increasing and decreasing functions in Figure 3.30. Notice that each of the lines you drew has a positive or negative slope, respectively.

▶ FIRST REFLECTION

The slope of a tangent line at a point on the graph of a function is equal to the value of the derivative at that point. Use the limit definition of the derivative to explain why the slope at a point where the function is increasing should be nonnegative.   ◀

The above considerations lead to the following general method for determining whether a function is increasing or decreasing on a given interval.

TESTS FOR INCREASING AND DECREASING

Suppose that f is continuous on $[a, b]$ and differentiable on (a, b). If $f' > 0$ at each point of (a, b), then f increases on $[a, b]$. If $f' < 0$ at each point of (a, b), then f decreases on $[a, b]$.

The derivative actually gives more information than just whether the function is increasing or decreasing. It also tells how fast the change is occurring. This can be critical information (as in the case of our heat shield, where the extreme rate of change of the temperature can eliminate a broad range of materials that might otherwise be suitable).

A second, less obvious (but still important) property of a function is concavity. Concavity is the rate at which the rate of change of the function is changing. More simply, it's the rate at which the first derivative changes.

Graph the function $y = \frac{1}{3}x^3 - 2x^2 + 3x$ with xmin=-1, xmax=2, ymin=-2, and ymax=2. The plot of the function on this interval is "bent downward." (See Figure 3.31.) In precise mathematical terms, we say the curve is *concave down*.

Think and Share

Break into groups. Use your For...EndFor statement from Exercise 32 of Section 3.1 to draw tangent lines on your graph. Notice that the slopes of the tangent lines are decreasing, moving from left to right. Since y' computes these slopes, y' must be decreasing.

Extend your graph by changing the value of xmax to 5. Notice that the new part of the curve is concave up. How are the slopes of the tangent lines changing over this part of the curve?

Our investigation leads us to the following operational definition of concavity.

CONCAVITY OF A DIFFERENTIABLE FUNCTION

The graph of a differentiable function $y = f(x)$ is **concave up** on an interval where y' is increasing ($y'' > 0$) and **concave down** on an interval where y' is decreasing ($y'' < 0$). (See Figure 3.32.)

FIGURE 3.31 $y = \frac{1}{3}x^3 - 2x^2 + 3x$ on $[-1, 2] \times [-2, 2]$

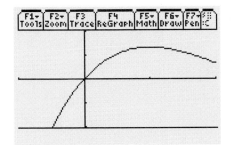

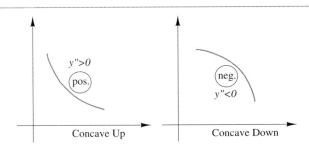

FIGURE 3.32   Two Decreasing Functions and Their Concavity

$y'' > 0$ (pos.)

(neg.) $y'' < 0$

Concave Up   Concave Down

The **second derivative function** is the derivative of the first derivative function. There are several notations for the second derivative, some of which are

$$y''$$
$$\frac{d^2y}{dx^2}$$
$$f''(x)$$
$$\frac{d}{dx}\left(\frac{d}{dx}f(x)\right) = \frac{d^2}{dx^2}f(x)$$

The second derivative measures the rate of change of the first derivative function.

Since y'' is the derivative of y', our test for increasing and decreasing tells us that if $y'' > 0$ then y' is increasing, and if $y'' < 0$ then y' is decreasing. In light of our definition for concavity, we are led to the following test.

SECOND DERIVATIVE TEST FOR CONCAVITY

Let f be a function that is twice differentiable on an interval I. If $y'' > 0$ on I, the graph of f over I is concave up. If $y'' < 0$ on I, the graph of f over I is concave down.

Notice that this test is precisely analogous to using the second differences to determine concavity at a point of a sequence. When the second differences are positive, the first differences are increasing, just as when the second derivative is positive, the slopes of the tangent lines are increasing. We illustrate this point in Figure 3.33 where we have drawn line segments between selected points as well as tangent lines at those points. Notice that both sets of slopes are increasing, and the curve is concave up. Similarly, when a curve is concave down, both the second differences between points and the second derivatives at the points are negative.

FIGURE 3.33 Line Segments and Tangent Lines

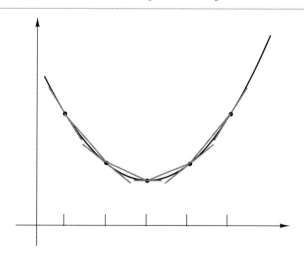

▶ SECOND REFLECTION

Determine the precise value of x at which the concavity of $y = \frac{1}{3}x^3 - 2x^2 + 3x$ changes from down to up. Justify your solution. ◀

Once again, the second derivative says more than just whether the function is concave up or concave down.

▶ **THIRD REFLECTION**

Graph $y = x^2$ and $y = x^4$ on the interval $[0, 1]$. From the graphs, which function do you think should have greater concavity at $x = \frac{1}{2}$? Justify your response. Compute the second derivatives of the two functions and evaluate them at $x = \frac{1}{2}$. Do the computations support your intuition?   ◀

Local Extreme Values: Maxima and Minima.   Maximum and minimum points are an important feature of a function. In economics, such points are used to maximize profits or minimize costs. In physics, they can be used to find equilibrium points in an energy system and classify their stability. In chemistry, maxima and minima correspond to equilibrium points in a chemical reaction.

It's not hard to approximate the location of maximum and minimum points on the graph of a function. Maximums occur at the "top of humps" and minimums occur at the "bottom of valleys." The function $y = \frac{1}{3}x^3 - 2x^2 + 3x$ (considered earlier) has a maximum point near $x = 1$. This maximum point is higher than any of the points near it, but it is not higher than the point at $x = 5$. We use the word "local" to indicate that we are only considering the behavior of the function near the point.

The point where the local maximum occurs is higher than any of the points near it, because the output of f at that point is greater than the output at any of the nearby points. A similar discussion works for minimum points, and we are led to the following definition.

FIGURE 3.34   Relative Extrema

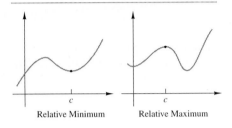

Relative Minimum   Relative Maximum

LOCAL OR RELATIVE EXTREMA

A function f has a **local** or **relative maximum value** at an interior point c of its domain if $f(c) \geq f(x)$ for all x in some open interval containing c. A function f has a **local** or **relative minimum value** at an interior point c of its domain if $f(c) \leq f(x)$ for all x in some open interval containing c.

Notice in Figure 3.34 that to the left of the local maximum the function is increasing, while to the right the function is decreasing. Thus, $f' > 0$ to the left of the local maximum; $f' < 0$ to the right. Similarly, $f' < 0$ to the left of the local minimum; $f' > 0$ to the right. We can summarize these observations in the following test.

FIRST DERIVATIVE TEST FOR LOCAL EXTREMA

Suppose that f is a continuous function on an interval containing c. The function f has a local maximum value at c if there is an interval $[a, b]$ containing c such that $f'(x) \geq 0$ for $a < x < c$ and $f'(x) \leq 0$ for $c < x < b$.

The function f has a local minimum value at c if there is an interval $[a, b]$ containing c such that $f'(x) < 0$ for $a < x < c$ and $f'(x) > 0$ for $c < x < b$.

Finding Local Extreme Values

How do we find local extreme values efficiently? Consider the graphs in Figure 3.35. For both a local minimum and a local maximum, the first derivative

FIGURE 3.35 Extreme Values

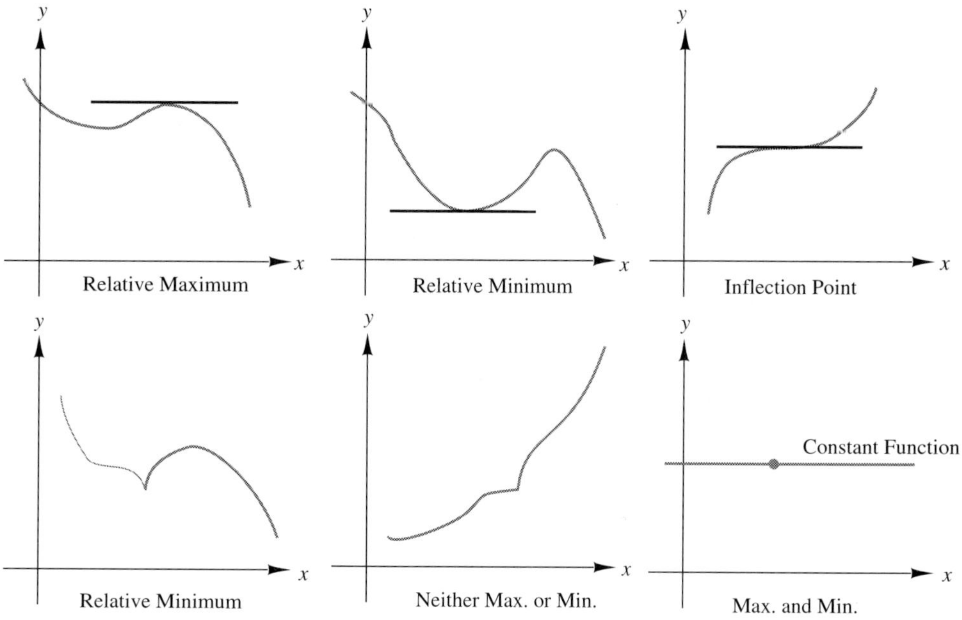

Relative Maximum

Relative Minimum

Inflection Point

Relative Minimum

Neither Max. or Min.

Constant Function

Max. and Min.

must be either zero or undefined. We call such points *critical points*. Local extrema, if they exist, must occur at the critical points.

CRITICAL POINT

An interior point c of the domain of a function f where f' is zero or undefined is a **critical point** of f.

We see from the illustrations that a critical point is not necessarily a relative maximum or relative minimum. On the other hand, relative extreme values only occur at critical points. Therefore, we begin our search for local extrema by examining the critical points. Those critical points where the first derivative changes sign must be local extrema. We formalize this idea as a strategy.

STRATEGY FOR LOCATING RELATIVE EXTREMA

1. Find the critical points, inputs to the function where the derivative is 0 or undefined.
2. For each critical point, evaluate the first derivative a little to the left and a little to the right.
3. Apply the first derivative test for local extreme values.

FIGURE 3.36   Graph of $g(x) = \frac{1}{3}x^3 - 2x^2 + 3x$

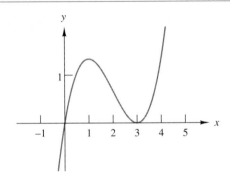

For those critical points at which the first derivative is zero, we can use the second derivative to verify our findings. In particular, if the second derivative is positive when the first derivative is zero, then the function has a minimum at the critical point. Alternatively, if the second derivative is negative when the first derivative is zero, then the function has a maximum at the critical point. If the second derivative is zero at a critical point, no conclusion may be inferred. The function could have a maximum, a minimum, or neither.

FIGURE 3.37   Concavity at Relative Extrema

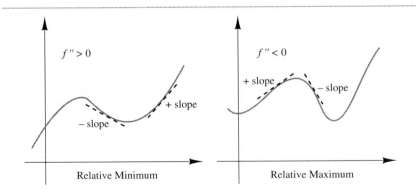

The Second Derivative Test formalizes the use of concavity in finding local extrema.

SECOND DERIVATIVE TEST FOR LOCAL EXTREMA

If $f'(c) = 0$ and $f''(c) < 0$, then f is concave down and has a local maximum at $x = c$.

If $f'(c) = 0$ and $f''(c) > 0$, then f is concave up and has a local minimum at $x = c$.

Think and Share

Break into groups. If we know that $f'(c) = 0$, we may use the concavity at c to determine if a relative maximum or minimum exists at c. Consider once more your graph of the function $y = \frac{1}{3}x^3 - 2x^2 + 3x$. What is the slope of the tangent line at the local maximum point? What is the slope of the tangent line at the local minimum point? What are the values of f' at these two points as shown in Figure 3.36? What is the concavity at these two points? What are the values of f'' at these two points? The Second Derivative Test for local extreme values uses the idea of concavity and the sign of the second derivative. Sketch the graph of a function that is concave down to the left of a maximum point and concave up to the right of the maximum point. What can you say about the first derivative at your maximum point? What can you say about the second derivative at your maximum point? Does the Second Derivative Test hold for this function? Sketch the graph of a function that is concave up on both sides of a maximum point. Why must we be careful to insist that $f'(c) = 0$?

Inflection Points

We defined inflection points for sequences as points where a change in concavity took place. Inflection points for continuous functions are defined analogously.

INFLECTION POINT

A point on the graph of a function where there is a tangent line and the concavity changes (as you move from left to right) is called an **inflection point.**

Since the second derivative is positive when the function is concave up and negative when the function is concave down, we have the following test.

TEST FOR INFLECTION POINTS

Suppose that f is continuous on an interval containing c. The function f has an inflection point at c if there exists an interval $[a, b]$ containing c such that $f'' > 0$ for $a < x < c$ and $f'' < 0$ for $c < x < b$ or $f'' < 0$ for $a < x < c$ and $f'' > 0$ for $c < x < b$.

FIGURE 3.38   Inflection Point at a Point with a Horizontal Tangent Line

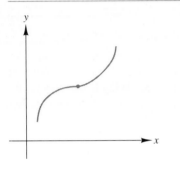

As with our other tests, this test is analogous to the situation for sequences where changes in the sign of the second differences corresponded to changes in concavity.

Summary of Properties of the Graph of a Continuous Function

To summarize the properties of a graph of a continuous function as characterized by the values of its derivatives, we create a table similar to the one we built for a sequence. In Table 3.3, we use the notation developed to describe one-sided limits. The symbol a^- means values of x to the left of a; a^+ means values to the right.

▶ FOURTH REFLECTION
How does Table 3.3 compare with the table of Difference Tests in Section 2.1? ◀

TABLE 3.3   Derivative Tests

Property at $x = a$	Derivative Tests
Increasing	$f'(a) > 0$
Decreasing	$f'(a) < 0$
Relative Maximum	$f'(a^-) > 0$ to $f'(a^+) < 0$
or	$f'(a) = 0$ **and** $f''(a) < 0$
Relative Minimum	$f'(a^-) < 0$ to $f'(a^+) > 0$
or	$f'(a) = 0$ **and** $f''(a) > 0$
Concave Up	$f''(a) > 0$
Concave Down	$f''(a) < 0$
Inflection Point	$f''(a^-) > 0$ to $f''(a^+) < 0$ **or**
	$f''(a^-) < 0$ to $f''(a^+) > 0$

Back to the Drawing Board

We have invested considerable energy in developing the above tests for determining properties of continuous functions from symbolic derivatives. Let's see how they can help us in analyzing the heat shield function.

EXAMPLE 3 **The Heat Shield Revisited**

Recall that the function

$$u(x) = \frac{2}{\pi} e^{-0.01\pi^2} \sin(\pi x) + \frac{2}{\pi} e^{-0.04\pi^2} \sin(2\pi x) + \frac{2}{3\pi} e^{-0.09\pi^2} \sin(3\pi x),$$

for $0 \leq x \leq 1$, describes the temperature, $u(x)$ (in thousands of degrees), at any depth, x (in centimeters), of the shield at a particular instant in time, in this case 0.01 seconds after reentry. You have been asked the following questions:

- What is the maximum temperature, and where does it occur?
- What is the greatest change in temperature per unit depth?
- Where is the temperature changing the fastest?

We compute the first and second derivatives, knowing they will help find important features of the function:

$$u'(x) = 0.8227 \cos(3\pi x) + 2.6953 \cos(2\pi x) + 1.8120 \cos(\pi x)$$
$$u''(x) = -7.75412 \sin(3\pi x) - 16.9351 \sin(2\pi x) - 5.6927 \sin(\pi x)$$

To find the maximum value of $u(x)$, we find the critical points. Our function and its derivatives are always defined, so we only need concern ourselves with zeros. We solve $u'(x) = 0$ for x on the interval [0, 1] using the following `solve` command:

$$\text{solve(d(y1(x),x,0.001)=0.0,x)| x>=0 and x<=1} \quad \textbf{(3.7)}$$

We find that the solutions are $x = 0.277145$ and $x = 0.915263$. Referring back to the graph of $u(x)$ in Figure 3.29, we expect that a relative maximum occurs at $x = 0.277145$.

▶ FIFTH REFLECTION

Use the first derivative test to verify that $x = 0.277145$ is a maximum. Use the second derivative test to confirm this result. Why is 0.95 too far away from 0.277145 to be used in the first derivative test? ◀

From our graph, we know that $x = 0.277145$ is also a global maximum on [0, 1]. We have answered the first question: the maximum temperature is $u(0.277145) = 0.907952$, or about 908°.

Where is the change in temperature the greatest? Since the first derivative is the rate of change of the function, we need to find where the continuous function $u'(x)$ reaches a maximum. To find the critical points of $u'(x)$, we set its derivative $u''(x) = 0$. We solve this equation using a slightly altered version of the `solve` command in Equation (3.7) with the 2 in the d(function indicating the second derivative.

$$\text{solve(d(y1(x),x,2)=0.0,x)|x>=0 and x<=1}$$

The *hypercritical* points (points where the second derivative is zero or undefined) are $x = 0$, $x = 0.48159$, and $x = 1$. The slopes at these three points are $u'(0) = 5.33008$, $u'(0.48159) = -2.71458$, and $u'(1) = 0.060527$. Confirming our analysis with our graph, we conclude that the greatest change in temperature occurs at $x = 0$, where the temperature is changing at a rate of 5330.08 degrees per centimeter. Now our materials engineers can determine whether or not a material exists that can withstand such a drastic change in temperature.

▶ SIXTH REFLECTION
Why does it make sense that the greatest change in the temperature of the heat shield occurs when $x = 0$? ◀

The Graphical Perspective

The information in our summary table (Table 3.3) for continuous functions is very general. It can be used to understand the graph of any continuous function. Consider the following example.

A quartic function is defined by a fourth degree polynomial. Such a function is continuous everywhere and can "turn" as many as three times. Hence, there can be a number of relative maximums, minimums, and inflection points.

EXAMPLE 4   **Estimating Extreme Values Graphically**

Consider the function $y = x^4 + 2x^3 - 7.75x^2 - 11.75x + 7.5$ on the domain $[-3.2, 2.4]$. We begin our analysis by graphing the function on the domain. Using the Trace feature of the TI calculator, we estimate the local maximum value of f to be 11.494 at $x = -.659$ and the two local minimum values of f to be -3.806 at $x = -2.588$ and -16.706 at $x = 1.741$.

▶ SEVENTH REFLECTION
What are the results of using the Maximum, Minimum options in the Math menu of the Graph window? ◀

EXAMPLE 5   **Finding Extrema and Inflection Points Exactly**

We now use the first and second derivatives to locate the exact local extrema and inflection points and to insure that we have located them all. We will make a script of our method. You should clear the HOME screen and clear 1-character variables. We begin by defining f.

```
define f(q)=q^4+2q^3-7.75q^2-11.75q+7.5
```

FIGURE 3.39
$y = x^4 + 2x^3 - 7.75x^2 - 11.75x + 7.5$
on $[-3.25, 2.25] \times [-20, 15]$

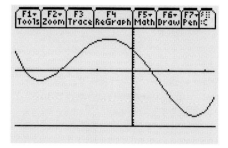

The graph of this function is displayed in Figure 3.39. We then use the differentiation operator to find the first and the second derivatives, storing them to `d1f(x)` and `d2f(x)` respectively, using the commands

$$d(f(x),x) \rightarrow d1f(x),$$

$$d(f(x),x,2) \rightarrow d2f(x).$$

We now find our critical points by solving `d1f(x)=0` for x.

$$solve(d1f(x)=0,x)$$

The critical points are $x = 1.727$, $x = -0.663$, and $x = -2.564$. We apply the first derivative test for local extrema by evaluating the first derivative on either side of each critical point as shown in Figure 3.40.

FIGURE 3.40   Critical Points and the First Derivative Test

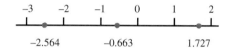

$$\left. \begin{array}{l} d1f(-3) = -19.25 \\ d1f(-1) = 5.75 \end{array} \right\} \quad \begin{array}{l} \text{a local minimum value of } -3.815 \\ \text{at } x = -2.564 \end{array}$$

▶ **EIGHTH REFLECTION**

Apply the first derivative test to determine what happens at the other two critical points. What are the other maximum and minimum values? Do your results confirm the results obtained earlier?   ◀

We continue in our investigation by finding the inflection points. We solve `d2f(x)=0` and evaluate `d2f(x)` on either side of these hypercritical points.

▶ **NINTH REFLECTION**

What are the coordinates of the inflection points?   ◀

We could have used the second derivative test for local extrema by evaluating the second derivative at each critical point. Since `d2f(-2.564)` is greater than zero, we know the graph is concave up and there is a relative minimum at $x = -2.564$.

▶ **TENTH REFLECTION**

`d2f(2)` is also positive. Is there a relative minimum at $x = 2$? Explain why or why not.   ◀

▶ **ELEVENTH REFLECTION**

Use the second derivative test to verify the other two extrema.   ◀

Create a Script. Open the Tools menu and save a copy of your HOME screen with the name `maxminIP`. Open the text file and you have your script. (Review the scripting process in Section 3.2 if necessary.)

EXERCISE SET 3.4

For Exercises 1–4, graph each function. Compute the first and second derivatives. Find all points where $y'(x) = 0$. Classify each critical point as a relative maximum, relative minimum, or neither. Find all inflection points. Justify your answers.

1. $y(x) = x^5 - 7x^4 + 7x^3 + 21x^2 - 30x + 25$

2. $y(x) = x^2 \left(\dfrac{7}{8}\right)^x$

3. $y(x) = x^3 \left(\dfrac{3}{4}\right)^x$

4. $y(x) = 3x^4 - 6x^2 + 12x - 8$

5. Find extreme values and inflection points for the following function.

$$y(x) = \left(\frac{1}{2}\right)^x \sin\left(\frac{x}{2}\right), \quad -2\pi \le x \le 3\pi$$

Be sure to explain and justify your results.

6. For the heat shield example, we just received a new model which changes the property constraints for the material required. The new model modifies the old by adding a term

$$u_{new}(x) = u(x) + \frac{1}{2\pi}e^{-.16\pi^2}\sin(4\pi x).$$

We are still interested in the properties at 0.01 seconds after reentry, and the temperature is in thousands of degrees. The questions remain the same:

   a. What is the maximum temperature and where does it occur?

   b. What is the greatest change in temperature per unit length?

   c. Where is the temperature changing the fastest?

7. Enter the following table of data (believed to be cubic in nature) into the `DataMatrix` editor.

   a. Build a difference table to determine the relative minimum, relative maximum, and inflection points.

   b. Use the `CubicReg` (a least-squares fit to a cubic polynomial) to calculate a cubic model for the data. Store the regression polynomial in $y1(x)$. Use the first and second derivatives of $y1(x)$ to determine the relative minimum, relative maximum, and inflection point of the polynomial model.

   c. Compare the values obtained from the two methods.

8. Enter the following table of data (believed to be cubic in nature) into the `DataMatrix` editor.

TABLE 3.5

x	−0.5	0.0	0.5	1.0
$f(x)$	−0.11764	2.7758	2.5412	2.1649
x	1.5	2.0	2.5	3.0
$f(x)$	−0.18861	−1.4583	−2.8114	−2.5487

   a. Build a difference table to determine the relative minimum, relative maximum, and inflection point.

   b. Use the `CubicReg` (a least-squares fit to a cubic polynomial) to calculate a cubic model for the data. Store the regression polynomial in $y1(x)$. Use the first and second derivatives of $y1(x)$ to determine the relative minimum, relative maximum, and inflection point of the polynomial model.

   c. Compare the values obtained from the two methods.

9. Enter the following table of data (believed to be quartic in nature) into the `DataMatrix` editor.

TABLE 3.4

x	−3	−2	−1	0
$f(x)$	48.305	14.050	0.17151	0.13577
x	1	2	3	4
$f(x)$	8.2406	18.011	24.113	20.208

TABLE 3.6

x	−1.0	−0.5	0.0	0.5
$f(x)$	−5.2567	0.19922	2.3342	3.0186
x	1.0	1.5	2.0	2.5
$f(x)$	1.7645	0.15838	−1.5475	−2.7529

a. Build a difference table to determine the relative minimum, relative maximum, and inflection point(s).
b. Use the `QuartReg` (a least-squares fit to a quartic polynomial) to calculate a quartic model for the data. Store the regression polynomial in $y1(x)$. Use the first and second derivatives of $y1(x)$ to determine the relative minimums, relative maximums, and inflection points of the polynomial model.
c. Compare the values obtained from the two methods.

10. Enter the following table of data (believed to be quadratic in nature) into the `Data/Matrix` editor.

TABLE 3.7

x	−1.5	−1.0	−0.5	0.0
$f(x)$	−7.7302	0.05941	3.0982	8.2868
x	0.5	1.0	1.5	2.0
$f(x)$	14.782	19.714	21.144	25.989

a. Build a difference table to determine the relative minimum, relative maximum, and inflection point.
b. Use the `QuadReg` (a least-squares fit to a quadratic polynomial) to calculate a quadratic model for the data. Store the regression polynomial in $y1(x)$. Use the first and second derivatives of $y1(x)$ to determine the relative minimums, relative maximums, and inflection points of the polynomial model.
c. Compare the values obtained from the two methods.

11. Find a cubic polynomial with an inflection point at $x = -2$.
12. Find a cubic polynomial for which $(2, 4)$ is an inflection point.
13. Find a cubic polynomial with a relative maximum at $(\frac{2}{3}, 5)$.
14. Find a cubic polynomial with a relative maximum at $(-\frac{4}{3}, 6)$ and a relative minimum at $(4, -8)$.
15. Explain why a cubic polynomial must have an inflection point.
16. Find conditions on the coefficients a, b, c, d with $a \neq 0$ that will guarantee that

$$f(x) = ax^3 + bx^2 + cx + d$$

is an increasing function. Hint: you will need to use the quadratic formula.

For each of the functions in Exercises 17–19 below,
a. Find the zeros of the function.
b. Find an equation for the asymptote of the function.

c. Find all relative minimums and maximums (give both coordinates).
d. Find all inflection points (give both coordinates).
e. Decide whether there are any other values of x at which the concavity of the function changes sign.

17. Analyze the function:

$$f(x) = \frac{x^3 - 2x - 2}{x^2 - 2x}$$

18. Analyze the function:

$$f(x) = \frac{x^3 - 11x + 19}{x^2 - 4x + 4}$$

19. Analyze the function:

$$f(x) = \frac{7x^3 + x^2 - 12x - 15}{x^3 - 4x}$$

20. The graph of the derivative of a mystery function is provided in Figure 3.41 below. The graphing window is $[-4, 4]$ by $[-10, 10]$ with tick marks at one-unit intervals. Discuss the concavity of the graph of the mystery function. Does the graph of the mystery function have a point of inflection? If so, where does it occur? Explain.

21. Sketch a plausible graph of the mystery function whose derivative is sketched in Figure 3.41.

22. The graph of the derivative of a mystery function is provided in Figure 3.42. The graphing window is $[-4, 4]$ by $[-10, 10]$ with tick marks at one-unit intervals. Discuss the concavity of the graph of the mystery function. Does the graph of the mystery function have a point of inflection? If so, where does it occur? Explain.

23. Sketch a plausible graph of the mystery function whose derivative is sketched in Figure 3.42.

FIGURE 3.41

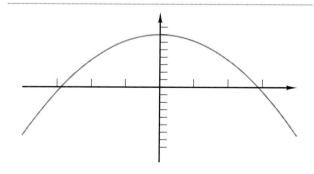

24. *Group Exercise* Develop conditions on the coefficients a, b, c, and d with $a \neq 0$ that will guarantee that the cubic polynomial

$$ax^3 + bx^2 + cx + d$$

has three distinct real roots.

FIGURE 3.42

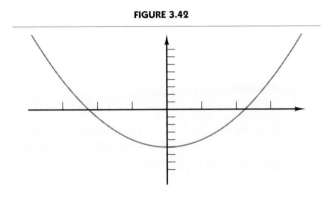

3.5

Properties of Continuous Functions

Continuous functions have two important properties that discrete functions do not have. The first property assures us that we can always find a zero of a continuous function between two input values whose output values have opposite signs. The second property guarantees us that, on any closed interval in the domain of a continuous function, there are input values whose output values are the maximum value and the minimum value for the function on the interval. We will begin the section with a banking example that demonstrates the use of the first property.

Two Properties of Continuous Functions

E X A M P L E   1   **Interest Paid by a Bank**

A bank offers a certificate of deposit (or CD, as it is sometimes called) that pays 8% interest every year. If the CD is cashed at any time during the year, no interest is accrued for that year. The value of the account for the first 20 years can be represented with a sequence. According to Table 3.8, if the account

TABLE 3.8   Value of a Certificate of Deposit

n	0	1	2	3	4	. . .	20
a_n	1000	1080	1166.40	1259.71	1360.49	. . .	4660.96

is cashed when it is one day short of being three years old, it will be worth $1166.40. Notice that the certificate is never worth exactly $1050 or any other value that is in between the values in the table.

On the other hand, another bank offers a savings account at an interest rate of 8% with interest compounded continuously. If $y(t)$ is the value of the account at time t, with t in years, then

$$y(t) = 1000e^{0.08t}, 0 \leq t \leq 20$$

computes the value of an initial investment of $1000 at time t. Let's compute $y(0)$ and $y(1)$, the value of the account initially and at the end of one year. First we evaluate $y(0)$ and get $y(0) = 1000e^{0.08(0)} = 1000$ as expected. Similarly, $y(1) = 1000e^{0.08(1)} \approx 1083.29$.

▶ First Reflection

What is the value of the savings account after 20 years? How does it compare with the CD? Speculate as to why there is a difference. Why would the bank compound interest continuously? ◀

If we want to know the exact time at which the savings account will be worth $1050 we must solve the equation $1000e^{0.08t} = 1050$. You can use the `solve` feature of the TI calculator to solve this equation. But, because $\ln(e^x) = x$, it is not too difficult to see that

$$t = \frac{1}{0.08} \ln\left(\frac{1050}{1000}\right) \approx 0.609877.$$

Recalling that t is in years and assuming there are 365 days in the year, we see that the account will become worth $1050 sometime during the 223rd day that it is in the account.

▶ Second Reflection

Pick a value between $2000 and $4660.96. How long will it be before the value of the savings account reaches the value you picked? Will the continuously compounded savings account reach every value between $1000 and $4660.96 during the first 20 years? What about the CD? Are there any values that the savings account will not reach during the first 20 years? What about $10,000? ◀

The Intermediate Value Property. The CD values are computed with a discrete function, while the savings account values are computed with a continuous function. The difference between the two functions that we have been exploring is due to a property of continuous functions known as the *intermediate value property* and it is important enough to warrant a formal statement.

THE INTERMEDIATE VALUE PROPERTY

If f is a continuous function on the closed and bounded interval $[a, b]$ and y is any number between $f(a)$ and $f(b)$, then there is an input value c between a and b such that $f(c) = y$. (See Figure 3.43.)

An interval is bounded if it is contained in a larger interval $[-M, M]$ for some real number M. The interval $[a, \infty)$ is not bounded.

The intermediate value property can be used to find zeros of a continuous function. Consider the arbitrary continuous function $y = f(x)$ on the closed interval $[a, b]$, shown in Figure 3.44. Since $f(a)$ is positive and $f(b)$ is negative, the number 0 lies between $f(a)$ and $f(b)$. Therefore, the intermediate value property guarantees us that there is a value c somewhere in the interval $[a, b]$, such that $f(c) = 0$, that is, f has at least one zero in $[a, b]$. The intermediate value property does not tell us where the zero is or even if there is more than one, but it does guarantee there is at least one.

FIGURE 3.43 The Intermediate Value Property　　　　**FIGURE 3.44** A Zero from the Intermediate Value Property

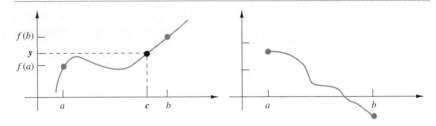

Think and Share

Break into groups. The intermediate value property can be used as the basis of a procedure to locate the zeros of a continuous function. Consider the function,

$$f(x) = x + \sin(x) - 3.$$

Note that

$$f(1) = \sin(1) - 2 < 0,$$

while $f(5) = \sin(5) + 2 > 0$. Hence, f must have a zero on the interval $[1, 5]$. Compute a decimal approximation for $f(3)$. Use your result to determine a smaller interval that must contain a zero of f. Devise a procedure to find the root as accurately as one wishes. Explain your procedure.

The Extreme Value Property. The second important property of continuous functions is the *extreme value property*, which says that if a function is continuous on a closed and bounded interval, then it takes on a maximum and minimum value somewhere on the interval. This property can't apply to discrete functions, since discrete functions are never defined on intervals, even though *finite* sequences have maximum and minimum values. Once again, the property does not tell us where or how often the function assumes its maximum and minimum values. It only guarantees that there is one maximum and one minimum value. This property is also important enough to warrant a formal statement.

THE EXTREME VALUE PROPERTY

FIGURE 3.45 $f(x) = \frac{1}{x-1}$

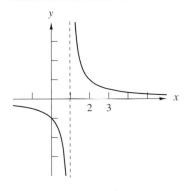

If f is a continuous function on the closed and bounded interval $[a, b]$, f must achieve a maximum and a minimum value on the interval.

The extreme value property is probably easier to understand if we consider a function which does not meet all of the conditions. The function

$$f(x) = \frac{1}{x-1}$$

for example, does not have a minimum value on the interval $(0, 1)$. The function is continuous on this interval, but the interval itself is not closed. On the other hand, if we consider the closed interval $[0, 1]$, the function f is not continuous (f is not defined at 1) and again it has no minimum. The function f satisfies both conditions of the extreme value property on the interval $[0, 0.5]$ and it does have both a maximum and a minimum value on this interval.

▶ **THIRD REFLECTION**

Graph $f(x) = \frac{1}{x-1}$ on the interval $[0, 0.5]$. What are the maximum and minimum values on this interval? Where do they occur? Does the extreme value property apply to the function

$$f(x) = \frac{1}{x-1}$$

on the interval $[3, \infty)$? If it does, give the domain value where your minimum value occurs. If not, explain why the extreme value property does not apply. ◀

▶ **FOURTH REFLECTION**

Observe the graph of

$$f(x) = \frac{1}{x-1}$$

FIGURE 3.46 Extreme Value Property

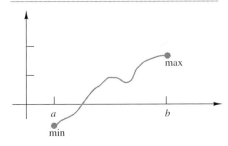

in Figure 3.45 to explain the meaning of the term "asymptotic behavior." Determine equations for the vertical and horizontal asymptotes. ◀

The extreme value property of continuous functions is important when we are trying to solve optimization problems that require us to find the largest or smallest value a function assumes over an appropriate domain. The function sketched in Figure 3.46 is continuous on the interval [a,b]. Its extreme values are labeled.

We use graphs in the following example to find approximations for the extreme values of a function.

EXAMPLE 2

Find the maximum and minimum values of the function $f(x) = -x^2 + 2x + \pi$ on the interval $[0, 3]$ and on the interval $[0, 1)$.

The graphs of the function on the indicated intervals appear in Figure 3.47 and Figure 3.48. From these graphs, we see that the maximum and minimum values of the function on $[0, 3]$ occur at 1 and 3, respectively, and are $f(1)$ and $f(3)$. On the interval $[0, 1)$, the minimum value occurs at 0 and is $f(0)$ and there is no maximum value since the function is increasing on an interval that is not closed at 1, its right point. The extreme value property does not apply to the second non-closed interval, $[0, 1)$, and thus does not require that f have a maximum on that interval.

FIGURE 3.47　$f(x) = -x^2 + 2x + \pi$ on $[0, 3] \times [-1, 5]$

FIGURE 3.48　$f(x) = -x^2 + 2x + \pi$ on $[0, 1) \times [2, 5]$

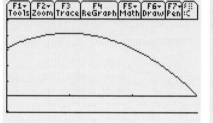
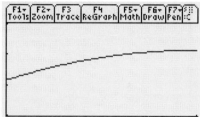

▶ FIFTH REFLECTION

Where is the first derivative of $f(x) = -x^2 + 2x + \pi$ positive, negative, and zero? Can you determine the maximum value of this differentiable function on a closed interval by just looking at input values where the derivative is 0?　◀

You will use graphs to help you identify important features of a continuous function in the next Think and Share.

Break into groups of four and then into pairs. Each first pair should use use graphs to find the maximum and minimum values of the function $f(x) = x^2 - 3x + \sqrt{5}$ on

1. On the interval $[-2, 0]$.
2. On the interval $[0, 2]$.
3. On the interval $[2, 3]$.
4. On the interval $[2, 3)$.

Think and Share

Each pair should compare there results with the other pair in their group. The groups should discuss where extreme values occur: always at endpoints, only at points that are not endpoints, a combination of endpoints and not endpoints. Each group should discuss why there is no maximum in (4) in light of the extreme value property.

Summary

In this section, we introduced the intermediate value property and the extreme value property for continuous functions. We then investigated how the extreme value property is applied using graphs of functions. And finally, we discussed when the extreme value property could not be applied.

EXERCISE SET 3.5

In Exercises 1–10, find the maximum and minimum values of the function on the given interval.

1. $f(x) = 3\sin(x - 2)$ on $[0, 2]$.
2. $g(x) = x^3 - 3x + 5$ on $[0, 5]$.
3. $h(x) = x^4 - 9x + 31$ on $[-3, 3]$.
4. $f(x) = 3\sin\left(\frac{x}{2}\right)$ on $[0, 2]$.
5. $g(x) = \tan^{-1}(x)$ on $[-2, 2]$
6. $h(x) = \tan^{-1}(x)$ on $(-\infty, \infty)$.
7. $f(x) = |x - 4|$ on $[-2, 1]$.
8. $g(x) = |2x - 3|$ on $[0, 20]$.
9. $h(x) = \cos(2x)$ on $(-\infty, \infty)$
10. $f(x) = \tan\left(\frac{x}{2}\right)$ on $(-\pi, \pi)$.
11. Does the extreme value property apply in Exercise 10 to f on the interval $(-\pi, \pi)$?
12. Is there one value of x for which the maximum value of h is obtained in Exercise 9? Does this contradict the extreme value property?
13. In Exercise 3, how many values of x are there for which the minimum values is obtained? Does this contradict the extreme value property.
14. The function $f(x) = \cos(x)$ has a maximum and minimum value on the interval $(-20, 20)$. Does this contradict the extreme value property?
15. Give an example of a function that has no maximum value on the interval from $[-2, 2]$. Can your function be continuous?

16. Give an example of a continuous function that is increasing but has no maximum value on some interval. Can your interval be a closed interval?
17. Give an example of a continuous function that is decreasing but has no minimum value on some interval. Can your interval be a closed interval?
18. A continuous increasing function on a closed interval has both a maximum and a minimum by the extreme value property. Where will the maximum occur on the interal?
19. A continuous decreasing function on a closed interval has both a maximum and a minimum by the extreme value property. Where will the maximum occur on the interval? For each of the continuous functions in Exercises 5 and 6:
    i. Graph the function using a window large enough to display the function's asymptotic behavior.
    ii. Determine an equation for the function's asymptote.
    iii. Describe the long-term behavior of the first differences.
    iv. Explain your result in terms of your equation for the function's asymptote.
    a. $y(x) = 5 - 4\left(\frac{1}{2}\right)^x$
    b. $y(x) = \dfrac{2x^3 + 2x + 100}{x^2 + 1}$
20. *Group Exercise* The intermediate value property indicates that we could find a root of a continuous function if we knew that the function values at the end points of some interval had opposite signs. Design an efficient method for finding a root on an interval. Provide a written description of your procedure (usually called an *algorithm* in mathematics).

3.6

Optimization: Finding Global Extrema

Many problems involve finding the greatest or smallest value (or values) that a function assumes over a domain of interest. Such values are called **global**

extrema. These problems often require that we determine the appropriate domain of interest ourselves. In this section, we will use an elementary example and two more involved examples to illustrate a general strategy for solving such problems.

Maximizing a Volume

Many problems involve maximizing (or minimizing) a function of one variable. Let's look at an example.

EXAMPLE 1   A Hamburger Box for Dave

The Challenge: Windy's is bringing back Dave's Deluxe! Dave has issued a challenge to design a box for his new hamburger accentuating the "voluminousness" of his flagship hamburger. He wants the box to be made from a piece of cardboard of size $8\frac{1}{2}$ inches by 11 inches. Free "frosties" for a year go to the person who designs the box with the largest volume. The design must be simple for ease of fabrication. Six squares of width x are to be cut from the cardboard that will then be folded into a box of height "x" (see diagram in Figure 3.49). The problem is to determine the size of x that will maximize the volume of the box.

FIGURE 3.49   Dave's Box

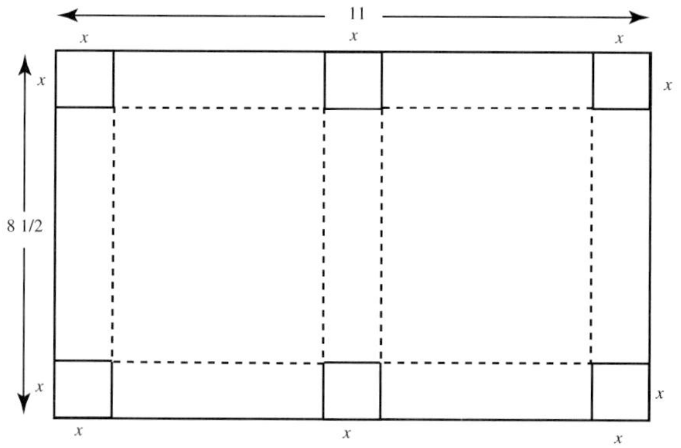

We wish to develop a strategy for meeting Dave's challenge. First, we give a precise definition for the type of quantity we seek.

GLOBAL OR ABSOLUTE EXTREMA

Let f be a function defined on the interval I. The function f has a global or absolute maximum value on I at c if $f(c) \geq f(x)$ for all x in I, and a global or absolute minimum value at c if $f(c) \leq f(x)$ for all x in I.

Notice the difference between global and local extrema. For local extrema we only consider points which are "nearby." For absolute extrema, we consider all points in the domain. Solving Dave's problem amounts to finding a global maximum on an appropriate domain.

Recall that the extreme value property, from Section 3.5, guarantees that if we are considering a continuous function on a closed interval then there will be a global maximum and there will be a global minimum, but the property says nothing about how to find these points. In this section, we will learn how to locate these global extreme points.

Consider the graphs in Figure 3.50. The domain D is defined to be the closed interval [a,b]. Notice that a global maximum value can occur at the left endpoint, the right endpoint, or at a critical point. Furthermore, a graph could even have two global maximum values.

FIGURE 3.50 Global Extrema

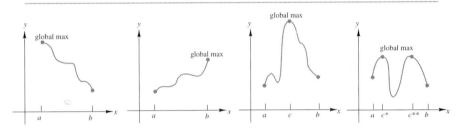

Every global extreme point that is interior to the domain of a function must also be a local extreme point. After all, a point that is higher than every point in the domain must be higher than any nearby points. From the graphs in Figure 3.50, we see that global extremes can also occur at endpoints. Hence, a global extreme point is either a critical point or an endpoint, and we have the following strategy for finding global extrema.

STRATEGY FOR LOCATING GLOBAL EXTREMA

To find the global maximum and minimum values of a *continuous* function f on the closed interval $[a, b]$,

1. Calculate the values of f at the critical points of f.
2. Calculate the values of $f(a)$ and $f(b)$.
3. The largest of the values from Steps 1 and 2 is the global maximum value; the smallest is the global minimum value.

Figures 3.51 and 3.52 show the critical points at c and d and the associated function values for the endpoints and these critical points for a function f on an interval $[a, b]$. The global maximum for the function is $f(b)$ that occurs at the endpoint b. The global minimum is $f(d)$ and occurs at the critical point d.

FIGURE 3.51   Critical Points and Endpoints   **FIGURE 3.52**   Maximum and Minimum Values

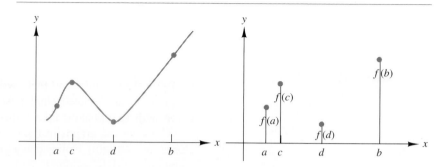

Let's reduce Dave's problem to a form that will allow us to apply our strategy. The strategy requires two elements: a continuous function and a closed interval. From Figure 3.49, we see that the length of the folded-up box is $\frac{11-3x}{2}$, the width of the box is $8\frac{1}{2} - 2x$, and the height of the box is x. So the volume of the box, which is the product of length, width and height, is given by the function:

$$V(x) = \frac{11 - 3x}{2}\left(8\frac{1}{2} - 2x\right)x = 3x^3 - 23.75x^2 + 46.75x.$$

We still have to determine an appropriate domain for our problem. Clearly x cannot be negative, so the left endpoint for x is zero. Since three cuts of size x must be made on the side that is 11 inches long, x cannot be greater than $\frac{11}{3}$. Thus, from these practical considerations, we determine the domain for x to be $\left[0, \frac{11}{3}\right]$.

▶ FIRST REFLECTION

If we consider two cuts on the short side of the box, the maximum value for x is $8.5/2 = 4.25$. Why did we choose $\frac{11}{3}$ and not 4.25 as the endpoint of the domain?   ◀

We now have the elements required to employ our strategy for finding a global maximum on the domain.

1.   The critical points are found by solving the equation $V'(x) = 0$.

$$V'(x) = 9x^2 - 47.5x + 46.75 = 0$$

$$x = 3.96904 \text{ and } 1.30874$$

The only critical point to consider in the domain is $x = 1.30874$, where $V(1.30874) = 27.2294$

2.   Evaluating $V(x)$ at the boundaries of the domain we have $V(0) = 0$ and $V(11/3) \approx 0$.

3.   For a global maximum, we choose the largest of the values at the endpoints and the values at the critical point(s). Thus, the global maximum on the interval is 27.2294 which is achieved at $x = 1.30874$.

Interpretation: The height of the box is $x = 1.30874$ inches; the length is 3.53689 inches; the width is 5.88252 inches; and the volume is 27.2294.

▶ SECOND REFLECTION
Why is the volume zero at both endpoints of the domain?   ◀

Think and Share

Break into groups. Figure 3.53 displays a graph of $V(x)$ over the domain [0, 4.25]. One contestant said that the domain of the function was [0, 4.25]. This domain includes the second critical point found above. Use the TI calculator to graph $V(x)$ over the domain [0, 4.25]. Analyze the concavity over this domain. Can you justify a global maximum at either critical point by your results? Use your calculator to confirm your analysis.

FIGURE 3.53 $V(x)$ on $[0, 4.5] \times [-2, 30]$

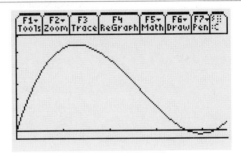

Storing Computers: A Single Warehouse

In the next example, we will apply what we have learned about derivatives to a significant problem: ordering computers for warehouse inventory.

EXAMPLE 2   **An Inventory Problem: A Single Warehouse**

Company Silicon has hired you to reduce costs for delivering and storing computers at their retail warehouses. They wish to know how many computers they should ship with each delivery and how frequently the deliveries should occur. You decide to focus on one warehouse. Each delivery to the franchise costs $1300. The accounting department is also able to determine that the franchise incurs a storage cost of $600 per thousand computers stored each day and that the franchise is selling 3400 computers every four weeks. You decide to find a delivery schedule that will minimize the average daily cost of delivering and storing a sufficient number of computers at the retail site to meet consumer demand.

If the company delivers a full year's supply of computers to the warehouse, they will have only one delivery charge. Of course, such a large delivery will result in very large storage costs. On the other hand, if they reduce their storage costs by making small deliveries, the deliveries will have to be made more often, thus running up their delivery costs. The problem can be simplified by thinking in terms of inventory cycles.

What is an *inventory cycle*? (Consider carefully Figure 3.54.) On the y-axis we plot the number of computers at any time t at the retail site; on the t-axis is time in days. The cycle begins with a delivery of Q computers. We let r be the number of computers purchased by customers each day, so our inventory is decreased at the constant rate r. In Figure 3.54, we show a cycle that begins with a delivery and ends with no computers in the inventory, at which time a delivery is made to begin another cycle. We let T be the length of the cycle; that is, the number of days between deliveries. The objective then is to find an *optimal inventory cycle* (one that will minimize total costs), if one exists. That is, we want values for Q and T which minimize the average daily cost of delivery and storage.

We seek an expression for the average daily cost. We begin by considering the *cost of an inventory cycle* of length T days. This cost can be computed from the formula

$$cost\ per\ cycle = delivery\ costs + storage\ costs.$$

The delivery costs are the constant amount $1300, since only one delivery is made at the beginning of the cycle. To compute the storage costs, consider Figure 3.54. The *average* daily inventory is $\frac{1}{2}Q$ computers. From the accounting information, the storage cost per computer is $0.60 per day. Thus, the average daily storage cost is $0.3Q$, and the total storage cost for a cycle of T days is $0.3QT$. We obtain the equation,

$$Total\ cost\ per\ cycle = 1300 + 0.3QT.$$

▶ THIRD REFLECTION

Using Figure 3.54, justify the assertion in the text that the average daily inventory is $\frac{1}{2}Q$.   ◀

The average daily cost is the total cost over the cycle divided by T, the length of the cycle. Letting c stand for the average daily cost, we obtain

$$c = \frac{1300}{T} + 0.3Q.$$

As we might have expected, the average costs depend on both the size of the delivery, Q, and the length of the inventory cycle, T. According to Figure 3.54, however, these two variables are related. In a single inventory cycle, the amount delivered equals the amount sold. Since the computers are being sold at a rate of $\frac{3400}{28}$ per day, the total number sold equals $\frac{3400}{28}T$. Thus, $Q = \frac{3400}{28}T$ and our average cost equation becomes

$$c = \frac{1300}{T} + 0.3\frac{3400}{28}T.$$

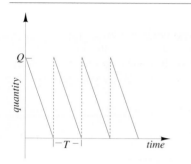

FIGURE 3.54   Inventory Cycles

▶ FOURTH REFLECTION

Find the inventory cycle length, T, which minimizes c. Use the equation relating T and Q to find the corresponding size for each delivery.   ◀

Storing Computers: The General Problem

We can consider the inventory problem in a more general setting.

EXAMPLE 3   **An Inventory Problem: The General Case**
_____

We have found a solution for a particular warehouse, but Company Silicon is not satisfied. In the first place, they have a number of other warehouses. Furthermore, delivery costs, storage costs, and demand rates are likely to change in the future. They are not willing to rehire a consultant every time that happens.

Fortunately, it is not difficult to convert our particular solution into a general solution. The difference from one warehouse to another is nothing more than a change in parameters. (Recall from Section 1.3 that parameters in a problem tend to be those quantities that are constant in a given context but will vary from context to context.) We need to assign letters to each parameter that we are using. The parameters for our problem were the specific constants provided for the particular warehouse: delivery cost, sales rate, and storage costs per computer. For the general problem, we let d be the cost in dollars for one delivery, r be the number of computers sold per day, and s be the storage cost per computer per day.

The total costs for one cycle are still computed from the formula

$$cost\ per\ cycle = delivery\ costs + storage\ costs.$$

For the general case, the delivery costs are the constant amount d. The average daily inventory is still $\frac{1}{2}Q$ computers, and, since s is the daily storage cost per computer, the average daily storage cost is $\dfrac{sQ}{2}$. We compute the total storage cost for a cycle of T days as $\dfrac{sQ}{2}T$, and we obtain the total cost equation

$$Total\ cost\ per\ cycle = d + \frac{sQ}{2}\,T.$$

Just as in the particular case, we divide by T to obtain the average cost equation

$$c = \frac{d}{T} + \frac{sQ}{2}.$$

Keep in mind that d and s are parameters. Thus, just as in the case of a single warehouse, the equation expresses the average costs as a function of Q and T. We continue to assume that the amount delivered equals the amount sold.

FIGURE 3.55   Daily Cost

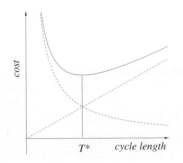

cost

$T*$ cycle length

Since the computers are being sold at a rate of r per day, the total number sold in T days equals rT. Hence, $Q = rT$ and our average cost equation becomes

$$c = \frac{d}{T} + \frac{sr}{2}T, \; T > 0.$$

▶ FIFTH REFLECTION

Why does the equation include the restriction $T > 0$?   ◀

Note that c is the sum of a hyperbola and a linear function. The average daily cost is graphed in Figure 3.55.

To find the value of T that minimizes cost, we begin by finding the critical points of $c(T)$. First, we differentiate c with respect to T:

$$c' = -\frac{d}{T^2} + \frac{sr}{2}.$$

Solving the equation

$$-\frac{d}{T^2} + \frac{sr}{2} = 0$$

for T gives two critical points. The positive critical point (since $T > 0$) is

$$T^* = \sqrt{\frac{2d}{sr}}.$$

We evaluate the second derivative at the positive critical point and find

$$c''\left(\sqrt{\frac{2d}{sr}}\right) = \frac{2d}{T^3}.$$

Since c'' is always positive for positive values of T, our critical point is the *relative minimum* that we expected from the graph in Figure 3.55. Also, from the graph in Figure 3.55, we know that $\sqrt{\dfrac{2d}{sr}}$ is a *global minimum*. Further, by using symbols for our parameters,

$$T^* = \sqrt{\frac{2d}{sr}}$$

solves all inventory problems for any s and r. The optimal amount to deliver is derived by knowing $Q = rT$, giving

$$Optimal \; delivery \; size = r\sqrt{\frac{2d}{sr}}.$$

As expected, the optimal time between deliveries and optimal amount to deliver increases as d increases, and decreases as s increases. Yet, the details of this relationship would have been difficult to discern without calculus.

▶ SIXTH REFLECTION

Show analytically that the optimal delivery time occurs at the intersection of the hyperbola and the linear function as suggested in Figure 3.55.   ◀

FIGURE 3.56 Buffer Stock

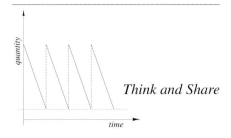

Think and Share

Break into groups. This Think and Share is about the notion of "Buffer Stock." Note from Figure 3.54 that the inventory cycle ends with no computers stored. Regardless of how good our estimates of demand are, we are likely to run out of inventory before a delivery is made. Many firms hedge against this situation by providing a *buffer stock* to help avoid stockouts and unsatisfied customers. The situation is depicted in Figure 3.56.

What is the effect of a buffer stock on the optimal inventory strategy we found above? Using Q_b as the size of the buffer stock, find an equation for the cost per cycle. What effect does this have on the location of the critical point? What is the optimal time between deliveries and optimal amount to deliver with a buffer stock? What is the real cost of adding the buffer stock?

In this section, we have learned how to optimize in general, and we developed an inventory strategy. To develop the mathematical model, we had to make several simplifications and approximations. Once we developed a continuous function to represent the cost as a function of a single variable, we were able to apply the principles we learned earlier in this chapter to find the best solution. The process developed in this section generalizes to many applications. Building models of this detail is generally best done by working in teams. You should work in teams to practice this process with different applications in the exercise set.

EXERCISE SET 3.6

1. Find the maximum and minimum values of each of the following functions on the given interval.
   a. $f(x) = 9x^4 - 16x^3 - 30x^2 + 24x + 2$ on $[-1, 2.5]$.
   b. $g(x) = -7x^4 + 12x^3 - x^2 - 48x + 5$ on $[-3, 4]$.
   c. $h(x) = x^5 - 7x^4 + 7x^3 + 21x^2 - 30x + 25$ on $[0, 5]$.
   d. $g(x) = x^3 \left(\dfrac{3}{4}\right)^x$ on $[-3, 3]$.
   e. $h(x) = 3x^4 - 6x^2 + 12x - 8$ on $[-2.5, 2]$.

2. Find extreme values and inflection points for the following function.
   $$f(x) = \left(\frac{1}{2}\right)^x \sin\left(\frac{x}{2}\right), \quad -\pi \le x \le 2\pi.$$

   Be sure to explain and justify your results.

3. Dave plans to sell stuffed pizzas and package them in the same style box he uses for Dave's Deluxe. The box will be constructed from a cardboard sheet that is 40 cm wide and 80 cm long. He will, of course, be looking to make the box with the largest volume. What are the dimensions of this box? Will Dave be offering a deep dish or thin crust pizza?

4. A Norman-style window consists of a rectangle topped by a semicircle as shown in Figure 3.57. For the entry of a new house you are designing, the clients have requested a Norman window consisting of a clear glass rectangle capped by a rose-colored glass semicircle. In order to minimize the cost of the window, you have recommended that the teak wood trim around the circumference of the window be kept to less than 25 ft. The clear glass allows 90% of the light to enter the room (absorbing the other 10%), while the rose glass allows only 65% of the light to enter. What are the dimensions of the window that will fit your recommendation and allow the maximum amount of light into the house?

FIGURE 3.57 A Norman Window

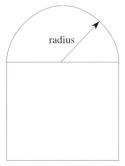

5. For centuries, the farmers of Northern China have dismantled parts of the Great Wall to use as construction materials. One enterprising young farmer, He Xin, is going to adapt this peaceful custom by using part of the existing wall as the back of a pigpen and dismantling another section of the wall to use as fencing for the other three sides. He Xin can dismantle enough of the wall to build 100 meters of fence. What is the largest number of pigs he can keep in the pen if each pig requires about two square meters of space?

6. He Xin's neighbor, Mai Wone, needs to build three farrowing pens for her sows. She, too, will back the pens against the Great Wall and separate the pens with fencing made from stones from the Wall. She is able to move 20 meters of fencing. Give plans and dimensions for the largest pens (in area) she can build.

7. A high school in Iowa has decided to construct a 440-yard track with space for a football field within the track. The track will be made of two straightaways and two semicircles. The football field must be 120 yards long and 50 yards wide. The athletic department requires a space of at least two yards from the sidelines to the edge of the track. Is the proposed design feasible? If it is feasible, find the dimensions of the track that will enclose the largest area.

8. A rectangle is inscribed in an ellipse so that its sides are parallel to the major and minor axes. Find the dimensions of the largest such rectangle that can be inscribed in a ellipse of major axis a and minor axis b.

9. Consider the parabola $y = x^2$.
   a. Find the point P, on the parabola, closest to the point $(3, 2)$.
   b. What is the slope of the tangent to the parabola at P?
   c. What is the slope of the line between P and $(3, 2)$? How is it related to the slope of the parabola at P?

10. Since the 1800s the United States Postal Service has offered parcel post, a low-cost service for shipping small packages. The criterion for what constitutes a "small" package was designed to be easy to implement with only a yardstick and a piece of string. Packages are assumed, for the most part, to be either cylinders or rectangular boxes. The longest axis of the package is designated as the "length" of the package. A string is wrapped around the middle of package perpendicular to the length. This length of string is the "girth" of the package. A package is acceptable for parcel post if the sum of the girth and length does not exceed 108 inches.
   a. What is the volume of the largest cubic package that can be sent parcel post?
   b. What is the volume of the largest mailing tube that can be sent parcel post?

c. Some boxes have a square base. What is the volume of the largest such box that can be sent parcel post?

11. The weekly cost in dollars of running a small production line is roughly modeled by the function

$$c(t) = 5500 + 1200t^2 - 65t^2 + t^3,$$

where $0 \leq t \leq 50$ is the number of hours the line is in operation during the week.
   a. Find the lowest weekly cost this model predicts.
   b. Find the lowest average cost this model predicts.
   c. What is the significance of the constant term in the model?
   d. What features of this model are realistic? What features are unrealistic?

12. The annual cost, $Cost$, of using an MRI scanner as a function of the number of cases, n, is graphed in Figure 3.58. The average cost of n cases can be interpreted graphically as the slope of the line through $(0, 0)$ and $(Cost, n)$.
   a. In Figure 3.58 mark the point on the curve at which the average cost is least.
   b. Describe this point in terms of its tangent line.

FIGURE 3.58   Cost vs. Number of Cases

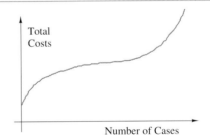

13. Economists have independently invented many of the ideas of the calculus. The economists' notion of "marginal cost at n" is essentially what we would mean by "instantaneous cost." Economists have stated the Average/Marginal Cost Rule: "The average cost of production is least when the average cost is equal to the marginal cost." Make sense of this principle in terms of derivatives and graphs of functions.

14. Graphs of the equations $3x + 2y = A$ form a family of parallel lines. For large values of A, the lines fail to intersect the parabola $y = x \cdot (5 - x)$. For smaller values of A the lines intersect the parabola twice. One line in the family meets the parabola in a single point. That line is tangent to the parabola at the point of intersection. Use this observation to determine the largest value assumed by the function

$F(x, y) = 3x + 2y$ for points (x, y), on or below the parabola, $y = x \cdot (5 - x)$.

15. You are maneuvering a heavy 20 foot length of pipe down a seven-foot wide hallway. You come to a (right angle) corner. Can you get the pipe around the corner while keeping it horizontal? (See Figure 3.59.) Fully explain and justify your answer.

FIGURE 3.59 Moving a Pipe

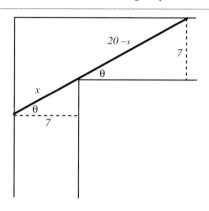

16. Can a 17 foot length of pipe be maneuvered around a corner joining a seven-foot hall and a five-foot hall? (See Figure 3.60.)

FIGURE 3.60 What are we doing with pipes in the hall anyway?

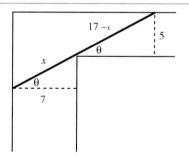

17. You have been supporting yourself in college working in the science storeroom. Your job is to collect and assemble equipment for biology and physics laboratories and deliver the equipment to various rooms in the science building. Over summer break you have some extra time and decide to make a dolly to help make your deliveries easy. Naturally, you

want to make the dolly as large as feasible. The narrowest halls you need to travel are five feet across, so you figure that four feet is the widest you could construct the top of the dolly. The tightest corner you need to negotiate is the one where the five-foot-wide hall meets the major eight-foot-wide corridor. How long can you make the top of the dolly and still negotiate this corner?

18. The top corner of a sheet of $8\frac{1}{2} \times 11''$ sheet of paper is folded to just touch the opposite side as shown in Figure 3.61. The crease extends across two adjacent sides. What is the shortest such crease? That is, minimize L in Figure 3.61.

FIGURE 3.61 Folded Paper

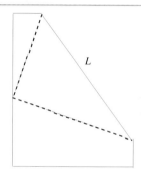

Exercises 17–21 involve the single warehouse model.

19. Describe the behavior of the optimal delivery size and the minimum cost in the inventory model if s, the storage cost, increases from $0 to $3 per day. The delivery cost is $1300 and the number of computers sold per day is 3400/28.

20. Describe the behavior of the optimal delivery size and the minimum cost in the inventory model if r, the number of computers sold per day, increases from zero to 20. The delivery cost is $1300 and the storage cost is $600 per thousand computers per day.

21. Describe the behavior of the optimal delivery size and the minimum cost in the inventory model if d, the cost in dollars for one delivery, increases from $0 to $1000. The storage cost is $600 per thousand computers per day and the number of computers sold per day is 3400/28.

22. Explain why the number of computers sold per day, Q, can be eliminated from our average daily cost equation.

23. Explain why the average daily inventory cost is $\frac{1}{2}Q$.

All of the following six extended exercises should be done in groups of three or four people.

24. Find the optimal time between deliveries and the optimal amount to deliver if $d = 2.1$ thousand dollars per delivery, $s = 0.7$ thousand dollars per thousand computers stored each month, and $r = 5$ thousand computers per month. (Hint: Measure time in months, not days.)

25. Consider an industrial situation where it is necessary to set up an assembly line. Suppose that each time the line is set up a cost, c, is incurred. Assume c is in addition to the cost of producing any items and is independent of the amount produced. Suggest submodels for the production rate. Now, assume a constant production rate, k, and a constant demand rate, r. What assumptions are implied by the model in Figure 3.62? Next, assume a storage cost of s (in dollars per unit per day) and compute the optimal length of the production run, P^*, in order to minimize the costs. List all of your assumptions. How sensitive is the average daily cost to the optimal length of the production run?

FIGURE 3.62 Determine the optimal length of the production run.

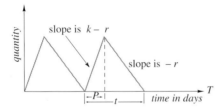

26. Consider a company that allows back-ordering. That is, the company notifies customers that a temporary stock outage exists and that their order will be filled shortly. What conditions might argue for such a policy? What effect does such a policy have on storage costs? Should costs be assigned to stock outages? Why? How would you make such an assignment? What assumptions are implied by the model in Figure 3.63? Suppose a *loss of goodwill* cost of w dollars per unit per day is assigned to each stock outage. Compute the optimal order quantity Q^* and interpret your model.

27. In the inventory model discussed in the text, we assumed a constant delivery cost that is independent of the amount delivered. Actually, in many cases, the cost varies in discrete amounts depending on the size of the truck needed, the number of platform cars required, and so forth. How would you modify the model in the text to take into account these changes? We also assumed a constant cost for the raw materials. However, often bulk-order discounts are given. How would you incorporate these discount effects into the model?

FIGURE 3.63 An Inventory Strategy that Permits Stock Outages

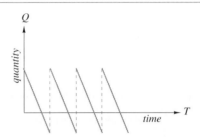

28. Discuss the assumptions implicit in the two graphical models depicted in Figures 3.64 and 3.65. Suggest scenarios in which each model might apply. How would you determine the submodel for demand in Figure 3.65? Discuss how you would compute the optimal order quantity in each case.

FIGURE 3.64 An Inventory Submodel

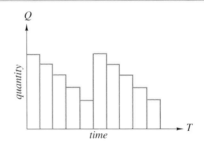

FIGURE 3.65 Another Inventory Submodel

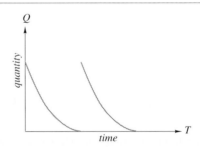

29. What is the optimal speed and safe following distance that allows the maximum flow rate (cars per unit time)? The solution to the problem would be of use in controlling traffic in tunnels, for roads under repair, or for other congested

FIGURE 3.66 Schematic

areas. In the schematic in Figure 3.66, l is the length of a typical car and d is the distance between cars. Justify that the flow rate is given by

$$f = \frac{\text{velocity}}{\text{distance}}.$$

Let's assume a car length of 15 ft. Assume the safe stopping distance is given by $d = 1.1v + 0.054v^2$ where d is measured in feet and v in miles per hour. Find the velocity in miles per hour and the corresponding following distance d that maximizes traffic flow. How sensitive is the solution to changes in v? Can you suggest a practical rule? How would you enforce it?

3.7

Implicit Differentiation and Its Applications

In this section we will use a new differentiation technique to shed light on several examples involving slope. This new method is extremely powerful and will prove quite useful later in the course. This new differentiation method and the method of parametric differentiation in the next section will allow us to solve problems that we could not solve otherwise even with the help of the TI calculator.

FIGURE 3.67 A Tetherball

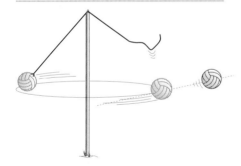

Flying off on a Tangent

A ball swings around at the end of a four-foot rope so that its trajectory can be described by a circle. (See Figure 3.67.) Suppose the rope breaks and the ball flies off its path along a line tangent to the circle (this is quite likely the origin of the figurative expression "flying off on a tangent"). How do we find an equation for this tangent line? More particularly, how do we find the slope of this line since we will already know a point (the location of the ball when the rope breaks)?

We will solve this problem in two ways. First, we will apply the differentiation techniques we already know to the equation of the circle in rectangular coordinates. We will then learn a new technique, *implicit differentiation*, which will lead to a simpler solution.

E X A M P L E 1 **Slope: the Old-Fashioned Way**

From geometry, the standard equation for a circle of radius four in rectangular coordinates is $x^2 + y^2 = 16$. To make the problem concrete, let's assume that the ball is at the point $(3, \sqrt{7})$ when the rope breaks. We can solve this problem using calculus by finding the value of $\frac{dy}{dx}$ when $x = 3$. This seems simple enough, but we are stopped immediately by one big problem. Our equation does not give y explicitly as a function of x. Hence, our first step is to solve the equation for y,

$$y = \pm\sqrt{16 - x^2}$$

where the positive square root represents the upper half of the circle, and the negative square root represents the lower half.

▶ **FIRST REFLECTION**

Solve $x^2 + y^2 = 16$ for y on the TI calculator. Is the solution given by the TI calculator equivalent to the one given above? Why do you think the calculator generates such a strangely written answer?   ◀

Since $(3, \sqrt{7})$ is on the upper half of the circle, we differentiate $y = \sqrt{16 - x^2}$ with respect to x,

$$\frac{dy}{dx} = \frac{1}{2}(16 - x^2)^{-1/2}(-2x) = \frac{-x}{\sqrt{16 - x^2}}.$$

(Compare this answer with the answer given by the TI calculator.)

Thus the slope of the tangent line to the circle at $(3, \sqrt{7})$ is

$$\frac{dy}{dx}\bigg|_{(3,\sqrt{7})} = \frac{-3}{\sqrt{16 - (3)^2}} = \frac{-3}{\sqrt{7}}.$$

The equation of the tangent line is

$$y - \sqrt{7} = \frac{-3}{\sqrt{7}}(x - 3).$$

We have solved the problem, but our solution includes an unsettling feature. At the very first step, we were required to solve our equation for y. Many equations involving both x and y cannot be solved for y explicitly.

IMPLICIT VERSUS EXPLICIT

The equation $x^2 + y^2 = 16$ implies that y depends upon x, because if we restrict our attention to either the top or the bottom of the circle, specifying a value for x determines a value for y. We call the equation an **implicit** relationship and say that y is **implicitly** defined as a function of x. The equation $y = \sqrt{16 - x^2}$, on the other hand, **explicitly** defines y in terms of x (notice that the equation will only generate points on the top half of the circle). We call this second equation an **explicit** relationship and say y is explicitly defined as a function of x.

EXAMPLE 2 **Slope: the Implicit Way**

Let's begin again using a new method called *implicit differentiation* to find $\frac{dy}{dx}$ without solving for y explicitly.

Try the differentiation operator on $x^2 + y^2 = 16$. The TI calculator treats y^2 as a constant when differentiating with respect to x. (Recall Section 3.3 where we first introduced the differentiation operator.) Therefore, the command

$$d(x\texttt{\textasciicircum}2 + y\texttt{\textasciicircum}2 = 16, x)$$

results in the answer $2x = 0$ as displayed in Figure 3.68. This result is correct if x and y are not related but this is decidedly not the case for our example.

Alter the Entry line to alert the TI calculator that y is to be treated as a function of x (refer to Figure 3.68) by using $(y(x))\texttt{\textasciicircum}2$ instead of $y\texttt{\textasciicircum}2$ and reenter the command. The result is also shown in Figure 3.68 and is

$$2 \cdot y(x) \cdot \frac{dy(x)}{dx} + 2 = 0 \quad \text{or} \quad 2 \cdot y \cdot \frac{dy}{dx} + 2 = 0. \tag{3.8}$$

We would like to solve the resultant equation for $\frac{d}{dx}(y(x))$. We could do this by hand (and it is not particularly difficult), but let's see if we can get the TI calculator to do the work for us. The most logical approach is to use the `solve` command. Unfortunately, the TI calculator will not solve for an expression like $\frac{d}{dx}(y(x))$. It requires a single variable name (try it). To get around this difficulty, we replace the expression $\frac{d}{dx}(y(x))$ with the single variable *slope* (a reasonable choice since $\frac{d}{dx}(y(x))$ is just another way to write $\frac{dy}{dx}$). You can edit the last answer that appears in the History area in the middle of the screen by following the instructions in *Editing an Earlier Output* in the margin. (See also Figure 3.69 and 3.70.)

FIGURE 3.68 Implicit Differention

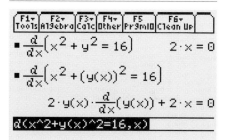

EDITING AN EARLIER OUTPUT

Clear the HOME screen Entry line. Highlight the desired output in the History area and press ENTER to transfer the output to the Entry line. To enter a "with" command, first put the cursor at the right end of the Entry line. Enter the "with" symbol (vertical bar), followed by `d(y(x),x) = slope`, and ENTER. Caution: be sure to use the differentiation operator.

Use the "with" operator (the vertical bar). Then enter `d(y(x),x)=slope`.

FIGURE 3.69 Substituting `slope` for
`d(y(x),x)`

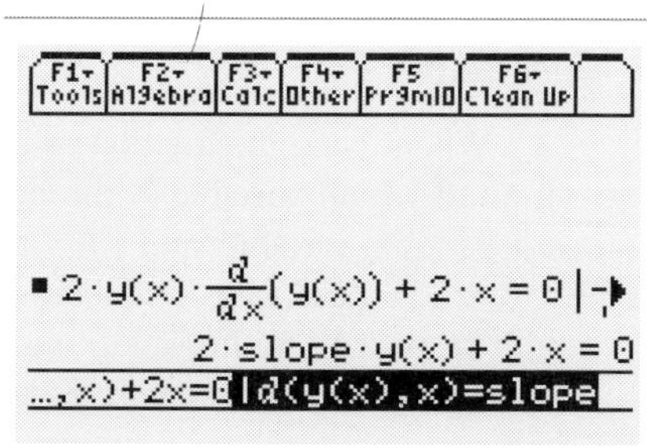

FIGURE 3.70 Solving for `slope`

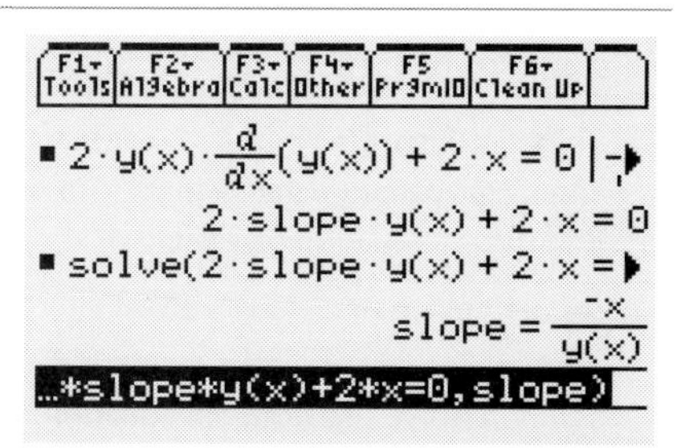

Solving this equation for *slope* and replacing *slope* with $\dfrac{dy}{dx}$, we find

$$\frac{dy}{dx} = \frac{-x}{y(x)}. \tag{3.9}$$

Notice that our result includes both x and y. This is common when we use implicit differentiation.

With pencil and paper, abbreviating $y(x)$ as y, we have

$$x^2 + y^2 = 16$$

$$2x + 2y\frac{dy}{dx} = 0$$

$$2y\frac{dy}{dx} = -2x$$

$$\frac{dy}{dx} = \frac{-x}{y}.$$

Here, we have abbreviated $y(x)$ with y.

Notice that this result matches our results in Figure 3.70 and Equation (3.9).

Let's check our results at the point where $x = 3$ and $y = \sqrt{7}$. Bring the `slope =` statement to the Entry line and type "with" `x=3 and y(x) = √(7)`. (See Figure 3.71.)

Try resetting memory if you get weird error messages.

FIGURE 3.71   Evaluating `slope` at a Point

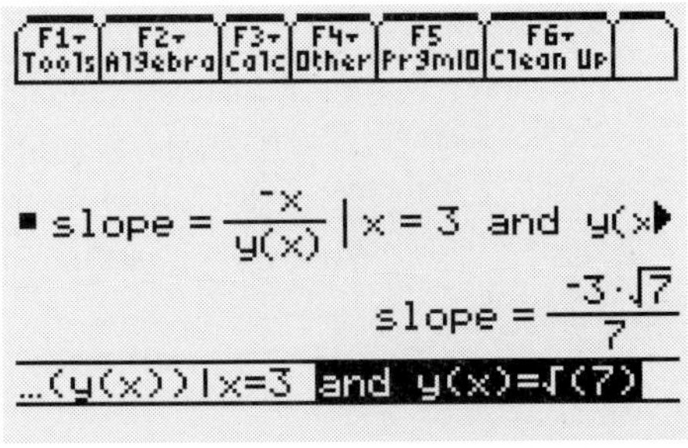

TECHNOLOGY NOTE

To find a derivative implicitly, differentiate the equation that defines y implicitly as a function of x using `y(x)` for `y`. Replace $\dfrac{d}{dx}(y(x))$ with the single variable *slope*, using the "with" symbol. Solve the resultant equation for *slope*.

When using the method of implicit differentiation, it is important to keep in mind which variable is the independent variable. In our example, x is the independent variable, and y is viewed as a function of x. When we differentiate x^2 with respect to x we obtain $2x$, as expected. When we differentiate y^2 with respect to x, however, we view y as a function of x so we are really differentiating $(y(x))^2$ with respect to x. This derivative requires the chain rule with $y(x)$ as the inner function. Consequently, when we differentiate y^2 with respect to x, our result is $2y\dfrac{dy}{dx}$.

Think and Share

Break into groups. Find $\frac{dy}{dx}$ if $x^2y + 3y^4 = 8x^3$ on the TI calculator and with pencil and paper. Which was easier for you? Could you find dy/dx explicitly this time or did you have to use implicit differentiation? Explain.

Related Rates Problems

We use implicit differentiation to solve *related rate* problems where we know the rate of change of one variable, and we want to know the rate of change of another, related variable. We can think of an equation as equating two functions whose derivatives will also be equal. When we create an equation relating the variables, we can differentiate that equation to generate another equation that relates the known and unknown rates of change. The two variables are implicit functions of time; and, when we differentiate the equation, we differentiate both sides of it with respect to time. In complicated related rate problems, more than two variables may be involved. An example will clarify this solution process.

FIGURE 3.72   Flying a Kite

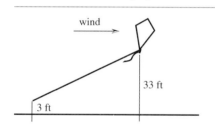

E X A M P L E   3   **The Student and the Kite**

A student flies a kite in a lovely park on a windy day in March. The kite is blown due east at 5 ft/sec while maintaining a constant height of 33 feet above the ground. How fast is the distance between student and the kite changing when the kite is 100 feet away from the student? Assume the student is holding the string three feet above the ground.

  We begin by drawing a diagram that shows a fixed position of the student and the kite. We try to indicate that the situation is not static as shown in Figure 3.72.

  Figure 3.72 shows the direction of the wind and the constant distance (33 ft) the kite is above the ground. We indicate the direction the wind is blowing with an arrow, but we have not shown the speed of the wind. If the kite has traveled 90 ft with respect to the ground, we can write an equation that relates 90 and 30 and the fixed distance s using the Pythagorean Theorem as displayed in Equation (3.10).

FIGURE 3.73   Flying a Kite with Variables

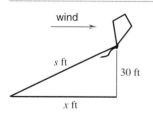

$$90^2 + 30^2 = s^2 \qquad \textbf{(3.10)}$$

But there is no way to relate the rate of change of the position of the kite with respect to the ground (the wind speed) and the rate of change of the distance between the kite and the student by implicit differentiation. All the numbers are constant and implicit differentiation with respect to time will give us that $0 = 0$. (Why?)

  We need to have distances that vary. Variable distances x and s and the constant 30 ft are displayed in the triangle in Figure 3.73. In this figure, the wind is blowing, s is the distance from the student to the kite, and x is the distance from the student to a point directly below the kite. We can now state symbolically the rate of change that we know and the rate of change that we do not know. For the ground speed of the kite (the rate at which x is changing as the kite is blown due east), we use dx/dt. For the rate of change of the distance between the student and the kite, we use ds/dt. Both of these rates of change (derivatives) are in ft/sec.

The variables we need are the functions we are differentiating with respect to t, namely, x and s. These variables are related by the Pythagorean Theorem in Equation (3.11).

$$x^2 + 30^2 = s^2 \tag{3.11}$$

We can differentiate Equation (3.11) with respect to t to obtain an equation where the rate of change for x and the rate of change for s are related as shown in Equation (3.12).

$$2x\frac{dx}{dt} = 2s\frac{ds}{dt}. \tag{3.12}$$

We know the derivative dx/dt is equal to 5 ft/sec, so

$$x \cdot 5 = 2s\frac{ds}{dt} \tag{3.13}$$

$$2s\frac{ds}{dt} = 10x$$

$$\frac{ds}{dt} = \frac{10x}{2s} = \frac{5x}{s}. \tag{3.14}$$

We have the rate of change of the distance between the student and the kite, ds/dt, but in terms of the variable x. We wanted to know what this rate will be when the kite is 100 ft from the student (along the kite string). We need to know what the value of x is when s is 100 ft. We can use an equation like Equation (3.10) where all the values were constant to do this.

$$x^2 + 30^2 = 100^2$$

$$x^2 = 10000 - 900$$

$$x^2 = 9100$$

$$x = \sqrt{9100} \approx 95.4 \text{ ft}$$

We want to know the rate of change for a fixed value of s (and x) based on the definition of the rate of change as a function of s and x given in Equation (3.13). We make the substitutions as displayed in Equation (3.15).

Thus, using this value of x, we have

$$\frac{ds}{dt} = \frac{5x}{s} \tag{3.15}$$

$$\frac{ds}{dt} \approx \frac{5 \cdot 95.4}{100} = 4.77 \text{ ft/sec} \tag{3.16}$$

Thus, the distance beteen the student and the kite is increasing at the rate of 4.77 ft/sec when the student is 100 ft from the kite.

Suppose that the student is standing at the edge of a cliff where there is an up draft that causes the kite to rise at the rate of 2 ft/sec as well as be blown due east a 5 ft/sec. You wish to determine how fast the distance between the student and the kite is changing when the kite has moved horizontally 100 ft. Assume the spool of kite string is held at the very edge of the cliff.

Do the following.

Think and Share

1. Draw a diagram of this situation involving a triangle.
2. Determine the number of variables you will need and indicate them on your diagram.
3. Write down the given rates of change symbolically (in Leibniz notation).
4. Write down the unknown rate of change symbolically (in Leibniz notation).
5. Use the Pythagorean Theorem to develop an equation that has 3 variables in it.
6. Find the required rate of change (derivative) in terms of the other variables using implicit differentiation.
7. Find the values of the three distances when the student is 100 feet from the kite.
8. Determine the desired rate of change in ft/sec.

Some related rate problems are based on a simple interpretation of a formula in two or more variables. The next example shows such a situation.

E X A M P L E 4 **Inflating a Balloon**

A spherical weather balloon is inflated at the rate of 2 ft^3/sec. How fast is the radius of the balloon expanding when the radius is 4 ft?

The formula we know (and will be able to derive with the tools in Chapters 4 and 5) for the volume of a sphere is

$$V = \frac{4}{3}\pi r^3. \qquad (3.17)$$

The rate at which the balloon is being inflated is the rate at which the volume is changing. And so, we know that $dV/dt = 2\text{ft}^3/\text{sec}$. We would like to know the rate of change of the radius. We can write that as dr/dt. Fortunately, the formula for the volume of the sphere (Equation (3.17)) relates the function $V(t)$ whose derivative we know and the function $r(t)$ whose derivative we want to find. Here we have written the two variables, V and r, as functions of time (even though we don't know what those explicit functions are). Thus, if we differentiate Equation (3.17) with respect to t, the independent variable for both functions, we will be able to relate these two rates. So, differentiating implicitly (thinking of the variables as functions), we have

$$\frac{d}{dt}(V) = \frac{d}{dt}\left(\frac{4}{3}\pi r^3\right)$$

$$\frac{dV}{dt} = 4\pi r^2 \frac{dr}{dt}$$

Solving for $\dfrac{dr}{dt}$, we have

$$\frac{dr}{dt} = \frac{dV/dt}{4\pi r^2}$$

We can now evaluate dr/dt since we know that the value of dV/dt is 2 ft^3/sec for all values of r and we have a value for r, namely 4 ft. Thus, the value of dr/dt is

$$\frac{dr}{dt} = \frac{2\text{ft}^3/\text{sec}}{4\pi \cdot 4^2\text{ft}^2} = \frac{1}{32\pi}\text{ft/sec} \approx 0.0099\text{ft/sec}.$$

Thus, the radius of the spherical balloon is increasing at the rate of 0.0099 feet per second when the balloon is being inflated at 2 ft^3/sec and the balloon's radius is 4 ft.

Conic Sections [Optional]

We can investigate the graphs of conic sections using implicit differentiation and the concept of derivative as slope. You can use the calculator tools we have developed to make the computations indicated in the discussion that follows if you wish.

Hyperbolas.   From previous mathematics courses, you know that the equation

$$\frac{x^2}{2^2} - \frac{y^2}{1^2} = 1 \tag{3.18}$$

represents a hyperbola. We can attempt to graph this hyperbola by computing the derivative of y with respect to x and then creating a graphical representation of the slopes of hyperbolas that have that derivative.

We begin by assuming that y is a function of x, that is, the variable y in Equation (3.18) can be written as $y(x)$. We then take the derivative of each side of this equation with respect to x assuming that each side is a function and that the derivatives of two equal functions are equal. Thus,

$$\frac{x^2}{2^2} - \frac{y^2}{1^2} = 1$$

$$\frac{d}{dx}\left(\frac{x^2}{2^2} - \frac{y^2}{1^2}\right) = \frac{d}{dx}(1)$$

$$\frac{2x}{4} - 2y\frac{d}{dx}y(x) = 0$$

We now solve for $\dfrac{d}{dx} y(x)$.

$$-2y\frac{d}{dx}y(x) = -\frac{x}{2}$$

$$\frac{d}{dx}y(x) = \frac{\frac{x}{2}}{2y} = \frac{x}{4y} \tag{3.19}$$

Now we would like to show the slopes of the hyperbola at various points on a coordinate plane based on this derivative $(dy/dx = x/4y)$. We could create a grid of points, compute the derivative with the x- and y-coordinates of each point, and then draw little slope lines at each of these grid points based on these computed derivatives. This would be a time consuming process by hand. Fortunately, we can do this on a TI-89 or TI-92 Plus calculator using the instructions in *Drawing a Slope Field Using the Implicit Derivative*.

DRAWING A SLOPE FIELD USING THE IMPLICIT DERIVATIVE

To draw a grid of slope lines on a coordinate plane, first change the `Graph...` mode to `DIFF EQUATIONS`. Next, in the `Y=` editor, enter the derivative expression for `y1'` using `t` for the independent variable and `y1` for the dependent variable. For example,

$$\frac{d}{dx}y(x) = \frac{x}{4y}$$

would be entered as

$$\mathtt{y1'=t/(4y1)}.$$

Leave the `yi1` (initial value for `y1`) line blank. While in the `Y=` editor, press `F1` and then `9` to format the Graph Formats `Fields...` item to `SLPEFLD` (short for SLOPEFIELD). Now select an appropriate graphing window using the WINDOW editor. Finally, press GRAPH to display the set of slope lines.

Display the slope lines for the derivative that you computed in Equation (3.19) using the process given in *Drawing a Slope Field Using the Implicit Derivative*, entering the derivative as `y1'=t/(4y1)`. Notice that the slopes seem to indicate hyperbolas centered at the origin as displayed in Figure 3.74. You can see the upper branch of the hyperbola using the Initial Value feature by pressing `F8`, moving the cursor to about $(3, 2)$, and pressing ENTER as displayed in Figure 3.75.

▶ SECOND REFLECTION

Display a field of slopes for the function $11x^2 + 10xy + y^2 = -16$. Is the hyperbola's axis of symmetry parallel to the x=axis? ◀

FIGURE 3.74   Slopes of an Hyperbola on   **FIGURE 3.75**   The Upper Branch of One
[−10, 10] × [−5, 5]   Hyperbola

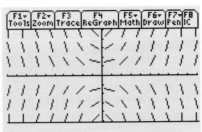
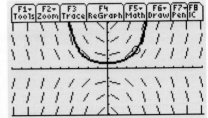

FIGURE 3.76   Slopes of an Ellipse on
[−10, 10] × [−5, 5]

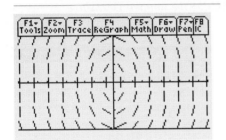

Ellipses.   Recall that the equation

$$\frac{x^2}{3^2} + \frac{y^2}{5^2} = 1$$

is the equation of an ellipse. You can determine the derivative of $y(x)$ as we did for the hyperbola above using implicit differentiation. The derivative is

$$\frac{d}{dx}y(x) = \frac{-50x}{18y}.$$

The graph of the slopes looks like Figure 3.76.

▶ THIRD REFLECTION
Display the field of slopes for the equation

$$20x^2 + 17xy + 5y^2 = 16$$

Is the axis of symmetry of the ellipse parallel to the x-axis?   ◀

EXERCISE SET 3.7

In Exercises 1–4: (a) Find dy/dx by implicit differentiation, (b) find dy/dx by solving the equation explicitly for y and then differentiating, (c) show that your answers to parts (a) and (b) are consistent.

1. $x^2 + 7x - 5xy = 3$
2. $x^3 - 5x^2y + 3x = 7$
3. $x^2 - y^2 = 1$
4. $\frac{y^2}{16} + \frac{x^2}{9} = 1$

In Exercises 5–12, find dy/dx:

5. $x^2 + 7xy + y^3 = 10$
6. $x^4 + y^4 - 5xy^3 = 4$
7. $\sin(y) + \cos(x) - xy = 2$
8. $\sin(xy) + y^2 = 5 - x$

9. $x^2 + y^2 + 16 = 0$
10. $y^5 = x$
11. $\cos^2(x) = y - \sin^2(x)$
12. $\sin(y) = x$

13. Are there any points that satisfy the equation n Exercise 9? Are there any tangent lines to the graph of the equation?
14. Derive the formula for the derivative of $f(x) = \sqrt[5]{x}$ from your result in Exercise 10.
15. Why is the slope of the tangent line always zero in Exercise 11?
16. Find the derivative of $f(x) = \sin^{-1}(x)$ from your result in Exercise 12.

In Exercises 17–20, (a) find the SLOPEFIELD for the given equation, (b) find a point that satisfies the equation, (c) use that point to draw a representation of the graph of the equation on your TI-92 Plus or TI-89 calculator, and (d) sketch the graph on paper.

17. $x^2 + y^2 + 2x - 4y - 21 = 0$
18. $x^2 - y^2 + 2x - 4y - 21 = 0$
19. $x^2 + y^2 + 4x - 6y - 17 = 0$
20. $x^2 - y^2 + 6x - 4y - 17 = 0$

Exercises 21–24 are related rate problems.

21. A spherical weather balloon is being inflated at 3 ft³/sec. Determine the rate ate which the diameter is changing when the diameter is 2 ft.

22. Sand is poured on a flat surface creating a conical sand pile. As the sand is poured, the the radius of the base of the cone is always twice its height. If the height of the cone is changing at the rate of 3 m/min, what is the rate of change of the radius of the cone? $\left[V_{cone} = \frac{1}{3}\pi r^2 h \right]$.

23. Water is poured into a cylindrical tank whose base is 5 meters at the rate of 2 m³/min. How fast is the depth of the water increasing when the water is 3 ft deep? $\left[V_{cylinder} = \pi r^2 h \right]$.

24. Two bicyclists ride from due north and due east toward an intersection. If the first bicyclist is riding at 20 ft/sec and the second bicyclist is riding at 30 ft/sec, how fast is the distance between them changing when they are both 100 ft from the intersection.

Explain, in your own words, the meaning of each of the following terms. Write your explanations in clear and complete sentences. Include annotated diagrams whenever appropriate.

25. explicit function
26. implicit function
27. flying off on a tangent
28. implicit differentiation

3.8

Modeling Motion with Parametric Equations

We will begin this section by recreating a model originally developed to resolve a question about motion that was of practical military significance in the late sixteenth century.

Question: What is the path followed by a cannonball?

To model motion in the plane (the change in distance over time), we need to represent horizontal and vertical distances in terms of time. Equations which express x and y in terms of a third variable or parameter are called **parametric equations** and serve as a perfect tool with which to model motion.

We will conclude this section with a discussion of slopes of tangent lines involving curves represented parametrically.

▶ FIRST REFLECTION

The numbers m and b are parameters in the linear equation $f(x) = mx + b$. How does this use of the word "parameter" differ from its use in the paragraph above? ◀

Parametric Equations

Parametric equations are often used to represent a point on a line at a given time t. For example, the parametric representation of the function $f(x) = 2x + 1$ given in Equations (3.20 and 3.21)

$$x(t) = 2t \qquad (3.20)$$
$$y(t) = 4t + 1 \qquad (3.21)$$

can be used to create the values in Table 3.9.

Parametric equations can describe straight lines and simple curves, as well as much more complicated curves, like the cycloid discussed at the end of this section.

▶ SECOND REFLECTION

Show that the parametric equations

$$x(t) = t + 1$$
$$y(t) = 3t + 8$$

describe the line given in Cartesian coordinates by $y = 3x + 5$. (Hint: Plot the points that result from $t = 0, 1,$ and 2.) Substitute the parametric value of x into the Cartesian equation. Describe what happens. ◀

Projectiles — Parametric Equations of Flight

There are an enormous number of factors that, potentially and in practice, affect the flight of a projectile like a cannonball. We will need to know the position from which the projectile is launched, and its initial velocity (both speed and direction) if we are to have any hope of predicting its flight. Gravity and wind will alter the path of the projectile. The resistance of air will slow the object. The effects of this wind resistance will depend on the size, shape, orientation, surface, and weight of the projectile. Baseball pitchers know that spin affects the path of a thrown ball. For long trajectories (such as those followed by ballistic missiles), the rotation and curvature of the earth are important. The process of modeling, however, is a process of simplification. For the flight of our cannonball, we ignore all factors except the first three: position, velocity, and gravity.

TABLE 3.9   A Parametric Table for $f(x) = 2x + 1$ with Parameter t

t	$x(t) = 2t$	$y(t) = 4t + 1$
0	0	1
1	2	5
2	4	9
3	6	13
4	8	17
5	10	21

Cannon

EXAMPLE 1 **The Cannonball Question**

Creating the Model

FIGURE 3.77 Components of the Vector v

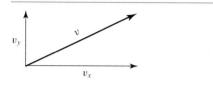

We assume that the flight of the cannonball is confined to a plane so that we need only track how two coordinates change as time advances. We describe the flight in terms of a horizontal coordinate x and a vertical coordinate y. A third variable, time, will be denoted by t, with $t = 0$ being the instant at which the projectile is launched. The initial position of the projectile is denoted by (x_0, y_0). (This use of zero as a subscript is a reflection of our decision to assign $t = 0$ to the moment the projectile is launched.) Further, the projectile's initial horizontal velocity is denoted by v_x, and its initial vertical velocity by v_y.

Traditionally, time is viewed as a parameter, and the equations and graphs of our model are called *parametric equations* and *parametric graphs*. Typically we measure t in seconds, or minutes, and measure x and y in meters, feet, yards, or kilometers, as dictated by the specific problem.

The velocity can be represented by an arrow in a given direction with a specific length. The length of the arrow is what we often call speed. It is also called the magnitude of the velocity. We can represent the velocity as the sum of its horizontal and vertical components as shown in Figure 3.77.

We build the model by adding in our three factors one at a time.

Building the Model

Position Alone. With no initial velocity and without the effect of gravity, the path of the cannonball is not all that interesting. Its position as a function of time is described by the constant parametric equations

The graph of these parametric equations is the single point (x_0, y_0).

$$x(t) = x_0 \tag{3.22}$$

$$y(t) = y_0 \tag{3.23}$$

for $0 \le t < \infty$. The cannonball simply sits there in the cannon. Perhaps the cannoneer has found something better to do (like running away).

Position with Velocity. Since for a constant rate of speed distance traveled is *rate · time*, and we are ignoring all forces that might alter the projectile's path, the horizontal distance traveled in time t is $v_x \cdot t$. Likewise the vertical distance traveled (remember we are ignoring gravity at the moment) is $v_y \cdot t$. Adding the effect of an initial velocity to Equation (3.22) and (3.23), our model becomes:

$$x(t) = x_0 + v_x \cdot t \tag{3.24}$$

$$y(t) = y_0 + v_y \cdot t \tag{3.25}$$

In the MODE *dialogue box, change* Graph ... *to* PARAMETRIC. *In the* Y= *editor, enter* xt1(t)=-5+10*t *and* yt1(t)=1+15*t

The WINDOW *settings for both Figure 3.78, and, later, Figure 3.79 are* tmin=-5, tmax=5, tstep=0.2, xmin=-10, xmax=10, xscl=1, ymin=-10, ymax=10, yscl=1.

Galileo established his principle of free fall by rolling a ball down a gently tilted board and measuring its progress with his newly invented pendulum clock. He found that the distances traveled in equal time intervals stood in ratios 1:3:5:7:9: . . . . His conclusion follows since the sum of the distances (the total distance traveled in n time intervals) is proportional to the sum of the first n odd integers $(1 + 3 + 5 + \cdots (2n - 1) = \sum_{k=1}^{n}(2n - 1) = n^2)$, *that is, a multiple of* n^2.

The "constant" g varies with altitude and position on the earth, but at sea level it may be taken as 32.2 ft/sec^2 in the American (ft-lbs-sec) system or 9.8 m/sec^2 in the metric system.

for $0 \leq t < \infty$. The graph of this parametric system is a straight line,[†] provided at least one of v_x and v_y is not zero.

In Figure 3.78 we have plotted Equation (3.24) and (3.25) with the totally arbitrary assignments of $x_0 = -5$, $y_0 = 1$, $v_x = 10$ and $v_y = 15$.

Our cannoneer would be quite surprised to see the cannonball follow this path, leaving the cannon and heading off in a straight line without ever returning to earth.

FIGURE 3.78   The line $x(t) = -5 + 10t$, $y(t) = -1 + 15t$

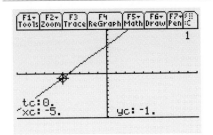

▶ THIRD REFLECTION

What is the slope of the line graphed in Figure 3.78?   ◀

Position, Velocity, and Gravity.   Adding the effects of gravity to our model requires an understanding of free fall that originally took Galileo nearly a decade to work out. He determined that the distance traveled by a body falling under the influence of gravity is proportional to the square of the time the body falls. The fraction $g/2$ is customarily (if somewhat mysteriously) used for this constant of proportionality. Gravity acts on the projectile only in the vertical direction; thus, we do not need to modify the horizontal component of our model in Equation (3.24). Since the effect of gravity is directed downward and acts to lessen the height (in fact gravity defines the notion "down") and g is customarily taken to be positive, we subtract the term for effect of gravity from the vertical component of Equation (3.25), obtaining

$$x(t) = x_0 + v_x t \qquad \textbf{(3.26)}$$
$$y(t) = y_0 + v_y t - \frac{g}{2}t^2$$

for $0 < t < \infty$. The system used in Figure 3.78, modified by subtracting $\frac{32.2}{2}t^2$ for the vertical component, is graphed in Figure 3.79. Here is a flight path our beleaguered cannoneer might recognize. Do you?

[†]This is a reflection of Newton's Second Law: Objects at rest or in uniform motion remain at rest or in motion unless acted upon by an outside force.

► **FOURTH REFLECTION**

What sort of curve does it look like? ◄

Now that we have developed this parametric approach to motion, we can apply it in many situations.

In the Y= *editor, enter* xt2=-5+10*t *and* yt2=yt1(t)-32.2/2*t^2

FIGURE 3.79 The graph $x(t) = -5 + 10t$, $y(t) = -1 + 15t - \frac{32.2}{2}t^2$ on $[-10, 10] \times [-10, 10]$

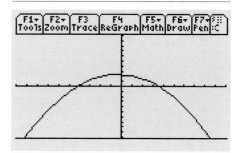

E X A M P L E 2 **The Path of an Arrow in Flight — The Arrow Example**

An archer releases an arrow from a height of four feet with an initial velocity of 150 ft/sec at an angle of 15° above the horizontal. Will the arrow strike a circular target three feet in diameter whose center is five feet above the ground 100 yards distant across a level field?

A Solution. First, we adapt our system of parametric equations to this situation. We take the feet of our archer as the origin of our coordinate system. (That seems as good a place as any, and puts ground level at height zero.) We orient the positive x-axis so that it points in the direction of the center of the target and assume that the archer aims the shot so that it travels in the coordinate plane, for the most part. Finally, we make sure all measurements are in feet and seconds.

With these conventions established, the initial position for the arrow is $x_0 = 0$ and $y_0 = 4$. As we can see from Figure 3.80, the initial horizontal and vertical components of the velocity are

$$v_x = 150\cos(15°) \text{ and } v_y = 150\sin(15°).$$

Your TI calculator can read angles in degrees as well as radian measure. Go to the MODE *dialogue box and set* Angle... *to* DEGREE.

Combining position, velocity, and gravity, our model becomes

$$x(t) = 0 + 150\cos(15°) \cdot t$$
$$y(t) = 4 + 150\sin(15°) \cdot t - \frac{32.2}{2}t^2.$$

FIGURE 3.80   Resolving Velocity into Horizontal and Vertical Components

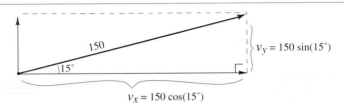

$v_y = 150 \sin(15°)$

$v_x = 150 \cos(15°)$

▶ **FIFTH REFLECTION**

Graph the flight of the arrow. According to your graph, will the arrow hit the target? Hint: Find the height of the arrow, $y(t)$, when it has traveled horizontally 300 feet (that is, $x(t) = 300$).   ◀

Parametric Equations Involving Circular Functions

A circle can be represented by parametric equations that involve circular functions.

Circles.   The *circular functions*, sine and cosine, can be defined as the vertical and horizontal components of points on the unit circle $x^2 + y^2 = 1$. If t is the length of the arc obtained by tracing in a counterclockwise direction from $(1, 0)$ along the circumference of the circle to (x, y) then $x = \cos(t)$ and $y = \sin(t)$ (see Figure 3.81). Thus, the unit circle can be described by the rectangular equation

$$x^2 + y^2 = 1 \tag{3.27}$$

or the parametric equations

$$x(t) = \cos(t) \tag{3.28}$$

$$y(t) = \sin(t) \tag{3.29}$$

Be sure to change your
MODE Angle... *to* RADIAN.

where $0 \le t \le 2\pi$.

▶ **SIXTH REFLECTION**

Complete each of the following:

1. Substitute the parametric expressions for x and y from Equations (3.28) and (3.29) into Equation (3.27) to obtain a well-known identity.
2. Graph the parametric Equations (3.28) and (3.29). Use WINDOW settings tmin=0,tmax=2π,tstep=0.1,xmin=-2,xmax =2,xscl=1,ymin=-2, ymax =2,yscl=1.
3. What does the graph look like?
4. Go to the Zoom menu and select ZoomSqr. Now what does the graph look like?
5. In the margin, record the parametric equations for a circle of radius two centered at the origin.

FIGURE 3.81 The unit circle with a central angle of measure t radians. The length of the arc in bold is subtended by the angle in t units.

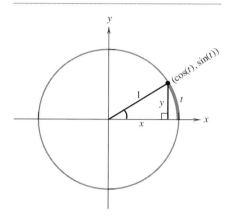

6. Confirm your equations by graphing them. ◀

Parametric Differentiation

A curve represented parametrically may have tangent lines and we can compute the slopes of these tangent lines. For example, given a parametric representation of the path of a cannonball in the plane, we can determine the direction of the cannonball (slope of the tangent line) at any time t.

E X A M P L E   3   **Slope: the Parametric Way**

We would like to find the slope of the tangent line to the circle of radius 4 at the point in time where $t = 0.7$ seconds. The parametric equations for our circle of radius 4 centered at the origin, are $x = 4\cos(t)$ and $y = 4\sin(t)$ (see the Sixth Reflection).

We can still view y as a function of x and we have explicit representations of both x and y as functions of t. The chain rule tells us that

$$\frac{dy}{dt} = \frac{dy}{dx} \cdot \frac{dx}{dt}.$$

Solving for $\dfrac{dy}{dx}$ leads to a method for differentiating functions with parametric representations called *parametric differentiation*.

$$\frac{dy}{dx} = \frac{\dfrac{dy}{dt}}{\dfrac{dx}{dt}}$$

In this equation, it appears that $\dfrac{dy}{dt}$ and $\dfrac{dx}{dt}$ are being treated as fractions by cancelling the dt from numerator and denominator to obtain $\dfrac{dy}{dx}$. This is *not* the case (see the discussion of the chain rule in Section 3.3). On the other hand, this is an easy way to remember the equation.

PARAMETRIC DIFFERENTIATION

Given parametric equations $x = f(t)$ and $y = g(t)$, compute $\dfrac{dx}{dt}$ and $\dfrac{dy}{dt}$. Form the ratio $\dfrac{dy}{dt}$ divided by $\dfrac{dx}{dt}$ to obtain $\dfrac{dy}{dx}$.

Let's try this technique on our example. Since $x = 4\cos(t)$ and $y = 4\sin(t)$, we find $\dfrac{dy}{dt} = 4\cos(t)$ and $\dfrac{dx}{dt} = -4\sin(t)$ so that

$$\frac{dy}{dx} = \frac{dy/dt}{dx/dt} = \frac{4\cos(t)}{-4\sin(t)} = -\cot(t).$$

We can compute the slope of our tangent line directly by substituting 0.722734 for t.

▶ **Seventh Reflection**

Explain how we know that the point $(3, \sqrt{7})$ corresponds to the t-value of 0.722734. What point on the circle corresponds to the t-value zero? Is there a tangent line when $t = 0$? Can you compute $\frac{dy}{dx}$ when $t = 0$? ◀

E X A M P L E   4   **Finding the Slope of Parabolic Curve Parametrically**

Find the slope of the tangent line to the curve given by the parametric equations

$$x(t) = 2t + 1 \tag{3.30}$$
$$y(t) = t^2 - 3t + 5 \tag{3.31}$$

at $t = 4$.

Since

$$\frac{dy}{dx} = \frac{\frac{dy}{dt}}{\frac{dx}{dt}},$$

Differentiating $y(t)$ and $x(t)$ with respect to t.

we have

$$\frac{dy}{dx} = \frac{\frac{dy}{dt}}{\frac{dx}{dt}} = \frac{2t - 3}{2}$$

so that

$$\frac{dy}{dx}\bigg|_{t=4} = \frac{5}{2}.$$

To check our result, we can write y as a function of x and differentiate explicitly. Since $x = 2t + 1$, we see that $t = (x - 1)/2$. Evaluating $y(t)$ for this t value in Equation (3.31), we have

$$y = \left(\frac{x-1}{2}\right)^2 - 3\left(\frac{x-1}{2}\right) + 5$$

and

$$y = \frac{x^2}{4} - 2x + \frac{27}{4}$$

which we recognize as an equation for a parabola.

Therefore,

$$y'(x) = \frac{x}{2} - 2.$$

Since $x = 9$ when $t = 4$, we evaluate $y'(x)$ for $x = 9$ and have

$$y'(9) = \frac{5}{2}.$$

We have verified that the slope of the tangent line to the parametric curve given by Equation (3.30) at the point where $t = 4$ is $5/2$.

Summary

In this section we have learned about the parametric representation of curves and about parametric differentiation. We have also learned about finding tangent lines to a curve represented by parametric equations using parametric differentiation.

EXERCISE SET 3.8

Exercises 1–4 refer to the curve described by the parametric equations

$$x(t) = 1 + 3\cos(t) \text{ and } y(t) = \sin(t).$$

1. Write down a description of the curve.

2. Find the points on the curve and the corresponding values of t where the tangent line is horizontal.

3. Are there any points on the curve where the tangent line is vertical? Justify.

4. Find the values of t for which x is increasing while y is decreasing. How do the values of dx/dt and dy/dt support your conclusion?

Exercises 5–7 refer to the model for the flight of the arrow in Example 2:

$$x(t) = 0 + 150\cos(15°t)$$
$$y(t) = 4 + 150\sin(15°t) - \frac{32.2}{2}t^2.$$

It seemed natural to express the angle in degrees when we first looked at this model. In order to take derivatives, however, we need to convert degree measure to radian measure. Since $30°$ corresponds to $\frac{\pi}{6}$ radians, we see immediately that $15°$ corresponds to $\frac{\pi}{12}$. For more difficult conversions, use your calculator.

5. Determine, from a graph, how high the arrow can get.

6. Use calculus to determine how high the arrow can get. Hint: what is the slope of the tangent line to the curve when it is at its highest point?

7. When is the arrow traveling the fastest? Explain, showing your reasoning and your computations.

In Exercises 8–11, explain, in your own words, the meaning of each of the following terms. Write your explanations in clear and complete sentences. Include annotated diagrams whenever appropriate.

8. parameter

9. parametric equations

10. speed

11. parametric differentiation

12. Give a system of parametric equations (and restriction of the parameter) that describes the line segment from $(-2, 5)$ to $(4, -4)$.

13. Eliminate the parameter t to find the point slope form of the line described by the parametric equations $\{x(t) = 2 + 15t, y(t) = 14 - 35t\}$.

14. Give two different systems of parametric equations that describe the line $y = \frac{3}{7}x + 5$.

15. Find the point of intersection of the lines $\{x(t) = 5 + t, y(t) = 4 - 7t\}$ and $\{x(t) = 4 + 2t, y(t) = 5 - 3t\}$. Check your work graphically on your calculator.

16. Show that the system of equations

$$x(t) = x_0 + v_x \cdot t$$
$$y(t) = y_0 + v_y \cdot t$$

is a straight line by giving the $y = mx + b$ equation of the same line. Explain the computation in a way that another student might be willing to read.

17. Show that the cannonball model predicts that the path of a projectile is a parabola by modifying the system

$$x(t) = x_0 + v_x t$$
$$y(t) = y_0 + v_y t - \frac{g}{2} t^2$$

so that it is in the form $y = ax^2 + bx + c$. Give the values of a, b and c in terms of x_0, y_0, v_x, v_y and g.

Exercises 18–25 refer to the Arrow Example.

18. At what time after the arrow in Example 2 is released will the arrow be five feet above the ground? How far back would the target in the example need to have been in order for the arrow to strike the center?

19. What is the maximum height attained by the arrow? At what time is this height achieved?

20. Assuming an elevation of $15°$ in the archer's aim for the arrow, what initial velocity would result in the arrow striking the center of the target?

21. How strong a head wind is needed in order for the arrow to hit the center of the target?

22. Assuming an initial velocity of 150 ft/sec for the arrow, at what angle should the archer aim in order to hit the center of the target? (There are two answers to this question.)

23. Assuming an elevation of $15°$ in the archer's aim for the arrow in the example, what range of velocities allows the archer to hit the target at all?

24. Assuming an initial velocity of 150 ft/sec for the arrow in the example, what range of angles allows the archer to hit the target at all?

25. If the archer is shooting at a target 75 yards away, find an angle of elevation and initial velocity that will allow the

arrow to strike the center of the target. Reasonable velocities range from 50 ft/sec to 175 ft/sec.

26. Give a system of parametric equations for the following circles.
    a. The circle of radius three centered at the origin.
    b. The circle of radius three centered at (0, 4).
    c. The circle of radius five centered at (−3, 4).
    d. The circle whose Cartesian relation is

    $$(x + 7)^2 + (y - 1)^2 = 81.$$

27. Give a system of parametric equations for the circle whose Cartesian relation is

    $$(x - a)^2 + (y - b)^2 = r^2.$$

28. a. Graph and describe the system of parametric equations $\{x(t) = 3\cos(t), y(t) = 4\sin(t)\}$ for $0 \le t \le 2\pi$.
    b. Use the trigonometric identity $\cos^2(\theta) + \sin^2(\theta) = 1$ to obtain the Cartesian relation (an equation in x and y) for this figure.

29. a. Graph and describe the system of parametric equations $\{x(t) = 7\cos(t) + 8, y(t) = 5\sin(t) - 3\}$ for $0 \le t \le 2\pi$.
    b. Use the trigonometric identity $\cos^2(\theta) + \sin^2(\theta) = 1$ to obtain the Cartesian relation (an equation in x and y) for this figure.

†30. Use the trigonometric identity $\cos^2(\theta) + \sin^2(\theta) = 1$ to obtain a Cartesian equation for the figure described by the parametric equations

    $$x(t) = a \cdot \cos(t) + c$$
    $$y(t) = b \cdot \sin(t) + d$$

    for $0 \le t \le 2\pi$. Explain your computations in a form that another student might be willing to read.

31. Use the trigonometric identity $\tan^2(\theta) + 1 = \sec^2(\theta)$ to obtain a Cartesian equation for the figure described by the parametric equations:

    $$x(t) = a \cdot \tan(t) + c$$
    $$y(t) = b \cdot \sec(t) + d$$

    for $0 \le t \le 2\pi$. Explain your computations in a form a fellow student might be willing to read.

3.9

Partial Derivatives

Functions of a Single Variable

So far, we have considered only functions of one variable, most often of the form $y = f(x)$. We interpreted the derivative, $f'(x)$, as the rate of change of the output variable with respect to the input variable. The graph of $y = f(x)$ was some sort of curve in the xy-plane and $f'(x)$ could also be interpreted as the slope of the tangent line to the curve $y = f(x)$ at any point $(x, f(x))$ on the curve.

The derivative tells us how y changes as x increases. If $f'(2) = -6$, we know that when $x = 2$, y decreases six units for each one unit increase in x. If we were to draw a tangent line to the curve at $x = 2$, we would see that the slope of this line is -6. (See Figure 3.82.)

FIGURE 3.82 Derivatives as Slopes of Tangent Lines

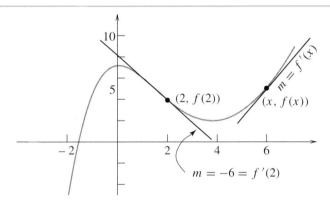

FIGURE 3.83 The Tangent Line to the Unit Circle at $t = \pi/4$

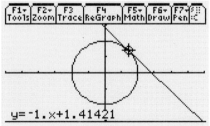

We have also considered functions defined parametrically, where x and y were both defined in terms of a third variable or parameter. When we sketched the graphs, we still obtained a curve in the xy-plane, and we could still find $\frac{dy}{dx}$ although the method of differentiation was different (recall parametric differentiation from Section 3.6). The graph in Figure 3.83 shows the tangent line drawn to the graph of a circle when $t = \pi/4$.

Functions of More Than One Variable

It is quite reasonable to expect that a quantity might depend upon more than one variable. The area of a triangle depends upon the length of its base as well as its height. The volume of a gas is dependent upon both pressure and temperature. In this section we will consider how a function of more than one variable changes as the variables change.

We turn our attention now to functions of two variables whose graphs are surfaces in three-space.

Just as a plane is divided into four quadrants by the coordinate axes, three-dimensional space is divided into eight octants by the three coordinate axes. Figure 3.84 shows a point in three-space.

The equation of the xy-plane is $z = 0$. Similarly, the equations of the xz- and yz-planes are $y = 0$ and $x = 0$, respectively.

▶ FIRST REFLECTION

What plane is represented by the equation $x = 2$? What is the equation of the plane parallel to the xz-plane that passes through the point $(1, -3, 4)$?   ◀

FIGURE 3.84   The Point $(x, y, z) = (4, 2, 3)$ in Three-Space

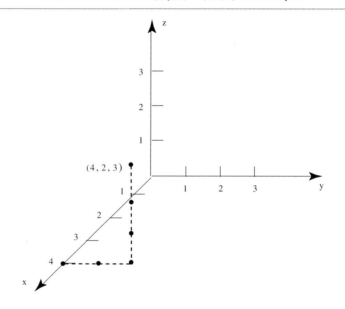

EXAMPLE 1   **The Sphere Example**

One of the simplest surfaces to visualize is the top half of a sphere. The equation for a sphere centered at the origin with radius two is

$$x^2 + y^2 + z^2 = 4.$$

Solving for z yields

$$z = \pm\sqrt{4 - x^2 - y^2}.$$

If we consider only the top half of the sphere, we see that z is a function of the two variables x and y. We use the notation $z = f(x, y)$ to indicate this functional relationship.

CREATING A 3D GRAPH

To obtain the graph in Figure 3.85, set Graph in the MODE dialogue box to 3D. Go to the Y= editor and enter $z1 = \sqrt{4 - x^2 - y^2}$. While in the Y= editor, choose Format in the Tools menu and set:

Coordinates...	to	RECT
Axes...	to	AXES
Labels...	to	ON
Style...	to	HIDDEN SURFACE

Press GRAPH to graph the function.

We note that the domain of f is $\{(x, y) | x^2 + y^2 \le 4\}$ or the set of points (x, y) such that $x^2 + y^2 \le 4$. The domain of a function of two variables is a region in the plane.

▶ SECOND REFLECTION

What is the range of f? ◀

FIGURE 3.85 The Top Half of a Sphere on $[-2, 2] \times [-2, 2] \times [0, 2]$

FIGURE 3.86 Graph Window with zmin=0 and zmax=2

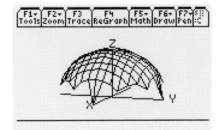
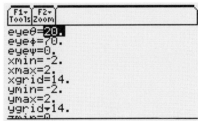

Observe the graph in Figure 3.85. Imagine a series of vertical planes all parallel to the yz-plane. These planes cut the surface in the parallel curves that look rather like parts of semicircles, opening downward from left to right across the screen. The planes are obtained by keeping x fixed or constant.

▶ THIRD REFLECTION

What planes produce the other set of parallel curves in Figure 3.85? ◀

Consider the surface $z = f(x, y)$ in Figure 3.87. The point $Q(a, b, f(a, b))$ on the surface corresponds to the point $P(a, b)$ in the xy-plane. We can no longer ask ourselves the question, "What is the slope of a tangent line to the surface at the point Q," because we see that there are any number of directions our tangent line might take.

FIGURE 3.87   Graph of $z = f(x, y)$ near (a, b) **FIGURE 3.88**   The Graph with Tangents at Q

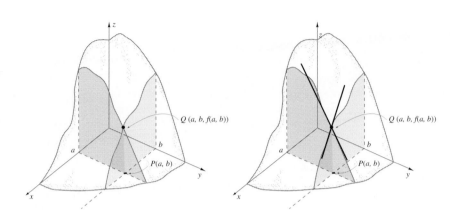

We restrict our attention to the two particular tangent lines, one in a plane parallel to the yz-plane and one in a plane parallel to the xz-plane. (See Figures 3.87 and 3.88.) When we consider the tangent line through the point

FIGURE 3.89   Derivative in the y Direction at   **FIGURE 3.90**   Derivative in the x Direction at
$y = b$ $x = a$

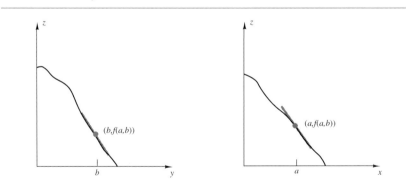

$(a, b, f(a, b))$ in the yz-plane (the x "direction"), we hold y fixed and differentiate with respect to x.

For example, if

$$z = f(x, y) = 5x^2 y^3,$$

the derivative of z with respect to x when $y = 5$ is actually the derivative of

$$z = f(x, 5) = g(x) = 5x^2 5^3$$

with respect to x. When $y = 2$, the derivative of z with respect to x is the derivative of

$$z = f(x, 2) = h(x) = 5x^2 2^3$$

with respect to x. In each case, we turn a function of two variables into a function of one variable. Thus, we can take the derivative with respect to x as if y were fixed (treating y as if it were a constant). Then the derivative of z with respect to x is $10xy^3$. On the other hand, if we hold x constant in the same way and differentiate z with respect to y, the derivative of z with respect to y is $5x^2(3y^2) = 15x^2y^2$.

These special rates of change in the positive x and y directions are called *partial derivatives* and special symbols are used to designate them.

PARTIAL DERIVATIVES

If $z = f(x, y)$, the partial derivative of z with respect to x is designated by

$$f_x(x, y) \text{ or } \frac{\partial z}{\partial x}.$$

Similarly, the partial derivative of z with respect to y is designated by

$$f_y(x, y) \text{ or } \frac{\partial z}{\partial y}.$$

▶ FOURTH REFLECTION

When we found $f_x(x, y)$ for $f(x, y) = 5x^2y^3$, why didn't we use the product rule? ◀

FINDING PARTIAL DERIVATIVES

To find partial derivatives on your TI calculator, define $f(x, y)$ and then use the differentiation operator, $d($. You must specify the variable of differentiation. Thus, $\dfrac{\partial f(x, y)}{\partial y}$ is entered as $d(f(x,y),y)$.

Fortunately, all the differentiation rules of Sections 3.2 and 3.3 still hold. There are no new rules, but we must remember when to apply the old ones.

Think and Share

Working in pairs, let's play *Can you beat the machine?* For each of the following functions, one player should compute the partial derivatives using the TI calculator and the other player should compute them by hand using the differentiation rules in Sections 3.2 and 3.3. The first player done wins.

$$z = \sin(x + y) \qquad z = e^{xy}$$

$$z = \sin(xy) \qquad z = x\sin(y)$$

$$z = \frac{x}{y} \qquad z = e^{xy}\sin(xy)$$

E X A M P L E   2   **A Bug on the Sphere**

Suppose a bug is located on a sphere of radius two directly above the point $(1, \frac{1}{2})$. How fast does the bug move if it travels in the direction of the positive y-axis? We see in Figure 3.91 the portion of the sphere in octant one. We pass a vertical plane parallel to the y-axis through the surface at the point $(1, \frac{1}{2}, \frac{\sqrt{11}}{2})$ (see Figure 3.92).

Recall from Example 1 that $f(x, y) = \sqrt{4 - x^2 - y^2}$. Keeping x constant, we calculate

$$f_y(x, y) = \frac{-y}{\sqrt{-(x^2 + y^2 - 4)}} = \frac{-y}{\sqrt{4 - x^2 - y^2}}.$$

FIGURE 3.91   The Sphere of Radius 2 with the Point Above $(1,1/2)$

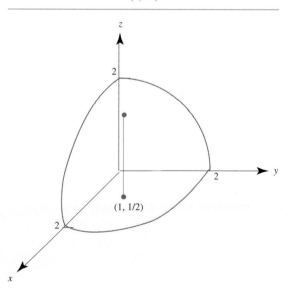

FIGURE 3.92   A Plane through the Point at $(1,1/2,\sqrt{11}/2)$

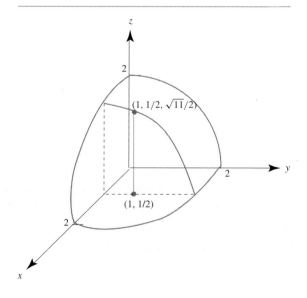

We then edit the command line, adding the "with" operator to obtain $f_y(1, \frac{1}{2}) = \dfrac{-\sqrt{11}}{11} \approx -0.30$. Thus, the bug is moving at a rate of 0.30 units in the negative y-axis direction.

▶ **FIFTH REFLECTION**

Is the bug ascending or descending? Justify your answer.   ◀

EXERCISE SET 3.9

In Exercises 1–6, find $\dfrac{\partial z}{\partial x}$ and $\dfrac{\partial z}{\partial y}$.

1. $z = 3x^2 - 7xy + y^3$.
2. $z = \cos(3x - 5y)$.
3. $z = x\sqrt{y}$.
4. $z = \dfrac{\sqrt{y}}{x}$.
5. $z = \sin(x^2 + y^2)$.
6. $z = \dfrac{xy}{x^2 + y^2}$.

7. If $z = \ln(e^x + e^y)$, show that $\dfrac{\partial z}{\partial x} + \dfrac{\partial z}{\partial y} = 1$.

8. If $z = x^5 y - \sin(y)$, find $\dfrac{\partial^2 z}{\partial x \partial y} = \dfrac{\partial}{\partial x}\left(\dfrac{\partial z}{\partial y}\right)$. (Note: First differentiate with respect to y, and then with respect to x.) Is $\dfrac{\partial}{\partial x}\left(\dfrac{\partial z}{\partial y}\right) = \dfrac{\partial}{\partial y}\left(\dfrac{\partial z}{\partial x}\right)$?

9. $V = k\dfrac{T}{P}$ where V represents the volume of gas, P is the pressure on it, and T is the temperature. Find and interpret $\dfrac{\partial V}{\partial T}$ and $\dfrac{\partial V}{\partial P}$. ($k$ is a constant.)

10. Consider the bug in Example 2. How fast is it moving in the direction of the positive x-axis when it is located at the point $(1, 1/2, \sqrt{11}/2)$?

11. Would it be possible to have a surface so that a bug would ascend in the positive x direction but descend in the positive y direction? Explain.

Summary

This chapter is about the derivative and how to use the derivative as a tool in measuring and describing change in continuous functions. We have used the TI calculator to perform the symbolic calculations (that is, do the measuring) for us, and we have therefore had more time to spend on the concepts.

In Chapter 2, we developed the notion that the instantaneous rate of change of a function at a point was equivalent to the slope of the unique tangent line at that point. In Chapter 3, we introduced the limiting process in order to develop the formal definition of the derivative as a refinement of the notion of instantaneous change. We developed rules for finding symbolic expressions for derivatives of functions given explicitly, implicitly, and parametrically. Then we extended the notion of derivative to functions of two variables (for example, $f(x, y) = \sin(xy) + 2x^2 y^3$).

We have gained a good conceptual understanding of the derivative as a function which measures change. We have also developed the differentiation concepts and used these concepts and the skills needed to find derivatives in investigating the complex behavior of real-world phenomena.

SUPPLEMENTARY EXERCISES

1. Use complete sentences to explain the concept of continuity. Provide original examples of functions that are continuous and functions that are not.

2. Explain the derivation of the definition of the derivative in terms of limits.

3. We have frequently mentioned the terms "smooth" and "continuous" in this chapter. Explain what this means in complete sentences. Provide original examples of functions that are smooth and functions that are not.

4. Agree or disagree with the following statements; support your position with a well-constructed argument. "Differentiability implies continuity; if a function is differentiable everywhere on its domain, then it must be continuous everywhere. Likewise, continuity implies differentiability; if a function is continuous everywhere on its domain, then it must be differentiable everywhere."

5. Use proper notation to write out a formal statement of the Mean Value Theorem. Explain what the theorem means, and provide original examples and nonexamples. Illustrate.

Exercises 6–11 refer to the graphs in Figure 3.93 and Figure 3.94. Explain all answers to the following exercises.

FIGURE 3.93

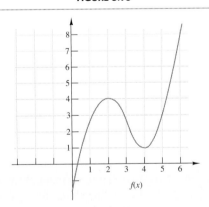

f(x)

6. Estimate the average rate of change of f over the interval [1,6].

7. Estimate the instantaneous rate of change of f at $x = 3$.

8. Estimate the slope of the secant line through the points $(1, g(1))$ and $(5, g(5))$.

FIGURE 3.94

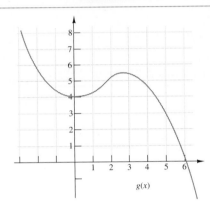

g(x)

9. Estimate the slope of the tangent line to the graph of g at $x = 4$.

10. Let $h(x) = f(x) \cdot g(x)$. Estimate $h'(3)$. Is h increasing or decreasing at $x = 3$?

11. Let $j(x) = f(g(x))$. Estimate $j'(3)$. Is j increasing at $x = 3$?

12. Analyze the function $f(x) = (x^2 - 2x - 3)\left(\frac{3}{4}\right)^x$. In particular,

    a. Identify all roots of f.

    b. Give both coordinates of all minimums and maximums of f.

    c. Give both coordinates of all inflection points of f, and record the slope of f at each inflection point.

13. Find a cubic polynomial with a relative maximum at $(-1/6, 5)$ and a relative minimum at $x = 4$.

14. Write an equation of the tangent line to the graph of

$$y = \frac{100 \sin(x)}{x^2 + 1}$$

at its point of inflection.

15. The parametric equations

$$y = \sin(3t)$$

and

$$x = \sin(2t)$$

describe a curve known as a Lissajous curve. Find the rate of change of y, with respect to x, at any time t. Is the tangent line to the curve ever horizontal? If so, where?

16. The graph of a function and its derivative is shown in Figure 3.95. Which is the graph of the derivative function? Explain why.

FIGURE 3.95 A Function and its Derivative

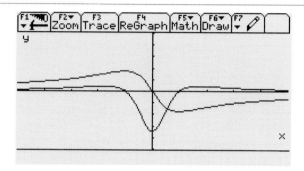

Exercises 17–19 refer to the function:

$$j(x) = \begin{cases} -4x & \text{if } x < 1 \\ x^2 + 1 & \text{if } x \geq 1 \end{cases}$$

17. Use the Y= editor to enter

$$\text{when}(x)=1, x^2 + 1, -4x)$$

as

$$y1(x)$$

and graph the function. Does the calculator draw the graph correctly? Explain.

18. Find $\lim\limits_{x \to 1^-} j(x)$ in three ways: numerically, graphically, and symbolically. Explain your process.

19. Find $\lim\limits_{x \to 1^+} j(x)$ in three ways: numerically, graphically, and symbolically. Explain your process.

20. The Orange Freight Company pays its long-haul drivers $21.75 per hour while they are on the road. The fuel costs for a fully loaded rig traveling at 45 miles per hour are $0.47 per mile. For each one-mile-per-hour increase in speed, there is a 1.5% increase in fuel costs. At what speed should the company instruct the drivers to travel so as to minimize their total cost per mile? Should the company's speed policy change for truckers that are deadheading (traveling without a load)?

21. An offshore docking facility for supertankers is being constructed two miles off the coast of Santa Teresa. Unfortunately, due to the presence of some environmentally sensitive wetlands, the docking facility cannot be located directly opposite the oil refinery it services. The refinery, as shown in Figure 3.96, is five miles east of the docking facility. Consequently, you must design a pipeline that goes from the station to the shore and then along the shore to the refinery. The pipeline will require a pumping station. The pumping station is to be placed where the underwater

FIGURE 3.96 The Santa Teresa Coastline

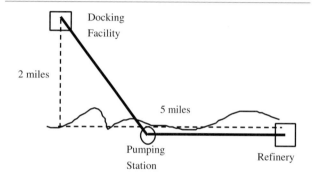

portion of the pipeline reaches the shore. You are, as yet, unable to get reliable estimates for the costs of building a pipeline under water or through the marsh, However, the best available information indicates the per-foot cost of building a pipeline underwater is 1.5 times the per-foot cost of building a pipeline through the marsh.

a. Where should you position the pumping station? Explain your reasoning and justify your answer.

b. A recent seismic survey discovers a previously unknown branch of the San Andreas fault that the underwater portion of the pipeline will have to cross. This will double the cost of the underwater portion of the pipeline. How much does this change your plans?

22. John Smith is responsible for periodically buying new trucks to replace older trucks in his company's fleet of vehicles. He is expected to determine the time a truck should be retained so as to minimize the average cost of owning the truck. Assume the purchase of a new truck is $9000. Also, assume the maintenance cost (in dollars) per truck for t years can be expressed analytically by the following empirical model:

$$C(t) = 640 + 180(t + 1)t$$

where t is the time in years that the company owns the truck.

a. Determine $E(t)$, the total cost function for a single truck retained for a period of t years.

b. Determine $E_A(t)$, the *average* annual cost function for a single truck that is kept in the fleet for t years.

c. Graphically depict $E_A(t)$ as a function of t. Justify the shape of your graph.

d. Analytically determine t^*, the optimal period that a truck should be retained in the fleet. Remember John Smith's objective is to minimize the average cost of owning a truck.

e. Suppose we have to round t^* to the nearest whole year. In general, would it be better to round up or round down? Justify your answer.

Project One: Flying off on a Tangent—Again!

This project is offered in a spirit of fun and exploration. The only prerequisites for success are an active imagination and the ability to interpret the equation of a tangent line.

Luke Skywalker is in trouble on the outskirts of the Delta Epsilon system where his spaceship has been caught in an elliptical orbit by an Empire tractor beam. R2-D2 has determined that his flight path under the influence of the beam is given by the parametric equations: $x(t) = \cos(t)$, $y(t) = 2\sin(t)$ where t is time in minutes. The little droid has also located the nearest rebel base at coordinates $(0, -4)$ on the planet Ebeohp.

Luke's only hope is to use the power of the FORCE to cancel all artificial power in the entire sector. The young Jedi can call on the force in a plane parallel to shut off the tractor beam but it will simultaneously shut off his ship's engines. He must invoke his power at the exact instant when his flight along the resultant tangent line will take him to safety at the rebel base.

1. When should Luke shut down the tractor beam?

2. Find an equation for the path he will follow after that time.

3. Exerting this level of power is extremely fatiguing and Luke can only keep it up for three minutes. Will he be able to keep the tractor beam shut down until he reaches the rebel base?

Project Report. Write a report on your investigations. Address your report to your classmates, assuming the material is new to them. Describe your reasoning and include all necessary background information. A minimal project report must include:

1. A description of Luke Skywalker's escape attempt.

2. Written responses to the three items above.

3. Justification and explanations for your calculations.

4. Suggestions for possible extensions of your observations and further questions for exploration.

Project Two: Refraction

You may have noticed that when you see an object partially submerged in water, it appears to be bent. This problem is known as refraction. Our understanding of refraction is based on a principle due to Pierre Fermat, a famous seventeenth

FIGURE 3.97   Where is the Coin?

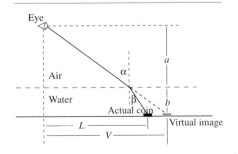

century mathematician. (Fermat was actually a lawyer who enjoyed mathematics as a hobby.) According to Fermat's principle, when light travels through one or more homogeneous media, say air or water, it follows the path that requires the least amount of total time.

In this project you will study refraction experimentally and then use calculus to discover a result called Snell's Law. The materials you need for this project are a container of water (preferably with a clear bottom like a glass baking dish), a common copper penny, and a ruler.

Experiment. Set up your experiment as shown in Figure 3.98. $h = 14$ inches, $b = 2$ inches, and $L = 13$ inches. You are to measure the distance between the two images of the penny (dimension $V - L$ below). Your project partner can use a ruler to make sure your eye is observing the images at the correct distances.

FIGURE 3.98 The Experimental Setup

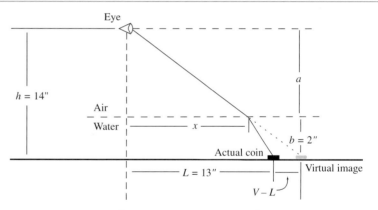

$v_{air} = 11.81 \times 10^9$ in/sec and $v_{water} = 8.89 \times 10^9$ in/sec

FIGURE 3.99 Snell's Law

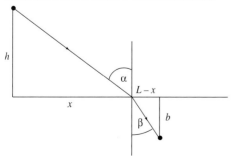

2. To mathematically represent Fermat's principle, we first need to find an expression for the total time light takes to travel from the coin to the eye. You should be familiar with the relationship *velocity = distance/time* or *time = distance/velocity*. Use these relationships to obtain *total time* as a function of x (see Figure 3.98). Write time as a function of h, b, x, L, v_1, and v_2, instead of the numbers from your data. Find the minimum value for your function using the values of h, b, x, L, v_1, and v_2 from your collected data.

3. Now for Snell's Law: in Figure 3.99 we use α and β to represent the angles made by the light rays as they travel from the coin to the eye. From Figure 3.99, write an expression for $\sin(\alpha)$ and $\sin(\beta)$ in terms of h, a, b, x, and L. Use these expressions and your first derivative set equal to zero to come up with a simple expression of Snell's Law.

4. Finally check your measured data to see how closely your results obey Snell's Law.

Project Report.   Write a report on your investigations. Address your report to your classmates, assuming the material is new to them. Describe your reasoning and include all necessary background information. A minimal project report must include:

1. A narrative of your experiment and a record of the data recovered for each of your and your partner's measurements. Be sure to record any difficulties you encountered.
2. Justification and explanations for your calculations and derivations in items 2 and 3 above.
3. A discussion of the possible sources of error that might account for any disagreement in your results in item 4 above.
4. Suggestions for possible extensions of your observations and further questions for exploration.

Project Three: Surveying

You are an engineer working for the Peace Corps in a country in Central America. Your group has just finished building a flight strip that will be used to bring supplies to a village. There is another village 100 km away that is inaccessible by vehicle from the flight strip (see Figure 3.100).

FIGURE 3.100   An Air Strip

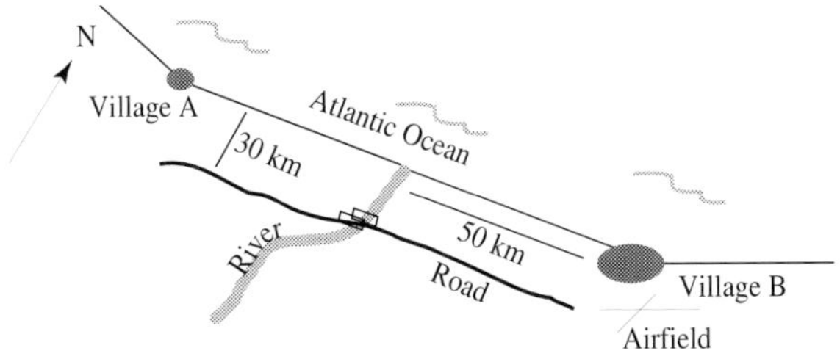

Your group has been asked to build a road connecting the two villages. Currently, there is an unimproved dirt road 30 km south of both villages, which has the only crossing of a river separating the two villages. The terrain north of that bridge is a riverbed that is swampy and impassable by vehicle. You want to build the most cost-effective road connecting the two villages.

1. After surveying the road and the terrain between the two villages, you estimate that the cost of materials and equipment to improve the existing road is $150K/km, whereas to construct a new road, the cost is $250K/km. Determine the most cost-effective route by which to connect the two villages with a road.

2. After a period of heavy rainfall, you determine that the terrain up to seven km to the west of village with the flight strip has become much more difficult to clear due to soil conditions. You estimate that it will now cost $1200K/km to build the road through this terrain. Based on this new information, determine the best route and estimate the cost of this road.

Project Report. Write a report on your investigations. Address your report to your classmates, assuming the material is new to them. Describe your reasoning and include all necessary background information. A minimal project report must include:

1. Discussion and computations to determine the most cost effective route in the first scenario above.
2. Discussion and computations to determine the most cost effective route in the second scenario above.
3. A policy recommendation as to which route to use. You must include justification for your discussion and may suggest ways to obtain additional information on which the decision may be based.
4. Suggestions for possible extensions of your observations and further questions for exploration.

4

THE DEFINITE INTEGRAL: ACCUMULATING CHANGE

4.1

Rate and Distance

4.2

Sums of Products

4.3

Error Bounds for the Left and Right
Endpoint Methods

4.4

The Definite Integral

4.5

Other Methods and Their Error Bounds

4.6

Applications

4.7

The Fundamental Theorem of Calculus,
Part I

Summary

We can estimate populations, distance, area, and other quantities that arise in fields such as ecology, business, and engineering when we know the rates at which such quantities are accumulating over time. In this chapter, we will develop methods for estimating the total accumulation of such quantities from their rates of accumulation. Our technique for computing estimates of total accumulation involves sums of products. We will then consider methods of finding exact values for such total accumulations.

4.1

Rate and Distance

Previously, we investigated how to determine velocities at various points in a journey using tables of distance versus time or formulas. In this section, we will study the reverse problem of determining the distance traveled over a fixed time period knowing the velocity at various points during that period.

FIGURE 4.1   Route of Third Stage

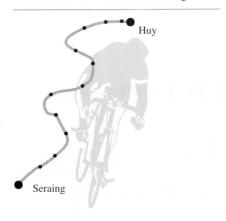

Huy

Seraing

TABLE 4.1   Third Stage Velocity Data

Time (min)	Rate (km/h)
2	36
31	32
74	40
173	41
224	35
301	20
387	26
452	35
507	35
591	33
637	21
702	29

The Tour de France

Every year hundreds of athletes gather from around the globe for the most important bicycle race in the world, the Tour de France. The race is run over a 23 or 24-day period and each year the course is changed. How would today's racers do against those of the past? Is the Tour as long today as it once was or has it slowly shrunk over time? How does the current Tour compare to the one run in 1930? These questions are harder to answer than one might think. The course does not go on standard roads, and some of the roads used in the past have been destroyed by time, urban development, and war. Data about the most recent races can be found in newspapers or on the Internet, but how do we find corresponding information for pre-war races?

The only data we do have from those early races are the times and rates at which each racer passed through the various villages along the course. Let's focus on the third stage of the 1930 race (each day's race is called a stage) as shown in Figure 4.1. This stage began in the center of Huy and passed through 12 villages, ending in the center of Seraing. The times and rates for one racer on this stage are given in Table 4.1.

Race officials recorded the times (in minutes) and the rates (in kilometers per hour) as the riders passed by their stations at the village limits with the last official being posted at the finish in the center of Seraing. How can we determine the length of the stage? Keep in mind that just because it may have taken the riders longer to complete the third stage in 1930 than it took them in 1996, that is no reason to believe they were slower riders. Perhaps the stage was longer.

The length of the stage can be determined if we know the length of each leg (portion of the course between two checkpoints). The length of a leg between two checkpoints equals the racer's average speed on the leg multiplied by the elapsed time for the leg. We can obtain elapsed times from Table 4.1. These elapsed times have been converted from minutes to hours in Table 4.2.

TABLE 4.2   Elapsed Time in Hours

Elapsed Time (h)	0.03	0.48	0.72	1.65		1.28	1.43	1.08	0.92	1.40	0.77	1.08
Rate (km/h)	36	32	40	41	35	20	26	35	35	33	21	29

▶ FIRST REFLECTION

How did we obtain the elapsed time entries in Table 4.2? Fill in the missing time in the table.   ◀

The average rate of our racer over one leg of the stage equals the length of the leg divided by the elapsed time for the leg. Thus, if we know the racer's average rate, we can compute the length of the leg exactly with the formula:

$$distance = average\ rate \times time$$

since we know

$$rate \times time = \frac{miles}{hour} \times hours = miles \text{ (a distance)}.$$

Rate on Each Leg.   Unfortunately, we do not know the racer's average rate over any of the legs. Instead, we know the racer's instantaneous rate at each checkpoint. How can we use these instantaneous rates to estimate the average rate over each leg?

Look Back Approach.   As a first try, let's use an approach where we assume that the racer's speed at any check point equals the average speed over the previous leg. We'll call this the *right endpoint method*. Using this assumption, we obtain our estimates of the distances between villages by multiplying the numbers in the second row of our table (speed at check point) by the numbers in the first row (time on previous leg). For example, we estimate the distance from the center of Huy to the first offical checkpoint to be $36 \times 0.03 = 1.08$ kilometers. Summing our distances, we obtain an approximation for the total length of the third stage:

$$(36 \times 0.03) + (32 \times 0.48) + (40 \times 0.72) + (41 \times 1.65) +$$
$$(35 \times 0.85) + (20 \times 1.28) + (26 \times 1.43) + (35 \times 1.08) +$$
$$(35 \times 0.92) + (33 \times 1.40) + (21 \times 0.77) + (29 \times 1.08)$$
$$= 369.11.$$

Look Ahead Approach.   There are several other ways to estimate the average rate for each leg. We could use what we will call the *left endpoint method* by assuming that the rate recorded at any check point will be the racer's average rate over the upcoming leg. For example, the estimate for the last leg is the product of the rate at the next-to-last checkpoint and the elapsed time on the last leg

$$21 \times 1.08 = 31.32 \text{ kilometers}$$

This method requires that we come up with a rate for the first two minutes as the racer goes from the center of Huy to its outer limits. If we assume that the stage got underway slowly with many racers, an estimated rate of 10 k/h might be appropriate. On the other hand, the racer might have been at the front of the pack, in which case we would do better to assume a speed of 34 or 35 k/h. Thus, the racer might have gone anywhere between one-third of a kilometer and 1.2 kilometers in the first two minutes of the stage. Let's assume our racer is ahead of the pack and compute another approximation for the total length of the third stage:

$$(35 \times 0.03) + (36 \times 0.48) + (32 \times 0.72) + (40 \times 1.65) + (41 \times 0.85) +$$
$$(35 \times 1.28) + (20 \times 1.43) + (26 \times 1.08) + (35 \times 0.92) +$$
$$(35 \times 1.40) + (33 \times 0.77) + (21 \times 1.08) = 372.99.$$

▶ **SECOND REFLECTION**

Explain how we determined the individual products in the Look Ahead (left endpoint) method. What is the basic difference between the left endpoint and right endpoint methods? What remains the same in each method?   ◀

An Averaging Approach.   We can obtain a third estimate by averaging the rates at the two ends of a leg as an estimate of the racer's average rate over the entire leg. We will call this the *averaging method*. If we assume that the speed at the center of Huy is zero kilometers per hour, the first average speed would be (0+36)/2. We compute

$$\left(\frac{(0+36)}{2} \times 0.03\right) + \left(\frac{(36+32)}{2} \times 0.48\right) + \left(\frac{(32+40)}{2} \times 0.72\right) +$$

$$\left(\frac{(40+41)}{2} \times 1.65\right) + \left(\frac{(41+35)}{2} \times 0.85\right) + \left(\frac{(35+20)}{2} \times 1.28\right) +$$

$$\left(\frac{(20+26)}{2} \times 1.43\right) + \left(\frac{(26+35)}{2} \times 1.08\right) + \left(\frac{(35+35)}{2} \times 0.92\right) +$$

$$\left(\frac{(35+33)}{2} \times 1.40\right) + \left(\frac{(33+21)}{2} \times 0.77\right) + \left(\frac{(21+29)}{2} \times 1.08\right)$$

$$= 370.53.$$

Why do we only get estimates? We know exactly how long it took our rider to complete each leg of the stage. If we knew the exact average rate over each leg, we could compute the exact distance. Unfortunately, we do not know these average rates and we are forced to estimate them. The idea behind all three methods is the same: we estimate the length of each leg by estimating the average rate for the racer on that leg and multiplying by the known elapsed time for that leg. In all three cases the distance traveled, "added-up," or "accumulated" by the racer during the race is the sum of the estimated lengths of each leg.

The only important difference between the three methods is in the way we estimate the racer's average rate over each leg. We know, by the Mean Value Theorem, that at some time during the leg the rider's instantaneous speed was equal to the average speed. Maddeningly, we don't know exactly when.

▶ **THIRD REFLECTION**

Describe features of the Tour that might conspire to make our estimates of the average rate of speed on a leg completely inaccurate.   ◀

Think and Share

Break into groups. Estimate the distance a car travels in one hour for the data recorded in Table 4.3. Assume the car travels in a straight line.

1.   Use the averaging method to estimate the total distance traveled.
2.   Use the left endpoint method to estimate the distance traveled by the car.

Your two estimates should be close because they are based on much more regular data than was available in the Tour de France situation. The computations do not have to take into account varying time intervals and incomplete information at the beginning of the race. Furthermore, drivers are more able to maintain speed than are bicyclists; thus, it is likely that the estimates for the average rate are much more accurate.

Think and Share

TABLE 4.3 Time vs. Velocity Data

Time(min)	Time(h)	Vel(km/h)
0	0	60.3
10	1/6	54.8
20	2/6	52.1
30	3/6	50.3
40	4/6	52.5
50	5/6	55.2
60	6/6	59.7

Modeling Data

E X A M P L E   1   **Distance a Car Travels**

We can use the TI calculator to create a model of the data from the previous Think and Share. Enter the times and velocities listed in Table 4.3 as a data table in the Data/Matrix editor with the Variable name carspeed (See Appendix A or B if you need help). Name the first two columns of your data table Time and Velocity. Use the Data/Matrix editor to plot the data as shown in Figure 4.2.

FIGURE 4.2   Data Plot $[-0.1, 1.1] \times [49, 62]$

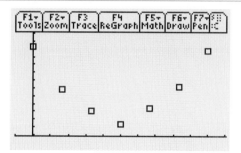

It is clear from the plot that the data is not linear. It does, however, appear that a quadratic function might fit the data well.

Calculate the quadratic regression function for Table 4.3 and store the function in y1(x). The quadratic function should be

$$y_1(x) = 36.428571x^2 - 36.557143x + 60.109524$$

as is shown in Figures 4.3 and 4.4.

Use this function and the detailed directions in the first and second steps of *Summing Sequences* below to create Table 4.4 for the estimated velocities of the car at five-minute intervals (the velocities have been rounded to two decimal places).

The estimated distance traveled in an hour using the right endpoint method is

$$57.32 \times \frac{1}{12} + 55.03 \times \frac{1}{12} + 53.25 \times \frac{1}{12} + 51.97 \times \frac{1}{12} +$$

FIGURE 4.3 Calculate Screen   **FIGURE 4.4** Stat Vars Dialogue Box

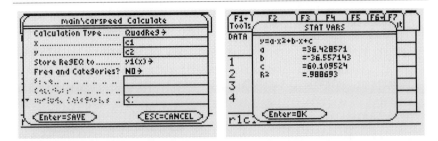

$$51.20 \times \frac{1}{12} + 50.94 \times \frac{1}{12} + 51.18 \times \frac{1}{12} + 51.93 \times \frac{1}{12} +$$

$$53.18 \times \frac{1}{12} + 54.94 \times \frac{1}{12} + 57.21 \times \frac{1}{12} + 59.98 \times \frac{1}{12} =$$

$$= 54.01 \text{ miles.}$$

TABLE 4.4   Velocities from y1(x)

Time (hr)	Velocity (km/h)
0	60.11
1/12	57.32
2/12	55.03
3/12	53.25
4/12	51.97
5/12	51.20
6/12	50.94
7/12	51.18
8/12	51.93
9/12	53.18
10/12	54.94
11/12	57.21
12/12	59.98

SUMMING SEQUENCES

The TI calculator can compute our sum automatically. Go to the HOME screen and type

- seq(k,k,1/12,1,1/12)

on the Entry line. The TI calculator interprets this command as instructions to create a list of times like those in Table 4.4 (from "5" to "60" minutes in increments of "5" minutes starting at the right endpoint of the first time interval). Press Enter to create a list of times. Edit this command to read

- y1(seq(k,k,1/12,1,1/12))

and press Enter to get the function values at each of these times (that is, the velocities). These values should match the values in the Table 4.4 starting at 57.32. Edit this line to read

- y1(seq(k,k,1/12,1,1/12))*(1/12)

and press Enter to obtain a list of the estimated distances for each five-minute (1/12 hour) interval using the right endpoint method. Finally, to sum these distances and obtain the total distance traveled, edit this line to read

- sum(y1(seq(k,k,1/12,1,1/12))*(1/12))

and press Enter.

▶ FOURTH REFLECTION

Explain the effect of each bulleted command in the *Summing Sequences* box above. ◀

► **FIFTH REFLECTION**

Edit the `seq` command in the last step of *Summing Sequences* to halve the size of the increment (increasing the number of intervals). Change the fraction at the end of the expression to reflect the decrease in interval length in hours. What is the estimated distance traveled?   ◄

Improving the Estimate

If we decrease the length of our time subintervals *Distance a Car Travels*, then there is no time for the car to vary its speed. This should improve the accuracy of using speeds at the ends of the interval to estimate the average rate of speed over the entire interval. Using one-minute intervals should give better estimates of the average speed over the interval than using five-minute intervals. As long as the quadratic function `y1` fits the data well, we should expect the estimate of the distance traveled to improve if we increase the number of time intervals which simultaneously decreases their length.

Think and Share

Break into groups. Use the function y_1 from the example above with smaller and smaller time intervals to complete Table 4.5. Edit the `seq` command to change the size of the increment (increasing the number of intervals). Then change the fraction at the end of the expression `sum(` to reflect the decrease in interval length in hours.

Notice that the curve does not go directly through all of the data points. How is the estimate of the distance traveled affected by this lack of precision? The original data was recorded in 10-minute intervals. What might have happened on the car trip during those intervals that could cause inaccuracies in our estimate of the distance traveled? Imagine at least three different possible circumstances and speculate as to their likelihood. How would the likelihood of your events be affected if the data were recorded more frequently?

TABLE 4.5 Distance Traveled Using Decreasing Time Interval Lengths

Time Interval (min)	3	2	1	1/2	1/3	1/6	1/10
Distance Traveled (k)	53.99					54.13	

Getting the Big Picture: Accumulation as Area

We can use technology to display our computations. Graph the velocity function, `y1`, on the interval from 0 to 60 minutes (one hour) with window settings `xmin=0`, `xmax=1`, `xscl=1/6`, `ymin=0`, and `ymax=65`. Draw a vertical line on the screen from the point `(1/6, 0)` to the point `(1/6, y1(1/6))` as shown in *Using the Line Command*.

USING THE LINE COMMAND

Go to the HOME screen. On the command line type `line 1/6, 0, 1/6, y1(1/6)` and press ENTER. Be sure to include the commas. The four numbers are the x and y coordinates of the points $(1/6, 0)$ and $(1/6, y1(1/6))$. The TI calculator will automatically go to the GRAPH screen.

The length of this vertical line is the difference of the two y-coordinates `y1(0)` and `y1(1/6)`. Thus, this length is `y1(1/6)`, the velocity when $x = 1/6$. You can continue the process, drawing vertical lines at 10-minute intervals to obtain a graph like the one in Figure 4.5. For example, the next vertical line is drawn with the command:

$$\texttt{line 2/6,0,2/6,y1(2/6)}$$

The velocity of the car at 10-minute intervals is represented graphically by the lengths of the vertical line segments.

The length of the intervals between each of these line seqments is one-sixth of an hour (10 minutes). Draw a horizontal line from the point $(0, y1(1/6))$ to the point $(1/6, y1(1/6))$ to obtain the graph in Figure 4.6.

FIGURE 4.5 Graph and Vertical Lines   **FIGURE 4.6**   First Rectangle

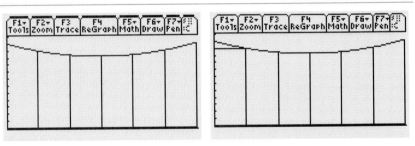

On the road

The distance traveled in miles over the first 10 minutes is the average speed in miles per hour over that interval times one-sixth of an hour. When we use the speed at the end of 10 minutes as an estimate for the average speed over the entire 10 minutes, we estimate the distance traveled in the first time interval as $y1(1/6)$ times the length of the interval. Thus, $y1(1/6) \cdot (1/6)$ estimates the distance traveled in the first 10 minutes. On the other hand, $y1(1/6) \cdot (1/6)$ equals the area of the rectangle that appears on the screen. The area of the rectangle graphically represents our estimate of the distance traveled in the first time interval using the right endpoint method. You can continue to draw horizontal lines to obtain the rectangles whose areas correspond to each of the terms in the right endpoint estimate as shown in Figure 4.7 (you will have to extend two of the vertical lines).

FIGURE 4.7   An Approximation Based on Six Rectangles

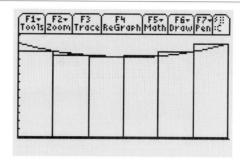

We have created a visual image of our estimate of the total distance traveled as the area of the six rectangles shown in Figure 4.7.

Think and Share

Break into groups. Calculate the areas of the rectangles in Figure 4.7. How does this compare with the results of estimating total distance with six equal time intervals? What is the relationship between the sum of the products of rate and time with the areas of these rectangles?

Putting it All Together.   In this section we have used two basic ideas. First, the total distance traveled for a trip of many legs is the sum of the distances traveled over each leg. Second, the distance over any one leg can be estimated by multiplying the time to complete the leg by an estimated average rate for that leg.

$$average\ rate \times time = \frac{miles}{hour} \times hours = miles\ \text{(a distance)}$$

Thus, the length of each leg is estimated by a product, and the total length is estimated by a sum of products. The usefulness of the idea of the sum of products will be further explored in the next section.

EXERCISE SET 4.1

In Exercises 1–6, estimate the distance traveled from the time and velocity tables given. The time is in hours, minutes, or seconds while the velocity is in some appropriate units like mph.
  a. Use the left endpoint method.
  b. Use the right endpoint method.
  c. Use the averaging method.

1. Use Table 4.6

TABLE 4.6

Time(secs)	0	1	2	3	4	5	6	7
Velocity(fps)	0	2	5	9	13	15	18	20

2. Use Table 4.7

TABLE 4.7

Time(secs)	0	1	2	3	4	5	6
Velocity(fps)	22	23	24	25	26	27	28

3. Use Table 4.8

TABLE 4.8

Time(h)	0	0.15	0.30	0.45	0.60	0.75
Velocity(mph)	31	33	44	32	28	29

4. Use Table 4.9

TABLE 4.9

Time(mins)	0	0.2	0.4	0.6	0.8	1.0
Velocity(fpm)	0	110	122	115	116	109

5. Use Table 4.10

TABLE 4.10

Elapsed Time(h)	0.1	0.1	0.1	0.1	0.1	0.1
Velocity(mph)	2	2.5	3.1	2.5	3.3	3.2

6. Use Table 4.11

TABLE 4.11

Elapsed Time(h)	1	1	1	1	1	1	1
Velocity(mph)	12	15	21	30	42	57	75

In Exercises 7–12,

   a. Fit the data in Exercises 1–6 with an appropriate model: linear, quadratic, or exponential. Justify your choice of model.

   b. Use your model, the `sum(seq(` command, and any one of the three estimating methods to find another estimate of the total distance traveled. Is this estimate better or worse than the original estimate? Justify.

   c. Double the number of time intervals and repeat part b.

   d. Multiply the number of time intervals by 10 and repeat part b.

7. Use Table 4.6.

8. Use Table 4.7.

9. Use Table 4.8.

10. Use Table 4.9.

11. Use Table 4.10.

12. Use Table 4.11.

13. The following figure gives the rate at which a student was bicycling on a straight street from her home to school. Write a story that interprets this information.

FIGURE 4.8

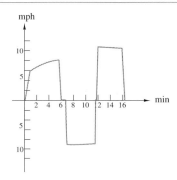

14. In Exercise 13, estimate the distance from the student's home to school.

15. In Exercise 13, estimate the distance traveled while the graph of the function is negative. Explain how you arrived at this estimate.

16. Explain what it means for the graph to be zero in Exercise 13. For what length of time was the student stopped? How far did the student go while the graph of the function was zero?

17. The following figure gives the rate at which a horse traverses a short steeplechase course with five hurdles. Write a story that interprets this information.

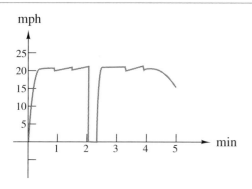

FIGURE 4.9 Steeplechase rate graph

18. At what time did the horse cross the first hurdle in Exercise 17? How far had the horse gone when it crossed the first hurdle? Explain how you estimated this distance.

19. At what time did the horse stop in Exercise 17? Explain why you know this from the graph. How far did the horse go while the graph of the function was zero?

†**20.** Create a scatter plot of the following table of velocities and times for a racing car from a standing start. Fit an exponential and a linear function to the data. Calculate the distance traveled for both functions using 32 time intervals. Explain why you think one of these estimates for the distance is better than the other.

TABLE 4.12 Racing Car Data

Time(sec)	0	3	6	9	12	15	18	21	24	27	30
Velocity(mph)	0	36	58	79	115	130	154	195	239	250	261

4.2

Sums of Products

In the last section, we viewed the total distance traveled by a bicyclist or a motorist as an accumulation of distances traveled over short segments of their trips. Estimates of those shorter distances were then used to produce a sum of products estimating the total distance. Estimating the total accumulation of a quantity by viewing it as the sum of smaller accumulations is a fundamental idea of calculus. In this section, we will continue to use this idea to investigate the accumulation of oil pumped through a pipeline and the accumulated power consumption of a town.

Pumping Oil Across the Desert

Oil companies transport oil from their wells to their refineries using pipelines. This method is extremely cheap, especially when compared to the other standard methods of transporting oil with trucks or tanker ships. Pipelines, however, suffer from a major disadvantage: leaks. When a truck or tanker leaks, it is immediately obvious. The problem is easily corrected (although in some cases not in time to avoid major environmental damage). Pipelines are a different story.

E X A M P L E   1   **The Pipeline Example**

Pipelines are often hundreds of miles long. Their enormous length increases their tendency to leak and complicates locating leaks; even determining if a leak exists can be difficult. The rate at which the oil flows through the

An Oil Pipeline Meter

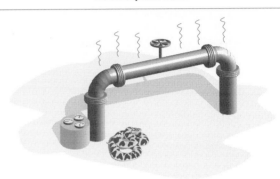

pipeline depends upon the oil's viscosity (thickness), which in turn depends upon temperature. The higher the temperature, the lower the viscosity and the more rapidly the oil will flow in the pipeline. If oil is pumped across a desert where the temperature varies from 2°C to 42°C, decreases in volume at the refinery due to leaks are impossible to distinguish from decreases due to variation in temperature.

To facilitate detecting leaks, an oil company makes two types of measurements. First, they measure the amount of oil pumped into the tanks at the refinery. Second, meters are installed along the pipeline to measure the rate at which oil is flowing at various strategic locations. This latter measurement allows the company to estimate the amount of oil that should have flowed across the desert and into the tanks. Comparing these two measurements can indicate the presence of a leak, and aid in locating it.

Every four hours over a 24-hour period, a technician observes the rate in barrels per hour at which oil is flowing through the pipeline at a specific meter. Her observations are recorded in Table 4.13. Note: the times are *not* elapsed times.

TABLE 4.13   Recorded Observations

Time (h)	0	4	8	12	16	20	24
Rate of Flow (b/h)	32	31	37	50	52	42	33

▶ FIRST REFLECTION
What is the elapsed time between observations? ◀

Even if the meter permits more frequent readings, a technician might well decide to look at this sample of the data to get a rough idea if there might be a problem. How can the flow rate measured in barrels per hour be used to estimate the number of barrels of oil that were pumped through the pipeline at this location?

This question has the same components as the Tour de France problem from the previous section. There, the cyclist's rate of speed was used to estimate his accumulation of miles (that is, distance traveled). In this problem, we want to use the flow rate of the oil to estimate the accumulation of barrels of oil. In both situations, the average rate of accumulation (speed in kilometers per hour for the cyclist or flow rate in barrels per hour for the oil) over a time interval multiplied by the length of the time interval equals the total accumulation (barrels or miles) for that interval. This can be written symbolically for the oil as

average number of barrels per hour $\times$ *hours = total number of barrels.*

Both problems have the same general underlying structure:

average rate of accumulation $\times$ *time = accumulated quantity.*

As we saw in the last section, the rate recorded at particular instants can be used to estimate the average rate of change in several different ways. Using the right endpoint method, we use the flow rate at the later time in each time interval to estimate the average flow rate for the entire interval and obtain

$$31 \times 4 + 37 \times 4 + 50 \times 4 + 52 \times 4 + 42 \times 4 + 33 \times 4 = 980$$

as an estimate for the total number of barrels of oil pumped through this particular location over the 24-hour period.

▶ SECOND REFLECTION
Estimate the total number of barrels of oil pumped using the left endpoint method. ◀

We don't know for sure whether these estimates are too high or too low. We do know that the products in these computations equal the areas of rectangles. By graphing these estimating rectangles in Figures 4.10 and 4.11 we obtain a visual representation of the computations.

▶ THIRD REFLECTION
What are the units of the area of each rectangle? Barrels/hours? Hours? or Barrels? ◀

FIGURE 4.10   Right Endpoint Method   **FIGURE 4.11**   Left Endpoint Method

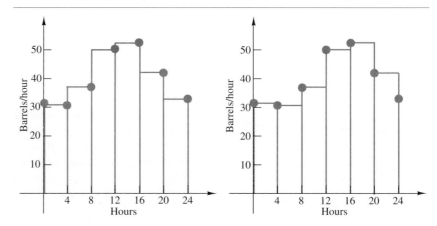

▶ FOURTH REFLECTION

Which of the graphs represents the left endpoint method? The following computation represents which endpoint method?

$$32 \times 4 + 31 \times 4 + 37 \times 4 + 50 \times 4 + 52 \times 4 + 42 \times 4$$

◀

Comparing the heights of the rectangles for the left endpoint and right endpoint methods, we see that although the right endpoint method gives a lower estimate for the oil accumulated over the first four-hour time period, it gives a much higher estimate over the second time period.

To be conservative, we would like an underestimate or *estimated lower bound* for the total number of barrels which passed by our meter for comparison with the number of barrels which actually arrived at the refinery. Neither of our two estimates appears to be a good lower bound since neither one gives the lowest estimate on every interval. On the other hand, if we use the smaller of the two estimates on each interval, we do obtain a good lower bound. In this minimum method, we would use the right endpoint estimate for the first four hours, the left endpoint estimate for the second four hours, etc.

▶ FIFTH REFLECTION

Think carefully about which endpoint of each subinterval generates the minimum rate for the subinterval. Then compute the underestimate for the barrels pumped through the pipeline at the meter as described in the text.   ◀

Break into groups. We can mimic the calculations the technician might perform as a quick check for a leak.

Think and Share

1. The technician will look more closely at the situation if the underestimate is more than 2% greater than the actual measurement of the oil accumulated at the refinery. Will the technician look more closely if the measurement at the refinery is 890 barrels? What could the technician do to get a better idea if there is a leak?

2. Compute an overestimate for the barrels pumped through the pipeline at this location and an estimate based on the average of the flow rates at the beginning and end of each time interval. Compare these estimates with the estimates obtained above. Does the 2% rule used by the technician seem reasonable?

3. Can you pin down the location of a leak if you have flow rate information for many locations? Explain.

4. What are the possible difficulties in these methods of estimating the number of barrels of oil that flow through the pipeline?

Energy Use in A Town

The flow of electricity through wires can be viewed very much like the flow of oil through a pipe. In both cases, we can use flow rates to estimate accumulation.

E X A M P L E 2 **Energy Use**

Suppose a town that owns its own electric power plant plans to join a consortium of electric power producers. The plant manager knows the monthly energy use from billing data, but the consortium members need to know the maximum amount of energy the town uses in a typical weekday. They also require an average weekday energy use profile (energy use recorded at four-hour intervals). The plant manager uses watt-meters to sample the rate of energy use of 5% of the homes and businesses in the town. The results of the survey are projected for the entire town to obtain Table 4.14 and the graph in Figure 4.12 (which predicts rate of energy use in megawatts at two-hour intervals).

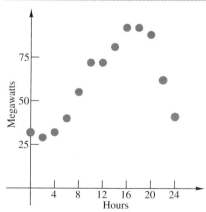

FIGURE 4.12 Rate of Energy Use

TABLE 4.14 Time vs. Power on Each Interval

Time(hrs)	0	2	4	6	8	10	12	14	16	18	20	22	24
Power(mw)	34	29	32	40	55	72	72	81	92	92	88	62	41

Energy use over a given time period equals the product of the rate of energy use (power) over the period multiplied by the length of the period.

$$\text{rate of energy use (megawatts)} \times \text{time (hours)}$$
$$= \text{energy (in megawatt-hours)}$$

The consortium prefers an overestimate or estimated *upper* bound of the energy use to plan for the optimum use of electricity in peak periods. Thus, the manager uses the maximum rate of energy use (power) for each two-hour interval to obtain an upper bound.

An estimated upper bound for the peak energy use is computed as the sum of products

$$(34 \times 2) + (32 \times 2) + (40 \times 2) + (55 \times 2) + (72 \times 2) + (72 \times 2) +$$
$$(81 \times 2) + (92 \times 2) + (92 \times 2) + (92 \times 2) + (88 \times 2) + (62 \times 2)$$
$$= 1624.$$

The consortium wants the weekday energy use profile to reflect typical rather than peak use. Consequently, the plant manager decides to use an averaging method. In Table 4.15, the average of the power consumption rates at the two ends of each time interval are used to estimate the average use over the interval.

TABLE 4.15   Average Power on an Interval

Time Interval	0-2	2-4	4-6	6-8	8-10	10-12	12-14	14-16	16-18	18-20	20-22	22-24
Power(mw)	31.5	30.5	36	47.5	63.5	72	76.5	86.5	92	90	75	51.5

Converting to four-hour periods, as required by the consortium, we obtain the data in Table 4.16. The graphical display of the energy profile for the town is given in Figure 4.13.

FIGURE 4.13   Energy Profile

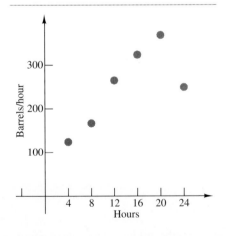

TABLE 4.16   Averaging Method Energy Use Profile Computation

Time(h)	Calculation	Energy Use(mw-h)
4	$31.5 \times 2 + 30.5 \times 2$	124
8	$36 \times 2 + 47.5 \times 2$	167
12	$63.5 \times 2 + 72 \times 2$	271
16	$76.5 \times 2 + 86.5 \times 2$	326
20	$92 \times 2 + 90 \times 2$	364
24	$75 \times 2 + 51.5 \times 2$	253

Think and Share

Break into groups. Estimate the minimum amount of energy consumption on a typical day. What situation might call for the use of this number? How would you go about checking if the energy consumption number was reasonable? Should you expect an overestimate or underestimate and why? What assumptions should be made about weekends in your calculations? How different from the monthly billing can the calculations be before you question their validity? What might you do to obtain better estimates?

We have introduced the notion and importance of accuracy in estimating quantities by using sums of products. We now develop a formal definition of these sums of products.

Riemann Sums

All of the examples in this and the previous section have involved estimating the accumulation of a quantity over a time interval. In all cases, we looked at the time interval in terms of smaller intervals. To avoid confusion, we will refer to the smaller intervals as *subintervals*. In the right endpoint and left endpoint methods we formed products of the type

$$f(t_k) \times (length\ of\ subinterval),$$

where t_k was chosen to be one of the endpoints of the subinterval. Our estimate of the total accumulation was obtained by summing these products. There is nothing particularly special about choosing t_k as an endpoint. We could obtain other estimates, perhaps as good or better, by using a value for t_k somewhere between the two endpoints of the subinterval.

Such sums of products are the foundation of integral calculus and are called **Riemann sums**. They need not refer to time as the independent variable.

▶ SIXTH REFLECTION

In Figure 4.14, label the tick mark between x_{k-1} and x_k as t_k and use it to construct the rectangle whose area represents the value of the kth term in the Riemann sum for f on the interval $[c, d]$. What is the length of the kth subinterval? Does the area of the rectangle you have drawn overestimate or underestimate the area under the graph and above the x-axis between x_{k-1} and x_k? ◀

In the next section, we will investigate various methods for generating Riemann sums to see if we can learn about the accuracy of our estimates.

Calculating Riemann Sums.   We can also compute a Riemann sum for a function f using the TI calculator $\sum$(command. We have already learned how to compute Riemann sums with the sum(and seq(commands. We can therefore compute $\sum_{k=1}^{6} f(t_k)\Delta x$ for right endpoints on the interval [3,11] with the interval length $\Delta x = 2$ in two ways:

```
sum(f(seq(k,k,5,11,2))*2)   or   ∑(f(3+k*2)*2,k,1,4)
```

RIEMANN SUM

Let the interval $[a, b]$ be subdivided into n subintervals by the points

$$a = x_0 < x_1 < x_2 < \ldots < \underbrace{x_{k-1} < x_k}_{k\text{th subinterval endpoints}} \ldots < x_{n-1} < x_n = b$$

The points $x_0, x_1, x_2, \ldots, x_n$ are called a partition. A **Riemann sum** for the function f on the interval $[a, b]$ is a sum of the form

$$R_n = \sum_{k=1}^{n} f(t_k)\Delta x_k$$

where t_k is some point in the kth subinterval $\left[x_{k-1}, x_k\right]$, and $\Delta x_k = x_k - x_{k-1}$ is the length of the kth subinterval. For now, it is assumed that the subdivision or partition is *regular*; that is, Δx_k is the same for all k and, hence, equals $\dfrac{b-a}{n}$.

FIGURE 4.14   Constructing the kth Term in a Riemann Sum

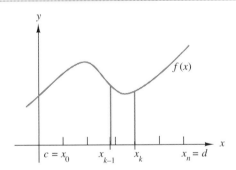

Notice that we compute the points, the functions values at the points, the products, and then the sum of the products using the sum(and seq(approach. In using the $\sum$(command, we must have a formula based on the index k for the endpoints and determine each of the products as the first argument of $\sum$(command. From now on, we will use the $\sum$(command for Riemann sum computations.

Use DelVar f *to set* f *as a general function.*

▶ SEVENTH REFLECTION

Show that the two ways of computing a Riemann sum described above are equivalent. Then write two ways to compute the Riemann sum

$$\sum_{k=1}^{5} f(t_k)\Delta x$$

for right endpoints on the interval [4,6] with $\Delta x = 2/5$. How can you change your $\sum ($ command to sum 100 products? ◀

1. Water flows through a pipe containing a flow measurement device. The measurements are given below in gallons per minute at various times. Estimate the total number of gallons of water that flow through the pipe in 20 minutes. Describe your method.

TABLE 4.17

Time(min)	0	5	10	15	20
Gallons/min	12	10	8	5	2

2. The flow of water in a river in Northern California changes with the seasons. The table below gives the rate in thousands of acre-feet per month at which the water is flowing every two months during the year. Estimate the total amount of water that flowed down the river for the entire year. Describe your method.

TABLE 4.18

Time(months)	0	2	4	6	8	10	12
Acre-feet/month	300	500	700	200	100	200	300

3. An artificial lake is created as a resevoir for agricultural irrigation. The rate at which the water flows out of the lake is regulated by a state agency. The table below gives the rate in hundreds of cubic feet per day at which the water is released every two days for 10 days. Estimate the total amount of water that is released in that period. Describe your method.

TABLE 4.19

Time(days)	0	2	4	6	8	10
Cubic feet/day	120	110	100	80	50	20

4. Grain is emptied from a grain elevater at various rates depending upon how much grain is in the elevator. The table below gives the rates in bushels per minute at which the grain is flowing for an eight-hour day. Estimate the total amount of grain that flows in this period. Describe your method.

TABLE 4.20

Time(min)	1	2	3	4	5	6	7	8
Bushels/min	150	200	170	250	150	100	200	250

5. The electrical energy from an electrical power plant is measured in volt–ampere–hours or kilowatt hours. The voltage on the main electrical carrier, which is rated at 2000 amperes from a power plant in Rhode Island, varies during the summer when there is a major drain on electricity due to air conditioning. The table below records the voltages in thousands of volts at one-hour intervals for four hours during a hot afternoon. Estimate the amount of energy that passed through this carrier. Describe your method.

TABLE 4.21

Time(hrs)	0	1	2	3	4
Volts	150	200	210	240	270

6. Electricity is generated by windmills in the Altamont Pass area of California at variable rates depending upon the wind velocity. The measured voltages in a wire rated at 70 amperes connected to a windmill are used to determine the amount of energy that the windmill produces. The table below contains the voltage measurements over a one-day period at four-hour intervals. Estimate the total amount of energy generated by this windmill on this day. Describe your method.

TABLE 4.22

Time(hrs)	0	4	8	12	16	20	24
Volts	200	170	250	230	280	300	210

In Exercises 7–12, create a function to model the data in each table in Exercises 1–6. Use the $\sum$ command and at least 10 times the original number of subintervals for each exercise.

7. Estimate the water flow in the pipe using a linear function.

8. Estimate the water flow in the river using a quadratic function.

9. Estimate the amount of water for irrigation using a linear function.

10. Estimate the number of bushels of grain taken from the grain elevator using a cubic function.

11. Estimate the volt-amperes that were generated by the plant using a linear function. Explain why this is not a good approximation.

12. Estimate the volt-amperes that were generated by the windmill farm using a linear function.

†**13.** The Electric Company claims that your windmill farm has generated 30,000 kilowatt hours of energy over the last month (30 days). You have recorded the following information from the electric power line (1000 ampere) that leads off your farm. Decide if the Electric Company's estimate of your energy production is correct. Give a rationale for your decision. (A watt is one volt-ampere.)

TABLE 4.23

Time(days)	0	5	10	15	20	25	30
Volts	20	25	20	15	12	30	0

4.3

Error Bounds for the Left and Right Endpoint Methods

In the first two sections, we solved problems using several methods for estimating the accumulation of a quantity. Each of the methods employed the use of a particular type of sum of products, known as a Riemann sum. Estimates are useless if we have no idea how close they are to the actual value. In this section, we will investigate the accuracy of each method.

Left and Right Endpoint Methods

Each of the methods developed in Sections 4.1 and 4.2 used the instantaneous rate of change of a quantity at a finite number of specified times. We used these rate values to generate estimates for the average rate of change of the quantity on the intervals between the specified times. We multiplied each estimate by the length of the corresponding interval, and summed these products to produce an estimate of the total accumulation of the quantity.

Often we have a function whose output is the rate of change of a quantity at any point on an interval. We can still use our same methods for estimating the total accumulation of the quantity over the interval simply by picking our own finite set of specified times. A specific example will help make this clearer.

EXAMPLE 1   **Distance Traveled by a Car**

Suppose that $v(t) = -40t^2 + 80t$ gives the velocity of a car at time t measured in hours. We wish to estimate the total accumulation of miles (that is, the total distance traveled) during the first hour.

Our function allows us to specify rates at any time we wish. Using the rates at 10-minute intervals, we obtain the following time versus rate table where each interval is 10 minutes (1/6 hours) long.

TABLE 4.24

Time (hrs)	0	1/6	2/6	3/6	4/6	5/6	6/6
Rate (mph)	0	12.222	22.222	30	35.556	38.889	40

Using the right endpoint method we obtain the Riemann sum

$$v\left(\frac{1}{6}\right)\times\frac{1}{6}+v\left(\frac{2}{6}\right)\times\frac{1}{6}+v\left(\frac{3}{6}\right)\times\frac{1}{6}+v\left(\frac{4}{6}\right)\times\frac{1}{6}+$$

$$v\left(\frac{5}{6}\right)\times\frac{1}{6}+v\left(\frac{6}{6}\right)\times\frac{1}{6}=\sum_{k=1}^{6}v\left(0+k\times\frac{1}{6}\right)\times\frac{1}{6}=29.815\text{miles},$$

where $v(1/6)=12.222$ from the table.

▶ FIRST REFLECTION

Use the right endpoint method to generate an estimate for the accumulated distance the car traveled over the first hour in Example 1 above by selecting rates at five-minute intervals. Write the Riemann sum notation.   ◀

RIGHT ENDPOINT METHOD

Let the interval $[a, b]$ be subdivided into n subintervals by the points

$$a=x_0<x_1<x_2<\ldots<\underbrace{x_{k-1}<x_k}_{k\text{th subinterval endpoints}}\ldots<x_{n-1}<x_n=b$$

and let $f(x)$ be a function that specifies the rate of change of a quantity over the interval $[a, b]$. The **right endpoint method** estimates the total accumulation of the quantity over the interval $[a, b]$ by

$$R_n=\sum_{k=1}^{n}f(x_k)\Delta x_k$$

where Δx_k is the length, x_k-x_{k-1}, of the kth subinterval, and $x_k=a+k\Delta x$ is the right endpoint of that subinterval.

RIGHT ENDPOINT SUM COMPUTATION

You can use the TI calculator to compute the Riemann sum of f on the interval $[a, b]$ using the right endpoint of each subinterval for n subintervals with the following command:

$$\Sigma(f(a+k*h)*h,k,1,n)$$

where a is the left endpoint of the interval $[a, b]$, $a+k*h$ is the right endpoint of the kth subinterval, and h is $(b - a)/n$.

The left endpoint method estimates the total accumulation by

$$L_n = \sum_{k=1}^{n} f(x_{k-1})\Delta x.$$

LEFT ENDPOINT SUM COMPUTATION

You can use the TI calculator to compute the Riemann sum of f on the interval $[a, b]$ using the left endpoint of each subinterval for n subintervals with the following command:

$$\Sigma(f(a+(k-1)*h)*h,k,1,n)$$

where a is the left endpoint of the interval $[a, b]$, $a+(k-1)*h$ is the left endpoint of the kth subinterval, and h is $(b - a)/n$.

FIGURE 4.15　Estimates Using Four Left Endpoints

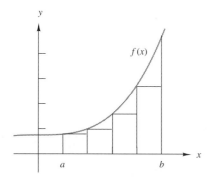

▶ SECOND REFLECTION

Use the left endpoint method to generate an estimate for the accumulated distance the car traveled over the first hour in Example 1 by selecting rates at five-minute intervals.　　◀

Error Bounds

How accurate are the estimates of total accumulation that are generated by the right and left endpoint methods? If the function is increasing over the interval (like the one in Example 1), we can determine a bound for the error associated with each of these estimates. A graph for an increasing function f on the interval $[a, b]$ with a graphical representation of the total accumulation using left endpoints is displayed in Figure 4.15.

The area of each rectangle corresponds to one of the computations made in the left endpoint method. Notice that for this increasing function the rate at the left endpoint, represented by the height of the rectangle, is smaller than the rate at any other point on the subinterval. This is easy to see from the graph simply because each of the rectangles lies entirely below the curve. Thus, the value of the rate function at the left endpoint of the subinterval must be less than the actual average value of the function over the whole subinterval.

Hence, for an increasing function, the left endpoint method underestimates the accumulation over each subinterval. By the same token, the right endpoint method will overestimate the accumulation over each subinterval. The difference between the two provides a way to bound the error.

Figure 4.16 gives a graphical representation of this idea. In this figure, the

FIGURE 4.16   Error Bound (Shaded)

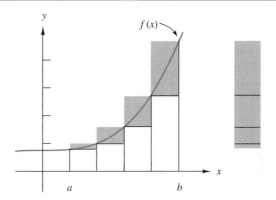

difference between the areas of the rectangles generated by the right and left endpoint methods in each subinterval is shaded. The area of each unshaded rectangle is the underestimation of the accumulation over the subinterval obtained with the left endpoint method. This area, combined with the shaded rectangle, produces the overestimation of the right endpoint method for increasing funtions using a regular partition. The true accumulation lies between these two values. Thus, the rectangle based on the left endpoint method will be in error by at most an amount equal to the area of the corresponding shaded rectangle. These shaded rectangles are combined together as a stack of rectangles at the right of the graph, and produce one shaded rectangle whose area is an upper bound for the total error in the estimate. This method of finding a bound on the error is called the *Stack Method Error Bound*.

Each one of the rectangles in the stack of rectangles represents an upper bound for the error for the corresponding subinterval. The area of each small rectangle in the stack equals the right endpoint overestimate minus the left endpoint underestimate for that subinterval. Notice that the stack has a width equal to Δx, the standard width of each subinterval. The height of this stack of rectangles is $f(b) - f(a)$, the maximum value of the function minus the minimum value of the function on $[a, b]$. Thus, the total error estimate equals the area of these shaded rectangles and can be computed as $(f(b) - f(a)) \cdot \Delta x$. First, break up the interval $[a, b]$ into n equal subintervals of length $\Delta x = (b - a)/n$. Then, letting E_{Left} be the error in the left endpoint method, we have

Stack Method Error Bound

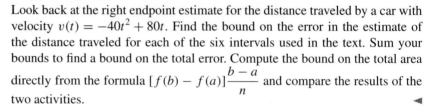

$$|E_{Left}| \le |f(b) - f(a)|\Delta x = |f(b) - f(a)|\frac{b-a}{n}. \tag{4.1}$$

Decreasing the Error. Increasing n, the number of subintervals, does not change the interval, $[a, b]$, or the height of the stack, $f(b) - f(a)$. Since n is in the denominator, increasing the value of n will decrease the size of our bound on the error. Thus, the accuracy of the left endpoint method will improve as we increase the number of subintervals (the number of estimating rectangles). We can make the computation as accurate as we like by choosing a sufficiently large number of subintervals.

Surprisingly, the same stack of rectangles represents an upper bound of the error in the overestimation produced by the right endpoint method. The computation of the error for a decreasing function will be similar.

▶ THIRD REFLECTION

Look back at the right endpoint estimate for the distance traveled by a car with velocity $v(t) = -40t^2 + 80t$. Find the bound on the error in the estimate of the distance traveled for each of the six intervals used in the text. Sum your bounds to find a bound on the total error. Compute the bound on the total area directly from the formula $[f(b) - f(a)]\dfrac{b-a}{n}$ and compare the results of the two activities.   ◄

Estimating the Area Under a Curve

The graphical representations for the left and right endpoint methods suggests another interpretation for these estimates. Notice that, for an increasing function that is nonnegative, the rectangles representing the right endpoint method completely cover the area under the curve on the interval $[a, b]$. Each of the rectangles contains extra space that is not below the curve. (See Figure 4.17.)

We see from the graph in Figure 4.16 that this extra space is always contained in one of the shaded rectangles used to bound the error in our estimate. We can make the size of this bound as small as we like simply by increasing the number of subintervals; similarly, we can make the size of the extra space as small as we like.

Therefore, we are justified in interpreting the Riemann sum generated by the right endpoint method as an estimate of the area under the curve. The computations remain the same whether the Riemann sum is interpreted in terms of area or accumulation. Hence, the error bound remains the same.

Furthermore, since all of our methods are estimating the same accumulation as the right endpoint method, all of them must also be estimating the area under the curve. We will use this idea freely for the rest of this section, and will revisit it in more detail in Section 4.6.

FIGURE 4.17   A Right Endpoint Approximation for an Increasing Function

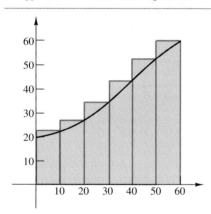

A More General Error Bound

The error bound determined above holds for functions that are strictly increasing or strictly decreasing on the interval $[a, b]$. To get a feel for what the error bound depends on in the general case, we will explore a few situations.

A Constant Function.   Consider the constant function $f(x) = c$ where c is any positive real number (see Figure 4.18 and 4.19). Now subdivide the interval

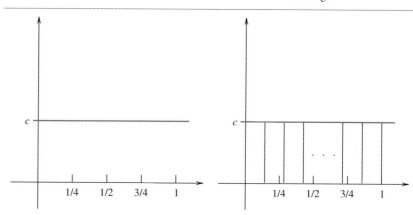

FIGURE 4.18   Constant Function   **FIGURE 4.19**   Constant Function with Rectangles

$[a, b]$ into n equal subintervals and use the left endpoint method to approximate the area under the curve of $f(x)$. There is no error in this case and the approximation is exact since the subinterval rectangles exactly fill up the entire rectangle.

A Linear Function.   Now consider the linear function $f(x) = mx$. In particular, let's look at the graphs of $f(x) = x$ and $f(x) = 4x$ on the interval $[0, 1]$ as shown in Figures 4.20 and 4.21.

Notice what happens when we use the left endpoint method with four equal subintervals (see Figures 4.22 and 4.23). The error (shaded area) certainly appears greater for the steeper line. The steepness of a line is, of course, determined by the slope of the line which, in turn, is the first derivative of the function (assuming it has a derivative). It is not surprising, therefore, to learn that, for any such *positive* function $f(x)$, an error bound for the area under the curve determined by f using the left endpoint method depends on the first derivative of $f(x)$ on $[a, b]$. If E_{Left} represents the error in the left endpoint

Derivative Method Error Bound   method, then one such bound is given by

$$\left| E_{Left} \right| \leq M \frac{(b - a)^2}{2n} \tag{4.2}$$

FIGURE 4.20 $f(x) = x$

FIGURE 4.21 $f(x) = 4x$

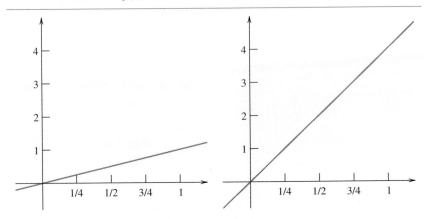

FIGURE 4.22   Area Using Left Endpoints

FIGURE 4.23   Area Using Left Endpoints

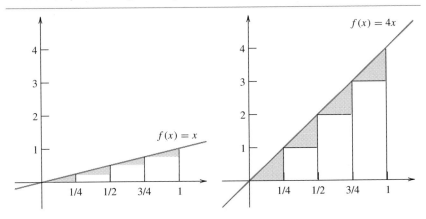

FIGURE 4.24   Area Using Left Endpoints

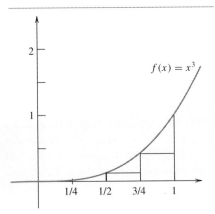

where M is the maximum value of the absolute value of the function's first derivative, $|f'(x)|$, on $[a, b]$ and n is the number of equal subintervals into which the interval $[a, b]$ is divided. It turns out that, as before, the same error bound holds for the right endpoint method.

▶ FOURTH REFLECTION

Why do we need to consider the absolute value of $f'(x)$ when finding error bounds?   ◀

EXAMPLE 2   **The Area Under a Curve**

Approximate the area under the curve of the function $f(x) = x^3$ on the interval $[a, b] = [0, 1]$ (see Figure 4.24) using four equal subintervals. Then compute an upper bound on the error involved.

Solution:

The length of each subinterval is

$$\frac{b-a}{n} = \frac{1-0}{4} = \frac{1}{4}.$$

The area under the curve is approximated by summing the areas of the rectangles using left endpoints. Thus, the approximated area is given by

$$f(0) \times \frac{1}{4} + f\left(\frac{1}{4}\right) \times \frac{1}{4} + f\left(\frac{2}{4}\right) \times \frac{1}{4} + f\left(\frac{3}{4}\right) \times \frac{1}{4}$$

$$= \left[(0)^3 + \left(\frac{1}{4}\right)^3 + \left(\frac{2}{4}\right)^3 + \left(\frac{3}{4}\right)^3\right] \times \frac{1}{4}$$

$$= [0^3 + 1^3 + 2^3 + 3^3] \times \frac{1}{4^4} \approx 0.141.$$

To obtain an error bound for this calculation, we use the formula

$$\left|E_{Left}\right| \le M\frac{(b-a)^2}{2n}.$$

The first derivative of $f(x) = x^3$ is $3x^2$ and the maximum of the first derivative on $[0, 1]$ occurs at the right endpoint. Therefore, we can use $M = 3(1)^2 = 3$ to obtain an error bound of

$$\left|E_{Left}\right| \le 3\frac{(1-0)^2}{2(4)} = \frac{3}{8} = 0.375.$$

In Sections 4.4 and 5.3, we will learn methods for computing the exact value of the area under the curve $f(x) = x^3$ on the interval $[0, 1]$ and above the x-axis. The exact area is 0.25. Hence, the error we make in using the left endpoint method with 4 equal subintervals is $|0.25 - 0.141| = 0.109$, clearly less than the error bound 0.375.

EXERCISE SET 4.3

In Exercises 1–6:

a. Reproduce a rough sketch of the graph of the given function f over the given interval. Divide the interval into four equal subintervals and draw the rectangles associated with a right endpoint sum.

b. Use the indicated method to estimate the total accumulation of the quantity whose rate of change is given by $f(x)$ over the specified interval using 8, 16, and 32 equal subintervals.

c. Determine whether the estimate is an overestimate or an underestimate.

d. Find an error bound using the stack method error bound for increasing or decreasing functions, and $n = 32$.

e. How many subintervals are required to guarantee that the estimate is within 0.005 of the total accumulation?

1. $f(x) = \sin(x)$ on $[0, \frac{\pi}{2}]$ using left endpoints.

2. $f(x) = x^2 + 3x - 5$ on $[2, 5]$ using right endpoints.

3. $f(x) = \cos(x)$ on $[0, \frac{\pi}{2}]$ using left endpoints.

4. $f(x) = \dfrac{1}{x-4}$ on $[1, 3]$ using right endpoints.

5. $f(x) = \tan(x)$ on $[-\frac{\pi}{4}, \frac{\pi}{4}]$ using left endpoints.

6. $f(x) = \cos^{-1}(x)$ on $[0, 1]$ using right endpoints.

In Exercises 7–10

   a. Use the left endpoint method to estimate the total accumulation of the quantity whose rate of change is given by $f(x)$ over the specified interval using 16 and 32 equal subintervals.

   b. Use the given exact value to determine the actual error made in the estimation in part (a).

   c. Use the derivative method error bound to find an upper bound for the error when $n = 32$. (Hint: You need to determine the maximum value of $|f'(x)|$ for the given function on the given interval.)

7. $f(x) = -\frac{1}{2}(x - 2)^2 + 3$ on $[0, 4]$. The exact accumulation is $\frac{28}{3}$.

8. $f(x) = \sin(x)$ on $[0, \pi]$. The exact accumulation is 2.

9. $f(x) = e^x$ on $[0, 1]$. The accumulation (accurate to five decimals) is 1.71828.

10. $f(x) = x^3 - 4x^2 + 5x + 10$ on $[-1, 4]$. The exact area is $\frac{775}{12}$.

In Exercises 11–14, find the value of n that will guarantee that your estimate of the total accumulation of the given function on the given interval will be accurate to within 0.001 using the stack method error bound.

11. $f(x) = x^2$ on $[2, 5]$

12. $g(x) = \sin(x)$ on $[\pi/2, \pi]$

13. $h(x) = \arctan(x)$ on $[1, 5]$

14. $k(x) = \ln(x)$ on $[1, 4]$

In Exercises 15–18, find the value of n that will guarantee that your estimate of the total accumulation of the given function on the given interval will be accurate to within 0.001 using the derivative method error bound.

15. $f(x) = x^2$ on $[2, 5]$

16. $g(x) = \sin(x)$ on $[\pi/2, \pi]$

17. $h(x) = \arctan(x)$ on $[1, 5]$

18. $k(x) = \ln(x)$ on $[1, 4]$

19. Discuss the relationshipbetween the values for n in Exercise 11 and in Exercise 15. How would the severity of the bending of the curve affect the values of n?

20. Discuss the relationship between the values for n in Exercise 12 and in Exercise 16. How would the severity of the bending of the curve affect the values of n?

21. Discuss the relationship between the values for n in Exercise 13 and in Exercise 17. How would the severity of the bending of the curve affect the values of n?

22. Discuss the relationship between the values for n in Exercise 14 and in Exercise 18. How would the severity of the bending of the curve affect the values of n?

23. Is it possible to find a function for which each estimation gives the exact value of the total accumulation? If so, do so. If not, explain why not.

24. Discuss how these methods can be used to evaluate the area under the graph of the function $f(x) = [x^2]$ on the interval $[0, 2]$, where $[u]$ is the greatest integer in u and is written int(u).

4.4

The Definite Integral

In previous sections, we have studied various methods for estimating the total accumulation of a quantity over a time interval from information on the rate of change of that accumulation. In this section, we will define such an accumulated quantity to be a *definite integral*. We will see how the error bounds found in the last section allow us to compute the *definite integral* exactly.

Visit to a Small Planet

A new, extremely small planet has recently been discovered on the outer edges of the solar system. On this planet, objects that are dropped off cliffs fall much more slowly than on the earth. A huge ore deposit has been discovered on the

Coalrus Mining Equipment

planet and the Coalrus Mining Company is making plans to set up operations there.

EXAMPLE 1 **The Coalrus Project**

Preliminary data from the planet show that the velocity of a dropped object in feet per second as a function of time in seconds is given by the equation

$$v(t) = 2t.$$

The Coalrus Company wants to use this formula to see how their equipment might have to be modified for operation on the planet.

In particular, it is easy to see that after three seconds, a lump of coal falls with a velocity of $v(3) = 2 \cdot 3 = 6$ feet per second, but *how far* has the lump fallen in that time? This appears to be just the sort of problem we've discussed in previous sections.

We begin by estimating the distance the coal has fallen using the right endpoint method. First, we subdivide the interval of time from zero to three seconds into small time intervals, say tenths of a second. Second, we use the value of the velocity function at the right endpoint of each subinterval to estimate the average velocity over the interval. For each interval, we take the product of our estimated average velocity with the length of the interval to estimate the distance the object has fallen during that interval. Finally, we estimate the total distance by summing the estimated distances for each of the 30 time subintervals. We compute

$$Distance = 2(.1)(.1) + 2(.2)(.1) + 2(.3)(.1) + \cdots + 2(2.9)(.1) + 2(3.0)(.1)$$
$$= 9.3 \, \text{ft}$$

as our estimate for the distance traveled by the falling object. We can even determine that our estimate is within

$$(v(3) - v(0)) \cdot \frac{3 - 0}{30} = 0.6 \, \text{ft}$$

Using the stack error bound formula.

of the actual distance.

▶ FIRST REFLECTION

Why are we justified in using this bound on the estimate?   ◀

Unfortunately, this is not good enough. The equipment used by Coalrus is extremely precise and the distance computation must be exact.

We know from our analysis of the error bound that we can increase the accuracy of the estimate by increasing the number of time intervals used. Can we use this fact to obtain an exact answer?

If we divide the interval into n subintervals, the length of each subinterval is $\Delta t = \dfrac{3 - 0}{n}$. The right endpoints of the first three subintervals are shown

FIGURE 4.25   Right Endpoints of Subintervals

in Figure 4.25 below the number line. Each right endpoint will be of the form $0 + k\Delta t$. Our estimate becomes

$$\sum_{k=1}^{n} 2(0 + k\Delta t)\Delta t.$$

The error for this estimate is equal to

$$(v(3) - v(0)) \cdot \frac{3 - 0}{n} = \frac{18}{n} \text{ ft.}$$

We can make the error as small as we like by choosing a large enough value for n.

▶ SECOND REFLECTION

Estimate the total distance the lump of coal falls using the first formula in *The Limit of a Sum of Products* below and the error for $n = 100$ and $n = 1000$. ◀

To obtain an exact answer, all we need to do is compute the limit of our sum as $n \to \infty$. We can represent this limit symbolically as

$$\lim_{n \to \infty} \sum_{k=1}^{n} 2(k\Delta t)\Delta t.$$

Although we will use h *for* Δt *in what follows, you can work with* Δt *by using* 2nd CHAR *to access the Greek character* Δ.)

THE LIMIT OF A SUM OF PRODUCTS

Using h for Δt for convenience, we define h to to be $\frac{3}{n}$ using the STO command, 3/n → h. Compute the sum of the products over n subintervals using right endpoints of each subinterval with the Σ command,

$$Σ(2(0+k*h)*h,k,1,n).$$

Finally, edit this Σ(command (or cut and paste) to obtain

$$\text{limit}(Σ(2(0+k*h)*h,k,1,n),n,\infty).$$

Again, we have used h for Δt.

FIGURE 4.26   Riemann Sum with h Used for
Δt

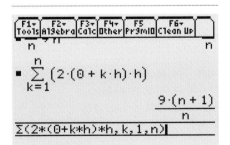

The first part of the computation is shown in Figure 4.26. The second step of the computation uses the formula for the sum of the first n integers, namely, $\frac{n(n + 1)}{2}$. It makes sense that the sum is a function of n since we know our estimates will be different if we use a different number of subintervals.

Also, we can make $\frac{n + 1}{n}$ as close as we like to one simply by choosing large values for n, so the final limit calculation, nine, is correct.

We can report to the Coalrus Company with confidence that the lump of coal will fall exactly nine feet in three seconds.

Definition of the Definite Integral

For the first time, we have computed the exact accumulation of a quantity over an interval using the function which represents the rate of that accumulation. This exact quantity, which we have only been estimating up to this point, is called the *definite integral*.

Recall that when all of the subintervals are the same size, $\dfrac{b-a}{n}$, we say the Riemann sum is being taken over a *regular partition*. We will soon give a more general definition for the definite integral by relaxing this regularity condition. For the time being, however, the stated definition is more than adequate for our purposes.

THE DEFINITE INTEGRAL: REGULAR PARTITIONS

The **definite integral** of a continuous function, f, on the interval $[a, b]$, written symbolically as

$$\int_a^b f(x)\,dx,$$

is defined to be the limit as $n \to \infty$ of a Riemann sum, $\sum_{k=1}^{n} f(t_k)\Delta x$, where t_k is an arbitrary point in the kth subinterval $\left[x_{k-1}, x_k\right]$ of $[a, b]$, and Δx is the length $x_k - x_{k-1}$ of the kth subinterval. Symbolically,

$$\int_a^b f(x)\,dx = \lim_{n\to\infty} \sum_{k=1}^{n} f(t_k)\Delta x.$$

The subintervals are taken to be of equal length, $\Delta x = \dfrac{b-a}{n}$.

The function f is called the **integrand** while a and b are the endpoints of the interval and are called the **lower** and **upper limits of integration**, respectively.

In the language of definite integrals, the Coalrus Company was asking us to compute the definite integral of $v(t) = 2t$ on the interval $[0, 3]$. Symbolically, we computed

$$\int_0^3 v(t)\,dt$$

using the limit as $n \to \infty$ of the right endpoint Riemann sum.

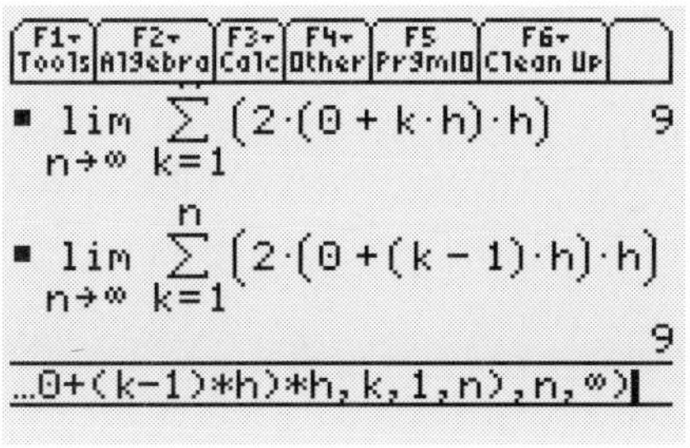

Computing the Definite Integral. To obtain the definite integral from the left endpoint Riemann sum instead of the right endpoint Riemann sum, we simply replace k in the formula with $k - 1$ as shown in Figure 4.27. Notice that the result is still nine.

FIGURE 4.27 Computing the Definite Integral

Think and Share

Break into groups. Write the midpoint Riemann sum for computing the definite integral of $v(t) = 2t$ on the interval $[0, 3]$. Input your sum in your TI calculator and compute its limit as $n \to \infty$.

We can choose the left endpoint, right endpoint, or midpoint for t_k in each subinterval. These choices generate Riemann sums that differ only in the value that is computed for the average rate of change ($f(t_k)$) on the subinterval. However, the limits of these Riemann sums are the same.

We can also use the value of our continuous function *at any point* in a subinterval as an estimate of the average rate of change over that subinterval. Moreover, the choice in each subinterval can be made in completely different ways. We could use the midpoint for the first subinterval, the left endpoint for the second subinterval, a random choice for the third subinterval, etc. Our choice can be arbitrary because for very large values of n, the subintervals are very small, and so all of the outputs of our continuous function on a particular subinterval must be very close together.

Again, we can use *any* point t_k in the kth subinterval. We can illustrate this property of the definite integral using the TI calculator `rand()` function. This function generates a random number between zero and one. Thus, `rand()·h` generates a random number between zero and `h` (Δt). If we compute `k*h - rand()h`, we obtain a random number in the kth subinterval of $[0, 3]$.

▶ THIRD REFLECTION

Write a Riemann sum for computing the definite integral of $v(t) = 2t$ on the interval $[0, 3]$ which uses a randomly generated value for t_k. Input your sum into the TI calculator and compute its limit as $n \to \infty$. How does your result compare to the previously computed solutions?   ◀

Limitations of the Limit.   Despite the power of the definite integral, we should not feel overly confident about our ability to compute values for them. After all, the function we were using, $v(t) = 2t$, was quite simple. A crucial step in the calculation was the following computation performed by the TI calculator.

$$\sum_{k=1}^{n} 2(k \, \Delta t) \Delta t = \frac{9(n+1)}{n}.$$

This is a surprisingly difficult computation by hand for such a simple function. It requires that you know

$$\sum_{k=1}^{n} k = \frac{n(n+1)}{2}.$$

EXAMPLE 2   **A More Complicated Integrand**

We now attempt to use the same approach to find

$$\int_0^\pi \sin(t)\, dt = \lim_{n \to \infty} \sum_{k=1}^{n} \sin(0 + k\,\Delta t)\Delta t.$$

Using the TI calculator, enter

π/n →h   **(4.3)**

limit(Σ(sin(0+k*h)*h,k,1,n),n,∞).

Figure 4.28 shows the results of this computation. Notice that π/n is split between the numerator and the denominator in the last answer.

Notice that the limit has not been carried out.

FIGURE 4.28   Limit Computation for $\int_0^\pi \sin(t)\, dt$

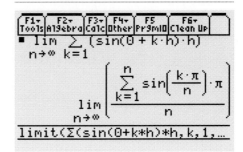

Just as was often the case for derivatives, an algebraic simplification is required before the limit can be computed. For polynomials, this simplification is accomplished with the use of formulas like

$$\sum_{k=1}^{n} k^1 = \frac{n(n+1)}{2},$$

$$\sum_{k=1}^{n} k^2 = \frac{n(n+1)(2n+1)}{6}, \text{ and}$$

$$\sum_{k=1}^{n} k^3 = \frac{n^2(n+1)^2}{4}.$$

The sine function requires a different simplification which is evidently not programmed into the TI calculator. We will revisit this integral in Chapter 5.

Think and Share

Break into groups. Estimate the value of $\int_0^\pi \sin(t)\,dt$ by evaluating a right endpoint Riemann sum with $n = 10, \ 100, \ 1000$. You can use the "with" ($\mathbf{I}$, the vertical bar) command to make your life easier. Conjecture a value for $\int_0^\pi \sin(t)\,dt$ based upon the results of your Riemann sum calculations. Write down your conjecture for future use. We will return to this integral as promised.

The error bounds for our Riemann sums with regular subdivisions lead to another subtle confusion regarding the definite integral. The bounds tell us that sums using a large number of subdivisions will all be close to the definite integral and so must be close to each other. It is easy to misinterpret this to mean that if we have sums that are close together, then the error must be small and they must have values close to the definite integral. For example, if the sum using 16 subintervals is within 0.01 of the sum using 17 subintervals it seems natural to think that the error in the estimate using 17 subintervals is at most 0.01.

Unfortunately, this is not always the case. Two Riemann sums can be closer together than the error bound for either one. We point this out because, in actual practice, the closeness of estimates is often used to determine error bounds.

Summary: Riemann Sums and the Definite Integral

In this section, we studied how to determine the exact amount accumulated over time by a rate function. We saw that the Riemann sum used in previous sections to estimate this quantity can be made exact because the bound on the error of this estimate can be made arbitrarily small. In the limit, our error bounds go to zero; hence, our estimates become exact. This idea of estimating a quantity with a Riemann sum so that a limit of the estimates is exact can be exploited in many situations, some of which we will explore in a later section.

EXERCISE SET 4.4

In Exercises 1–3, find the definite integral by computing the limit of a Riemann sum using the left or right endpoint sums.

1. $\int_{-2}^5 4\,dx$. If the integrand, 4, is a rate function, does your answer agree with your intuition? Explain.

2. $\int_{-2}^5 8\,dx = \int_{-2}^5 2 \cdot 4\,dx$. Compare your result with that of Exercise 1.

3. If $\int_{-1}^4 f(x)\,dx = 17$, conjecture a value for $\int_{-1}^4 kf(x)\,dx$, where k is any real number. Justify.

4. $\int_5^{-2} 4\,dx$. (If the upper limit is less than the lower limit, Δx will be negative.) Explain the relationship between the result of this exercise and the result of Exercise 1.

5. If $\int_{-1}^4 f(x)\,dx = 17$, conjecture a value for $\int_4^{-1} f(x)\,dx$. Justify.

6. Graph $f(x) = x^2 - 3x - 4$. Explain why $\int_0^4 x^2 - 3x - 4\,dx$ is negative.

7. Must $\int_{-1}^4 g(x)\,dx$ be positive if $g \geq 0$ on [-1,4]? Justify, using a Riemann sum argument.

8. Can $\int_{-5}^3 f(x)\,dx$ be positive if $f \leq 0$ on [-5,3]? Explain, using a Riemann sum argument.

9. Is $f(x) = x^3 + 5x$ an even or odd functon? (Recall: f is *even* if $f(-x) = f(x)$ and f is *odd* if $f(-x) = -f(x)$.) Graph $f(x)$.

10. Estimate $\int_{-5}^5 x^3 + 5x\,dx$ using Riemann sums with $n = 4$, 16, and 32. When is the definite integral of an odd rate function or integrand zero, as it is in this case? Explain.

In Exercises 11–22, build tables of *exact* values for definite integrals with a fixed lower bound and different upper bounds using the methods of this section and compare your results. The problems are arranged in groups of three.

11. $\int_0^b 4x \, dx$

b	0	1	2	3	4	5
$\int_0^b 4x \, dx$	0					

12. $\int_1^b 4x \, dx$

b	0	1	2	3	4	5
$\int_1^b 4x \, dx$	−2					

13. Compare the values in Exercises 10 and 11. Write a sentence describing the relationship between the two integrals. Support your conclusion with an illustration.

14. $\int_0^b 3x^2 \, dx$

b	0	1	2	3	4	5
$\int_0^b 3x^2 \, dx$						

15. $\int_2^b 3x^2 \, dx$

b	0	1	2	3	4	5
$\int_2^b 3x^2 \, dx$						

16. Compare the values in Exercises 14 and 15. Write a sentence describing the relationship between the two integrals. Support your conclusion with an illustration.

17. $\int_0^b 5x^2 \, dx$

b	0	1	2	3	4	5
$\int_0^b 5x^2 \, dx$						

18. $\int_1^b 5x^2 \, dx$

b	0	1	2	3	4	5
$\int_1^b 5x^2 \, dx$						

19. Compare the values in Exercises 17 and 18. Write a sentence describing the relationship between the two integrals. Support your conclusion with an illustration.

20. $\int_0^b \frac{x^2}{10} \, dx$

b	0	1	2	3	4	5
$\int_0^b \frac{x^2}{10} \, dx$						

21. $\int_1^b \frac{x^2}{10} \, dx$

b	0	1	2	3	4	5
$\int_1^b \frac{x^2}{10} \, dx$						

22. Compare the values in Exercises 20 and 21. Write a sentence describing the relationship between the two integrals. Support your conclusion with an illustration.

23. Use estimation methods to complete the table of values for the definite integral $\int_0^b \sin\left(\frac{\pi}{2}x\right) \, dx$.

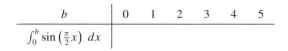

b	0	1	2	3	4	5
$\int_0^b \sin\left(\frac{\pi}{2}x\right)\, dx$						

24. Use estimation methods to complete the table of values for the definite integral $\int_1^b \sin\left(\frac{\pi}{2}x\right)\, dx$.

b	0	1	2	3	4	5
$\int_1^b \sin\left(\frac{\pi}{2}x\right)\, dx$						

25. Compare the values in Exercises 23 and 24. Write a sentence describing the relationship between the two integrals.

26. Investigate, through a trial and error procedure, the types of functions for which your calculator can compute the limit of a Riemann sum. You might try classes of functions such as polynomial, rational, exponential, logarithmic, radical, and trigonometric. Write your conclusion in a systematic manner.

4.5

Other Methods and Their Error Bounds

The left endpoint and right endpoint methods provide us with a straightforward method for estimating the value of a definite integral. These methods involve sums of products $f(t_k)\Delta x$, where $f(t_k)$ remains constant on each subinterval. We can illustrate each of these methods by drawing rectangles of height $f(t_k)$ and length Δx. As we have seen, we can decrease the error by increasing the number of subintervals.

▶ FIRST REFLECTION

What are the stack and derivative error bound formulas? When must we use the derivative method?   ◀

Although these estimation techniques are simple to visualize and to compute, they are not very efficient, often requiring extremely large values of n which, in turn, can take a long time to compute on the TI calculator.

In this section, we will develop more elaborate methods for estimating the value of definite integrals using a linear approximation for the function on each subinterval for two of the methods, and a quadratic approximation for the third. The accuracy of these new methods will be compared with the accuracy of left and right endpoint methods.

The Trapezoid Method

One method for estimating total accumulation in Sections 4.1 and 4.2 used the average value of the function at the left and right endpoints of each subinterval as the estimate for the average rate of change over the subinterval. So far, we have not given a graphical representation for this "averaging method." Figure 4.29 displays the trapezoids formed by joining successive points on the graph of a function at the subinterval endpoints with what is called a **secant**

line of the function. The area of each of these trapezoids is the average of the lengths of the sides of the trapezoid (which correspond to the function values at the endpoint) times the width of the trapezoid (Δx). Our computation for the estimated accumulation over a subinterval using the averaging method is exactly equal to the area of the corresponding trapezoid. For this reason, the method is called the *trapezoid method*.

You may recall that the area of a trapezoid is $A = \dfrac{b_1 + b_2}{2} h.$

FIGURE 4.29 One Trapezoid

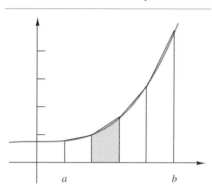

Dripping Coffee Urn

FIGURE 4.30 One Trapezoid

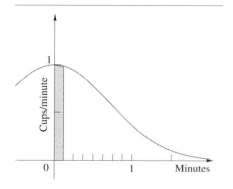

E X A M P L E   1   **Dripping Coffee via the Trapezoid Method**

Coffee is dripping from a large coffee urn at a rate given by the function $f(x) = e^{-x^2}$ where x is the time (in minutes) the coffee has been dripping and $f(x)$ is measured in cups per minute. We use the trapezoid method to estimate the amount of coffee that has dripped out of the urn during the first minute using eight subintervals of equal length.

shows the graph of the function $f(x) = e^{-x^2}$ on the interval from 0 to 1 with eight subintervals. The area of the trapezoid over the first subinterval from 0 to $\frac{1}{8}$ is

$$\frac{f(0) + f\left(\frac{1}{8}\right)}{2} \Delta x.$$

Figure 4.30 The area of the second trapezoid on the subinterval from $\frac{1}{8}$ to $\frac{2}{8}$ is

$$\frac{f\left(\frac{1}{8}\right) + f\left(\frac{2}{8}\right)}{2} \Delta x.$$

Since the subintervals are all the same length $\left(\Delta x = \frac{1}{8}\right)$, the general form for the area of the kth trapezoid, is given by

$$\frac{f\left(\frac{k-1}{8}\right) + f\left(\frac{k}{8}\right)}{2} \frac{1}{8}.$$

The estimate of the accumulation using the trapezoid method and eight subintervals is given by

$$\sum_{k=1}^{n} \frac{f(\frac{k-1}{8}) + f\left(\frac{k}{8}\right)}{2} \frac{1}{8}.$$

Having written our estimate in summation notation, it is fairly straight forward to find the corresponding TI command.

Finding this general form allows us to use the $\Sigma($ command on the TI calculator to compute our estimate.

THE TRAPEZOID METHOD ON THE TI CALCULATOR

To find the trapezoid method estimate of the accumulation due to a rate function f on the interval $[a, b]$, first define f and store values of a, b, and n. For example,

```
Define f(q)=e^(-q^2)
0→a
1→b
8→n
(b-a)/n→h
```

(Remember to use the e^x key in entering e^{-q^2}.) Type the command

```
Σ((f(a+(k-1)*h)+f(a+k*h))/2*h,k,1,n)
```

where n is the number of subintervals of equal length.

We obtain 0.745866 as the approximate value of our trapezoid method estimate for the coffee accumulated over the first minute by pressing ◇ and then Enter ($\approx$).

Error Discussion. How good is the estimate? Figure 4.31 leads us to suspect that the trapezoids are better estimators of the area under the curve when the curve is less severely bent.

FIGURE 4.31 Concavity and Error

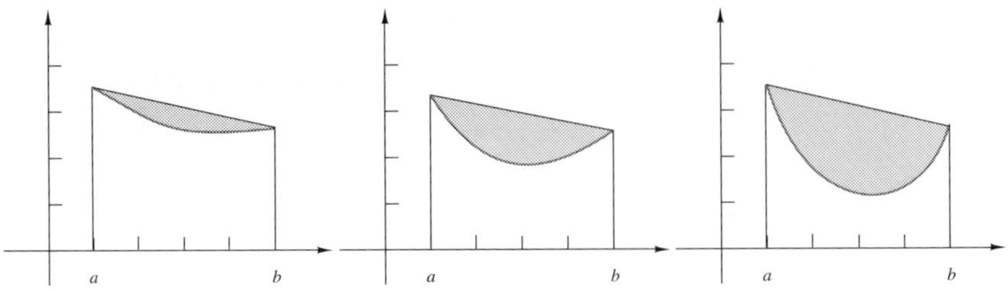

In Chapter 3, we saw that the second derivative measures the "bend" or "concavity" of a function. Consequently, we should not be surprised if a bound on the error for the trapezoid method depends on the second derivative. If E_T represents the error in the trapezoid method estimate, then one such bound is given by

$$|E_T| \leq M \frac{b-a}{12} \Delta x^2 = M \frac{(b-a)^3}{12n^2} \tag{4.4}$$

where M is the maximum value of the absolute value of the function's second derivative on the interval $[a, b]$. How good is our estimate of the coffee accumulation over the first minute?

The second derivative of the function $f(x) = e^{-x^2}$ is

$$f''(x) = 4x^2 e^{-x^2} - 2e^{-x^2}$$

on $[0, 1]$. To find the maximum value of the absolute value of $f''(x)$, we observe that the graph of $|f''(x)|$ is biggest when $x = 0$.

▶ SECOND REFLECTION

Trace your graph of $|f''(x)|$ to find a value for M. Is it the same as the maximum of $f''(x)$?   ◄

The maximum value of the absolute value of the second derivative on the interval $[0, 1]$ computed at $x = 0$ is

$$|f''(0)| = |-2| = 2.$$

Using this as the value of M in the formula for the error bound and T_n to represent the trapezoid method using n subintervals, we find

$$\left| E_{T_8} \right| \leq M \frac{(b-a)^3}{12 \cdot n^2} = 2 \cdot \frac{1}{12 \cdot 8^2} = 0.002604.$$

Thus, the actual amount of coffee (in cups) accumulated after one minute is within 0.002604 of 0.745866 or 0.745866 ± 0.002604.

Think and Share

Break into groups. Calculate an estimate of the coffee accumulated after one minute using the trapezoid method with 64 subintervals. Compute a bound on the accuracy of your estimate. Compute the ratio of the bound found above (0.002604) divided by your bound. How does increasing the number of subintervals by a factor of eight affect the bound?

Midpoint Method

Yet another method for estimating accumulation from a rate of change function uses the midpoint of each subinterval to generate the estimated average rate of change over the subinterval. As before, the length of each subinterval is Δx.

Thus, the estimating product for each subinterval using the *midpoint method* is computed as

$$\text{product} = f(\text{midpoint}_k)\Delta x$$

$$= f\left(a + (k-1)\Delta x + \frac{\Delta x}{2}\right)\Delta x$$

where the interval under consideration is $[a, b]$ and $\Delta x = \dfrac{b-a}{n}$. The expression $a + (k-1)\Delta x$ gives the left endpoint of the kth subinterval; $\Delta x/2$ is half the subinterval length. The sum of these two numbers gives the midpoint of the kth subinterval. If we store Δx as h and the number of subintervals as n, then the total accumulation is estimated by summing these products,

$$\sum_{k=1}^{n} f\left(a + (k-1)*h + \frac{h}{2}\right)h.$$

▶ **THIRD REFLECTION**

Write the TI command to compute the above sum.   ◀

Once again we need a bound on our estimate. Figure 4.32 shows the effect

FIGURE 4.32   Concavity and Midpoint Method Errors

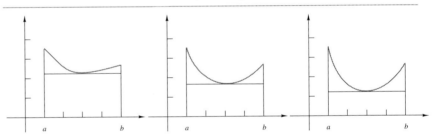

of bend on the midpoint method for several functions. As was the case for the trapezoid method, the more the function is bent, the more erroneous is the midpoint method. (This is easiest to see from the graph if you think in terms of areas.) As with the trapezoid method, a formula to estimate the size of the error once again depends on the maximum value of the absolute value of the second derivative on the interval. Letting E_{Mid} represent the error in the midpoint method on the interval $[a, b]$, the error is given by

Midpoint Method Error Bound

$$|E_{Mid}| \le M\frac{b-a}{24}\Delta x^2 = M\frac{(b-a)^3}{24n^2} \tag{4.5}$$

where M is the maximum value of the absolute value of the function's second derivative on the interval $[a, b]$. Notice that this error bound is half the error bound for the trapezoid method.

EXAMPLE 2   **Dripping Coffee via the Midpoint Method**

Using the midpoint method to estimate the total accumulation generated by the rate function $f(x) = e^{-x^2}$ with $n = 8$ on the interval $[0, 1]$ we find

$$\sum_{k=1}^{8} f\left(0 + (k-1)\cdot\frac{1}{8} + \frac{1}{8}\cdot\frac{1}{2}\right)\cdot\frac{1}{8} = 0.747304.$$

The bound on the error for this estimate is computed as

$$\left|E_{Mid_8}\right| \leq 2\frac{1}{24\cdot 8^2} = 0.001302. \tag{4.6}$$

▶ FOURTH REFLECTION

Use the midpoint method to estimate the total accumulation generated by the rate function $f(x) = e^{-x^2}$ on the interval $[0, 1]$ with $n = 64$ and $n = 128$. Determine the accuracy of your estimates. Do the results seem reasonable? ◀

   We see from Figures 4.33 and 4.34 that the trapezoid method overestimates

FIGURE 4.33   Concave Up Error Comparison   **FIGURE 4.34**   Concave Down Error Comparison

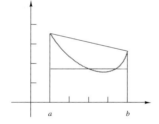
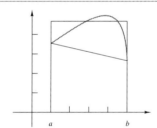

the area under the function when the midpoint method underestimates that area, and vice versa. This relationship is always true for functions that are always concave up or always concave down on the interval.

▶ FIFTH REFLECTION

In Figures 4.33 and 4.34, shade in the trapezoid method estimates and then the midpoint method estimates. How do the shaded areas compare?   ◀

   The area of each rectangle in the midpoint method turns out to be the same as the area under the tangent line drawn at the midpoint of each subinterval. The tangent line is always below the curve, if the curve is concave up; and always above, if the curve is concave down. As a result, when the curve is concave up, the midpoint method underestimates; and when the curve is concave down, the midpoint method overestimates in intervals where the function is always concave up or always concave down. Figure 4.35 shows such tangent line area approximations for the functions in Figures 4.33 and 4.34.

FIGURE 4.35   Comparison of Trapezoid and Midpoint (Tangent Line) Methods

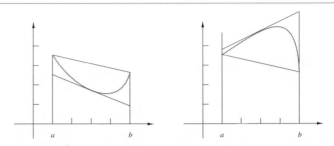

▶ SIXTH REFLECTION

What can you do if the concavity changes from up to down (or vice versa) in the interval? Illustrate your answer with a graph.   ◀

▶ SEVENTH REFLECTION

In Figure 4.35, shade in the trapezoid method estimates and then the midpoint method estimates. How do they compare?   ◀

Simpson's Method

From the error bound formulas in Equations (4.4) and (4.5), we see that the bound on the error for the trapezoid method is twice the bound on the error from the midpoint method. A natural way to exploit these observations is to create a weighted average of the two estimates. We hope to produce an estimate that is more accurate than either one used alone. Since the error bound of the trapezoid method is the larger, we use less of its estimate than we use of the midpoint method's estimate. For example, if a function is increasing and concave down on an interval, we use two parts of the midpoint method overestimate to one part of the trapezoid method underestimate. The errors will tend to balance out since the error of the trapezoid method is the same size as twice the error of the midpoint method and opposite in sign. This estimate of the area under the curve is called Simpson's method.

$$\text{Simpson's method error} = \frac{2 \cdot \text{midpoint} + \text{trapezoid}}{3}$$

where we divide by three since we are using three estimates and we wish to obtain their average.

E X A M P L E   3   **Dripping Coffee via Simpson's Method**

We have already computed the trapezoid and midpoint estimates for the total accumulation generated by the rate function $f(x) = e^{-x^2}$ with $n = 8$ on the interval $[0, 1]$. Applying Simpson's method to these estimates we obtain the following:

$$\text{Simpson's Method} = \frac{2 \cdot \text{midpoint} + \text{trapezoid}}{3} = \frac{2 \times 0.747304 + 0.745866}{3}$$
$$= 0.746825.$$

The error bound formula for Simpson's methodindexSimpson's method error is given by

$$|E_S| \leq M\frac{b-a}{180}\Delta x^4 = M\frac{(b-a)^5}{180n^4} \tag{4.7}$$

where M is the maximum value of the absolute value of the function's fourth derivative on the interval $[a, b]$.

Think and Share

Break into groups. Use Simpson's method along with your estimates from previous reflections to estimate the total accumulation generated by the rate function $f(x) = e^{-x^2}$ with $n = 64$ on the interval $[0, 1]$. Find an upper bound for the error.

You will consider the error in Simpson's method in the exercises.

EXERCISE SET 4.5

In Exercises 1–6, (a) use the indicated method to estimate the total accumulation of the given rate function over the specified interval using 8, 16, and 32 equal subintervals, (b) determine whether the estimate is an overestimate or an underestimate, and (c) find an error bound for each estimate.

1. $f(x) = x^2 - 3x - 5$ on $[3, 7]$ using the midpoint method.
2. $f(x) = \sin(x)$ on $[0, \pi]$ using the trapezoid method.
3. $f(x) = 5x^2 - 3x + 2$ on $[-1, 4]$ using the trapezoid method.
4. $f(x) = x^3 - 3x^2 - 5x + 2$ on $[3, 5]$ using Simpson's method.
5. $f(x) = \cos(x)$ on $[-5, 5]$ using the midpoint method.
6. $f(x) = \sin(x)$ on $[0, \pi]$ using Simpson's method.

7. Let $f(x) = \cos(x)$ on the interval $[0, \frac{\pi}{2}]$. Let $n = 25$ and rank the estimates L_{25}, T_{25}, R_{25}, Mid_{25}, and S_{25}.

8. Suppose $f'(x) > 0$ and $f''(x) > 0$ on the interval $[0, 10]$. Which is bigger, Mid_{2000} or R_{100}? Justify.

In Exercises 9–14, find the value of n that will guarantee that your estimate of total accumulation of the given rate function on the given interval will be accurate to within 0.001 using the error bound formulas.

9. $f(x) = x^2 - 3x - 5$ on $[3, 7]$ using the midpoint method.
10. $f(x) = \sin(x)$ on $[0, \pi]$ using the trapezoid method.
11. $f(x) = 5x^2 - 3x + 2$ on $[-1, 4]$ using the trapezoid method.
12. $f(x) = x^3 - 3x^2 - 5x + 2$ on $[3, 5]$ using Simpson's method.
13. $f(x) = \cos(x)$ on $[-5, 5]$ using the midpoint method.
14. $f(x) = \sin(x)$ on $[0, \pi]$ using Simpson's method.

In Exercises 15–18, find the value of n required to guarantee an accuracy of 0.001 for the midpoint, trapezoid, and Simpson's estimates of total accumulation of the given rate function on the given interval. Indicate which is the most efficient method.

15. $f(x) = \sin(x)$ on $[0, \pi]$.
16. $f(x) = \cos(x)$ on $[-5, 5]$.
17. $f(x) = \cos(x^2)$ on $[0, \pi/2]$.
18. $f(x) = x^4 - 3x^2 - 5x + 2$ on $[3, 5]$ using Simpson's method.

19. Based on your computations from Exercises 15–18, discuss which method is most effective and justify your result. (Hint: Compare the error bound formulas.)

20. For the function $f(x) = \sin^2(2000\pi x)$, find the Simpson's method estimate for $\int_0^2 \sin^2(2000\pi x)$, $n = 50$ and 100. From your computations, is the value of the integral zero? Is the function always positive or zero? Discuss your computed values and the area under the graph of this function.

21. Find the value

$$\int_0^\pi \sin(x^2)\, dx$$

with an accuracy of 0.000001 for the midpoint, trapezoid, and Simpson's methods. Indicate which method requires the smallest number of subintervals.

22. From your computations in Exercise 21, discuss which of these methods you prefer for the following qualities.
   a. Ease of entry.
   b. Length of time to a value.
   c. Ease of finding accuracy.

23. What method is programmed into your calculator to compute definite integrals? (Hint: You may wish to consult your *User's Guide*.)

4.6

Applications

In the last section, we saw that if we can estimate a quantity with a Riemann sum, then we can represent the exact value for the quantity with a definite integral. The definite integral is computed by taking the limit of a Riemann sum as $n \to \infty$. So far, we have only applied this idea to determining total accumulation given a rate of change. In this section, we will see that Riemann sums can be used as an estimation tool in a much wider class of problems. For each of these problems, the definite integral will provide the exact value of the quantity or measurement in question.

Volume

An interesting use of the definite integral is to compute volumeindexVolume.

EXAMPLE 1 **The Cone Example**

We take our first example from geometry. You may already be familiar with the formula for the volume of a right circular cone of height h and radius r (if you're not, don't worry). Have you ever wondered where the formula comes from? We will use the definite integral to derive this formula.

As a first step, we place the cone in a coordinate system so that we can conveniently represent the aspects of the cone analytically (that is, with formulas). We choose to orient the cone sideways with the x-axis running along the axis of the cone, as in Figure 4.36.

The top edge of the cone is a straight line through the points $(0, r)$ and $(h, 0)$. The slope of the line is $\dfrac{-r}{h}$ and the y-intercept is r; hence, we can

FIGURE 4.36   Oriented Cone

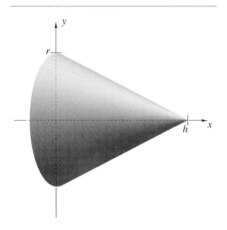

FIGURE 4.37   Slices of a Cone

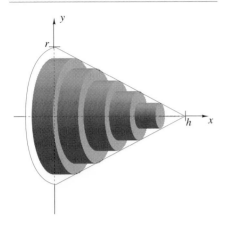

represent the line with the formula $f(q) = \dfrac{-r}{h}q + r$. Define this function on your TI calculator.

Our next step is to find an estimate for the volume of the cone. Suppose we slice the cone vertically into six pieces and approximate each slice with a cylinder as in Figure 4.37. The volume of each of the six pieces can be estimated by a cylinder (see Figure 4.38). Notice that the radius of the first cylinder equals

FIGURE 4.38   Volume of a Slice is the Volume of a Cylinder

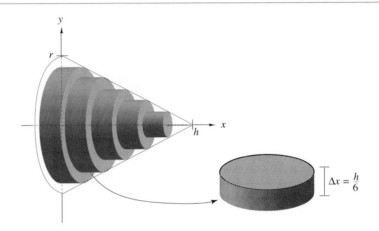

The volume of a cylinder is base × height = $\pi r^2 h$.

$f(\frac{h}{6})$. The height of the cylinder (which corresponds to its width when it is standing on end as in the figure) also equals $\frac{h}{6}$. Thus, the volume of the first cylinder equals $\pi \left(f(\frac{h}{6}) \right)^2 \left(\frac{h}{6} \right)$. Each of the cylinders has a similar formula for its volume and we can use the sum of these volumes to estimate the total volume of the cone.

▶ FIRST REFLECTION

Is the estimate of the volume of the cone given by our cylinders too big or too small? Use the figure to explain your answer.   ◀

Compute this estimate using the TI calculator by computing

$$\sum_{k=1}^{6} \pi \left(f \left(k \cdot \frac{h}{6} \right) \right)^2 \frac{h}{6}$$

where $k \cdot \dfrac{h}{6}$ is the right endpoint of the kth subinterval. Be aware that the estimate you obtain depends upon both the height and the radius. This makes sense because, if we change the height or radius of the cone, we expect that the volume will also change.

▶ SECOND REFLECTION

Look up the formula for the volume of a cone, if you don't already know it. How does the estimate of volume computed by the TI calculator compare to the volume obtained by using that formula? Does this make sense in light of your answer to the last reflection? ◀

Our estimate differs from the actual volume of the cone because of extra space; that is, parts of the cone which are not contained in any cylinder. We can reduce this extra space by using more cylinders. Replacing every $h/6$ in our sum with an h/n, we obtain a formula for the estimate of the volume of the cone when n cylinders are used,

$$\sum_{k=1}^{n} \pi \left(f\left(k \cdot \frac{h}{n} \right) \right)^2 \frac{h}{n}.$$

Use your TI calculator to compute this sum.

Our sum is the Riemann sum generated from the right endpoint method. To see this, let $x_k = k \cdot \frac{h}{n}$, $\Delta x = \frac{h}{n}$, and $g(x) = \pi \left(f(x) \right)^2$ to obtain

$$\sum_{k=1}^{n} g(x_k) \Delta x.$$

You may wish to take a moment to verify this transformation. Moreover, if we take enough cylinders, we can make the volume of the extra space as small as we like. Thus, the exact volume of the cone can be obtained by computing the limit of our sum as $n \to \infty$. As previously defined, this limit is the definite integral where $a = 0$ and $b = h$. In symbols,

$$Volume\ of\ cone = \lim_{n \to \infty} \sum_{k=1}^{n} \pi \left(f(x_k) \right)^2 \Delta x = \int_0^h \pi \left(f(x) \right)^2 dx.$$

▶ THIRD REFLECTION

Compute this limit. How does your answer compare to the formula you looked up (or knew) for the volume of a cone? ◀

▶ FOURTH REFLECTION

In Figure 4.38, divide the second subinterval in half and draw the approximating disks based on the two newly created subintervals. Does this new partition using right endpoints improve the estimate? ◀

This example demonstrates a general technique for using the definite integral to find exact values of unknown quantities. All we need to do is to find a Riemann sum which is a "good" approximator of that quantity. To be a "good" approximator for the quantity, we must be able to make the error in the Riemann sum as small as we like by using large values of n in approximating the volume of the cone. Notice that the Riemann sum does not have to be at all close to the actual value of the quantity when n is small.

Population Estimates

Every 10 years the United States Census Bureau tries to count the population of the United States. The count is used in a variety of important ways which include determining the number of representatives each state has in Congress and the amount of federal aid to which cities and states are entitled.

Recently, the Census Bureau numbers have been the subject of frequent attacks. Many statisticians claim there are serious flaws in the way the Bureau collects its data. The next example describes the sort of study a statistician might undertake to justify such claims.

FIGURE 4.39 A Metropolitan Area

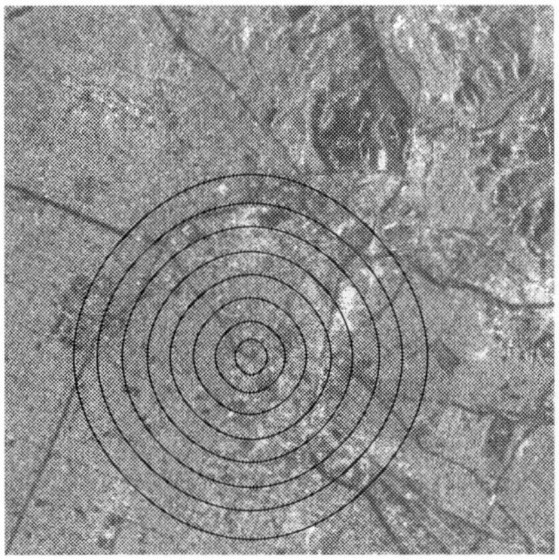

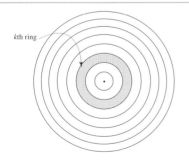

FIGURE 4.40 Population Density Rings

*k*th ring

E X A M P L E 2   **Estimating a City's Population**

A large metropolitan area in the Midwest has a rapidly growing population. Our statistician wishes to determine the accuracy of the United States Census Bureau count of 510,000 for the current population of the area.

An analysis of the Bureau's data reveals that the population measured in people per square mile is a function of the distance from the center of the major city. In particular, as one moves away from the center, the density decreases. As a result, the statistician decides to partition a map of the city into eight rings starting from the center of the city (including the center circle of radius one mile), as shown in Figure 4.39.

The first ring, which is actually a circle, includes the part of the city that is within one mile of the city center. The second ring includes the part of the city that is between one and two miles from the city center. The kth ring includes the part of the city between $k - 1$ and k miles from the city center.

The statistician samples the number of people living in 40 city blocks by randomly selecting five blocks in each of the eight rings. The collected data is displayed in Table 4.25, where the distance is measured in miles and the population density is an average of the densities from the sample blocks in people per square mile.

TABLE 4.25   Distance from City Center vs. Population Density

Distance from center (mi)	1	2	3	4	5	6	7	8
Population density (p/mi^2)	8300	6100	4800	3800	2650	1720	1500	800

The population in the kth ring is the area of that ring multiplied by the population density for that region. From Figure 4.40, the area of the ring is the difference of the areas of two circles. The bigger circle has area πk^2, the smaller has area $\pi(k - 1)^2$; hence, the ring area equals $\pi k^2 - \pi(k - 1)^2$. If the population densities in the table are used as estimates of the population density in the corresponding ring, we can compute an estimate for the total population of the city as

$$\{\pi(1)^2 8300\} + \{(\pi(2)^2 - \pi(1)^2)6100\} + \{(\pi(3)^2 - \pi(2)^2)4800\}$$
$$+ \{(\pi(4)^2 - \pi(3)^2)3800\} + \{(\pi(5)^2 - \pi(4)^2)2650\}$$
$$+ \{(\pi(6)^2 - \pi(5)^2)1720\} + \{(\pi(7)^2 - \pi(6)^2)1500\}$$
$$+ \{(\pi(8)^2 - \pi(7)^2)800\}$$
$$= 475, 857.$$

The estimate seems to indicate that the census figures are too high, but our statistician is still not satisfied. Using the population densities from the table to estimate the density of the ring may be extremely inaccurate. A plot of the data reveals what appears to be more like an exponential decline.

FIGURE 4.41   Population Density Plot on $[0, 9] \times [0, 9000]$

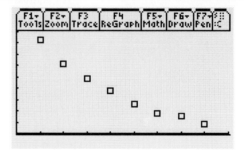

$$p(x) = 12221.6(0.72659)^x \qquad \textbf{(4.8)}$$

The TI calculator gives us Equation (4.8) as an exponential fit for the data where $p(x)$ represents the population density function and x is the distance from the center of the city (see Section 2.2 if you need to review exponential fits).

▶ FIFTH REFLECTION

Graph the exponential function with the data points. Does the curve appear to fit the data? Does our model produce an overestimate or an underestimate? Explain.   ◀

Our continuous exponential function models the discrete data sample. We can use this function to obtain better estimates by subdividing the city more finely than the original nine one-mile rings. (You should be thinking "Riemann sums.")

We suppose that the city is subdivided into n rings where the kth ring includes the part of the city which is between x_{k-1} and x_k miles from the city center and $x_0 = 0$, $x_n = 8$. As before, to determine the number of people in the kth ring, we compute the product of the area of the ring with an estimate of the population density for the ring. Once again, the area of the ring is the difference of the areas of two circles or $\pi x_k{}^2 - \pi x_{k-1}{}^2 = \pi(x_k^2 - x_{k-1}^2)$. We use $p(x_k)$, the population density at the far edge of the kth ring, as our estimate for the population density over the entire ring. Thus, the population estimate for the kth ring equals

$$p(x_k)\pi(x_k^2 - x_{k-1}^2).$$

The sum of these populations is an estimate of the total population,

$$\sum_{k=1}^{n} p(x_k)\pi(x_k^2 - x_{k-1}^2).$$

The limit of these sums, as the width of each ring approaches zero (with each ring the same width), provides our best estimate for the population using this model.

$$Total\ Population \approx \lim_{n \to \infty} \sum_{k=1}^{n} p(x_k)\pi(x_k^2 - x_{k-1}^2)$$

We say "approximately equal" in this case because $p(x)$ is a model, not an exact density function. It appears that all we need to do is recognize this limit as a definite integral and compute. Unfortunately, our sum is not a Riemann sum!

▶ SIXTH REFLECTION
Compare the above sum to the definition of a Riemann sum. Why isn't our sum a Riemann sum? ◀

What to do? The expression $x_k^2 - x_{k-1}^2$ can be factored.

$$(x_k^2 - x_{k-1}^2) = (x_k + x_{k-1})(x_k - x_{k-1}) = (x_k + x_{k-1})\Delta x$$

The limit can now be rewritten

$$\lim_{n \to \infty} \sum_{k=1}^{n} p(x_k)\pi(x_k + x_{k-1})\Delta x.$$

The sum in this limit is still not a Riemann sum because it includes both x_k and x_{k-1} and is not a function of just one point in the kth subinterval.

Our decision to use $p(x_k)$, the population density at the outer edge of the ring, was arbitrary. We could use the value of p at any point between x_{k-1} and x_k. In particular, we could have chosen the midpoint, $\dfrac{x_k + x_{k-1}}{2}$. With this choice and just a bit more algebra, we rewrite our limit as the limit of a Riemann sum,

$$\lim_{n \to \infty} \sum_{k=1}^{n} p\left(\frac{x_k + x_{k-1}}{2}\right)\pi\frac{2(x_k + x_{k-1})}{2}\Delta x$$

$$= 2\pi \lim_{n\to\infty} \sum_{k=1}^{n} p(t_k) t_k \Delta x$$

$$= 2\pi \int_{0}^{8} x p(x)\, dx$$

where $t_k = \dfrac{x_k + x_{k-1}}{2}$ is the midpoint of the kth ring.

We have written our estimate as a definite integral. Unfortunately, the TI calculator cannot compute the limit of this Riemann sum. The exact computation of this integral will have to wait until we develop additional techniques in Chapter 5. We can, however, compute an estimate for the population using a left or right Riemann sum with $n = 100$.

▶ SEVENTH REFLECTION

Compute an approximation for the population with $n = 100$. Would a larger value of n give a better approximation in this situation?   ◀

The Center of Mass

The center of mass is useful in the design of airplanes, automobiles, and three-wheeled vehicles. It is also crucial in physics in understanding planetary motion and quantum mechanics. We can build the tools and concepts needed to deal with such complicated applications by studying simple ones.

E X A M P L E   3   The Seesaw
_____

Susan and her younger sister, Jodi, have been playing on a seesaw (a board supported in the middle by a bar or fulcrum) in the park. Susan weighs 30

FIGURE 4.42   Susan and Jodi on a Seesaw
_____

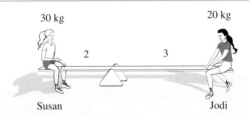

kilograms and Jodi weighs 20 kilograms. Jodi has noticed that no matter where she sits on the seesaw, Susan can always find a place to sit on the other side so that the seesaw is perfectly balanced. The two children decide to record their distances from the fulcrum when the seesaw is balanced. Susan notices that the seesaw is balanced whenever the product of her weight and her distance from

TABLE 4.26   Balancing Distances in Meters

Susan	2	1.66	1.33
Jodi	3	2.5	2

the fulcrum equals the product of Jodi's weight and Jodi's distance from the fulcrum. The product of Susan's mass times her distance from the fulcrum is called her **moment of mass** about the fulcrum.

▶ EIGHTH REFLECTION

What is Jodi's moment of mass when she sits one meter from the fulcrum? ◀

The board they are using is 10 meters long but very light. Jodi asks Susan if she can figure out how they can place the board so that they can each sit on the two ends. Susan reasons that if the fulcrum is $\bar{x}$ meters from her end of the seesaw it will be $10 - \bar{x}$ meters from Jodi's end (see Figure 4.43). To be balanced, Susan's moment of mass about $\bar{x}$ must equal Jodi's moment of mass about $\bar{x}$ which leads to the equation,

$$30\bar{x} = 20(10 - \bar{x}).$$

FIGURE 4.43 The Seesaw on the Number Line

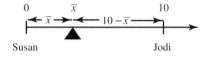

▶ NINTH REFLECTION

What is the answer to Jodi's question? ◀

A standard principle of physics states that for the seesaw to balance, the fulcrum must be placed so that the moments of mass on one side of the fulcrum equal the moments of mass on the other side. The point where the fulcrum must be placed to balance the seesaw is called the **center of mass**. If Jodi were a physicist, she would have phrased her question, "Where is the center of mass if the two point masses are placed on the opposite ends of the rigid structure?"

E X A M P L E 4 **Three on a Seesaw**

Jodi's friend, Eddy, arrives at the park and sits on the same side of the seesaw as Jodi, but two meters closer to the center (Susan and Jodi are still on the ends). Eddy weighs 18 kilograms. Now Susan wants to know where she should put the fulcrum so that she can balance both Jodi and Eddy.

Susan decides to use mathematics to solve her problem. With a piece of chalk, she marks a number line on the seesaw at one-meter intervals. The center of the seesaw is at zero; Eddy is sitting at 3, Jodi is at 5, and Susan is at -5. Susan can see right away that placing the fulcrum at the center will not work, but what if she places the fulcrum at 1? In that case, her moment of mass about the fulcrum, also called her moment of mass about 1, is $-(30)(6)$. The minus sign indicates she's to the left of 1. Eddy's moment of mass about 1 is $(18)(2)$

FIGURE 4.44 Three on a Seesaw

and Jodi's moment of mass about 1 is $(20)(4)$. Summing the three moments of mass, Susan computes

$$-(30)(6) + (18)(2) + (20)(4) = -64.$$

Once again, the seesaw will not balance. To balance, the sum must be zero. Susan notices that she can compute all three moments of mass about the fulcrum with the same expressions $m(x - 1)$, where m is the person's mass and x the person's location on the seesaw. The expression even gives the proper signs, since the expression will be negative for anyone sitting to the left of 1.

Susan gets an idea. If, as before, she lets $\bar{x}$ be the location of the unknown center of mass, she can write down an equation and solve it for $\bar{x}$. Her equation is

$$30(-5 - \bar{x}) + 18(3 - \bar{x}) + 20(5 - \bar{x}) = 0.$$

Susan solves her equation and places the seesaw so that she can exactly balance Eddy and Jodi.

▶ TENTH REFLECTION

Solve Susan's equation to find $\bar{x}$, the center of mass. Does your result make sense in this situation?   ◀

In principle, any number of children can be used to balance the seesaw, but this is not recommended for safety reasons.

EXAMPLE 5   **Center of Mass: General Case of n Point Masses**

Often, as in the example, we wish to locate the center of mass using the location of objects along the number line (seesaw). Since we don't know the location of the center of mass, we can't place it at zero on the number line. On the other hand, if the center of mass is at $\bar{x}$ on the number line then the moment of mass for an object at location x with mass m can be computed as $m(x - \bar{x})$. This computation will assign the proper negative value to objects to the left of $\bar{x}$ and the proper positive value to objects to the right of $\bar{x}$. If we have n objects, then their moments of mass about $\bar{x}$ equal zero, which we can write symbolically as

$$\sum_{i=1}^{n} m_i(x_i - \bar{x}) = 0,$$

where x_i and m_i are the location and mass of the ith object. We can solve this equation for $\bar{x}$.

$$\sum_{i=1}^{n} m_i(x_i - \bar{x}) = 0$$

$$\sum_{i=1}^{n} m_i x_i - \sum_{i=1}^{n} m_i \bar{x} = 0$$

$$\sum_{i=1}^{n} m_i x_i = \sum_{i=1}^{n} m_i \bar{x}$$

$$\sum_{i=1}^{n} m_i x_i = \bar{x} \sum_{i=1}^{n} m_i$$

$$\bar{x} = \frac{\displaystyle\sum_{i=1}^{n} m_i x_i}{\displaystyle\sum_{i=1}^{n} m_i}$$

▶ ELEVENTH REFLECTION

Can we cancel the m_i in the equation for $\bar{x}$ and say

$$\bar{x} = \sum_{i=1}^{n} x_i ?$$

Justify. Remember, we only need one counterexample to prove a statement to be false.   ◀

The quantity $m_i x_i$ is the moment of mass of the ith object about the origin. Therefore, the numerator of our expression equals the sum of the moments of mass about the origin for the point masses. The denominator of the expression is the sum of all the masses, and so equals the total mass. We can summarize what we've learned with a definition.

CENTER OF MASS

The **center of mass** of a set of n point masses distributed on a horizontal line is the location for a fulcrum at which the moments of mass to the left are exactly balanced by the moments of mass to the right. This location equals the sum of the moments of mass about the origin for the point masses divided by the total mass of the system. Symbolically,

$$\bar{x} = \frac{\displaystyle\sum_{i=1}^{n} m_i x_i}{\displaystyle\sum_{i=1}^{n} m_i}. \tag{4.9}$$

The mass of the seesaw can be ignored if we assume a thin supporting structure (filament) for the set of discrete point masses. We neglected the mass of the seesaw because its mass is distributed uniformly rather than being located at a few specific points.

Although a beam cannot be said to consist of a finite number of point masses, we can modify our general point mass solution to solve this new problem. In the next example, we look at the center of mass of a tapered beam.

EXAMPLE 6   The Tapered Beam

An architect wishes to use a beautiful tapered oak beam as the support for a ceiling. The beam is 6.5 m long and tapers from a diameter of 1 m to 0.5 m. For maximum stability, a crossbeam must be placed at the oak beam's center of mass.

How can we apply our understanding of center of mass for a finite set of point masses to this continuous problem? Calculus is perfect for just this sort of process; that is, going from the discrete to the continuous. We need only find a continuous model which is a good approximator of our discrete situation.

We begin by noticing that our beam is actually just a cone with its top cut off. Thus, we can use many of the same ideas we used when we computed the volume of a cone. As with the cone, we orient the beam so that the x-axis runs through its center, as in Figure 4.45 with the origin at the left end of the beam. We divide the beam into n vertical slices (one of which is shown in

FIGURE 4.46   The kth Slice

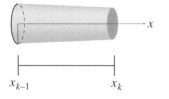

x_{k-1} x_k

FIGURE 4.45   A Beam

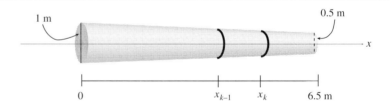

1 m   0.5 m

0   x_{k-1} x_k 6.5 m

Figure 4.45), just as we did with the cone, and analyze the kth slice. (Don't worry. The slices are imaginary; we don't actually cut the beam.) The kth slice and its approximating cylinder are shown in Figures 4.46 and 4.47, respectively.

FIGURE 4.47   The kth Approximation

▶ TWELFTH REFLECTION

What geometric shape is the cross section of each slice of the beam?   ◀

At this point, we introduce mass, using density. The density of an object is its mass per unit volume. We will assume that our oak beam has uniform density ρ. (For most oak beams this is a fairly reasonable assumption.) The mass of an object of uniform density equals the product of its density with its volume.

For our kth slice we write $m_k = \rho V_k$, where m_k is the mass of the slice and V_k is its volume. The total mass of the beam is

$$Total\ Mass = \sum_{k=1}^{n} \rho V_k.$$

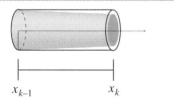

$mass = density \times volume$

$$kg = \frac{kg}{m^3} \times m^3$$

To use our discrete definition of center of mass, we must be able to compute moments of mass. That computation requires that the mass of each slice has a location. Eventually, we plan to create a definite integral by taking a limit as the width of the slices goes to zero. The definition of the definite integral tells us we can choose any location on the intersection of the x-axis with our slice as the location of the slice's mass. We use the right endpoint method and assume that the mass of each of the slices is located at the right end of the slice. With this assumption, we use the sum of the moments of mass of our slices about the origin to approximate the total moment of the beam about the origin,

$$\text{total moment about origin} \approx \sum_{k=1}^{n} x_k \rho V_k.$$

We cannot use our two equations until we find an expression for V_k, the volume of the kth slice, but this is exactly what we had to do in the first example of this section when we computed the volume of a cone.

Think and Share

Break into groups. The top edge of the beam is a line through the points $(0, .5)$ to $(6.5, 0.25)$. Find a function f whose graph between 0 and 6.5 runs along this edge of the beam. Assume that each slice has the same width which we represent as Δx. The volume of the kth slice can be approximated with a cylinder of height Δx and base area $\pi(f(x_k))^2$. Write a formula for V_k using the expression for f you found above. Rewrite the formulas for the *total mass* and the *total moment* using the expression you found which approximates V_k. Put $\lim_{n\to\infty}$ in front of each of your sums and then rewrite them as integrals. Be sure to put the proper limits of integration. Since all of the slices have the same width, $x_k = \dfrac{6.5}{n} \cdot k$ and $\Delta x = \dfrac{6.5}{n}$. Compute the *total mass* and the *total moment* and take their ratio as the center of mass of the beam. Explain in terms of the oak beam why ρ does not appear in the final solution.

In Chapter 5, we will use our notion of a one-dimensional center of mass as a basis for studying a two-dimensional center of mass.

FIGURE 4.48 Area of Region

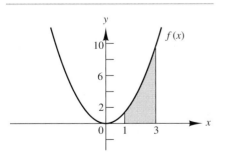

Area

The subsection, "Estimating the Area Under a Curve" Section 4.3 included a brief discussion of interpreting a Riemann sum and the associated definite integrals in terms of area. Such interpretations can provide valuable graphical insights into challenging questions. The next example clarifies this idea.

E X A M P L E 7 **Find the Area of a Region: A Graphing Approach**

Find $\int_1^3 x^2\,dx$, the area of the region shown in Figure 4.48. Using the definition of definite integral, we can rewrite this integral as the limit of a Riemann sum:

FIGURE 4.49 Right Endpoint Rectangle

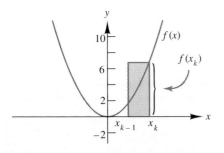

$$\int_1^3 x^2\, dx = \lim_{n \to \infty} \sum_{k=1}^{n} f(t_k) \cdot \Delta x$$

$$= \lim_{n \to \infty} \big(f(t_1) \cdot \Delta x + f(t_2) \cdot \Delta x$$

$$+\, f(t_3) \cdot \Delta x + \cdots + f(t_{n-1}) \cdot \Delta x + f(t_n) \cdot \Delta x \big)$$

where $f(x) = x^2$ and t_k is any point in the subinterval $[x_{k-1}, x_k]$.

Each of the products in the Riemann sum is represented graphically by the area of a rectangle whose base is on the x-axis. The width of this rectangle is Δx and its height is $f(t_k)$, the value of the function at the left or right endpoint or somewhere in-between. A rectangle corresponding to the right endpoint choice is shown in Figure 4.49. We would like to draw n rectangles representing each product in the sum moving left to right.

Creating a Script for a Riemann Sum. To do this with our calculators, we use x for x_{k-1} and x + h for x_k (don't try this yet, keep reading). The general rectangle with baseline segment from x to x + h is displayed in Figure 4.50. The heights of the right rectangles are based on the function value at the right endpoint. In this case, the value is f(x+h). This rectangle can be drawn with three line segments starting at $(x, 0)$ using the Line command. (See *Graphing the First Product.*)

GRAPHING THE FIRST PRODUCT

Clear the HOME screen. Define $f(x)$. (For our example, use Define f(x)=x^2.) Use the WINDOW editor to set the xmin and xmax values at the lower and upper limits of integration (1 and 3 for our example). The values for ymin and ymax should be -1 and 10, respectively. Graph $f(x)$ by entering Graph f(x) on the HOME screen. Return to the HOME screen and store the value for n, the number of subintervals, using the STO command. (We will begin with $n = 4$.) Store xmin to x.

Compute the width of the subinterval and store the value to h:

$$(\text{xmax-xmin})/n \rightarrow h$$

since xmin and xmax are the left and right endpoints of the interval we are sub-dividing. Finally, draw the first rectangle by entering the three Line commands given below. Return to the HOME screen after each command.

```
Line x,0,x,f(x+h)
Line x,f(x+h),x+h,f(x+h)
Line x+h,f(x+h),x+h,0
```

FIGURE 4.50 Right Endpoint Rectangle Construction

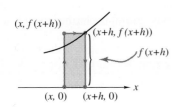

After completing the steps in *Graphing the First Product*, your HOME screen should look something like Figure 4.51. Convert this HOME screen into a script named riemsum using the Save Copy As... command from the

FIGURE 4.51   Drawing a Rectangle

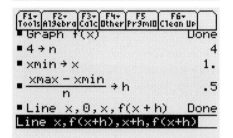

You will need to use F2 *to enter the command symbol* C: *at the beginning of a line.*

FIGURE 4.52   Riemann Sum Representation Script

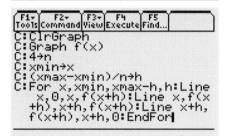

Toolbox menu (refer to Section 2.5 if you need to review creating a script). The statement Define f(x) = x^2 is off the screen (remember to use q for x in the definition if you are using a TI-92).

The For. . . EndFor Command.   We can alter riemsum so that it will draw all of the rectangles for our Riemann sum at once using the For. . . EndFor statement. This process is described in *Graphing a Riemann Sum Representation*.

GRAPHING A RIEMANN SUM REPRESENTATION

Use the Text editor in the APPS menu to edit the riemsum script to look like Figure 4.52. To do this, insert the clrgraph command just above the command Graph f(x) by first moving the cursor down one line to just before the G in Graph and pressing ENTER. Second, press F2 and ENTER to make the Graph line a command with a C: before it. Then move the cursor up one line and type clrgraph.

Now we need to create the For . . . EndFor command. In a script, the commands inside a For. . . EndFor statement must be separated by colons and must all be on the same line. The syntax of the For· · · Endfor command is

> For variable, start, end, increment
>
> *Lines or other commands*
>
> EndFor.

Now, without pressing ENTER, move the cursor to just before the L in the first Line command and edit the Line commands to create the For . . . EndFor command.

Our particular command starts the value of x at xmin and continues by increments of h to xmax. The colons (:) separate the parts of the For· · · EndFor structure. Use the backspace key to place all the Line commands together and then insert colons between them, making them part of the For· · · EndFor statements as shown in Figure 4.52.

▶ THIRTEENTH REFLECTION

Edit riemsum to obtain the graphical representation of the Riemann sum when n is 10, 25, 100. How do the areas of the rectangles compare to the area under the curve? Which method for computing a Riemann sum is riemsum using? ◀

Computing the Approximate Area.   Now that we have a script that will draw the rectangles for us, let's add a Σ(command so that we can find the sum of the areas of these rectangles. Add it as the next command after your For. . . EndFor statement. The command

$$\Sigma(f(xmin+k*h)*h, k, 1, n)$$

will compute the sum of the areas of n right endpoint rectangles.

▶ **FOURTEENTH REFLECTION**

Edit `riemann` to draw the rectangles and compute the sum of the areas of 100 left endpoint rectangles.   ◀

Think and Share

Break into groups of two. In this Think and Share, you will create a script to graphically represent a definite integral on the TI calculator. When you have completed the steps in *Graphing the First Product* and in *Graphing a Riemann Sum Representation*, use the script to draw rectangles for $n = 5, 20,$ and 50. Answer the following questions.

1. Is the sum of the area of the rectangles a better approximation for the area as n gets larger?
2. Alter `riemsum` to use the left endpoint method and display the rectangles for $n = 20$. How do the two graphical representations (left and right endpoint methods) compare?

Work

In physics the work done on an object is defined to be the product of the force applied to the object times the distance the object moves while the force is applied.

$$work = force \times distance$$

where the force is constant and the distance the object moves is along a straight line. For example, the force done in lifting an object weighing 30 pounds from the ground to a ledge 20 feet in the air is

$$work = force \times distance$$
$$= 30 \text{ lbs} \times 20 \text{ feet} = 60 \text{ ft-lbs}$$

The length of an interval is the right endpoint minus the left endpoint. This is true even for vertical intervals. In particular, the distance between the points −5 and 3 on the y-axis is $3 - (-5) = 3 + 5 = 8$ as shown in Figure 4.53.

FIGURE 4.53   Vertical Interval Length

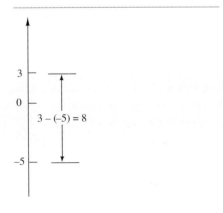

EXAMPLE 8   **Pumping Water**

How much work is done in pumping all of the water in a cylindrical wading pool to the top of the pool? The pool is six feet in diameter and four feet deep and water weighs 62.4 lbs/ft^3.

Following the major idea from this section, we imagine slicing the water into small cylindrical discs all of the same height, and estimate the work done pumping each slice to the top of the pool. Figure 4.54 displays the kth slice of water. We let Δy be $y_k - y_{k-1}$, the height of the slice. The force required to move the kth slice equals its weight and is computed as *weight = volume* × 62.4. Thus, the work done on the kth slice can be approximated.

$$work_{k\text{th slice}} = weight \times distance \approx \pi 3^2 \Delta y \times 62.4 \times (4 - t_k)$$

where $4 - t_k$ represents the approximate distance the kth slice of water is moved.

You will complete this example in the following Reflection.

FIGURE 4.54 Wading Pool

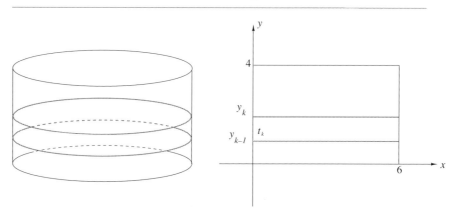

▶ FIFTEENTH REFLECTION

Write a Riemann sum which approximates the work done for n slices. Write an integral that represents the work done exactly. Compute the value of your integral by computing the $\lim_{n \to \infty}$ of your sum. How would this computation change if the tank were the same height but had a square bottom six feet on a side? ◀

EXERCISE SET 4.6

1. Orient a right circular cone of base radius r and height h sideways as done in Example 1, but place its vertex at the origin.
   a. Find a formula for the top edge of the cone through the points $(0,0)$ and (h, r). Define this function on your TI calculator.
   b. Slice the cone vertically into six slices. Draw a picture to show the kth element analysis.
   c. Find an expression for the volume of the kth slice.
   d. Write the sum to estimate the volume of the cone with n slices, and write the associated definite integral.
   e. Find the volume of the cone by finding the limit as n approaches infinity of the approximating sum.

2. Orient a right circular cone of base radius r and height h in a vertical position, with the vertex at the origin, so that the y-axis runs along the cone's axis.
   a. Slice the cone horizontally into six slices. Draw a picture to show the kth element analysis. Observe that the thickness of each slice is Δy.
   b. Find a formula for the edge of the cone through the points $(0,0)$ and (r, h). Solve for x in terms of y, and define this function on your TI calculator.
   c. Find an expression for the volume of the kth slice.

   d. Write the sum to estimate the volume of the cone with n slices, and write the associated definite integral.
   e. Find the volume of the cone by finding the limit as n approaches infinity of the approximating sum.

3. Using the kth element analysis, write the approximating sum and its associated definite integral to calculate the volume of a right circular cylinder of height eight feet and radius three feet positioned so that the x-axis runs through the axis of the cylinder. Include an annotated diagram and calculate the volume of this cylinder.

4. Orient a rectangular pyramid with square base of length b and height h so that the x-axis runs along the axis of the pyramid, and the top of the pyramid points to the right.
   a. Find a formula for the top edge of the pyramid through the points $(0, b/2)$ and $(h, 0)$. Define the function in your TI calculator.
   b. Slice the pyramid vertically into six slices. Draw a picture to show the kth element analysis. Observe that each cross section is a square with thickness Δx. What is the length of each side of the square?
   c. Find an expression for the volume of the kth slice.
   d. Write a sum to estimate the volume of the pyramid with n slices, and write the associated definite integral.

   e. Find the volume of the pyramid by finding the limit as n approaches infinity of the approximating sum.

5. Compute the population density in Example 2 using a cubic function to approximate the data. Should the city demand a recount?

In Exercises 6 and 7, use the following data table where D represents distance from the center of the city and P represents the population density.

TABLE 4.27

D (mi)	1	2	3	4	5	6	7	8	9
P (p/mi²)	7200	5100	3900	2600	1650	720	450	800	100

6. Compute the population of a city that is on the edge of a large lake (like Lake Michigan) from the data in the table above using an exponential function to approximate the data. Clearly, there are no individuals living in the lake!

7. Compute the population of a city that is on the edge of a large lake (like Lake Michigan) from the data in the table above using a cubic polynomial function to approximate the data. Clearly, there are no individuals living in the lake!

8. Compute the population of an ancient city whose population density doubles each mile starting at the city walls and moving towards the center. The city walls are uniformly five miles from the center of the city. The population density at the edge of the city is 200 inhabitants per square mile.

9. Find the center of mass of a rod of uniform material whose density is ρ, whose length is 30 cm, and whose diameter tapers from two centimeters to one centimeter. Indicate the units of ρ.

10. Find the center of mass of a rod of uniform material whose density is ρ, whose length is 15 in, and whose diameter tapers from 0.5 in to 0.25 in. Indicate the units of ρ.

11. Find the center of mass of a compound rod made up of two metals that have density ρ_1 and ρ_2. The compound rod consists of 10 cm of the first metal and 20 cm of the second metal welded together.

12. Find the center of mass of a compound rod made up of two metals that have density ρ_1 and ρ_2. The compound rod consists of 5 in of the first metal and 7 in of the second metal welded together.

In Exercises 13–28, show the kth slice or element analysis including the Riemann sum, write down the resulting definite integral, and give an estimate or exact value for the integral. Be sure to indicate whether your result is an estimate or an exact value.

13. Find the area above the x-axis and below the graph of the function $f(x) = x^2 + 5$ between $x = 2$ and $x = 4$.

14. Find the area above the x-axis and below the graph of the function $g(x) = 2e^x$ between $x = -2$ and $x = 7$.

15. Find the area above the x-axis and below the graph of the function $g(x) = \cos(2x - 3)$ between $x = 1$ and $x = 2$.

16. Find the area above the x-axis and below the graph of the function $h(x) = x^2 - 4x - 5$ between $x = -7$ and $x = -5$.

17. Find the area above the x-axis and below the graph of the function $f(x) = \ln(x)$ from $x = 1$ to $x = 3$.

18. Find the area above the x-axis and below the graph of the function $f(x) = e^{-x^2}$ from $x = -1$ to $x = 1$.

19. Find the area above the x-axis and below the graph of the function $f(x) = 7 - x^2$ from $x = 0$ to $x = 2$.

20. Find the area above the x-axis and below the graph of the function $f(x) = \sin(3x + 1)$ from $x = -2$ to $x = -1.5$.

21. An underground gasoline storage tank is a cylinder lying horizontally on its side and must be replaced for environmental reasons. The radius of the tank is 1 meter, its length is 3 meters, and its top is 3 meters below the ground. The tank is three quarters full. Find the total amount of work done in pumping the gasoline to a point one meter above the ground?

22. Find the work done in pumping the water out (just up to the top) of a cylindrical tank full of water. The tank is five feet deep and eight feet in diameter.

23. A one-ton block of ice in the shape of a cube must be lifted 50 ft, and it is melting at a rate of 2 lb per minute. If it is lifted at the rate of one foot every minute, find the work done in moving the block of ice the 50 ft.

24. Find the work done in pumping the water out (just up to the top) of a cylindrical tank that is six feet deep and nine feet in diameter (a) if the tank is full and (b) if the tank is only half full.

25. Find the work done in pumping the water out (just up to the top) of a conical tank that is five feet high and eight feet in diameter (a) if the tank is full and (b) if the tank is only half full. Be sure to draw a diagram.

26. Find the work done in pumping the water a distance of six feet above the top of a conical tank that is five feet high and eight feet in diameter (a) if the tank is full and (b) if the tank is only half full. Be sure to draw a diagram.

27. A small paper plant in Oregon discharged carbon tetrachloride (CCl_4) into the Klamath River. The EPA learned of the situation in 1994 when the company was discharging CCl_4 at the rate of 16 m³/year. The EPA ordered that within 2 years a filtering system be installed that would slow and finally stop the discharge of CCl_4 from the paper plant. The company completed the installation of the filtering system in 18 months. From the time the filtering system was installed until the time the discharge was stopped, the rate of flow was approximated by

$$\text{Rate (in m}^3\text{/year)} = t^2 - 8t + 16$$

where t is time measured in years since the company received the order.

a. Draw a graph showing the rate of CCl_4 discharge into the river as a function of time, beginning when the company received the order.

b. How many years elapsed between the time the company received the order and the time the discharge stopped entirely?

c. How much CCl_4 was discharged into the river during the time shown in your graph in part (a)?

28. During the 1900s, the yearly electrical energy use (in megawatt-hours) in the US has been increasing at an exponential rate of 7% per year. Assuming that the electricity consumption in 1900 was 1.4 million megawatt-hours and that the exponential trend continues throughout the 1900s,

a. Write down the rate of electricity consumption as a function of time, t, measured in years since 1900 ($t = 0$) and sketch a graph of this rate function.

b. What was the total electrical energy use during the 1900s given this model?

c. What was the average yearly electrical consumption throughout the 1900s given this model?

d. The consumption of electricity was closest to this average in what year(s)?

e. By looking at your sketch in part (a), could you have predicted in which half of the century the answer to part (c) would occur? Explain.

4.7

The Fundamental Theorem of Calculus, Part I

Introduction

Integration, the art of evaluating or computing integrals, has a very broad range of applications. Yet, the initial introduction of the concept of integration in the early 1600s was not immediately followed by a burst of new results. A moment's reflection on our work in Section 4.4 indicates a possible explanation. The algebra required to compute the limit of a Riemann sum can be quite demanding, even for a simple function. Recall that for the function $v(t) = 2t$, we were required to know

$$\sum_{k=1}^{n} 2(k\Delta x)\Delta x = \frac{9(n + 1)}{n}.$$

One of the advantages of technology is that it enables us to compute such limits. On the other hand, even the technology is limited. The TI calculator, for example, was unable to compute

$$\int_0^{\pi} \sin(x) \, dx = \text{limit}(\Sigma(\sin(k*h)*h, k, 1, n), n, \infty)$$

where $h = \dfrac{\pi - 0}{n}$.

In the second half of the 1600s, a powerful method for computing integrals was discovered. It is not a coincidence that this discovery was followed shortly by Isaac Newton's groundbreaking work in the physics of planetary motion.

In the ensuing years, Newton's work was followed by a flood of discoveries, not only in physics, but in a host of different fields, showing the broad range of its applications in areas from economics to population dynamics. Without a convenient method for computing integrals, the pace of these discoveries would certainly have been slowed if not halted altogether.

A word of warning before we begin our study of this method; the method will allow us to compute the definite integral for a very large class of functions, but not all functions. In most cases, we will be freed from the necessity of using limits to compute integrals. However, these computations are useless unless we understand their meaning. To translate a problem from an English description to a mathematical description using integrals, as we did in Section 4.6, requires understanding the definition of the integral as the limit of a Riemann sum.

Computing the Integral of the Sine from 0 to π

In Section 4.4, we were unable to compute

$$\int_0^\pi \sin(x) \, dx$$

because the TI calculator could not evaluate $\lim_{n \to \infty} \sum_{k=1}^n \sin(kh)h$. More generally, we can rewrite our integral as

$$\lim_{n \to \infty} \sum_{k=1}^n \sin(t_k) \Delta x \qquad \textbf{(4.10)}$$

where t_k is any number in the interval $[x_{k-1}, x_k]$, $x_0 = 0$, $x_n = \pi$, and $\Delta x = \dfrac{\pi - 0}{n}$. The Riemann sum in this expression is only an approximation to the definite integral because the expression $\sin(t_k) \Delta t$ is only an approximation to the accumulation of the sine function on the interval $[x_{k-1}, x_k]$. The inaccuracy is caused by the use of $\sin(t_k)$ to approximate the average value of $\sin(t)$ on the interval.

We can use antiderivatives (introduced in Chapter 3 and formally defined below) and the Mean Value Theorem to eliminate this inaccuracy.

AN ANTIDERIVATIVE

The function F is called an **antiderivative** of the function f if $F'(x) = f(x)$.

We begin by noting that $-\cos(x)$ is an antiderivative of $f(x) = \sin(x)$. If we use the label F for this function $-\cos(x)$, then $F(x) = -\cos(x)$. As a result of this labeling convention, $\sin(x) = F'(x)$ is the rate function for F. Thus, $\int_0^\pi \sin(x) \, dx = \int_0^\pi F'(x) \, dx$ is the total accumulation of this newly named function F on the interval $[0, \pi]$.

According to the Mean Value Theorem, there is a number, c_k, in the kth subinterval $[x_{k-1}, x_k]$, such that

$$F'(c_k) = \sin(c_k) = \frac{F(x_k) - F(x_{k-1})}{x_k - x_{k-1}}.$$

Clearing the denominator, we find $\sin(c_k)(x_k - x_{k-1}) = F(x_k) - F(x_{k-1})$. In other words, $\sin(c_k)(x_k - x_{k-1})$ equals the *exact* accumulation of F on the interval $[x_{k-1}, x_k]$. To obtain the total accumulation over the interval $[0, \pi]$ we simply sum these exact accumulations (no need to take the limit) on each of the subintervals,

$$\int_0^\pi \sin(x)\, dx = total\ accumulation \tag{4.11}$$

$$= \sum_{k=1}^n \sin(c_k)(x_k - x_{k-1})$$

$$= \sum_{k=1}^n (F(x_k) - F(x_{k-1}))$$

$$= (F(x_1) - F(x_0)) + (F(x_2) - F(x_1))$$
$$+ (F(x_3) - F(x_2)) + \ldots + (F(x_{n-1})$$
$$- F(x_{n-2})) + (F(x_n) - F(x_{n-1})).$$

The sum on the right of this equation is an example of what is called a *telescoping sum*. All of the terms of this sum cancel (add to zero) except $F(x_n)$ and $F(x_0)$. No matter what partition we use for our Riemann sum, x_0 is always the left endpoint and x_n is always the right endpoint. For our example, $x_0 = 0$ and $x_n = \pi$; hence, our Equation (4.11) reduces to

$$total\ accumulation = \sum_{k=1}^n \sin(c_k)(x_k - x_{k-1}) = F(x_n) - F(x_0)$$

$$= F(\pi) - F(0)$$

$$= (-\cos(\pi)) - (-\cos(0)).$$

Our original definite integral also equals the total accumulation and we have found that

$$\int_0^\pi \sin(x)\, dx = (-\cos(\pi)) - (-\cos(0)) = 2. \tag{4.12}$$

Think and Share

Break into groups. Use the TI calculator to compute the average rate of change of $-\cos(t)$ on the intervals: $\left[0, \frac{\pi}{4}\right], \left[\frac{\pi}{4}, \frac{\pi}{2}\right], \left[\frac{\pi}{2}, \frac{3\pi}{4}\right], \left[\frac{3\pi}{4}, \pi\right]$. For each of the four intervals, find numbers c_1, c_2, c_3, c_4 such that $\sin(c_k)$ equals the average rate of change on the corresponding interval. Compute $\sum_{k=1}^4 \sin(c_k)\frac{\pi}{4}$ using your choices for c_k. How does your result compare to $\int_0^\pi \sin(x)\, dx$ computed above?

The Fundamental Theorem of Calculus, Part I

The technique we used to compute $\int_0^\pi \sin(x)\, dx$ can be used to compute

$$\int_a^b f(x)\, dx$$

for any function f which is continuous on $[a, b]$ and whose antiderivative is known. Suppose that F is an anti-derivative of f and we wish to compute the definite integral

$$\int_a^b f(x)\, dx = \lim_{n\to\infty} \sum_{k=1}^n f(t_k)\Delta x.$$

Let $a = x_0, x_1, \ldots, x_n = b$ be a partition of $[a, b]$. According to the Mean Value Theorem, there exists a number c_k in $\left[x_{k-1}, x_k\right]$ such that

$$F'(c_k) = f(c_k) = \frac{F(x_k) - F(x_{k-1})}{x_k - x_{k-1}}.$$

Thus, $f(c_k)(x_k - x_{k-1}) = F(x_k) - F(x_{k-1})$. We sum these n equalities to obtain

$$\sum_{k=1}^n f(c_k)(x_k - x_{k-1}) = \sum_{k=1}^n \left(F(x_k) - F(x_{k-1})\right) = F(b) - F(a). \quad \textbf{(4.13)}$$

If we recognize that $\Delta x = x_k - x_{k-1}$, the expression on the left of this equation is a Riemann sum. To obtain the integral, we take the limit of the left expression as n goes to infinity. The expression on the right, $F(b) - F(a)$, does not depend on n, so, it remains constant as n goes to infinity. Therefore,

$$\int_a^b f(x)\, dx = \lim_{n\to\infty} \sum_{k=1}^n f(c_k)\left(x_k - x_{k-1}\right)$$
$$= F(b) - F(a).$$

We summarize our observations in the following theorem.

FUNDAMENTAL THEOREM OF CALCULUS, PART I

If f is a continuous function on $[a, b]$, and F is an antiderivative on $[a, b]$ then

$$\int_a^b f(x)\, dx = F(b) - F(a).$$

▶ FIRST REFLECTION

Explain why f has to be continuous. Expand the second sum in Equation (4.13) and verify that it is a telescoping sum that results in $F(b) - F(a)$.   ◀

The Significance of the Fundamental Theorem of Calculus.   The Fundamental Theorem of Calculus (the first part of which we have dealt with here) has deep theoretical and practical significance. Using the concept of an antiderivative, the Fundamental Theorem ties together the two basic operations of calculus: differentiation and integration. The practical significance of the Fundamental Theorem is equally important.

The theorem (Part I) tells us that we can compute a definite integral on an interval simply by taking the difference of the values of an antiderivative at the two endpoints. Perhaps this seems to beg the question, since the theorem does not tell us how to find an antiderivative. We will develop methods for finding antiderivatives of many functions in Chapter 5. For the moment, we will rely on your knowledge of derivatives to generate antiderivatives.

We begin with a simple application of the Fundamental Theorem of Calculus, Part I.

E X A M P L E   1   **Applying the Fundamental Theorem, Part I**

Find

$$\int_0^2 3t^2 \, dt$$

Solution:

First, recall that we know that the function $F(t) = t^3$ is an antiderivative of $f(t) = 3t^2$. Thus,

$$\int_0^2 f(t) \, dt = F(2) - F(0) = 2^3 - 0^3 = 8$$

by the Fundamental Theorem of Calculus.

This exact calculation (based on knowing an antiderivative) is much easier than the calculation that we were required to perform in Sections 4.4 and 4.5 to obtain an exact value for definite integrals.

In the not-too-distant past, entire books were published which were filled with tables of antiderivatives. The advent of technology has ended this practice. Your TI calculator contains its own huge list of antiderivatives, and you can access that list using the $\int$ integrate feature in the Calc menu or using the $\int$ key.

We will use the first part of the Fundamental Theorem of Calculus to compute exact definite integral values frequently. It would, therefore, be helpful if there were a compact notation for $F(b) - F(a)$, namely, $F(x)\Big|_a^b$, where

$F(x)$ is an antiderivative of $f(x)$ and a and b are the lower and upper limits of integration. We will use a standard convention for this expression,

$$\int_a^b f(x)\,dx = F(x)\Big|_a^b = F(b) - F(a).$$

COMPUTING INTEGRALS WITH THE TI CALCULATOR

You compute the definite integral $\int_0^\pi \sin(t)\,dt$ using the following commands:

∫(sin(t), t, 0, π).

You compute an antiderivative of $\sin(t)$ using either of the following commands:

d(sin(t),t,-1) or ∫(sin(t), t).

The following example demonstrates the Fundamental Theorem in action. Try computing the definite integrals using paper and pencil and then compare your solutions to the ones given in the text.

EXAMPLE 2   **Applying the Fundamental Theorem**

Find each of the following definite integrals by guessing an antiderivative and using the Fundamental Theorem of Calculus.

1. $\displaystyle\int_1^5 (x^2 - 3x - 4)\,dx$

2. $\displaystyle\int_{-\pi/2}^{\pi/2} \cos(x)\,dx$

3. $\displaystyle\int_0^1 \frac{1}{(x-2)^3}\,dx$

4. $\displaystyle\int_0^5 \frac{1}{(x-2)^3}\,dx$

Solution:

1.
$$\int_1^5 x^2 - 3x - 4\,dx = \overbrace{\frac{x^3}{3} - 3\frac{x^2}{2} - 4x}^{\text{Guess (check by differentiating)}}\Bigg|_1^5$$
$$= \frac{5^3}{3} - 3\cdot\frac{5^2}{2} - 4\cdot 5 - \left[\frac{1^3}{3} - 3\cdot\frac{1^2}{2} - 4\cdot 1\right]$$
$$= \frac{125}{3} - 3\cdot\frac{25}{2} - 20 - \left[\frac{1}{3} - 3\cdot\frac{1}{2} - 4\right]$$
$$= -32/3$$

2.
$$\int_{-\pi/2}^{\pi/2} \cos(x)\, dx = \sin(x)\Big|_{-\pi/2}^{\pi/2}$$

$$= \sin\left(\frac{\pi}{2}\right) - \sin\left(\frac{-\pi}{2}\right) = 1 - (-1) = 2$$

3.

$$\int_0^1 \frac{1}{(x-2)^3}\, dx = \int_0^1 (x-2)^{-3}\, dx = \overbrace{\frac{(x-2)^{-2}}{-2}}^{\textit{Guess (check)}}\Big|_0^1 = \frac{1}{-2(x-2)^2}\Big|_0^1$$

$$= \frac{1}{-2(1-2)^2} - \left(\frac{1}{-2(0-2)^2}\right)$$

$$= \frac{1}{-2} - \left(\frac{1}{-8}\right) = -\frac{1}{2} + \frac{1}{8} = -\frac{3}{8}$$

4. The integrand in $\int_0^5 \frac{1}{(x-2)^2}\, dx$ is undefined at $x = 2$. The values of the integrand approach ∞ both from the left and the right at $x = 2$. Thus, this definite integral cannot be evaluated. This is an example of what we call an *improper* integral. What condition of the Fundamental Theorem is not true for this integral?

Possible Confusion.   The Fundamental Theorem is so powerful and so often used that students often confuse it with the definition of the definite integral. After computing dozens of definite integrals of the form $\int_a^b f(x)\, dx$ simply by computing $F(b) - F(a)$ where F is an antiderivative of f, it is easy to begin to believe that $\int_a^b f(x)\, dx$ is defined to be $F(b) - F(a)$.

There are two very good reasons to avoid this confusion. First, knowing how to compute a definite integral is no help at all in converting a problem into mathematical form. The analysis that led to the representation of the volume of a cone in terms of a definite integral is completely independent of the Fundamental Theorem. The same statement is true for all the examples in Section 4.6, and it will continue to be true whenever you use the definite integral to solve problems.

The second reason is that finding a formula using basic functions for an antiderivative can be difficult or even impossible. The function $f(x) = \sin(x^2)$ has no elementary antiderivative and thus the formula

$$\int_0^2 \sin(t^2)\, dt = F(2) - F(0)$$

is of no use because we cannot find an expression for F.

▶ **SECOND REFLECTION**

Explain how to evaluate $\int_0^2 \sin(t^2)\, dt$ without being able to find an antiderivative of $\sin(t^2)$.   ◀

Properties of the Definite Integral

There are properties for the definite integral just as there are properties for the derivative. Several important properties of the definite integral are listed below for your reference.

PROPERTIES OF THE DEFINITE INTEGRAL

If f and g are continuous functions on an interval I which contains a, b, and c, then the following are true.

1. $\displaystyle \int_a^a f(t)\, dt = 0$

2. $\displaystyle \int_a^b f(t)\, dt = -\int_b^a f(t)\, dt$

3. $\displaystyle \int_a^b kf(t)\, dt = k\int_a^b f(t)\, dt$

4. $\displaystyle \int_a^b [f(t) + g(t)]\, dt = \int_a^b f(t)\, dt + \int_a^b g(t)\, dt$

5. $\displaystyle \int_a^c f(t)\, dt + \int_c^b f(t)\, dt = \int_a^b f(t)\, dt$

Proving the Properties.   The Fundamental Theorem of Calculus is important in proving properties 3 and 5.

Proof: Assume that F is an antiderivative of f, and G is an antiderivative of g; both f and g are defined on the interval I.

3. From the Fundamental Theorem of Calculus, Part I:

$$\int_a^b kf(t)\, dt = kF(b) - kF(a)$$
$$= k[F(b) - F(a)]$$
$$= k\int_a^b f(t)\, dt$$

where $F(x)$ is an antiderivative of $f(x)$.

5. From the Fundamental Theorem of Calculus, Part I:

$$\int_a^c f(t)\, dt + \int_c^b f(t)\, dt = F(c) - F(a) + F(b) - F(c)$$
$$= F(b) - F(a) = \int_a^b f(t)\, dt$$

where $F(x)$ is an antiderivative of $f(x)$.

Think and Share

Break into groups. Use the Fundamental Theorem of Calculus, Part I to prove Property 4.

The other two properties can be proved in a similar manner and are left for you in the exercises. These properties are often used to simplify or transform definite integrals into simpler, more manageable integrals.

We will study the Fundamental Theorem of Calculus, Part II as well as applications of the Fundamental Theorem in Chapter 5.

EXERCISE SET 4.7

In Exercises 1–10, using the Fundamental Theorem of Calculus and an antiderivative you find by guessing (and checking), evaluate the following definite integrals by hand. Check your results with your calculator.

1. $\int_0^3 2x \, dx$

2. $\int_1^4 x^2 \, dx$

3. $\int_{-1}^2 x^2 - 3x + 5 \, dx$

4. $\int_2^7 -x^2 - 3x + 4 \, dx$

5. $\int_1^3 \frac{1}{x^2} \, dx$

6. $\int_1^3 \frac{1}{x^3} \, dx$

7. $\int_{-1}^1 \frac{1}{x^2} \, dx$

8. $\int_{-\pi/2}^{\pi/2} \cos(x) \, dx$

9. $\int_{\pi/6}^{\pi/3} \sec^2(x) \, dx$. Hint: What is the derivative of $\tan(x)$?

10. $\int_{\pi/6}^{\pi/3} \sec(x)\tan(x) \, dx$. Hint: What is the derivative of $\sec(x)$?

11. Explain in clear and complete sentences, and without using mathematical symbols, the meaning of each of the properties of the definite integral.

In Exercises 12–17, use the properties of the definite integral to rewrite the given definite integral(s). Write down each property as you use it. *Do not evaluate them.*

12. $\int_{-5}^5 7\sin^2(x) \, dx$

13. $\int_2^5 7\tan^5(x) \, dx + \int_5^{11} 7\tan^5(x) \, dx$

14. $\int_2^4 \pi \sin(x) \, dx$

15. $\int_{-3}^4 \sin(\pi) \cdot x \, dx$

16. $\int_2^5 3 \cdot x^2 \, dx$

17. $\int_1^2 11 \cdot 2t \, dt$

In Exercises 18 and 19, evaluate the integral.

18. $\int_2^2 (3x^2 - 5) \, dx$

19. $\int_5^5 e^{-x^2} \, dx$

20. How are the values of the following integrals related?
  a. $\int_5^1 x^{25} \, dx$
  b. $\int_1^5 x^{25} \, dx$

21. Describe the meaning of the definite integral $\int_a^b f(t) \, dt$ in terms of rate and accumulation functions.

22. Prove Property 2 using the Fundamental Theorem of Calculus, Part I.

23. Prove Property 3 using the Fundamental Theorem of Calculus, Part I.

24. If $F(x) = \int_a^x f(t) \, dt$, does $F(a+b) = F(a) + F(b)$? Explain.

Summary

The major concept of this chapter is that we are often able to evaluate an unknown quantity exactly by appropriately estimating it over smaller and smaller subintervals. An understanding of this concept is critical to building problem-solving skills transferable to new situations and other fields of study.

This idea is widely used, as we have seen through our examples. There are many, many more situations in which you can apply this approach. The central idea is that when considering a situation where a basic principle cannot be applied globally (over the entire interval), we can often apply the principle locally (on small pieces or subintervals of the whole). Summing these small estimates gives us an estimate for the solution we are seeking. And finally, the limit process gives us the exact solution. We will use the notion of work to demonstrate this important process which we call *kth element analysis*.

Work is defined as the product of a constant force acting on an object and the distance the object is moved along a straight line. What happens, for instance, if an object is being moved from point a to point b by a force that is not constant (see Figure 4.55)?

FIGURE 4.55   An Object Moved by a Variable Force

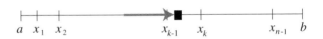

By subdividing the path from a to b into small pieces, you can consider the force over a small distance (from x_{k-1} to x_k) to be constant. That would seem reasonable, since it would normally be difficult for a varying force to dramatically change in size over a short distance. Now you can approximate the work over this short distance using the principle *Work = Force x Distance*. You continue by using this principle for each of the small pieces and then add up the work done over all the small pieces to obtain an estimate for the total work done by the variable force. It again seems reasonable that the smaller the pieces (distances), the better the estimate, since the assumption that the force is constant gets better as the distance grows smaller on each piece.

The theoretical development is, as always, important to justify applying the theory to a particular situation and to its potential use in other situations. The central theory in this chapter is this: we consider a continuous function on a closed interval $[a, b]$; subdivide the interval into n subintervals; analyze a typical subinterval, say the kth subinterval; and form the product of the function value at some point $(t_k, f(t_k))$ with the length of the subinterval, Δx_k (see Figure 4.56).

The arbitrary point t_k is in the kth subinterval $[x_{k-1}, x_k]$ and Δx_k is the length, $x_k - x_{k-1}$. Summing these products leads to a Riemann sum,

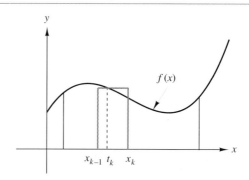

FIGURE 4.56 Random Rectangle Area for the kth Interval

$\sum_{k=1}^{n} f(t_k)\Delta x_k$. This sum represents an estimate for the solution we are seeking. The exact solution is the limit of the Riemann sum,

$$\lim_{n \to \infty} \sum_{k=1}^{n} f(t_k)\Delta x_k.$$

This limit results in the definite integral of the function $f(x)$ on the interval $[a, b]$, and is written

$$\lim_{n \to \infty} \sum_{k=1}^{n} f(t_k)\Delta x_k = \int_a^b f(x)dx.$$

The focus of the entire chapter centers on this idea. All the examples and problems use this idea. Time spent in studying and solving the problems with the goal of connecting the various problems to the one central idea will result in deeper understanding. If you set out to gain as much insight as possible into this central idea, you will maximize your problem-solving ability in situations where this theory applies.

SUPPLEMENTARY EXERCISES

In Exercises 1–6, be sure to use complete sentences.

1. Explain how to use a right endpoint or left endpoint sum to estimate total accumulation from a given data set. Include an annotated diagram.

2. Explain how to apply the midpoint rule. Include a diagram.

3. Explain how to apply the trapezoid rule. Include a diagram.

4. Explain the meaning of Riemann sum. Include a diagram which illustrates how the products in a Riemann sum can be interpreted as areas of rectangles.

5. Explain the definition of the definite integral as a limit of Riemann sums. Write out the definition and explain the meaning of each symbol.

† **6.** Describe the kth element analysis used in this chapter.

Exercises 7–21 refer to the table below which lists a car's velocities at 10-minute intervals.

TABLE 4.28

Time(min)	0	10	20	30	40	50	60
Velocity(mph)	0	48	65	62	57	55	40

7. Use the left endpoint method to estimate the total distance traveled.

8. Write the Riemann sum associated with the left endpoint method, with $n = 6$.

9. Draw a picture to show how the six products involved in the Riemann sum of Exercise 8 can be represented by the areas of six rectangles. Write a clear explanation of your drawing.

10. Use the trapezoid method to estimate the total distance traveled. Draw a picture to show your trapezoids. Does this method overestimate or underestimate the total distance traveled if the velocity function is concave down? Justify.

11. Use the midpoint method to estimate the total distance traveled. Draw a picture to show your three midpoint rectangles. Does this method overestimate or underestimate the total distance traveled if the velocity function is concave down? Justify.

12. Use Simpson's method to estimate the total distance traveled. Does this method overestimate or underestimate the total distance traveled if the velocity function is concave down? Justify.

13. Use the Data/Matrix editor to find a continuous function with which to model the data. Produce a copy of the data plot and the plot of your continuous function. Provide the equation for your function.

14. Use your continuous model to estimate the total distance traveled with $\Delta t = 1$ using a left endpoint method. Compare this result with the result obtained using the discrete data. Which result do you think is better? Justify.

15. Use your continuous model to estimate the total distance traveled with $\Delta t = 1$ using the trapezoid method. Compare this result with the result obtained using the discrete data. Which result do you think is better? Find an error bound for your result. Justify.

16. Use your continuous model to estimate the total distance traveled with $\Delta t = 1$ using the midpoint method. Compare this result with the result obtained using the discrete data. Which result do you think is better? Find an error bound for your result. Justify.

17. Use your continuous model to estimate the total distance traveled with $\Delta t = 1$ using Simpson's method. Compare this result with the result obtained using the discrete data. Which result do you think is better? Find an error bound for your result. Justify.

18. Use your continuous model to estimate the total distance traveled with $\Delta t = 1$ using a right endpoint method. Compare this result with the result obtained using the

discrete data. Which result do you think is better? Find an error bound for your result. Justify.

19. Write down the definite integral defined by the limit of the Riemann sum associated with the left endpoint method of Exercise 14. Show how to change the sum so that you have a right endpoint sum. How does the form of the definite integral associated with the right endpoint sum differ from the integral associated with the left endpoint sum? Explain.

20. Write down the definite integral defined by the limit of the Riemann sum associated with the trapezoid method used in Exercise 15.

21. Write down the definite integral defined by the limit of the Riemann sum associated with the midpoint method used in Exercise 16.

22. Let f be a function that is decreasing, continuous, and concave up on the interval $[a, b]$. Rank the left, right, midpoint, trapezoid and Simpson's methods for f on $[a, b]$ for six subintervals; that is, rank the values L_6, R_6, Mid_6, T_6, and S_6 from smallest to largest. What can be said about the ranking of L_{50}, R_{50}, Mid_{50}, T_{50}, and S_{50} for this function?

23. Follow the procedure of Example 1 in Section 4.6 to find the volume of a cylinder of radius R and height H.

24. Use kth element analysis to find a definite integral to measure the length of a curve on an interval $[a, b]$. (Hint: Let Δs_k estimate the length of the curve on the kth subinterval. Use the Pythagorean Theorem to find an expression for Δs_k in terms of Δx_k and Δy_k. Find an approximating sum, and then manipulate your sum algebraically to obtain a Riemann sum.) Take the special limit, and voilà—you should have your definite integral.

25. Evaluate the following definite integrals.
    a. $\int_1^2 x^2 - 5x - 11\, dx$
    b. $\int_{-2}^5 \sin(y)\, dy$
    c. $\int_0^\pi \cos(x)\, dx$
    d. $\int_{-3}^{-2} \frac{1}{t}\, dt$

26. Evaluate the following definite integrals.
    a. $\int_0^2 2x^2 + 4x - 3\, dx$
    b. $\int_{-3}^5 \sin(x^2)2x\, dx$
    c. $\int_0^{\frac{\pi}{4}} \sec^2(y)\, dy$
    d. $\int_3^5 \frac{1}{p}\, dp$

27. Find an antiderivative of the function $f(x) = \sec(7x)\tan(7x)$.

28. Find an antiderivative of the function $g(t) = \sec^2(5t)$.

P R O J E C T S

Project One: The Volume of a Sphere

The purpose of this project is to find the volume of a sphere using the methods of this chapter.

1. Write the equation of a circle centered at the origin with radius r.
2. Discuss how this equation represents many functions and which function (including its domain) you will use in your computations.
3. Draw a careful sketch of your function with a compass and ruler and sketch in a three-dimensional sphere.
4. Sketch, on this drawing, several slices of the sphere including the kth slice or element you use in creating a Riemann sum.
5. Describe the kth element analysis you use in creating a Riemann sum.
6. **a.** Complete the following table of approximating sums for the volume of a sphere of *radius one* for the given number of subintervals n.

TABLE 4.29

n	16	64	256	1024	4096
Approximate volume					

    b. Discuss the accuracy of the last ($n = 4096$) of these computations.
    c. Divide each of these numbers by π. How many multiples of π does this sequence of numbers seem to be approaching?
7. Create a definite integral by taking the limit of your Riemann sum as $n \to \infty$.
8. Compute the Simpson's method approximation for this definite integral and record your results in the table below.

TABLE 4.30

n	16	32	64	128	256	
Approximate volume						

9. Discuss the accuracy of the last ($n = 256$) approximation in your table.
10. Divide each of these numbers by π. What rational number (3/5, 7/4, for example) multiple of π does this sequence of numbers seem to be approaching?
11. Compare the accuracy of Simpson's method to the accuracy of the method you used in approximating the Riemann sum above.
12. Use the TI calculator integration function ($\int$ ⟨) to compute the value of your definite integral. Explain how you know that the TI calculator result is exact or an approximation.

13. Compare your value with the known formula for the volume of a sphere that you may have learned in your high school geometry course.

Project Report.   Write a report on your investigations. Address your report to your classmates, assuming the material is new to them. Describe your reasoning and include all necessary background information. A minimal project report must include:

1.   Written responses to each of the thirteen items above.
2.   Justification and explanations for your calculations.
3.   Suggestions for possible extensions of your observations and further questions for exploration.

Project Two: Simpson's Method Revisited

The purpose of this project is to show the equivalence of Simpson's definite integral estimation method and an estimation based on quadratic or parabolic functions. Both methods use the two endpoints and the midpoint of each subinterval used in the estimation.

In this project, you should complete the following tasks:

1.   Write down

$$Simpson's\ method = \frac{2 \times midpoint + trapezoid}{3}$$

for the integral $\int_{-h}^{h} f(x)\,dx$ using one interval and function notation. Your result should contain $f(-h)$, $f(h)$, and $f(0)$.

2.   Define $f(x) = ax^2 + bx + c$ and evaluate your expression for Simpson's method with this definition of f (the general quadratic or parabolic function).

3.   Find the definite integral $\int_{-h}^{h} f(x)\,dx$ for $f(x) = ax^2 + bx + c$.

4.   Compare the results from (2) and (3).

5.   Explain why three non-collinear points determine a parabola.

6.   Discuss the accuracy of Simpson's methods for linear functions.

7.   Show that

$$Simpson's\ Method = \frac{2 \times midpoint + trapezoid}{3}$$

estimates $\int_{1}^{5} 3x^2 + 2x - 5\,dx$ exactly. Determine if Simpson's method estimates every definite integral of a quadratic function exactly.

8.   Explain why Simpson's method is a parabolic or quadratic estimation method. Use graphs to illustrate your explanation.

9.   Show that Simpson's method estimates the following integral exactly.

$$\int_{0}^{2} x^3 + 2x^2 + 3\,dx$$

10. Does Simpson's method exactly estimate every definite integral of a cubic function? Justify.

Project Report. Write a report on your investigations. Address your report to your classmates, assuming the material is new to them. Describe your reasoning and include all necessary background information. A minimal project report must include:

1. Written responses to each of the ten items above.
2. Justification and explanations for your calculations.
3. Suggestions for possible extensions of your observations and further questions for exploration.

Project Three: Finding Definite Integrals Using the Monte Carlo Method

Monte Carlo techniques are used to solve numerous problems in industry like designing the lead vest used in dentists' offices to protect you from harmful radiation. These problems are often complex but the general method itself is fairly straightforward. For example, if you randomly select 100 points in a region of known area, then the number of points n (on average) that fall within a subregion of unknown area allows you to approximate that unknown area with:

$$\frac{n}{100} \approx \frac{\text{the area of the unknown region}}{\text{the area of the known region}}$$

.

1. Use the Monte Carlo method to approximate the area of a unit circle. The TI calculator's random number function $rand()$ generates values between 0 and 1. By counting the proportion of points of the form $(rand(), rand())$ all of which lie inside the square with corners at $(0, 0)$, $(1, 0)$, $(1, 1)$, $(0, 1)$ that fall interior to the circle $x^2 + y^2 = 1$, you can estimate the area of one quarter of the unit circle.

2. Use the Monte Carlo method to find approximate values of the following integrals. You will need to use the random number generator of your TI calculator tailored to the intervals involved in the problem.

a. $\displaystyle\int_0^1 e^{-x^2}\, dx$

b. $\displaystyle\int_0^1 e^{\sin(x)}\, dx$

c. $\displaystyle\int_0^{\sqrt{\pi}} \sin(x^2)\, dx$

d. $\displaystyle\int_1^2 \frac{\sin(x)}{x}\, dx$

e. $\displaystyle\int_2^e \frac{1}{\ln(x)}\, dx$

Project Report.   Write a report on your investigations. Address your report to your classmates, assuming the material is new to them. Describe your reasoning and include all necessary background information. A minimal project report must include:

1. A list of the commands and functions you used to estimate each of the areas and integrals above. This should include the initial value of `RandSeed` you used.
2. Data from the Monte Carlo experiments you have run.
3. If you have used multiple experiments to estimate a single integral, include a statistical analysis of the results.
4. Justification and explanations for your calculations.
5. Suggestions for possible extensions of your observations and further questions for exploration.

FIGURE 4.57   Novocaine in the Body

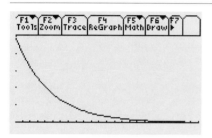

FIGURE 4.58   A Secant Line and the Tangent

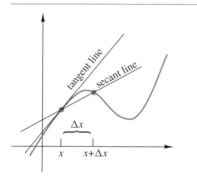

FIGURE 4.59   Approximating a Cone

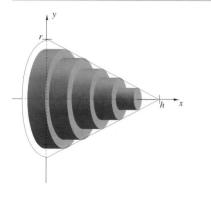

INTERLUDE:

The Truth About Limits

Throughout the text, we have used algebra and arithmetic to model geometric or physical processes. We modeled drug dosage, populations and rumor with difference equations. We have used differences and fractions to understand speed and acceleration, slope and concavity. Areas, volumes, and the dripping of coffee urns have both been modeled by sums of products. Repeatedly, we have seen in our models limiting values that we could approximate numerically but which were difficult to find algebraically. The decay of a single dose of novocaine to zero, the maximum concentration of amoxicillin in a patient's system, the "sum" of an infinite geometric series, the instantaneous velocity of a car, the slope of a tangent line, and the convergence of Riemann sums are all examples of these limits.

Each time we encountered a limit, we used some special algebraic, numeric, or geometric device to investigate it. At times, we have appealed to arithmetic experience. In Chapter 1, we remarked that if $-1 < a < 1$, then the limit of a^n, "as n grows to infinity," is zero. In Chapter 2, we invoked our numerical intuition less formally and used nDeriv(with $h = .001$ as our tool to investigate the derivative. In Chapter 4, we found the area of the region under an increasing function $f(x)$ for $a \leq x \leq b$, by trapping the area between left and right endpoint sums. Seeing that the sums differed by only $\frac{f(b)-f(a)}{n}$, we reasoned that the area must be the common limit of the left and the right sums.

Throughout the text, we use limits to give a definite value in situations where algebra or arithmetic alone fail to supply an answer. We use three related but distinct types of limits. They are:

1. The limit of a sequence $\{a_n\}_{n=1}^{\infty}$ as n goes to infinity.
2. The limit of a function $f(x)$ as x goes to infinity.
3. The limit of a function $f(x)$ as x goes to a point p.

The Awful Truth

The time has come for us to examine more closely how we determine limits and to collect together some elementary facts about them. First, we need to make clear our own limitations with limits. Calculating limits is a hard problem, not unlike the problem of solving equations. Finding the root of a linear equation is reasonably easy, and we can always find a solution. Finding the roots of a quadratic equation is harder, but the quadratic equation gives us a simple and guaranteed process to find the roots or see that there are none. For equations of degree three and four, there are formulas (not simple) guaranteed to produce the roots of the equation. But for higher degree equations and for equations involving trigonometric or exponential functions, life holds no such guarantees. There are no simple rules that always work for solving equations. The problem of finding limits is much the same. There are no rules guaranteed to compute the limit of every sequence or function.

As we shall see in Chapter 9, we sometimes have to be satisfied with knowing that there is a limit.

Limits and Approximations

Where does that leave us? Continuing the analogy to equations, we can always tell when we have a root, just plug it into the equation and see if it works. Is the same possible with limits? After all, a limit must fit seamlessly into an overall pattern. Happily, the answer is "yes," we can at least identify a limit once we have it. The key to this identification is found in our discussion of integrals. We *defined the value* of an integral as the limit of sums. Then, when these values prove tricky to find, we used these very sums (left, right, midpoint, random or trapezoidal) to *approximate the value* of the integral. This interplay between *value* and *approximation* is how we identify a limit once we have one.

LIMITS RECOGNIZED

For any positive *error*:

1. We say $\lim\limits_{n \to \infty} a_n = A$ whenever for all indices m sufficiently large

$$A - error < a_m < A + error.$$

2. We say $\lim\limits_{x \to \infty} f(x) = A$ whenever for all values of x sufficiently large

$$A - error < f(x) < A + error.$$

3. We say $\lim\limits_{x \to p} f(x) = A$ whenever for all values of x sufficiently close to p

$$A - error < f(x) < A + error.$$

The first statement tells us that for A to be the limit of the sequence $\{a_n\}$ all of the a_n's with sufficiently large indices must be good approximations to A. The third says that when A is the limit of $f(x)$ at $x = p$, any of the values $f(x)$ where x is close enough to p will supply a close approximation to A.

▶ **FIRST REFLECTION**

How can you paraphrase the second statement on limits of $f(x)$ as x goes to infinity? ◀

Think and Share

Break into groups of three. Have one of the members of your group use the limit recognition principle above to explain why the limit of the constant sequence $a_n = L$ is L. Then have the other two members of the group explain why if $f(x) = L$ then $\lim_{x \to \infty} f(x) = L$ and $\lim_{x \to p} f(x) = L$.

Known Limits

Throughout the first four chapters, we have worked algebraic magic whenever we needed to find a limit. We identified some limits as rest points of the sequences generated by the recursion relations $u_n = F(u_{n-1})$, and then found them among the solutions of $x = F(x)$. When we used limits to calculate the derivatives of polynomials, we first found a formula for the difference quotient in terms of Δx. Then we *very carefully* simplified the formula *before* setting $\Delta x = 0$. When we wanted to calculate the derivative of $\sin(x)$, and could not eliminate Δx from the denominator, we had to rely on graphical data.

Clever observations and algebraic insights are always part of a mathematician's toolkit. However, tricks are not the whole story. There are a number of well-known limits that we have already used. For reference, we list them below:

1. $\lim\limits_{n \to \infty} \dfrac{1}{n} = 0.$

2. $\lim\limits_{n \to \infty} \left(1 + \dfrac{1}{n}\right)^n = e.$

The first and third limits can be attributed to Archimedes. The other three are due to Leonard Euler.

3. $\lim\limits_{x \to \infty} \dfrac{1}{x} = 0.$

4. $\lim\limits_{x \to \infty} \left(1 + \dfrac{1}{x}\right)^x = e.$

5. $\lim\limits_{x \to 0} \dfrac{\sin(x)}{x} = 1.$

▶ **SECOND REFLECTION**

Go to a Trigonometry, Precalculus, or Math Analysis text, and look up the definition of $\arctan(x)$. Explain why $\lim\limits_{n \to \infty} \arctan(x) = \dfrac{\pi}{2}$ ◀

New Limits From Old

These five or six basic limits constitute a modest stock of examples. Happily, we can use them to calculate more complicated limits. We record some properties that we have, to some degree, already used.

PROPERTIES OF LIMITS OF SEQUENCES

1. If $\lim_{n\to\infty} a_n = A$ then, for any real number c,

$$\lim_{n\to\infty} (c \cdot a_n) = c \cdot A.$$

2. If $\lim_{n\to\infty} a_n = A$ and $\lim_{n\to\infty} b_n = B$, then

$$\lim_{n\to\infty} (a_n + b_n) = A + B.$$

3. If $\lim_{n\to\infty} a_n = A$ and $\lim_{n\to\infty} b_n = B$, then

$$\lim_{n\to\infty} (a_n \cdot b_n) = A \cdot B.$$

4. If $\lim_{n\to\infty} a_n = A$ and $\lim_{n\to\infty} b_n = B$, then, provided $B \neq 0$,

$$\lim_{n\to\infty} \left(\frac{a_n}{b_n} \right) = \frac{A}{B}.$$

5. If A is in the domain of the continuous function $g(x)$ and $\lim_{n\to\infty} a_n = A$, then

$$\lim_{n\to\infty} g(a_n) = g(A).$$

PROPERTIES OF LIMITS OF FUNCTIONS AT INFINITY

1. If $\lim_{x\to\infty} f(x) = A$, then for any real number c

$$\lim_{x\to\infty} (c \cdot f(x)) = c \cdot A.$$

2. If $\lim_{n\to\infty} f(x) = A$ and $\lim_{x\to\infty} g(x) = B$, then

$$\lim_{x\to\infty} (f(x) + g(x)) = A + B.$$

3. If $\lim_{x\to\infty} f(x) = A$ and $\lim_{x\to\infty} g(x) = B$, then

$$\lim_{x\to\infty} (f(x) \cdot g(x)) = A \cdot B.$$

4. If $\lim_{x\to\infty} f(x) = A$ and $\lim_{x\to\infty} g(x) = B$, then, provided $B \neq 0$,

$$\lim_{x\to\infty} \left(\frac{f(x)}{g(x)} \right) = \frac{A}{B}.$$

5. If A is in the domain of the continuous function $g(x)$ and $\lim_{x\to\infty} f(x) = A$,

$$\lim_{x\to\infty} g(f(x)) = g(A).$$

PROPERTIES OF LIMITS OF FUNCTIONS AT A POINT

1. If $\lim\limits_{x \to a} f(x) = A$, then for any real number c,

$$\lim_{x \to a} (c \cdot f(x)) = c \cdot A.$$

2. If $\lim\limits_{n \to p} f(x) = A$ and $\lim\limits_{x \to p} g(x) = B$, then

$$\lim_{x \to a} (f(x) + g(x)) = A + B.$$

3. If $\lim\limits_{x \to p} f(x) = A$ and $\lim\limits_{x \to p} g(x) = B$, then

$$\lim_{x \to p} (f(x) \cdot g(x)) = A \cdot B.$$

4. If $\lim\limits_{x \to p} f(x) = A$ and $\lim\limits_{x \to p} g(x) = B$, then, provided $B \neq 0$,

$$\lim_{x \to p} \left(\frac{f(x)}{g(x)} \right) = \frac{A}{B}.$$

5. If A is in the domain of the continuous function $g(x)$ and $\lim\limits_{x \to p} f(x) = A$, then

$$\lim_{x \to p} g(f(x)) = g(A).$$

E X A M P L E   1   **Finding the Limit of a Sequence**

Use the Properties of Limits to determine $\lim\limits_{n \to \infty} u_n$, if $u_n = \dfrac{n+2}{3n+1}$.

Of course, if you are in a hurry, you may consult your TI calculator.

Solution:

We start with an algebraic manipulation that will introduce the fraction $1/n$. Factoring n out of both the numerator and denominator of u_n, gives

$$u_n = \frac{n}{n} \cdot \frac{1 + \frac{2}{n}}{3 + \frac{1}{n}} = \frac{1 + 2 \cdot \frac{1}{n}}{3 + \frac{1}{n}}.$$

Analyzing the numerator, we reason that since $\lim\limits_{n \to \infty} \dfrac{1}{n} = 0$, it follows that $\lim\limits_{n \to \infty} 2 \cdot \dfrac{1}{n} = 2 \cdot 0 = 0$. Consequently, we see that $\lim\limits_{n \to \infty} 1 + 2 \cdot \dfrac{1}{n} = 1 + 0 = 1$. Similar reasoning tells us that the limit of the denominator is 3. Finally, we find

$$\lim_{n \to \infty} u_n = \frac{\lim\limits_{n \to \infty} \left(1 + 2 \cdot \dfrac{1}{n} \right)}{\lim\limits_{n \to \infty} \left(3 + \dfrac{1}{n} \right)} = \frac{1}{3}$$

E X A M P L E 2 **Finding a Limit of a Function at Infinity**

Use the Properties of Limits to determine $\lim\limits_{x \to \infty} f(x)$, if $f(x) = \left(1 + \frac{.08}{x}\right)^x$.

Solution:

If we set $\dfrac{0.08}{x} = \dfrac{1}{y}$, we find that $x = 0.08y$. Rewriting $f(x)$ in terms of y gives us

$$\left(1 + \frac{0.08}{x}\right)^x = \left(1 + \frac{1}{y}\right)^{0.08y} = \left(\left(1 + \frac{1}{y}\right)^y\right)^{0.08}.$$

We complete the analysis as follows. As x goes to infinity, so does $y = 0.08x$. Thus, the limit we seek can be evaluated as

$$\lim_{y \to \infty} \left(\left(1 + \frac{1}{y}\right)^y\right)^{0.08} = \left(\lim_{y \to \infty} \left(1 + \frac{1}{y}\right)^y\right)^{0.08}$$
$$= (e)^{0.08}$$

▶ **THIRD REFLECTION**

What property of limits justifies moving the limit inside the power in the above equations? ◀

E X A M P L E 3 **Finding the Limit of a Function at a Point**

Use the Properties of Limits to determine $\lim\limits_{x \to 0} \dfrac{1 - \cos(x)}{x}$.

Solution:

Since the one limit we have for trigonometric functions involves $\sin(x)$, we transform $\dfrac{1 - \cos(x)}{x}$ by multiplying both the numerator and denominator by $1 + \cos(x)$.

$$\frac{1 - \cos(x)}{x} \cdot \left(\frac{1 + \cos(x)}{1 + \cos(x)}\right) = \frac{1 - \cos^2(x)}{x \cdot (1 + \cos(x))}$$
$$= \frac{\sin^2(x)}{x \cdot (1 + \cos(x))}$$
$$= \frac{\sin(x)}{x} \cdot \frac{\sin(x)}{(1 + \cos(x))}$$

Now we can use the Properties of Limits to evaluate this expression.

$$\lim_{x \to 0} \frac{1 - \cos(x)}{x} = \lim_{x \to 0} \frac{\sin(x)}{x} \cdot \lim_{x \to 0} \frac{\sin(x)}{(1 + \cos(x))}$$

$$= \lim_{x \to 0} \frac{\sin(x)}{x} \cdot \frac{\lim\limits_{x \to 0} \sin(x)}{\lim\limits_{x \to 0} (1 + \cos(x))}$$

$$= 1 \cdot \frac{0}{2}$$

$$= 0$$

EXERCISE SET 4.7

1. Use algebra and the Properties of Limits to show

$$\lim_{n \to \infty} \frac{3n^2 + n - 7}{2n^2 + 5n} = \frac{3}{2}.$$

2. Use algebra and the Properties of Limits to show

$$\lim_{n \to \infty} \left(1 + \frac{2}{n}\right)^n = e^2.$$

3. Use algebra and the Properties of Limits to show

$$\lim_{n \to \infty} \frac{a_n \cdot x^n + a_{n-1} \cdot x^{n-1} + \cdots + a_0}{b_n \cdot x^n + b_{n-1} \cdot x^{n-1} + \cdots + b_0} = \frac{a_n}{b_n}.$$

4. Use algebra and the Properties of Limits to show

$$\lim_{n \to \infty} \left(1 + \frac{r}{n}\right)^n = e^r.$$

5. Use algebra and the Properties of Limits to show

$$\lim_{x \to \infty} \left(1 + \frac{1}{x}\right)^{x/2} = \sqrt{e}.$$

6. Use algebra and the Properties of Limits to show

$$\lim_{x \to 0} \frac{1 - \cos(x)}{x^2} = \frac{1}{2}.$$

7. Use algebra and the Properties of Limits to show

$$\lim_{x \to 0} \frac{\tan(x)}{x} = 1.$$

8. Use algebra and the Properties of limits to show

$$\lim_{x \to 0} \frac{\sin(2x)}{3x} = \frac{2}{3}.$$

9. Use algebra and the Properties of Limits to show

$$\lim_{x \to 0} \frac{\sin(a \cdot x)}{b \cdot x} = \frac{a}{b}.$$

10. Show that there is no number A for which

$$\lim_{x \to 0} \sin\left(\frac{1}{x}\right) = A.$$

5

THE INTEGRAL: THEORY, APPLICATIONS, AND TECHNIQUES

5.1
Rate, Accumulation, and the Fundamental
Theorem of Calculus, Part II

5.2
The Antiderivative Concept

5.3
Finding Antiderivatives Using Properties
and Formulas

5.4
Applications

5.5
Using the Chain Rule in Finding
Antiderivatives

5.6
Techniques of Integration

Summary

In the first four chapters we have studied the two fundamental operations of calculus: differentiation and integration. Differentiation computes the instantaneous rate at which change is occurring. Integration computes the accumulation based on the rate at which change is occurring. These two operations are related in profound and important ways. The Fundamental Theorem of Calculus, Part I, which you saw in Chapter 4, allows us to easily evaluate the definite integral of a function whose antiderivative we know. In this chapter, we will develop the Fundamental Theorem of Calculus, Part II which assures us that an antiderivative exists for all continuous functions. Unfortunately, it does not tell us how to find such antiderivatives.

5.1

Rate, Accumulation, and the Fundamental Theorem of Calculus, Part II

The rate at which distance, money, area, or oil accumulates is a function of time. If we have a rate function, then we can generate an estimate for how much of a quantity has accumulated in a given time period. To obtain the estimate, we partition the time interval into smaller subintervals, take an actual rate or an average of actual rates on each subinterval as an estimate of the average rate on that subinterval, multiply by the length of the subinterval, and sum the resultant products. We can make the estimate as accurate as we like by decreasing the lengths of all the subintervals. Using the limit idea we write the exact total accumulation as a definite integral.

Accumulation Function

If we assign zero to the time at which the accumulation begins, then the amount accumulated over a time period of length T is given by the definite integral, $\int_0^T f(t)\, dt$. The integrand $f(t)$ is the instantaneous rate of change of accumulation at time t; the definite integral $\int_0^T f(t)\, dt$ is the total amount accumulated over the time interval $[0, T]$.

The total amount of the quantity that accumulates depends on T, the length of time that the accumulation lasts. If we know how long the accumulation lasts then we can compute the accumulation using the definite integral. Defining $A(T)$ to be the total accumulation over the interval $[0, T]$, we have a formula for $A(T)$,

$$A(T) = \int_0^T f(t)\, dt.$$

Table 5.1 is an input/output table for $A(T)$. All of the outputs are definite integrals. Their numerical values can be computed once we specify the rate function, $f(t)$.

TABLE 5.1   Input/Output for $A(T)$

T	1	2	3	4	5
A(T)	$\int_0^1 f(t)\, dt$	$\int_0^2 f(t)\, dt$	$\int_0^3 f(t)\, dt$	$\int_0^4 f(t)\, dt$	$\int_0^5 f(t)\, dt$

We have used the rate function and the definite integral to generate a new function, the accumulation function, over the interval $[0, T]$. In this section, we will begin to study this function.

A Small Planet Revisited

Recall the rate function for the small planet we visited in Chapter 4. The velocity of an object dropped from a cliff on that small planet in ft/sec was given by $v(t) = 2t$. The accumulation function for this situation equals the distance the object travels from the time it was dropped at time $t = 0$ to time $t = T$. Using $D(T)$ to denote this function, we have the formula $D(T) = \int_0^T v(t)\, dt = \int_0^T 2t\, dt$. The $\int_0^T v(t)\, dt$ notation here might be confusing because time appears both as an instantaneous time with the variable of integration (t) and also as a fixed time with the length of the time interval (T). The definite integral $\int_0^T v(t)\,dt$ measures the accumulation of the rate function $v(t)$ over the fixed set of values for t given by $0 \leq t \leq T$.

Think and Share

Break into groups. Use the graph of $v(t) = 2t$ in Figure 5.1 and the notion that the integral of a nonnegative function can represent the area under a curve to find the rest of the distances in Table 5.2 for the given values of T. You may use *Computing Limits of Riemann Sums* below as a computation method. Confirm at least two of your entries by a second method of computation.

FIGURE 5.1   The Graph of $v(t) = 2t$

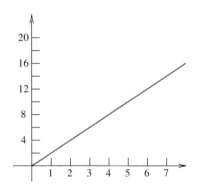

TABLE 5.2 $v(t) = 2t$ Starting at 0

T	0	1	2	3	4	5	T
D(T)	$\int_0^0 2t\,dt$	$\int_0^1 2t\,dt$	$\int_0^2 2t\,dt$	$\int_0^3 2t\,dt$	$\int_0^4 2t\,dt$	$\int_0^5 2t\,dt$	$\int_0^T 2t\,dt$
Distance	0	1			16		

COMPUTING LIMITS OF RIEMANN SUMS

To find the value of $\int_a^b f(t)\,dt = \lim\limits_{n\to\infty} \sum\limits_{k=1}^n f(t_k)\Delta t$, you first define $f(t)$ by entering `define f(q)= formula in q`. Then store the value of Δt to h by entering `(b-a)/n → h`. The following commands give the limit of our Riemann sum for t_k as either the left or right endpoint of the kth subinterval,

$$\texttt{limit(}\Sigma\texttt{(f(a + k*h) * h , k, 1, n), n,} \infty\texttt{)} \qquad (5.1)$$

$$\texttt{limit(}\Sigma\texttt{(f(a + (k-1)*h) * h, k, 1, n), n,} \infty\texttt{)} \qquad (5.2)$$

where we assume each of the n subintervals has equal length Δt.

From the table we see

$$\text{distance}(T) = D(T) = \int_0^T 2t\,dt = T^2.$$

Evidently, on our small planet, the distance traveled by an object is the square of the time it travels. During the first second, over the interval $[0, 1]$, the object falls one foot. During the second second, over the interval $[1, 2]$, the object falls an additional three feet. Clearly, the average rate at which our object is accumulating distance is increasing. How is the instantaneous rate of change of that accumulation changing?

Explain the terms *average* rate and *instantaneous rate.* ◀

We can use the derivative to determine the instantaneous rate at which our object is accumulating distance. The function $D(T)$ is the accumulation at time T, so, the derivative of $D(T)$ with respect to T is the instantaneous rate of change of accumulation at time T. Using our observation that $D(T) = T^2$, we compute this derivative,

$$\frac{d}{dT}(D(T)) = 2T.$$

Hence, the derivative of our accumulation function at time T equals $2T$. Our original function, $v(t) = 2t$, represented the rate at which the object accumulated distance at time t. It should come as no real surprise, therefore, that $v(T)$ equals $2T$ since both expressions represent the instantaneous rate at which the object accumulates distance at time T.

If we replace $D(T)$ in the previous formula with its integral representation, we can write

$$\frac{d}{dT}\int_0^T 2t \; dt = 2T.$$

The definite integral $\int_0^T 2t \; dt$ is a function whose derivative is the doubling function. Thus, $D(T) = T^2$ is an antiderivative of the function defined by $2T$. (Recall, an antiderivative of f is a function whose derivative is f.)

▶ **SECOND REFLECTION**
Suppose you jump off a 25-foot cliff on the small planet. How long does it take you to reach the ground? How fast are you traveling when you hit? ◀

Break into groups.

Think and Share

1. Find an expression for a function whose derivative is $f(t) = 4t$ by completing Table 5.3 using a graph of f or by computing limits of Riemann sums.

2. Describe to each other the rate function given by the expression $4t$ and the accumulation function given by $\int_0^T 4t \; dt$.

TABLE 5.3 $v(t) = 4t$ Starting at 0

T	0	1	2	3	4	5
$\int_0^T f(t)\, dt$	$\int_0^0 4t\, dt$	$\int_0^1 4t\, dt$	$\int_0^2 4t\, dt$	$\int_0^3 4t\, dt$	$\int_0^4 4t\, dt$	$\int_0^5 4t\, dt$
Value	0	2			32	

We graph the rate function $v(t) = 4t$ as shown in Figure 5.2. Figure 5.3 shows $D(T)$, the accumulation function.

The slope of the tangent line to $D(T)$ at each point $(T, 2T^2)$ is

$$\frac{d}{dt}D(T) = \frac{d}{dt}\left(\int_0^T 4t\ dt\right)$$

$$= \frac{d}{dt}(2T^2)$$

$$= 4T.$$

This is the value of the rate function at $t = T$ which is $4T$.

FIGURE 5.2 Graph of $v(t) = 4t$ **FIGURE 5.3** Graph of D(T)

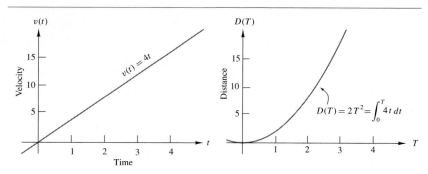

The rate at which the object's distance is changing after two seconds is the value of $v(t)$ when $t = 2$, as shown in Figures 5.4 and 5.5. As in the first case,

FIGURE 5.4 Function Value when t is 2 for the   **FIGURE 5.5** Tangent Line to the Graph of
Function $v(t) = 4t$ $D(T)$ at $t = 2$

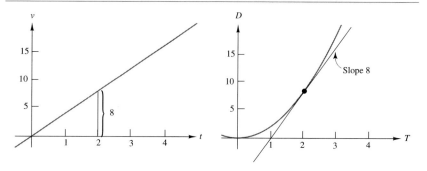

we see that $\int_0^T 4t \, dt$ is an expression whose derivative is $4T$, the original rate function, since $f(T) = 4T$ is the same function as $f(t) = 4t$.

The Rate Function versus the Derivative of the Accumulation Function.   To make this point even clearer, graph the accumulation function, the rate function, and the derivative of the accumulation function together as shown in *Graphing Accumulation and Derivative Functions* below. How many graphs do you see? Use the Trace feature to convince yourself that the derivative of the accumulation function is the rate function.

GRAPHING ACCUMULATION AND DERIVATIVE FUNCTIONS

In the Y= editor, enter

y1=4x   the rate function
y2=∫(4t,t,0,x)   the accumulation function
y3=∂(y2(x),x)   the derivative of the accumulation function

The integral operator is found above the 7 on the numeric keypad. Set xres to 5 in the WINDOW editor since plotting the accumulation function takes a long time.

A Different Starting Place

Table 5.4 shows the distance traveled starting at time $t = 0$ by an object dropped from a cliff on a small planet where velocity is $v(t) = 4t$. If we ask what

TABLE 5.4 $v(t) = 4t$ Starting at 0

T	0	1	2	3	4	5	T
D(T)	$\int_0^0 4t \, dt$	$\int_0^1 4t \, dt$	$\int_0^2 4t \, dt$	$\int_0^3 4t \, dt$	$\int_0^4 4t \, dt$	$\int_0^5 4t \, dt$	$\int_0^T 4t \, dt$
Distance	0	1			16		

distance the object has traveled starting at time $T = 1$, we construct Table 5.5 (fill in the missing entries using a graph of the velocity function) to describe the accumulation at various times T.

TABLE 5.5 $v(t) = 4t$ Starting at 1

T	1	2	3	4	5	T
$\int_1^T f(t)\,dt$	$\int_1^1 4t\,dt$	$\int_1^2 4t\,dt$	$\int_1^3 4t\,dt$	$\int_1^4 4t\,dt$	$\int_1^5 4t\,dt$	$\int_1^T 4t\,dt$
Value	0	6			48	

▶ **THIRD REFLECTION**

From the values in Table 5.5, find an accumulation function to represent $\int_1^T f(t)\,dt$. Graph this accumulation function. How is this graph related to the graph of $D(T) = \int_0^T 4t\,dt$ in Figure 5.3? ◀

The distance function starting at zero in Table 5.3 differs by a constant amount from the distance function starting at one in Table 5.5 (if we look at the same values of T). Both distance functions have the same derivative, $4T$. Thus, the instantaneous rate of change for both accumulation functions is the same. This makes sense since the rate at which an object falls should not depend on when we start watching it.

Break into groups.

Think and Share

1. Complete Table 5.6 for the earth where the velocity function is $v(t) = 32t$ like the one completed earlier for the small planet.

2. Describe to each other the rate function given by the expression $32t$ and the rate of change of accumulation function given by $\dfrac{d}{dT} \displaystyle\int_0^T 32t\,dt$.

TABLE 5.6 $v(t) = 32t$ Starting at 0

T	0	1	2	3	4	5
$\int_0^T f(t)\,dt$	$\int_0^0 32t\,dt$	$\int_0^1 32t\,dt$	$\int_0^2 32t\,dt$	$\int_0^3 32t\,dt$	$\int_0^4 32t\,dt$	$\int_0^5 32t\,dt$
Value	0	16			256	

Putting it All Together

We have used a continuous rate function and the definite integral concept to generate a new function, the accumulation function, over the interval $[a, T]$. We have seen that this new function has as its derivative the original rate function. Symbolically,

$$\frac{d}{dT} \int_a^T rate(t)\,dt = rate(T).$$

This equation is the Fundamental Theorem of Calculus, Part II and is formally defined below.

FUNDAMENTAL THEOREM OF CALCULUS, PART II

Let f be a continuous function. The function $F(x) = \int_a^x f(t)\, dt$ is an antiderivative of f. That is, $F'(x) = \dfrac{d}{dx} \int_a^x f(t)\, dt = f(x)$

EXERCISE SET 5.1

In Exercises 1–4, $f(t)$ represents a rate function. Complete each table as follows:

a. Write each integral as the limit of a Riemann sum.
b. Find the limit of this Riemann sum to complete the table.
c. Find a formula for the accumulation function.
d. Differentiate the accumulation function.
e. Compare the derivative of the accumulation function with the original rate function.

1.

TABLE 5.7

T	0	1	2	3	4
$\int_0^T f(t)\, dt$	$\int_0^0 6t\, dt$	$\int_0^1 6t\, dt$	$\int_0^2 6t\, dt$	$\int_0^3 6t\, dt$	$\int_0^4 6t\, dt$
Value	0	3			

2.

TABLE 5.8

T	0	1	2	3	4	5
$\int_0^T f(t)\, dt$	$\int_0^0 3\, dt$	$\int_0^1 3\, dt$	$\int_0^2 3\, dt$	$\int_0^3 3\, dt$	$\int_0^4 3\, dt$	$\int_0^5 3\, dt$
Value	0					

3.

TABLE 5.9

T	1	2	3	4	5
$\int_1^T f(t)\, dt$	$\int_1^1 8t\, dt$	$\int_1^2 8t\, dt$	$\int_1^3 8t\, dt$	$\int_1^4 8t\, dt$	$\int_1^5 8t\, dt$
Value	0				

4.

TABLE 5.10

T	0	1	2	3	4
$\int_0^T f(t)\, dt$	$\int_0^0 3t^2\, dt$	$\int_0^1 3t^2\, dt$	$\int_0^2 3t^2\, dt$	$\int_0^3 3t^2\, dt$	$\int_0^4 3t^2\, dt$
Value	0				

In Exercises 5–9, graph the accumulation function starting at $t = a$ for each $f(t)$ (a rate function) and write a formula for each accumulation function.

5. $f(t) = 3t^2$ $a = 0$
6. $f(t) = 3t^2$ $a = 1$
7. $f(t) = 3t^2$ $a = 2$
8. $f(t) = 3t^2$ $a = -1$
9. $f(t) = 3t^2$ $a = -2$

10. Evaluate $\int_0^a 3t^2\, dt$ for $a = 1, 2, 3, 4$. Describe the relationship between the values of $\int_0^T 3t^2\, dt$, $\int_a^T 3t^2\, dt$, and $\int_0^a 3t^2\, dt$ using the values of $\int_0^T 3t^2\, dt$ in Exercise 4.

11. Verify the equality $\int_0^3 3t^2\, dt + \int_3^5 3t^2\, dt = \int_0^5 3t^2\, dt$. Give a graphical representation of this equality.

12. Verify the equality $\int_0^3 2t\, dt + \int_3^5 2t\, dt = \int_0^5 2t\, dt$. Give a graphical representation of this equality.

Exercises 13–21 prepare you for work that you will be doing later in the chapter. Use your TI calculator to differentiate each function. Describe the derivative as a sum or product of the functions displayed on the screen. Decide which derivatives are easier to write down directly using the differentiation rules and which require the use of technology.

13. $f(x) = \cos(x^3)$
14. $g(x) = x^2 - 3x + 5$
15. $h(x) = \sec(3x^3 + 5x - 2)$
16. $f(x) = x^4 \cdot \cos(x)$
17. $g(x) = \dfrac{2x}{\cos(x)}$
18. $h(x) = \tan(3x^3 + 5x - 2) + 5x$
19. $f(x) = e^{3x^2 - 5}$
20. $g(x) = \sin(x)x$
21. $h(x) = \cos(x)\tan(x)$

22. Graph the rate function $f(t) = 2t - 5$, graph its accumulation function starting at $t = 1$; graph the tangent line to the accumulation function at $t = 3$. Explain how this is a specific case of the Fundamental Theorem of Calculus, Part II.

23. Let $G(x) = \int_{-1}^{x} f(t)\, dt$ where f is the function whose graph is shown in Figure 5.6. The tick marks are one unit apart in both the horizontal and vertical directions. The graphing window is $[-8, 7]$ by $[-5, 10]$.
    a. Estimate $G(-2)$. Justify.
    b. For which values of x is $G(x)$ increasing? Justify.
    c. Is $G(x)$ ever concave up? Justify.
    d. Describe $G(x)$ and $f(x)$ in terms of rate function and accumulation function and explain how these functions are related.
    e. Find $G'(x)$.

FIGURE 5.6   Graph of $f(x)$

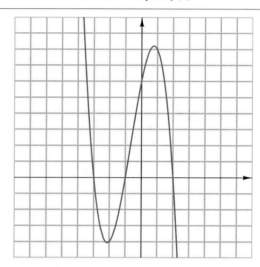

5.2

The Antiderivative Concept

In Chapter 3, we saw that the derivative of a function defined on an interval is itself a function. We also introduced the concept of antiderivative. In the previous section, we used the definite integral and a function, $f(t)$, to create yet another function, namely an antiderivative for $f(t)$. This connection between the definite integral and an antiderivative warrants further investigation.

Families of Antiderivatives

We begin by restating the formal definition of antiderivative.

AN ANTIDERIVATIVE

The function F is called an **antiderivative** of the function f if $F'(x) = f(x)$.

▶ **FIRST REFLECTION**

1. Write down the derivative of each of the following functions:

    $$f(x) = x^2 + 2$$
    $$g(x) = x^2 + 8$$
    $$h(x) = x^2 - 10$$
    $$j(x) = x^2 + \pi$$
    $$k(x) = x^2 - 31$$

    a. Is $f(x)$ an antiderivative of $g'(x)$?

    b. Write the formula for another function with $2x$ as its derivative.

    c. Explain what happens to the constant when you differentiate the functions you wrote down.

2. Discuss what it might mean to find *the* antiderivative of $2x$.   ◀

Clearly, *the* antiderivative of a function cannot be a single function. Each one of the functions you just differentiated is an antiderivative of the function $r(x) = 2x$. Consequently, none of them can be called *the* antiderivative of $r(x)$ although each of them is *an* antiderivative of $r(x)$.

When we refer to *the* antiderivative of a function f, we will mean the collection or family of all functions that are antiderivatives of f. The functions in such a class are closely related, as we can see from the following theorem.

THEOREM

> Given two functions $F(x)$ and $G(x)$ such that $F'(x) = G'(x)$, then $F(x) = G(x) + C$ where C is any real number.

Proof: We know that $F'(x) = G'(x)$. Thus, $F'(x) - G'(x) = 0$ and, from the *Sum of Two Functions Rule* for differentiation, $F'(x) - G'(x) = (F - G)'(x)$. So, $(F - G)' = 0$. The only functions that have zero as derivatives are the constant functions which can be represented as $h(x) = C$. Thus, $(F - G)(x) = C$, $F(x) - G(x) = C$ and $F(x) = G(x) + C$.

Our "proof" has a hole. The statement, "the only functions that have zero as derivatives are the constant functions," while true, requires a proof you will provide in the following Reflection.

▶ **SECOND REFLECTION**

Suppose $H'(x) = 0$ (everywhere) on the interval $[a, b]$. Let x be a number in $[a, b]$. By the Mean Value Theorem in Section 2.6, there exists a number c in (a, x) such that

$$H'(c) = \frac{H(x) - H(a)}{x - a}.$$

FIGURE 5.7 Antiderivative of $r(x) = 2x$

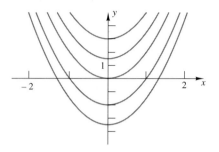

What is the value of $H'(c)$? Substitute the value for $H'(c)$ into this equation and solve for $H(x)$. How does $H(x)$ compare to $H(a)$? Since x was an arbitrary number in $[a, b]$, what can you conclude about the output values of H on the interval $[a, b]$? ◄

According to the theorem, the antiderivative of a function is a family of functions that differ only by a constant. In particular, the antiderivative of $r(x) = 2x$ is the class of all functions that can be written in the form $x^2 + C$ where C is a real number.

Adding a constant to a function will result in a vertical shift of its graph which leads to a graphical interpretation of the antiderivative. The graph of the general antiderivative is a set of parallel curves. Some of the functions in the family of the antiderivative of $r(x) = 2x$ are graphed in Figure 5.7.

► **THIRD REFLECTION**

In Figure 5.7, draw the vertical line $x = \frac{1}{2}$. Sketch a tangent line at the point of intersection of the vertical line with each of the graphed functions. What do you notice about the slopes of these tangent lines? ◄

Break into pairs.

Think and Share

1. Write down several specific antiderivatives of $3x^2$.
2. Write down the general antiderivative for $3x^2$.
3. **a.** Write down a function whose antiderivative you know.
   b. Have your partner write down several specific antiderivatives of your function and the general antiderivative.

The indefinite integral of a function is the antiderivative of that function.

The conventional symbol for the general antiderivative, a class of functions, is related to the symbol for the definite integral. The antiderivative is also called the *indefinite integral* and is written as follows:

$$\int f(x) \, dx$$

If F is an antiderivative of f, then

$$\int f(x) \, dx = F(x) + C$$

If $f(x)$ is the derivative of $F(x)$, then $F(x)$ is an antiderivative of $f(x)$. Thus, whenever you know a derivative, you know a corresponding antiderivative. Any table of derivatives is also a table of antiderivatives. This is similar to the notion that a table of squares is also a table of square roots. Table 5.12 on page 324 contains standard derivative rules and Table 5.13 contains some examples of specific derivatives with the accompanying antiderivative given for the first item, $\int 4x^3 \, dx$. In Table 5.13, each entry lists the rules from Table 5.12 needed to differentiate the function.

The point of studying these tables is *not* to memorize them. The TI calculator is programmed with all the information in these tables and more. Just as in Chapter 2 when we studied the derivative, we can understand antiderivatives and use them more effectively if we work with and understand their properties. The properties of antiderivatives, as you might suspect, are closely related to the properties of derivatives.

Break into groups of four and then into teams of two. In this Think and Share, we will explore the relationship between derivatives and antiderivatives for specific functions.

Think and Share

1. Differentiate the functions whose algebraic expressions are given below.

   a. $x^2 + 3x - 5$
   b. $2\sin(x)$
   c. $5\tan(x) + x^2$
   d. $x^{5/2} + e^x$
   e. $x^{1/2}$
   f. $\dfrac{x}{x+1}$
   g. $x^2 \cos(x)$

2. Your team should rewrite (scramble) the derivatives in (1) and give them to the other team to work out antiderivatives. For example, $x^2 + 3x - 5$ has derivative $2x + 3$. You could scramble this as $3 + 2x$ or $x + 2 + x + 1$.

3. Write down a partial table of antiderivatives by reversing every other derivative equation in Table 5.13. For example, $\dfrac{d}{dx}x^4 = 4x^3$ would be rewritten as $\displaystyle\int 4x^3\,dx = x^4 + C$.

4. Make up a function and differentiate it. Give your derivative to another pair of students for them to provide an antiderivative.

E X A M P L E   1   **Finding an Antiderivative**

Find $\int [5x^3 + \cos(x)]\,dx$.

Solution:

$$\int [5x^3 + \cos(x)]\,dx = \frac{5x^4}{4} + \sin(x) + C$$

We have used the constant times a function and the power rule to find the antiderivative of $5x^3$ and the sine rule for $\cos(x)$.

Graphing the Antiderivative

In Chapter 2, we saw that we could sketch the graph of the derivative of a function by estimating the slopes of tangent lines to the graph of the function. What about the reverse question? Can we sketch the graph of an antiderivative of a function from the graph of the function? Symbolically, from $F' = f$, can we use the graph of f to sketch a possible graph of F?

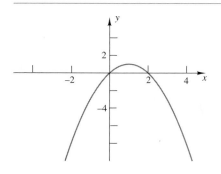

FIGURE 5.8   Graph of f

E X A M P L E   2   **Graphing a Function from its Antiderivative**

The graph of f is given in Figure 5.8. We wish to draw the graph of F, an antiderivative of f.

Solution:

The problem is complicated because there are many antiderivatives of f. Given the graph of one of them we can obtain the graphs of others simply by moving the given graph up and down, because adding a constant corresponds to a vertical shift. By such shifts, we can get an antiderivative to go through a specified point. In particular, we will sketch a possible graph of the antiderivative that passes through the point $(0, 2)$.

Recall from Chapter 3 that the function F is increasing when its derivative f is positive, and decreasing when f is negative. Also, a function has a relative extrema when it changes from increasing to decreasing or from decreasing to increasing. In Figure 5.8, the function f is zero at $x = 0$ and at $x = 2$. At $x = 0$ the function F will have a relative minimum since its derivative, f, is negative to the left of zero and positive to the right of zero. At $x = 2$, F will have a relative maximum.

Finally, a function is concave up when its derivative is increasing and concave down when its derivative is decreasing. Thus, F is concave up to the left of $x = 1$ and concave down to the right of $x = 1$.

Putting this information together we have sketched a possible graph of F in Figure 5.9

This graph only shows the basic shape since we do not know the relative maximum value at $x = 2$. To obtain a more accurate graph of F, we examine more closely the information given by the graph of f. The function value of f at x is the slope of the tangent line to F at $(x, F(x))$. Thus, we can estimate the slope of the tangent lines to the graph of F at several points by estimating the function values for f at the x-coordinate of those points. By knowing the slope, we know the "steepness" of the graph. Several of these slopes are given in Table 5.11. You may wish to complete the table.

TABLE 5.11   Slopes of Tangent Lines

x	-2	-1	0	1	2	3
$f(x)$	-6		0		0	

We have used these slopes to draw a more accurate sketch of the antiderivative in Figure 5.10

FIGURE 5.9   Sketch of the Graph of an   **FIGURE 5.10**   A More Accurate Sketch of F
                Antiderivative of f

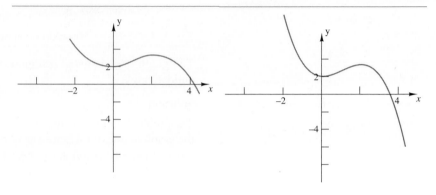

Putting It All Together

We can sometimes find antiderivatives from information we have about derivatives. We can always symbolically represent an antiderivative of $f(x)$ from our
work in Section 5.1 as

$$F(x) = \int_a^x f(t)\, dt$$

since the accumulation function is an antiderivative of the rate function. Thus,

$$\frac{d}{dx} \int_a^x f(t)\, dt = f(x).$$

As we recall, this is the Fundamental Theorem of Calculus, Part II, and says that
the derivative of an accumulation function is the rate at which the accumulation
occurred. In other words, the rate function is the derivative of its accumulation
function.

FUNDAMENTAL THEOREM OF CALCULUS, PART II

Let f be a continuous function. The function $F(x) = \int_a^x f(t)\, dt$ is an an-

tiderivative of f. That is, $F'(x) = \dfrac{d}{dx} \int_a^x f(t)\, dt = f(x)$

TABLE 5.12 Differentiation Rules

1. **The Power Rule** $\dfrac{d}{dx}x^n = nx^{n-1}$ $\qquad\qquad \int x^n\,dx = \dfrac{x^{n+1}}{n+1} + C$

2. **The Power Rule for Fractions** $\dfrac{d}{dx}x^{p/q} = \dfrac{p}{q}x^{(p/q)-1}$

3. **A Constant Times a Function** $\dfrac{d}{dx}kf(x) = k\dfrac{d}{dx}f(x)$

4. **The Sum of Two Functions** $\dfrac{d}{dx}(f(x) + g(x)) = \dfrac{d}{dx}f(x) + \dfrac{d}{dx}g(x)$

5. **The Product of Two Functions**
$$\dfrac{d}{dx}(f(x) \cdot g(x)) = f(x) \cdot g'(x) + g(x) \cdot f'(x)$$

6. **The Quotient of Two Functions**
$$\dfrac{d}{dx}\left(\dfrac{f(x)}{g(x)}\right) = \dfrac{g(x) \cdot f'(x) - g'(x) \cdot f(x)}{(g(x))^2}$$

7. **The Chain Rule** $\dfrac{d}{dx}f(g(x)) = f'(g(x)) \cdot g'(x)$

8. **Sine** $\dfrac{d}{dx}\sin(x) = \cos(x)$

9. **Cosine** $\dfrac{d}{dx}\cos(x) = -\sin(x)$

10. **Tangent** $\dfrac{d}{dx}\tan(x) = \sec^2(x)$

11. **Cosecant** $\dfrac{d}{dx}\csc(x) = -\csc(x)\cot(x)$

12. **Secant** $\dfrac{d}{dx}\sec(x) = \sec(x)\tan(x)$

13. **Cotangent** $\dfrac{d}{dx}\cot(x) = -\csc^2(x)$

14. **The Exponential Function** $\dfrac{d}{dx}e^x = e^x$

15. **The Natural Logarithm Function** $\dfrac{d}{dx}\ln(x) = \dfrac{1}{x}$

TABLE 5.13   Derivatives for Specific Functions

1. **Power Rule** $\dfrac{d}{dx}x^4 = 4x^3$ $\displaystyle\int 4x^3\,dx = x^4 + C$

2. **Power Rule with a Negative Exponent** $\dfrac{d}{dt}t^{-6} = -6t^{-7}$

3. **Power Rule with a Fractional Exponent** $\dfrac{d}{dx}x^{1/2} = \dfrac{1}{2}x^{-1/2}$

4. **Power Rule with a Negative Fractional Exponent**
$\dfrac{d}{dx}x^{-4/3} = -\dfrac{4}{3}x^{-7/3}$

5. **A Constant Times a Function** $\dfrac{d}{dx}5x^4 = 5\dfrac{d}{dx}x^4 = 20x^3$

6. **A Constant Times a Function** $\dfrac{d}{dx}3\sec(x) = 3\sec(x)\tan(x)$

7. **Sum of Functions**
$\dfrac{d}{dx}(x^4 + 3x^2 - 4x - 5) = 4x^3 + 6x - 4$

8. **Sum of Functions and Power Rule and Sine**
$\dfrac{d}{dx}(x^{-3} - \sin(x)) = -3x^{-4} - \cos(x)$

9. **Sum of Functions, Constant Times a Function, Exponential, and Cosine**
$\dfrac{d}{dx}(3e^x - \cos(x) - 4) = 3e^x - (-\sin(x))$

10. **Sum of Functions, Logarithm Function, and Power Rule**
$\dfrac{d}{dx}(\ln(x) + 5x^{1/3}) = \dfrac{1}{x} + \dfrac{5}{3}x^{-2/3}$

11. **Product Rule, Sine, and Power Rule**
$\dfrac{d}{dx}x^2 \cdot \sin(x) = x^2 \cdot \cos(x) + \sin(x) \cdot 2x$

12. **Product Rule, Exponential Function, and Power Rule**
$\dfrac{d}{dx}\left(e^x \cdot x^{-2/3}\right) = e^x \cdot \left(-\dfrac{2}{3}\right)x^{-5/3} + x^{-2/3} \cdot e^x$

13. **Quotient Rule, Power Rule, and Cosecant**
$\dfrac{d}{dx}\left(\dfrac{x^3}{\csc(x)}\right) = \dfrac{\csc(x) \cdot (3x^2) - (-\csc(x)\cot(x)) \cdot x^3}{\csc^2(x)}$

14. **Sum of Functions, Chain Rule, and Power Rule**
$\dfrac{d}{dx}(5x^2 - 3x + 7)^{11} = 11(5x^2 - 3x + 7)^{10} \cdot (10x - 3)$

15. **Chain Rule, Sine, and Power Rule** $\dfrac{d}{dx}\sin(x^4) = \cos(x^4) \cdot 4x^3$

EXERCISE SET 5.2

In Exercises 1–10, find each antiderivative. Describe the rules or process you used to determine each antiderivative. Check your antiderivatives using the TI calculator by differentiating each of your antiderivatives.

1. $\int (3x^2 - 5x + 4)\, dx$

2. $\int 5\cos(t)\, dt$

3. $\int 2x\sin(x^2)\, dx$

4. $\int (x^3 - 15x^2 + 2)\, dx$

5. $\int 5x^4 \sec(x^5)\tan(x^5)\, dx$

6. $\int \frac{25}{3}u^{2/3}\, du$

7. $\int \frac{5}{3x}\, dx$

8. $\int 20x(5x^3 - 8)(x^5 - 4x^2 - 11)^4\, dx$

9. $\int 66x^2 e^{2x^3}\, dx$

10. $\int \frac{-49}{2}z^{-9/2}\, dz$

In Exercises 11–14, graph the accumulation function from the given rate function.

11. A rate function.

FIGURE 5.11

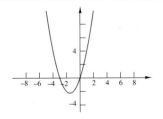

12. A rate function.

FIGURE 5.12

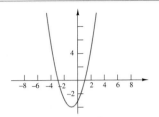

13. A rate function.

FIGURE 5.13

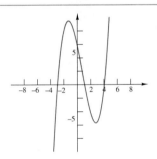

14. A rate function.

FIGURE 5.14

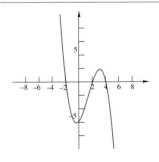

In Exercises 15–20, explain how you would find the indefinite integral of the given function.

15. $f(x) = 2x\sin(x^2)$

16. $g(x) = 3x^2\cos(x^3)$

17. $h(x) = (2x - 7)(x^2 - 7x + 2)^5$

18. $f(x) = (7 + 6x)(-3x^2 - 7x + 5)^4$

19. $g(x) = (5 - x)\left(5x - \dfrac{x^2}{2}\right)$

20. $h(x) = \cos(\cos(x^2)) \cdot (-\sin(x^2)) \cdot 2x$

21. Let $G(x) = \int_{-3}^{x} f(t)\,dt$ where f is the function whose graph is shown in Figure 5.15. The tick marks are one unit apart in both the horizontal and vertical directions. The graphing window is $[-8, 7]$ by $[-5, 10]$.

   a. Estimate $G(-2)$. Justify.

   b. For which values of x is $G(x)$ increasing? Justify.

   c. Is $G(x)$ ever concave up? Justify.

   d. Describe $G(x)$ and $f(x)$ in terms of rate function and accumulation function and explain how these functions are related.

   e. Find $G'(x)$.

FIGURE 5.15   Graph of $f(x)$

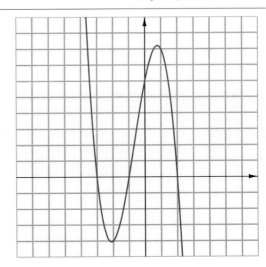

5.3

Finding Antiderivatives Using Properties and Formulas

Antiderivatives are important in evaluating definite integrals as we know from the Fundamental Theorem of Calculus, Part I in Chapter 4. In previous sections, we have guessed (and checked) some antiderivatives for a number of simple functions like polynomials. In this section, we will develop our ability to find antiderivatives using the properties of antiderivatives and a few basic antiderivative formulas.

Properties and Rules of Antiderivatives

The following properties of antiderivatives can be verified by differentiation. The first two are similar to properties of definite integrals.

1. **The Sum of Functions Property**

$$\int [f(x) + g(x)]\,dx =$$

2. **A Constant Times a Function Property**

$$\int kf(x)\,dx =$$

3. **The Power Rule**

$$\int x^n \, dx = \frac{x^{n+1}}{n+1} + C, n \neq -1$$

4. **The Power Rule for Functions**

$$\int [f(x)]^n f'(x) \, dx = \frac{[f(x)]^{n+1}}{n+1} + C, n \neq -1$$

5. **The Chain Rule in Reverse**

$$\int f'(g(u))g'(u) \, du =$$

We can directly apply the first two properties to polynomial functions.

E X A M P L E 1   **Integrating a Polynomial**

Find $\int (4x^2 - 3x + 7) \, dx$.

Solution:

We will use the sum of functions and the constant times a function properties of indefinite integrals.

$$\int (4x^2 - 3x + 7) \, dx = \int 4x^2 \, dx - \int 3x \, dx + \int 7 \, dx$$

$$= 4 \int x^2 \, dx - 3 \int x \, dx + 7 \int 1 \, dx$$

$$= 4 \frac{x^3}{3} - 3 \frac{x^2}{2} + 7x + C$$

An indefinite integral or antiderivative is a family of functions. The addition of C indicates this family of functions.

► **FIRST REFLECTION**

Why can't $n = -1$ in the Power Rule? Give an example.   ◄

E X A M P L E 2   **The Power Rule for Functions**

Find $\int (x^2 + 3)^4 \cdot 2x \, dx$.

Solution:

We notice that the derivative of $(x^2 + 3)$, a part of the integrand, is $2x$ which also occurs as part of the integrand. From this, we see that the Power Rule for Functions can be applied.

$$\int (x^2 + 3)^4 \cdot 2x \, dx = \frac{(x^2 + 3)^5}{5} + C$$

Sometimes the derivative that we need in order to apply the Power Rule for Functions is missing a constant multiplier. The next example shows how to deal with such integrals.

E X A M P L E 3   **A Missing Constant Multiplier**

Find $\int (3x^2 - 4)^5 \cdot 2x \; dx$.

Solution:

The derivative of $3x^2 - 4$ is $6x$. The factor that exists in the integrand is $2x$. From our knowledge of rational expressions, we can multiply $2x$ by 3 as long as we divide the integrand by 3 as well.

$$\int (3x^2 - 4)^5 \cdot 2x \; dx = \int (3x^2 - 4)^5 \cdot \frac{6x}{3} \; dx$$

$$= \frac{1}{3} \int (3x^2 - 4)^5 \cdot 6x \; dx$$

$$= \frac{1}{3} \frac{(3x^2 - 4)^6}{6} + C$$

▶ SECOND REFLECTION

Complete the following table of basic formulas for antiderivatives. You can check any antiderivative or indefinite integral by differentiation. As we know, the derivative of an indefinite integral should be its integrand.   ◀

We can use these basic formulas and the properties of indefinite integrals to find a wider variety of indefinite integrals or antiderivatives.

E X A M P L E 4   **Integrating Term by Term**

Find $\int (\sin(x) + \sec^2(x)) \; dx$

Solution:

We notice that the integrand is a sum of two functions whose antiderivatives we know.

$$\int (\sin(x) + \sec^2(x)) \; dx = \int \sin(x) \; dx + \int \sec^2(x) \; dx$$

$$= -\cos(x) + \tan(x) + C$$

The Chain Rule for Derivatives' pattern involves a composite function of the form $f(g(x))$ and the derivative of the "inside" function $g'(x)$ as a product, $f(g(x)) \cdot g'(x)$. Your experience with differentiation will allow you to recognize this pattern. It occurs in numerous integrals.

TABLE 5.14 Table of Basic Formulas for Integrals

1. **Power Rule**

$$\int x^n \, dx =$$

2. **Sine Function**

$$\int \sin(x) \, dx =$$

3. **Cosine Function**

$$\int \cos(x) \, dx =$$

4. **Secant Squared Function**

$$\int \sec^2(x) \, dx =$$

5. **The Product of Secant and Tangent Functions**

$$\int \sec(x) \tan(x) \, dx =$$

6. **The Reciprocal Function**

$$\int \frac{1}{x} \, dx =$$

7. **The Exponential Function**

$$\int e^x \, dx =$$

8. **The Power Rule for Functions**

$$\int [f(x)]^n f'(x) \, dx = \frac{[f(x)]^{n+1}}{n+1} + C, n \neq -1$$

E X A M P L E 5 **Recognizing the Chain Rule**

Find $\int \sin(x^2) \cdot 2x \, dx$.

Solution:

The derivative of the "inside" function x^2 is $2x$. Thus,

$$\int \sin(x^2) \cdot 2x \, dx = -\cos(x^2) + C$$

Notice that our integral $\int \sin(x^2) \cdot 2x \, dx$ fits the form $\int \sin(f(x)) \cdot f'(x) \, dx$ whose antiderivative is $-\cos(f(x)) + C$. All the basic formulas given above can be extended in this way to handle a wider variety of antiderivatives.

▶ **THIRD REFLECTION**
Write a similar expression for $\int \cos(f(x)) \cdot f'(x) \, dx$.   ◀

E X A M P L E   6   **Recognizing the Integrand**

Find $\int \sec(2x) \tan(2x)\, dx$.

Solution:

The integrand of this indefinite integral reminds us of the Basic Formula (5) above. We would have to start with $\sec(2x)$ to arrive at this integrand through the use of the Chain Rule. Unfortunately, the derivative of $\sec(2x)$ is $\sec(2x) \tan(2x) \cdot 2$. Fortunately, we can deal with the constant 2 as follows:

$$\int \sec(2x) \tan(2x)\, dx = \int \sec(2x) \tan(2x) \frac{2}{2}\, dx$$

$$= \frac{1}{2} \int \sec(2x) \tan(2x) \cdot 2\, dx$$

$$= \frac{1}{2} \sec(2x) + C$$

Think and Share

Break into pairs. Work out each integration problem. Explain your technique to your partner, or get help from your partner. Check by differentition.

1. $\int \left(3x^4 - 7x^2 + \dfrac{5}{x}\right) dx =$

2. $\int (x^3 - 1)^{1/2} \cdot 3x^2\, dx =$

3. $\int (x^3 - 1)^4 \cdot x^2\, dx =$

4. $\int (\cos(x) + \sec(x)\tan(x))\, dx =$

5. $\int \sec^2(3x)\, dx =$

E X E R C I S E   S E T   5 . 3

In Exercises 1–20, find the indicated antiderivative and state the properties and formulas you used.

1. $\int (x^2 + 3x - 7)\, dx$

2. $\int (3x^5 - 4x^2 + 5)\, dx$

3. $\int (5x^4 - x + 7)\, dx$

4. $\int (11x^3 - 2x^2 - 3x - 4)\, dx$

5. $\int (x^2 + 3x)^5 \cdot (2x + 3)\, dx$

6. $\int (3x^2 - 5x)^7 \cdot (6x - 5)\, dx$

7. $\int (5x^2 - \sin(x) + 2)\, dx$

8. $\int (\cos(x) - 5x^3)\, dx$

9. $\int (\sec^2(x) - \sec(x)\tan(x) + 5)\, dx$

10. $\int \dfrac{2}{x} - x^2 + \sin(x)\, dx$

11. $\int \sin^3(x) \cos(x)\, dx$

12. $\int \cos^3(x) \sin(x)\, dx$

13. $\int \cos(x^2) \cdot 2x\, dx$

14. $\displaystyle\int \sec^2(x^3) \cdot 3x^2 \, dx$

15. $\displaystyle\int \sec^2(x^3) \cdot x^2 \, dx$

16. $\displaystyle\int \sin^4(x) \cos(x) \, dx$

17. $\displaystyle\int \cos^4(x) \sin(x) \, dx$

18. $\displaystyle\int \sin(x^3) \cdot x^2 \, dx$

19. $\displaystyle\int \sin^3(x) \cos(x) + 4 \, dx$

20. $\displaystyle\int \, dx$

In Exercises 21–40, find the indicated antiderivative.

21. $\displaystyle\int \sin(x^4) \cdot 4x^3 \, dx$

22. $\displaystyle\int \cos(x^4) \cdot 4x^3 \, dx$

23. $\displaystyle\int x^3 \sin(x^4) \, dx$

24. $\displaystyle\int x^3 \cos(x^4) \, dx$

25. $\displaystyle\int \sin(x) \cos(x) \, dx$

26. $\displaystyle\int (x^3 + 2x - 5)^4 \cdot (3x^2 + 2) \, dx$

27. $\displaystyle\int (x^5 + 3x^3 - 7x - 3)^5 \cdot (5x^4 + 9x^2 - 7) \, dx$

28. $\displaystyle\int \sec^2(3x) \, dx$

29. $\displaystyle\int x \sin(x^2) \, dx$

30. $\displaystyle\int (x^2 + \sec(x)\tan(x)) \, dx$

31. $\displaystyle\int (2x + 3)(x^2 + 3x - 11) \, dx$

32. $\displaystyle\int (x + 1)(x^2 + 2x + 5)^6 \, dx$

33. $\displaystyle\int \frac{1}{x} + x \cos(x^2) \, dx$

34. $\displaystyle\int \left(\frac{3}{x^3} + 2\right) \, dx$

35. $\displaystyle\int \sec^4(x)\tan(x) \, dx$

36. $\displaystyle\int \sec^5(x)\tan(x) \, dx$

37. $\displaystyle\int x \sin(x^2) \cos(x^2) \, dx$

38. $\displaystyle\int x^2 \sin(x^3) \cos(x^3) \, dx$

39. $\displaystyle\int \tan(x) \, dx$ Hint: $\tan(x) = \frac{\sin(x)}{\cos(x)}$

40. $\displaystyle\int \cot(x) \, dx$ Hint: $\cot(x) = \frac{\cos(x)}{\sin(x)}$

In Exercises 41–46, find the value of the definite integral.

41. $\displaystyle\int_0^3 (x^2 + 2x - 5) \, dx$

42. $\displaystyle\int_0^{\sqrt{\pi}} x \sin(x^2) \, dx$

43. $\displaystyle\int_0^1 (\sec(x)\tan(x) + x^2) \, dx$

44. $\displaystyle\int_{-1}^5 (2x + 5)(x^2 + 5x - 7)^4 \, dx$

45. $\displaystyle\int_1^3 \frac{4}{x^2} \, dx$

46. $\displaystyle\int_1^5 \left(\frac{1}{x} + 3x^2 + \sin(x)\right) \, dx$

Find the general antiderivative for each of the following. Check each answer by differentiation.

47. $\displaystyle\int (x^3 + 5x^2 - 7x + 1) \, dx$

48. $\displaystyle\int \left(x^4 - 7x^3 + \frac{3}{x}\right) \, dx$

49. $\displaystyle\int (\cos(3x) + \sin(5x) \, dx$

50. $\displaystyle\int \sec^2(2x) \, dx$

51. $\displaystyle\int (3x^2 + 1)^4 \cdot x \, dx$

52. $\displaystyle\int (2x^4 - 6)^3 \cdot x^3 \, dx$

53. $\displaystyle\int \tan(3x)\sec(3x) \, dx$

54. $\displaystyle\int \left(e^x + \frac{1}{x}\right) \, dx$

55. Is it possible to obtain two different (and both correct) antiderivatives for one function f? Explain.

56. Find $f(x)$ if $\int f(x) \, dx = x^4 - 7x^2 + 1$

57. Is it always possible to find an antiderivative for a continuous function f? Explain.

58. Is it always possible to evaluate a definite integral using Part I of the Fundamental Theorem of Calculus? Explain.

5.4

Applications

Area Between Two Curves

Determining the area under a curve is a specific example of the more general question of determining the area between two curves.

E X A M P L E   1   **Area Between Curves**

Compute the area of the region bounded by the two functions $f(x) = (x + 1)^3$ and $g(x) = x^2 - 25$ and the lines $x = 1$ and $x = 4$.

Our first step is to graph the two functions as shown in Figure 5.16. As before in Chapter 4, we slice up the region between them into n strips of equal width (Δx), perpendicular to the x-axis, and analyze the kth strip. Using the right endpoint method, we approximate the area of the kth strip with a rectangle based on the function

FIGURE 5.16   Area Between Graphs   **FIGURE 5.17**   Area of One Rectangle

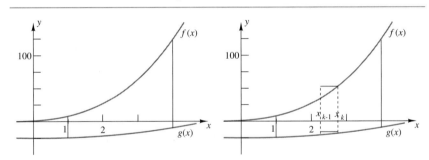

values of f and g. The height of the rectangle is $f(x_k) - g(x_k)$ and the width is Δx (see Figure 5.17), so we can compute its area with the formula

$$\text{Area}_{rectangle} = (f(x_k) - g(x_k))\Delta x.$$

The total area of the region between the curves is estimated by the sum of the areas of the n rectangles,

$$\textit{area of the rectangles} = \sum_{k=1}^{n}(f(x_k) - g(x_k))\Delta x. \qquad \textbf{(5.3)}$$

Think and Share

Break into groups. Use the `Save Copy As ...` option in the `Toolbox` menu to rename the `riemsum` script as `fgdif`. Open this new script and alter it to obtain a graph of the approximating rectangles between the two functions when $n = 10$. Find the sum of the areas of the 10 rectangles.

Fortunately, the sum of the areas of the rectangles is a Riemann sum. The exact area of the region between the curves equals the associated definite integral and is computed by taking the limit of this Riemann sum.

Equation (5.3) is a Riemann sum for the single function $h(x) = f(x) - g(x)$.

Symbolically,

$$\text{Area}_{region} \approx \sum_{k=1}^{n}(f(x_k) - g(x_k))\Delta x$$

$$\text{Area}_{region} = \lim_{n \to \infty}\sum_{k=1}^{n}(f(x_k) - g(x_k))\Delta x$$

$$= \int_{1}^{4}[f(x) - g(x)]\,dx = \int_{1}^{4}[(x+1)^3 - (x^2 - 25)]\,dx$$

$$= \left[\frac{(x+1)^4}{4} - \frac{x^3}{3} + 25x\right]\Bigg|_{1}^{4}$$

$$= \left[\frac{(4+1)^4}{4} - \frac{4^3}{3} + 25 \cdot 4\right] - \left[\frac{(1+1)^4}{4} - \frac{1^3}{3} + 25 \cdot 1\right]$$

$$= \frac{825}{4}.$$

Of course, we could compute this value directly on the TI calculator using the command

$$\int((x+1)^3-(x^2-25),x,1,4).$$

You can use the $\int($ command to find a numerical value for definite integrals that are not evaluated in the text.

The discussion in the previous example will work for any f and g as long as $f(x) \geq g(x)$ on the interval in question. We make the following definition.

THE AREA OF A PLANE REGION

The area of the region bounded by the functions $f(x)$ and $g(x)$ ($f(x) \geq g(x)$) and the vertical lines $x = a$ and $x = b$ is given by

$$\text{Area} = \lim_{n \to \infty}\sum_{k=1}^{n}[f(t_k) - g(t_k)]\Delta x = \int_{a}^{b}[f(x) - g(x)]\,dx$$

where t_k is an arbitrary point in the kth subinterval of $[a, b]$, and $[a, b]$ is partitioned into subintervals of equal length.

FIGURE 5.18 *k*th Slice Rectangle

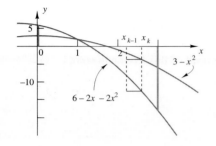

▶ **FIRST REFLECTION**

Why is finding the area under a curve a special case of finding the area between two curves? ◀

The definition can be used even when the graph of one function is not always above the graph of the other. We just split the interval into intervals over which only one function is on top. The next example illustrates this idea.

EXAMPLE 2 **Intersecting Graphs**

Find the area of the region bounded by the functions $f(x) = 6 - 2x - 2x^2$ and $g(x) = 3 - x^2$ and the lines $x = 0$ and $x = 3$.

Graph the two functions, slice up the region into n strips of equal width (Δx), and construct the kth strip rectangle, as shown in Figure 5.18. Notice that $f > g$ to the left of $x = 1$ (the point of intersection of the two functions) and $g > f$ to the right of $x = 1$. For a strip on the left of $x = 1$, the height of the kth rectangle is $f(x_k) - g(x_k)$, while a strip on the right of $x = 1$ has height $g(x_k) - f(x_k)$. In both cases, the width is denoted by Δx. Thus, rectangles on the left have area

$$\text{Area}_{rectangle} = (f(x_k) - g(x_k)) \cdot \Delta x$$

while rectangles on the right have area

$$\text{Area}_{rectangle} = (g(x_k) - f(x_k)) \cdot \Delta x$$

The total area of the region is the sum of the area to the left of $x = 1$ and the area to the right of $x = 1$. These two areas are estimated by the two Riemann sums,

$$\text{Area}_{left} \approx \sum_{k=1}^{n} \left(f(x_k) - g(x_k) \right) \Delta x \qquad \text{Area}_{right} \approx \sum_{k=1}^{n} \left(g(x_k) - f(x_k) \right) \Delta x$$

where $x_0 = 0$ and $x_n = 1$ for Area_{left}, and $x_0 = 1$ and $x_n = 3$ for Area_{right}. The limit of these two Riemann sums are definite integrals which equal the exact area between the curves to the left and right of $x = 1$, respectively.

$$\text{Area}_{left} = \lim_{n \to \infty} \sum_{k=1}^{n} \left(f(x_k) - g(x_k) \right) \Delta x$$

$$= \int_0^1 (f(x) - g(x))\, dx = \int_0^1 \left((6 - 2x - 2x^2) - (3 - x^2) \right)\, dx$$

$$\text{Area}_{right} = \lim_{n \to \infty} \sum_{k=1}^{n} \left(g(x_k) - f(x_k) \right) \Delta x$$

$$= \int_1^3 (g(x) - f(x))\, dx = \int_1^3 \left((3 - x^2) - (6 - 2x - 2x^2) \right)\, dx$$

$$\text{Area}_{region} = \text{Area}_{left} + \text{Area}_{right} = \frac{37}{3}$$

You should carry out the evaluation of Area $_{left}$ and Area $_{right}$ to convince yourself that the answer is correct.

Think and Share

Break into groups. Find the area bounded by the two functions defined by $f(x) = x^3 - 8x + 15$ and $g(x) = 4x^2 + 3x - 15$ by completing steps 1, 2, and 3.

1. Graph the two functions.
2. Find their points of intersection.
3. Find the area between the two graphs.

Is $\int_a^b (f(x) - g(x))\, dx$ the area between the two graphs where a and b are the outside intersection points. Explain.

FIGURE 5.19   Region

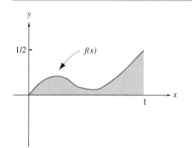

Volumes of Revolution

Figure 5.19 shows a region bounded by the function $f(x)$, the x-axis, and the line $x = 1$. Solid objects can be generated by rotating such regions about a line as long as the line does not pass through the interior of the region. For example, a solid object can be generated by revolving the region in the figure about the x-axis, as shown in Figures 5.20 and 5.21. We call such an object a *volume of revolution*.

FIGURE 5.20   Region Rotated   **FIGURE 5.21**   Object in Vertical Position

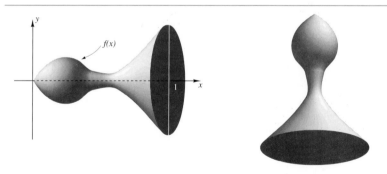

Figure 5.20 shows the solid as generated. We can move the solid and look at it from any angle. Figure 5.21 shows the object repositioned vertically.

▶ SECOND REFLECTION

Describe the object that would be generated if the region in Figure 5.19 were rotated about the line $y = -1$. Describe three physical objects which could be viewed as volumes of revolution.   ◀

▶ **THIRD REFLECTION**

Can you describe the volume that would be generated if the region in Figure 5.19 were rotated about the line $y = 1/2$? Why is the condition that "...the line does not pass through the interior of the region..." included? Can you describe a region for which the volume generated by rotating it about a line through the interior of the region would generate a "reasonable" volume? ◀

Many beautiful works of art can be viewed as volumes of rotation. Their symmetry makes them mathematically elegant.

E X A M P L E   3   **A Volume of Revolution**

We wish to find the volume of the object formed by revolving the region bounded by $f(x) = x^2$, the x-axis, and the line $x = 1$ about the x-axis. The region is drawn in Figure 5.22.

FIGURE 5.22   Region   **FIGURE 5.23**   Region Rotated

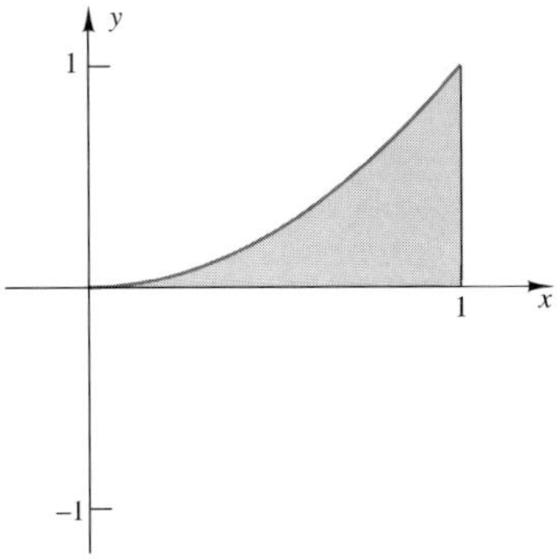

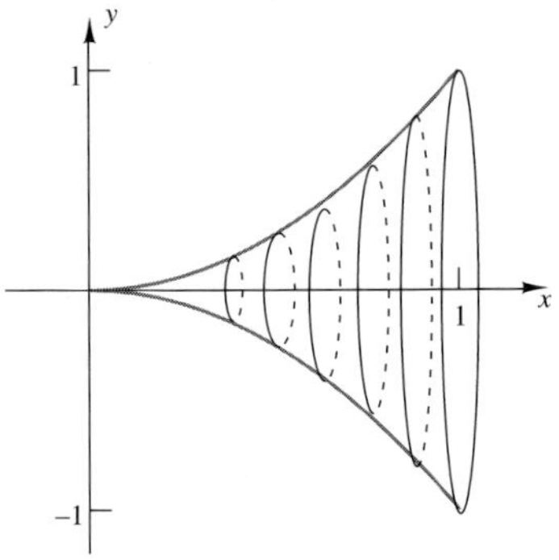

We slice the solid object into n pieces of equal thickness and let Δx be the thickness. Each slice is made parallel to the y-axis and perpendicular to the x-axis, as shown in Figure 5.23. To take advantage of the symmetry, we usually slice a solid of revolution perpendicular to the axis of revolution.

Using a kth element (in this case the kth slice) analysis, we see that the kth slice can be approximated by a disk (right circular cylinder), as shown in Figure 5.24. The disk has volume equal to the area of its base (a circular disk) times its thickness, Δx. Notice that in Figure 5.24, the cylinder is standing on end, so its base is perpendicular to the x-axis. Also, from Figure 5.25, we see that the radius of the circular base is $f(x_k)$, the vertical distance from the right endpoint of the kth subinterval to the graph of f. The volume of the kth disk is

$$\text{Volume}_{k \text{ th disk}} = \pi \left[f(x_k) \right]^2 \Delta x.$$

FIGURE 5.24   One Disk   **FIGURE 5.25**   Radius of kth Disk

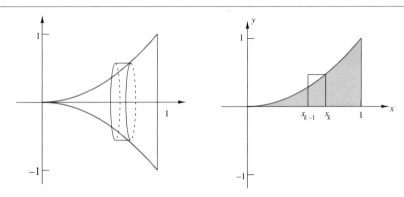

The volume of the kth cylinder approximates the volume of the kth slice. Thus, the total volume of the object, which equals the sum of the volumes of the slices, is approximated by the sum of the volumes of the cylinders,

$$\text{Volume}_{\text{Solid of Revolution}} \approx \sum_{k=1}^{n} \pi \left[f(x_k) \right]^2 \Delta x.$$

We find the exact value for the volume of the solid of revolution by taking the limit of this Riemann sum

$$\text{Volume}_{\text{Solid of Revolution}} = \lim_{n \to \infty} \sum_{k=1}^{n} \pi \left[f(x_k) \right]^2 \Delta x = \int_{0}^{1} \pi [x^2]^2 \, dx.$$

▶ **FOURTH REFLECTION**

What is the volume of this solid? What features of the original region led to the upper and lower limits of integration?   ◀

Population Projections

Population projections permeate discussions of resource allocations and man's impact on the environment. The issues are very real since the earth's resources are finite. What level of population can the earth comfortably support, and how soon will that level be reached (assuming we haven't reached it already)?

The first question is very difficult to answer because people differ in their definitions of "comfort." For any particular level of population, however, we can determine fairly accurately how soon that level will be reached.

In the case of a single nation, birth and death rates can be combined to estimate the population growth rate. We can create an accumulation function that indicates this estimated population growth beginning at some fixed point in time, say 1910. By adding the population growth to the population in 1910, we can predict the nation's population at any specific time in the future (although the prediction becomes less accurate the farther we project into the future because these rates change over time).

EXAMPLE   4   **Population Prediction**

Table 5.15 displays a nation's combined birth and death rates (in thousands per year). Using this information, we will determine a population growth rate function and use that function to estimate the population of the nation in 2000.

FIGURE 5.26 Graph of Population Data on [0, 100] × [0, 220]

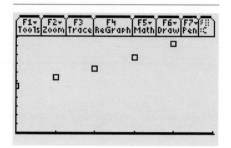

TABLE 5.15   Birth Rate Minus Death Rate (Thousands per year)

Year	1910	1930	1950	1970	1990
Birth rate−Death rate	106	126	148	175	206

First we convert the table to years starting from 1910 as shown in Table 5.16. As a second step, plot this data on your TI calculator. The plot of the data looks like Figure 5.26. Population growth rate is usually exponential,

TABLE 5.16   Birth Rate Minus Death Rate

Elapsed Time(years)	0	20	40	60	80
Birth rate−Death rate	106	126	148	175	206

We can assume that this rate function is valid through the year 2000.

so we will assume an exponential model. Use your calculator to compute an exponential fit. You should obtain the following growth rate function.

$$rate(t) = 106.3(1.008321)^t$$

▶ FIFTH REFLECTION

Graph the data and the exponential function in the same window. How closely does the function appear to fit the data?   ◄

We can compute the population each year based on this growth rate function. In the kth year, we use the growth rate at the beginning of the year as an estimate of the average rate of population growth for the entire year. With this in mind, we estimate the increase in population for the kth year as

$$population\ growth_{k\text{th year}} = growth\ rate(t_{k-1})\Delta t.$$

We estimate the total population growth by summing our estimates of each year's growth.

$$total\ population\ growth \approx \sum_{k=1}^{n} growth\ rate(t_{k-1})\Delta t$$

Better estimates for the population growth are obtained by using shorter and shorter subintervals of time.

The process we're describing should be very familiar to you by now, and you should not be surprised when, in our next step, we take the limit of our sum as the number of time subintervals increases (forcing the length of each subinterval to approach zero). Our result turns out to be the limit of a Riemann sum.

$$total\ population\ growth = \lim_{n \to \infty} \sum_{k=1}^{n} growth\ rate(t_{k-1})\,\Delta t$$

$$= \int_{0}^{90} growth\ rate(t)\ dt$$

▶ **SIXTH REFLECTION**

Why are the limits of integration from 0 to 90? Use the TI calculator or the first part of the Fundamental Theorem of Calculus to determine the value of this integral. This number is the increase in population from 1910 to 2000. Add this value to the population in 1910 (92,200,000) to find the projected population of the nation in 2000. ◀

Think and Share

Break into groups. Discuss and list the possible problems that might result from this approach to population projection. Are births and deaths the only means by which a nation's population can change? Can you really assume that the population rate of change is purely exponential over almost a century? Are there other issues that should be considered in projecting the population of this country? Would you trust this model to predict this nation's population in 2050? Explain why or why not. What nation might this be?

Two-Dimensional Center of Mass

In Section 4.6, we studied the concept of center of mass in a one-dimensional case. We used a seesaw to demonstrate the discrete situation and then used an oak beam to generalize to the continuous case. The ideas we developed will form the foundation for our investigation of a two-dimensional situation.

FIGURE 5.27

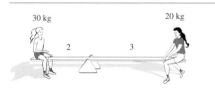

30 kg   20 kg

2   3

E X A M P L E 5 **A Question of Balance**

FIGURE 5.28 Tray with Cups and Glasses

Sharing a milk shake at an outdoor cafe with her sister, Susan begins to wonder if she could get a job as a waitress. Jodi notices a waitress clearing a table and carrying the tray with only one hand (see Figure 5.28).

"You couldn't do that," says Jodi. "You're not strong enough."

"It's not a question of strength," Susan replies. "The tray doesn't weigh that much. It's a question of balance."

Susan decides to see if she can figure out where the balance point of the tray should be, using what she already knows about seesaws.

To begin with, she simplifies the problem by imagining that the tray is sitting on the edge of a flat board (see Figure 5.29).

Sitting on the board, the tray acts like a seesaw. Each of the cups and glasses on the tray tend to make it tip one way or the other depending on their masses and their distances from the board.

Susan realizes she must assign precise locations to the cups and glasses. She imagines two perpendicular lines drawn through the center of the tray which will serve as her x- and y-axes (see Figure 5.30). To each of the cups and glasses she assigns coordinates which correspond to their centers (which fortunately are all fairly symmetric) as illustrated in Figure 5.31.

FIGURE 5.29   Tray Sitting on the Edge of a Flat Board

FIGURE 5.30   Tray with Axes

FIGURE 5.31   Axes with Tray

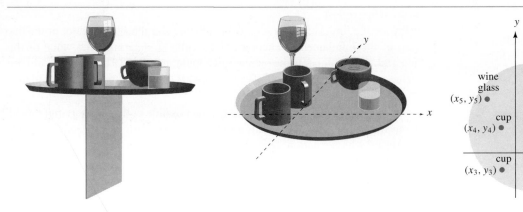

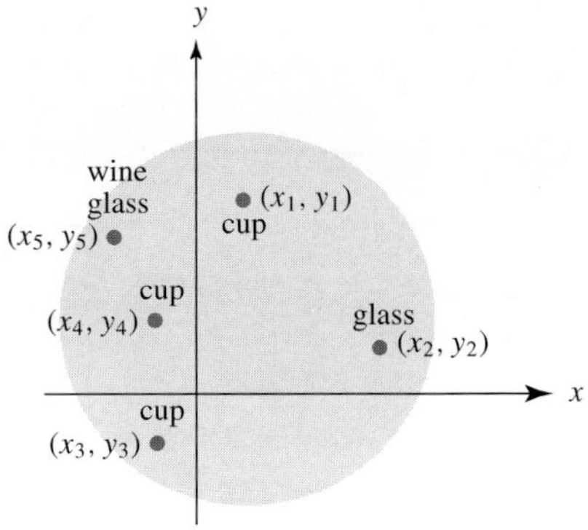

FIGURE 5.32   Tray with Axes and $x = \bar{x}$ Line

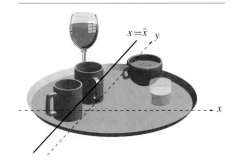

Susan imagines that her board is placed parallel to the y-axis so that the tray is balanced. Using her coordinate system, she knows it will run along a line with equation $x = \bar{x}$ where $\bar{x}$ is some constant (see Figure 5.32).

A cup located at (x_i, y_i) will be $x_i - \bar{x}$ units away from the board. If the cup's mass is m_i, then its **moment of mass about the board** (the line $x = \bar{x}$) will be $m_i(x_i - \bar{x})$. Susan realizes that this is exactly the same expression she got for the seesaw, so the same formula must give her the value of $\bar{x}$; that is,

$$\bar{x} = \frac{\displaystyle\sum_{i=1}^{n} m_i x_i}{\displaystyle\sum_{i=1}^{n} m_i}.$$

Here, $\dfrac{\displaystyle\sum_{i=1}^{n} m_i x_i}{\displaystyle\sum_{i=1}^{n} m_i}$ is called the total moment of mass about the y-axis.

Susan also realizes that if she repeats the analysis using a board which runs along a line parallel to the x-axis with equation $y = \bar{y}$ she will find a similar formula,

$$\bar{y} = \frac{\sum_{i=1}^{n} m_i y_i}{\sum_{i=1}^{n} m_i}$$

FIGURE 5.33   An Irregular Shape Sitting on Its Center of Mass

where $\dfrac{\displaystyle\sum_{i=1}^{n} m_i y_i}{\displaystyle\sum_{i=1}^{n} m_i}$ is called the total moment of mass about the x-axis.

To be balanced, the tray must be balanced in both the x and y directions. Thus, the balance point or *center of mass* of the tray must lie on the line $x = \bar{x}$ and on the line $y = \bar{y}$. The only point that will do this is their point of intersection, $(\bar{x}, \bar{y})$.

▶ **SEVENTH REFLECTION**

Place the lid of your calculator flat along your index finger in such a way that your finger is parallel to one side of the lid, and the lid is balanced. Repeat with your finger perpendicular to its original location. Estimate where the two lines intersect. Try to balance the lid on the tip of your finger at the point of intersection.   ◄

E X A M P L E   6   **The Center of Mass of a Thin Plate**

There is a problem with Susan's analysis. She has neglected to consider the weight of the tray itself. The farther her computed center is from the center of the tray, the greater the error caused by her omission. Computing the center of mass for the tray, we need to modify Susan's argument from the discrete to the continuous.

The circular tray is so symmetric that it is easy to guess correctly that its center of mass is simply the center of the circle, but suppose we wished to determine the center of mass for a plate like the one in Figure 5.33.

This figure looks like a piece of modern furniture; if you were designing such a table, you would want to place the single leg at the balance point or the center of mass. To simplify the problem, we will assume the plate has uniform density and we will look for $\bar{y}$, the y-coordinate of the center of mass. Just as we have done before, we compute $\bar{y}$ by dividing the total moment of the plate about the x-axis by the total mass of the plate.

Following Susan's example to find the center of mass of the plate, we introduce a coordinate system so that all the points on the plate have precise locations (see Figure 5.34). By orienting our plate properly, we can let the

FIGURE 5.34 Plate with Coordinate System

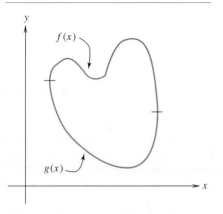

FIGURE 5.35 Plate Partitioned

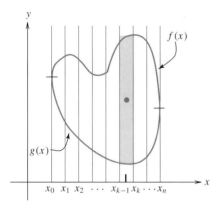

function $f(x)$ represent its top boundary and the function $g(x)$ represent its bottom boundary. This can be done in several ways by subdividing the object into several parts. Just as in the discrete case, if we can balance the plate on a board that runs parallel to the x-axis, the line will have equation $y = \bar{y}$.

To obtain a discrete approximation of our plate, we partition the x-axis into n subintervals, and construct n rectangles, one of which is shown in Figure 5.35. In this figure, $n = 8$ and $k = 6$. We have shaded the sixth slice. The ellipsis (...) indicates the missing subinterval endpoints. You need to realize that there could be any number (n) of slices; we focus on the shaded kth slice. The rectangles partition the x-axis into the subintervals,

$$[x_0,\ x_1],\ [x_1,\ x_2],\dots,[x_{k-1},\ x_k],\dots,[x_{n-1},\ x_n],$$

where $x_0 = a$ and $x_n = b$. The width of the kth rectangle is $x_k - x_{k-1}$ or Δx, if we assume all the rectangles have the same width. From the drawing, we see that the height of the kth rectangle (at the midpoint of the interval) is $f\left(\dfrac{x_k + x_{k-1}}{2}\right) - g\left(\dfrac{x_k + x_{k-1}}{2}\right)$. To simplify our expressions, we set $\bar{x}_k$ equal to $\dfrac{x_k + x_{k-1}}{2}$. With this convention, the area of the rectangle is $\left(f\left(\bar{x}_k\right) - g\left(\bar{x}_k\right)\right)\Delta x$. We use ρ for the density of the plate, so m_k, the mass of the kth rectangle, can be computed by $m_k = \rho\left(f\left(\bar{x}_k\right) - g\left(\bar{x}_k\right)\right)\Delta x$.

Since the plate has uniform density, the center of mass of each rectangle is at the center point of the rectangle; that is, the point with coordinates $\left(\bar{x}_k,\ \dfrac{f\left(\bar{x}_k\right) + g\left(\bar{x}_k\right)}{2}\right)$. To compute the moment of mass about the x-axis, we will treat the rectangle as if all of its mass were located at its center. (Physicists use the term *point mass* for this assumption.)

The distance from the center of the rectangle to the x-axis is the y-coordinate of the center point. Thus, the moment of mass of the rectangle about the x-axis is given by

$$moment\ of\ mass_{k\text{th rectangle}} = mass \cdot distance\ from\ the\ x\text{-axis}$$

$$= m_k \cdot \frac{f\left(\bar{x}_k\right) + g\left(\bar{x}_k\right)}{2}$$

$$= \rho\left(f\left(\bar{x}_k\right) - g\left(\bar{x}_k\right)\right)\Delta x \cdot \frac{f\left(\bar{x}_k\right) + g\left(\bar{x}_k\right)}{2}.$$

Summing the moments of mass about the x-axis for each of the n rectangles gives an estimate for the total moment about the x-axis.

$$M_x \approx \sum_{k=1}^{n} \rho\left(f\left(\bar{x}_k\right) - g\left(\bar{x}_k\right)\right)\Delta x \cdot \frac{f\left(\bar{x}_k\right) + g\left(\bar{x}_k\right)}{2}$$

Since $\bar{x}_k$ is in the interval $[x_{k-1}, x_k]$ we see this is a Riemann sum. Thus, we can compute the exact moment of mass about the x-axis by taking the limit as $n \to \infty$.

$$M_x = \lim_{n \to \infty} \sum_{k=1}^{n} \rho \frac{[f(\bar{x}_k)]^2 - [g(\bar{x}_k)]^2}{2} \Delta x$$

$$= \int_a^b \rho \frac{[f(x)]^2 - [g(x)]^2}{2} \, dx$$

The total mass is computed by summing up the masses of each rectangular strip and again taking the limit as $n \to \infty$:

$$M = \lim_{n \to \infty} \sum_{k=1}^{n} \rho \left(f(\bar{x}_k) - g(\bar{x}_k) \right) \Delta x$$

$$= \int_a^b \rho \left(f(x) - g(x) \right) \, dx.$$

The y-coordinate of the center of mass is obtained by dividing M_x, the total moment of mass about the x-axis, by M, the total mass:

$$\bar{y} = \frac{M_x}{M} = \frac{\displaystyle\int_a^b \rho \frac{[f(x)]^2 - [g(x)]^2}{2} \, dx}{\displaystyle\int_a^b \rho(f(x) - g(x)) \, dx}.$$

▶ **EIGHTH REFLECTION**

Can you divide out the variable ρ in the answer for $\bar{y}$ above? Explain.　◀

Think and Share

Break into groups. Determine the distance from the center of mass of the kth rectangle to the y-axis. Find an expression for the moment of mass of the kth rectangle about the y-axis. By summing the moments of mass about the y-axis for the n rectangles, obtain a sum which approximates the moment of mass of the plate about the y-axis. Take the limit as $n \to \infty$ of your sum and rewrite it as an integral which equals M_y, the exact moment of mass of the plate about the y-axis. Divide your result by the integral representation of the mass of the plate and use your expression to obtain the equation for $\bar{x}$, the x-coordinate of the center of mass.

$$\bar{x} = \frac{M_y}{M} =$$

One appropriate window is xmin=-1, xmax=4, ymin=-3, ymax=5.

Think and Share

Graph the functions

$$f(x) = 2\sqrt{-(x^2 - 3x - 1.75)}$$
$$g(x) = 0.75(x - 1.5)^4 - 2.75(x - 1.5)^2 + 1$$

in an appropriate window so that you can clearly see the region bounded by the two curves. Assume your picture shows a plate of uniform density and find the coordinates of its center of mass. Plot the point you find. Does it look reasonable?

EXERCISE SET 5.4

In Exercises 1–9, sketch the region, shade your kth rectangle, write the Riemann sum and evaluate its associated integral to find the indicated area.

1. Find the area between the x-axis and the graph of the function $f(x) = x^2 - 4x - 5$.
2. Find the area between the x-axis and the graph of the function $f(x) = -x^2 + 4x + 5$.
3. Find the area between the graphs of $f(x) = x^2 - 2x + 5$ and $g(x) = 3x - 4$ between $x = -2$ and $x = 3$.
4. Find the area between the graphs of $f(x) = x^2 + 5$ and $g(x) = 2^x$ between $x = -1$ and $x = 3$.
5. Find the area between the graphs of $f(x) = x^2 + 3$ and $g(x) = \sin(x) + 2$ between $x = -3$ and $x = 3$.
6. Find the area between the graphs of $f(x) = x^2 - 2x + 5$ and $g(x) = -3$.
7. Find the area between the graphs of $f(x) = -x^2 - 2$ and $g(x) = -2x - 5$ between $x = 0$ and $x = 3$.
8. Find the area between the graphs of $f(x) = x^3 - 2x^2 - 6x + 20$ and $g(x) = 2x^2 + 5x - 10$.
9. Find the area between the graphs of $f(x) = x^3 - 10x - 8$ and $g(x) = -x^2 + 7x - 8$.

In Exercises 10–13, show your kth element analysis, write the Riemann sum, and evaluate its associated integral to find the volume of the solid of revolution determined by rotating the given region about the given line.

10. The region enclosed by $f(x) = x^3$, the x-axis, and the line $x = 1$ about the x-axis.
11. The region enclosed by $f(x) = \sin(x)$ and the x-axis between $x = 0$ and $x = \pi/2$ about the x-axis.
12. The region enclosed by $f(x) = -x^2 + 4x$ and the x-axis about the x-axis.
13. The region enclosed by $f(x) = e^{-x}$, the x-axis, the y-axis and the line $x = 1$ about the x-axis.

Exercises 14 and 15 are about population projections.

14. The table below shows the population increase (in thousands) for the given years. Predict the population increase from 1980 to 1985 and the total population in 1985 if the population in 1920 was 13,000,000.

TABLE 5.17   Population Increase in Thousands

Year	1920	1940	1960	1980
Birth rate–Death rate	99	115	134	156

15. The table below shows the population increase (in thousands) based on birth and death rates for the given years. Predict the population increase from 1990 to 1995 and the total population in 1995 if the population in 1950 was 61,350,000.

TABLE 5.18   Population Increase in Thousands

Year	1950	1960	1970	1980	1990
Birth rate–Death rate	577	635	699	770	848

In Exercises 16–19, find the center of mass of the thin plate occupying the specified region. Assume that the thin plate is of uniform density ρ.

16. The region bounded by $f(x) = 2x$, the x-axis, and $x = 3$.
17. The region bounded by $g(x) = \sin(x)$ and the x-axis between $x = 0$ and $x = \pi$.
18. The region bounded by $h(x) = x^2 + 1$ and the line $y = 5$.
19. The region bounded by $f(x) = \frac{1}{x}$ and the x-axis between $x = 1$ and $x = 5$.

In Exercises 20–26, find the the volume of the solid of revolution determined by rotating the given region about the given line.

20. The region in Exercise 1 rotated about the line $y = 1$.

21. The region in Exercise 2 rotated about the line $x = 5$.

22. The region in Exercise 3 rotated about the line $y = -10$.

23. The region in the first quadrant enclosed by $f(x) = x^2$, the y-axis, and the line $y = 4$ about the line $x = 2$.

24. The region in the previous exercise rotated about the line $x = 5$.

25. The region enclosed by the functions $f(x) = x^3 + 1$, $g(x) = x^2 + 3x - 3$, $x = -1$ and $x = 1/2$ rotated about the line $y = 2$.

26. The region enclosed by the functions $f(x) = \cos(x)$ and $g(x) = \dfrac{x^2}{4}$ between $x = -1$ and $x = 1$ rotated about the line $y = 1$.

You can approach the situations in Exercises 27–31 using the kth element analysis method. However, these situations have not been introduced in the text.

27. Population studies often look at age groups or cohorts. In this exercise, we will consider the age group of women over 45 years of age who give birth. The number of live births each year from 1950 through 1988 to women in this age group in the U.S.A. is given in Table 5.19[†].

TABLE 5.19 Live Births

Year	1950	1955	1960	1965	1970	1975
Births	5322	5430	5182	4614	3146	1628

Year	1980	1985	1986	1987	1988
Births	1200	1162	1257	1375	1427

a. Find a model that fits this data. Is this a rate of change model? Explain.

b. Using this model, find an expression for the rate of change of annual births per year as a function of time (in years).

c. Using this model, find an expression for the total number of live births since 1950 as a function of time since 1950 ($t = 0$).

d. Determine the number of live births from 1970 to 1980 and from 1980 to 1990 assuming this model.

[†]*Statistical Abstracts of the U.S.*, Brenan Press, 1992.

28. Businesses spend large sums of money each year on advertising in order to stimulate sales of products. Table 5.20 shows the approximate increase in sales (in thousands of dollars) that an additional $100 spent on advertising, at various levels, can be expected to generate for CMM, a retail store.

TABLE 5.20 Advertising Expenditures versus Revenue

Advertising Expenditures (in hundreds of dollars)	25	50	75	100	125	150	175
Revenue Increase Due to an Extra $100 in Advertising (in thousands of dollars)	5	60	95	105	104	79	34

a. Fit a model to these data.

b. Use the model to determine an expression for the total sales revenue $R(x)$ as a function of the amount x spent on advertising. Their revenue is $900,000 when $5000 is spent on advertising.

c. Graph the revenue function $R(x)$ and estimate the point where returns begin to diminish for revenue as a function of advertising.

29. A major tire manufacturer began marketing an all weather tire in 1968 and sales began to skyrocket in 1972. Table 5.21 shows the rate of change of sales every four years since 1968:

a. Fit a model to the data based on an examination of a scatter plot of the data.

b. Use the model to produce an expression that gives the total sales function $S(t)$ as a function of t. The total sales in 1976 were $30,000.

c. Find the change in sales from 1976 through 1986 and from 1986 through 1996.

TABLE 5.21 Natural Gas Prices: 1980-1990

Year	1980	1981	1982	1983	1984	1985
Price	3.68	4.29	5.17	6.06	6.12	6.12

Year	1986	1987	1988	1989	1990
Price	5.83	5.54	5.47	5.64	5.77

30. Table 5.21 shows the price in dollars per 1000 cubic feet of natural gas used by residential consumers in the U.S. from 1980 to 1990.

a. Fit a model to the price data.

b. Determine the average price per 1000 cubic feet of natural gas from 1980 to 1990.

c. Sketch, on the same axes, a graph of the price as a function of time and a constant function whose value is the average price.

d. Are the areas under these two curves the same? How could you compute the average price using a definite integral?

5.5

Using the Chain Rule in Finding Antiderivatives

The Fundamental Theorem of Calculus is a powerful tool for computing definite integrals. It requires, however, that we be able to find a usable formula for an antiderivative of the function being integrated. Many practical situations are modeled by functions for which a usable antiderivative cannot be found.

One standard approach to alleviate these difficulties is to transform the integral into a simpler form. In the next two sections, we will be learning three different techniques for making such transformations (there are many more). In the examples in these sections, we will see that we can actually find the resulting simpler integral. Keep in mind also that the simplification itself can be a big help in understanding, even when it doesn't lead directly to a suitable antiderivative.

The Chain Rule Revisited

Our first transformation technique comes from our old friend, the chain rule. Although the chain rule is usually stated in terms of differentiation, it can also be viewed in terms of integration. Recall that every derivative formula is also an integration formula.

By the chain rule,

$$\frac{d}{dx}\sin(x^2) = \cos(x^2)2x.$$

It follows that

$$\int \cos(x^2)2x \, dx = \sin(x^2) + C.$$

Picking apart the integrand of this indefinite integral can be illuminating. We see that the integrand has two factors: a composite function, $\cos(x^2)$, and $2x$. The derivative of x^2, the "inside" part of the composite function, equals $2x$, the other factor in the integrand. On the right side, the indefinite integral is, of course, the composite function we started with, plus a constant.

More generally, the chain rule says $(f(g(x)))' = f'(g(x))g'(x)$. Rewriting this equation in terms of integrals, we have

$$\int (f(g(x)))' \, dx = \int f'(g(x))g'(x) \, dx \qquad \textbf{(5.4)}$$

or

$$f(g(x)) + C = \int (f'(g(x))g'(x)) \, dx.$$

or

$$\int (f'(g(x))g'(x)) \, dx = f(g(x)) + C.$$

This equation tells us that if we can recognize the derivative of the inner function of a composite function as a factor of our integrand, then we can usually write down the antiderivative directly or at least simplify the integral.

Break into pairs.

Think and Share

1. Create or make up a composition of two functions as in the discussion above. (You may use polynomial, radical, rational, exponential, trigonometric, and logarithmic functions.) Differentiate it. Give the derivative to your partner to guess the antiderivative. Be sure to tell your partner that the antiderivative will be a family of functions, not just a single function.
2. Repeat this process twice, varying the classes of functions that you use.

Finding Antiderivatives by Guessing and Checking

To integrate using the chain rule, we study the integrand looking for a factor (an expression) whose derivative also appears as a factor (or nearly a factor) of the integrand. If we find such an expression as a factor, we use it to create a likely guess for the antiderivative. We check this educated guess by differentiating this proposed antiderivative. The "check" part is very important. Often, even when the guess is not correct, we can fix it with a minor adjustment.

We will demonstrate this approach in the following examples. For each example, make your own guess before reading the one in the text.

E X A M P L E   1   **The Guess and Check Method**

We will find the following indefinite integrals based on our knowledge of the chain rule using the "guess and check" method.

1. $\int (5x^4 - 8x^3 - 4)^7 (5x^3 - 6x^2) \, dx$

2. $\int \dfrac{3x^2 - 5}{(x^3 - 5x + 7)^2} \, dx$

3. $\int \sec^2(4x^3 - 3x^2 + 5)(2x^2 - x) \, dx$

4. $\int \sec(5x^2 - 2x + 4) \tan(5x^2 - 2x + 4)(5x - 1) \, dx$

5. $\int x^3 \sqrt{4 - x^2}\, dx$

Solution:

1. The indefinite integral

$$\int (5x^4 - 8x^3 - 4)^7 (5x^3 - 6x^2)\, dx$$

has two factors. Only the first is a composite function: x^7 composed with $5x^4 - 8x^3 - 4$. The expression $\dfrac{x^8}{8}$ is an antiderivative of x^7 which suggests $(5x^4 - 8x^3 - 4)^8$ as a first guess for an antiderivative. Differentiating our guess with respect to x gives

$$\frac{8(5x^4 - 8x^3 - 4)^7}{8} \cdot (20x^3 - 24x^2) = (5x^4 - 8x^3 - 4)^7 \cdot 4(5x^3 - 6x^2)$$

$$= 4(5x^4 - 8x^3 - 4)^7 (5x^3 - 6x^2).$$

Our first guess was wrong. Its derivative is four times the original integrand. This is where a minor adjustment can be made.

We recall the constant multiplier rule: " the derivative of a constant times a function equals the constant multiplied by the derivative of the function." According to this rule, we can get the derivative of our "guess" to be divided by four simply by dividing the "guess" itself by four. We do this and see that the derivative of $\dfrac{(5x^4 - 8x^3 - 4)^8}{32}$ is our original integrand. Thus,

$$\int (5x^4 - 8x^3 - 4)^7 (5x^3 - 6x^2)\, dx = \frac{(5x^4 - 8x^3 - 4)^8}{32} + C.$$

2. Rewrite (transform) the indefinite integral as

$$\int \frac{3x^2 - 5}{(x^3 - 5x + 7)^2}\, dx = \int (x^3 - 5x + 7)^{-2}(3x^2 - 5)\, dx,$$

where $(x^3 - 5x + 7)^{-2}$ is a composite function: x^{-2} composed with $x^3 - 5x + 7$. Noting that $(3x^2 - 5)$ is the derivative of the "inside" of the composite function, we guess $\dfrac{(x^3 - 5x + 7)^{-1}}{-1}$ as our antiderivative since we know an antiderivative of x^{-2} is $\dfrac{x^{-1}}{-1}$. Our "check,"

$$\frac{d}{dx}(x^3 - 5x + 7)^{-1} = -1((x^3 - 5x + 7)^{-2}(3x^2 - 5) = \frac{-(3x^2 - 5)}{(x^3 - 5x + 7)^2},$$

shows we guessed correctly. Thus,

$$\int \frac{3x^2 - 5}{(x^3 - 5x + 7)^2} \, dx = -(x^3 - 5x + 7)^{-1} + C.$$

3.   The composite function for the indefinite integral

$$\int \sec^2(4x^3 - 3x^2 + 5)(2x^2 - x) \, dx$$

must be $\sec^2(4x^3 - 3x^2 + 5)$. To apply our guess and check method, we must be able to guess an antiderivative for the outer function of our composite function. In this case, we must recall that the derivative of $\tan(x)$ is $\sec^2(x)$. Also, a mental computation shows that the derivative of the input $(4x^3 - 3x^2 + 5)$ to the composite function is $12x^2 - 6x$ which at least has the same degree as $(2x^2 - x)$. We take

$$\tan(4x^3 - 3x^2 + 5)$$

as our first guess at an antiderivative. Differentiating and checking we obtain

$$\sec^2(4x^3 - 3x^2 + 5) \cdot (12x^2 - 6x) = \sec^2(4x^3 - 3x^2 + 5) \cdot 6(2x^2 - x).$$

Notice that we did not completely factor the expression $(12x^2 - 6x)$. We're not trying to simplify; we're trying to make the expression look as much like the original integrand as possible. What we have almost matches the integrand, but not quite. Just as in the first item of our example, we are off by a constant factor (in this case the constant factor is six). Thus, we divide our original antiderivative by six and obtain

$$\int \sec^2(4x^3 - 3x^2 + 5)(2x^2 - x) \, dx = \frac{\tan(4x^3 - 3x^2 + 5)}{6} + C.$$

4.   Our first impression of the indefinite integral

$$\int \sec(5x^2 - 2x + 4) \tan(5x^2 - 2x + 4)(5x - 1) \, dx$$

is that things look bleak. There are three factors in the integrand, two of which are composite functions! If we choose either one to actually be our composite function, the rest of the expression is far too complicated. (Remember, the factor outside of the composite function should be the derivative of the composite function's inner function.) We are saved by noticing that both of the composite functions have the same inner function, $5x^2 - 2x + 4$. Thus, we are led to consider their product as the outer function, with $5x^2 - 2x + 4$ as the inner function. Reflecting on derivatives of trigonometric functions, we recall that $\sec(x) \tan(x)$ is the derivative of $\sec(x)$. We therefore guess

$$\sec(5x^2 - 2x + 4)$$

as our antiderivative. We check our result by differentiation and find that the derivative

$$\sec(5x^2 - 2x + 4)\tan(5x^2 - 2x + 4) \cdot (10x - 2)$$

is off by a constant factor of two. Once again the discrepancy is easily handled and we obtain our solution,

$$\int \sec(5x^2 - 2x + 4)\tan(5x^2 - 2x + 4)(5x - 1)\,dx =$$

$$\frac{\sec(5x^2 - 2x + 4)}{2} + C.$$

Check your result with your graphing calculator.

You must use

$$\frac{1}{\cos(5x^2 - 2x + 4)} \quad \text{instead of} \quad \sec(5x^2 - 2x + 4)$$

because the secant function is not available on your calculator unless you define it. You should find that our result checks with your calculator.

5. The integrand of $\int x^3 \sqrt{4 - x^2}\,dx$ is $x^3\sqrt{4 - x^2}$. Probing this integrand for functions whose derivative is some portion of the integrand, we find that x^2 is almost the derivative of x^3. However, x^2 is not a factor in the integrand, but rather a term in a square root. The derivative of $4 - x^2$ is not x^3. The derivative of $\sqrt{4 - x^2}$ is complicated and also doesn't appear as a factor of the integrand. We are out of possibilities for this method at this point. Other techniques that we have not developed might work. Some of these techniques are discussed in the Interlude titled *Techniques of Integration*. Guess and check does not work. You might try this integral on your TI calculator.

Differentials

When the guess and check method does not work, we must often turn to another methodknown as simple substitution or change of variable. We must digress just a bit, however, to explain "differential notation" because differentials are needed when we use a change of variable in finding antiderivatives. The tangent line is a linear approximation to the function $y = f(x)$ at the point $(x, f(x))$. If we move some short distance Δx away from x, then the difference between the value of the linear function approximation (tangent line) at $x + \Delta x$ and the output value of f at x is dy (see Figure 5.36). The symbol dy is called the *differential* of y and represents this difference or change in y.

If the slope of a linear function is m, then the values of the function at x + 1, x + 2, and x + 3 are given by
$$f(x) + m \cdot 1,$$
$$f(x) + m \cdot 2,$$
and f(x) + m · 3,
respectively, as shown in Figure 5.37.

FIGURE 5.36   The Differential dy

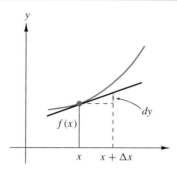

FIGURE 5.37   Slope Times Increment

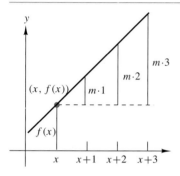

Since slope is the change in y over the change in x, we know $f'(x) = \frac{dy}{dx}$. We write

$$dy = f'(x)\Delta x.$$

If we use dx (the differential with respect to x) for Δx, then we can write

$$dy = f'(x)\, dx. \tag{5.5}$$

We will frequently make use of equation (5.5) in finding antiderivatives symbolically.

Simple Substitution

Guessing and checking is often a useful strategy in finding an antiderivative. In order to work, however, it requires that we find a function which is an antiderivative for the outer function of the composite function. This is often difficult, and is sometimes impossible. On the other hand, guess and check is the basis of an important transformation technique—*simple substitution.*

As we stated at the beginning of this section, we often transform a complicated integral to a simpler form just to increase our understanding of the underlying problem situation. Simple substitution is one type of transformation that can sometimes help us transform an integral to a simpler form. Simple substitution is merely cosmetic: it changes the appearance of the integrand. But a "pretty" integrand is worth its weight in comprehension.

E X A M P L E 2   **The Simple Substitution Method**

We will use simple substitution to transform

$$\int \cos(x^2 + 4x - 5) \cdot (x + 2)\, dx \tag{5.6}$$

into a simpler integral.

We recognize that the composite function is $\cos(x^2 + 4x - 5)$ in the integrand. We also note the derivative of the inner function is $(2x + 4)$. This is not quite what we wanted, but it's close enough that we take a chance anyway. The derivative of the expression $x^2 + 4x - 5$ is almost part of the integrand. From our work in finding antiderivatives with the chain rule, we substitute a single letter, say u, for the expression $x^2 + 4x - 5$ in x. Thus, we say

$$u = x^2 + 4x - 5.$$

With this substitution and using differentials, we find

$$du = (2x + 4)\, dx$$
$$= 2(x + 2)\, dx.$$

Solving for dx yields

$$dx = \frac{du}{2(x + 2)}.$$

Replacing $x^2 + 4x - 5$ with u, and dx with $\dfrac{du}{2(x + 2)}$, Equation (5.6) becomes

$$\int \cos(u)(x + 2) \cdot \frac{du}{2(x + 2)} = \int \cos(u)\frac{1}{2}\, du.$$

Notice we have arrived at an integral we know and can write its antiderivative.

$$\int \cos(u)\frac{1}{2}\, du$$
$$= \frac{1}{2}\int \cos(u)\, du$$
$$= \frac{1}{2}\sin(u) + C$$
$$= \frac{1}{2}\sin(x^2 + 4x - 5) + C.$$

Check it!

In the last step, we have rewritten the final answer in terms of the variable x using our original u-substitution $u = x^2 + 4x - 5$.

We could just as well have guessed the antiderivative based on our previous experience. The point of the method, however, is to simplify the integrand, not necessarily to find an antiderivative. On the other hand, simple substitution or *u-substitution*, as it is sometimes called, may allow you to find some indefinite integrals for which the guess and check method is very difficult.

Formal Notation for Simple Substitution

From Equation (5.4), we have

$$\int f'(g(x)) \cdot g'(x)\, dx = \int [\, f(g(x))]'\; dx = f(g(x)) + C.$$

This equation is generally the one we have in mind when we use the guess and check method. Using the guess and check method to find indefinite integrals is a great way to develop the ability to recognize patterns in symbolic expressions. You will find that this ability is very useful in doing mathematics, and we recommend that you use it with the guess and check method as often as you can.

In the method of u-substitution, we set $u = g(x)$, so $f(g(x))$ becomes $f(u)$, $g'(x)\, dx = du$, and $f'(g(x)) = f'(u)$. Making these substitutions in the previous equation, we see that

$$\int f'(g(x)) \cdot g'(x)\, dx = \int f'(u)\, du$$
$$= f(u) + C \qquad\qquad \textbf{(5.7)}$$
$$= f(g(x)) + C.$$

Even if we don't know an antiderivative for the outer function of our composite function, we can still make the simple substitution transformation given by Equation (5.7).

$$\int f'(g(x)) \cdot g'(x)\, dx = \int f'(u)\, du \qquad\qquad \textbf{(5.8)}$$

Notice that you substitute

$$du \text{ for } g'(x)\, dx \quad \text{as well as} \quad u \text{ for } g(x).$$

A substitution is successful if we can eliminate x from the integral on the right. This last step may involve algebra or even further u-substitutions.

EXERCISE SET 5.5

In Exercises 1–21, use guess and check or simple substitution to find the indefinite integral. Explain the process you used for each integral. Check your results with your calculator after you have written down your answer.

1. $\displaystyle\int 5\left(x^2 - 2x + 3\right)^4 (2x - 2)\, dx$

2. $\displaystyle\int 7\left(x^3 - 3x^2 + 5x - 7\right)^6 \left(3x^2 - 6x + 5\right)\, dx$

3. $\displaystyle\int \sec(e^x)\tan(e^x)e^x\, dx$

4. $\displaystyle\int 2\sec(2x - 3)\tan(2x - 3)\, dx$

5. $\displaystyle\int 2xe^{x^2}\, dx$

6. $\displaystyle\int x^2 e^{x^3}\, dx$

7. $\displaystyle\int \frac{2\ln(x)}{x}\, dx$

8. $\displaystyle\int \frac{4x}{2x^2 - 3}\, dx$

9. $\displaystyle\int \frac{2x}{\sqrt{2x^2 + 5}}\, dx$

10. $\displaystyle\int \frac{2x}{\left(3x^2 - 2\right)^{2/3}}\, dx$

11. $\int 3\,(\sin(x))^2\cos(x)\,dx$

12. $\int 3\,\csc(x^3)\cot(x^3)x^2\,dx$

13. $\int \dfrac{1}{x\ln(x)}\,dx$

14. $\int 4\cos(x^2)x^2 + 2\,\sin(x^2)\,dx$

15. $\int x\sec^2(x^2)\,dx$

16. $\int (\sec(x))^2\tan(x)\,dx$

17. $\int \left(6x^2 - 3\right)\left(2x^3 - 3x + 11\right)^3\,dx$

18. $\int \dfrac{1}{x\sqrt{\ln(x)}}\,dx$

19. $\int 5\dfrac{e^x}{e^x - 1}\,dx$

20. $\int (\tan(x))^2\,dx$

21. $\int e^{3x}\sin(x)\left(3e^{3x}\sin(x) + e^{3x}\cos(x)\right)\,dx$

In Exercises 22–25, to find the indefinite integral, let $u =$ the function under the radical in the simple substitution method and solve the resulting equation for x. Then rewrite the integral in terms of u. Find an antiderivative and write your final result

in terms of x. Compare your result with that of your calculator. Are the results equivalent?

22. $\int x\sqrt{x - 1}\,dx$

23. $\int \sqrt{1 + x^2}\,x^3\,dx$

24. $\int x\sqrt{x + 1}\,dx$

25. $\int \sqrt{1 + x^2}\,x^5\,dx$

Exercises 26–29 involve the composition of three functions. Find the given integrals. Use simple substitution to transform each integral to a form $\int f'(u)\,du$. Show your choice of u.

26. $\int \cos(e^{x^2+1})xe^{x^2+1}\,dx$

27. $\int \dfrac{\cos(\ln(x^2 + 1))x}{x^2 + 1}\,dx$

28. $\int (\csc(x))^2\cot(x)e^{(\csc(x))^2}\,dx$

29. $\int \cos(\cos(x^2))\sin(x^2)x\,dx$

30. Working with another person, discuss the strategy each of you uses to find integrals. Write down a strategy that incorporates both of your approaches.

31. Make up an original example of an indefinite integral that can be solved by simple substitution. Write out your solution process in a manner acceptable as a textbook example.

5.6

Techniques of Integration

In the previous section, we used simple substitution as a technique for transforming integrals into a simpler form. In this section, we develop two more transformation techniques: integration by parts and integration using partial fractions.

We begin with integration by parts, which is based upon the product rule for differentiation. We will use this method to find the indefinite integral of certain special products of functions.

Caution!

The antiderivative of the product of two functions is *not* the product of the antiderivatives of the two functions.

Integration by Parts

Just as simple substitution was based on the chain rule for differentiation, integration by parts is based on the product rule. Recall that the derivative of the product of two functions equals the product of the derivative of the first function times the second added to the product of the first times the derivative of the second function. Symbolically,

$$[f(x) \cdot g(x)]' = f'(x) \cdot g(x) + f(x) \cdot g'(x). \tag{5.9}$$

Notice that on the right side, the roles of f and g are interchanged from one term to the next. We can use this interchange idea to transform one indefinite integral into another. If we're careful, or lucky, the "transformed" integral will be simpler than the original integral.

Let's see how this transformation occurs by antidifferentiating both sides of Equation (5.9) to obtain an integral form of the product rule.

$$\int [f(x) \cdot g(x)]' \, dx = \int f'(x) \cdot g(x) \, dx + \int f(x) \, dx \cdot g'(x) \, dx$$

Notice that $f(x) \cdot g(x)$ is an antiderivative of the integrand on the left. Making this substitution and solving for $\int f(x)g'(x) \, dx$ we have

$$\int f(x) \cdot g'(x) \, dx = f(x) \cdot g(x) - \int f'(x) \cdot g(x) \, dx$$

$$= f(x) \cdot g(x) - \int g(x) \cdot f'(x) \, dx.$$

We have transformed the problem from finding an antiderivative of

$$\int f(x) \cdot g'(x) \, dx$$

into finding an antiderivative of

$$\int g(x) \cdot f'(x) \, dx.$$

Our hope is that the second indefinite integral is one we know how to do or at least one that is simpler to understand.

E X A M P L E   1   **Integration by Parts Method**

Find the antiderivative of $\int xe^x \, dx$.

Solution:

Let $f(x) = x$ and $g'(x)dx = e^x \, dx$. Then $f'(x) = 1$ and $g(x) = e^x$ and

$$\int xe^x \, dx = xe^x - \int e^x \cdot 1 \, dx$$

$$= xe^x - e^x + C.$$

Notice that the antiderivative of the product of two functions $(g(x) \cdot f'(x))$ is *not* the product of the antiderivatives of these two functions.

A Simpler Notation for Integration by Parts

We can write this transformation more simply using symbol substitution (an idea you used with the chain rule in Section 5.5). We will use u for $f(x)$ and v for $g(x)$, to transform the equation

$$\int f(x) \cdot g'(x) \, dx = f(x) \cdot g(x) - \int g(x) \cdot f'(x) \, dx$$

into the more compact form

$$\int u \, dv = uv - \int v \, du.$$

▶ **FIRST REFLECTION**

What part of the integrand was replaced with dv?   ◀

Just as with simple substitution, the integration by parts method requires that the integrand be a product of two factors. We choose one factor or "part" of the product to be "dv" and choose what's left to be "u" (the other factor or "part"). (The part used for "dv" is a differential and must include the dx.) Usually, we try to pick "dv" to be the most complicated expression we can antidifferentiate (easily) since finding an antiderivative is more difficult than finding a derivative.

We demonstrate this transformation in the following examples. For each example, make your own choices of u and dv before reading the one in the text.

E X A M P L E   2   **More Integration by Parts**

Find the following antiderivatives or indefinite integrals.

1. $\int x \cdot \cos(x) \, dx$
2. $\int x^2 \ln(x) \, dx$

Solution:

1. We know an antiderivative for $\cos(x)$ and for x but the $\cos(x)$ antiderivative is more complicated, so we choose

    $$u = x \quad \text{and} \quad dv = \cos(x) \, dx.$$

    Differentiating x and picking an antiderivative for $\cos(x)$, we find

    $$du = 1 \, dx \quad \text{and} \quad v = \sin(x).$$

Making these substitutions, we obtain

$$\int x \cdot \cos(x) \, dx = x \cdot \sin(x) - \int \sin(x) \, dx.$$

Check it!

Notice that the integral on the right is simpler than the integral on the left, simple enough that we can compute it to obtain

$$\int x \cdot \cos(x) \, dx = x \sin(x) - (-\cos(x)) + C$$

$$= x \sin(x) + \cos(x) + C.$$

2. In this case, both factors become simpler when differentiated, but we do not relish the thought of finding an antiderivative for $\ln(x)$. Thus, we make the substitution

$$u = \ln(x) \quad \text{and} \quad dv = x^2 \, dx$$

$$du = \frac{1}{x} \, dx \quad \text{and} \quad v = \frac{x^3}{3} \qquad \text{(differentiating } \ln(x) \text{ and integrating } x^2\text{)}.$$

Making the substitution

Check it!

$$\int x^2 \ln(x) \, dx = \int \ln(x) \cdot x^2 \, dx = \ln(x) \cdot \frac{x^3}{3} - \int \frac{x^3}{3} \cdot \frac{1}{x} \, dx$$

$$= \frac{x^3}{3} \cdot \ln(x) - \frac{1}{3} \int x^2 \, dx$$

$$= \frac{x^3}{3} \cdot \ln(x) - \frac{x^3}{9} + C,$$

once again, the "transformed" integral was easily integrated.

There are other choices you could make for u and dv. You should think about the pair of functions you choose in terms of integrating and differentiating. Functions that can be easily integrated are candidates for dv; the remaining portion of the integrand becomes the choice for u. The choice of dv through integration gives you v; the choice of u through differentiation gives you du. Like the u-substitution method, integration by parts is a mechanical guess and check process. Unlike the u-substitution method, you (usually) cannot analyze the integrand to guess its antiderivative directly.

Think and Share

Break into pairs. Discuss the effect of various choices of u and dv that are possible in using the integration by parts method with

$$\int u \, dv = uv - \int v \, du$$

on

$$\int x^3 \cos(x) \, dx.$$

Think and Share

For example, one choice might be $u = x^3$ and $dv = \cos(x)\, dx$. Explore this and other choices.

Integration Using Partial Fractions: Population Growth

Our final technique for transforming integrals is designed to simplify indefinite integrals of rational functions (quotients of polynomials). Although the technique is very general, we will only consider simple rational functions with constants in the numerator and products of linear functions in the denominator.

In the 1830s, P.F. Verhulst proposed the following equation to model population growth
indexPopulation growth model

$$\frac{dP}{dt} = kP\left(1 - \frac{P}{C}\right) \tag{5.10}$$

where P is the size of the population, and k and C are positive constants. If the population is small, so that P is close to zero, the model predicts that dP/dt, the rate of growth of the population, will also be small. The model also predicts that dP/dt will be small if the population is close to C, 0 if the population is C, and negative if the population is greater than C. Thus, the model includes an upper bound for the population, C, which is often called the *carrying capacity* of the environment. You have already seen examples of this sort of growth in Chapters 1 and 2.

A standard technique for analyzing such a formula is to rewrite it in the form

$$\frac{1}{P \cdot (C - P)} \frac{dP}{dt} = \frac{k}{C}. \tag{5.11}$$

This technique is called *separation of variables* and you will learn more about it in Section 7.1.

▶ **SECOND REFLECTION**
Show how to rewrite Equation (5.10) in the form given in Equation (5.11). Are both side of the equation functions of t? Can you find the anti-derivative of both sides? ◀

The analysis continues by integrating both sides of the equation. The left side leads us to consider

$$\int \frac{1}{P \cdot (C - P)}\, dP.$$

This is exactly the sort of integral that our final technique is designed to simplify, because the integrand is a ratio of two polynomials in P. The method of integration using partial fractions simplifies this integral by splitting the integrand into two simpler fractions. The whole process amounts to the reverse of adding fractions.

▶ THIRD REFLECTION
What is the degree of the polynomial in the numerator above?   ◀

We begin by assuming there are two simpler fractions and writing the equation

$$\frac{1}{P \cdot (C - P)} = \frac{A}{P} + \frac{B}{C - P}.$$

We now need to determine values of A and B that will make the equation true. In the following algebraic manipulation, we multiply both sides of the equation by $P \cdot (C - P)$ and simplify.

$$\frac{1}{P \cdot (C - P)} = \frac{A}{P} + \frac{B}{C - P}$$
$$1 = A(C - P) + B \cdot P \text{ (multiplying through by } P(C - P))$$
$$1 = AC - AP + BP \text{ (using the Distributive Law)}$$
$$1 = AC + (-A + B)P$$

We know from our study of polynomials that two polynomials in P are equal if the coefficients of like degree are equal. Thus, $AC = 1$, so that $A = \frac{1}{C}$. Also, $(-A + B) = 0$ (since there is no P term on the left side), so that $B = A$. Thus, the original expression can be rewritten as the sum of two terms whose denominators are linear polynomials, and we obtain a simplification of our integral,

$$\int \frac{1}{P \cdot (C - P)} \, dP = \int \left[\frac{A}{P} + \frac{B}{C - P} \right] dP$$
$$= \int \left[\frac{\frac{1}{C}}{P} + \frac{\frac{1}{C}}{C - P} \right] dP$$
$$= \int \frac{\frac{1}{C}}{P} \, dP + \int \frac{\frac{1}{C}}{C - P} \, dP.$$

When the population is small, P is "close" to zero, and $\int \frac{\frac{1}{C}}{P} \, dP$ is much larger than $\int \frac{\frac{1}{C}}{C - P} \, dP$, and therefore dominates the model of the population. On the other hand, when P is "close" to C, $\int \frac{\frac{1}{C}}{C - P} \, dP$ is the dominating term. Thus, the two integrals can be used to enhance our understanding of the population growth.

▶ FOURTH REFLECTION
Carry out the integration $\int \frac{1}{P \cdot (C - P)} \, dP$ using the partial fractions developed above. Does your result match the result you obtained with your TI calculator? ◀

EXERCISE SET 5.6

In Exercises 1–10, find the integral using the integration by parts transformation, if possible, and indicate your choices for u and dv. Indicate which of these exercises could also be done by simple substitution.

1. $\int x \sin(x) \, dx$

2. $\int_1^4 x \cdot e^x \, dx$

3. $\int x \ln(x) \, dx$

4. $\int x^3 \sqrt{9 - x^2} \, dx$

5. $\int x \sin(9 - x^2) \, dx$

6. $\int x \cdot (x + 2)^8 \, dx$

7. $\int_0^3 x \cdot e^{x^2} \, dx$

8. $\int x \sqrt{x + 1} \, dx$

9. $\int x^2 \sin(x) \, dx$

10. $\int x^3 \ln(x) \, dx$

In Exercises 11–16, find the integral using the partial fraction transformation.

11. $\int \frac{7}{(x - 2)(x + 3)} \, dx$

12. $\int \frac{7}{(x - 1)(2x - 3)} \, dx$

13. $\int \frac{3}{x \cdot (x + 1)} \, dx$

14. $\int \frac{2}{1 - x^2} \, dx$

15. $\int \frac{5}{(x + 1)(x + 6)} \, dx$

16. $\int \frac{-1}{x^2 + x - 2} \, dx$

In Exercises 17 and 18, find the integral.

17. $\int e^x \sin(x) \, dx$

18. $\int \sec^3(x) \, dx$

In Exercises, 19–25 write down the partial fraction expansion of each of the integrands using the expand command on your TI calculator. Find the integrals using your calculator. Indicate which term in your result is associated with each term of the partial fraction expansion.

19. $\int \frac{x^2 + x - 1}{(3x^2 + 1)(x - 1)} \, dx$

20. $\int \frac{x + 3}{x^2 + 3x + 2} \, dx$

21. $\int \frac{4x - 2}{x^3 - x^2 - 2x} \, dx$

22. $\int \frac{4}{x^4 - 1} \, dx$

23. $\int \frac{2x}{(x^2 + 1)(x + 1)^2} \, dx$

24. $\int \frac{x + 1}{x^2 + 2x + 2} \, dx$

25. $\int \frac{5x^2 - 3}{x^3 - x} \, dx$

26. Determine which of the previous seven integrals can be done by the chain rule guess and check method. In the exercise(s) you can do using the chain rule method, which function has its derivative as part of the integrand? Explain why, using pencil and paper techniques, you would try the guess and check method before the Partial Fraction method in integrating a rational function.

27. Examine the partial fraction decompostition of each integrand and explain why one of the following antiderivatives contains an arctan(x) ($\tan^{-1}(x)$) function and the other does not.

   a. $\int \frac{2t^2 - 8t - 8}{(t - 2)(t^2 + 4)} \, dt$

   b. $\int \frac{4t^2 + 2t + 8}{t(t^2 + 4)} \, dx$

Summary

The major ideas in this chapter are contained in the Fundamental Theorem of Calculus, in two parts. Part I has practical significance. It gives us a simpler way to evaluate a definite integral, as long as we can find an antiderivative expression (closed form). Part II of the Fundamental Theorem, on the other hand, shows the connection between differentiation and integration. It is theoretically significant because it guarantees that every continuous function has an antiderivative, even if we cannot find an expression for it.

The Fundamental Theorem of Calculus, Part I.   The Fundamental Theorem of Calculus, Part I states that if F is any particular antiderivative of a function f, then $\int_a^b f(x)\,dx = F(b) - F(a)$.

This initially seems like a great benefit. If we can find any particular antiderivative $F(x)$ of $f(x)$ on the interval $[a, b]$, then all we need to do is evaluate $F(x)$ at the endpoints of the interval and subtract their values in the appropriate order. Our TI calculator (or any symbolic algebra package) can even do the work for us. The TI calculator can evaluate the definite integral exactly for every function for which an antiderivative exists. In this case, we do not have to use the limit definition of the definite integral.

It is imperative that we remember, however, that Part I of the Fundamental Theorem can only be applied when an antiderivative exists in closed form. If this antiderivative does not exist in closed form, we must apply the definition of the definite integral. We subdivide things into small parts, use approximations on these small parts, and then add up these approximations and take the limit of their sum as the number of parts increases. The definition of the definite integral as the limit of a Riemann sum was the major concept covered in Chapter 4, and is the basis for understanding the Fundamental Theorem of Calculus, the major concept of Chapter 5.

The Fundamental Theorem of Calculus, Part II.   The function $F(x) = \int_a^x f(t)dt$ is called an accumulation function and is obtained by integrating a rate function. Part II of the Fundamental Theorem of Calculus reverses this process, and states that the derivative of the accumulation function gives the rate function. With the antiderivatives of f represented in the form

$$F(x) = \int_a^x f(t)dt,$$

where $f(t)$ is a continuous function on the interval $[a, x]$, Part II of the Fundamental Theorem of Calculus states that

$$F'(x) = \frac{d}{dx} \int_a^x f(t)\,dt = f(x).$$

This part of the theorem shows the strong connection between the derivative and the integral.

The General Antiderivative or Indefinite Integral of a Function.   It is important to understand that the general antiderivative or indefinite integral of a function $f(x)$ is a family of functions which differ only by a constant, and their graphs are parallel curves. In symbols, $\int f(x)\,dx = F(x) + C$ is a general antiderivative of $f(x)$. $F(x)$ is any particular antiderivative of $f(x)$, and C is any real number.

As a consequence, whenever you know a derivative of a function, you always know a corresponding antiderivative. For example, since we know that $\frac{d}{dx}(x^4) = 4x^3$ we immediately know that $\int 4x^3\,dx = x^4 + C$.

Methods of Integration.   The sections in which we introduce several methods of integration might falsely be considered obsolete when viewed with the capability of the TI calculator. However, the techniques of finding antiderivatives using the chain rule, integration by parts, and integration using partial fraction are meant to develop a deeper insight and understanding of the integration process. In addition, the skills inherent to successfully transform one integrand into another are valuable skills which transfer to other areas of mathematics and science. Transformations, more than rote manipulation, give value to the techniques of integration studied in this chapter. Always remember that you are in charge of your symbol manipulation package, whether it is on a computer or a handheld calculator. You must understand the process involved in finding antiderivatives so that you can correctly interpret calculator or computer output.

Applications.   Lastly, the applications in Section 5.4 continue to emphasize the left side of the Fundamental Theorem of Calculus, Part I, where the subdividing, approximating, summing, and limit-taking provide the solutions to a variety of problems. You should continue to search for the conditions under which problem solutions can be achieved with this method and place them in your resources for future use.

SUPPLEMENTARY EXERCISES

1. Use complete sentences to explain, in your own words, the meaning of each of the following.
   a. antiderivative
   b. definite integral
   c. integrand
   d. upper limit of integration
   e. accumulation function
   f. volume of revolution
   g. center of mass
   h. telescoping sum

2. Explain the practical significance of Part I of the Fundamental Theorem. In other words, what does this part of the theorem allow us to do? What are its limitations?

3. Explain the theoretical significance of Part II of the Fundamental Theorem. In other words, what does this part of the theorem guarantee?

4. Explain the difference between a definite integral and an indefinite one by remarking upon the results of each operation.

5. How are the derivative and the integral connected? Explain in complete sentences.

6. Use your calculator to find a general antiderivative for $\dfrac{x^2 - 5x + 3}{\sqrt{x + 3}}$. Graph the specific antiderivatives for $C = 0$ and $C = -3$. How are the two graphs related?

7. Graph the accumulation functions $F(x) = \int_1^x \sin(t)\,dt$ and $G(x) = \int_{-2}^x \sin(t)\,dt$. Let the TI calculator do the hard work for you by defining $F(x)$ as $y1(x) = \int(\sin(t), t, 1, x)$. Define $G(x)$ in a similar manner. Change $xres$ to 5, or even 10, to speed up the graphing. How are the accumulation functions related? Explain.

8. Find $F'(x)$ and $G'(x)$ for the functions defined in the previous problem. Explain why the two derivatives are equal.

9. Graph the accumulation function $F(x) = \int_1^x t + 5\,dt$ and the rate function $f(t) = t + 5$ on the same axes. Identify which function is the derivative function. Find intervals over which F is increasing and decreasing, concave up or

concave down, by observing the graph of its derivative. (Use the vocabulary of Chapter 3.) Does F have a maximum value? If so, where? Justify.

10. Provide an original integration problem that can be done by the guess and check method. Show how you go about guessing and checking.

11. Do the integration problem in Exercise 10 by simple substitution, if possible. Carefully explain the method. If

    it is not possible to use the method of simple substitution, explain why it is not.

12. Explain why $\int (x^2 - 7)^5 \, dx$ cannot be done by simple substitution.

13. Provide an original integration problem that can be done by the integration by parts method. Write out the solution process.

14. Do the integration problem in Exercise 13 by simple substitution, if possible. Carefully explain the method. If it is not possible to use the method of simple substitution, explain why it is not.

15. Make up a partial fraction integration problem with denominator $(x - 7)(x + 3)$. You may create your own numerator. Find the general antiderivative with or without your calculator. Indicate which term in your antiderivative is associated with each term of the partial fraction expansion.

16. Can all definite integrals be evaluated using Part I of the Fundamental Theorem? Explain.

PROJECTS

Project One: Finding Integrals: An Adventure

The purpose of this project is to integrate functions that cannot be integrated directly using the $\int \!\!\!\!\!\!\int$ command. (Software improvements may allow the TI calculator to perform integrations that it could not do in 1997, when this text was written.) Before using transformations, be sure to check first that the TI calculator integrator fails.

1. Find and then describe how you found

$$\int \frac{1}{x(x + 1)(x + 2)(x + 3) \cdots (x + n)} \, dx.$$

2. Find and then describe how you found

$$\int \frac{\sin(x)}{\cos(2x)} \, dx.$$

3. Find and then describe how you found

$$\int \frac{\sin(x)}{\sin(2x)} \, dx$$

4. Find and then describe how you found

$$\int \frac{\cos(\ln(\tan^{-1}(\sqrt{x})))}{2(\cos(\sin(\ln(\tan^{-1} \sqrt{x}))))^2 \tan^{-1}(\sqrt{x})\sqrt{x}(x + 1)} \, dx.$$

5. Find and then describe how you found

$$\int_0^\infty \frac{\ln(u)}{1+u^2}\, du.$$

(Hint: The value is *not* 4.5E^{-13}.)

Project Report.   Write a report on your investigations. Address your report to your classmates, assuming the material is new to them. Describe your reasoning and include all necessary background information. A minimal project report must include:

1.   Written narratives for each of the five integrals above. Your description should include the methods you tried that failed to transform the integral into an integral or integrals that you can integrate. You should give a detailed description of how you successfully integrated the function. You might wish to consult the Interlude following Chapter 5 entitled *Further Techniques of Integration.*
2.   The differentiation process that proves your result is correct in each case.
3.   Suggestions for possible extensions of your observations and further questions for exploration.

Project Two: The Definite Integral of Functions that Are Not Continuous

The purpose of this project is to investigate definite integration of functions on intervals where the functions may not be continuous.

We will look at four definite integrals in this project.

1.   *I.* $\displaystyle\int_2^4 [x]\, dx$, where $[x]$ is the greatest integer in x function

2.   *II.* $\displaystyle\int_1^3 g(x)\, dx$ where $g(x) = \begin{cases} x^2 + 1, & x \le 2 \\ x + 3, & x > 2 \end{cases}$

3.   *III.* $\displaystyle\int_0^4 h(x)\, dx$ where $h(x) = \begin{cases} x^2 + 1, & x \le 2 \\ x - 4, & x > 2 \end{cases}$

4.   *IV.* $\displaystyle\int_{-1}^1 \frac{1}{x^2}\, dx$

You can find these integrals using the steps outlined below.

1.   *I.*   For the integral $\int_2^4 [x]\, dx$ (with $[x]$ defined using the `int` statement), step through the following investigation:

   **a. *A.*   Use the Simpson's method approximation to complete the following table of approximations.

TABLE 5.22 Simpson's Method Approximations

n	16	32	64	128
Simpson's Approximation				

b. *B.* Find the definite integral for the two halves of the interval [2, 4].

c. *C.* Compare these two approximations.

d. *D.* Use the TI calculator integration function ($\int$ ⟨) to find this integral. Does the result from the TI calculator make sense in light of your computations?

2. ***II.*** For the integral $\int_1^3 g(x)\,dx$, step through the following investigation:

a. *A.* Enter $g(x)$ using the When statement on your TI calculator.

b. *B.* Use the Simpson's method approximation to complete the following table of approximations.

TABLE 5.23 Simpson's Method Approximations

n	16	32	64	128
Simpson's Approximation				

c. *C.* Finding the definite integral for the two halves of the interval [1, 3].

d. *D.* Compare these two approximations.

e. *E.* Use the TI calculator integration function ($\int$ ⟨) to find this integral. Does the result from the TI calculator make sense in light of your computations?

3. ***III.*** For the integral $\int_0^4 h(x)\,dx$, step through the following investigation:

a. *A.* Enter $h(x)$.

b. *B.* Use the Simpson's method approximation to complete the following table of approximations.

TABLE 5.24 Simpson's Method Approximations

n	16	32	64	128
Simpson's Approximation				

c. *C.* Finding the definite integral for the two halves of the interval [0, 4].

d. *D.* Compare these two approximations.

e. *E.* Use the TI calculator integration function ($\int$ ⟨) to find this integral. Does the result from the TI calculator make sense in light of your computations?

4. *IV.* For the integral $\int_{-1}^{1} \frac{1}{x^2} \, dx$, step through the following investigation:

a. *A.* Use the Simpson's method approximation to complete the following table of approximations.

TABLE 5.25   Simpson's Method Approximations

n	16	32	64	128
Simpson's Approximation				

b. *B.* Finding the definite integral for the two halves of the interval $[-1, 1]$.

c. *C.* Compare these two approximations.

d. *D.* Use the TI calculator integration function ($\int \langle$) to find this integral. Does the result from the TI calculator make sense in light of your computations?

Project Report.   Write a report on your investigations. Address your report to your classmates, assuming the material is new to them. Describe your reasoning and include all necessary background information. A minimal project report must include:

1. Written narratives of each of the four investigations above.

2. Justification and explanations for your calculations.

3. From your experience with these four integrals, a description of the characteristics a function should have so that the definite integral would exist. Be sure to include examples.

4. Suggestions for possible extensions of your observations and further questions for exploration.

Project Three: Hydrostatic Pressure

You may have experienced a discomfort in your ears from diving down into water. You may also be able to appreciate that lifting a 2 ft by 2 ft metal plate in water is more difficult if you lift it horizontally versus vertically. These effects have to do with the fact that the pressure on an object varies with its depth in the water.

The physics principle involved states that the **pressure** acting on an object located in an open container at a certain depth in a liquid is the product of the density of the liquid and the depth.

$$p = \text{density} \times \text{depth}$$
$$= \rho \cdot h$$

where ρ is the density of the liquid and h is the depth of the object below the surface of the liquid.

The **total force** acting on a thin plate located horizontally at a depth h in a liquid is defined as the area of the plate times the pressure ($F = pA = \rho hA$). These principles are involved in the solution of such problems as designing submarines and dams. The problem of computing the total force on the vertical face of a dam would use these principles along with the idea of subdividing an object into small parts, analyzing a typical subinterval (kth), summing up the products, and taking the limit of this sum.

In the following exercises, use the pressure and total force principles to find the hydrostatic (total) force on the given objects. Leave your answers in terms of the density ρ of the given liquid.

1. The vertical ends of a water trough are equilateral triangles whose sides are 2 ft long. Find the hydrostatic force on an end panel when the trough is

   a. full of water
   b. half full of water

2. The vertical ends of a water trough are semicircular of radius 3 ft. Find the hydrostatic force on an end panel when the trough is

   a. full of water
   b. three quarters full of water

Project Report. Write a report on your investigations. Address your report to your classmates, assuming the material is new to them. Describe your reasoning and include all necessary background information. A minimal project report must include:

1. Descriptions and calculations for each of the two scenarios above.
2. Justification and explanations for your calculations.
3. Suggestions for possible extensions of your observations and further questions for exploration.

Further Techniques of Integration

We have already studied the two major integration techniques, integration by parts and integration using partial fractions, in Section 5.6. They are essential in solving population models and for theoretical purposes. In this Interlude, we will study these techniques more completely as well as additional techniques that can help us find antiderviatives of functions that exist in closed form.

We already know that we can find antiderivatives for a large class of functions, provided they exist, using our TI calculator. What is the purpose of this Interlude then? Perhaps the major benefit is the potential to sharpen our problem-solving skills in the sense that, faced with an indefinite integral to evaluate, we learn to ask the following questions: what resources do we need to evaluate it? When can we be reasonably sure an antiderivative in closed form of a given function doesn't exist?

To this end, we will extend our knowledge of the two techniques we have already studied: integration by parts and integration using partial fractions. We will also study a few additional methods: trigonometric combinations, trigonometric substitution, and completing the square. We start with a strategy that encompasses an overall approach to finding antiderivatives.

A Strategy for Finding Antiderivatives

The strategy we will use in looking for an antiderivative of a particular function has as its first two priorities; (1) does the integrand fit a *standard* form, or (2) does the integrand fit an *extended standard* form. The standard and extended standard forms are listed in Table 5.1

TABLE 5.1   Integration Strategy

Standard Form	Extended Form

Power Form

$$\int x^n \, dx = \frac{x^{n+1}}{n} + C, \quad n \neq -1$$

Power Form

$$\int [f(x)]^n f'(x) \, dx = \frac{[f(x)]^{n+1}}{n+1} + C, \quad n \neq -1$$

Logarithm Form

$$\int \frac{1}{x} \, dx = \ln|x| + C$$

Logarithm Form

$$\int \frac{1}{f(x)} f'(x) \, dx = \ln|f(x)| + C$$

Exponential Form

$$\int e^x \, dx = e^x + C$$

Exponential Form

$$\int e^{[f(x)]} f'(x) \, dx = e^{[f(x)]} + C$$

Trigonometric Forms

$$\int \sin(x) \, dx = -\cos(x) + C$$

$$\int \cos(x) \, dx = \sin(x) + C$$

$$\int \sec^2(x) \, dx = \tan(x) + C$$

$$\int \sec(x)\tan(x) \, dx = \sec(x) + C$$

$$\int \csc^2(x) \, dx = -\cot(x) + C$$

$$\int \csc(x)\cot(x) \, dx = -\csc(x) + C$$

Trigonometric Forms

$$\int \sin[f(x)]f'(x) \, dx = -\cos[f(x)] + C$$

$$\int \cos[f(x)]f'(x) \, dx = \sin[f(x)] + C$$

$$\int \sec^2[f(x)]f'(x) \, dx = \tan[f(x)] + C$$

$$\int \sec[f(x)]\tan[f(x)]f'(x) \, dx = \sec[f(x)] + C$$

$$\int \csc^2[f(x)]f'(x) \, dx = -\cot[f(x)] + C$$

$$\int \csc[f(x)]\cot[f(x)]f'(x) \, dx = -\csc[f(x)] + C$$

There are nine standard and nine extended standard forms in the table. This number would increase if we decided to include integrals whose antiderivatives involve inverse trigonometric and/or hyperbolic and inverse hyperbolic functions. We will leave it at these nine for now. In addition, we need to realize that the integrals of all polynomials are to be considered in standard form. For example, the integral $\int (5x^2 - 2x - 3) \, dx$ is equal to $5 \int x^2 \, dx - 2 \int x \, dx - 3 \int \, dx$, which is now three standard integrals.

E X A M P L E   1   **A Standard Form**

Find the antiderivative of $\int \sec^2(x) \, dx$.

This is a standard integral and so, we can write its antiderivative.

$$\int \sec^2(x) \, dx = \tan(x) + C$$

E X A M P L E   2   **An Extended Standard Form**

Find the antiderivative of $\int x e^{x^2}\, dx$.

From our recent studies, we might consider this a good candidate for integration by parts. However, we need to first check if this is a standard form. If not, then we check to see if it's an extended standard form. Since the integrand contains a form of e^x, we should definitely check the exponential form. We can see it is not the standard exponential form since the integrand is not just e^x. Therefore, we check the extended standard exponential form. We note that the exponent of e is x^2, which we take to be $f(x)$. In addition, $f'(x) = 2x$ which, except for the multiplier 2, is the remaining factor of our integrand. We can easily adjust for the missing multiplier 2. We have identified the integral as an extended standard exponential form and can proceed to write the antiderivative.

$$\int e^{f(x)} f'(x)\, dx = e^{f(x)} + C$$

$$\int x e^{x^2}\, dx = \frac{1}{2} \int e^{x^2} 2x\, dx$$
$$= e^{x^2} + C$$

E X A M P L E   3   **A Power Form**

Find the antiderivative of $\int (6x^2 - 10x)\sqrt{2x^3 - 5x^2 + 8}\, dx$.

Since the integrand is not a polynomial, it cannot be a standard form. Next we consider if this is an extended standard power form that fits the pattern $\int [f(x)]^n f'(x)\, dx$. Rewriting the integrand and continuing, we obtain

$$\int (6x^2 - 10x)\sqrt{2x^3 - 5x^2 + 8}\, dx = \int (2x^3 - 5x^2 + 8)^{1/2}(6x^2 - 10x)\, dx$$

where we have anticipated that $f(x) = 2x^3 - 5x^2 + 8$. Since the derivative of this function is $6x^2 - 10x$, we have an exact match for the extended standard power form. This allows us to write

$$\int (2x^3 - 5x^2 + 8)^{1/2}(6x^2 - 10x)\, dx = \frac{(2x^3 - 5x^2 + 8)^{3/2}}{3/2} + C.$$

As usual, if we wish to learn mathematics well, we need to become thinkers and observers. For example, if the factor $(6x^2 - 10x)$ in the example above had been $(3x^2 - 5x)$ or $(12x^2 - 20x)$, we need to easily see that it still fits the extended standard power form. All that is needed is a small adjustment by a constant, since $2(3x^2 - 5x) = 6x^2 - 10x$ and $\frac{1}{2}(12x^2 - 20x) = 6x^2 - 10x$. Remember that we check for a standard form first; if that fails, we check for an extended standard form. The rest of this Interlude focuses on the procedure we follow if the integrand is neither standard nor extended standard.

► **FIRST REFLECTION**

Since we can always determine if we have found a correct antiderivative by taking its derivative, convince yourself that the extended standard power form

$$\int [f(x)]^n f'(x)\, dx = \frac{[f(x)]^{n+1}}{n+1} + C$$

is correct by taking the derivative of the right-hand side. ◄

Extending Integration by Parts

The technique of integration by parts should not be considered unless and until the integral has been found to fit neither the standard forms nor extended standard forms. The main types of integrals that lend themselves to the integration by parts technique are those that have a mix of functions, integrands containing the $ln(x)$ function, or inverse trigonometric functions in some form. A mix of functions means integrands containing algebraic and trigonometric functions, algebraic and exponential functions, trigonometric and exponential functions, etc. The technique of integration by parts may also need to be used more than once to arrive at a solution. Finally, this technique can be used in other situations that are difficult to categorize. Recall that the integration by parts technique is $\int u\, dv = uv - \int v\, du$.

EXAMPLE 4 **A Mix Requiring Integration by Parts**

Find the antiderivative of $\int e^x \sin(x)\, dx$.

After deciding the integral does not fit either a standard or extended standard form, we decide to try integration by parts since it fits the mix of functions pattern. We try $dv = \sin(x)$ and $u = e^x$ first. This gives $du = e^x\, dx$ and $v = -\cos(x)$. Continuing on, we obtain

$$\int e^x \sin(x)\, dx = -e^x \cos(x) + \int e^x \cos(x)\, dx.$$

The new integral did not become more complicated, so this is a good sign. We apply integration by parts again with choices $dv = \cos(x)$ and $u = e^x$ (if we switched choices now, we would undo the advantage we gained). This choice results in $du = e^x\, dx$ and $v = \sin(x)$, which in turn gives

$$\int e^x \sin(x)\, dx = -e^x \cos(x) + e^x \sin(x) - \int e^x \sin(x)\, dx.$$

We notice that the latest integral matches the integral on the left, so we can combine them and complete the solution as follows:

$$2 \int e^x \sin(x)\, dx = -e^x \cos(x) + e^x \sin(x)$$

$$\int e^x \sin(x)\, dx = \frac{1}{2}[-e^x \cos(x) + e^x \sin(x)] + C.$$

Think and Share

Try switching the choices after the first application of the intregation by parts technique to see what happens. Be careful with your signs.

▶ **SECOND REFLECTION**

What change in the integral $\int e^x \sin(x)\,dx$ would cause it to be an extended exponential form? An extended trigonometric form?　　◀

EXAMPLE 5
_____

Find the antiderivative of $\int \ln(x)\,dx$.

It takes little time to recognize that this integral does not fit either a standard or extended standard form. We try integration by parts since it contains the ln function. Since we cannot integrate $\int \ln(x)\,dx$ directly, we have only one choice for dv; namely, dx. We then choose $u = \ln(x)$, resulting in $du = \dfrac{1}{x}\,dx$ and $v = x$. These choices give us

$$\int \ln(x)\,dx = x\ln(x) - \int x \cdot \frac{1}{x}\,dx$$

$$= x\ln(x) - \int 1\,dx$$

$$= x\ln(x) - x + C.$$

Trigonometric Combinations

There are a number of approaches we can take when looking for an antiderivative of a function containing **only** trigonometric functions. As you might guess, this depends on the particular makeup of the function. We will study two of these approaches.

Using Trigonometric Identities. When we are faced with an integrand that contains only trigonometric functions and we have determined that it doesn't fit either the standard or extended standard form, trigonometric identities might be a viable option.

EXAMPLE 6
_____

Find the antiderivative of $\int \tan(x)\,dx$.

This integral does not fit a standard trigonometric form (review them if necessary). In consdering using trigonometric identities, we recall that $\tan(x) = \dfrac{\sin(x)}{\cos(x)}$.

$$\int \tan(x)\,dx = \int \frac{\sin(x)}{\cos(x)}\,dx$$

We reexamine our procedure so far. Suspecting it might be an extended logarithm form, we write so as to help us determine if it is this form.

$$\int \frac{\sin(x)}{\cos(x)}\, dx = \int \frac{1}{\cos(x)} \cdot \sin(x)\, dx$$

We now can decide if it fits the extended logarithm form by considering $f(x) = \cos(x)$ and noticing that, aside from a negative sign, its derivative adjacent to it. Thus, it is an extended logarithm form and by adjusting for the negative sign, we can carry out the integral.

$$\int \frac{1}{f(x)} f'(x)\, dx = \ln|f(x)| + C$$

$$\int \frac{1}{\cos(x)} \cdot \sin(x)\, dx = -\int \frac{1}{\cos(x)} \cdot [-\sin(x)]\, dx$$
$$= -\ln|\cos(x)| + C$$

▶ **THIRD REFLECTION**

Try to recall the double angle identity for the cosine function (look it up, if necessary) and use one of its three forms to transform $\sin^2(x)\, dx$ into one that eliminates the squared term, then carry out the integral $\int \sin^2(x)\, dx$. ◀

E X A M P L E 7 **Another Extended Standard Power Form**

Find the antiderivative of $\int \tan^2(x) \sec^2(x)\, dx$.

Study the integral and satisfy yourself that it does not fit a standard trigonometric form. Our strategy tells us we should try the extended standard power form. We know that the derivative of the tangent function contains the secant function, so we attempt to put it into this form.

$$\int \tan^2(x) \sec^2(x)\, dx = \int [\tan(x)]^2 \sec^2(x)\, dx$$

We can now see that it fits the extended standard power form and so we can write down the antiderivative.

$$\int [\tan(x)]^2 \sec^2(x)\, dx = \frac{[\tan(x)]^3}{3} + C = \frac{\tan^3(x)}{3} + C$$

You may have already seen the extended standard power form before the rewrite. Whether you did or not is unimportant at this time; what is important is that you will be able to easily spot various patterns as you gain more experience.

▶ **FOURTH REFLECTION**

Break into pairs. Determine what category of extended standard form the integral $\int \sin(x) \cos^3(x)\, dx$ falls into, then find its antiderivative. ◀

Think and Share

Think about the approach you might use to integrate

$$\int \frac{1 + \sin^2(x)}{\cos^2(x)}$$

and then find the antiderivative. Use of trigonometric identities is one technique that should come to mind. You may need to use some algebraic manipulations first.

Special Trigonometric Combinations. Finding antiderivatives of certain combinations of sine/cosine products and tangent/secant products lend themselves to techniques we will study next. The sine/cosine combinations we will study here are capable of being transformed into standard trigonometric or extended standard trigonometric form and/or standard power or extended standard power form.

The Sine/Cosine Combination. We start with the easiest sine/cosine combinations.

E X A M P L E 8 **The Power Form Pattern**

Find the antiderivarive of $\int \cos^3(x) \sin^2(x) \, dx$.

Whenever the sine or cosine has an odd, natural number exponent greater than one, the integral fits a convenient pattern. If the exponent is one, then it already fits a standard trigonometric form or an extended power form. We can take all but one of the functions with the odd, natural number exponent and change them to the other function using the identity $\cos^2(x) + \sin^2(x) = 1$. In doing so, we obtain

$$\int \cos^3(x) \sin^2(x) \, dx = \int \cos^2(x) \sin^2(x) \cos(x) \, dx$$

$$= \int [1 - \sin^2(x)] \sin^2(x) \cos(x) \, dx$$

$$= \int \sin^2(x) \cos(x) \, dx - \int \sin^4(x) \cos(x) \, dx$$

$$= \int [\sin(x)]^2 \cos(x) \, dx - \int [\sin(x)]^4 \cos(x) \, dx.$$

We can now write down the antiderivative upon recognizing that both integrals are extended standard power forms.

$$\int [\sin(x)]^2 \cos(x) \, dx - \int [\sin(x)]^4 \cos(x) \, dx = \frac{\sin^3(x)}{3} - \frac{\sin^5(x)}{5} + C$$

$$\int [f(x)]^n f'(x) \, dx = \frac{[f(x)]^{n+1}}{n + 1} + C$$

E X A M P L E 9 **The Double Angle Pattern**

Find the antiderivative of $\int \sin^2(x) \cos^2(x)\, dx$.

Whenever both sine and cosine functions in the integrand are raised to even, whole number powers, we use the two equivalent forms of the double angle identity for the cosine function below.

$$\cos(2x) = 2\cos^2(x) - 1$$
$$= 1 - 2\sin^2(x)$$

For instance, we can solve for $\cos^2(x)$ in the identity above to obtain

$$\cos(2x) = 2\cos^2(x) - 1$$
$$1 + \cos(2x) = 2\cos^2(x)$$
$$\frac{1 + \cos(2x)}{2} = \cos^2(x).$$

Using the other form of the cosine double angle identity, we can solve for $\sin^2(x)$ to obtain

$$\sin^2(x) = \frac{1 - \cos(2x)}{2}.$$

Using these results to transform the original integral, we obtain

$$\int \sin^2(x) \cos^2(x)\, dx = \int \frac{1 - \cos(2x)}{2} \cdot \frac{1 + \cos(2x)}{2}\, dx$$
$$= \frac{1}{4} \int \left[1 - \cos^2(2x) \right]\, dx$$
$$= \frac{1}{4} \left[\int 1 \cdot dx - \int \cos^2(2x)\, dx \right]$$
$$= \frac{1}{4} \left[x - \int \cos^2(2x)\, dx \right].$$

The integral remaining to be evaluated contains an even, whole number power of the cosine function and so causes us to use the cosine double angle identity once again.

$$\int \cos(f(x)) f'(x)\, dx = \sin(f(x)) + C$$

$$\frac{1}{4}\left[x - \int \cos^2(2x)\, dx \right] = \frac{1}{4}\left[x - \int \left[\frac{1 + \cos(4x)}{2}\, dx \right] \right]$$
$$= \frac{1}{4}\left[x - \frac{1}{2}\left[\int 1 \cdot dx + \int \cos(4x)\, dx \right] \right]$$
$$= \frac{1}{4}\left[x - \frac{1}{2}\left[x + \frac{1}{4}\int \cos(4x) \cdot 4\, dx \right] \right]$$
$$= \frac{1}{4}\left[\frac{x}{2} - \frac{1}{8}\sin(4x) \right] + C$$

The Tangent/Secant Combinations.   The tangent/secant combinations are slightly more complicated than the sine/cosine combinations in that they include one more possibility than those for sine/cosine.

E X A M P L E   1 0   **The Power Form for Tangent**

Find the antiderivative of $\int \tan^2(x) \sec^4(x)\, dx$.

As always, we examine the integral to see if it fits a standard or extended standard form. Whenever an integrand contains an even power of the secant function in the tangent/secant combination, we can fit this into one or more power or extended power forms. $\sec^2(x)$ is the derivative of $\tan(x)$. Hence, after removing two powers of $\sec(x)$, we can replace the remaining even powers of secant by tangents using the identity $1 + \tan^2(x) = \sec^2(x)$. That is, we will fit the pattern $\int [\tan(x)]^n \sec^2(x)\, dx$. The result follows.

$$
\int \tan^2(x)\sec^4(x)\, dx = \int \tan^2(x)\sec^2(x)\sec^2(x)\, dx
$$

$$
= \int \tan^2(x)[1 + \tan^2(x)]\sec^2(x)\, dx
$$

$$
= \int \tan^2(x)\sec^2(x)\, dx + \int \tan^4(x)\sec^2(x)\, dx
$$

$$
= \int [\tan(x)]^2 \sec^2(x)\, dx + \int [\tan(x)]^4 \sec^2(x)\, dx
$$

$$
\int [f(x)]^n f'(x)\, dx = \frac{[f(x)]^{n+1}}{n+1} + C
$$

We now can fit the extended power form by considering $f(x) = \tan(x)$ and write the solution as

$$
\int [\tan(x)]^2 \sec^2(x)\, dx + \int [\tan(x)]^4 \sec^2(x)\, dx = \frac{\tan^3(x)}{3} + \frac{\tan^5(x)}{5} + C.
$$

E X A M P L E   1 1   **The Power Form for Secant**

Find an antiderivative of $\int \tan^3(x)\sec^3(x)\, dx$.

Whenever an integrand contains an odd power of the tangent function in the tangent/secant combination, we can also fit it into a power or extended power form. $\sec(x)\tan(x)$ is the derivative of $\sec(x)$. Hence, after removing the expression $\sec(x)\tan(x)$, we can replace the remaining even powers of tangent by secants to fit the pattern $\int [\sec(x)]^n \sec(x)\tan(x)\, dx$. The sequence of steps follow.

$$
\int \tan^3(x)\sec^3(x)\, dx = \int \tan^2(x)\sec^2(x)\sec(x)\tan(x)\, dx
$$

$$
= \int [\sec^2(x) - 1]\sec^2(x)\sec(x)\tan(x)\, dx
$$

$$
= \int \sec^4(x)\sec(x)\tan(x)\, dx - \int \sec^2(x)\sec(x)\tan(x)\, dx
$$

$$\int [f(x)]^n f'(x)\, dx = \frac{[f(x)]^{n+1}}{n+1} + C$$

$$= \int [\sec(x)]^4 \sec(x) \tan(x)\, dx - [\sec(x)]^2 \sec(x) \tan(x)\, dx$$

The extended power form should now be fairly easy to see, giving the result

$$\int [\sec(x)]^4 \sec(x) \tan(x)\, dx - \int [\sec(x)]^2 \sec(x) \tan(x)\, dx$$

$$= \frac{\sec^5(x)}{5} - \frac{\sec^3(x)}{3} + C.$$

The case involving the tangent raised to an even power and the secant raised to an odd power requires the technique of intergration by parts.

▶ **FIFTH REFLECTION**

Explain how integrals with combinations of cotangent/cosecant funtions are similar to the tangent/secant combinations. What are the minor differences? ◀

Trigonometric Substitution

There are certain forms that can appear in an integrand that lend themselves to a technique called **trigonometric substitution**. The integrands that benefit most from this technique are those containing one of the forms $\sqrt{a^2 - x^2}$, $\sqrt{x^2 + a^2}$, or $\sqrt{x^2 - a^2}$, where a is a positive constant. You use a trigonometric substitution to eliminate the square root. For example, if the integrand contains the factor $\sqrt{a^2 - x^2}$, what type of trig substitution would eliminate the square root? Recalling that $1 - \sin^2(x) = \cos^2(x)$, let's examine the substitution $x = a \sin(\theta)$. This substitution gives

$$\sqrt{a^2 - x^2} = \sqrt{a^2 - a^2 \sin^2(\theta)}$$

$$= \sqrt{a^2[1 - \sin^2(\theta)]}$$

$$= a\sqrt{1 - \sin^2(\theta)} \quad \text{since } a \text{ is positive}$$

$$= a\sqrt{\cos^2(\theta)}.$$

We know that $\sqrt{\cos^2(\theta)} = |\cos(x)| = \pm \cos(x)$ (not just $\cos(x)$). However, this does not affect our search for an antiderivative, so we will disregard the absolute value sign for the time being while we focus on finding antiderivaives. Be sure, however, to keep this fact in mind as **it is important** in the final result. In addition, it is equally important with trigonometric substitutions to find the differential of x in order to complete the change in variable.

EXAMPLE 12 **An Integral That Can Use Trigonometric Substitution**

Find the antiderivaive of $\int x^3 \sqrt{4 - x^2}\, dx$.

Since the integrand contains a factor of the form $\sqrt{a^2 - x^2}$, we attempt the following trigonometric substitution:

$$\text{let} \quad x = 2\sin(\theta)$$

$$\text{then} \quad dx = 2\cos(\theta)\,d\theta$$

The substitution results in the transformation

$$\int x^3\sqrt{4 - x^2}\,dx = \int 8\sin^3(\theta)\sqrt{4 - 4\sin^2(\theta)}\,2\cos(\theta)\,d\theta$$

$$= 16\int \sqrt{4[1 - \sin^2(\theta)]}\,\sin^3(\theta)\cos(\theta)\,d\theta$$

$$= 16\int 2\sqrt{1 - \sin^2(\theta)}\,\sin^3(\theta)\cos(\theta)\,d\theta$$

$$= 32\int \sqrt{\cos^2(\theta)}\,\sin^3(\theta)\cos(\theta)\,d\theta$$

$$= 32\int \sin^3(\theta)\cos^2(\theta)\,d\theta \quad \text{neglecting absolute value.}$$

At this point, we see that this latest integral fits into the pattern of special trigonometric combinations. Continuing on, we find

$$\int \sin^3(\theta)\cos^2(\theta)\,d\theta = \int \sin^2(\theta)\cos^2(\theta)\sin(\theta)\,d\theta$$

$$= \int [1 - \cos^2(\theta)]\cos^2(\theta)\sin(\theta)\,d\theta$$

$$= \int \cos^2(\theta)\sin(\theta)\,d\theta - \int \cos^4(\theta)\sin(\theta)\,d\theta$$

$$= \frac{\cos^3(\theta)}{3} - \frac{\cos^5(\theta)}{3} + C$$

FIGURE 5.32 Information triangle

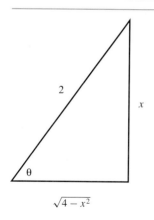

However, we were asked to find the antiderivative of a function in x and we have one in terms of θ. This is always the case when we change variables, so we look to where we made the substitution to obtain the solution. Since we let $x = 2\sin(\theta)$, then $\sin(\theta) = x/2$. Looking at the triangle in Figure 5.32, we see that $\cos(\theta) = \dfrac{\sqrt{4 - x^2}}{2}$. Thus, our final solution is

$$\int x^3\sqrt{4 - x^2}\,dx = \frac{(4 - x^2)^{3/2}}{3} - \frac{(4 - x^2)^{5/2}}{5} + C.$$

▶ SIXTH REFLECTION

Find two different changes in the integrand of the example above that make it fit the extended power form. ◀

Break into pairs. The example above can also be solved using the integration by parts technique. Discuss what choices you would make for dv and u. Then carry out the integration and compare your results with those of the example. Would integraion by parts help if the integral was $\int x^2 \sqrt{4 - x^2}\, dx$? Explain.

EXAMPLE 13

Find the antiderivative of $\int \dfrac{1}{1 + x^2}\, dx$.

We notice that this integral does not fit the extended logarithm form. Although it also does not contain a square root, the trigonometric substitution technique can be considered. Recalling that $1 + \tan^2(x) = \sec^2(x)$, we make the substitution

$$\text{let} \quad x = 1 \cdot \tan(\theta)$$
$$\text{then} \quad dx = \sec^2(\theta)\, d\theta.$$

Making the transformation yields

$$\int \frac{1}{1 + x^2}\, dx = \int \frac{1}{1 + \tan^2(\theta)} \sec^2(\theta)\, d\theta$$
$$= \int \frac{1}{\sec^2(\theta)} \sec^2(\theta)\, d\theta$$
$$= \int 1\, d\theta$$
$$= \theta + C.$$

We look to our substitution to express the result in terms of the original variable x. Since we let $x = \tan(\theta)$, then $\theta = \tan^{-1}(x)$, so

$$\int \frac{1}{1 + x^2}\, dx = \tan^{-1}(x) + C.$$

▶ **SEVENTH REFLECTION**

What change in the integrand in the example above would cause the integral to be an extended logarithm form? ◀

▶ **EIGHTH REFLECTION**

One can see from the two examples above that an integrand containing the third form $\sqrt{x^2 - a^2}$ would require a different trigonometric substitution. Actually, the substitution would be to let $x = a \sec(\theta)$. Convince yourself that this substitution in this case would eliminate the square root. ◀

As mentioned before, the trigonometric substitution technique has the most benefit when the integrand contains an expression of the form $\sqrt{a^2 - x^2}$,

$\sqrt{x^2 + a^2}$, or $\sqrt{x^2 - a^2}$. The substitutions we make in these cases are $x = a\sin(\theta)$, $x = a\tan(\theta)$, and $x = a\sec(\theta)$, respectively. These same substitutions should be considered when the integrand contains an expression of the form $a^2 - x^2$, $a^2 + x^2$, or $x^2 - a^2$, even when they are raised to the 3/2 power, 5/2 power, or even natural number powers. The same technique should be considered in extended forms such as an integrand containing an expression like $\sqrt{8 - (3x + 5)^2}$. The parallel is to let $3x + 5 = \sqrt{8}\sin(\theta)$, take the differential of both sides, and continue on. These should be considered after checking for standard and extended standard forms and, perhaps, partial fractions and integration by parts.

Completing the Square

Another technique that can help us evaluate integrals when other techniques fail involves the method of completing the square of quadratic expressions.

EXAMPLE 14

Find the antiderivative of $\displaystyle\int \frac{1}{\sqrt{6 + 4x - x^2}}\, dx$.

We use our strategy developed so far to determine that none of the prior techniques work. We, therefore, try our latest technique of completing the square.

$$\int \frac{1}{\sqrt{6 + 4x - x^2}}\, dx = \int \frac{1}{\sqrt{-x^2 + 4x + 6}}\, dx$$

$$= \int \frac{1}{\sqrt{-(x^2 - 4x + 4) + 6 - 4}}\, dx$$

$$= \int \frac{1}{\sqrt{2 - (x - 2)^2}}\, dx$$

We are now in a position to try trigonometric substitution. The integrand seems to indicate using the sine function.

$$\text{let} \quad x - 2 = \sqrt{2}\sin(\theta)$$
$$\text{then} \quad dx = \sqrt{2}\cos(\theta)\, d\theta$$

Hence,

$$\int \frac{1}{\sqrt{2 - (x - 2)^2}}\, dx = \int \frac{1}{\sqrt{2 - 2\sin^2(\theta)}}\sqrt{2}\cos(\theta)\, d\theta$$

$$= \int \frac{1}{\sqrt{2}\sqrt{1 - \sin^2(\theta)}}\sqrt{2}\cos(\theta)\, d\theta$$

$$= \int \frac{1}{\sqrt{\cos^2(\theta)}}\cos(\theta)\, d\theta$$

$$= \int \frac{1}{\cos(\theta)} \cos(\theta) \, d\theta$$

$$= \int 1 \, d\theta$$

$$= \theta + C$$

where we again leave the absolute value part until it is necessary. Since

$$x - 2 = \sqrt{2} \sin(\theta),$$

then

$$\theta = \sin^{-1}\left(\frac{x - 2}{\sqrt{2}}\right)$$

and

$$\int \frac{1}{\sqrt{2 - (x - 2)^2}} \, dx = \sin^{-1}\left(\frac{x - 2}{\sqrt{2}}\right) + C$$

Think and Share

Break into pairs. Complete the square in x for the general quadratic expression $ax^2 + bx + c$, $a \neq 0$. Then consider the various signs for a, b, and c and describe which of the three types of trigonometric substitution you would use for each inusing this technique.

Extending Integration Using Partial Fractions

We studied partial fraction integration in Section 5.6 for the case with a constant numerator and products of linear functions in the denominator. To fully cover this topic, we need to examine three more possible cases. First, however, we can dispense with a constant numerator and allow it to be any polynomial. The three additional cases have denominators with products of (1) linear functions, one or more of which repeat; (2) quadratic functions not factorable over the reals; and (3) quadratic functions not factorable over the reals, one or more of which repeat.

We start by stating that we can factor every nth degree polynomial in one variable into n linear factors provided we allow complex number coefficients. This fact comes from the Fundamental Theorem of Algebra. If we want real number coefficients, some of the factors may have to be quadratic (e.g. $x^2 + 4$). In addition, we consider only rational expressions with numerators that are polynomials of degree less than the degree of the denominator. If the degree of the numerator is greater than the degree of the denominator, long division will always help the problem and, in most cases, will involve the integral of a rational function (the remainder).

One last point: the method of partial fractions reverses the process of combining fractions. For example, to write the fraction $\dfrac{1}{x^2 - x - 2}$ equivalently as two fractions, we would start as follows:

$$\frac{1}{x^2 - x - 2} = \frac{1}{(x - 2)(x + 1)} = \frac{A}{x - 2} + \frac{B}{x + 1}.$$

From here, we would use algebra to solve for the constants A and B. However, since our focus is on finding antiderivatives, we will give our answers in terms of these constants. The integral of this particular rational function would look like the following:

$$\int \frac{1}{x^2 - x - 2}\, dx = \int \frac{1}{(x - 2)(x + 1)}\, dx$$

$$= \int \frac{A}{x - 2}\, dx + \int \frac{B}{x + 1}\, dx$$

$$= A \int \frac{1}{x - 2}\, dx + B \int \frac{1}{x + 1}\, dx$$

$$= A \ln|x - 2| + B \ln|x + 1| + C.$$

$$\int \frac{1}{f(x)} f'(x)\, dx = \ln|f(x)| + C$$

Although other forms of rational functions may require more complicated calculations for the constants involved in the separated fractions, they are, nonetheless, algebraic calculations, so we will skip them. Our focus is the integration process.

We now consider the three remaining cases of integrals of rational functions with the various denominators.

Linear Factors With Only One or More Repeating. When a rational expression contains a repeated linear factor in the denominator, then the partial fraction decomposition for the repeated factor will require as many fractions as the **number of times** the factor is repeated. The idea is demonstrated in the following example.

EXAMPLE 15

Find the antiderivative of $\displaystyle\int \frac{x^2 - 3x + 5}{(x - 5)(x + 2)^3}\, dx.$

$$\int \frac{x^2 - 3x + 5}{(x - 5)(x + 2)^3}\, dx$$

$$= \int \frac{A}{x - 5}\, dx + \int \frac{B}{x + 2}\, dx + \int \frac{C}{(x + 2)^2}\, dx + \int \frac{D}{(x + 2)^3}\, dx$$

$$= A \int \frac{1}{x - 5}\, dx + B \int \frac{1}{x + 2}\, dx + C \int \frac{1}{(x + 2)^2}\, dx + D \int \frac{1}{(x + 2)^3}\, dx$$

$$= A \ln|x - 5| + B \ln|x + 2| - \frac{C}{x + 2} - \frac{D}{2(x + 2)^2} + E$$

The technique here enabled us to break the original integral into four integrals that fell into the categories of extended standard logarithm form and extended standard power form. Notice that the linear factor $x + 2$, repeated three times, required three separate fractions with the degree of this factor going up from one to three. Had the linear factor been repeated five times, then there would have been five fractions with the degree of the factor going from one to five, and so on for n repeats.

Denominators Containing Quadratic Factors With No Repeats.   The technique for integrals of rational functions with denominators containing quadratic factors is very similar to the technique used for linear factors. The main difference is that the numerators of the separate fractions have to be linear, not constant. Remember, such quadratic factors cannot be factored over the real numbers.

E X A M P L E   1 6

Find the antiderivative of $\displaystyle\int \frac{3x + 5}{(2x - 3)(x^2 + 25)}\, dx$

We begin by breaking the fraction into partial fractions.

$$\int \frac{3x + 5}{(2x - 3)(x^2 + 25)}\, dx$$

$$= \int \frac{A}{2x - 3}\, dx + \int \frac{Bx + C}{x^2 + 25}\, dx$$

$$= A \int \frac{1}{2x - 3}\, dx + B \int \frac{x}{x^2 + 25}\, dx + C \int \frac{1}{x^2 + 25}\, dx$$

$$= \frac{A}{2} \int \frac{1}{2x - 3} \cdot 2\, dx + \frac{B}{2} \int \frac{1}{x^2 + 25} \cdot 2x\, dx + C \int \frac{1}{x^2 + 25}\, dx$$

$$= \frac{A}{2} \ln|2x - 3| + \frac{B}{2} \ln(x^2 + 25) + C \int \frac{1}{x^2 + 25}\, dx$$

All that remains is to evaluate $\displaystyle C \int \frac{1}{x^2 + 25}\, dx$. Using our strategy, we eventually come to the conclusion that a trigonometric substitution might work.

$$\text{let} \quad x = 5 \tan(\theta)$$

$$\text{then} \quad dx = 5 \sec^2(\theta)\, d\theta$$

This trigonometric substitution transforms our integral into

$$C \int \frac{1}{x^2 + 25}\, dx = C \int \frac{1}{25 \tan^2(\theta) + 25} 5 \sec^2(\theta)\, d\theta$$

$$= C \int \frac{1}{25[\tan^2(\theta) + 1]} 5 \sec^2(\theta)\, d\theta$$

$$= \frac{C}{5} \int \frac{1}{\tan^2(\theta) + 1} \sec^2(\theta)\, d\theta$$

$$= \frac{C}{5} \int \frac{1}{\sec^2(\theta)} \sec^2(\theta)\, d\theta$$

$$= \frac{C}{5} \int 1\, d\theta$$

$$= \frac{C}{5}\theta.$$

We recall that the original substitution allows us to write the answer in the original variable. Since we let $x = 5\tan(\theta)$, then $\theta = \tan^{-1}(\frac{x}{5})$. We can now write the complete answer.

$$\int \frac{3x + 5}{(2x - 3)(x^2 + 25)}\, dx = \frac{A}{2} \ln|2x - 3| + \frac{B}{2} \ln(x^2 + 25) + \frac{C}{5} \tan^{-1}\left(\frac{x}{5}\right) + D$$

The important point here is to remember that a linear factor like $Bx + C$ is needed in the numerator when the denominator of a rational fraction is quadratic and not factorable over the real numbers.

Denominators Containing Quadratic Factors, One or More of Which Repeat.

Had the integral in the previous example been $\int \dfrac{3x + 5}{(2x - 3)(x^2 + 25)^2}\, dx$, then, following the lead of the associated linear case, we would write the partial fractions as

$$\int \frac{3x + 5}{(2x - 3)(x^2 + 25)^2}\, dx = \int \frac{A}{2x - 3}\, dx + \int \frac{Bx + C}{x^2 + 25}\, dx + \int \frac{Dx + E}{(x^2 + 25)^2}\, dx$$

and proceed from there. The difference is in the last integral. We can break it up into an integral that will fit an extended standard power form and the other looks like a good candidate for a trigonometric substitution.

▶ NINTH REFLECTION

Think about how you would do the last integral above, $\int \dfrac{Dx + E}{(x^2 + 25)^2}\, dx$. Then work out the solution.　　　◀

Finishing up the technique of finding antiderivatives using partial fractions involves a complete quadratic factor in the denominator that is not factorable over the real numbers. For example, if factors like $2x^2 + 3x + 5$ appear in the denominator of an integral that will require the use of partial fractions, then completing the square should help us.

Putting It All Together

The topic of techniques of integration does a reasonable job of helping us increase our problem-solving skills. It provides a discipline of using a workable strategy combined with a thought process required to consider an appropriate approach that is dictated by the particular form of the integral. A summary of the strategy follows.

- Always start by examining the integral to see if it is a standard form. Naturally, this means we need to know the standard forms.
- If the integral does not fit a standard form, we next check to see if it fits an extended standard form.

If the integral does not fit these first two patterns, then we look to one of the techniques below.

1. If the integrand is composed of trigonometric functions only, then

    a. Use identities to see if you can transform the integral into a recognizable integral.

    b. See if it fits one of the special sine/cosine or tangent/secant combinations to integrate it.

2. Use integration by parts if the integral is a mix of functions or contains the ln function or inverse trigonometric functions. This technique can also work for other forms as well.

3. Use partial fractions if the integral contains a rational function. Be sure to use long division first if the degree of the numerator is greater than or equal to the degree of the denominator.

4. Use trigonometric substitution if the integrand contains one of the expressions $\sqrt{a^2 - x^2}$, $\sqrt{a^2 + x^2}$, or $\sqrt{x^2 - a^2}$, $a > 0$. Also, consider using trigonometric substitution even when the inside expressions are raised to other powers.

5. Complete the square on expressions of the form $ax^2 + bx + c$.

Recall that the techniques listed in 1-5 above should be considered only after you have examined the integral and find it does not fit either the standard or extended standard form.

Again, we know that the TI calculator can integrate most of the integrals we will encounter in a calculus course. However, thinking about and working with techniques of integration helps sharpen our problem-solving skills.

EXERCISE SET 5.7

Since the main objective of this Interlude is to enhance your problem-solving skills, you should make a first pass at the following exercises by looking at the integral and mentally determining and writing what category it falls into. That is, standard (power, log, etc.) form, extended (power, log, etc.) form, parts, partial fractions, etc. Then go back over the problems and try the category you chose. Finally, if necessary, use your TI calculator but first, review the strategy.

In Exercises 1–36, first write down the category of each integral and the form. Then find the antiderivative.

1. $\displaystyle\int \sin^{5/2}(x) \cos(x) \, dx$

2. $\displaystyle\int \frac{dx}{\sqrt{1 + 9x^2}}$

3. $\displaystyle\int \frac{6x}{\sqrt{6 + x^2}} \, dx$

4. $\displaystyle\int x^2 e^{3x} \, dx$

5. $\displaystyle\int \frac{4x - 7}{x^2 + 3x} \, dx$

6. $\displaystyle\int \tan^2(2x)\, dx$

7. $\displaystyle\int \frac{\sin(x)}{2 - \cos(x)}\, dx$

8. $\displaystyle\int \frac{\ln^2(x)}{x}\, dx$

9. $\displaystyle\int \frac{\ln(x^2)}{x}\, dx$

10. $\displaystyle\int \frac{3}{\sqrt{25 - x^2}}\, dx$

11. $\displaystyle\int \frac{dx}{\sqrt{x^2 - 5x}}$

12. $\displaystyle\int \sec^3(x) \tan^3(x)\, dx$

13. $\displaystyle\int x \sin(2x^2)\, dx$

14. $\displaystyle\int 4x^2 e^{5x^3}\, dx$

15. $\displaystyle\int \tan(1 - 2x) \sec(1 - 2x)\, dx$

16. $\displaystyle\int x \ln^2(x)\, dx$

17. $\displaystyle\int x \ln(x^2)\, dx$

18. $\displaystyle\int \frac{2x^2 - 7}{(1 - 3x)(5x^2 + 2)}\, dx$

19. $\displaystyle\int \sqrt{6x - 2x^2}\, dx$

20. $\displaystyle\int \sin^4(3x) \cos^2(3x)\, dx$

21. $\displaystyle\int \frac{dx}{8x^2 - 1}$

22. $\displaystyle\int \frac{7x}{8x^2 - 1}\, dx$

23. $\displaystyle\int x \sec^2(x^2)\, dx$

24. $\displaystyle\int x^3 \sec^2(x^2)\, dx$

25. $\displaystyle\int \frac{1}{\sqrt{x^2 - 2}}\, dx$

26. $\displaystyle\int \frac{3x^2}{\sqrt{4 + x^4}}\, dx$

27. $\displaystyle\int \sin^5(x) \cos^3(x)\, dx$

28. $\displaystyle\int \frac{x^2}{(16 - x^2)^{3/2}}\, dx$

29. $\displaystyle\int \frac{\sqrt{1 - 4x^2}}{x^3}\, dx$

30. $\displaystyle\int \frac{x}{\sqrt{x^2 - 2x + 5}}\, dx$

31. $\displaystyle\int \frac{x^2 - 3x - 12}{x^2 - 2x - 8}\, dx$

32. $\displaystyle\int \frac{3x^2 - 5x + 2}{(3x - 2)(x^2 + 1)^2}\, dx$

33. $\displaystyle\int \frac{e^{2x}}{\sqrt{1 + e^{2x}}}\, dx$

34. $\displaystyle\int \sec^2(5x) \tan^2(5x)\, dx$

35. $\displaystyle\int x^3 \sqrt{x^2 + 4}\, dx$

II

MODELING WITH CALCULUS

Change is a fundamental characteristic of the universe. Objects change location, size, direction of motion, temperature, mass, charge, and so on. The mathematics that existed before the invention of calculus, such as algebra and geometry, were hindered in describing the universe by their static nature. Calculus, the mathematics of change, is specifically designed to create mathematical models of a changing universe.

Building a mathematical model is an iterative process. A modeler must keep many activities and ideas in mind during the process of creating and checking a model. Often, puzzling and confusing situations arise which force the modeler to retrace and rethink previous steps in the process.

As a good first step, we identify the situation to be studied and collect information about that situation. We then study the information to discover meaningful patterns. Using those patterns and our understanding of the universe, simplifying assumptions are made that lead to the creation of a model. Often, though not always, the model is mathematical. If the model cannot be analyzed with the tools at hand, we make further simplifications. Such simplifications include treating some variables as constants, neglecting some variable, gathering together others by assuming simple relationships between them, or further restricting the phenomenon under investigation.

The creation of the model does *not* end the process. The model must then be checked for agreement with the collected information *and* for its accuracy in making predictions (that is, its agreement with subsequent observations). If, as is often the case, the model is not sufficiently accurate, then we must refine it and check again. Sometimes, the refinements amount to minor tinkering with one or more parameters. Other times, the fundamental assumptions of the model must be reexamined. The process of simplification and refinement determine the generality, realism, and precision of the model and constitutes the art of modeling .

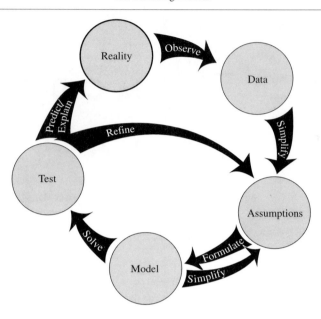

Now that we are familiar with the basic concepts of calculus, we can begin to participate, at a deeper level, in the exciting process called modeling. This part of the book is designed to develop model-building abilities. In Chapter 6, we will investigate simple models representing the change in a single quantity over time. In the process, we will move from the discrete to the continuous and back. In Chapter 7, we will provide various tools, both symbolic and numeric, for analyzing the models we've created. These tools are extremely important, both in checking the validity of a given model and in using the model to understand the particular phenomenon being studied. In Chapter 8, we expand our study to include more complicated situations that involve the interaction of two or more quantities that are changing over time.

C H A P T E R

6

MODELING

WITH

THE DERIVATIVE

6.1
One Day in the Life of a Modeler

6.2
Warming and Cooling

6.3
Population Modeling

6.4
Euler's Method

6.5
Slope Fields

6.6
Errors in the Model Construction

Terminology

Summary

6.1

One Day in the Life of a Modeler

The simplest models represent changes in a single quantity over time. Such models require an independent variable to represent time and a dependent variable to represent the quantity of interest. In some cases, we can construct a satisfactory model from observed values of the quantity using the discrete paradigm,

$$future = present + change.$$

In other cases, circumstance may lead us to prefer a model which is continuous. Then we would need to develop a continuous paradigm to construct a model. The two types of models are closely linked, as we shall see.

Modeling

EXAMPLE 1 **Modeling a Chemical Reaction**

You are a freelance analyst consulting for a variety of companies. Monday morning, a senior laboratory chemist in Research and Development at CMM Chemical Corporation sends you an email message.

FIGURE 6.1   Marie's Email Message

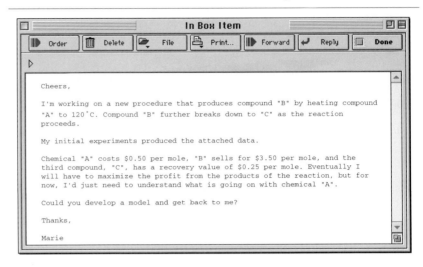

To get a quick look at the situation, you graph the data for *A*. Create the data matrix `Reaction` with the data for time in `c1` and *A* in `c2`. Clearly, the

FIGURE 6.2   Reaction Diagram

TABLE 6.1   Reaction Data (Attachment)

Time (min)	0	2	6	10	20	30	50	70	90	120	150	200
A (mole)	1.00	0.88	0.69	0.53	0.28	0.15	0.043	0.012	0.00	0.00	0.00	0.00
B (mole)	0.00	0.12	0.29	0.42	0.56	0.57	0.46	0.33	0.22	0.12	0.06	0.02
C (mole)	0.000	0.003	0.030	0.050	0.16	0.28	0.50	0.66	0.78	0.88	0.94	0.98

FIGURE 6.3   Amount of Chemical *A* vs. Time   **FIGURE 6.4**   Divided Differences vs. Time

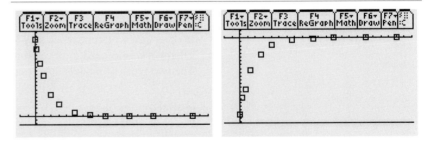

graph is decreasing; but how fast is it decreasing? Since data was not collected at regular intervals, you check the first divided differences, $\Delta A / \Delta t$, to obtain information about the rate of decrease. Recall that for a sequence of values the forward differences are defined as

$$\Delta a_n = a_{n+1} - a_n.$$

In the HOME *screen, define the forward difference function (as in Chapter 2) as*

```
Define fd(q) = shift(q,1) - q
```

In the Data/Matrix *editor define the third column to be the divided differences with*

```
c3= fd(c2)/fd(c1)
```

► FIRST REFLECTION

Why are the Δa's negative?　◄

Once again, you look to a graph for insight by plotting the values for the divided differences against time.

This graph looks suspiciously familiar. If it were reflected through the x axis, it would look very much like the original graph of the data. You decide to reflect this plot and compare the reflected image to the original data side by side. Make a column for the reflected differences:

```
c4= -c3
```

Instead of plotting the two data sets c2 and c4 on the same axes, since the values are quite different in magnitude, you make two separate plots for Marie. Follow the direction in the technology note, *Split Screen Dual Graphing*, to create the graphs in Figure 6.5.

► SECOND REFLECTION

What will happen if you do plot both data sets on one set of axes?　◄

Use ZoomData on both plots and then set the horizontal scales the same without changing the vertical scale. The two plots have the same shape! The vertical scales are very different, but the horizontal scales are the same suggesting that the divided differences are proportional to the amount of chemical A left in the solution.

FIGURE 6.5 Comparing A and $\Delta A/\Delta t$

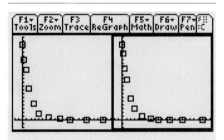

SPLIT SCREEN DUAL GRAPHING

Go to page 2 of the MODE dialog and set:

Split Screen...... Left-Right.

Split 1 App....... Graph

Split 2 App....... Graph

Number of Graphs.. 2

and press Enter. This arrangement sets the display to handle graphs with each having its own Y= and WINDOW editor. Press 2nd APPS to switch between windows. In the left window, use the Y= editor to turn Plot 1 on and Plot 2 off. Switch to the right window, then turn Plot 1 off and Plot 2 on.

► **THIRD REFLECTION**

Can you determine a scaling factor k that you can use to transform the c3 data with the formula

$$c4=kc3$$

so that the graphs for c2 and c4 have the same size as well as shape?   ◄

The symbol $\propto$ is used to indicate direct proportionality.

These observations lead us to the following model. Take the proportionality constant, k, to be negative and include the initial value of A,

$$\frac{\Delta A}{\Delta t} \propto A \qquad \text{(6.1)}$$
$$A(0) = 1.$$

Having spent the morning working on the problem, you send an email message to Marie explaining your results and go to lunch.

► **FOURTH REFLECTION**

How should you respond to Marie? Write an email message that includes a comment on the graphic effect of taking k to be negative. Also, include an explanation as to why the graphs suggest that $\Delta A/\Delta t$ is proportional to A.   ◄

Returning from lunch you find the following email message waiting on your computer.

FIGURE 6.6   Marie's Email Reply

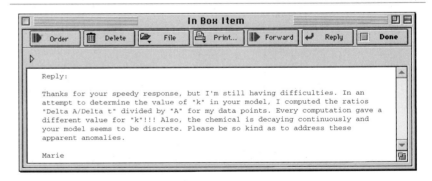

► **FIFTH REFLECTION**

Go to the Data/Matrix editor and create a fifth column which displays the chemist's computations of the estimate for the constant of proportionality, k. Do your computations match Marie's? How do these values compare with the scale factor from Third Reflection.   ◄

The original model is too simple and needs to be refined. The chemical decays continuously, so the model should also be continuous. The interpretation of divided differences as an approximation of the derivative of A over the interval Δt suggests that the derivative itself is proportional to the amount of chemical A. By analogy to your discrete model, you propose a continuous model $a(t)$ for the change in A,

$$\frac{da}{dt} \propto a$$

$$a(0) = 1,$$

(6.2)

where the new constant of proportionality, κ, will need to be determined from the data.

▶ **SIXTH REFLECTION**

Compose a second email message to Marie explaining the continuous model. Specify which k value from Marie's computations gives the most reasonable approximation for κ and justify your choice.   ◀

By 3:00 p.m. you've finished sending your response to Marie and you've forwarded an invoice for your standard $2,000 fee to accounting. You decide to knock off for the rest of the day, but just as you're about to go out the door the phone rings. The plant manager for Mountain View Mercury Processing is on the line with a problem.

The concentration of pollutant in a water solution can be reduced by pumping the polluted water into a large holding pond of pure water, mixing, and pumping the diluted water out. At first the polluted water is diluted by the large quantity of pure water in the pond. Over time, as the water in the pond becomes more and more polluted, it will still dilute the concentration of the pollutant, but with less and less effect.

E X A M P L E   2   **Mercury Pollution**

"We've just received the latest OSHA guidelines," he says, "and I'm worried that one of our older treatment ponds may not be in compliance with the latest standards. The pond contains 10,000 gallons of water. We're filling with a solution that contains 0.1 grams of mercury per gallon at the rate of three gallons per minute and we have polluted water flowing out of the pond also at three gallons per minute. We've cut back on the amount of mercury we are pumping into the pond since when we took over the operation the pond was way out of compliance. The concentration of mercury in the pool was 0.7 grams per gallon, but I'm afraid that we have not brought the mercury down low enough. What I need is a model that will let me figure out how much mercury is in the pond at any time."

The problem seems simple enough and you agree to consider it. You decide to represent the quantity of mercury in the pond at time t with a differentiable (and hence continuous) function $p(t)$. You realize that while the rate of change in the amount of mercury in the tank varies with time t, it can be described as the difference between the rate of change of mercury in the inflow and that in the outflow. The rate of mercury flowing into the tank is constant and can be computed as

$$r_{in}(t) = 0.1 \frac{\text{g}}{\text{gal}} \times 3 \frac{\text{gal}}{\text{min}} = 0.3 \frac{\text{g}}{\text{min}}.$$

The rate of mercury flowing out varies with time, t, but can be described in terms of $p(t)$

$$r_{out}(t) = \frac{p(t)}{10,000} \frac{g}{gal} \times 3 \frac{gal}{min} = 0.0003 \, p(t) \frac{g}{min}.$$

▶ **SEVENTH REFLECTION**

Put the *in* and *out* rates together to get an equation that models the amount of mercury in the pond as

$$\frac{dp}{dt} = r_{in}(t) - r_{out}(t) = ?$$

◀

The manager is asking, "Does the long-term concentration of mercury in the holding pond meet government standards?" By the time you work out the model, it's 7:00 p.m., so you decide to go home. There will be time to look up the government standards and answer the manager's question tomorrow.

We'll pick this example up at a later time; for now, let's step back and reflect on the process of modeling we've used in the these examples.

Putting It All Together

Two models for the change over time in the amount of chemical A were developed: the discrete model,

$$\frac{\Delta A}{\Delta t} \propto A \tag{6.3}$$

$$A(0) = 1 \tag{6.4}$$

and the continuous model,

$$\frac{da}{dt} \propto a(t) \tag{6.5}$$

$$a(0) = 1. \tag{6.6}$$

*We use the term **differential equation** for any equation that relates a function to one or more of its derivatives.*

By choosing a constant value for Δt — which then determines the constant of proportionality k — we can write our discrete model as a difference equation:

$$A_n = A_{n-1} + change$$
$$= A_{n-1} + k \cdot A_{n-1}$$
$$A_0 = 1$$

By determining a value for the constant of proportionality, κ, in the continuous model, we obtain the *differential equation*

$$\frac{da}{dt} = \kappa \cdot a(t)$$
$$a(0) = 1.$$

The two models are quite similar. This should not be surprising in light of our understanding of the derivative. The left sides of the two Equations (6.3) and (6.5) are related by

$$\lim_{\Delta t \to 0} \frac{\Delta A}{\Delta t} = \frac{da}{dt}.$$

The differential equation model is sometimes described as the limit (as Δt tends zero) of the family of difference equation discrete models. Consequently, for small values of Δt, we expect the corresponding difference equation and differential equation to give similar information.

In our second example, the information we were given to work with was phrased in terms of rates, so it was natural for us to go directly to a description involving a derivative.

$$\frac{dp}{dt}(t) = 0.3 - 0.0003p(t) \qquad \textbf{(6.7)}$$

$$p(0) = 0. \qquad \textbf{(6.8)}$$

Summary

The art of developing a mathematical model involves formulation, testing, reformulation, and retesting. We approach the building of models by analyzing change. The first questions we ask in constructing a model are: "What is changing?" and "How fast is it changing?" We use two fundamentally different but related ways to model rate of change. The first uses the *average rate of change* and gives rise to difference equations tracking change discretely at fixed intervals of time. The second uses *instantaneous rate of change* and gives rise to differential equations tracking change continuously in time. Since instantaneous rate of change, the derivative, is the limit (as Δt goes to zero) of the average rates of change, the relation between these two approaches is deep.

We often start the modeling process by examining data, which is necessarily collected at discrete moments in time. Consequently, a discrete model is frequently, but not always, the first to be developed. Having constructed a promising discrete model for a continuous quantity, we may then use analogy to build a continuous model. The test of any model is how far it advances our understanding of the situation being modeled. While elegance and simplicity are useful guides, models fall or stand by the insights and predictions they supply. If modelers can create a mathematical model, then they can bring the power of axiomatic mathematics to bear while often creating interesting mathematical questions. Both mathematics and science have long been enriched by the interplay of ideas that go into mathematical modeling.

Thomas Huxley, the eminent 19th century biologist bemoaned, "The great tragedy of Science: the slaying of a beautiful hypothesis by an ugly fact."

EXERCISE SET 6.1

In Exercises 1–5, the given difference equations model behavior over small time intervals. Write the analogous differential equations or initial value problem.

1. $\Delta P_n = 0.3(50 - P_n)$ for $n = 0, 1, 2, \ldots$

2. $\Delta V_n = -32$ for $n = 0, 1, 2, \ldots$

3. $\left.\begin{array}{rcl} \Delta V_n &=& -32 - 0.5 V_n \\ V_0 &=& 32 \end{array}\right\}$ for $n = 0, 1, 2, \ldots$

4. $\left.\begin{array}{rcl} X_{n+1} &=& X_n + 0.05 X_n \\ X_0 &=& 1000 \end{array}\right\}$ for $n = 0, 1, 2, \ldots$

5. $\left.\begin{array}{rcl} X_{n+1} &=& X_n + 0.05 X_n + 500 \\ X_0 &=& 1000 \end{array}\right\}$ for $n = 0, 1, 2, \ldots$

6. $\left.\begin{array}{rcl} Y_{n+1} &=& Y_n^2 + Y_n - 3 \\ Y_0 &=& 5 \end{array}\right\}$ for $n = 0, 1, 2, \ldots$

7. Determine whether an exponential growth or decay model is appropriate for chemical B by examining the data in Table 6.1.

8. Determine whether an exponential growth or decay model is appropriate for chemical C by examining the data in Table 6.1.

Write the associated difference equation for the following differential equations or initial value problems.

9. $\dfrac{dx}{dt} = 16000 - t^2$.

10. $\dfrac{dp}{dt} = 0.1p$ and $p(0) = 100$.

11. $\left.\begin{array}{rcl} \dfrac{dy}{dx} &=& y^2 + x \cdot y + 2 \\ y(0) &=& 1 \end{array}\right\}$ for $y \geq 0$.

12. $\left.\begin{array}{rcl} \dfrac{dp}{dt} &=& 0.1p(10000 - p) \\ p(0) &=& 1000 \end{array}\right\}$ for $y \geq 0$.

13. A soft drink can is taken from a refrigerator. The drink is at $40°$F and is left on a table by a swimming pool. It's $90°$F outside. Make a model for the temperature of the soft drink.

14. The average student breathes 20 times per minute, exhaling 100 in^3 of air containing 4% CO_2 (carbon dioxide). Model the amount of CO_2 in the air of a closed classroom that has dimensions 30 ft. $\times$ 35 ft. $\times$ 8 ft. The classroom originally has fresh air containing 1% CO_2 as 30 students enter. Make a model for the amount of CO_2 in the air.

15. Identify two types of errors in the modeling process. Explain the sources of these errors and describe their impact.

Explain, in your own words, the meaning of each of the following terms or phrases. Write your explanations in clear and complete sentences. Include annotated diagrams whenever appropriate.

16. mathematical model

17. validation of a model

18. model simplification

19. model refinement

†20. In the first example we arrived at the differential equation model $\left\{\begin{array}{rcl} \dfrac{da}{dx} &\propto& a \\ a(0) &=& 1 \end{array}\right\}$. Use the fact that for the class of exponential functions, $f(x) = b^x$, $f(x)$ is proportional to $f'(x)$ to find a continuous function that models the decay of Chemical A. Which of the values Marie proposed for k gives the most reasonable fit?

21. Develop a difference equation model for the amount of chemical C created in the reaction described in Table 6.1. Estimate numerical values for any parameters in your model and test your model against the data.

6.2

Warming and Cooling

In this section, we will examine the relation between discrete and continuous models. Additionally, we will see how the mathematics of one model can lead to a refinement of that model or even an altogether new approach.

Warming Beer

A mathematician measured the temperature of beer as it warmed over an evening spent contemplating ways to improve the performance of calculus students. The measurements are recorded in Table 6.2. The original time data was recorded in hours, minutes, and seconds, but we also provide the decimal forms. The mathematician noted that the temperature of the kitchen was 20.5°C. Our objective is to develop one or more models that give us insight into the processes that govern warming and cooling. We begin as usual by entering the

TABLE 6.2 Time and Temperature

Time(h:m:s)	6:45:45	6:48:45	6:52	6:55	6:57:30	7:00:15	7:05:30
Time (decimal)	6.7625	6.8125	6.8667	6.9167	6.9583	7.0042	7.0917
Temperature	7.70	8.60	9.44	10.0	10.5	11.0	11.6
Time (h:m:s)	7:10	7:15:25	7:21	7:29	7:35	7:42	7:50
Time (decimal)	7.1667	7.2569	7.3500	7.4833	7.5833	7.7000	7.8333
Temperature	12.3	13.1	13.8	14.6	15.1	15.7	16.3
Time (h:m:s)	8:00	8:12	8:30	8:50	9:09	9:30	10:00
Time (decimal)	8.0000	8.2000	8.5000	8.8333	9.1500	9.5000	10.0000
Temperature	16.9	17.3	18.1	18.6	19.1	19.6	20

FIGURE 6.7 Temperature Plotted Against Time

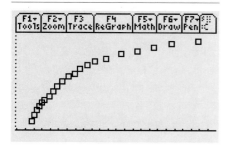

In the cell containing c_3 in the third column, enter the formula $c_3=c_1-c_1[1]$. The c_3 column now begins with the number 0. Each successive entry in the c_3 column is the time elapsed from 6:45:45 or 6.7625.

The temperature of the room is called the "ambient" temperature.

data in a Data table called beertemp and looking at a plot of temperature versus time. Enter the time in decimal form in c_1 and temperature data in column c_2. From the Data/Matrix editor, use the Plot Setup Dialogue box to Define a scatter plot with c_1 as x and c_2 as y. Graph the data using ZoomData as shown in Figure 6.7.

▶ FIRST REFLECTION

How would you transform 6:45:45 into decimal form? ◀

Having seen the shape of the data we wish to model, we decide to simplify our task by converting actual (clock) time into *elapsed time* (from the first measurement). To obtain elapsed time, we subtract 6.7625 from each entry in c_1 and place the results in column c_3. (See margin note.)

▶ SECOND REFLECTION

By transforming clock time into elapsed time, we are discarding some information. Are we making any assumptions about the way beer warms by discarding this information? If so, what are they? Are these assumptions warranted? ◀

The graph of the temperature versus elapsed time is given in Figure 6.8. Looking at the graph, we make the the following observations. First, the temperature is increasing with time. That makes sense, beer warms in a warm room. Second, the data clearly shows a downward concavity. The rate of increase is itself decreasing. And third, not only is the graph concave down, but it appears to be leveling off as time passes. From our experience we expect the beer to warm up *to* but *not beyond* the temperature of the room. What determines the

FIGURE 6.8 Temperature Plotted Against
Elapsed Time

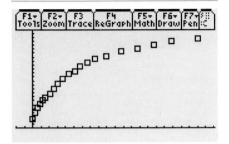

*We will examine the effects of choosing
different numerical values for Δt later.*

*The virtue of simplicity has long been
recognized and is known variously as
"Occam's Razor," "The Law of
Economy," or "The Law of Parsimony."*

rate at which the beer is warming cannot be the temperature of the beer itself,
but rather the difference between temperature of the beer and the room.

▶ **THIRD REFLECTION**
How would the temperature of a can of beer change in a sauna? If the can were
put into a freezer?   ◀

A Family of Discrete Models.   We have identified the quantities we will use in
our model: elapsed time and the difference between the beer's temperature the
ambient temperature. Now we are in a position to construct a variety of models.
Our starting point is:

$$future = present + change. \tag{6.9}$$

Our first objective will be to use a model to construct a sequence of values
for the beer's temperature measured at constant time intervals and compare
our data with the original temperatures. We denote the length of the constant
time interval by Δt. The elements of the sequence, are $u_0, u_1, u_2, \ldots$, where
u_n represents temperature when the elapsed time is equal to $n \cdot \Delta t$. Relate the
sequence to Equation (6.9) by identifying $future = u_n$ and $present = u_{n-1}$.

Our earlier analysis that the change in temperature is governed by the
difference between the current temperature and the ambient temperature tells
us that *change* will be a function of the quantity $\left(20.5 - u_{n-1}\right)$. The sign
of *change* is positive as long as u_{n-1} is less than 20.5. The size of *change*
grows smaller as u_{n-1} grows closer to 20.5. There are literally infinitely many
mathematical relations fitting this bill that we could use to construct our model.
However, simplicity is a virtue, so we take the simplest. We assume that the
change in temperature is directly proportional to $\left(20.5 - u_{n-1}\right)$.

$$change \propto \left(20.5 - u_{n-1}\right)$$

We take $u_0 = 7.7$ as our initial condition (why?). We have a family of difference
equations as our model.

$$u_n = u_{n-1} + k\left(20.5 - u_{n-1}\right) \tag{6.10}$$
$$u_0 = 7.7$$

Equation (6.10) describes a family rather than a single model since we have
not yet specified a value for Δt. Each different choice of time interval yields a
model with a different numerical value for the constant of proportionality k.

▶ **FOURTH REFLECTION**
How would you expect the numerical values of k for a Δt of ten minutes and
a Δt of thirty seconds to compare?   ◀

To see if we have produced a viable model, we need to pick a value for
Δt and then estimate the corresponding value of k. We choose a six minute
time interval, $\Delta t = 0.1$ hr. Examining our table of data, we find there is one

time interval of six minutes. We can use the data from this interval to estimate a value for k. From

$$15.1 = 14.6 + k \cdot (20.5 - 14.6)$$

we find $k \approx 0.085$ as our estimate of the constant of proportionality. The difference equation corresponding to a six minute time interval is

$$u_n = u_{n-1} + 0.085 \left(20.5 - u_{n-1} \right)$$
$$u_0 = 7.7$$

To compare the results of our difference model with the original data graphically, we will plot the points $\left(n \cdot \Delta t, u_n \right)$ and the data together. To see the full three plus hours, we set $nmin = 0$ and $nmax = 33$ in the Window menu. The resulting image is shown in Figure 6.10. Apparently we are on to something with this model!

▶ **FIFTH REFLECTION**

Can you improve the fit by adjusting the numerical value for k? ◀

Set your calculator to sequence *mode. Enter* u1(n) = 0.1 n *and* ui1=0 *to generate the sequence of times, and* u2(n)=u2(n-1)+0.085(20.5-u2(n-1)) *and* ui2 = 7.7 *to generate the sequence of temperatures. Go to the* Axes *menu and set up* CUSTOM *axes, as shown in Figure 6.9.*

FIGURE 6.9   The Custom Axes Menu

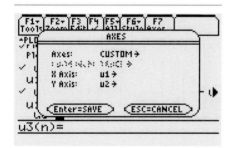

FIGURE 6.10   A Difference Equation Model

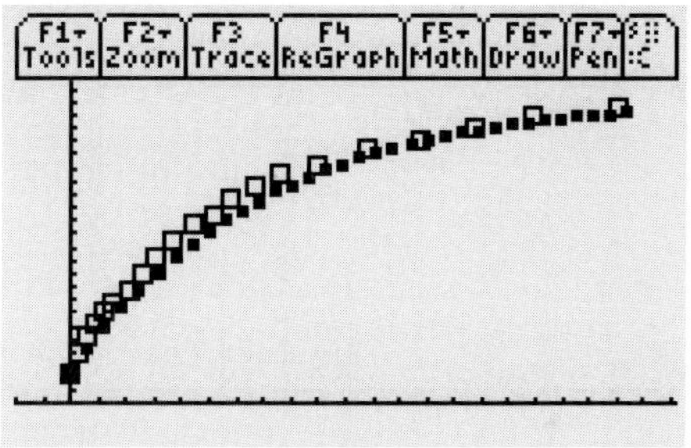

Think and Share

Form into groups of three. In each team, establish difference equation models for time intervals Δt of twelve minutes, three minutes, and one minute. Produce a graph for each of these models. Report the values of k that give the best fit for each model to the class as a whole.

Making New Models From Old

Our discrete approach to modeling the beer warming data has proven successful. But intuitively, temperature is a continuous quantity, so we may wish to establish a continuous model. Recall that our model assumed that *for a fixed time interval* Δt, the change in temperature per time interval was proportional to the difference between the ambient temperature and the temperature of the beer at the start of the interval. Let $f(t)$ represent the temperature of the warming beer at *elapsed time t*. Since the rate of change of temperature from time t to time $t + \Delta t$ is

$$\frac{f(t + \Delta t) - f(t)}{\Delta t},$$

our observation above can be written as

$$\frac{f(t + \Delta t) - f(t)}{\Delta t} \propto (20.5 - f(t)).$$

A big — and occasionally unwarranted — leap of faith.

Now comes a leap of faith. We assume that f will not only be continuous, but differentiable, and the proportionality relationship between $(f(t + \Delta t) - f(t))/\Delta t$ and $(20.5 - f(t))$ survives passing to the limit as Δt goes to zero. We propose the differential equation model

$$\frac{df}{dt} = \kappa \cdot (20.5 - f(t)) \qquad \textbf{(6.11)}$$

$$f(0) = 7.7. \qquad \textbf{(6.12)}$$

We can use algebra and calculus to manipulate and understand the model of Equation (6.11). Both our discrete and continuous models were developed by recognizing the key role of the quantity $(20.5 - f(t))$. We can highlight the role of this quantity in Equation (6.11) by setting

$$g(t) = (20.5 - f(t)).$$

Substituting the formula for g into Equation (6.11) yields

$$\frac{dg}{dt}(t) = -\kappa \cdot g(t)) \qquad \textbf{(6.13)}$$

$$g(0) = 12.8 \qquad \textbf{(6.14)}$$

▶ **Sixth Reflection**
Carry out the steps to derive Equations (6.13) from Equations (6.11).   ◀

We can make use of our knowledge of derivatives. Equation (6.13) tells us that the derivative of g is directly proportional to g itself. Recalling that the differentiation rule for exponential functions is

$$\frac{d}{dt}b^t = \ln(b) \cdot b^t,$$

we can identify $g(t)$ as an exponential function of the form

$$g(t) = 12.8 \cdot b^t. \qquad \textbf{(6.15)}$$

▶ **Seventh Reflection**
Where did the 12.8 in Equation (6.15) come from?   ◀

Using one of our data points, in this case $t = 1.2375$ and $f(t) = 16.9$, in Equation (6.15), we arrive at the explicit formulas $g(t) = 1.28 \,(0.3878)^t$ and

$f(t) = 20.5 - 12.8\,(0.3878)^t$. Figure 6.11 shows the function f graphed with the transformed data.

▶ **EIGHTH REFLECTION**

Carry out the algebra necessary to derive the formulas for g and f. ◀

Think and Share

Form into groups of three. Have each person use a different data point to estimate the parameter b in the formula for g. How great is the variation in your values? Report your results to the class as a whole. What are the largest and smallest values reported? List one or more ways you might use to determine what value of b is "best."

FIGURE 6.11 A Function Model

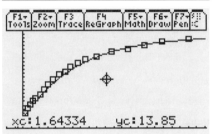

xc:1.64334   yc:13.85

Summary

We have seen how a difference equation model can give rise to a related differential equation model. We accomplished this by interpreting the change from one sequence element to the next, $u_n - u_{n-1}$, as a rate of change. Whenever we have a family of difference equations

$$u_n = u_{n-1} + F(u_{n-1}, k)$$

that successfully model a continuous quantity $f(t)$ over fixed time intervals Δt, by interperting the difference $u_n - u_{n-1}$ as a rate of change, $(f(t + \Delta t) - f(t))/\Delta t$, we can produce a continuous model. Since

$$\frac{df}{dt} = \lim_{\Delta t \to 0} \frac{f(t + \Delta t) - f(t)}{\Delta t},$$

we can reasonably hope that the differential equation

$$\frac{df}{dt} = F(f(t), \kappa)$$

will supply a viable continuous model. The test of any model is whether or not it adds to our understanding of the real world situation we are modeling. Differential equation models have been particularly attractive since differential equations are amenable to algebraic manipulation. As we have seen, algebraic manipulation can lead to further insights and new classes of models.

The differential equation we established for the temperature of an object over time was

$$\frac{dT}{dt} = \kappa \cdot (A - T)$$

where T is the temperature of the object at time t, A is the ambient temperature and κ is a (positive) constant or proportionality. This model is known as **Newton's Law of Cooling**. Newton's Law of Cooling has proven to be a very useful model of reality.

We have also seen that the relations we need to construct a model are not always apparent from the original data collected. We may need to transform the data in order to uncover useful information and insight.

FIGURE 6.12 Newton: 1642–1727

EXERCISE SET 6.2

1. Build a differnce equation model for the untransformed (clock) data from our example.

2. Translate the function f to obtain a model of the original untransformed (clock) data.

† **3.** In the data file `beertemp` define the fourth column to be the difference between the ambient temperature and the temperature of the warming beer with `c4= 20.5-c2`. Use the `ExpReg` command to find a different expression for $g(t)$. How does this function compare with the formula obtained in the text?

4. Use the model of Equation (6.10) to create a table of the temperature of the beer in the example every fifteen minutes from 6:30 until 10:00. Describe the ways in which this data is more useful than the data actually collected.

5. Suppose the temperature of the room in the example had been 22°C. Find a function that models the data under this condition.

6. Suppose the thermometer that was used to measure the temperature of the beer was off by 1°C but that the temperature of the room was correctly measured. Find a function that models the data under this condition.

7. Sketch a graph of the cooling of beer from room temperature to the temperature of ice. Explain why your graph accurately reflects how beer (or another liquid) cools.

8. Determine a model for the cooling of coffee at 130° F to a room temperature of 72° F.

9. Determine a model for the cooling of tea at 131° F to a room temperature of 54° F.

10. Use your knowledge of derivatives to find a function $f(x)$ that satisfies the equation $\dfrac{df}{dx} = x^2 + x$.

11. Use your knowledge of derivatives to find a function $f(x)$ that satisfies the equation $\dfrac{df}{dx} = 2e^{2x}$.

12. Use your knowledge of derivatives to find a function $f(x)$ that satisfies the equation $\dfrac{df}{dx} = 2f(x)$.

13. Use your knowledge of derivatives to find a function $f(x)$ that satisfies the equation $\dfrac{d^2 f}{dx^2} = \cos(x) + \sin(x)$.

14. Use your knowledge of derivatives to find a function $f(x)$ that satisfies the equation $\dfrac{df}{dx} = -f(x)$.

15. What is the significance of the phrase *Non sunt multiplicanda entia praeter necessitatem*?

6.3

Population Modeling

In the previous sections, we used difference equations and differential equations to create models. We build such models whenever studying a quantity which changes over time. As a particular example, biologists often study the growth and decline of various populations. What light can mathematical models shed on their investigations?

We first saw discrete logistic models used to describe population growth in Section 1.3 (and Section 2.3). In contrast to Chapter 1, our focus for building the model is now on rate of change. Before, we experimented by choosing values for the parameters and checking the fit to Carlson's yeast data in order to validate the model. Now, we will accept the model and concentrate on developing techniques to estimate the model's parameters from collected data.

TABLE 6.3 Yeast Culture Growth Data

Time t	Size of Biomass P_t
0	9.6
1	18.3
2	29.0
3	47.2
4	71.1
5	119.1
6	174.6
7	257.3
8	350.7
9	441.0
10	513.3
11	559.7
12	594.8
13	629.4
14	640.8
15	651.1
16	655.9
17	659.6
18	661.8

FIGURE 6.13 The Growth of a Yeast Culture

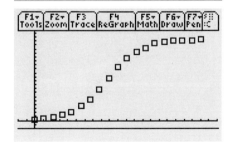

In Chapter 1, we tacitly ignored the size of Δt by assuming it was equal to $\Delta t = 1$.

Pearl's Yeast Experiment

An experiment[†] measuring the growth of yeast in a culture generated the data in Table 6.3. The population size is measured in units of biomass and time in hours. We wish to build a model to predict the size of the biomass t hours after beginning the experiment with an initial biomass of 9.6.

When considering data generated by observation, we cannot always expect patterns in the numbers to be immediately apparent. Errors in observations and the effects of independent factors may obscure important trends. A well-designed experiment limits the effects of such unwanted influences, but the representation of the data may still prevent us from recognizing underlying patterns. Numerical tables are particularly good at concealing patterns since humans are not well suited to making sense out of long strings of numbers. Graphs, on the other hand, are an excellent tool for recognizing trends (so good, they sometimes indicate the existence of trends that aren't there). Enter the data in the Data/Matrix editor as `yeast` and plot it. Figure 6.13 shows a plot of the yeast culture data.

The Discrete and Continuous Logistic Models

In Chapter 1, modeling very similar data from the Carlson experiment, we developed the family of difference equations

$$u_n = u_{n-1} + r \cdot u_{n-1} - p \cdot u_{n-1}^2 \qquad (6.16)$$
$$u_0 = B.$$

where u_n represents the biomass of the colony at time $n \cdot \Delta t$, and p and r were parameters such that $0 < p < r$ (where p is very much less than r). The numerical values of p and r depend on the choice of time interval Δt. This family of difference equations is known as the *Discrete Logistic Model* for limited growth. Subtracting u_{n-1} from both sides and replacing $u_n - u_{n-1}$ by Δu_n, we see that the average change in biomass per time interval Δt, is given by

$$\Delta u_n = r_{\Delta t} \cdot u_{n-1} - p_{\Delta t} \cdot u_{n-1}^2. \qquad (6.17)$$

▶ **FIRST REFLECTION**

Why is it appropriate to use the subscript Δt in $r_{\Delta t}$ and $p_{\Delta t}$ in Equation (6.17). ◀

As we saw in the previous sections, a difference equation can be used to build a differential equation model. In this case, we let the function $f(t)$ represent the mass of the yeast colony at time t. Interpreting the right side of Equation (6.17) as the average rate of change of the mass per time interval Δt

---

[†]This data is from R. Pearl, 1927. The growth of a population. *Quart. Rev. Biol.* 2:532-548. This seminal article popularized logistic models and was the first in a 20 year series of papers culminating in Pearl and Reed's 1940 logistic fit of the U. S. population.

of the yeast colony, we conjecture that the instantaneous rate of change will have the same relation to the mass of the colony. This reasoning leads to the differential equation

$$\frac{df}{dt}(t) = r \cdot f(t) - p \cdot (f(t))^2 \qquad \textbf{(6.18)}$$
$$f(0) = B$$

as a continuous model for the growth of the colony. This equation is known as the *Continuous Logistic Model*.

Estimating the Parameters I

Relative average rate of change, $\Delta u_n / u_{n-1}$, in the discrete model and relative instantaneous rate of change $f'(t)/f(t)$ in the continuous model.

Both of the logistic models make an interesting prediction. They imply that the relative rate of change of the biomass of the colony will be a linear function of the mass of the colony itself! We see this relation after dividing both sides of the Equation (6.17) by u_{n-1} and Equation (6.18) by $f(t)$.

$$\frac{\Delta u_n}{u_{n-1}} = r_{\Delta t} - p_{\Delta t} \cdot u_{n-1} \text{ and}$$
$$\frac{df/dt}{f(t)} = r - p \cdot f(t)$$

Recall that fd(q)=shift(q,1)-q.

Plot the relative first differences of our data versus biomass to see how linear the resulting plot looks. Set c3=fd(c2)/c2 and plot c2 versus c3 on the *x*- and *y*-axes respectively as in Figure 6.14. Using the LinReg command

FIGURE 6.14 Mass vs. Relative Differences   **FIGURE 6.15** The Linear Fit

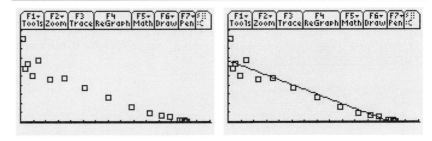

From the Data/Matrix *editor, you can find* LinReg *under* Calculation Type *in the dialog box from the* Calc *menu. Set* c2 *as the x and* c3 *as the y variables.*

to fit a line to the data as in Figure 6.14. As we can see in Figure 6.15, the resulting linear function,

$$0.66935 - 0.001053x,$$

fits the data reasonably well. We accept the coefficients as estimates for the parameters and take $p = 0.001053$ and $r = 0.66935$, respectively.

Think and Share

To see how successfully the linear regression estimated the parameters, define $p = 0.001053$ and $r = 0.66935$ in the HOME screen and use the difference equation

$$u_n = u_{n-1} + r \cdot u_{n-1} - p \cdot u_{n-1}^2$$
$$u_0 = 9.6$$

to complete Table 6.4.

TABLE 6.4 Observed and Predicted Values

Time $t = n$ hours	Observed Biomass	Predicted Biomass u_n
0	9.6	9.6
1	18.3	15.9
2	29.0	
3	47.2	
4	71.1	
5	119.1	
6	174.6	
7	257.3	
8	350.7	
9	441.0	
10	513.3	
11	559.7	
12	594.8	
13	629.4	
14	640.8	
15	651.1	
16	655.9	
17	659.6	
18	661.8	

Graph the predicted data along with the actual data to check our work visually. Your results should look similar to Figure 6.16.

Estimating the Parameters II

In Section 1.3, we noted that Equation (6.16) had two rest points 0 and r/p. These values must be of biological importance. Once a population falls to 0 or grows to r/p, both of the logistic models predict that there will be no

FIGURE 6.16   Observations and Predictions

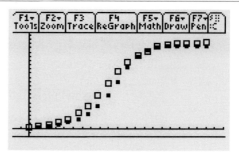

Why is no further change predicted?

further change in the population. The biological meaning of a population of 0 is obvious, but the significance of r/p is not immediately apparent. A quick examination of the algebraic signs in the discrete logistic equation

$$\Delta u_n = p \cdot u_{n-1} \cdot \left(\frac{r}{p} - u_{n-1} \right)$$

leads to the following observation. For positive values of u_{n-1} whenever the population, u_{n-1}, is less than r/p, Δu_n is positive and the population increases toward the rest value r/p. On the other hand, if u_{n-1} is greater than r/p, then Δu_n is negative and the population will decline, again toward r/p.

▶ **SECOND REFLECTION**

What does the Continuous Logistic model predict when $f(t) > r/p$? When $0 < f(t) < r/p$? When $f(t) = r/p$?   ◀

Since Equation (6.16) predicts that a population greater than r/p will decline towards r/p and that a smaller population will rise towards r/p. The value $r/p = M$ is called the *carrying capacity*. In Figure 6.16, the yeast population appears to be approaching a limiting value, the carrying capacity.

These observations provide a different approach to approximating the parameters p and M. We estimate from the graph that the limiting value of the population is (roughly) 665. That is, we take $M = r/p = 665$. (That's an "eyeball" or visual approximation.) Putting this value into Equation (6.17), we find that the model now has only the single parameter p.

$$\Delta u_n = p u_n (665 - u_n), \qquad n = 0, 1, 2 \ldots.$$

Solving the equation for p, we find

$$p = \frac{\Delta u_n}{u_n (665 - u_n)}.$$

To find an approximation for p, we can calculate and examine values for $\Delta u_n / \left(u_n (665 - u_n) \right)$ given by the original data.

Think and Share

Break into groups. Complete the Table 6.5. Use the `OneVar` statistic command from the `Calc` menu to determine a numerical value for p. Use your

Think and Share

value of p in the difference equation

$$u_n = u_{n-1} + p \cdot u_{n-1} \left(665 - u_{n-1}\right)$$
$$u_0 = 9.6$$

to predict the biomass every hour between 0 and 18. Overlay your predictions and observations on the same plot. Do you think your model does a reasonable job of capturing the trend of the data?

TABLE 6.5   Estimating p

Time, t	Size of Biomass	Change in Biomass, Δu_n	$p = \dfrac{\Delta u}{u_n(665 - u_n)}$
0	9.6	8.7	0.00138
1	18.3	10.7	0.0009
2	29.0		
3	47.2		
4	71.1		
5	119.1		
6	174.6		
7	257.3		
8	350.7		
9	441.0		
10	513.3		
11	559.7		
12	594.8		
13	629.4		
14	640.8		
15	651.1		
16	655.9		
17	659.6		
18	661.8		

Summary

Equation (6.20) was first introduced by the Belgian mathematical biologist Pierre-Francois Verhulst (1804-1849). We will return to Verhulst's equation again in Section 7.1 where we symbolically determine a function $f(t)$ which satisfies the continuous model. The model of Equation (6.19) gives rise to a wide variety of different behaviors for different values of M and p. This model has been extensively studied and has many applications beyond the field of biology. You will be asked to explore several of these alternative applications in the exercises.

Discrete Logistic Model:

$$\Delta u = p(M - u_{n-1})u_{n-1}, \quad t = 0, 1, 2, \dots \tag{6.19}$$
$$u_0 = B$$

Continuous Logistic Model:

$$\frac{df}{dt}(t) = r \cdot (M - f(t)) \cdot f(t), \qquad t \geq t_0 \tag{6.20}$$
$$f(t_0) = B$$

EXERCISE SET 6.3

1. The data in Table 6.6 were obtained for the growth of a sheep population, P_t, introduced into a new environment on the island of Tasmania.
   a. Make an estimate of the maximum possible population size M by graphing P_t.
   b. Use Equation (6.19) to model the growth.
2. Use Equation (6.19) to model the U.S. Housing Density data in Table 6.7. Estimate M and r. Assume you are making the prediction in 1990 using previous census. Estimate the housing density in 2000 using your model.
3. Determine whether a logistic model is appropriate for the U.S. Land Area data in Table 6.8.
4. Use Equation (6.19) to model the U.S. population for 1790 to 1940 using the data in Table 6.9. Estimate M and r. Assume you are making the prediction in 1941 using previous census. Estimate the population in 1941 using your model. How well did you do? What factors other than those considered in the logistic model determine the U.S.

TABLE 6.6   Sheep Population in Tasmania[*]

Year	1814	1824	1834	1844	1854	1864
Sheep	125	275	830	1200	1750	1650

[*] Source: Davidson, J. 1938. On the growth of the sheep population in Tasmania. *Trans. Roy. Soc. S. Australia* 62:342-346.

TABLE 6.7   U.S. Housing Density (in units/km²), 1940 to 1990[*]

Year	1940	1950	1960	1970	1980	1990	2000
Count	4.1	5.0	6.4	7.5	9.6	11.2	???

[*] Source: U.S. Bureau of the Census.

TABLE 6.8   U.S. Land Area (in miles²/1000's), 1790 to 1990.[*]

Year	1790	1800	1810	1820	1830	1840
Count	865	865	1682	1749	1749	1749
Year	1850	1860	1870	1880	1890	1900
Count	2940	2970	3541	3541	3541	3547
Year	1910	1920	1930	1940	1950	1960
Count	3547	3547	3557	3557	3552	3541
Year	1970	1980	1990	2000
Count	3537	3539	3536	???

[*] Source: U.S. Bureau of the Census.

population? Compare your result to the fit found in the 1940 paper of Pearl, Reed, and Kish, "The logistic curve and the census count of 1940." *Science* 92, 486–488. Also compare your result with the fit given in 1997 by Bauldry, "Fitting logistics to the U. S. population," *MapleTech* **4** No 3, 73–77.

TABLE 6.9   U.S. Population (in 1000's), 1790 to 1940[*]

Year	1790	1800	1810	1820	1830	1840
Count	3929	5308	7240	9638	12866	17069
Year	1850	1860	1870	1880	1890	1900
Count	23192	31442	38558	50189	62980	76212
Year	1910	1920	1930	1940
Count	92228	106022	123203	132165

[*] Source: U.S. Bureau of the Census.

5. Adjust your model from Exercise 4 to include the U.S. population data in Table 6.10. Estimate the new M and r. Assume you are making the prediction in 1991 using previous census. Estimate the population in 2000 using your model. How confident are you in your prediction? Return to class in September of 2001 to check your estimate.

TABLE 6.10   U.S. Population (in 1000's), 1950 to 1990*

Year	1950	1960	1970	1980	1990	2000
Count	151326	179323	203302	226542	248710	???

* Source: U.S. Bureau of the Census.

6. Consider the spread of a highly communicable disease on an isolated island with population size N. A portion of the population travels abroad and returns to the island infected with the disease. You would like to predict the number of people x who will have been infected as of time t. Consider the following model:

$$\frac{dx}{dt} = kx(N - x)$$

a. List two major assumptions implicit in the preceding model. How reasonable are the assumptions you list?

b. Graph dx/dt versus x. At what population level is the disease spreading the fastest?

7. Assume we are considering the survival of whales; if the number of whales falls below a minimum survival level T, called the *threshold*, the species will become extinct. Also assume the population is limited by the carrying capacity M of the environment. That is, if the whale population is above M, then it will experience a decline because the environment cannot sustain that large a population level.

a. Discuss the following model for the whale population,

$$\frac{dw}{dt} = k(M - w)(w - T)$$

where $w(t)$ denotes the whale population at time t and k is a positive constant.

b. Graph dw/dt versus w and graph w versus t.

c. Assuming the model reasonably estimates the whale population, what implications are suggested for fish populations and the fishing industry? What controls would you suggest?

8. Sociologists recognize a phenomenon called "social diffusion," which is the spreading of a piece of information, a technological innovation, or a cultural fad among a population. The members of the population can be divided into two classes: those who have the information, and those who do not. In a fixed population whose size is known, it is reasonable to assume the rate of diffusion is proportional to the number who have the information times the number yet to receive it. Let $x(t)$ denotes the number of individuals who have the information at time t in a population of N people. Then a mathematical model for social diffusion is given by $dx/dt = kx(N - x)$, where t represents time and k is a positive constant.

a. At what time is the information spreading fastest?

b. How many people will eventually receive the information?

9. Name 10 factors that affect the growth of a human population not considered in either the Malthusian or logistic models. Would you expect attempts to predict future human populations to be very precise?

Explain, in your own words, the meaning of each of the following terms or phrases. Write your explanations in clear and complete sentences. Include annotated diagrams whenever appropriate.

10. discrete logistic model

11. continuous logistic model

12. carrying capacity

13. parameter estimation

6.4

Euler's Method

Differential equations, that is equations that involve the derivative of a function, arise naturally in models of continuous change. As we have seen, it is sometimes possible to identify a function from its derivative. When this can be done, we can replace a differential equation model with a compact "closed form" formula for the function described by the differential equation. Unfortunately, for many important examples, this is not possible, no closed form formula exists. Consequently, we must develop numerical and graphical representations of the model. In this and the following section, we develop such techniques.

Solutions for Differential Equations

One way of "solving" a differential equation with an initial condition is to produce a "solution curve" that satisfies both the differential equation and the given initial condition. A solution can be presented symbolically as a formula, numerically as a table, or a graphically as a curve. The oldest method for generating numerical solutions for a differential equation was developed by the great eighteenth century Swiss mathematician, Leonhard Euler. His is one of a number of similar techniques which fall into the branch of mathematics known as *numerical analysis*. Euler's method is not widely used in modern practice because it requires a very large number of calculations to be effective. Nevertheless, Euler's method is still worth studying for the geometric intuition it provides and since more sophisticated modern approaches are based on similar ideas.

Euler's method is simple: follow the tangent line. Euler reasoned that a function that satisfies a differential equation must be locally linear. Thus, a short step along its tangent line takes one to very nearly the same point as would a short step along the function itself. By repeatedly stepping along tangents, Euler's Method creates a sequence of discrete values closely approximating the function.

*A differential equation together with an initial condition is called an **Initial Value Problem**.*

FIGURE 6.17　Euler's Tangent Line Method

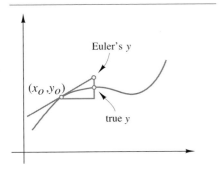

Mercury Pollution Revisited

In Section 6.1, we developed a model for the amount of mercury in a treatment pond.

$$\frac{dp}{dt} = 0.3 - 0.0003\,p(t) \qquad\qquad (6.21)$$
$$p(0) = 7000.$$

Let's use Euler's method to estimate how the quantity of mercury in the pond evolves. In order to calculate the amount of mercury twice a day, set $\Delta t = 12$. Let y_n represent our approximations of $p(t_n)$. First, find the slope-intercept equation $y = y_1 + m \cdot (t - t_1)$ for the line tangent to $p(t)$ at the point $(t_1, y_1) = (0, 7000)$. The differential equation gives the slope of this tangent line as $m = 0.3 - 0.0003 \cdot 7000 = -1.8$. Step to the second point, (t_2, y_2), using the tangent's equation. We find that $(t_2, y_2) = (12, 6978.4)$, as follows.

$$t_2 = t_1 + \Delta t$$
$$= 0 + 12$$
$$= 12$$

$$y_2 = y_1 + m(y_1) \cdot \Delta t$$
$$= 7000 + (-1.8) \cdot 12$$
$$= 6978.4.$$

As we continue the procedure, keep in mind that the y_i's approximate the total mercury $p(t_i)$ in the pool at time t_i and are produced by Euler's method. To estimate the level of mercury at 24 hours, start from the point $(12, 6978.4)$ and repeat the process above. Begin by computing the slope of the tangent line at $(12, 6978.4)$ as $m = 0.3 - 0.0003 \cdot 6978.4 = -1.79352$. Then we find $(t_3, y_3) = (24, 6956.88)$ from

$$t_3 = t_2 + \Delta t$$
$$= 12 + 12$$
$$= 24$$
$$y_3 = y_2 + m(y_2) \cdot \Delta t$$
$$= 6978.4 + (-1.79352) \cdot 12$$
$$= 6956.88.$$

At the nth stage, we compute the slope of the tangent line using the expression given by the differential equation evaluated at (t_{n-1}, y_{n-1}). To simplify the notation, `Define` the function m by `m(q)=0.3-0.0003q`. Let's use this slope function in computing the Euler estimate for the amount of mercury at 36 hours.

$$t_4 = t_3 + \Delta t$$
$$= 36$$
$$y_4 = y_3 + m(y_3) \cdot \Delta t$$
$$= 6935.43$$

▶ **FIRST REFLECTION**
Calculate (t_5, y_5).   ◀

Observe that in each case we generated the next point (t_n, y_n) from the current point (t_{n-1}, y_{n-1}) by using the difference equations

$$t_n = t_{n-1} + \Delta t$$
$$y_n = y_{n-1} + m(y_{n-1}) \cdot \Delta t.$$

The expression $m(y_{n-1}) \cdot \Delta t$ is the change in y along the tangent line between two data points. Our computations follow the familiar pattern

$$future = present + change.$$

for both t and y. We are using the change along the tangent line to estimate the actual change in p.

The first four points don't show much. On the other hand, if we generate the first *seven hundred and thirty* points, we can see how the amount of mercury behaves over the first year of operation of Mountain View Mercury Processing. Performing the calculations required to compute so many points would be extremely tedious.

Let's use our calculators to implement Euler's method. To avoid clashing with system variables and to shorten typing, we'll use h for Δt. In the HOME screen enter

$$\texttt{Define h=12}$$
$$\texttt{Define m(q)=0.3-0.0003q}$$

FIGURE 6.18   Euler's Method Data Table

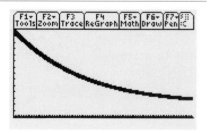

n	u1	u2
1.	12.	6978.4
2.	24.	6956.88
3.	36.	6935.43
4.	48.	6914.07
5.	60.	6892.77

n=1.

In Figure 6.19, n *runs from 1 to 730. The plot window is* [0, 730] × [0, 7000].

In the the MODE dialogue box set $\texttt{Graph...}$ to $\texttt{SEQUENCE}$ and enter t_n as $\texttt{u1}$ and y_n as $\texttt{u2}$ in the Y= editor.

$$\texttt{u1 = u1(n-1) + h}$$
$$\texttt{ui1 = 0}$$
$$\texttt{u2 = u2(n-1) + h} * \texttt{m(u2(n-1))}$$
$$\texttt{ui2 = 7000}$$

Set both $\texttt{tblStart}$ and Δ $\texttt{tbl}$ to 1 in the $\texttt{TblSet}$ dialog box. Look at the table generated for the first several values of t and y. Since it's much easier to see trends visually than from data tables, we turn to a graph. Set the $\texttt{Axes}$ in the Y= editor to $\texttt{Custom}$ with $\texttt{u1}$ on the x-axis and $\texttt{u2}$ on the y-axis to obtain Figure 6.19.

▶ **SECOND REFLECTION**

Determine the minimum value for the level of mercury in the pool using the rest points of Equation (6.21).   ◀

Notice that both approximations for p as well as the table and the plot were produced without obtaining a closed form formula. The power of Euler's method (and of more advanced numerical methods) lies in the ability to derive information directly from a differential equation without the need for explicit formulas for the function. In calculus terms this is, "If we know the rate of change, then we know the function." This method can be used on any differential equation that can be put in the form $dy/dt = f(t, y)$.

FIGURE 6.19   Mercury Pollution

E X A M P L E   1   **Purification Tank**

A purification tank contains 100 gallons of water. We pump brine containing two pounds of salt per gallon into the tank at a rate of two gallons per minute. A mixed solution is drawn from the tank at the rate of three gallons per minute. Since the rate flowing out is larger than the rate flowing in, the volume $V(t)$ of brine in the tank is $V(t) = 100 + 2t - 3t = 100 - t$ gallons after t minutes. Model the concentration of salt in the tank.

The rate at which salt enters the tank equals the rate at which the brine enters the tank multiplied by the concentration of salt in the brine. Letting r_{in}

FIGURE 6.20 Brine Tank

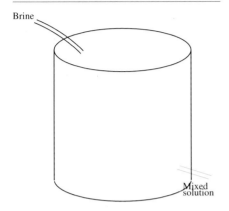

FIGURE 6.20 Brine Tank

Brine

Mixed solution

Think and Share

represent the rate at which salt enters the tank we obtain

$$r_{in} = 2\,\frac{lb}{gal} \times 2\,\frac{gal}{min} = 4\,\frac{lb}{min}.$$

We let $x(t)$ represent the amount of salt in the tank at time t minutes after the pumping begins. Thus, the concentration of salt in the tank equals x lb/$(100 - t)$ gal. The concentration of salt in the outflow is the same as that of the tank. Letting r_{out} represent the rate at which salt leaves the tank, we have

$$r_{out} = \frac{x}{100 - t}\,\frac{lb}{gal} \times 3\,\frac{gal}{min} = \frac{3x}{(100 - t)}\,\frac{lb}{min}.$$

Break into groups. The rate of change of the amount of salt in the tank is equal to the rate of salt entering minus the rate of salt leaving. Write the differential equation for the instantaneous rate of change of the amount of salt:

$$\frac{dx}{dt} = r_{in} - r_{out} = ?$$

The expression on the right side of your equation defines the slope function m and should depend on both t and x. Use the differential equation to define m(t,x)=.... Also define h=0.5. Since the slope function m now has two arguments, we need to change the definition of u2.

```
u1 = u1(n-1) + h

ui1 = 0

u2 = u2(n-1) + h * m(u1(n-1),u2(n-1))

ui2 = 0
```

According to your model, when does the tank run dry? What is the maximum salt concentration? When is the maximum concentration achieved? Check the table to verify your estimations.

Euler's Method Abstractly

Euler's method approximates the solution curve to an initial value problem by following tangent lines generated by using the differential equation to determine slopes. If we have a differential equation in the form

$$\frac{dy}{dt} = f(t, y)$$
$$f(t_1) = y_1,$$

then the slope of a line tangent to the solution curve at the point (t_1, y_1) is found from

$$m_1 = f(t_1, y_1).$$

The equation of the tangent line is

$$y = y_1 + m_1 \cdot (t - t_1).$$

Taking a step of size Δt from t_1 along the tangent line to get to (t_2, y_2) where

$$t_2 = t_1 + \Delta t$$
$$y_2 = y_1 + m_1 \cdot \Delta t.$$

The third point is generated by following the same procedure.

$$t_3 = t_2 + \Delta t$$
$$y_3 = y_2 + m_2 \cdot \Delta t$$

where $m_2 = f(t_2, y_2)$ is the slope of the tangent line through (t_2, y_2). Writing these relations in the general form gives us Euler's method.

EULER'S METHOD

Euler's method generates the sequence of points (t_k, y_k) that approximates the solution curve for an initial value problem. For the differential equation $dy/dt = f(t, y)$ with $y = y_1$ at $t = t_1$, set

$$t_k = t_{k-1} + \Delta t$$
$$y_k = y_{k-1} + m_{k-1} \Delta t$$

where $m_{k-1} = f(t_{k-1}, y_{k-1})$ for $k = 2, 3, \ldots$.

▶ THIRD REFLECTION

What should we choose the value of Δt to be if we want to use Euler's method to generate a sequence of points starting at $t_1 = 2$ and going to $t_n = 5$ in 45 steps?   ◀

Summary

For locally linear functions,

$$\lim_{\Delta t \to 0} \frac{\Delta y}{\Delta t} = \frac{dy}{dt}.$$

Thus, for small values of Δt, we know that $\Delta y / \Delta t \approx dy/dt$. Multiplying both sides of this approximation by Δt, we obtain the relation that lies at the heart of Euler's Method,

$$\Delta y \approx \frac{dy}{dt} \Delta t.$$

This estimate is quite similar to the approximations we used in Chapters 4 and 5 when we developed the definite integral. We expect that taking more and shorter steps, that is, using a smaller Δt, should result in better estimates. In Chapter 7, we will study the effects of choosing a smaller Δt.

Euler's method and other numeric techniques provide us with extremely powerful tools for approximating the solutions of differential equations for continuous models. However, because they only generate approximations, these methods must be handled with care. Since Euler's method produces only one solution from a given initial value, might it miss other solutions as a result? Also, Euler's method ignores features that occur between estimated points: What if something unusual, like a discontinuity, occurs? These and other important questions will be addressed in Chapter 7.

EXERCISE SET 6.4

Use Euler's method with step size $h = 0.1$ to find the indicated value in Exercises 1–4.

1. Find $y(2)$ where $y(x)$ is determined by

$$y' = 5 - x$$
$$y(0) = 1.$$

2. Find $y(2)$ where $y(x)$ is determined by

$$y' = 5 - y$$
$$y(1) = 0.$$

3. Find $y(3)$ where $y(x)$ is determined by

$$y' = x^2 + 1/x$$
$$y(1) = 0.$$

4. Find $y(2.5)$ where $y(x)$ is determined by

$$y' = \frac{y^2 + 1}{y}$$
$$y(1) = 0.$$

Use Euler's method to produce a graph for solutions of the differential equations given in Exercises 5 and 6.

5. The function $y(x)$ is determined by

$$y' = 3^{-x^2}$$
$$y(0) = 2.$$

6. Graph, for t in $[0, 2\pi]$, the function $x(t)$ determined by

$$x' = \sin(t^2)$$
$$x(0) = 1.$$

† **7.** *Group Exercise* A crucial question for fish farms is "How many fish do we harvest, and how often?" Consider a

government-managed trout farm where the harvested fish will be used to stock lakes throughout the state. Before we can consider harvesting strategies, we need to model the long-term population. We assume the main factors affecting population growth for our model will be the birth rate, death rate, and availability of resources such as food, space, and so forth. These assumptions lead to the logistic differential equation describing the number of trout, P, at time t months,

$$\frac{dP}{dt} = k P (M - P),$$

where k is a positive constant of proportionality and M is the maximum population or carrying capacity. From long-term data collection, the manager of the trout farm has estimated that $k = 3 \cdot 10^{-6}$ and $M = 77,000$. The manager plans to start the population with a batch of $5,000$ trout fry (young). Use Euler's method with a Δt of one month to model the long term trout population without harvesting. What is the population at the beginning of the second year? The third year? The fourth year? What does the model predict for the trout population in the long run? Produce tables and graphs to support your conclusions. Alter the logistic model by subtracting a constant harvesting term k, the number of fish harvested per month:

$$\frac{dp}{dt} = rp (M - p) - k.$$

Apply Euler's method to the modified trout model to generate the sequence form of the differential equation. Is it feasible to harvest:

a. 500 fish per month?

b. 1000 fish per month?

c. 1100 fish per month?

8. A soft drink can is taken from a refrigerator. The drink is at 40°F and is left on a table by a swimming pool. It's 90°F outside.

  a. Make a model for the temperature of the soft drink.

  b. Use Euler's method to produce a graph of the drink's temperature.

  c. Describe the long-term trend you observe.

9. The average student breathes 20 times per minute, exhaling 100 in^3 of air containing 4% CO_2 (carbon dioxide). Model the amount of CO_2 in the air of a closed classroom that has dimensions 30 ft × 35 ft × 8 ft. The classroom originally has fresh air containing 1% CO_2 as 30 students enter.

  a. Make a model for the amount of CO_2 in the air.

  b. Use Euler's method to produce a graph and a table of values.

  c. Describe the long-term trend you observe.

10. Explain why the function $y(t)$ solving $y' = y$ with $y(0) = A$ must be increasing when $A > 0$.

Explain, in your own words, the meaning of each of the following terms or phrases. Write your explanations in clear and complete sentences. Include annotated diagrams whenever appropriate.

11. Euler's method.

12. Continuous logistic model with harvesting.

6.5

Slope Fields

Differential equations are used to model biological, physical, sociological, and chemical phenomena. A mathematical model can capture relationships that transcend any one particular situation. Now we will explore a graphical representation of a differential equation called a *slope field*[†] that can display important features of a model.

In the previous section, we used Euler's method to generate data tables and graphs that approximated a solution for a given model. While powerful and simple, Euler's method has the disadvantage of producing only one solution starting from any given initial point. Any change in the initial value, step size, or parameters requires that we rerun the entire procedure to obtain a new solution. Even if we run Euler's method for several variations, we may easily miss overall trends.

Consider the differential equation

$$\frac{dy}{dx} = f(x, y). \tag{6.22}$$

We could pick *any* point (x_i, y_i) as an initial value and recover a solution $y(x)$ that satisfies Equation (6.22). That solution $y(x)$ will have a slope of $f(x_i, y_i)$ at its initial point (x_i, y_i). We can illustrate this by drawing a short line segment centered at the point (x_i, y_i) with slope $f(x_i, y_i)$. By creating a grid of such points and drawing short line segments with the appropriate slopes, we obtain a global picture of the behavior of the differential equation. These graphs are called **slope fields**.

[†]Drawing slope fields is a feature built into the TI-89 and the TI-92 Plus Module. In Appendix C we present a simple program for use on a TI-92 to draw slope fields.

Interpreting Slope Fields

We will illustrate the process with the differential equation $dy/dx = 2x - 6$. We can create a slope field with a grid made up of points of the form (j, k), where j and k are integers from -1 to 10 and -10 to 10, respectively. Through each point (j, k), plot a short line segment having the slope $m = 2j - 6$. Through the point $(2, -3)$, draw a short segment with slope $2 \cdot 2 - 6 = -2$. Through the point $(3, -3)$, draw a short segment with slope $2 \cdot 3 - 6 = 0$. Continuing the procedure for all 252 points produces the slope field shown in Figure 6.21.

FIGURE 6.21   Slope Field of $y' = 2x - 6$

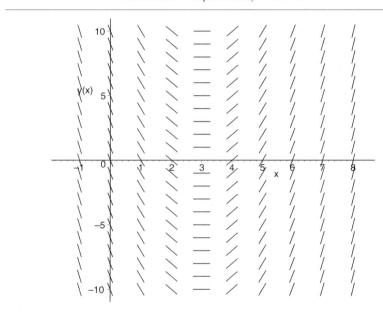

E X A M P L E   1   **A Simple Slope Field**

Use the Draw menu to add the function $y = x^2 - 6x + 1$ to the graph with the command $\texttt{DrawFunc x\^{}2-6x+1}$.

In the MODE *menu set the* Graph *mode to* DIFF EQUATIONS. *Enter the differential equation* $dy/dx = 2x - 6$ *in the* Y= *editor as* y1'(t) = 2t-6. *Leave the initial condition* yi1 *blank and select* ZoomStd *from the* Zoom *menu, to produce the image in Figure 6.22.*

FIGURE 6.22   Slope Field of $y' = 2x - 6$

FIGURE 6.23 $y' = 2x - 6$ with $y = x^2 - 6x + 1$

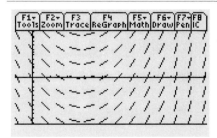

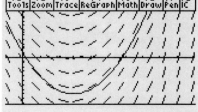

▶ **FIRST REFLECTION**

How does the new function relate to the differential equation? To the slope field? ◀

To get comfortable with slope fields, we'll revisit several of the models we've seen previously.

FIGURE 6.24   Slope Field for $dA/dt = \kappa A$

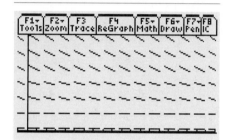

E X A M P L E   2   **The Chemical Reaction's Slope Field**

In "Modeling a Chemical Reaction" in Section 6.1, we developed a model using the differential equation

$$\frac{dA}{dt} = \kappa A$$

where κ is a negative constant. For this example use $\kappa = -0.064$. Enter `y1'(t)` `= -0.64y1` in the Y= editor, then set the graph window to $[-1, 20] \times [0, 1]$.

▶ **SECOND REFLECTION**

Where are the line segments drawn steepest? With reference to the differential equation, explain why your response makes sense. ◀

E X A M P L E   3   **Mercury Pollution, Again**

The "Mercury Pollution" example from Section 6.1 led to the differential equation

$$\frac{dp}{dt} = 0.3 - 0.0003p.$$

Enter this differential equation as `y1'(t) = 0.3-0.0003y1`.

FIGURE 6.25   Slope Field for $\frac{dp}{dt} = 0.3 - 0.0003p$

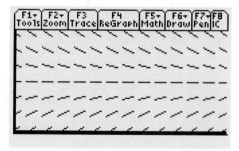

▶ **THIRD REFLECTION**

Describe the overall trends that you observe. ◀

EXAMPLE 4 **Yeast Data**

In Section 6.3 we developed the logistic model for the growth of population with limited resources and explored techniques for estimating the parameters. One method of estimating the parameters resulted in the differential equation

$$\frac{dy}{dt} = 0.6693y - 0.00105y^2$$

In Figure 6.26 we plot the slope field for this equation and the Pearl yeast data from Table 6.3 that gave rise to the parameters. The window is $[-1, 18] \times [-10, 900]$.

FIGURE 6.26 Slope Field and the Pearl Yeast Data

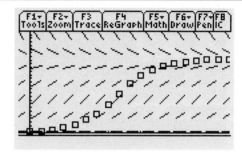

▶**Fourth Reflection**

From looking at the slope field, predict what would happen to the population with a initial biomass of 850? ◀

Slope Fields and Solution Curves

We can use slope fields and our calculator's numerical methods together to get very good pictures of the behavior predicted by a differential equation model. The slope field shows the global range of possibilities. By selecting a point $y(t_i) = B$ and adding the solution curve passing through that point, we can view features of particular interest.

EXAMPLE 5 **Logistic Solution Curve**

Use the IC menu to set the initial condition $t = 9$ and $y = 441$. A solution curve that passes through $(9, 441)$ is drawn on the slope field. Figure 6.27.

Think and Share

Break into groups. Choose different initial conditions from the yeast data of Table 6.3. Compare your results and report on which choice of starting point gives the best graphical fit of our model to the data. Can you offer reasons why some values work much better than others?

Your calculator will generate and graph a numerical approximation automatically. You can set it to use either Euler's method or a more sophisticated RK method – named for Runge and Kutta. In Chapter 7 we will develop one of the family of RK approximations.

FIGURE 6.27   Logistic Slope Field and Solution Curve

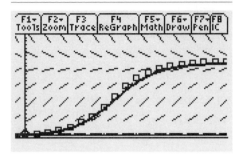

FIGURE 6.28 $\frac{dy}{dt} = \frac{t^2 - y^2}{2}$

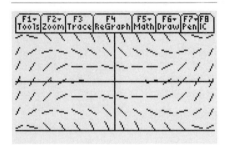

FIGURE 6.29   A Slope Field with Four Solution Curves

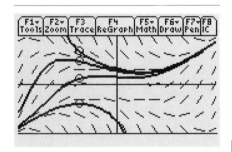

Think and Share

E X A M P L E   6

We will close with a slightly more complicated differential equation.

$$\frac{dy}{dt} = \frac{t^2 - y^2}{2}$$

Once a grid of points (t_i, y_i) is chosen, we know that the slope field for this differential equation is made of short line segments centered at (t_i, y_i) with slope $(t_i^2 - y_i^2)/2$. Define `y1' = (t^2-y1^2)/2` and graph the slope field in the window $[-3.1, 3.1] \times [-3, 3]$ as shown in Figure 6.28. The picture shows a good deal of information about the behavior of the family of solution curves.

▶ **FIFTH REFLECTION**

Describe regions where solution curves increase or decrease. Are there are regions or points that appear to attract or repel solutions?   ◀

We can investigate a differential equation by using the IC menu (F8) to add several solution curves as we have done in Figure 6.29. The circles indicate the initial values we have chosen. Notice the three upper curves in Figure 6.29. Even though these curves start with different initial values,they come together as t increases and appear to be converging. Contrast their behavior with the two lower curves that start out very close together, but quickly move apart as t increases, displaying radically different trends.

Break into groups. Add solution curves generated by the following initial values:

1. $y(0) = 9.6$
2. $y(0) = 150$
3. $y(0) = 700$
4. $y(0) = 750$

to the slope field for the differential equation we used to model the yeast data of Table 6.3.

Think and Share

$$\frac{dy}{dt} = 0.66935y - 0.001053y^2.$$

Use a y-axis range of $[0, 800]$. Solve the equation $\frac{dy}{dt} = 0$ for y. For which value of a does the line $y = a$ seem to attract solution curves? For which value of a does the line $y = a$ seem to repel solution curves?

Summary

Differential equations are used to model and describe complicated physical and biological processes. Not only can differential equations predict complex behaviors, but those predictions may vary greatly given different initial values. The slope field of a differential equation supplies an image that allows us to visualize the response of a differential equation to a wide range of initial conditions. When we use a numerical technique, such as Euler's method, to add representative solution curves to a slope field, we can frequently develop a greater understanding of a model. In many important situations it is possible to find elementary formulas for solutions. In Chapter 7 we will introduce two of the most powerful standard techniques for finding symbolic solutions.

EXERCISE SET 6.5

1. Produce and describe the slope field of

$$\frac{dy}{dx} = (y^2 + 1) \cdot (y - 2) \cdot (y + 3).$$

2. Produce and describe the slope field of

$$\frac{dy}{dx} = y^3 - 4y^2 - y + 4.$$

† **3.** Produce a slope field for

$$\frac{dy}{dx} = 2x.$$

Add several graphs from the family $y = x^2 + C$ and describe the relation between the slope field and the functions.

4. Produce a slope field for

$$\frac{dy}{dx} = 3x^2.$$

Add several graphs from the family $y = x^3 + C$ and describe the relation between the slope field and the functions.

5. Produce a slope field for

$$\frac{dy}{dx} = cos(x).$$

Add several graphs from the family $y = sin(x) + C$ and describe the relation between the slope field and the functions.

† **6.** Produce a slope field for

$$\frac{dy}{dx} = 2y.$$

Add several graphs from the family $y = C \cdot e^{2x}$ and describe the relation between the slope field and the functions.

7. The average student breathes 20 times per minute, exhaling 100 in³ of air containing 4% CO_2 (carbon dioxide). Model the amount of CO_2 in the air of a closed classroom that has dimensions 30 ft × 35 ft × 8 ft.

  a. Model the amount of CO_2 in the air.
  b. Draw a slope field for your model.
  c. Use Euler's method to add a solution curve to your slope field plot, assuming that the classroom originally has fresh air containing 1% CO_2 as 30 students enter.
  d. Describe the long term trend you observe.

A **nullcline** is a line on a slope field on which the slopes given by the differential equation are all zero. Horizontal nullclines are very useful for determining *rest points*; they are most easily found by solving the equation $dy/dx = 0$ for both x and y.

8. Sketch the nullclines on the slope field below.

FIGURE 6.30   Slope Field I

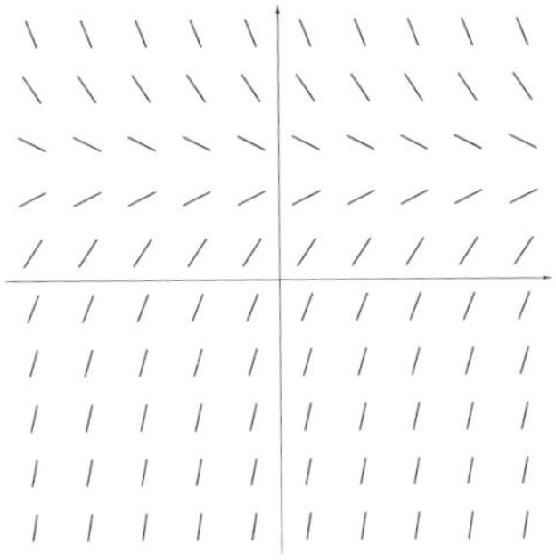

9. Sketch the nullclines on the slope field below.

FIGURE 6.31   Slope Field II

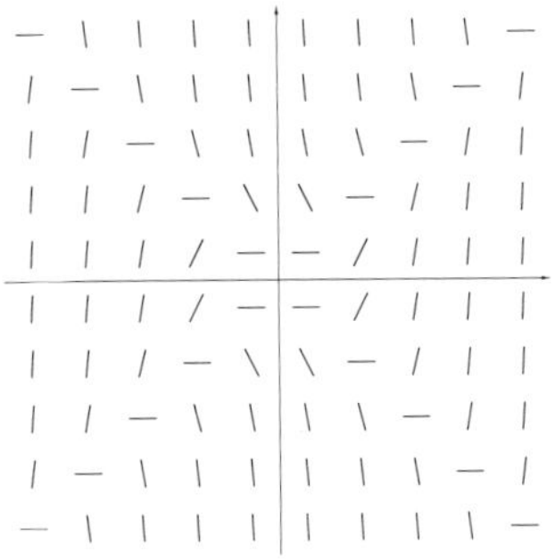

10. Sketch several solution curves on the slope field below.

FIGURE 6.32   Slope Field III

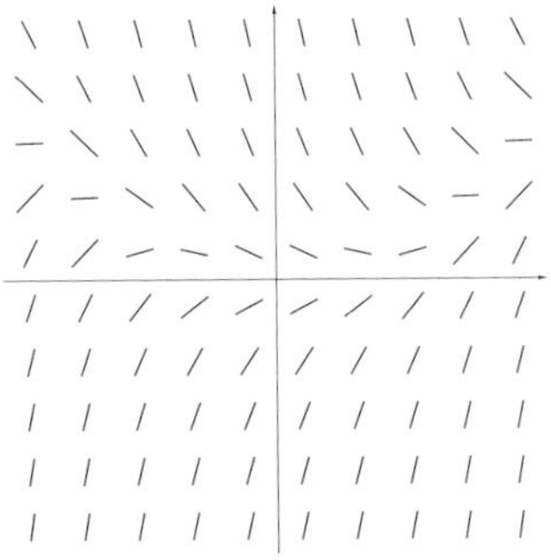

11. Sketch the nullclines and several integral curves on the slope field below.

FIGURE 6.33   Slope Field IV

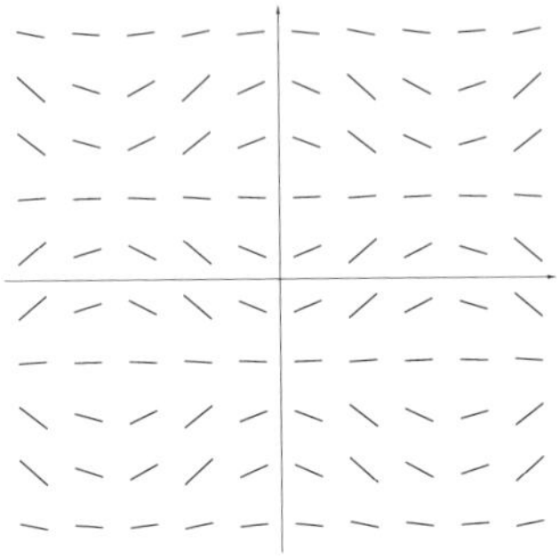

12. A horizontal nullcline on a slope field can *attract* or *repel* solution curves. Find one of each kind on a slope field for the logistic model

$$\frac{dp}{dt} = 0.1p \cdot (10 - p)$$

from a slope field using the window $[-2, 15] \times [-2, 15]$.

13. A horizontal nullcline can attract solution curves from one side and repel solution curves from the other. Find and analyze the nullcline for the differential equation

$$\frac{dy}{dx} = (y - 1)^2.$$

14. Research the terms *mathematical chaos* and *dynamical systems*. Report on their relation to differential equations and models.

Explain, in your own words, the meaning of each of the following terms. Write your explanations in clear and complete sentences. Include annotated diagrams whenever appropriate.

15. slope field

16. rest point

17. attracting nullcline

18. repelling nullcline

6.6

Errors in the Model Construction

Mathematical modeling is both a science and an art. Since the full complexity of reality can never be completely captured, when we fit mathematical descriptions to real phenomena, we make simplifying assumptions. These suppositions constitute one major source of error in the modeling process. Simple models have the virtue of being easily understood. On the other hand, simplified models may miss important features of the situation being studied. Moreover, even the simplest models are subject to errors in data collection.

Constructing a model requires careful consideration of the effects of errors. Examining discrepancies between a model and observed data can lead to greater understanding of the underlying reality. That understanding can then be applied to the creation of an improved model.

Projectiles Revisited, or Stop the Plane, I Want to Get Off!

Recall the Galilean model we developed in Section 3.6 for the flight of a projectile. The height of an object falling under the influence of gravity is

$$y(t) = y_0 + v_y t - \frac{1}{2} g t^2$$

where y_0 is the initial height, the parameter v_y is the initial vertical velocity, and g is the gravitational constant. Differentiating this function with respect to t, we can see that it is a solution to the differential equations

$$\frac{dy}{dt} = v_y - gt$$

Verify this!

with the initial conditions $y(0) = y_0$.

A number of assumptions were made when we developed the Galilean model and are inherent in the differential equation. The differential equation includes the term gt for the effect of gravity on vertical motion and reflects the assumption that there are no other forces acting in the vertical direction.

Let's apply the model above to a skydiver jumping from an airplane flying level at an altitude of 10,000 feet. An observer sets a camera on a tripod on the ground and photographs the jumper every half-second. By carefully examining the photographs, he arrives at the data shown in Table 6.11 for the height of the skydiver during the first five seconds of her flight.

FIGURE 6.34 Skydiver

TABLE 6.11   Skydiving Data

Time	0.0	0.5	1.0	1.5	2.0	2.5
Height	10000	9996.12	9984.91	9967	9942.96	9913.3

Time	3.0	3.5	4.0	4.5	5.0
Height	9878.52	9839.04	9795.26	9747.55	9696.25

The skydiver's initial height is $y_0 = 10,000$. Given that $g = 32\text{ft/sec}^2$, our model yields the equation

$$y(t) = 10000 - 16t^2$$

for the jumper's height at time t. Let's see how well the model fits the observed data.

FIGURE 6.35 Skydiving Data and the Galilean Model

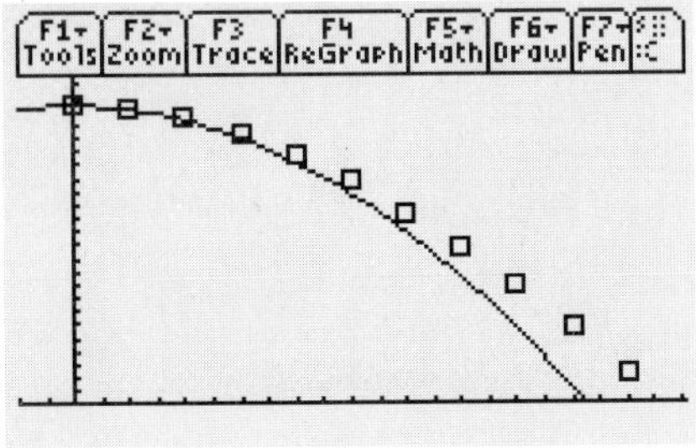

▶ **FIRST REFLECTION**

Create a data table with time in the first column and the height in the second. Define the function y1(x)=10000-16x^2 and add a new column to table c3= y1(c1). In the fourth column, compute the differences between the recorded data and the values predicted by the model c4= c3-c2. How far off is the model after one second? How far off is the model after five seconds? Describe how the error is changing. ◀

Plotting the observed heights together with the function y1 predicted by the model produces the graph in Figure 6.35. The plot shows that the skydiver is not falling as quickly as the model predicts. It's doubtful she will agree to fall more rapidly, so we need to revise our model.

Curve Fitting or Modeling?

One way to modify the model is to find a function which is a better approximation of the observed data. Many software packages, including those on your calculator, are capable of fitting different functions to data.

Think and Share

Form groups. In the Data/Matrix editor, use the Calc menu (F5) to fit your group's assigned *regression function* to the data. Save the RegEq in y2(x). Graph the function and the data together. How well does the function fit the data? Is the new function a good model of the skydiver's flight? Do you expect the new function to be a good predictor of the skydiver's flight over the first 10 seconds? Explain why or why not.

We use models to predict the future. Conversely, we use the accuracy of those predictions to evaluate of the model. Let's see how well the new models match further observations. Table 6.12 shows the height of our jumper over the second five seconds of flight.

TABLE 6.12 Additional Skydiving Data

Time	5.0	5.5	6.0	6.5	7.0	7.5
Height	9696.25	9641.65	9584.04	9523.68	9460.8	9395.61
Time	8.0	8.5	9.0	9.5	10.0	
Height	9328.31	9259.08	9188.08	9115.47	9041.37	

First, clear the heading in column `c4` *to avoid a* `Dimension mismatch` *error.*

We should be slow to abandon the Galilean model for projectiles — it's worked well for nearly 500 years!

▶ **SECOND REFLECTION**

Add the new observations to the data table and plot it along with y2. Describe how well $y2$ fits the data for the first 10 seconds of flight. Do you expect $y2$ to be a good predictor of the skydiver's height in the third five seconds? ◀

We could try to improve our model by repeating the process. That is, by finding a new fit to the larger set of data. However, this approach feels unsettling. How do we know our new model will be any good over the first 15 seconds of flight? Suppose we were to obtain data for the first 15 seconds of flight and that our new model matched it. Does that guarantee the new model will be valid over the first 20 seconds? We need to reexamine how we made the model.

One possible source of difficulties might be the assumption that we only need to fit the data better in order to obtain a good model for skydiving. Perhaps this situation cannot be properly modeled with one of the calculator's regression functions. Unfortunately, this leaves us with an even more difficult question. Of the many possible different types of functions, which should we use? There are an infinite number of possibilities even if we restrict ourselves to polynomials.

Some of the regression functions gave much better approximations to the observed data. What do we make of this in terms of the model? Replacing $y(t) = y_0 + v_y t - gt^2/2$ with a different function would be making strong statement about the physics of skydiving.

We are faced with a dilemma. Given any finite set of data, we can always find a function that approximates the data very closely. Unfortunately, fitting a given set of data well with some function does not guarantee that that function will be a good predictor. One of the most important applications of modeling is to make just such predictions.

There is only one way out of this predicament. We must develop a theory which justifies our model. Models without basis in reality are at best a convenient way of organizing data, and at worst, totally misleading.

The Theory Behind Jumping Out of a Plane

We've seen that the model generated by our theory does not match the observed data for the skydiver, but we haven't discovered the reason for the discrepancy. To obtain the original model, we simplified by considering only initial position, initial velocity, and gravity. If we radio our skydiver and ask her what additional factors she thinks our model should include, she will be unable to hear our question because of the loud noise of the air rushing past. If she could hear us, she would no doubt point out that air resistance must be incorporated into the model — it's a very important factor to skydivers.

▶ **THIRD REFLECTION**

Which direction will the force of air resistance be pushing? How does this force change over the course of the jump? How does this compare to our assumptions about the force of gravity? ◀

Newton's second law of motion states that the acceleration experienced by the skydiver is proportional to the sum of the acting forces. We can use the data to approximate acceleration by computing second divided differences.

Clear all the columns in your table except for the original data in columns $c1$ and $c2$. Define the third column to be the first divided differences by entering

Remember that fd(q)= shift(q,1)-q.

$$c3 = fd(c2)/fd(c1).$$

These divided differences represent the average velocity over each half second time interval. Define the fourth column to be the second divided differences by entering

$$c4 = fd(c3)/fd(c1).$$

The second divided differences approximate acceleration.

▶**FOURTH REFLECTION**

What trends do you see in the data? What do the trends indicate about the velocity and acceleration of the skydiver? What are the effects on the skydiver?◀

According to Newton's laws, gravity produces constant acceleration for a relatively light object like a skydiver close to the surface of the earth. If our skydiver were experiencing constant acceleration, the second differences in our table would all be nearly the same, but they are not. Given that the acceleration due to gravity is constant, we conclude that the acceleration due to air resistance must vary.

Here again, our skydiver helps us out. She tells us that the faster she falls, the harder the air pushes against her. That is, the acceleration due to air resistance varies directly with the velocity. Thus, we model the acceleration due to air resistance with

$$A(t) = k \cdot \frac{dy}{dt}$$

where $A(t)$ is the acceleration due to air resistance at time t, dy/dt is velocity, and k is the proportionality constant.

The Improved Model

Acceleration is the second derivative of position. Hence, adding our linear term for the acceleration due to air resistance and using -32 ft/sec^2 for the constant acceleration due to gravity, we obtain the model

$$\frac{d^2y}{dt^2} = -32 + k \cdot \frac{dy}{dt}. \tag{6.23}$$

▶**FIFTH REFLECTION**

Why does the model use -32 instead of $+32$ for the acceleration due to gravity? What will be the sign of dy/dt? How should the sign of the air resistance compare to the sign of the velocity? Should k be a positive or negative constant? ◄

The value for k depends on the skydiver. If she stretches out parallel to the ground, we would expect the air resistance to be greater than if she dives head first. We'll use our observed data to estimate the value of k. Begin by solving Equation (6.23) for k:

$$k = \frac{\frac{d^2 y}{dt^2} + 32}{\frac{dy}{dt}}.$$

Now define column c5 to be

c5= (c4+32)/c3.

Notice that the values are not all the same. This is a problem, since k is supposed to be a constant. The discrepancy stems in part from our necessary use of average velocities and accelerations to approximate the instantaneous quantities.

We choose the median of the data -0.18 as the value for k Our acceleration equation becomes

$$\frac{d^2 y}{dt^2} = -32 - 0.18 \cdot \frac{dy}{dt}.$$

A differential equation that involves a second order derivative and no higher is called a second order differential equation.

This is a new form of differential equation for us, it involves the second derivative. However, since acceleration is the *first derivative* of velocity as well as the *second derivative* of position, we can continue by making a simple substitution. We let $v(t)$ represent dy/dt, the vertical velocity at time t. The differential equation, with initial condition that the downward velocity of the skydiver was zero when she jumped out of the plane, becomes

$$\frac{dv}{dt} = -32 - 0.8 \cdot v \qquad \textbf{(6.24)}$$
$$v(0) = 0$$

FIGURE 6.36 The Improved Model on $[-1, 11] \times [9000, 10100]$

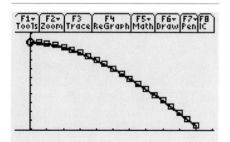

To graph this model set your calculator to DIFF EQUATIONS graphing mode. Enter y1′ = -32-0.18*y1, yi1=0, y2′=y1, y2i=10000. Under the Axes menu, select Custom axes with t for the horizontal and y2 for the vertical axes.

Data in the Real World

We have provided idealized data for the heights of the skydiver. Real experiments never look quite so "clean." Errors are made in collecting, measuring, and transcribing data. For instance, the level of a liquid in a pipette may be misread, the scale of an ohmmeter may not be fine enough, or two digits in a measurement may be transposed during transcription. Each of these types of error will affect the model that we develop in different ways.

▶ Sixth Reflection

Describe how you think each kind of error could affect a model. How might these errors be recognized?   ◀

In the graph shown in Figure 6.37, taken from a research article,[†] there are three curves that describe the activity of the enzyme *NADPH oxidase* in combination with proposed activators added at different times. The researchers will have a difficult time fitting continuous curves and models to such rough data.

FIGURE 6.37   Enzyme Activation

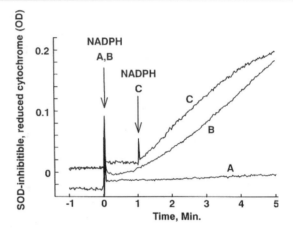

Conclusion

We construct models by making assumptions in order to create a simplified version of reality. Those assumptions imply differences between the model and the reality being represented. If the differences are too great, the model's predictions are not accurate and we are forced to reconsider the simplifications. Our skydiver discussion illustrates a simple example of this process.

Data collection introduces another source of error into the model construction process. Even if we manage to eliminate recording errors, we must still deal with the limitations of our instruments. Some of the discrepancies in Newton's model of the universe could not have been discovered while he was alive since they were beyond the accuracy of any instruments existing at that time.

[†] The graph is from Bauldry, S. A., Nasrallah, V. N., and Bass, D. A. 1992. Activation of NADPH Oxidase in Human Neutrophils Permeabilized with *Staphylococcus aureus* α-Toxin. *J. Biological Chemistry 267*-1:323–330.

Once a model has been constructed, we analyze it in the hope that understanding the model will help us better understand reality. Even after introducing air resistance, the skydiver model we obtained has a closed-form solution. Hence, we obtained a model which could be completely understood. When a model of an important process does not have a closed-form solutions, we perform our analysis using approximating techniques such as Euler's method. These techniques allow us to analyze a much broader class of problems than would otherwise be possible. However, we will see that these techniques introduce their own types of errors.

EXERCISE SET 6.6

1. Use the `QuadReg` feature of your calculator to find a quadratic function which fits the data points $(-3, 2)$, $(0, 1)$ and $(1, 0)$. Is the function a good fit? Use `CubicReg` to find a cubic function which fits the data points $(-3, 2)$, $(0, -3)$, $(1, 0)$ and $(2, 5)$. Is the cubic function a good fit? Suppose you are given n data points. What degree polynomial function should you expect to give a very good fit? The text states that, given any finite set of data, we will be able to find functions which approximate the data very closely. Is this statement true if we restrict ourselves to polynomial functions?

2. Explain why the velocity function satisfying the differential equation

$$\frac{dv}{dt} = -32 + k \cdot v$$

cannot be a nonconstant polynomial function. (Hint: think about degrees.)

3. The model in the text uses the simplification that the force of gravity is constant. Newton's theory actually predicts that the force of gravity is inversely proportional to the square of the distance between the jumper and the center of mass of the earth. Use the value of 4,000 miles for the radius of the earth and modify the differential equation to include this effect.

† 4. Verify that the function $v(t) = 177.778 \cdot (0.83527)^t - 177.778$ satisfies (approximately) the differential equation

$$\frac{dv}{dt} = -32 - 0.18 \cdot v$$

with the initial condition $v(0) = 0$. Plot the velocity for t from 0 to 100. Describe what the graph indicates regarding velocity.

† 5. Integrate the velocity function you found in the previous problem and use the initial condition $y(0) = 10,000$ to obtain the corresponding height function. Graph your height function along with the given data. The text states that the height data is "idealized." Based on your graph, explain why it is unlikely that the data could have been generated as described in the text (i.e. from photographs).

6. The skydiver's horizontal velocity is not affected by gravity. It is, however, affected by the air resistance. Given that the skydiver's initial horizontal velocity is 50 mph, determine a differential equation which models her horizontal motion. Suppose that the skydivers horizontal velocity after one second is 42 mph. Find a function which represents the horizontal velocity for the skydiver.

7. Suppose the skydiver pulls her ripcord at 2,000 feet. Use your results from the previous two questions to determine how far the model predicts she will have traveled horizontally when she pulls her ripcord.

8. The number -0.18 is a parameter of the model, called the coefficient of drag, and was estimated using divided differences. We can obtain a value for this parameter which matches the data even more accurately. Let d represent the coefficient of drag to obtain the differential equation

$$\frac{dv}{dt} = -32 + d \cdot v$$

Verify that the function $v(t) = \frac{32}{d}\left(1 - e^{d \cdot t}\right)$ satisfies both this differential equation and the initial condition $v(0) = 0$. Use integration to obtain a height function in terms of d. Use trial and error to obtain a value for d which matches the given data exactly.

When a skydiver falls with constant velocity we say the skydiver has reached terminal velocity (Note: this does **not** mean that the skydiver has died.) Exercises 9–12, investigate terminal velocity.

9. What will be the value of $\frac{dv}{dt}$ when the skydiver reaches terminal velocity? Use the differential equation

$$\frac{dv}{dt} = -32 - 0.18 \cdot v$$

to determine the terminal velocity predicted by the model. Use the velocity function you found in Exercise 4 to determine the time t at which the skydiver's velocity is within 0.1 foot per second of the terminal velocity. What is the skydivers height at this time?

10. Does the velocity function predict a time at which the skydiver's velocity equals the terminal velocity? Does this feature of the model seem realistic? Explain why or why not.

11. Suppose the skydiver is jumping downward from the belly of the plane and is given an initial shove, so her initial vertical velocity is -10 feet per second instead of 0. Still assuming the differential equation

$$\frac{dv}{dt} = -32 - 0.18 \cdot v.$$

and an initial height of 10,000 feet, find a height function for this new initial velocity. How does the terminal velocity in this case compare to the terminal velocity you found in the previous problem?

12. Let v_t represent the terminal velocity of a skydiver in feet per second. Determine a differential equation which models the flight of the skydiver in terms of v_t. Let v_0 represent the initial velocity of the skydiver and determine a velocity function for the skydiver in terms of v_t and v_0. Let h_0 represent the initial height of the skydiver and determine a height function for the skydiver in terms of v_t, v_0, and h_0.

13. Suppose the camera was bumped at some point in the process of photographing the skydiver. How might this mishap be recognized from a plot of the data?

14. The coefficient of drag will vary from one skydiver to the next. Table 6.13 contains idealized data for a second skydiver jumping from 11,000 feet. Estimate the coefficient

TABLE 6.13   Skydiving Data

Time	0.0	0.5	1.0	1.5	2.0
Height	11000	10995.87	10982.94	10960.34	10927.09
Time	2.5	3.0	3.5	4.0	4.5
Height	10882.16	10824.38	10752.46	10665.02	10560.49
Time	5.0				
Height	10437.2				

of drag for the second skydiver. Write a differential equation to model the second skydiver's fall. Solve your differential equation to obtain a velocity function and a height function

for the second skydiver. Assuming the second skydiver pulls his ripcord at 2,000 feet, how long does your model predict he will be in free fall?

15. When modeling the skydiver, we assumed a constant coefficient for the force of air resistance. This assumption must be altered when modeling the reentry of a space shuttle into the earth's atmosphere. As the shuttle descends, the density of the atmosphere increases which in turn increases the force of air resistance. Write a differential equation to model the flight of the shuttle under the assumption that the coefficient of drag is inversely proportional to the shuttle's height.

16. The force of air resistance on a falling leaf is proportional to the square of the velocity. Write a differential equation to model the flight of a falling leaf. Find the velocity function for the leaf in terms of the coefficient of drag. Obtain a leaf and record the time it takes to fall various distances. Use your data to obtain a value for the leaf's drag coefficient.

The following information will be used in Exercises 17–19. A mass of ten grams attached to a spring oscillates up and down after being pulled down 5 centimeters and released. Table 6.14 gives the height of the mass in relation to its rest position during the first second following its release. Letting $h(t)$ represent the height of the mass at time t, Hooke's law along with Newton's second law can be used to obtain the differential equation:

$$\frac{d^2h}{dt^2} = -k\,h(t).$$

as a model for the motion of the mass.

TABLE 6.14   Spring-Mass Data

Time	0.0	0.1	0.2	0.3	0.4
Height	-5	2.07033	3.23575	-4.72935	0.712973
Time	0.5	0.6	0.7	0.8	0.9
Height	4.01986	-4.09447	-0.66038	4.60061	-3.1561
Time	1.0				
Height	-1.94117				

17. Show that $h(t) = -5\cos(\omega t)$ is a solution for the differential equation.

18. Find a value for ω in the formula $h(t) = -5\cos(\omega t)$ so that h is a reasonably good match for the data given in the table.

19. According to the given differential equation the mass will continue to oscillate forever. In reality, the force of friction acts against the motion of the mass and will eventually bring it to a halt. Write a differential equation to model the motion of the mass which includes a term for friction.

Terminology

Differential equations have been studied since the derivative was first defined in the seventeenth century. Consequently a great deal of terminology has been developed. We have introduced a little of the notation as we needed it. Here we collect the terms and their definitions for easy reference.

DIFFERENTIAL EQUATION

An equation that involves the derivatives of a function is called a **differential equation**.

A differential equation that involves only the first derivative of the function is called a *first order* differential equation. If the equation involves the second derivative and no higher order derivatives, it is a *second order* differential equation.

A *solution* to a differential equation is a function or family of functions that satisfies the differential equation. For example, the function $f(t) = 3\,e^{2t}$ is a *particular solution* of the differential equation

$$\frac{dy}{dt} = 2y. \tag{6.25}$$

The family of functions $f(t) = C\,e^{2t}$ is called the *general solution* to Equation (6.25). If the function can be described as an explicit formula, we have a *closed form solution* for the differential equation.

SOLUTION CURVE

The graph of a particular solution to a differential equation is called a *solution curve*.

When using differential equations to model a phenomenon, we may be interested from the outset in particular solutions if in addition to the differential equation, we know a numerical *initial value*. Adding the requirement that $f(0) = 3$ to Equation (6.25) has the effect of specifying $f(t) = 3\,e^{2t}$ as the only solution.

The set of equations

$$\frac{dy}{dt} = g(t, y)$$
$$y(t_0) = y_0$$

is called a **first order initial value problem**. A set of equations of the form

$$\frac{d^2 y}{dt^2} = g(t, y, y')$$
$$y(t_0) = y_0$$
$$y'(t_0) = z_0$$

is a **second order initial value problem**.

Summary

Model building requires coordination of a number of diverse but interrelated activities. We define the problem to be studied and collect relevant data. By making simplifying assumptions and examining data in tables and graphs, we discover underlying patterns. We model the patterns of change with difference equations, or when the problem indicates a continuous model, with differential equations. Euler's method generates discrete data from a continuous model, which can then be compared to observed data for closeness of fit. Slope fields provide a global view of the model's behavior. These techniques, and many others, are used to analyze the accuracy and precision of the model. The results of our analysis often lead us to refine the model and restart the process. The test of a model is how good its predictions are and what insights into the problem it gives.

In the coming chapters, we will expand our techniques for analyzing models. We will learn to find explicit closed form formulas for solutions for certain types of differential equation models. When formulas cannot be found, we'll use approximations to analyze the model. We will investigate the error in our approximation techniques and consider fundamental difficulties that arise in their use. Eventually, we will see how to generalize our methods to more complicated models requiring systems of difference or differential equations. Watch for the blending of approaches, and closely observe how melding the information we get from different sources leads to deeper insights.

SUPPLEMENTARY EXERCISES

Find the associated differential equations for the difference equations in Exercises Exercises 1–5.

1. $x_n = 1.6x_{n-1}$

2. $\Delta F_n = F_{n-1}$

3. $\theta_n = \theta_{n-1}^2 + \theta_{n-1} + n + 1$

4. $\left\{ \begin{array}{rcl} p_n & = & \kappa\, p_{n-1} \\ p(0) & = & p_0 \end{array} \right\}.$

5. $\dfrac{\Delta P}{\Delta n} + nP = 0$

Find the associated difference equations for the differential equations in Exercises Exercises 6–10.

6. $\dfrac{dy}{dx} = e^x \sin(x)$

7. $\dfrac{dx}{dt} + t = x$

8. $\dfrac{dy}{d\theta} = \dfrac{e^{y+\theta}}{y+2}$

9. $x(y^3 + 1)\,dx - y^2\,dy = 0$

10. $\dfrac{dP}{dn} + nP = 0$

11. Use Euler's method to find $P(5)$ and $P(10)$ for

$$\frac{dP}{dt} = P \cdot (1 - P)$$
$$P(0) = 0.1$$

a. with a step size of $h = 1.0$

b. with a step size of $h = 0.5$

12. Use a logistic differential equation to model the Hawaii population data in Table 6.15.

a. Estimate M and r.

b. Estimate the population in 1990 and 2000 using your model. How well did you do?

c. What factors other than those considered in the logistic model determine the Hawaii population?

TABLE 6.15 Hawaii Population (in 1000's), 1900 to 1990*

Year	1900	1910	1920	1930	1940	1950
Count	154	192	226	368	423	500
Year	1960	1970	1980	1990	2000	
Count	633	770	965	1108	???	

* Source: U.S. Bureau of the Census.

Match the differential equations in Exercises Exercises 13–18 with the correct slope field in Figure 6.38.

13. $y' = y \cdot (2.5 - y)$ is slope field _____ .

14. $y' = \dfrac{(3 - x)(3 + x)}{6}$ is slope field _____ .

15. $y' = x^2 - y^2$ is slope field _____ .

16. $y' = -y$ is slope field _____ .

17. $y' = -\dfrac{x}{y}$ is slope field _____ .

18. $y' = \dfrac{(3 - y)(3 + y)}{6}$ is slope field _____ .

19. For the differential equation $\dfrac{dy}{dx} = \dfrac{2xy}{x^2 + 1}$:

a. Graph the slope field.

b. Use Euler's method to add solution curves through the points $(0, 4)$, $(0, 2)$, $(0, 0)$, $(0, -2)$, $(0, -4)$.

c. Discuss what you observe.

20. For the differential equation $\dfrac{dy}{dx} = \dfrac{2xy}{x^2 - 1}$:

a. Graph the slope field.

b. Use Euler's method to add solution curves through the points $(0, 4)$, $(0, 2)$, $(0, 0)$, $(0, -2)$, and $(0, -4)$.

c. Discuss what you observe.

21. For the inital value problem

$$y' = y^\alpha$$
$$y(0) = y_0,$$

produce slope fields, one each for $\alpha = -1, 0, 1$, that show the function family's behavior.

FIGURE 6.38 Slope Fields for Exercise 16

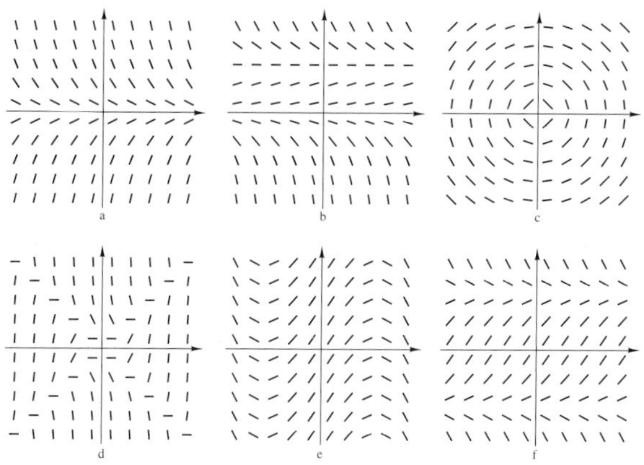

22. How could you use a slope field to check the validity of a solution curve generated by a numerical method?

PROJECTS

Project One: Predicting Populations

We've looked at several different population models in this and previous chapters. The main models are the Malthusian simple exponential growth, Verhulst logistic, and Gompertz. Each of these models is a statement about the rate of change of the population. In this project, you'll work with a simple population model where the rate of change depends on fixed percentages of the current populations.

FIGURE 6.39 The World

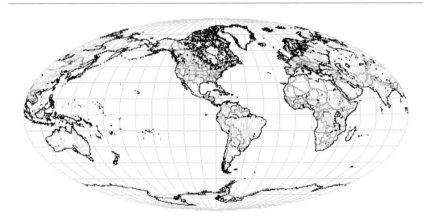

Assemble your project group and do the following:

1. Choose a country outside North America.

2. Go to a computer laboratory and use a World Wide Web browser to visit the Central Intelligence Agency's *World Fact Book* site. The 1996 Fact Book web page's address is:

`http://www.odci.gov/cia/publications/factbook/index.html`

You may need to search the web if the address has changed; search with the keywords "cia" and "factbook."

3. Find the following information for your group's country:

    a. current population

    b. birth rate

    c. migration rate

    d. death rate

4. Construct both differential and difference equation models for your country's population.

5. Predict next year's population value.

Project Report.   Write a report on your investigations. Address your report to your classmates, assuming the material is new to them. Describe your reasoning and include all necessary background information. A minimal project report must include:

1. Background and demographic information about the country.
2. Maps of your chosen country and it's place in the major geographic region to which it belongs.
3. Justification and explanations for your calculations.
4. Suggestions for possible extensions of your observations and further questions for exploration.

Project Two: Radioactive Pathways

Radioactive elements go through several stages, changing into different products, before they finally reach a stable end to their decay pathways. For example, the uranium isotope U^{238} goes through 13 intermediate changes before finally decaying to lead Pb^{206}. The half-lives of these intermediate elements range from 245,000 years for U^{234} to 160 microseconds for the polonium isotope Po^{214}.

In this project you will be tracking the amount of one isotope by knowing the initial amount, the decay rate of the isotope, and the decay rate of the previous element in the pathway. The rate of change of the amount of your chosen isotope is equal to the rate of the amount gained from the previous element minus the amount lost by decay of your chosen element.

Assemble your project group and do the following:

1. Choose an intermediate element in the Uranium decay pathway.
2. Determine the needed decay rates from the chart above.
3. Construct a model for your element assuming that you begin with ten grams of your isotope.
4. Use your model to make a plot for your isotope.

Project Report.   Write a report on your investigations. Address your report to your classmates, assuming the material is new to them. Describe your reasoning and include all necessary background information. A minimal project report must include:

1. Descriptions and calculations for decay scenario you have chosen.
2. Justification and explanations for your calculations.
3. Historical background on your chosen element, including the date and circumstances of its discovery.
4. Suggestions for possible extensions of your observations and further questions for exploration.

FIGURE 6.40   Marie Curie, Co-Discoverer of Plutonium and Radium

FIGURE 6.41   The Uranium Decay Pathway

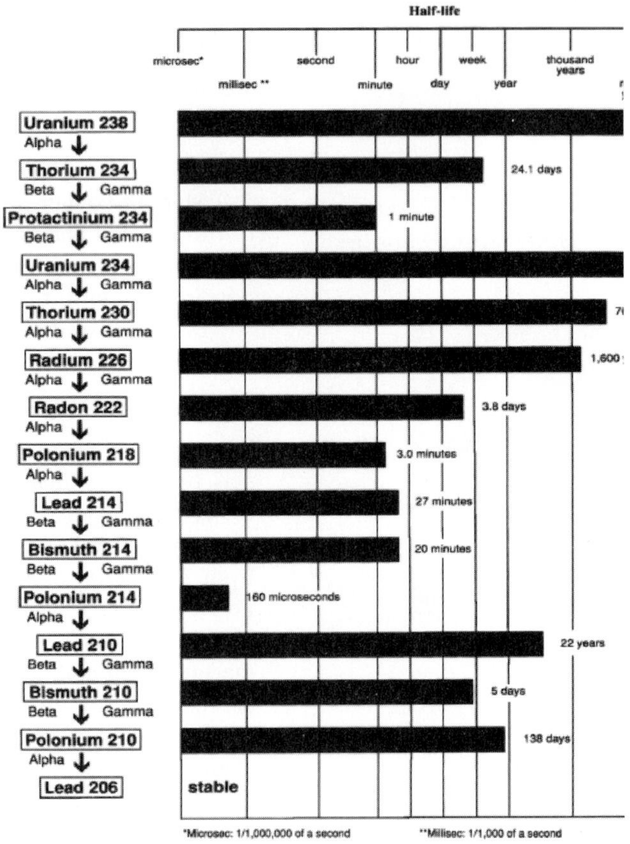

Project Three: Detecting Leaks in a Heating System[†]

Just as you're about to go out the door, the phone rings. The plant manager for Fargo Farms in North Dakota is on the line with a problem.

"I'm worried about a leak in the heating system for one of our hog far-rowing barns," he says. "The system works by pumping a salt solution through pipes embedded in a cement floor. I flushed the system with a 0.05 pound per gallon salt solution flowing at the rate of one gallon per minute and took some measurements of the salt concentration in the output. Could you come up with a model so I can figure out what's going on without tearing up the floor?"

You agree to consider the problem and the manager faxes you the data presented in Table 6.16.

[†]This project was inspired by the following article:

Aboufadel, E. and Tavener, S.J. 1996. Detecting a Leak in an Underground Storage Tank *The C*ODE*E Newsletter* 5:1.

TABLE 6.16 Concentration Measurements

t minutes	0	10	20
$C(t)$ lb salt per gal	0.20	0.18	0.16

If there is a leak you should be able to detect it by keeping track of the total amount of salt in the system, so you let $S(t)$ represent this quantity. You would like to know the amount of salt in the system when the flushing began but you don't. You would like to know the volume of liquid in the system when the flushing began but, again, you don't. You do know something about the rates at which liquid is entering and leaving the system so you decide to focus on the rate of change of $S(t)$.

Salt can go into the system only through the input pipe, and can leave the system only through the output pipe or through a leak. You know the rate of flow into and out of the system through the pipes is one gallon per minute but you don't know the rate of flow through the leak; so, you assign the parameter R_{leak} to that rate.

The rate at which salt enters the system equals the rate at which the salt solution enters the system times the concentration of salt in the solution. Since the salt solution is entering the system at one gallon per minute and has a concentration of 0.05 pounds per gallon, salt is entering the system at the rate of 0.05 pounds per minute.

Similarly, the rate at which salt leaves the system equals the rate at which liquid leaves the system times the concentration of salt in the system. You know the rate at which the liquid leaves equals $1 + R_{leak}$, but what about the concentration of salt? This concentration should be the total amount of salt in the system at time t divided by the volume of liquid in the system at time t. The flow through the input pipe equals the flow through the output pipe, so the volume of liquid in the system at time t must equal $V_0 - R_{leak} \cdot t$ where V_0 is the volume of liquid in the system at time 0. Thus, the concentration of salt in the system at time t equals

$$\frac{S(t)}{V_0 - R_{leak} \cdot t}$$

and the rate at which salt is leaving the system must equal

$$\frac{S(t)}{V_0 - R_{leak} \cdot t} (R_{out} + R_{leak}).$$

Assemble your project group and do the following:

1. Using the appropriate expressions from the preceding paragraphs, write an equation which represents dS/dt, the rate of change of salt in the system, as the difference between the rate at which salt enters the system and the rate at which it leaves. Write a difference equation model from your differential equation model.

2. From the table, the initial concentration of salt in the system is 0.20. Use that data to write an equation relating V_0 to S_0, the amount of salt in the system when the flushing begins.

3. Rewrite your model, replacing V_0 with its value in terms of S_0.

Project Report.   Write a report on your investigations. Address your report to your classmates, assuming the material is new to them. Describe your reasoning and include all necessary background information. A minimal project report must include:

1. Development of your model for the amount of salt in the heating system, and your estimate for the numeric values of all parameters.

2. An answer to the managers questions. Should the manager have the floor torn up to look for a leak?

3. Justification and explanations for your calculations.

4. Suggestions for possible extensions of your observations and further questions for exploration.

Existence and Uniqueness Theorems

In Chapters 6 and 7 we have found functions that solve differential equations with initial conditions (IVPs) using graphical, symbolic, and numerical techniques. If there are several functions that solve an IVP, how good are predictions made using one of them? A solution function needs to be unique to be useful. Does every IVP have a solution? If a solution does not exist, then searching for one with numerical or other techniques is useless. In this Interlude we will examine theorems that begin to answer these questions about the uniqueness and existence of solutions to IVPs.

Uniqueness

A Lipschitz Condition

There are different conditions that we can impose on an IVP to guarantee uniqueness. Among these is one called a *Lipschitz condition*.

LIPSCHITZ CONDITION

A function f is said to satisfy a Lipschitz Condition on the interval $[a, b]$ if there is a constant L such that for every two values x_1 and x_2 in $[a, b]$, it follows that

$$|f(x_2) - f(x_1)| \leq L\,|x_2 - x_1|$$

Can you show that "Lipschitz implies continuous" is true?

In words the Lipschitz condition is: given any two points of a function f, the distance between corresponding y's is always less than a fixed constant times the distance between the x's. A Lipschitz function is continuous but not necessarily differentiable.

The Uniqueness Theorem

The Uniqueness theorem of this section demonstrates that a Lipschitz condition is enough to give uniqueness of solution for an IVP. We'll use a standard technique in the proof: assume that we have two solutions and show that they're really the same.

UNIQUENESS THEOREM

If the function $\phi(y)$ is *Lipschitz* on the interval $[a, b]$, then the *autonomous initial value problem*

$$y' = \phi(y)$$
$$y(a) = y_a$$

has *at most* one solution.

Proof.   Suppose that we have two solutions $y_1(x)$ and $y_2(x)$ to our initial value problem. Define the new function

$$f = y_2 - y_1$$

Two properties of f are readily apparent:

1. $f(a) = 0$ since $y_1(a) = y_2(a) = y_a$
2. f is differentiable on $[a, b]$ and $f'(x) = y_2'(x) - y_1'(x)$ on $[a, b]$

Now it's time to use the Lipschitz condition of the hypotheses. Since $\phi(y)$ is Lipschitz, there is a nonnegative constant L so that

$$|\phi(y_2) - \phi(y_1)| < L\,|y_2 - y_1|$$

for all $x \in [a, b]$. Then combining this condition with the differential equation from the IVP

$$f'(x) = y_2'(x) - y_1'(x)$$
$$= \phi(y_2) - \phi(y_1)$$

gives us

$$|f'(x)| = |\phi(y_2) - \phi(y_1)|$$
$$\leq L\,|y_2 - y_1|$$

Since $f = y_2 - y_1$, the relation above is the same as

$$|f'(x)| \leq M\,|f(x)|$$

Next, we'll show that this relation forces f to be the zero function.

Choose $x_0 \in (a, b)$ such that

$$0 < x_0 - a < \frac{1}{L}$$

and let

$$M_0 = \max_{a \leq x \leq x_0} |f(x)|$$
$$M_1 = \max_{a \leq x \leq x_0} |f'(x)|$$

Since we know that $|f'(x)| \leq L\,|f(x)|$, then we have

$$M_1 \leq L\,M_0 \tag{7.29}$$

Recall that $f(a) = 0$ and f is differentiable on $[a, b]$. Choose any x with $a < x < x_0$ and apply the *Mean Value Theorem* to

$$f(x) = f(x) - f(a) = f'(\xi)(x - a)$$

and so, since $a < x < x_0$,

$$|f(x)| \leq |f'(x)|\,(x_0 - a)$$

In light of the definitions of M_0 and M_1, this inequality becomes

$$M_0 \leq M_1\,(x_0 - a)$$

When we use Inequality (7.29), then

$$M_0 \leq M_0\,L\,(x_0 - a)$$

Remember, we chose x_0 so that $L\,(x_0 - a)$ is less than 1, hence the only way M_0 can be less than $M_0\,L\,(x_0 - a)$ is

$$M_0 = 0$$

Since the maximum M_0 of $|f(x)|$ is zero, then $f(x)$ must be zero on $[a, x_0]$. We need only iterate the argument to show that $f(x) = 0$ on all of $[a, b]$. Thus, since $f = y_2 - y_1$, we have that

$$y_2(x) = y_1(x)$$

for all $x \in [a, b]$ and uniqueness is established.
QED.

Existence

Existence theorems use a variety of proof techniques. One method is to assume a solution doesn't exist and derive a contradiction. While this approach is logically sound and useful, it doesn't supply any insight into the process of *solving an IVP*. A different technique is to construct a general solution, much like deriving the quadratic formula from the general quadratic equation. Usually, only special cases fall to such a direct attack. Our first theorem is of this kind.

A Constructive Existence Theorem

CONSTRUCTIVE EXISTENCE THEOREM

If the functions $P(x)$ and $Q(x)$ are continuous on the interval $[a, b]$, then the *linear initial value problem*

$$y' + P(x)\, y = Q(x)$$
$$y(a) = y_a$$

has *at least* one solution.

If the left side of the IVP were the result of the *product rule* for derivatives, we could just integrate both sides as for a separable differential equation. The proof technique is to find a multiplier to make it so.

Proof. Multiply the left side of the IVP by $\mu(x)$ and consider the related differential equation

$$\mu(x)\, y'(x) + \mu(x)\, P(x)\, y(x) = \frac{d}{dx}\, (\mu(x)\, y(x))$$
$$= \mu(x)\, y'(x) + \mu'(x)\, y(x)$$

From this relation, we see that

$$\mu'(x) = P(x)\, \mu(x)$$

Thus, separating and integrating, we have

$$\mu(x) = e^{\int P(x)dx}$$

We are sure the integral exists since P is continuous by hypothesis.

Multiply the differential equation by $\mu(x)$ while applying the product rule (in reverse) to the left side to arrive at

$$\frac{d}{dx}\, (\mu(x)\, y(x)) = \mu(x)\, Q(x)$$

Since μ is an exponential of an integral of a continuous function and Q is continuous by hypothesis, $\mu\,Q$ is continuous on $[a, b]$ and, therefore, integrable. Now integrate both sides and divide by μ to obtain

$$y(x) = \mu^{-1}(x)\left(C + \int \mu(t)\,Q(t)dt\right)$$

Substituting $y(a) = y_a$ and determining C solves the initial value problem. For the sake of completeness, replace μ by its definition

$$y(x) = e^{-\int P(x)dx}\left(C + \int e^{\int P(t)dt}\,Q(t)dt\right)$$

QED.

Even though we have a formula for the solution of our IVP, in practice the indefinite integrals may likely prove difficult, if not impossible, to calculate.

An Algorithmic Existence Theorem

The technique used in the next theorem is quite novel. An approximate solution is improved to a better approximation. This approach is somewhat reminiscent of Newton's method for finding roots—from one guess, produce a better guess at the root. However, here we use integrals instead of tangent lines. This technique is called *Picard's Successive Approximations.* The proof requires several advanced concepts that we've not yet seen; these will be noted in the development.

ALGORITHMIC EXISTENCE THEOREM

If both the functions $\phi(y)$ and $\frac{d}{dy}\,\phi(y)$ are continuous on the interval $c \le y \le d$ containing y_a, then the *autonomous initial value problem*

$$y' = \phi(y)$$
$$y(a) = y_a$$

has *at least* one solution.

Proof. First translate the x-interval so that $x = a$ becomes $x = 0$. Now change the translated initial value problem to an *integral equation* by formally integrating both sides from 0 to x.

$$y(x) = y_0 + \int_a^x \phi(y(t))\,dt \tag{7.30}$$

Set our starting approximation of the solution to the initial value

$$y_0(x) = y_0 = 0$$

Define y_1 in terms of the integral Equation (7.30) and y_0.

$$y_1(x) = y_0 + \int_0^x \phi(y_0(t))\, dt$$
$$= y_0 + \phi(y_0)\, x$$

Continue this process, successive approximations, defining y_n by

$$y_n(x) = y_0 + \int_0^x \phi(y_{n-1}(t))\, dt$$

We claim that the sequence of functions $y_n(x)$ converges to a function $y(x)$ that is a solution of the IVP. We'll establish this claim with several steps.

First we'll collect several facts about ϕ and ϕ'.

1. Since ϕ is continuous on a closed interval, it has a maximum, say M_0, on the interval.
2. Since ϕ' is continuous on a closed interval, it has a maximum, say M_1, on the interval.
3. Since ϕ is differentiable on the closed interval, applying the Mean Value Theorem to two values z_1 and z_2 in the interval gives

$$|\phi(z_2) - \phi(z_1)| = |\phi'(\xi)|\, |z_2 - z_1|$$
$$\leq M_1\, |z_2 - z_1|$$

The proof now hinges on the statement

If $|x| < h$, then

$$|y_n(x) - y_{n-1}(x)| \leq \frac{M_0\, M_1^{n-1}\, |x^n|}{n!}$$
$$\leq \frac{M_0\, M_1^{n-1}\, h^n}{n!} = \frac{M_0}{M_1}\, \frac{(M_1 h)^n}{n!}$$

This assertion is verified by induction.

For $n = 1$, we see that

$$|y_1(x) - y_0| \leq \int_0^x |\phi(y_0)|\, dt$$
$$\leq M_0\, |x| \leq M_0 h$$

by fact 1 above and the integral inequality $|\int f(z)\, dz| \leq \int |f(z)|\, dz$.

Assume the relation holds for $n = k$, then consider

$$|y_{k+1}(x) - y_k(x)| \leq \int_0^x |\phi(y_k(t)) - \phi(y_{k-1}(t))|\, dt$$

An application of fact 3 gives

$$|y_{k+1}(x) - y_k(x)| \leq \int_0^x M_1\, |(y_k(t) - y_{k-1}(t)|\, dt$$

Now use the induction hypothesis that the relation holds for $n = k$, so that

$$|y_{k+1}(x) - y_k(x)| \leq M_1 \frac{M_0 M_1^{k-1}}{k!} \int_0^x |t^k| \, dt$$

$$\leq M_1 \frac{M_0 M_1^{k-1}}{k!} \frac{|x^{k+1}|}{k+1}$$

$$\leq \frac{M_0 M_1^k h^{k+1}}{(k+1)!}$$

The rest of the proof uses concepts from Chapter 9.

The statement then follows by the Principle of Mathematical Induction.

Now we look at the series

$$\sum_{k=1}^{n} \left(y_k(x) - y_{k-1}(x) \right) = y_n(x) - y_0$$

Hence

$$|y_n(x) - y_0| \leq \sum_{k=1}^{n} \left(y_k(x) - y_{k-1}(x) \right)$$

$$= \sum_{k=0}^{n} \frac{M_0}{M_1} \frac{(M_1 h)^k}{k!}$$

$$= \frac{M_0}{M_1} \sum_{k=0}^{n} \frac{(M_1 h)^k}{k!}$$

Since

$$\sum_{k=0}^{\infty} \frac{(M_1 h)^k}{k!} = e^{M_1 h}$$

converges, we have that the series $\sum_{k=1}^{n} (y_k(x) - y_{k-1}(x))$, and so $y_n(x)$, converges *absolutely and uniformly* to $y(x)$. (The significance of this type of convergence is that we may interchange limits and integrals.) Thence

$$y(x) = \lim_{n \to \infty} y_n(x)$$

$$= y_0 + \lim_{n \to \infty} \int_0^x \phi(y_{n-1}(t)) \, dt$$

$$= y_0 + \int_0^x \lim_{n \to \infty} \phi(y_{n-1}(t)) \, dt$$

and so,

$$y(x) = y_0 + \int_0^x \phi(y(t)) \, dt$$

Differentiating this last result shows that $y(x)$ is indeed a solution to the initial value problem.

QED.

Just as noted for linear differential equations, the integrations needed for the successive approximations may be quite difficult, or even impossible, to calculate. When symbolic solutions cannot be found, we must use numeric and graphical procedures.

The General Existence and Uniqueness Theorem

To prove a general existence theorem, even for first order equations is beyond the scope of this text. The mathematics required is much more advanced than what we have available[†]. Nonetheless, let's state the theorem.

GENERAL EXISTENCE AND UNIQUENESS THEOREM

If the function $\phi(x, y)$ is continuous for $x \in [a, b]$ and $y \in [c, d]$, if the derivative of $\phi(x, y)$ with respect to y (treating x as a constant) is Lipschitz for $x \in [a, b]$ and $y \in [c, d]$, and $a < x_0 < b$ and $c < y_0 < d$, then there is a subinterval $[a_1, b_1]$ of $[a, b]$ on which the *first order initial value problem*

$$y' = \phi(x, y)$$
$$y(x_0) = y_0$$

has *exactly* one solution.

Keep in mind that, although we've proved that a *solution* exists, this in no wise guarantees that a *formula* exists. Many important differential equations do not have closed form solutions. This lack caused the topic of Special Functions to be developed.

Here are two questions for you to ponder. What do you think continuity in terms of two variables involves? What do you think "differentiate y treating x as a constant" will do?

[†]For a proof of the general theorem, see Chapter 9 of *Elementary Differential Equations* by E. Rainville and P. Bedient.

EXERCISE SET 7.0

1. Investigate Lipschitz conditions
  a. Let $f(x) = |x|$ on $[-a, a]$.
    i. Show that f satisfies a Lipschitz condition on $[-a, a]$. What is L?
    ii. Is f continuous on $[-a, a]$?
    iii. Is f differentiable on $[-a, a]$?
  b. Let $f(x) = \sqrt{(x)}$ on $[0, a]$.

  i. Let $M > 0$. Let $0 < x_1 < x_2 < 4/M^2$. Show that

$$f(x_2) - f(x_1) > M(x_2 - x_1)$$

by "rationalizing the numerator" $f(x_2) - f(x_1)$. Thus f is not Lipschitz on $[0, a]$.
  ii. Is f continuous on $[0, a]$?
  iii. Is f differentiable on $[0, a]$?

c. Discuss the statement, "A Lipschitz condition is stronger than continuity but weaker than differentiability."

2. Use the methods of the constructive proof to solve the initial value problem

$$y' + y = 1$$
$$y(0) = 1.$$

3. Use the methods of the algorithmic proof to construct y_3 as an approximate solution to the initial value problem

$$y' + y = 1$$
$$y(0) = 1.$$

7

SOLVING DIFFERENTIAL EQUATIONS

7.1

Integration and Separation of Variables

7.2

Linear Differential Equations

7.3

Errors in Euler's Method

7.4

Improving Euler's Method

7.5

Advanced Numeric Techniques

Summary

In Chapter 6, we constructed discrete and continuous models. Each continuous model included a differential equation or an initial value problem. We analyzed initial value problems by using approximate solutions presented as tables or graphs. We could avoid the difficulties and uncertainties inherent in approximations by finding closed-form solutions to the differential equation. Our focus turns to symbolic solution techniques; we'll develop two methods for solving differential equations. The questions we need to address include: Can we find exact solutions? If so, how? If not, can we trust the approximate solutions? How close is Euler's approximation to the actual solution? Could there be more than one solution?

If we are to trust an approximate solution, we *must* have an idea of how far estimated values are from the actual value. We need to know the error in the approximation. Using the insights we developed in Chapter 4, we can bound this error. As with numerical integration, the bound can be made as small as we like simply by reducing the step size. However, a small step size brings with it two very practical problems. A small step size requires a large number of computations. The calculations may take too much time. Additionally, a small step size can give rise to rounding errors. To address these problems, we will make modifications to Euler's method that improve the technique.

7.1

Integration and Separation of Variables

Our first choice for understanding a given model represented by a differential equation would be to obtain a closed form formula for solutions. Because a formula uses familiar functions, it can quickly give us insight into the behavior of the model and by inference, the reality it portrays.

Antiderivatives and Differential Equations

We have already studied finding formulas for solutions to differential equations when we studied antidifferentiation in Chapter 5.

E X A M P L E 1

Consider the differential equation,

$$\frac{dx}{dt} = t^2.$$

To find a closed form formula for a solution to this differential equation, we must find a family of functions with independent variable t whose derivative is t^2. That is, we must find the family of antiderivatives for t^2. Antidifferentiating with respect to t we obtain $x(t) = t^3/3 + C$ as our formula where C is an arbitrary constant.

E X A M P L E 2

Suppose we desire a symbolic solution for the initial value problem

$$\frac{dy}{dx} = \cos(x) + \sin(x)$$
$$y(0) = 2.$$

First, we antidifferentiate (this time with respect to x), obtaining $y(x) = \sin(x) - \cos(x) + C$ for some constant C. Then, solving the equation $2 = \sin(0) - \cos(0) + C$ for C, we find $y(x) = \sin(x) - \cos(x) + 3$.

The whole subject of indefinite integrals can be viewed as simply a special subtopic of differential equations. We can find a formula for a solution to a differential equations of the type

$$\frac{dy}{dx} = f(x) \tag{7.1}$$

by antidifferentiating $f(x)$ with respect to x. Of course, even this fundamental case is not all that simple. We know that the many antiderivatives cannot be expressed in terms of elementary functions. (For example, the antiderivatives of $\sin(x^2)$, $\sin(x^4)$, or $\sin(x^6)$, ...).

Extending Antidifferentiation

Several times we have needed to solve a differential equation of the form

$$\frac{dp}{dt} = \kappa p(t), \qquad p(t) > 0. \tag{7.2}$$

Equation (7.2) describes a function whose derivative is proportional to the function itself. In Section 6.2, we determined that p was an exponential function of t from that observation. Recognizing a solution from some special property of the function or its derivative is not a very reliable technique. In the following example, we solve Equation (7.2) by a line of reasoning that is useful for solving a number of differential equations.

To obtain an explicit formula for p in terms of t from Equation (7.2) by antidifferentiation, we would need only integrate the right hand side with respect to t. Unfortunately, in order to do so we would need an explicit formula for p in terms of t. But that is our original problem. Let's use some algebra to break out of this circle.

Why can we assume p is never zero?

If we can assume p is never 0 we can divide both sides of Equation (7.2) by p to obtain

$$\frac{1}{p}\frac{dp}{dt} = \kappa. \tag{7.3}$$

If we stare at this equation long enough, we might notice a familiar form. The left side looks like the chain rule,

$$\frac{d}{dt}f(p) = \frac{df}{dp}\cdot\frac{dp}{dt}. \tag{7.4}$$

The right hand side of Equation (7.4) shows a function of p multiplied by the derivative of p with respect to t. This matches the left hand side of Equation (7.3). What is $f(p)$ if

$$\frac{df}{dp} = \frac{1}{p}$$

Why, in this case, can we use $f(p) = \ln(p)$ instead of $f(p) = \ln(|p|)$?

The function $f(p) = \ln(p)$ has derivative $1/p$. Then Equation (7.3) becomes

$$\frac{d}{dt}\ln(p(t)) = \kappa.$$

$\int \kappa\, dt = \kappa \cdot t$

We see that the derivative of $\ln(p(t))$ equals the constant κ. We certainly also know that κt has the same derivative, κ. The Fundamental Theorem of Calculus guarantees that since $\ln(p(t))$ and κt have the same derivative, they differ by at most a constant. That is

$$\ln(p) = \kappa t + C.$$

If we use exponentiation to "undo" the logarithm, we find

$$p = e^{\ln(p)} = e^{\kappa t + C} = e^{\kappa t} e^C = C_1 \cdot e^{\kappa t} \qquad \textbf{(7.5)}$$

where we let $C_1 = e^C$.

▶ **FIRST REFLECTION**

Show that the function p in Equation (7.5) solves the differential equation (7.2) by substituting it into equation.   ◀

The Method of Separation of Variables

The reasoning we used to solve Equation (7.2) applies to a special class, the *separable* differentiable equations.

SEPARABLE DIFFERENTIAL EQUATIONS

The differential equation $\frac{dy}{dt} = f(t, y)$ is **separable** if $f(t, y)$ can be factored into a product of a function of t alone, multiplied by a function of y alone. That is,

$$\frac{dy}{dt} = f(t, y)$$
$$= g(t) \cdot h(y)$$

Whenever we recognize a separable differentiable equation, we can adapt our reasoning above as a method to produce a closed form formula. We call the method *separation of variables*. The steps of the procedure are:

1. Factor $f(t, y)$ to obtain $dy/dt = g(t) \cdot h(y)$.
2. Assume $h(y)$ is never 0 and divide both sides by $h(y)$. (This "separates" the variables t and y.)
3. Find antiderivatives for both sides of the equation. The left-hand side will always look like an application of the Chain Rule.
4. Set the antiderivative on the left equal to the one on the right plus a constant, C. The Fundamental Theorem of Calculus gives us this equality.
5. Solve for y in terms of t and C.
6. Determine the arbitrary constant, C, by using the initial value, if one is given.

▶ **SECOND REFLECTION**

We have implied that $\frac{dp}{dt} = \kappa \cdot p$ is a separable differential equation. If so, what are the functions f, g, and h?   ◀

E X A M P L E   3   **A Symbolic Solution**

We illustrate the procedure used for separation of variables for the following initial value problem.

$$\frac{dy}{dt} = 1 + t + y^2 + t \cdot y^2 \tag{7.6}$$
$$y(0) = 0$$

Typically the flow and format of the computations are as follows.

Factor to get $f(t, y) = g(t) \cdot h(y)$.

$$\frac{dy}{dt} = (1 + t) \cdot (1 + y^2)$$

Divide through by $h(y)$.

$$\frac{1}{1 + y^2} \cdot \frac{dy}{dt} = 1 + t$$

Integrate both sides with respect to t.

$$\int \left(\frac{1}{1 + y^2} \cdot \frac{dy}{dt} \right) dt = \int (1 + t)\, dt$$
$$\int \frac{1}{1 + y^2}\, dy = \int (1 + t)\, dt$$

▶ **THIRD REFLECTION**

Explain how was the Chain Rule was used in the previous calculation.   ◀

Apply the Fundamental Theorem of Calculus.

$$\arctan(y) = t + \frac{1}{2} t^2 + C$$

Next, solve for y, and we have the general solution of the differential equation.

$$y(t) = \tan \left(t + \frac{1}{2} t^2 + C \right)$$

To find the particular solution, substitute $y(0) = 0$, and solve for C.

$$0 = \tan(C)$$
$$C = 0, \pm\pi, \pm 2\pi, \ldots$$

▶ **FOURTH REFLECTION**

Why do each of the formulas $\tan(t + t^2/2 + k \cdot \pi)$, $k = 0, \pm 1, \pm 2, \ldots$ represent the same function?   ◀

Verify this.

Finally, the solution to the initial value problem given in Equations (7.6) is

$$y(t) = \tan(t + 1/2 t^2)$$

Separation of variables is a clever and useful approach for finding closed form solutions to differential equations. There are, however, a number of obstacles to the method that can arise. To begin with, we may not be able to factor $f(t, y)$. Even if we can factor $f(t, y)$, we may not be able to justify the assumption that $h(y)$ is never 0. In which case, we cannot perform the division step. Assuming we can separate the variables, there may not be closed form expressions for one or both of $\int 1/h(y)dy$ and $\int g(t)dt$. Finally, we may not be able to solve for the equation that relates the two antiderivatives for y. A lot can go wrong. Nonetheless, separation of variables allows us to find closed form solutions for a large number of important differential equations.

E X A M P L E   4   **A Trout Farm**

In the exercises of Section 6.4, a logistic model was developed for a population of unharvested trout.

$$\frac{dx}{dt} = (3 \times 10^{-6}) \cdot x \cdot (77000 - x), \qquad x > 0$$
$$x(0) = 5000$$

where $x(t) \geq 0$ is the number of trout at time t. For convenience we set $r = 3 \times 10^{-6}$, $M = 77,000$, and $x_0 = 5000$. This notation gives us a "smaller" equation.

$$\frac{dx}{dt} = r \cdot x \cdot (M - x)$$
$$x(0) = x_0$$

We factor the right hand side of this equation into the constant function $g(t) = r$ and $h(x) = x \cdot (M - x)$. Dividing by $h(x)$ or $x \cdot (M - x)$, we obtain

$$\frac{1}{x \cdot (M - x)} \frac{dx}{dt} = r.$$

Integrating both sides with respect to t, we get

$$\int \left(\frac{1}{x \cdot (M - x)} \frac{dx}{dt} \right) dt = \int r \, dt$$
$$\int \frac{1}{x \cdot (M - x)} dx = \int r \, dt.$$

A closed form for the integral on the left can determined using partial fractions.

Check that $\frac{1}{x \cdot (M-x)} = \frac{1}{M} \left(\frac{1}{x} + \frac{1}{M-x} \right)$.

$$\frac{1}{M} \left(\int \frac{1}{x} dx + \int \frac{1}{M - x} dx \right) = \int r \, dt$$
$$\left(\int \frac{1}{x} dx + \int \frac{1}{M - x} dx \right) = M \cdot \int r \, dt$$
$$\ln(|x|) - \ln(|M - x|) = rMt + C$$

$$\ln(x) - \ln(M - x) = rMt + C$$

$$\ln\left(\frac{x}{M - x}\right) = rMt + C$$

Why can we drop the absolute values in this case?

Using the condition $x(0) = x_0$ we solve for C and obtain

$$C = \ln\left(\frac{x_0}{M - x_0}\right)$$

Substituting this value for C into the last equation yields

$$\ln\left(\frac{x}{M - x} = rMt + \ln\frac{x_0}{M - x_0}\right)$$

We now exponentiate both sides and solve for x.

$$\frac{x}{M - x} = e^{rMt} \cdot \frac{x_0}{M - x_0}$$

$$x \cdot (M - x_0) = (M - x) \cdot e^{rMt} \cdot x_0$$

$$x \cdot \left(x_0 \cdot e^{rMt} + M - x_0\right) = e^{rMt} \cdot Mx_0$$

$$x = \frac{e^{rMt} \cdot Mx_0}{x_0 \cdot e^{rMt} + M - x_0}$$

$$x = \frac{Mx_0}{x_0 + (M - x_0)e^{-rMt}}$$

Reinstating the values $r = 3 \times 10^{-6}$, $M = 77,000$, and $x_0 = 5000$, we arrive at a formula for the trout population.

$$x(t) = \frac{77000 \cdot 5000}{5000 + (77000 - 5000)\left(e^{-(3 \times 10^{-6})77000t}\right)} \tag{7.7}$$

▶ **FIFTH REFLECTION**

Find the limit of $x(t)$ as t goes to infinity. Is this what you expected? ◀

Break into groups. Modify Equation (7.7) to obtain a model for the trout population for each of the following initial trout populations.

Think and Share

1. $x_0 = 0$
2. $x_0 = 10,000$
3. $x_0 = 50,000$
4. $x_0 = 77,000$
5. $x_0 = 85,000$

Graph the each equation. Overlay the the graphs. What is the long-term behavior in each case? Recommend a starting value for the Trout farm. Explain to the manager the reasoning behind your recommendation.

EXAMPLE 5

Separation of variables can be used to find closed form solutions for a number of important problems. Unfortunately, some differential equations are beyond its reach.

In Section 6.3, we constructed the differential equation

$$\frac{dy}{dt} = 4 - \frac{3y}{100 - t}$$

as a model for the amount of brine in a tank. The right side of this equation cannot be factored as a function of t times a function of y. *Try it!* Consequently, this problem cannot be solved using separation of variables. If we are to find a function which satisfies the differential equation, we must find another method.

Summary

Separation of variables only works for equations we can algebraically transform so that:

1. the right side depends only on the independent variable, and
2. the left side looks like the result of differentiation using the chain rule.

The method depends on recognizing the chain rule, and so, it is analogous to the Method of Substitution used in Section 5.5 for integration. The chain rule is not the only rule we know for differentiation. In fact, in Section 5.6 we used the Product Rule for differentiation to develop Integration by Parts. Perhaps we can make use of the product rule to solve differential equations. We will pursue this idea in the next section.

EXERCISE SET 7.1

In Exercises 1–4, determine by substitution whether the following functions are symbolic solutions to the given differential equation.

1. $\dfrac{dy}{dx} = 3x^2 e^{-y}$, $y = \ln(x^3 + c)$

2. $\dfrac{dy}{dx} = 2(x + y^2 x)$, $y = \tan(x + c)$

3. $\dfrac{dy}{dx} = \dfrac{3y(x + 1)^2}{y - 1}$, $y - \ln(|y|) = (x + 1)^3 + c$

4. $\dfrac{dP}{dt} + P = Pte^t$, $\ln(|p|) = (1 - t)(1 - e^t)$

5. Solve $\dfrac{dy}{dx} = \dfrac{y - 2}{x + 2}$, $y(-2) = 0$.

6. Solve $\dfrac{dy}{dx} = y \sin(\theta)$, $y(\pi) = 2$.

7. Solve $x\, dx + y\, dy = 0$, $y(0) = a$.

8.

a. Solve $\dfrac{dy}{dx} - 2y(t - 1) = 0$, $y(0) = 4$.

b. Find the maximum value of the function you found in part (a).

9. Solve $y\dfrac{dy}{dx} = e^{-y^2}(x + 1)$, $y(0) = -1$.

10. Solve $\dfrac{dy}{dx} = 2 + y$, $y(0) = 1$.

11. Solve $\dfrac{dP}{dt} = 0.000005\, P(t)(500 - P(t))(P(t) - 150))$,
    $P(0) = 350$

12. Solve $\dfrac{dy}{dx} = 3x^2 e^{-y}$.

13. Solve $\dfrac{dy}{dx} = 2(x + y^2 x)$.

14. Solve $\dfrac{dy}{dx} = xe^x$.

15. Solve $\dfrac{dy}{dx} = \ln(x)$.

16. Solve $e^{-x}\dfrac{dy}{dx} = x.$

17. Solve $\dfrac{dy}{dx} = y^2 \tan^{-1} x.$

18. Solve $\dfrac{dy}{dx} = y \sin^{-1} x.$

19. Solve $\dfrac{dy}{dx} = y^2 \tan^{-1} x.$

20. Solve $\sqrt{2xy}\,\dfrac{dy}{dx} = 1.$

21. Solve $(\ln(x))\dfrac{dy}{dx} = xy.$

22. Solve $\dfrac{dy}{dx} = \dfrac{e^y}{xy}.$

23. As we saw in Chapter 1, the body removes a substance from the blood at a rate proportional to the amount present. If an antibiotic is given intravenously (IV), there is a constant amount added to the bloodstream continuously. When the IV is disconnected, the input immediately drops to zero. The model is

$$\frac{du}{dt} = -ku + \left\{ \begin{array}{ll} \alpha & \text{if } t < T \\ 0 & \text{if } t \geq T \end{array} \right\}$$

$$u(0) = 1.5$$

where k is the rate of removal and α represents the constant input which is terminated at time T. To avoid obscuring the ideas with computation, take $k = 2$, $\alpha = 1$, and $T = 1$.

a. Find the solution u_t

$$\frac{du}{dt} = -2u + 1$$

$$u(0) = 1.5.$$

b. Find the value of u at the point of discontinuity $t = 1$.

c. Solve

$$\frac{du}{dt} = -2u + 0$$

$$u(0) = u_1(1).$$

Explain how this choice of initial value connects the solutions.

d. Set

$$u(t) = \left\{ \begin{array}{ll} u_1(t) & \text{if } t < 1 \\ u_2(t) & \text{if } t \geq 1. \end{array} \right.$$

e. Graph $u(t)$. Is u continuous?

f. Find a t such that the amount of antibiotic in the blood stream is $u = 0.05$.

24. Use separation of variables to derive

$$x(t) = \frac{Mx_0}{x_0 + (M - x_0)e^{-rM(t-t_0)}}, \qquad 0 < x < M.$$

as the solution to the initial value problem

$$\frac{dx}{dt} = r \cdot x \cdot (M - x)$$

$$x(0) = x_0$$

under the assumption that $0 < x < M$. Derive closed form solutions for the same initial value problem if $x_0 > M$ and so $x > M$. If $x_0 < 0$.

Explain, in your own words, the meaning of each of the following terms. Write your explanations in clear and complete sentences. Include annotated diagrams whenever appropriate.

25. symbolic solution

26. numerical solution

27. separation of variables

7.2

Linear Differential Equations

In the last section, we saw antidifferentiation as the first technique that produces symbolic solutions to differential equations. We went on to develop a second method, aeparation of variables, that can produce closed form solutions. These techniques work by recognizing special algebraic forms of an equation. Consequently, they are effective for solving restricted — but important — classes of differential equations. Antidifferention requires that we have already have an explicit closed form formula. Separation of variables requires that we be able to rewrite (transform) the differential equation so that the right hand side can be antidifferentiated immediately and the left side is identifiable as a differentiation via the Chain Rule. In this section, we focus on another algebraic class of differential equations, *linear differential equations*. The special form of linear differential equations permits us to develop a solution technique, the *Method of Integrating Factors*, based on our ability to transform the equation's left hand into the result of differentiation using the product rule. Solution techniques for differential equations tend to build on one another. Separation of variables uses antidifferentiation. We will see that the Method of Integrating Factors uses both antidifferentiation and separation of variables.

LINEAR FIRST ORDER DIFFERENTIAL EQUATIONS

A **linear differential equation** is one that can be manipulated into the form

$$\frac{dy}{dt} + P(t) \cdot y = Q(t). \tag{7.8}$$

The technique for solving linear equations is based on the product rule for derivatives.

The Product Rule.

$$\frac{d}{dt}(u \cdot y) = u \cdot \frac{dy}{dt} + \frac{du}{dt} \cdot y. \tag{7.9}$$

Notice that the left side of Equation (7.8) resembles the right side of Equation (7.9). Each is the sum of two terms, one containing dy/dt, the other y. What is missing in Equation (7.8) is a factor u in the dy/dt term. We can, however, build one.

The Method of Integrating Factors

In the course of solving the original differential equation, we will need to solve two smaller differential equations. The first of these auxiliary equations is a separable equation, the second an antiderivative. The technique follows.

1. Bring the differential equation into linear form.

This equation is separable.

$$\frac{dy}{dt} + P(t) \cdot y = Q(t)$$

2. Determine a new function $\mu(t)$ that satisfies

The product rule is at the heart of the technique.

$$\frac{d\mu}{dt} = \mu(t) \cdot P(t).$$

3. Multiply both sides by $\mu(t)$.

$$\mu(t) \cdot \frac{dy}{dt} + \mu(t) \cdot P(t)\, y = \mu(t) \cdot Q(t),$$

and simplify

$$\frac{d}{dt}\left(\mu(t) \cdot y(t)\right) = \mu(t) \cdot Q(t).$$

$H(t) = \int \mu(t)Q(t)dt$

4. Determine an antiderivative, $H(t)$ for the right hand side. By the Fundamental Theorem of Calculus $H(t)$ differs by at most a constant from $\mu(t) \cdot y(t)$.

$$\mu(t) \cdot y(t) = H(t) + C$$

5. Use the initial condition, if there is one, to determine the constant of integration C.
6. Divide by $\mu(t)$, obtaining a closed form solution to Equation (7.8).

EXAMPLE 1   **The Brine Model Solved**

In section 6.4 we developed the initial value problem

$$\frac{dy}{dt} = \frac{400 - 4t - 3y}{100 - t} \tag{7.10}$$
$$y(0) = 0$$

to model the amount of salt (in pounds) in a 100 gallon tank as a function of time t measured in minutes as brine is pumped in and simultaneously drained out. The method of integrating factors allows us to find a closed form solution.

1. Bring the differential equation into linear form

$$\frac{dy}{dt} + P(t)\, y = Q(t)$$

Starting with Equation (7.10), we clear the fractions, bring the term containing y to the left hand side of the equation, and simplify the right side to arrive at

$$\frac{dy}{dt} + \left(\frac{3}{100 - t}\right) \cdot y = 4. \tag{7.11}$$

This equation is a linear differential equation with $P(t) = 3/(100 - t)$ and $Q(t) = 4$.
2. Determine a new function $\mu(t)$ that satisfies

$$\frac{d\mu}{dt} = \mu(t) \cdot P(t).$$

Transform the differential equation for μ

$$\frac{d\mu}{dt} = \mu \cdot \frac{3}{100 - t} \tag{7.12}$$

into

$$\frac{1}{\mu} \cdot \frac{d\mu}{dt} = \frac{3}{100 - t}$$

separating the variables. Integrating, we see that

$$\ln(\mu) = \int \frac{3}{100 - t} dt$$
$$= -3\ln(|100 - t|)$$
$$= -3\ln(100 - t)$$

and, hence,

$$\mu(t) = \frac{3}{100 - t}$$

▶ **First Reflection**

We have done some potentially dangerous algebra here: dividing Equation (7.12) by the function $\mu(t)$, replacing $\ln(|100 - t|)$ with $\ln(100 - t)$, suppressing the constant of integration when determining $\mu(t)$. How can each of these steps be justified? ◀

3. Multiply both sides of Equation (7.11) by $\mu(t)$ and simplify, keeping the product rule in mind.

$$\mu(t) \cdot \left(\frac{dy}{dt} + P(t) \cdot y \right) = \frac{1}{(100 - t)^3} \cdot 4$$

$$\mu(t) \cdot \frac{dy}{dt} + (\mu(t) \cdot P(t)) \cdot y = \frac{4}{(100 - x)^3}$$

$$\mu(t) \cdot \frac{dy}{dt} + \frac{d\mu}{dt} \cdot y = \frac{4}{(100 - x)^3}$$

$$\frac{d}{dt}(\mu(t) \cdot y(t)) = \frac{4}{(100 - x)^3}$$

4. Determine antiderivatives and replace μ.

$$\mu(t) \cdot y(t) = \int \frac{4}{(100 - t)^3} dt \tag{7.13}$$

$$\frac{1}{(100 - t)^3} \cdot y(t) = \frac{2}{(100 - t)^2} + C$$

5. Use the initial condition to determine the constant of integration C.
   Substituting in $y(0) = 0$ gives us $C = -0.0002$.

6. Divide by $\mu(t)$ to solve for $y(t)$.
   Substituting our value of C into Equation (7.13) and multiplying to clear the fraction, we have a closed form solution to Equation (7.10).

$$y(t) = 2(100 - t) - 0.0002(100 - t)^3. \tag{7.14}$$

▶ **SECOND REFLECTION**

Substitute $y(t) = 2(100 - t) - 0.0002(100 - t)^3$ back into Equation (7.10) to verify that it is a solution. ◀

The window is $[0, 110] \times [-50, 100]$.

FIGURE 7.1   Solution for the Brine Tank

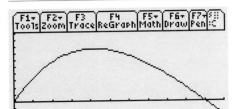

Think and Share

Break into groups. Compare Figure 7.1 with the numerical approximation computed in Section 6.4. Use Equation (7.14) to determine $y(0)$ and $y(100)$. Is this what you expected? What happens for $t > 100$? Why? Use Equation (7.14) to determine the maximum amount salt in the tank. When does it occur? Use Equation (7.14) to determine the maximum *concentration* of salt in the brine. When does it occur? Is that what you expected?

Let's apply the procedure to another example.

E X A M P L E 2 **Coffee at Sunrise**

We propose to model the temperature of a mathematician's morning coffee as it cools from from $180°$ in her mug. In Section 6.2, we developed the model *Newton's Law of Cooling* that we can apply now. Let $F(t)$ represent the temperature of the coffee as a function of time and A be the ambient temperature, then $F(t)$ satisfies the initial value problem

$$\frac{dF}{dt} = \kappa \cdot (A - F) \tag{7.15}$$
$$F(0) = 180.$$

When we used this model earlier, we assumed a constant ambient temperature. In this case, however, our mathematician pours her coffee outdoors at sunrise when the Central Texas temperature begins rising quite rapidly. The assumption of a constant ambient temperature is not valid. In particular, suppose that the outdoor temperature depends on t according to the simple linear equation

$$A(t) = 70 + 0.1\,t$$

where t is measured in minutes (giving an increase of six degrees every hour). Substituting this formula for A into Equation (7.15), we get

$$\frac{dF}{dt} = \kappa \cdot ((70 + 0.1\,t) - F) \tag{7.16}$$
$$F(0) = 180.$$

Manipulating the differential equation slightly, we identify it as linear.

$$\frac{dF}{dt} + \kappa \cdot F = \kappa \cdot (0.1\,t + 70).$$

▶ **THIRD REFLECTION**

What are the functions $P(t)$ and $Q(t)$?   ◀

The integrating factor μ is the solution of

$$\frac{d\mu}{dt} = \mu(t) \cdot \kappa.$$

This gives us $\mu(t) = e^{\kappa t}$. Multiplying by $\mu(t)$ as before, we get

$$e^{\kappa t} \cdot \frac{dF}{dt} + \kappa\, e^{\kappa t} \cdot F(t) = \kappa\, e^{\kappa t} \cdot (0.1\,t + 70).$$

Identifying the left hand side as the derivative of $e^{\kappa t} \cdot F(t)$ and antidifferentiating the right, we have

$$e^{\kappa t} \cdot F(t) = \int \kappa\, e^{\kappa t} \cdot (0.1\,t + 70)\, dt \tag{7.17}$$
$$e^{\kappa t} \cdot F(t) = e^{\kappa t} \cdot \left(70 - \frac{0.1}{\kappa} + 0.1\,t\right) + C.$$

Substitute the initial condition $F(0) = 180$ to get

$$C = 110 + \frac{0.1}{\kappa}.$$

Put the value of C into Equation (7.17) and solve for $F(t)$. We arrive at our solution to Equation (7.16).

$$F(t) = 70 - \frac{0.1}{\kappa} + 0.1\,t + e^{-\kappa t} \cdot \left(110 + \frac{0.1}{\kappa}\right) \tag{7.18}$$

The assumption of a linear temperature, while simple, is defensible. Remember, the point of the derivative is that any smooth function is locally linear. How long would the assumption be valid?

▶ **FOURTH REFLECTION**

What is $F(0)$? What is the long term behavior of this model? Justify your responses. ◀

Think and Share | Design and carry out an experiment to estimate the numerical value of the parameter κ. How does κ depend on the coffee mug?

Difficulties with Integrating Factors

As with the method of separation of variables, there are places where the Method of Integrating Factors can stall. Specifically, there are two stages at which the the Method of Integrating Factors requires the determination of a closed form antiderivative. The process can be stopped at either of these points, as in our next example.

E X A M P L E   3
_____

Find a solution for the differential equation

$$\frac{dy}{dt} + \sin(t^2)\, y = 1.$$

Finding μ proves to be a challenge, especially since $\int \sin(x^2)\, dx$ has no elementary antiderivative. Without μ, we cannot use the technique to obtain a closed form formula.

Summary

Linear equations are used to model many important behaviors, including pollution problems similar to the brine mixture considered in our first example. The mixture taking place may involve pollutants in the air or lakes, chemicals in compounds, or other mixing situations. Linear differential equations can also be used as first approximations to more complex behaviors.

We have explored techniques for finding closed form solutions to special types of differential equations. When available, symbolic solutions to differential equations can provide great insight into the model's behavior. Many important differential equation models do not have closed form solutions. To examine these models, we must rely on numerical methods or other approximation techniques. To use numeric techniques with confidence, we must have an idea of their accuracy. Although simple and theoretically important, in practice Euler's method lacks the accuracy and reliability of other more sophisticated techniques. In the next section, we will explore the error in Euler's method. In subsequent sections, we will see more stable and accurate numerical methods.

E X E R C I S E   S E T   7 . 2

In Exercises 1–4, find the solution to the initial value problem given.

1.
$$\frac{dy}{dx} + \frac{2y}{x} = -x^2 \sin(x)$$
$$y(\pi) = \pi^2$$

2.
$$\frac{dy}{dx} = 20e^{-10t} - 0.2r$$
$$r(0) = 20$$

3.
$$y\frac{dy}{dx} = e^{-y^2}(x + 1)$$
$$y(0) = -1$$

4.
$$\frac{dy}{dx} = 2 + \frac{y}{x}$$
$$y(x_0) = a$$

In Exercises 5–6, find the solution to the given differential equation.

5. $\dfrac{dy}{dx} = y + e^{3x}$

6. $x\dfrac{dy}{dx} + 2y = x^{-3}$

7. Consider
$$\frac{dy}{dx} = 2 + y$$
$$y(0) = 1.$$

  a. Determine the solution using separation of variables.
  b. Determine the solution as a linear differential equation.

8. Solve the differential equation
$$\frac{dP}{dt} = 0.000005\,P(t)(500 - P(t))(P(t) - 150)$$
$$P(0) = 350.$$

9. a. Show that the solution for
$$\frac{dy}{dx} + 2xy = 1$$
$$y(1) = 2$$

is given by

$$y(x) = e^{-x^2}\left(2e + \int_1^x e^{t^2}\,dt\right).$$

  b. Write a description of how you could attempt to calculate $y(2)$.

10. Giving examples, discuss this statement: "The family of solutions satisfying a differential equation may or may not be expressible with a single formula."

In Exercises 11 and 12, the following idea is used. A differential equation that can be put in the form

$$\frac{dy}{dx} + p(x)y = q(x)y^n$$

is called a *Bernoulli differential equation*. These equations can be made linear by a clever change of variables.

11. Find the solution to $\dfrac{dy}{dx} + \dfrac{1}{x}y = y^2$ using these steps.

  Step 1. Divide both sides by y^2.

  Step 2. Change variables to $v = y^{-1}$. Note that $v' = -\dfrac{\frac{dy}{dx}}{y^2}$.

  Step 3. Solve the new linear differential equation.

  Step 4. Change v back to y.

12. Find the solution to $\dfrac{dy}{dx} + xy = xy^5$ using these steps.

  Step 1. Divide both sides by y^5.

  Step 2. Change variables to $v = y^{-4}$. Note that $v' = -\dfrac{\frac{dy}{dx}}{y^5}$.

  Step 3. Solve the new linear differential equation.

  Step 4. Change v back to y.

13. Find the solution to $\dfrac{dy}{dx} + p(x)y = q(x)y^n$ using these steps.

  Step 1. Divide both sides by y^n.

  Step 2. Change variables to $v = y^{1-n}$. Note that $v' = -\dfrac{\frac{dy}{dx}}{y^n}$.

  Step 3. Solve the new linear differential equation.

  Step 4. Change v back to y.

Group Exercise. Many models require functions that are not continuous, yet we have only seen continuous functions in this section. In Exercise 14, we will handle a simple "on/off" function in an initial value problem (IVP).

14. As we saw in Chapter 1, the body removes a substance from the blood at a rate proportional to the amount present. If an antibiotic is given intravenously (IV), there is a constant amount added to the bloodstream continuously. When the IV is disconnected, the input immediately drops to zero. The model is

$$\frac{du}{dt} = -kn + \left\{ \begin{array}{ll} \alpha & \text{if } t < T \\ 0 & \text{if } t \geq T \end{array} \right\}$$

$$u(0) = 1.5$$

where k is the rate of removal and α represents the constant input which is terminated at time T. To avoid obscuring the ideas with computation, take $k = 2$, $\alpha = 1$, and $T = 1$.

a. *Step 1.* Find the integral curve $u_1(t)$ for the IVP:

$$\frac{du}{dt} = -2u + 1$$
$$u(0) = 1.5.$$

b. *Step 2.* Find the value of u_1 at the point of discontinuity $t = 1$.

c. *Step 3.* Find the integral curve $u_2(t)$ for the IVP:

$$\frac{du}{dt} = -2u + 0$$
$$u(0) = u_1(1)$$

Explain how this choice of "initial value" connects the integral curves.

d. *Step 4.* Set

$$u(t) = \left\{ \begin{array}{ll} u_1(t) & \text{if } t < 1 \\ u_2(t) & \text{if } t \geq 1. \end{array} \right.$$

e. Graph $u(t)$. Is u continuous?

f. Find a t such that the amount of antibiotic in the bloodstream is $u = 0.05$.

15. Apply the method of this section to the general linear equation

$$\frac{dy}{dt} + P(t)\, y = Q(t)$$

to derive

$$y = e^{-\int P\, dx} \left(\int e^{\int P\, dx}\, Q\, dx + C \right).$$

as its formal symbolic solution.

Explain, in your own words, the meaning of each of the following terms. Write your explanations in clear and complete sentences. Include annotated diagrams whenever appropriate.

16. separation of variables

17. linear differential equation

18. integrating factor

7.3

Errors in Euler's Method

Once a model is constructed, we must turn our attention to determining how well it represents reality. We compare values predicted by the model to actual observations. Thus, we must use the model to generate predictions. For a model with a closed form solution, calculating values is relatively easily. On the other hand, for a model with no closed form solution, we have to use numerical techniques to generate approximations of the predictions. These numerical methods, of course, introduce error. Even when we have a symbolic formula, computational errors can be introduced by the routines used internally in calculators or computers.

Euler's method is one of the simplest numerical techniques. Unfortunately, the method's simplicity generally results in its being impractical for professional applications. Nevertheless, the most commonly used "numerical differential equation solvers" are based on the same idea: stepping along a tangent. Studying the error in Euler's method provides a basis for understanding the error inherent in more sophisticated techniques.

Comparing Numeric and Closed Form Solutions. Since Euler's method is based on local linearity, a small change in the input will give a proportionally small change in the output. Thus, we expect that the values generated by Euler's method will be close to the true values of the solution. How close? We must answer this question if we are going to use Euler's method with any degree of confidence.

Let's look at a simple example where we can find the solution analytically in order to calculate the actual error.

E X A M P L E   1

Consider the initial value problem

$$\frac{dy}{dt} = (y-2)^2$$
$$y(1) = 1.$$

Enter the slope function `m(q)=(q-2)^2` and the stepsize `h=1.1`. Define Euler's method with sequences as we did in Section 6.3.

$$u1 = u1(n-1) + h$$
$$ui1 = 1.$$
$$u2 = u2(n-1) + h * m(u2(n-1))$$
$$ui2 = 1.$$

Set the `Graph` mode to `SEQUENCE`, use the `CUSTOM axes` menu to plot `u1` on the *x*-axis and `u2` on *y*-axis.

▶ **FIRST REFLECTION**
Plot the generated points for `nmin=1` to `nmax=16` in the window $[-0.1, 17] \times [-1, 10]$. Describe the behavior of the function from the plot.   ◀

This differential equation is separable, so we can find the symbolic solution $y(t) = 2 - 1/t$ directly.

▶ **SECOND REFLECTION**
Use separation of variables and the initial condition $y(1) = 1$ to find the symbolic solution.   ◀

Go to the HOME screen and define the solution as `f(x)`. Add the symbolic solution to your plot with the command

$$\text{DrawFunc } f(x).$$

The resulting graph is displayed in Figure 7.2. Notice that the error in Euler's

FIGURE 7.2   Euler's Method and the Symbolic Solution on $[-0.1, 17] \times [-1, 10]$

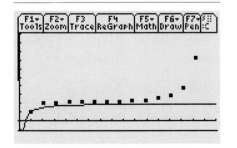

method remains reasonably small for a while and then suddenly becomes extremely large. In order to anticipate such difficulties, we must develop our understanding of the error occurs.

Analyzing Error in Euler's Method. Let's use Euler's method to generate an approximate solution for the initial value problem

$$\frac{dy}{dt} = f(y, t)$$
$$y(t_0) = y_0$$

on the interval $[t_0, t_N]$.

To avoid unnecessary complication, we consider the special case $y' = f(t)$. Using the sequence form of Euler's method from Section 6.3 gives

$$t_n = t_{n-1} + h \qquad \qquad \textbf{(7.19)}$$
$$y_n = y_{n-1} + h \cdot f(t_{n-1})$$

for n from 1 to N.

Subtracting y_{n-1} from both sides of Equation (7.19) leads to

$$y_n - y_{n-1} = h \cdot f(t_{n-1}).$$

The right side of this equation can be viewed as a one-term, left Riemann sum approximating an integral:

$$\int_{t_{n-1}}^{t_n} f(t) \, dt \approx f(t_{n-1}) \cdot (t_n - t_{n-1})$$
$$\approx f(t_{n-1}) \, h.$$

In Chapter 4, we determined that the error from left approximations is proportional to the width of the approximating rectangles. By applying the Mean Value Theorem to f and f', we can refine this result on the subinterval $[t_{n-1}, t_n]$ to obtain

$$\left| \int_{t_{n-1}}^{t_n} f(t) \, dt - f(t_{n-1}) \cdot h \right| \le M_n h^2$$

where M_n is a value larger than $\left| f'(x) \right|$ for all $x \in [t_{n-1}, t_n]$.

To extend our result over the entire interval $[t_0, t_N]$, we make a leap of faith and assume all the errors simply add together.[†] To find E_{Euler}, the total error

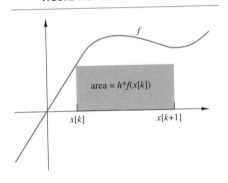

FIGURE 7.3 Area Based on Rate

area = $h*f(x[k])$

$x[k]$ $x[k+1]$

[†]We're ignoring that an error in y_1 affects y_2, which affects y_3, etc. A more detailed analysis, however, would show our total error estimate is correct. For our purposes, such analysis would add complexity without illuminating the concepts.

in Euler's method, overestimate each M_n with M, a number larger than $|f'|$ on the whole interval $[t_0, t_N]$, and compute the sum

$$E_{Euler} \leq \sum_{i=1}^{N} M h^2 = M h^2 \sum i = 1^N 1 = M h^2 N.$$

The step size h and the number of steps N are tied together according to the equation $N = (t_N - t_0)/h$. Thus,

$$E_{Euler} \leq M h^2 \frac{t_N - t_0}{h} = \left(M(t_N - t_0) \right) \cdot h.$$

Notice that the constant $M(t_N - t_0)$ depends on the interval $[t_0, t_N]$ and on the function f (through f'), but does not depend on the step size h. As a rule of thumb, we say that "The error in Euler's method is proportional to the step size."

Return to the initial value problem given at the beginning of this section:

$$\frac{dy}{dt} = (y - 2)^2$$
$$y(1) = 1.$$

Think and Share

Differentiate both sides implicitly to obtain $d^2 y/dt^2$ as a function of y. Consider the y values that were generated by Euler's method. What sort of y values should correspond to "good" Euler approximations? Compute the value of $d^2 y/dt^2$ for several of the y predicted values. What happens as the value of t increases? How does this relate to our formula for the total error?

Computation Error in Euler's Method

According to our analysis, we need only decrease h in order to increase the accuracy of Euler's method. Making the step size as small as necessary, however, only works if there are no limits to our calculations. Computers and calculators have restrictions that impose limitations on accuracy. For example, the size of computer memory and the speed of the processor limit the number of calculations that can be done. Also, the use of a fixed number of decimal places, *fixed precision*, in arithmetic operations results in errors. If our required step size is below these limits, we cannot generate a sufficiently accurate numeric solution.

Computational devices also incurr *round-off error* which results from their use of fixed decimal arithmetic. For example, using two-decimal-place arithmetic, the sum

$$\frac{1}{3} + \frac{1}{3} + \frac{1}{3} = 1$$

becomes

$$0.33 + 0.33 + 0.33 = 0.99 \neq 1.$$

As the step size goes down, the theoretical error in Euler's method also goes down. On the other hand, as the step size decreases, the round-off error goes up. Unfortunately, it is *very* difficult to determine round-off error, and its analysis requires statistical procedures. This situation is illustrated in Figure 7.4. Finding the optimal step size is an elusive goal.

FIGURE 7.4 Total Error vs. Step Size

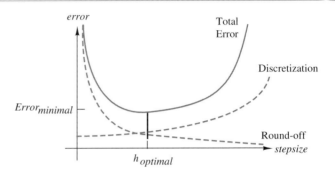

Summary

If we are to estimate a model's solution accurately, we must have some idea of how good (or bad) that approximation might be. In other words, we must understand the error in the model and in our calculations. Errors come from many different sources during the modeling process. The data may be inaccurate due to measuring devices which are badly calibrated or simply incapable of sufficiently fine resolution. The assumptions of the model may neglect important factors. These errors, for the most part, lie outside the realm of mathematics, although mathematics can shed light on their relative importance. Even if the data is sufficiently accurate and the assumptions of the model are reasonable, use of numeric techniques introduces additional sources of error. The fields of *Applied Mathematics* and *Numerical Analysis* investigate these errors. All numerical methods have theoretical limitations to their accuracy. In addition, the computational devices we use to implement such methods have their own practical limitations which create errors that are difficult to take into account and usually require statistical analysis to be adequately understood. Knowing how to control these errors allows us to use our models effectively.

EXERCISE SET 7.3

Use Euler's method with step size $h = 0.1$ to find the indicated value for the initial value problems given in Exercises 1–4. Find a symbolic formula and compare the values.

1. Find $y(2)$ where $y(x)$ is determined by

$$y' = 5 - x$$
$$y(0) = 1.$$

2. Find $y(2)$ where $y(x)$ is determined by

$$y' = 5 - y$$
$$y(1) = 0.$$

3. Find $y(3)$ where $y(x)$ is determined by

$$y' = x^2 + 1/x$$
$$y(1) = 0.$$

4. Find $y(2.5)$ where $y(x)$ is determined by

$$y' = \frac{y^2 + 1}{y}$$
$$y(1) = 0.$$

5. Calculate $y(1)$, choosing h to be 0.1, for the initial value problem

$$y' = y \sin(2x)$$
$$y(0) = 2$$

6. Given the initial value problem

$$\frac{dN}{dt} = N^2$$
$$N(0) = 1.$$

   a. Find a closed form formula for $N(t)$.
   b. Explain what happens at $N(1)$.

Chaotic Behavior Euler's method has problems that limits its use in practice. For Exercises 7–10, use the initial value problem

$$\frac{dN}{dt} = 3N - 0.01N^2$$
$$N(0) = 1$$

7. Use Euler's method with step size 0.1 to find $N(3)$. Graph your results with Style set to line.
8. Find the integral function of the IVP and graph it.
9. Describe the differences in your plots.

10. Repeat the analysis with a step size 0.05. Discuss your observations.

Euler's method is based on Riemann left sums. In Exercises 11–13, we'll develop a different refinement based on the Midpoint method

$$\int_a^{a+h} f(x)\, dx \approx \frac{f(a) + f(a + h)}{2} \cdot h.$$

11. Use the area approximation above to write a slope formula given the DE $y' = f(x, y)$.
12. Use your slope formula from the previous exercise to write a rule analogous to Euler's method.
13. Use your formula to find $y(2)$ for the initial value problem

$$y' = 5 - y$$
$$y(1) = 0.$$

Compare your answer to the symbolic result.

14. Report on the *Backward Euler's method* and discuss the results of the change the technique provides.

15. Let $y' = f(x)$ for $[t_1, t_N] = [0, 2]$. Suppose that we know

$$\max |f(x)| < \frac{1}{2}$$

Find the step size needed to keep the total error using Euler's method below 10^{-4}.

16. Let $y' = \sin(x \cdot y)$ for $[t_1, t_N] = [\pi, 8\pi]$. Find the step size needed to keep the total error using Euler's method below 10^{-3}.

Explain, in your own words, the meaning of each of the following terms. Write your explanations in clear and complete sentences. Include annotated diagrams whenever appropriate.

17. step size
18. discretization error
19. round-off error
20. fixed precision arithmetic

7.4

Improving Euler's Method

Many differential equations do not have closed form solutions. Thus, prior to the second half of this century, analyzing real world models constructed with differential equations proved to be a formidable task. Numerical methods, such as Euler's method, were known and well understood, but applying the techniques in practice was generally not feasible. The advent of computers has dramatically altered this situation. Executing 10,000 multiplications or other arithmetic operations was once considered out of the question; now, such a task is routine. The growth in the capabilities of technology has brought with it a corresponding growth in the importance of numerical techniques.

At the same time, the instruments used for obtaining data have also realized dramatic leaps in measurement accuracy. In Euler's method, the balance between the small stepsize needed for prediction precision and the large stepsize needed to avoid roundoff error is too fragile to allow the technique to be of practical use. Consequently, other more accurate methods are used. We saw in Section 7.3 that Euler's method is equivalent to the left sum method of approximating integrals. We can use other methods for estimating integrals as the basis for alternate numerical techniques.

Modifying Euler's Method

Consider the initial value problem

$$\frac{dy}{dt} = g(t, y) \tag{7.20}$$
$$y(t_0) = y_0$$

and suppose that we know it's solution is $y = f(t)$.

The Problem with Euler's Method.   We've seen that Euler's method generates a sequence, $y_1, y_2, \ldots$, that approximate $f(t)$ at a corresponding sequence $t_1, t_2, \ldots$. The process starts with choosing the step size Δt. Having chosen Δt, the values of t_n are given by the formula $t_n = t_0 + n \cdot \Delta t$. Each of the y_n's is generated from the previous point (t_{n-1}, y_{n-1}) using Equation (7.20) as shown below.

$$m_{n-1} = g(t_{n-1}, y_{n-1}) \tag{7.21}$$
$$y_n = y_{n-1} + m_{n-1} \cdot \Delta t \tag{7.22}$$

That y_n is close to $f(t_n)$ is ultimately justified from the observation that $\Delta y_{n-1} = m_{n-1} \cdot \Delta t$ is a "tangent line approximation" to $\Delta f = f(t_n) - f(t_{n-1})$. This works because tangent lines supply excellent estimators (over short intervals) to their functions. Tangent lines are not the functions themselves, but they do deviate from their functions in predictable ways. A tangent line lies above a function that is concave down and below a function that is concave up. On intervals where f is concave down, Euler's method will consistently overestimate Δf, on intervals where f is concave up, Euler's method will consistently underestimate Δf. This is responsible, in part, for the growing uncertainty in Euler's method as we use it to estimate values of f further away from the initial value.

Addressing the Problem.   We seek a procedure that will automatically decrease Euler's Δy_{n-1} when f is concave down and automatically increase Δy_{n-1} when f is concave up. To accomplish this, we'll use a strategy known as *prediction and correction*. Suppose, for the sake of argument, that f is concave down. Starting at (t_{n-1}, y_{n-1}), use Euler's method to calculate a preliminary estimate for $f(t_n)$, calling it z_n.

$$m_{left} = g(t_{n-1}, y_{n-1})$$
$$z_n = y_{n-1} + m_{left} \cdot \Delta t.$$

Since f is concave down, we expect $m_{left} \cdot \Delta t$ to overestimate $f(t_n) - f(t_{n-1})$. Now we change our perspective to the point (t_n, z_n). We calculate a new slope $m_{right} = G(t_n, z_n)$. Since f is concave down, looking backwards from (t_n, z_n), we expect $m_{right} \cdot \Delta t$ to underestimate $f(t_n) - f(t_{n-1})$.

▶ **First Reflection**

Explain why $\dfrac{dy}{dt}(t_{n-1}) \cdot \Delta t$ overestimates $f(t_n) - f(t_{n-1})$ and $\dfrac{dy}{dt}(t_n) \cdot \Delta t$ underestimates $f(t_n) - f(t_{n-1})$ when f is concave down. What is the situation when f is concave up?   ◀

The Method.   We would like the value $y_n - y_{n-1} = \Delta y_{n-1}$ to be as close to $f(t_n) - f(t_{n-1}) = \Delta f$ as possible. Given a known overestimate and a known underestimate, it is natural to average them. This idea gives us the following sequence of computations.

$$t_n = t_0 + n \cdot \Delta t$$
$$m_{left} = g(t_{n-1}, y_{n-1})$$
$$z_n = y_{n-1} + m_{left} \cdot \Delta t$$
$$t_n = t_0 + n \cdot \Delta t$$
$$m_{right} = g(t_n, z_n)$$

Average the two slopes, and then calculate the next point.

$$m = \frac{m_{left} + m_{right}}{2}$$
$$y_n = y_{n-1} + m \cdot \Delta t$$

Let's apply the new technique to the initial value problem of Example 1 from the last section and see whether it does perform better than Euler's method.

We did exactly this in Chapter 4 when we averaged the left and right Riemann sums to create the Trapezoid rule. The comparison is more than superficial.

E X A M P L E   1
_____

Use the calculations outlined above to approximate the solution to the initial value problem

$$\frac{dy}{dt} = (y - 2)^2 \qquad\qquad \textbf{(7.23)}$$

$$y(1) = 1.$$

In order to compare our new estimates with those from Section 7.3, we take Δt to be 1.1. Step from $(t_0, y_0) = (1, 1)$ to the next point, (t_1, y_1) by calculating

$$t_1 = 2.1$$
$$m_{left} = g(1, 1) = 1.0$$
$$z_1 = 1 + 1.0 \cdot 1.1 = 2.1$$
$$m_{right} = g(2.1, 2.1) = 0.01$$
$$m = \frac{1.0 + 0.01}{2} = 0.505$$
$$y_1 = 1.0 + 0.505 \cdot 1.1$$
$$= 1.5555$$

Verify the computation of y_1.

To find (t_2, y_2), repeat the process starting at $(t_1, y_1) = (2.1, 1.5555)$.

$$t_1 = 3.2$$
$$m_{left} = 0.19758$$
$$z_1 = 1.77284$$
$$m_{right} = 0.051602$$
$$m = 0.124591$$
$$y_1 = 1.69255$$

A graph showing the first 15 calculated points and the closed form solution is shown in Figure 7.5. This is a clear improvement on Euler's method as displayed in Figure 7.6.

FIGURE 7.5   Modified Euler's Method Plot and y1 Graph   **FIGURE 7.6**   Euler's Method Plot and y1 Graph

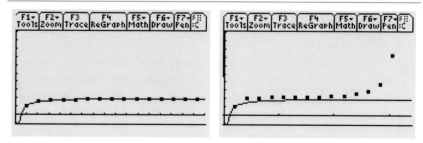

Modified Euler's Method on Your Calculator

Euler's method and Modified Euler's method only differ in the way they compute m. In Euler's method, m is the slope of a tangent line, while in Modified Euler's method, m is the average of the slopes of two tangent lines. Consequently, we can change to Modified Euler's method with just a slight change to our implementation of Euler's method. All we need to do is define a function that will calculate m. To define the slope function, `me`, for Modified Euler's method we need to use the `Program Editor`.

ENTERING A FUNCTION IN THE PROGRAM EDITOR

Choose `Program Editor` from the `APPS` menu. Select `New` and press ENTER. In the `NEW` dialog box, change the `Type` to `Function` and enter the name `me` in the `Variable` edit box. Press ENTER twice. The `Program` editor is now ready to accept your function. Use the cursor to insert the arguments (line 1) and then to move to line 3 (see Figure 7.7). In the `Program` editor, press ENTER to finish a line and go to the next line.

`:me(q,r)`	*Insert the arguments* `q,r`
`:Func`	*Indicates that we are editing a function*
`:Local m1,m2`	*Names the two variables as being temporary*
`:m(q,r) → m1`	*Calculate* `m(q,r)` *and STOre the result in* `m1`
`:m(q+h,r+m1*h) → m2`	*Calculate and STOre the result in* `m2`
`:Return (m1+m2)/2`	*Sends the result back*
`:EndFunc`	*Ends the function's definition*

The line

   `Local m1,m2`

indicates that the variables `m1` and `m2` are temporary, used internally by the function `me`, and erased when the function *returns* its calculated value.
Return to the HOME screen.

FIGURE 7.7   Entering the `Program` editor   **FIGURE 7.8**   The function `me` in the editor

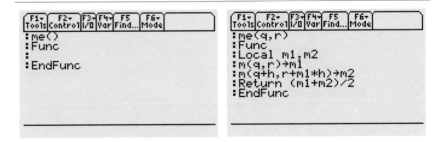

Think and Share

Use both Euler's and the Modified Euler's methods to solve the initial value problem in Equation (7.23).

```
u1(n) = u1(n-1) + h                                      time

  ui1 = 1

u2(n) = u2(n-1) + h * m(u1(n-1),u2(n-1))      Euler

  ui2 = 1.0

u3(n) = u3(n-1) + h * me(u1(n-1),u3(n-1))   Modified Euler

  ui3 = 1.0.
```

Think and Share

Enter the differential equation's slope function as m(q, r)=(r−2)^2 and set h to 1.1. Plot the solutions on the same screen by going to the Y= editor and setting Axes to CUSTOM. Set the x-axis to u1 and the y-axis to u. Remember to deselect u1.

Which method gives the better approximation of $f(12) = 11/12$? Try cutting the step size in half. Describe the result. After plotting, return to the Y= editor and reselect u1. Go to the Table screen to compare the values computed by the two methods (see Figure 7.9). Keep in mind that u1 is t, u2 is y_n from Euler's method, and u3 is y_n from the Modified Euler's method.

FIGURE 7.9 Values for Euler's Method and Modified Euler's Method

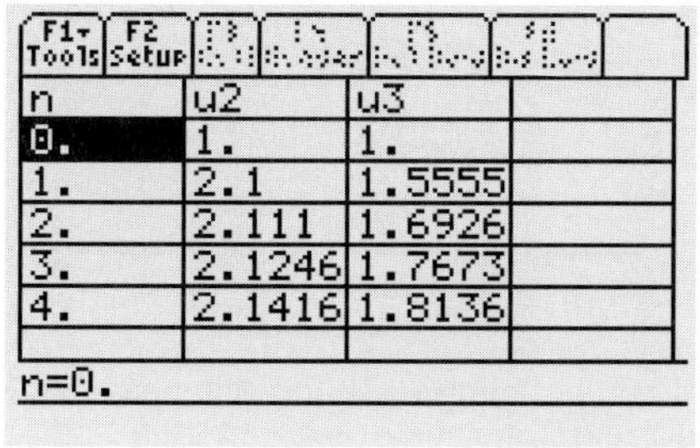

n	u2	u3	
0.	1.	1.	
1.	2.1	1.5555	
2.	2.111	1.6926	
3.	2.1246	1.7673	
4.	2.1416	1.8136	

n=0.

Error in Modified Euler's Method

In our error analysis, we compared Euler's method to a left endpoint integral approximation. Modified Euler's method can also be viewed as an integral approximation, the Trapezoid method. We leave the analysis as a project but state the result.

Let $a = t_0$ be the initial point, b be the right end point, and K_2 be a maximum of the second derivative of f throughout the interval $[a, b]$. The error for Modified Euler's, $E_{ModifiedEuler}$, with step size h is bounded according to the inequality

$$|f(t_n) - y_n| \leq \frac{1}{12} K_2 h^3 \frac{b - a}{h}$$

Collecting the constants together and adding the terms over the interval, we have

$$E_{ModifiedEuler} \leq C \cdot h^2$$

where C is a constant that depends on the interval and f, but not on h. In words, "the error is proportional to the square of the step size." Hence, halving the step size, decreases the error by a fourth. This reduction is quite an improvement over Euler's method and Modified Euler's Method requires only a little more computation.

► SECOND REFLECTION

Compare the error formula for Modified Euler's method to the error bounds given for the Trapezoid rule on page 265 of Section 4.5.   ◄

Summary

The error analysis for Euler's Method and the Modified Euler Method reveal that each method is closely related to one of the approximate integration methods developed in Chapter 4. Euler's method is analogous to a left endpoint Riemann sum, Modified Euler to the Trapezoid rule. That techniques to approximate solutions to initial value problems and to approximate integrals are closely related, should not be too surprising. In Sections 7.1 and 7.2, we saw that the symbolic solution of differential equations involved symbolic antiderivatives. The topics and histories of Integration and Differential Equations are intertwined. We know of two other methods for approximating integrals from Chapter 4: the Midpoint method and Simpson's rule. A quick check of the error in the Midpoint method reveals that it is a significant improvement over the Trapezoid method. Simpson's rule is another story. In the next section, we combine the predictor-corrector approach and Simpson's Rule to develop one of the most widely used numerical techniques for solving initial value problems.

EXERCISE SET 7.4

Use the Modified Euler's method with step size $h = 0.1$ to find the indicated value for the initial value problems given in Exercises 1–4.

1. Find $y(2)$ where $y(x)$ is determined by

$$y' = 5 - x$$
$$y(0) = 1.$$

2. Find $y(2)$ where $y(x)$ is determined by

$$y' = 5 - y$$
$$y(1) = 0.$$

3. Find $y(3)$ where $y(x)$ is determined by

$$y' = x^2 + 1/x$$
$$y(1) = 0.$$

4. Find $y(2.5)$ where $y(x)$ is determined by

$$y' = \frac{y^2 + 1}{y}$$
$$y(1) = 0.$$

5. Verify that

$$y = 2 e^{\left(\frac{1}{2} - \frac{1}{2} \cos(2t)\right)}$$

is the symbolic solution for the initial value problem

$$y' = y \sin(2x)$$
$$y(0) = 2.$$

6. Graph the solution of the following initial value problem over the interval $[0, 200]$ using a step size of $h = 4$.

$$\frac{dP}{dt} = 0.000005 \, P(t) \, (500 - P(t)) \, (P(t) - 150)$$
$$P(0) = 350.$$

Write a description of the behaviors and the differences observed using:
a. Euler's method.
b. Modified Euler's method.

7. Apply Euler's method to the initial value problem

$$y' = e^x \sin(2y)$$
$$y(0) = 4$$

to find the $y(6)$. What went wrong? Why?

8. Apply Modified Euler's method to the initial value problem

$$y' = e^x \sin(2y)$$
$$y(0) = 4$$

to find $y(6)$. What went wrong? Is it the same problem as in the previous exercise?

Chaos Revisited, Part I Both Euler's and Modified Euler's methods have problems that limit their use in practice. For Exercises 9–12, use the logistic initial value problem

$$\frac{dN}{dt} = 3N - 0.01N^2$$
$$N(0) = 1.$$

9. Use Euler's method with step size 0.1 to find $N(3)$. Graph the results with Style set to line.

10. Use Modified Euler's method with step size 0.1 to find $N(3)$. Graph the results with Style set to line.

11. Find the symbolic solution and graph it.

12. Describe the differences observed in the plots.

It is known that *round off error* is proportional to $1/stepsize$.

13. Compare roundoff error to the error in Euler's method.

14. Compare roundoff error to the error in Modified Euler's method.

The *cost* of calculating the next value, y_n, from y_{n-1} is the total number of arithmetic operations used. For Euler's method, $y_n = y_{n-1} + m \cdot h$, we have

$$cost(Euler) = one\ addition + one\ multiplication + one\ function$$
$$= 3\ arithmetic\ operations$$

where the "one function" is the slope calculation $m = f(x_{n-1}, y_{n-1})$.

15. Determine the cost of the Modified Euler's method.

16. Determine the cost of Euler's method.

17. Discuss the relative cost of each method in relation to its order of accuracy.

18. Report on *Huen's method*.

7.5

Advanced Numeric Techniques

Carle Runge and Martin Kutta were German mathematicians active at the turn of the twentieth century. We'll study the method developed by Runge in 1895 and extended to systems of differential equations by Kutta in 1901. Today an entire class of techniques fall under the heading of Runge–Kutta methods.

Simpson's Rule

We developed Simpson's rule to approximate an integral by taking a weighted average of the Trapezoid method and the Midpoint method. Simpson's Rule uses three points on each interval of length Δt to approximate the function: the left endpoint, the midpoint and the right endpoint.

On each interval $[t_{n-1}, t_{n-1} + \Delta t]$, Simpson's rule approximates the integral of the function g according to the formula

Runge was very active; at his 70th birthday party, he did handstands for his grandchildren!

$$\frac{1}{6}\left(g(t_{n-1}) + 4g\left(t_{n-1} + \frac{\Delta t}{2}\right) + g(t_{n-1} + \Delta t)\right) \cdot \Delta t. \qquad \textbf{(7.24)}$$

Runge–Kutta, the Method

Once again, we consider the initial value problem

$$\frac{dy}{dt} = g(t, y) \qquad \textbf{(7.25)}$$

FIGURE 7.11   Simpson's Rule with $h = \Delta t$ and $a = t_{n-1}$ for Convenience

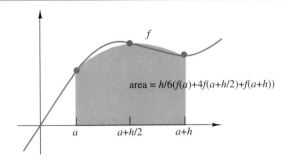

area = $h/6(f(a)+4f(a+h/2)+f(a+h))$

$$y(t_0) = y_0.$$

The outline we use for our computations is the same as we used for Euler's and the Modified Euler's methods. We pick a Δt which generates a sequence of values $t_n = t_0 + n \cdot \Delta t$. Starting from the point (t_0, y_0) we generate a sequence of estimates (t_1, y_1), (t_2, y_2), ...(t_n, y_n). We hope that each y_k approximates $f(t_k)$ acceptably well. We calculate the next estimated point, (t_n, y_n) from the preceding point (t_{n-1}, y_{n-1}) via:

$$y_n = y_{n-1} + m \cdot \Delta t.$$

We seek a way to determine m. Modeling our computation for m directly on Formula (7.24), we estimate the average slope of f on the interval $[t_{n-1}, t_n]$ as

$$m = \frac{1}{6} \left(m_{left} + 4\, m_{middle} + m_{right} \right) \qquad \textbf{(7.26)}$$

We have no difficulty computing the slope at the left endpoint, call it m_1.

$$m_1 = g(t_{n-1}, y_{n-1}).$$

However, we have a problem determining the midpoint and the right endpoint estimates for the slope. We employ a *predictor-corrector* process — calculating two values for the midpoint — in the determination of m_{middle}.

$$v_n = y_{n-1} + m_1 cdot \frac{\Delta t}{2}$$

$$m_2 = g \left(t_{n-1} + \frac{\Delta t}{2}, v_n \right)$$

$$w_n = y_{n-1} + \frac{\Delta t}{2} \cdot m_2$$

$$m_3 = g \left(t_{n-1} + \frac{\Delta t}{2}, w_n \right)$$

$$m_{middle} = \frac{1}{2} (m2 + m3)$$

To calculate an estimate for the slope at the right endpoint, we use our corrected estimate for the slope at the midpoint, m_3.

$$\begin{aligned} z_n &= y_{n-1} + m_3 \cdot \Delta t \\ m_{right} &= g(t_{n-1} + \Delta t, z_n) \\ m_4 &= g(t_{n-1} + \Delta t, z_n) \end{aligned}$$

Finally, we generate the next point in our sequence with

$$\begin{aligned} y_n &= y_{n-1} + \frac{1}{6}\left(m_{left} + 4 \cdot m_{middle} + m_{right}\right) \cdot \Delta t \\ &= y_{n-1} + \frac{1}{6}\left(m_1 + 2 \cdot m_2 + 2 \cdot m_3 + m_4\right) \cdot \Delta t. \end{aligned}$$

Runge-Kutta Method, via Sequences. Runge-Kutta is more complex than either Euler's method or Modified Euler's method. On the other hand, the complications boil down to a different technique for estimating m. We can define a function `rk` that computes m, and then Runge-Kutta can be implemented on your calculator along with the other two methods.

We define the `rk` function in the `Program` Editor as shown in Figure 7.12.

FIGURE 7.12 Program Editor

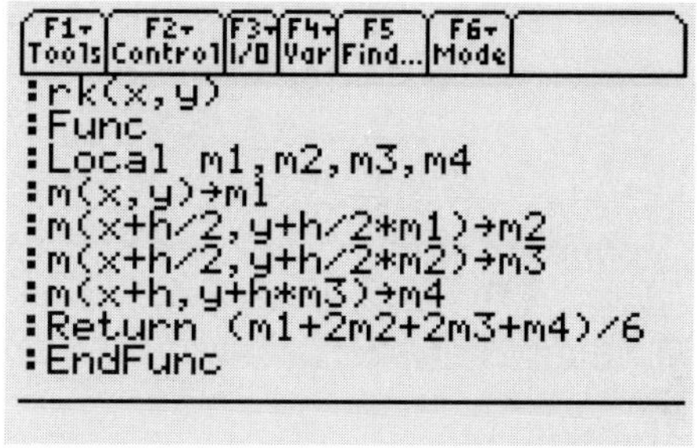
F1▾ Tools
F2▾ Control
F3▾ I/O
F4▾ Var
F5 Find...
F6▾ Mode
:rk(x,y)
:Func
:Local m1,m2,m3,m4
:m(x,y)→m1
:m(x+h/2,y+h/2*m1)→m2
:m(x+h/2,y+h/2*m2)→m3
:m(x+h,y+h*m3)→m4
:Return (m1+2m2+2m3+m4)/6
:EndFunc

EXAMPLE 1

In order to compare Runge–Kutta to the Euler methods, we return once more to our familiar example. Estimate y at $t = 12$ for the initial value problem.

$$\begin{aligned} \frac{dy}{dt} &= (y - 2)^2 \\ y(1) &= 1. \end{aligned}$$

First, define the derivative function `m(q,r)=(r-2)^2` and set the step size `h=1.1`. Add the sequence `u4` in the `Y=` editor to have

`u1(n)`	`=`	`u1(n-1)+h`	*time*
`ui1`	`=`	`1`	
`u2(n)`	`=`	`u2(n-1)+h*m(u1(n-1),u2(n-1))`	*Euler*
`ui2`	`=`	`1.0`	
`u3(n)`	`=`	`u3(n-1)+h*me(u1(n-1),u3(n-1))`	*Modified Euler*
`ui3`	`=`	`1.0`	
`u4(n)`	`=`	`u4(n-1)+h*rk(u1(n-1),u3(n-1))`	*Runge–Kutta*
`ui4`	`=`	`1.0`	

Go to the TABLE screen to compare the values generated by the three methods. The `u1` column is t, `u2` is generated by Euler's method, `u3` is from the Modified Euler's method, and `u4` is from Runge–Kutta. What approximations do the three methods give for y when $t = 12$ compared to the exact value $y(12) = 23/12$?

TABLE 7.1   First Steps

Actual $y(t_1)$	Euler y_1	Modified Euler y_1	Runge–Kutta y_1
$\dfrac{23}{12}$	2.1	1.5555	1.55029

Think and Share

Break into groups. Make three separate plots, one for each of the different methods, Euler, Modified Euler, and Runge–Kutta, timing how long each takes. Compare accuracy to speed. What do you think will happen with a complex equation over a long interval, say 0 to 100, with small steps? Make a differential equation and test your prediction.

Runge–Kutta Error

Each step of Runge-Kutta requires the computation of four slopes, which is four times the number of slopes computed for Euler's method and twice the number of slopes computed for Modified Euler's method. As might be expected, the increased number of calculations for each step does result in a significant improvement in accuracy. The analysis of error for the Runge-Kutta method can be done in the same fashion as for Euler's and Modified Euler's method.

Let $a = t_0$ be the initial point, b be right end point, and K_4 be an upper bound for the absolute value of the fourth derivative of f throughout the interval $[a, b]$. The error for the Runge-Kutta method, E_{rk}, is bounded according to the inequality

$$|y_n - f(t_n)| \le \frac{1}{180} K_4 h^5 \frac{b-a}{h}$$

Collecting the constants and adding the errors over all the subintervals, we have

$$E_{rk} \leq C \cdot h^4$$

where C is a constant that depends on the interval and f (via K_4), but not on h. The short form is "The error is proportional to the fourth power of the step size." With Runge–Kutta techniques, halving the step size reduces the error by a factor of 16. This is a considerable improvement!

► FIRST REFLECTION

Compare the error formula for the Runge–Kutta method to the error bounds given for Simpson's rule in Section 4.5. ◄

We can rate the efficiency and complexity of a numerical technique by summing the number of arithmetic operations used by the method. We count arithmetic computations as "adds" for addition or subtraction, "multiplies" for multiplication or division, "powers" for exponents and roots, and "functions" for others, such as evaluating $\sin(x)$ or just $f(x)$ in general.

Think and Share

Given the same Δt, determine the number of computations for one step of Euler, Modified Euler, and Runge-Kutta methods. How does halving Δt affect the number of computations? Compare the accuracy of Euler's method to the accuracy of Runge-Kutta when they both use the same number of computations.

A Rotating Skydiver

FIGURE 7.13 Rotating Skydiver

Let's return to our skydiver. By adding a term representing air resistance, we created a reasonable model for her jump. Nevertheless, we still ignored a host of other possibly important factors. Suppose, for example, that our skydiver slowly rotates as she falls. Her air resistance can no longer be modeled with a constant coefficient of drag.

A spotter determines by watching through a telescope that the skydiver makes one half turn every three seconds, and, from a quick calculation based on relative surface area, that her air resistance drops by 20% when she is turned sideways.

Replacing kv by $k \cdot (1 - \cos(2\pi t/6)/5) \, v$, our velocity equation becomes

$$\frac{dv}{dt} = -32 - 0.07 \left(1 - \frac{1}{5} \cos\left(\frac{\pi t}{3}\right)\right) v.$$

We've generated a linear differential equation for our model. Can we find a closed form solution? We recognize that the equation cannot be separated. To solve the equation using an integrating factor we must compute

$$\mu(t) = e^{\int 0.07\left(1 - \frac{1}{5} \cos\left(\frac{\pi t}{3}\right)\right) dt}.$$

► SECOND REFLECTION

There is a difficulty proceeding with the technique after multiplying by this integrating factor. What fails? ◄

Although our two methods for finding a closed form solution have failed, we can still analyze the new differential equation via numeric methods.

Think and Share

Break into groups. Use Runge–Kutta to generate an approximation for the flight of the skydiver. How do the rotating skydiver's heights compare to the data collected for the skydiver that wasn't spinning? Over how long a time period do you expect the approximations generated by Runge–Kutta to be reasonable? Generate a second model using Euler's method. Assuming the points generated by Runge–Kutta are more accurate, do you detect a trend in the error in Euler's method?

Summary

Modified Euler and Runge–Kutta both fall into a broad category of numeric techniques known as *predictor-corrector* methods. The common strategy of these procedures is first predicting a future point, and then using that point to correct the prediction. Runge–Kutta methods are particularly popular because they are easy to implement (on computers) and give a high degree of accuracy with relatively few calculations.

The formula

$$m = \frac{1}{6} \left(m_{left} + 2m_{mid,1} + 2m_{mid,2} + m_{right} \right)$$

can be viewed as a *weighted average* of four slopes. Advanced courses in numeric analysis investigate how to adjust the weights to achieve improved accuracy.

EXERCISE SET 7.5

Use the Runge–Kutta method with step size $h = 0.1$ to find the indicated value for the initial value problems given in Exercises 13–17. Find a symbolic solution and compare the values.

1. Find $y(1)$ where

$$y' = x^3 - x$$
$$y(0) = 0.$$

2. Find $y(2)$ where

$$y' = \sin(x) + x^2 - 1$$
$$y(0) = 1.$$

3. Find $y(1)$ where

$$y' = y^3 - y$$
$$y(0) = 1.$$

4. Find $y(1)$ where

$$y' = x \cdot \cos(y)$$
$$y(0) = \pi.$$

5. Find $y(3)$ where

$$y' = \frac{1}{y - 5}$$
$$y(0) = 0.$$

6. Compare the effects of changing the step size. Find $y(1)$ for the initial value problem

$$y' = 5 - y$$
$$y(0) = 4:$$

a. For $h = 0.1$.
b. For $h = 0.05$.
c. For $h = 0.01$.
d. For $h = 0.001$. (*If you have the time.*)

7. Graph the function determined by the initial value problem

$$\frac{dP}{dt} = 0.000005\, P(t)\, (500 - P(t))\, (P(t) - 150)$$
$$P(0) = 350$$

using the Runge-Kutta method.

8. Make additional graphs of numeric solutions determined by the initial value problem

$$\frac{dP}{dt} = 0.000005\, P(t)\, (500 - P(t))\, (P(t) - 150)$$
$$P(0) = a$$

using the Runge-Kutta method with:
a. $a = 700$.
b. $a = 400$.
c. $a = 100$.
d. $a = -100$.
Relate what you observe to a plot of

$$y = 0.000005\, x\, (500 - x)\, (x - 150).$$

9. Compare the effects of changing approximations to irrational initial values in

$$N' = N \cdot \left(N - \sqrt{12} \right)$$
$$N(0) = \sqrt{12}:$$

a. Use $N(0) = 3.4$ with a step size of 0.1.
b. Use $N(0) = 3.5$ with a step size of 0.1.

c. Find the symbolic solution directly — without integrating — by analyzing the original problem.

Chaos Revisited, Part II We saw that both Euler's and Modified Euler's methods have problems that limit their use in practice. For Exercises 22–25, use the logistic initial value problem

$$\frac{dN}{dt} = 3N - 0.01N^2$$
$$N(0) = 1.$$

10. Use the Runge-Kutta method with step size 0.1 to find $N(3)$. Graph the results with `Style` set to `line`.

11. Find the integral function of the initial value problem and graph it.

12. Describe the differences in the plots.

13. Compare the results with the those found above and in the previous section.

It is known that roundoff error is proportional to $1/stepsize$ and that truncation error for Runge–Kutta is proportional to $stepsize^4$.

14. Discuss how roundoff error changes in relation to truncation error when stepsize increases or decreases. Is this balance of errors more or less sensitive than in Euler's method?

15. What led Runge to study numerical methods of solving differential equations?

16. What led Kutta to study numerical methods of solving differential equations?

Summary

We have looked at different methods of solving differential equations that were developed as models. First, two formulas from differentiation were turned into methods that could be used. The chain rule allowed us to separate the variables into two integration problems. The product rule led us to producing an integrating factor that made a differential equation relatively easy to solve. The difficulties that are inherent in finding antiderivatives are the main limiting factors in these techniques. We desired symbolic formulas for solutions since we could use formulas more easily than tables of values to understand the behavior of the model.

Analyzing the model's properties required that we consider sources of error. There are several types of error that can arise in modeling. We investigated errors that are caused by assumptions made in developing a model, that come from data collection, and that are intrinsic to arithmetical calculation. The error due to calculation can be reduced by using more sophisticated methods of approximation; we viewed the Runge-Kutta methods as an enhancement of Euler's "follow the tangent line" technique. We saw that the choice of which method to use and what stepsize to take are made to balance truncation errors of the method with roundoff errors in the calculator or computer. Reducing error and enhancing the predictive value of a model are the genesis of the "test and refine" cycle in modeling.

Now that we know symbolic, numeric, and graphic methods of analysis, we'll continue in Chapter 8 by looking at more sophisticated problems.

SUPPLEMENTARY EXERCISES

Find the solution of the differential equations in Exercises 1–5.

1. $y' = e^x \sin(x)$

2. $\dfrac{dx}{dt} + t = x$

3. $\dfrac{dy}{d\theta} = \dfrac{e^{y+\theta}}{y+2}$

4. $x(y^3 + 1)\,dx - y^2\,dy = 0$

5. $\dfrac{dP}{dn} + nP = 0$

Find the indicated value of the solution curve for the initial value problems in Exercises 6–10.

6. Find $y(1)$ when

$$y' = t\,y^3$$
$$y(0) = 1.$$

7. Find $y(1)$ when

$$y' = t\,y^3$$
$$y(0) = -1.$$

8. Find y(4) when

$$\frac{dy}{dx} - \frac{y}{x-1} = x^2 + 1$$
$$y(2) = 1.$$

9. Find $y(1)$ when

$$y' = \frac{t}{y}$$
$$y(1) = -1.$$

10. Find P(10) when

$$P' = \frac{P}{10}(10 - P)$$
$$P(0) = 0.1.$$

In Exercises 11–14, use the indicated method to find $P(5)$ and $P(10)$ for the IVP:

$$P' = P \cdot (1 - P)$$
$$P(0) = 0.1.$$

11. a. Use Euler's method with a step size of $h = 1$.
   b. Use Euler's method with a step size of $h = 0.5$.

12. a. Use the Modified Euler's method with a step size of $h = 1$.
   b. Use the Modified Euler's method with a step size of $h = 0.5$.

13. a. Use the Runge–Kutta method with a step size of $h = 1$.
   b. Use the Runge–Kutta method with a step size of $h = 0.5$.

14. Discuss the results of the three methods.

15. Write a comparison of the way slope is chosen in Euler's, Modified Euler's, and the Runge–Kutta methods.

In Exercise 16, match the correct slope field to the differential equation.

16. a. $y' = y \cdot (2.5 - y)$ is slope field _____ .
   b. $y' = \dfrac{(3-x)(3+x)}{6}$ is slope field _____ .
   c. $y' = x^2 - y^2$ is slope field _____ .
   d. $y' = -y$ is slope field _____ .
   e. $y' = -\dfrac{x}{y}$ is slope field _____ .
   f. $y' = \dfrac{(3-y)(3+y)}{6}$ is slope field _____ .

17. *Group Exercise.* Given $\dfrac{dy}{dx} = \dfrac{2xy}{x^2 + 1}$:
   a. Graph the slope field.
   b. Use Euler's method to add solution curves through the points $(0, 4)$, $(0, 2)$, $(0, 0)$, $(0, -2)$, $(0, -4)$ to the plot.
   c. Use the Runge–Kutta method to add solution curves to a new slope field graph through the same points.

FIGURE 7.14   Slope Fields for Exercise 16

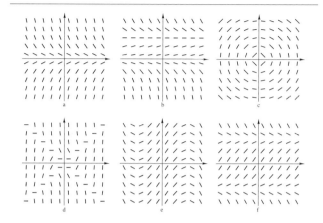

d. Solve the initial value problem symbolically. Add the function's graph to your plots.

e. Discuss what you observe.

18. *Group Exercise.* Given $\dfrac{dy}{dx} = \dfrac{2xy}{x^2 - 1}$:

a. Graph the slope field.

b. Use Euler's method to add solution curves through the points $(0, 4)$, $(0, 2)$, $(0, 0)$, $(0, -2)$, and $(0, -4)$ to the plot.

c. Use the Runge–Kutta method to add solution curves to a new slope field graph through the same points.

d. Find the solution symbolically. Add the function's graph to your plots.

e. Discuss what you observe.

19. For the inital value problem

$$y' = y^\alpha$$
$$y(0) = y_0 :$$

a. Find a closed form function family that satisfies the IVP.

b. Determine all critical values of α; that is, all values of α at which behavior abruptly changes.

c. Produce solution curves that show the generated function family's behavior.

d. Produce slope fields that show the generated function family's behavior.

20. Discuss the advantages Euler's method has over the Runge-Kutta method.

21. Discuss the advantages the Runge-Kutta method has over Euler's method.

22. How could you use a slope field to verify the solution curve generated by any numerical method?

23. Discuss Huxley's comment on theory and fact.

24. Discuss *Occam's Razor*.

P R O J E C T S

Project One: Torricelli's Law

Evangelista Torricelli was a Renaissance scientist and researcher in seventeenth century Italy. Although we'll be using his law of fluid velocity in this project, he is much better known for his work in atmospherics. He is especially remembered for having invented the barometer. Torricelli also worked on finding the value of the gravitational constant g.

Torricelli demonstrated that the velocity of water flowing from a hole in the bottom of a vessel is proportional to the free fall velocity of the *head* of the water. The **head** is the height of the water above the hole (see Figure 7.16).

It's a straightforward exercise to develop the expression for the free fall velocity from a height of h meters—do it!

Torricelli's law is

$$v_{aperture} = \sqrt{2gh},$$

where g is the force of gravity and h is the *head* of water.

FIGURE 7.16   Leaking Water

Shortly after Torricelli announced his law, the French researcher Jean Borda observed that the area (cross-sectional) of a stream of liquid is smaller than the area of the aperture through which the liquid flows. Borda's law is,

$$area_{liquid} = constant \times area_{aperture},$$

for some constant between zero and one that depends on the liquid's viscosity. This number is called *Borda's constant* and is usually between one-half and one. For water, Borda's constant is approximately 0.6. Putting these two results together gives the Torricelli–Borda law

$$v_{aperture} = b\sqrt{2gh}.$$

FIGURE 7.17   Clepsydra

Clepsydra comes from combining the words kleptein and hydor. What do these words mean?

Clepsydra.   In the ancient world, just as today, politicians and lawyers enjoyed having an audience. In the interest of fairness in court, it was necessary to monitor (and limit) the amount of time each side spoke. The Babylonians developed the *clepsydra*, a water clock, about 1400 B.C. Start with a vessel that has a hole in the bottom. Choose the shape of the vessel so that the water flows at a constant rate from the hole. Rules on the side of the clepsydra indicate how much time has passed as the water level drops. (There is only one city in North America that has a working clepsydra as the town clock. Look for its Web page to see a diagram of its water clock tower.)

Assemble your project group and do the following theoretical work:

1. Develop a first-order differential equation for the rate of change of the volume $V(t)$.
2. Solve your differential equation.
3. Show that your solution gives a constant rate of flow.
4. Choose an aperture size and find the distance between rulings that will mark five-minute intervals.

Experiment.   The second part of this project is an experiment with soapy water, which should probably be performed outdoors.

Assemble your project group.

Materials: For this procedure, you will need the following.

1. One-gallon plastic milk jug, clean and empty
2. One tablespoon of dish soap
3. 10d nail
4. Micrometer (The physics department undoubtedly has one.)
5. Paper strip
6. Yardstick
7. Stopwatch or a watch that displays seconds

Methods: Follow this procedure for the experiment.

1. Measure the nail with a micrometer. Push the nail *horizontally* into the milk jug approximately one inch above the bottom.
2. Put one tablespoon of dish soap into the jug.

FIGURE 7.18   Experimental Apparatus

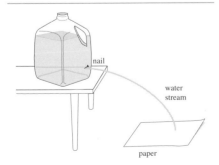

3. Fill the jug to just below the narrowing at the top with water, in order to have a tank of (relatively) constant width. Put the jug on a table.

4. Put paper on the ground and drop a plumb line (a string with a pointed weight hanging from it) to mark the point directly below the nail. Measure the height of the nail above the ground exactly.

5. At time $t = 0$, pull the nail from the jug.

6. At time $t = 0$ and at 10-second intervals, mark where the water stream hits the paper.

Analysis: Calculate the velocity of the stream from the position marks on the paper strip. Determine Borda's constant. Discuss why it did or did not change.

Project Report. Write a report on your investigations. Address your report to your classmates, assuming the material is new to them. Describe your reasoning and include all necessary background information. A minimal project report must include:

1. Development of your theoretical model for the Clepsydra.

2. A proposed experimental design to verify your model.

3. A narrative of your experiment, be sure to describe any difficulties encountered.

4. The data derived from your experiment and calculations to determine Borda's constant.

5. Justification and explanations for your calculations.

6. Suggestions for possible extensions of your observations and further questions for exploration.

Project Two: Extending the Logistic Model

In this project, you are going to enhance the logistic model. You will be looking at three refinements of the model and analyzing the results.

You will need to use the TI-92 program, `SlopeFld`, from Section 7.4.

Project 2: The Logistic Model. The logistic model can be extended easily. We developed the logistic model by observing that the growth rate in a population is jointly proportional to the current population, P, and the difference between the carrying capacity, M, of the environment and P. That is $P' = r \cdot P \cdot (M - P)$ with $P(0) = P_0$. Rewrite $M - P$ as $M \cdot (1 - P/M)$ and then set $k = rM$ so as to start with

$$\frac{dP}{dt} = k \cdot P \cdot \left(1 - \frac{P}{M}\right).$$

Use $k = 0.1$ and $M = 10$ for this project.

1. Plot a slope field for your logistic model. For your first WINDOW settings, use `xmin=0, xmax=40, ymin=-10`, and `ymax=25`.

2. Add solution curves to it that illustrate the behaviors.

3. Draw the rest points, that is, the nullclines on your graph.

We'll enhance this model three ways.

Adding a Threshold Term.   When a population dips below a critical level, the probability of a contact between breeding individuals becomes too low. The species becomes extinct. This model can explain the extremely rapid disappearance of passenger pigeons at the end of the nineteenth century. The pigeon population dipped below the threshold value, which was a surprisingly high number, and the pigeons became extinct *within 30 years*.

Include the threshold term $P/T - 1$ in your model.

$$\frac{dP}{dt} = k \cdot P \cdot \left(1 - \frac{P}{M}\right) \cdot \left(\frac{P}{T} - 1\right)$$

Use the threshold value $T = 7$.

1. Plot a slope field for your threshold model using the same WINDOW settings.
2. Add solution curves that illustrate the behaviors.
3. Draw the rest points of the threshold model on your graph.

Low-Level Harvesting.   There are two sorts of harvesters of fish: sport fishers and commercial fishers. Given a large enough fish population, the number of fish caught for sport is relatively constant. That is, sport fishing removes fish at a (relatively) constant rate. Modify the logistic model by subtracting h, the constant rate of harvest.

$$\frac{dP}{dt} = k \cdot P \cdot \left(1 - \frac{P}{M}\right) - h$$

Use the low-level harvest rate $h = 0.2$.

1. Plot a slope field for your low-level harvest model using the same WINDOW settings.
2. Add solution curves that illustrate the behaviors.
3. Draw the rest points of the low-level harvest model on your graph.

Intense Harvesting.   Commercial fishers are much more effective at removing fish from the population. They are highly concentrated and specialized. A more reasonable term to account for commercial fishing is a percentage of the fish population. Account for commercial fishing with a term of the form $-HP$.

$$\frac{dP}{dt} = k \cdot P \cdot \left(1 - \frac{P}{M}\right) - H \cdot P$$

Use the intense harvest rate percentage $H = 0.05$.

1. Plot a slope field for your commercial harvest model using the same WINDOW settings.
2. Add solution curves to it that illustrate the behaviors.
3. Draw the rest points of the commercial harvest model on your graph.

Extension.

1. Discuss the passenger pigeon extinction in light of the threshold model.

Project Report.   Write a report on your investigations. Address your report to your classmates, assuming the material is new to them. Describe your reasoning and include all necessary background information. A minimal project report must include:

1. Description of the long-term behavior for each of your models.
2. Results (tabular and graphical) of experiments you ran with each model.
3. Illustrative and annotated slope fields showing trajectories.
4. A discussion of any obvious shortcomings in the model.
5. A discussion as to how one might verify the validity of a model in either a natural or laboratory setting.
6. Justification and explanations for your calculations.
7. Suggestions for possible extensions of your observations and further questions for exploration.

Project Three: The Rotating Skydiver Model

In this project, you are going to further study the skydiving model. You will be looking at a refinement of the model, adding rotation, and analyzing the results.

Project Three: The Skydiving Model.   We developed the skydiving model and refined it by adding air resistance. At the end of Section 7.5, we further refined the model by considering a changing coefficient of drag based on observing that the skydiver was rotating as she fell. Our new differential equation was

$$\frac{dv}{dt} = -32 - 0.07\left(1 - \frac{1}{5}\cos\left(\frac{\pi t}{3}\right)\right)v.$$

1. Use Euler's method with an appropriate Δt to generate a table of values of the skydiver's velocity $v(t)$ for $t \in [0, 15]$.
2. Determine error bounds for your data.
3. Use the relation

$$\frac{\Delta h}{\Delta t} = v$$

    to find a recurrence equation for h_n.
4. Using $h_0 = 10,000$, construct a table for the skydiver's height. Compare your data to to the data in Table 7.2.

Project Report.   Write a report on your investigations. Address your report to your classmates, assuming the material is new to them. Describe your reasoning and include all necessary background information. A minimal project report must include:

1. Description of the long-term behavior of your models.

2. Results (tabular and graphical) of experiments you ran with each model.
3. Illustrative and annotated slope fields showing trajectories of the skydiver.
4. A discussion of any obvious shortcomings in the model.
5. Justification and explanations for your calculations.
6. Suggestions for possible extensions of your observations and further questions for exploration.

8

MODELING WITH SYSTEMS

8.1
Spirals of Change: You Are What You Eat

8.2
Modeling

8.3
Numerical Solutions: Iteration and Euler's Method

8.4
Symbolic Solutions of Systems of Differential Equations

8.5
Bungee Jumping

Summary

Many important problems in biology, economics, environmental science, and physics require understanding situations where two or more linked quantities are changing over time. In these situations we require two or more dependent variables, one for each of the quantities, as well as the independent variable, time. This use of more than one dependent variable leads to models involving more than one equation; that is, *systems* of equations.

To obtain a discrete model, we apply the simple paradigm

$$future = present + change$$

to each of the time dependent quantities that lead to systems of **difference equations**. If, on the other hand, we choose to create a continuous model, then we model the instantaneous rate of change of each time dependent quantity that leads to a system of **differential equations**.

As in previous chapters, we will use our models to predict the future . What long-term behavior is suggested? Are the predicted outcomes for the system sensitive to the starting conditions? If so, which starting conditions lead to which outcomes? Since models only approximate reality, we also will investigate the sensitivity of outcomes to changes in the constants (or parameters) of the model. What range of values for specific constants are associated with particular outcomes?

8.1

Spirals of Change: You Are What You Eat

Discrete systems of difference equations are extremely valuable for modeling complex interactions. In this section, we develop a discrete model for an important biological interaction.

Predators and Prey

Consider two species, one of which derives some or all of its nutrition by eating the other — who naturally isn't too crazy about this arrangement. Baleen whales eating krill, lynxes eating hares, wolves eating caribou, and parasitic wasps dining on their prey beetles are all examples of this predator (eater) and prey (eaten) relationship. For species linked as predator and prey, changes in one population will cause changes in the other. In order to visualize changes caused by these interactions, we will introduce new graphical techniques.

▶ FIRST REFLECTION
How would you expect a decrease in the number of prey to affect the size of the predator population? How would a decrease in the number of predators affect the size of the prey population? Explain your response.   ◀

Building the Model

We track the populations of predator and prey at fixed-time intervals. Ideally we choose these intervals to have a biological significance for both species of our model. The appropriate size of the interval varies with the species we are modeling. Whales reproduce on the scale of years, lynxes on the scale of months, and wasps, weeks. The units used for the two populations are likely to be different. It is not uncommon that the predator population is measured in number of individuals, while the prey population is measured in thousands or even millions of individuals. In some cases (yeast), we may simply track the total mass of each population rather than attempt a head count.

▶ SECOND REFLECTION
Suggest appropriate time intervals and units of population for the four examples mentioned in the opening paragraph (that is, baleen whales eating krill, lynxes eating hares, wolves eating caribou, and parasitic wasps eating beetles).   ◀

In order to construct a simple and manageable model, we make the following assumptions regarding sources of change in the two populations.

1. In the absence of the prey species, the predator population will decrease by a fixed proportion a during any of the tracking intervals.
2. In the absence of the predator species, the prey population will increase by a fixed proportion c during any of the tracking intervals.

3.   A certain proportion of the potential interactions between the two species result in an increase in the population of the predator and a decrease in the population of the prey during any of the tracking intervals.

▶ THIRD REFLECTION

Why does the predator population decrease in the absence of the prey species? That is, what is happening to the predator population?   ◀

The terms $b \cdot x_{n-1} \cdot y_{n-1}$ and $d \cdot x_{n-1} \cdot y_{n-1}$ each measure the effects of predator eating prey, which has a positive effect on the predator population and a negative effect on the prey population.

We measure our independent variable, time, by the number of intervals that have passed since we began tracking the two species. We have two dependent variables corresponding to the sizes of the two populations. After $\{n - 1\}$ time intervals, $\{x_n\}$ represents the size of the predator population and $\{y_n\}$ represents the size of the prey population. Since the data are discrete, we apply the discrete paradigm

$$future = present + change$$

to each of our two populations and obtain the system of equations

$$
\begin{aligned}
x_n &= x_{n-1} + (-a \cdot x_{n-1} + b \cdot x_{n-1} \cdot y_{n-1}) \\
y_n &= y_{n-1} + (c \cdot y_{n-1} - d \cdot x_{n-1} \cdot y_{n-1}) \\
x_1 &= Q \\
y_1 &= R.
\end{aligned}
\qquad \textbf{(8.1)}
$$

The values of the parameters, a, b, c, d, Q, and R, must be positive and depend on the particular species being studied. In addition, a and c must be between 0 and 1 because they are the proportions of a whole.

▶ FOURTH REFLECTION

Which of our assumptions would be violated if a were negative? Explain your choice.   ◀

Exploring the Model

Let's model a particular predator/prey interaction where yearly time periods are appropriate.

Arctic wolves hunt caribou in one of the most fascinating and well-studied examples of the predator/prey relation. Caribou in flight run at speeds of about 30 miles per hour, a full 10 miles per hour faster than the pursuing wolves. The smaller, slower wolves exploit their one physical advantage over the caribou, a slight edge in stamina. Wolves will run hour after hour until they finally overtake and bring down their exhausted quarry.

Caribou feed on vegetation. Without wolves to keep their numbers in check, the caribou population becomes so large that it destroys the available food source by over-grazing. With scarce food, the caribou fall victim to starvation and disease, their numbers dropping precipitously. Wolves hunt. If the population of the caribou becomes too small, the wolves are unable to make enough kills to sustain the pack

Given the annual reproductive periods for both animals, we choose to track their interaction with yearly time periods. We cannot easily obtain values for the parameters of our model. Many scientific studies are designed to estimate such parameters, and given the complexity of interaction, estimates are all that can be expected. We will assume a reasonable set of values for our parameters: $a = 0.2$, $b = 0.002$, $c = 0.5$, $d = 0.025$, $Q = 25$ and $R = 120$. To enter this model on your TI calculator, switch to SEQUENCE mode, go to the Y= editor and enter

```
u1(n)=(1-.2)*u1(n-1)+.002*u1(n-1)*u2(n-1)
ui1=25
u2(n)=(1+.5)*u2(n-1)-.025*u1(n-1)*u2(n-1)
ui2=120.
```

▶ **FIFTH REFLECTION**

What does the value of 0.5 assigned to c say about the caribou population? Explain the meaning of the values Q and R in terms of the caribou and wolves. ◀

Table 8.1 shows the sizes of the two populations (rounded to the nearest one-hundredth of an individual) predicted by the model over the first eight years.

TABLE 8.1   Wolves and Caribou

Year	1	2	3	4	5	6	7	8
Wolves	25	26.0	26.26	25.70	24.43	22.69	20.76	18.83
Caribou	120	105	89.25	75.28	64.56	57.42	53.55	52.53

▶ **SIXTH REFLECTION**

Compute $\{x_2\}$ and $\{y_2\}$ by hand and check that your results match the Year Two values in the table. Verify the rest of the table by viewing the Table screen on your TI calculator. ◀

Plot the two sequences for wolf and caribou populations, using the WINDOW settings

```
nmin=1,nmax=40,
plotstrt=1,plotstep=1,
xmin=0,xmax=40,xscl=10,
ymin=0,ymax=50,yscl=10.
```

Figure 8.1 shows the population of the wolves over the first 40 years.

Figure 8.2 shows the population of the caribou over the same time period. We can see that the populations vary, but the images do not tell us much

The WINDOW *settings for Figure 8.2 are the same as for Figure 8.1 except for* ymax=250.

FIGURE 8.1 Wolf Population   **FIGURE 8.2** Caribou Population

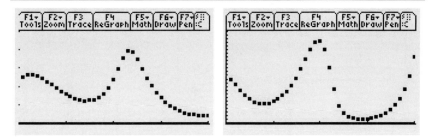

about how the two species interact. Plotting both populations on the same axes as in Figure 8.3 offers no new insight. On the contrary, the difference in scales between the wolf and caribou populations makes it harder to see the variation in the wolves. These graphs display the development of the two populations indi-

To obtain the graph in Figure 8.3 go to the Y= *editor. Highlight the formula for* u1, *open the* Style *menu and select* Line. *Repeat the process to assign the* Style Thick *to* u2.

FIGURE 8.3 Predator and Caribou Populations
Plotted Against Time

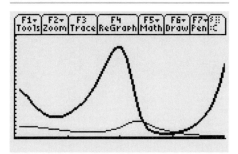

vidually over time. We would like a graph that explicitly reveals the interaction between the two species. Fortunately, we can produce just such a graph.

Visualizing the Model

To view the interaction between the wolves and the caribou, we plot the population of one against the population of the other. In Figure 8.4 we have plotted the points $\{(x_n, y_n) | n = 1 \ldots 40\}$ and connected them in sequential order, looking for a hidden image, just as in a child's connect-the-dots book.

There *is* a hidden image. The two populations are locked in an ever-widening spiral. Notice that the passage of time is reflected by the direction that the curve follows as it is drawn. In the next image, we have added arrows between successive stages to indicate the direction the populations are cycling as time increases. (For example, the first arrow starts at $(x_1, y_1) = (25, 120)$ and ends at $(x_2, y_2) = (26, 105)$.) This particular spiral is bad for both populations,

500   CHAPTER 8   MODELING WITH SYSTEMS

In the Y= *editor, from the* Axes ...
dialogue box choose CUSTOM *with* u1
on the X Axis *and* u2 *on the* Y Axis.
Assign the Style Thick *to both of* u1
and u2.

WINDOW *settings are*
nmin=1, nmax=40,
plotstrt=1, plotstep=1,
xmin=0, xmax=40, xscl=10,
ymin=-1, ymax=250, yscl=10.

FIGURE 8.4   Caribou Population (on the *y*-axis)
Plotted Against Wolf Population (on the *x*-axis)

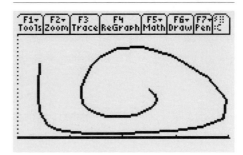

FIGURE 8.5   The Arrows Mark Transitions from Stage to Stage

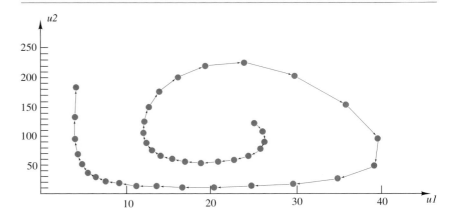

as we see by looking a little further into the future predicted by our model. In
year 47 the model predicts a negative population of prey. This is *very* bad for
the caribou and not good for the wolves.

The WINDOW *settings are*
nmin=1, nmax=47,
plotstrt=1, plotstep=1,
xmin=0, xmax=125, xscl=10,
ymin=0, ymax=700, yscl=100.

FIGURE 8.6   Extinction

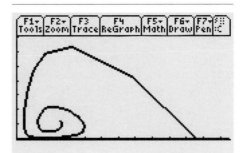

In the past, wolves and other large predators have dealt with this phenomenon by migrating to better hunting grounds. This strategy is becoming less and less viable by the intrusion of humans into natural habitats. The caribou and other prey species adopt a different strategy. They manipulate their own fertility in order to respond to changing conditions. If times are tough, animals may fail to come into heat (estrus requires a certain percentage of body fat), spontaneously abort (marsupials are particularly quick to abort a fetus), or practice selective or total infanticide (a sow will eat her own or another's young under conditions of stress).

Rapid Response.   Let's see how the situation changes if we change our model to lynxes eating hares. These animals reproduce more frequently than do wolves and caribou. Hence, we should make more frequent observations; that is, we should shorten our time interval. Accordingly, we alter our model to use quarterly (rather than yearly) time intervals but maintain the same annual rates of decline to facilitate comparisons. Thus, the lynxes suffer an annual 20% decline in the absence of hares, which translates to an estimated quarterly decline of $\frac{0.2}{4}$. Similarly, the 50% annual increase in the hares when no lynxes are present becomes a quarterly increase of $\frac{0.5}{4}$ for the hares. Our equations become

$$x_n = x_{n-1} + \left(-\frac{0.2}{4}\right) x_{n-1} + \left(\frac{0.002}{4}\right) x_{n-1} \cdot y_{n-1}$$

$$y_n = y_{n-1} + \left(\frac{0.5}{4}\right) y_{n-1} - \left(\frac{0.025}{4}\right) x_{n-1} \cdot y_{n-1}$$

$$x_1 = 25$$

$$y_1 = 120.$$

The next figure plots the interaction of these two populations over 50 years (two hundred quarters), that is, for $n = 1, \ldots 50 \times 4$.

The WINDOW settings for Figure 8.7 and Figure 8.8 are nearly the same as for Figure 8.4, only nmax *and* yscl *are changed. The settings are*
nmin=1,
(for Figure 8.7 use nmax=50*4
— for 8.8 use nmax=50*12*),*
plotstrt=1, plotstep=1,
xmin=0, xmax=40, xscl=10,
ymin=-1, ymax=250, yscl=100.

FIGURE 8.7   Tracking Prey and Predator Populations at Quarterly Intervals

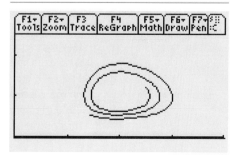

FIGURE 8.8   Tracking Prey and Predator
Populations at Monthly Intervals

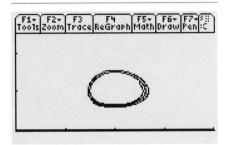

► **SEVENTH REFLECTION**

Write down the system of difference equations in the box below that would model our populations if we were to make monthly rather than yearly observations.   ◄

The spiraling remains, and apparently at the same rate. Although the amplitude is smaller, there seem to be about two and a third spirals in both images.

Form into groups of four, then separate into two pairs. One of these pairs will track predators; the other, prey. One member of each pair should create the graph shown in Figure 8.7, the other should construct the graph shown in Figure 8.8. For each graph, the predator (prey) team should use the `Trace` feature to determine the values of n that correspond to points where the curve stops moving right and starts moving left (stops moving up and starts moving down for the prey team) and to points where the curve stops moving left and starts moving right (stops moving down and starts moving up). These points correspond to times when the predator population is at a high point or a low point. Meanwhile, the prey team should use the `Trace` feature to determine the values of n that correspond to points where the curve stops moving up and starts moving down and to points where the curve stops moving down and starts moving up. These points correspond to times when the prey population is at a high point or a low point. Each team member should use their graph to answer the following questions and then share the results with their partner. The predator and prey teams should then compare their findings with each other.

Think and Share

- Suppose your species population is at a high point. How many years does the model seem to predict before the population will reach its next low point?
- How long will it take for your species population to be back at a high point?
- How does the change in the size of your species population when the other species is abundant compare to its change when the other species is scarce?
- Figure 8.7 or Figure 8.8 shows the prey population increasing toward a point on the curve and decreasing after a point on the curve. Determine a point on the curve where the prey population stops increasing and starts decreasing. Interpret that meaning of this point in terms of the predator population.
- Determine an interval of the curve where both the prey and predator populations are increasing. Interpret the meaning of this interval in terms of both the prey and predator populations.

Analyzing the Model

We can use algebra to understand the spiraling structures displayed by our model. The general predator-prey model, Equation (8.1), repeated below is

$$x_n = x_{n-1} + (-a \cdot x_{n-1} + b \cdot x_{n-1} \cdot y_{n-1})$$
$$y_n = y_{n-1} + (c \cdot y_{n-1} - d \cdot x_{n-1} \cdot y_{n-1})$$
$$x_1 = Q$$
$$y_1 = R,$$

where all parameters are positive and $0 < a < 1$, $0 < b < 1$. When considering models for the change in a single quantity in previous chapters, rest values for the quantity often illuminated the characteristics of the model. Here, we consider two quantities, both of which must remain fixed. The rest points for the two populations, if any, are points (x, y) for which the change in both populations is zero, or points (x, y) that satisfy the equations

$$0 = x \cdot (-a + b \cdot y)$$
$$0 = y \cdot (c - d \cdot x).$$

The TI-92 solve(*command will not solve systems of equations. The TI-89 and TI-92 Plus module* solve(*command will.*

The solutions, $(0, 0)$ and $(c/d, a/b)$, are a pair of rest points for the system of difference equations. What is their biological meaning? The point $(0, 0)$ corresponds to zero predators and zero prey and should be a rest point by anyone's criteria. As for $(c/d, a/b)$, we can at least note that in our example with lynxes and hares, $(c/d, a/b) = (40, 100)$ and that $(40, 100)$ is more or less at the center of the spiral.

▶ **EIGHTH REFLECTION**

Verify by direct computation that $(c/d, a/b)$ is a rest point for Equation (8.1). That is, plug in $x_{n-1} = c/d$ and $y_{n-1} = a/b$ and calculate x_n and y_n. Use (40,100) as a starting point for the model of lynxes and hares. What happens to the values of u1 and u2 as time increases?   ◀

Summary

In general, we can use a system of two or more difference equations to investigate situations that involve interaction between two or more dependent variables. Not surprisingly, complicated situations can be hard to analyze and often have patterns that are difficult to find. The choice of axes in our graphs may determine what we can see. The graphs of dependent variables plotted one against the other can reveal patterns that are not apparent when each is plotted against time.

EXERCISE SET 8.1

1. Put arrows on the graphs in Figure 8.1 and Figure 8.2 to indicate the direction of increasing time.

2. The graph in Figure 8.1 has a high point and then declines all the way to the horizontal axis. How are the two populations

changing during this decline? Explain why each changes as it does.

† **3. Group Exercise** Explain how the following three assumptions can be interpreted algebraically as the difference equations of Equation (8.1).

    a. In the absence of the prey species, the predator population will decrease by a fixed proportion a during any of the tracking intervals.

    b. In the absence of the predator species, the prey population will increase by a fixed proportion c during any of the tracking intervals.

    c. A certain proportion of the potential interactions between the two species result in an increase in the population of the predator and a decrease in the population of the prey during any of the tracking intervals.

4. For the recursion equations

$$x_n = 0.94x_{n-1} + 0.0006x_{n-1} \cdot y_{n-1}$$
$$y_n = 1.15y_{n-1} - 0.0075x_{n-1} \cdot y_{n-1},$$

    a. With the initial conditions $x_1 = 25$ and $y_1 = 0$, plot x_n against n. A plot of $\{x_n\}$ against n is known as a *time series*.

    b. With the initial conditions $x_1 = 0$ and $y_1 = 120$, plot y_n against n.

    c. Explain the resulting graphs in light of the first two assumptions of the preceding exercise.

5. Using your system of recursion formulas for a pair of species (Seventh Reflection) responding to changing conditions on a monthly basis, graph the trajectories over 480 months of populations that start with each of the following initial conditions. (Use the predator population as the horizontal axis, the prey population as the vertical axis.)

    a. $x_1 = 20$ and $y_1 = 100$

    b. $x_1 = 19$ and $y_1 = 100$

    c. $x_1 = 10$ and $y_1 = 100$

6. In many places and times, it has been customary to kill predators (wolves, coyotes, eagles) on sight. This harvesting of predators changes the interaction of predator and prey species. Likewise, hunting of prey by humans will also alter the interaction of the species. We used the system of equations

$$x_n = \left(1 - \frac{.2}{4}\right)x_{n-1} + \frac{.002}{4}x_{n-1} \cdot y_{n-1}$$
$$y_n = \left(1 + \frac{.5}{4}\right)y_{n-1} - \frac{.025}{4}x_{n-1} \cdot y_{n-1}$$
$$x_1 = 25$$
$$y_1 = 120$$

to model the interaction between a predator and its prey in each quarter.

    a. Modify the predator's equation to account for yearly losses of a single predator (an average of one-quarter individual per quarter) by hunters.

    b. Track the resulting populations over 200 intervals (50 years in our model).

    c. Add a term to the prey's equation that models a yearly hunt killing four individuals annually.

    d. Track the resulting populations if both predator and prey are hunted.

    e. Track the populations if predators are protected while prey are hunted at a rate of four individuals a year.

† **7.** *Group Exercise* A commercial lumber company detects an outbreak of beetles that threatens the productivity of one of the forests it manages. Your company supplies biological controls for various pests, and you raise a species of wasp that preys on this beetle. With the population of both species measured in thousands of individuals per square mile, your research indicates that the interaction of the two species is governed by the recursion equations

$$w_n = (0.999)w_{n-1} + 0.00001w_{n-1} \cdot b_{n-1}$$
$$b_n = (1.05)b_{n-1} - 0.0025w_{n-1} \cdot b_{n-1}$$

where n measures time in weeks, w_n represents the number of wasps at week n, and b_n represents the number of beetles at time n. The lumber company estimates that the current level of infestation is 120,000 beetles per square mile. At that level, the forest is suffering minor damage. The forest suffers no damage as long as the number of beetles remains below 50,000 per square mile. Moderate damage is suffered once the number of beetles exceeds 250,000 per square mile, with severe damage being sustained at concentrations above 400,000 beetles per square mile. You are asked to prepare an analysis of the lumber company's treatment options. The company will make a single treatment and intends to take no further control measures for five years. Your company can supply wasps at levels of 10,000, 15,000, and 20,000 individuals per square mile. Prepare an analysis of the advantages and disadvantages of each of the three treatments. Include estimates as to the level of damage the forest will suffer under each treatment and the length of time it will suffer that damage. Conclude by recommending a course of action.

8. In the vector field in Figure 8.9, trace the estimated trajectory of the populations from each of the indicated starting points, A, B, C, and D. Explain the meaning of the trajectories in your own words. The central point of the axes is a rest point.

The predator is positive up and the prey is positive to the right.

9. In the vector field in Figure 8.10, trace the estimated trajectory of the populations from each of the indicated starting points, A, B, C, and D. Explain the meaning of the trajectories in your own words. The central point of the axes is a rest point. The predator is positive up and the prey is positive to the right.

10. In the vector field in Figure 8.11, trace the estimated trajectory of the populations from each of the indicated starting points, A, B, C, and D. Explain the meaning of the trajectories in your own words. The central point of the axes is a rest point. The predator is positive up and the prey is positive to the right.

FIGURE 8.10

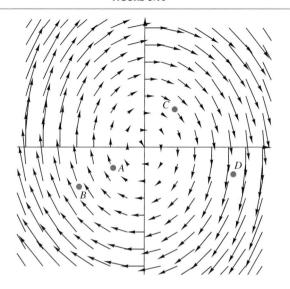

FIGURE 8.9

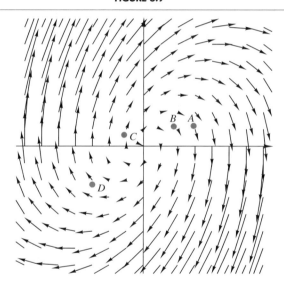

FIGURE 8.11

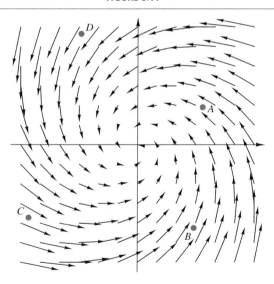

8.2

Modeling

The Modeling Construct

The construct

$$future = present + change$$

will lead to a *system of difference equations* if the change is taking place over **discrete** time periods. For example, consider the number of people voting Democrat or Republican in a two-party political system where the discrete time period is the time between elections. On the other hand, if the change is taking place **continuously** with respect to time, such as the flow of heat in a mechanical system, the construct will lead to a *system of differential equations*.

Difference and Differential Equations

Difference and differential equations are closely related. In this section, we derive systems of differential equations by first constructing a system of difference equations. In addition to modeling discrete behavior, difference equations are powerful, because when the time intervals are small they can be used to **approximate** a corresponding system of differential equations. If starting conditions are given, a system of difference equations . can be *iterated*, or evaluated recursively. Later, in Section 8.4 we will direct our attention to finding analytic, or symbolic, solutions to systems of differential equations where they can be found. We will begin by modeling with systems of difference equations and of differential equations.

FIGURE 8.12   Reaction Data Graph

E X A M P L E   1   **More Chemicals**

You are still an analyst for CMM Chemical Corps. (It's easy to hold a job when you know a little mathematics.) In Section 6.1 and 7.1, you modeled and analyzed the amount of *A* as a difference equation and as a differential equation. But if you remember, there were two more chemicals involved in the reaction. To quickly review, *A* produces *B* which produces *C* (see Figure 6.1). The chemist provided the data below and that *A* costs $0.50 per mole, *B* provides a profit of $3.50 per mole, and *C* provides a profit of $1.00 per mole. She wants to know when to stop the reaction to maximize the profit.

We begin our modeling process by first defining our variables for the discrete model

$$A_n = \text{the number of moles of chemical } A \text{ at the } n\text{th time period}$$
$$B_n = \text{the number of moles of chemical } B \text{ at the } n\text{th time period}$$
$$C_n = \text{the number of moles of chemical } C \text{ at the } n\text{th time period.}$$

TABLE 8.2 Reaction Data

Index	1	2	3	4	5	6	7	8	9	10	11	12
Time (min)	0	2	6	10	20	30	50	70	90	120	150	200
A (mole)	1.00	0.88	0.69	0.53	0.28	0.15	0.043	0.012	0.00	0.00	0.00	0.00
B (mole)	0.00	0.12	0.29	0.42	0.56	0.57	0.46	0.33	0.22	0.12	0.06	0.02
C (mole)	0.000	0.003	0.030	0.050	0.16	0.28	0.50	0.66	0.78	0.88	0.94	0.98

The variables for the continuous model are

$$a(t) \quad = \quad \text{the number of moles of chemical } A \text{ at time } t \text{ in seconds}$$

$$b(t) \quad = \quad \text{the number of moles of chemical } B \text{ at time } t \text{ in seconds}$$

$$c(t) \quad = \quad \text{the number of moles of chemical } C \text{ at time } t \text{ in seconds.}$$

You may have already modeled the amount of C in the exercises for Section 6.1. Whether or not you did, we will take a different approach here. If you remember our model of A, we said that the rate of change of A was proportional to the amount of A. Although the time intervals for our data are unequal, in order to construct a difference equation model, we must choose a fixed time interval Δt. This done, the difference equation model is $A_n = A_{n-1} - k_A A_{n-1}$. The corresponding differential equation model is $\frac{da}{dt} = -\kappa_A a$. In words, these models say that the change in A is proportional to the amount of chemical A present.

Let's look at C next.

▶ FIRST REFLECTION

Recall the data table `Reaction`. If it is not in your calculator, input the data from Table 8.2 with time in `c1` and A in `c2`. Input the data for B in `c3`, and C in `c4`. Label your columns so that you can keep track of what is where. Do a scatter plot of the amount of C versus time, or `c1` versus `c4`. What kind of growth is displayed? ◀

This scatter plot does not explain much other than we have a constrained growth situation. But looking at Figure 8.12, it appears that all of C comes from B. So now let's look at B.

▶ SECOND REFLECTION

Do a scatter plot of B. Describe what is happening to B in words. ◀

If C is dependent on B, the shape of B may explain the shape of C. Since we have established a relationship where C is dependent on B, let's try the change in C is proportional to B. For this assumption, our difference equation model is

$$C_n = C_{n-1} + k_B B_{n-1}.$$

CHAPTER 8   MODELING WITH SYSTEMS

Similarly, for the differential equation model, we would have

$$\frac{dc}{dt} = \kappa_B b.$$

At this point we have a model we know works for chemical A and one we think works for chemical C. So what about chemical B? In words, chemical B is growing from chemical A and decaying to chemical C. The question is at what rates? Since chemical A is decaying into chemical B, it makes sense that B is growing at the same rate that A is decaying. And since chemical C is growing from chemical B, then B must be decaying at the same rate that C is growing. This assumption gives the difference equation model

$$B_n = B_{n-1} + \left(k_A A_{n-1} - k_B B_{n-1}\right).$$

Similarly for the differential equation model, we have

$$\frac{db}{dt} = \kappa_A a - \kappa_B b.$$

A quick aside: looking at the differential equation model of chemical B, it reminds us of the mixture problems where that rate of change of b is equal to the rate in ($\kappa_A a$) minus the rate out ($\kappa_B b$).

Putting things together, a system of difference equations that models the chemical chain is

$$A_n = A_{n-1} - k_A A_{n-1}$$
$$B_n = B_{n-1} + k_A A_{n-1} - k_B B_{n-1}$$
$$C_n = C_{n-1} + k_B B_{n-1},$$

where $A_0 = 1.00$, $B_0 = 0$ and $C_0 = 0$. The analogous system of differential equations would be

$$\frac{da}{dt} = -\kappa_A a$$
$$\frac{db}{dt} = \kappa_A a - \kappa_B b$$
$$\frac{dc}{dt} = \kappa_B b,$$

where $a(0) = 1.00, b(0) = 0,$ and $c(0) = 0$. In order to use our family of difference equation models for a numerical simulation we need to fix a value for Δt and estimate the corresponding values for k_A, k_B. In order to use our continuous model we need values for κ_A, and κ_B. We may as well take Δt to be one minute as a simple place to start. Now we need $k_A, k_B, \kappa_A,$ and κ_B.

▶ THIRD REFLECTION

Compute a column of divided differences of A in c5. Compute a column of estimates for k_A in c6. What is your choice for k_A? ◀

► **Fourth Reflection**

Based on the information you have seen so far, what would you choose for κ_A (the proportionality constant for the differential equation for chemical A)? How would you decide if this choice was good or bad? How do these choices for k_A and κ_A compare with those in Chapters 6 and 7? ◄

Your helper, a future math major, has chosen $k_A = 0.0484$, and $\kappa_A = 0.0639$.

► **Fifth Reflection**

Why did she chose different values for k_A and κ_A? What values would you have chosen? ◄

Think and Share

Break into groups. Devise a method for finding k_B and κ_B, and then compute them. How will you determine if they are any good?

Again, your helper has chosen $k_B = 0.0221$ and $\kappa_B = 0.0214$. So our models are now

$$A_n = A_{n-1} - 0.0484 A_{n-1}$$
$$B_n = B_{n-1} + 0.0484 A_{n-1} - 0.0221 B_{n-1}$$
$$C_n = C_{n-1} + 0.0221 B_{n-1},$$

where $A_0 = 1$, $B_0 = 0$, $C_0 = 0$, and

$$\frac{da}{dt} = -0.0639a$$
$$\frac{db}{dt} = 0.0639a - 0.0214b$$
$$\frac{dc}{dt} = 0.0214b,$$

where $a(0) = 1.0$, $b(0) = 0$, and $c(0) = 0$.

Note that the system of difference equations is convenient for iterating, or repeated numeric evaluating. That is, knowing A_{n-1}, B_{n-1}, and C_{n-1}, we can compute A_n, B_n, and C_n. Also, the system of differential equations tells us that the rate of change in the amount of chemical A is proportional to the amount of A present, and the rate of change in C is proportional to the amount of B, and the rate of change in B is equal to the rate of change of A minus the rate of change of C. For this particular example, the system of differential equations seems simpler and much more intuitive, mainly because of the unequal step sizes. Yet, the above methodology for formulating systems of difference and differential equations to model interactive systems is quite powerful! Let's get some practice.

E X A M P L E   2   **A Two-Party System**

Let's consider a two-party political system consisting of Democrats and Republicans. Suppose pollsters have observed that about 30% of the Democrats register as Republicans before each election. Likewise, Republicans migrate to the Democratic party at the rate of 25% between elections. As time advances, how many voters belong to each party? Is the result sensitive to the starting strengths? We begin by defining variables:

$$D_n = \text{the number of Democrats for the } n\text{th election}$$
$$R_n = \text{the number of Republicans for the } n\text{th election.}$$

Using the construct *future = present + change*, we derive a system of difference equations to model a time period of one election period.

$$D_n = D_{n-1} - 0.30D_{n-1} + 0.25R_{n-1} = 0.70D_n + 0.25R_{n-1} \quad \textbf{(8.2)}$$
$$R_n = R_{n-1} + 0.30D_{n-1} - 0.25R_{n-1} = 0.30D_n + 0.75R_{n-1}$$

Assuming that the number of Democrats and Republicans can be represented as differentiable functions of time, we can construct a differential equation model. We reformat (8.2) as the differences $\Delta D_{n-1} = D_n - D_{n-1}$ and $\Delta R_{n-1} = R_n - R_{n-1}$ resulting in,

$$\Delta D_{n-1} = -0.30D_{n-1} + 0.25R_{n-1}$$
$$\Delta R_{n-1} = 0.30D_{n-1} - 0.25R_{n-1}.$$

Fixing notation for the continuous model with

$$d(t) = \text{the number of Democrats at any time } t$$
$$r(t) = \text{the number of Republicans at any time } t.$$

Interpreting the differences ΔD_{n-1} and ΔR_{n-1} as a rate of change per election cycle (the change in time in our difference equations (8.2)), the analogous system of differential equations is

$$\frac{dd}{dt} = -0.30d + 0.25r$$
$$\frac{dr}{dt} = 0.30d - 0.25r.$$

Here—for lack of any other way to estimate them —we have assumed that the constants of proportionality we derived for the difference equation model are valid for the differential equation model. We summarize the models for discrete and continuous change in future elections in Table 8.3.

E X A M P L E   3   **Species in Competition**

There are many ecological systems in which species compete against each other for some resource, such as food. Imagine a small pond that is mature enough

TABLE 8.3 Two-party Political System: Comparison of Difference and Differential Equations

System of Difference Equations	System of Differential Equations
Difference Equations	**Differential Equations**
$D_{n+1} = D_n - 0.30D_n + 0.25R_n$	$dd/dt = -0.30d + 0.25r$
$R_{n+1} = R_n + 0.30D_n - 0.25R_n$	$dr/dt = 0.30d - 0.25r$

to support wildlife. We desire to stock the pond with game fish, trout and bass. Is coexistence of the two species in the pond possible? If so, how sensitive are the population levels predicted by the model with respect to the initial stockage levels and to external influences on the environment, such as floods, disasters, and epidemics? Let's assume that in the absence of competition for the resources, either species would exhibit *unlimited* growth. That is, the growth, or increase in population, would be proportional to the amount present during the previous period. (In the problem set, we ask you to model logistic growth.) Fixing an interval of time Δt and letting X_n represent the number of trout present at the nth time period and Y_n, the number of bass, we write the difference equation for **growth in isolation**

$$X_n = X_{n-1} + aX_{n-1}$$
$$Y_n = Y_{n-1} + cY_{n-1}.$$

The values of the constants a and c represent the relative growth rates (for the fixed time interval, Δt).

We assume the species do not prey on one another. Rather, they compete with one another for living space and a common food supply. We assume that the intensity of the competition is roughly proportional to the number of possible interactions between the two species. If either species is scarce, there are fewer interactions. There are many ways to model this interaction, but one simple submodel assumes that the **decrease** in population is proportional to the product $X_n \cdot Y_n$. This leads to the system of difference equations from one season to the next,

$$X_n = X_{n-1} + aX_{n-1} - bX_n \cdot Y_{n-1}$$
$$Y_n = Y_{n-1} + cY_{n-1} - dX_{n-1} \cdot Y_{n-1}.$$

The constants b and d represent the intensity of the competition and the relative "hunting effectiveness" of the two species. All of a, b, c, and d are positive and have values that depend on Δt, the fixed time interval. The change in population size from one time period to the next is given by

$$\Delta X_{n-1} = X_n - X_{n-1} = aX_{n-1} - bX_{n-1} \cdot Y_{n-1}$$
$$\Delta Y_{n-1} = Y_n - Y_{n-1} = cY_{n-1} - dX_{n-1} \cdot Y_{n-1}.$$

Thus, the corresponding differential equation model is

$$\frac{dx}{dt} = ax - bx \cdot y$$
$$\frac{dy}{dt} = cy - dx \cdot y,$$

where $x(t)$ and $y(t)$ represent the population levels of trout and bass at any time t.

We have summarized our models for discrete and continuous change for future reference in Table 8.4:

TABLE 8.4   Species in Competition: Comparison of Difference and Differential Equations

System of Difference Equations	System of Differential Equations
System of Difference Equations	**System of Differential Equations**
$X_n = X_{n-1} + aX_{n-1} - bX_{n-1}Y_{n-1}$	$dx/dt = ax - bxy$
$Y_n = Y_{n-1} + cY_{n-1} - dX_{n-1}Y_{n-1}$	$dy/dt = cy - dxy$

▶ **SIXTH REFLECTION**

How are these models of competing hunters similar to and different from the predator-prey model in Section 8.1?   ◀

Putting it All Together

There are many advantages to modeling problems with systems of more than one dependent variable, both discretely and continuously. Chiefly from this section, we get insights from each model, whether a system of difference equations or a system of differential equations. Often one type of model makes more sense for the particular situation you are trying to clarify. In the next sections, as we seek to answer the questions that came to mind as we were modeling, we will learn how to attack these systems numerically, graphically, and symbolically. These solution techniques will extend some ideas from old chapters and expand our bevy of problem solving skills.

EXERCISE SET 8.2

1. We are interested in the variation of the price of corn. We have heard that there is a relationship between the supply (quantity) and demand (price) of corn. Since we are only interested in the iteration between price and quantity, we can reorder the data by decreasing price as shown in Table 8.5. When the price of corn is high, then more farmers supply corn the next season. Yet, increasing the quantity tends to drive the price down. The previous table of data has been reordered by decreasing price so that we can investigate this relationship. The basic idea behind the model is that the change in price (from one year to the next) is proportional to the quantity of corn the previous year, and the change in quantity (from one year to the next) is proportional to the price of corn the previous year.

a. From the data in Table 8.5, determine the proportionality constant k_p for the change in price and k_q for the change in quantity.

b. Construct a system of difference equations for price and quantity of corn with a time interval of one season.

c. Construct a system of differential equations that models the price and quantity of corn for time t in seasons.

d. Compare (b) and (c). Which seems more appropriate for this situation?

2. Consider the following economic model. Let P be the price of a single item on the market. Let Q be the quantity of the item available on the market. Both P and Q are functions of time. If one considers price and quantity as two interacting species, the following model might be proposed:

$$\frac{dP}{dt} = aP\left(\frac{b}{Q} - P\right)$$
$$\frac{dQ}{dt} = cQ\left(dP - Q\right),$$

where a, b, c, and d are positive constants. Justify and discuss the adequacy of the model. Hint: is there a quantity for which there is no change in price? Is there a price for which there is no change in quantity? Does this make sense? What are the values at which no change takes place? What do these values represent in terms of market prices?

3. The gross national product (GNP) represents the sum of consumption (purchases of goods and services) and gross private investment (which is the sum of increases in inventories plus buildings constructed and equipment acquired). Assume that the GNP is increasing at the rate of 3% per year and that the national debt is increasing at a rate proportional to the GNP. Construct systems of two difference and two differential equations modeling the interaction between the GNP and the national debt.

4. Develop a model for the growth of trout and bass, assuming that in isolation trout demonstrate exponential decay and the bass population grows logistically with a population limit of M. Use a system of difference equations to determine a corresponding system of differential equations.

5. Consider the model below where $x(t)$ represents the trout population and $y(t)$ represents the bass population.

$$\frac{dx}{dt} = a\left(1 - \frac{x}{k_1}\right)x - cxy$$
$$\frac{dy}{dt} = b\left(1 - \frac{y}{k_2}\right)y - dxy$$

a. What assumptions are implicitly being made about the growth of trout and bass in the absence of competition?

b. Interpret the constants a, b, c, d, k_1, and k_2 in terms of the stated problem.

6. Consider a 100-gal tank A and a 100-gal tank B. Tank A is initially filled with water in which 25 lbs of salt are dissolved. Tank A has 4 gal/min of salt-free water being input. Tank B is initially filled with salt-free water. The well-mixed solution from tank A is constantly being pumped into tank B at a rate of 6 gal/min, and the solution in tank B is constantly being pumped into tank A at 2 gal/min. The solution in tank B also exits the tank at the rate of 4 gal/min.

a. Draw a sketch of the tanks and flows.

b. From the description, describe what you believe will happen to the salt in the tanks.

c. Construct a model of a system of difference equations for the amount of salt in each tank with a time interval of one minute.

d. Construct a model of a system of differential equations for the amount of salt in each tank at time t in minutes.

e. Compare the two system models of parts (c) and (d). Are they the same? If so, how can this be? If not, which model would be most appropriate for this situation and why?

7. The two-compartment model shown in Figure 8.13 occurs frequently in drug kinetics. Suppose a concentration

FIGURE 8.13 Two Compartment Model Diagram

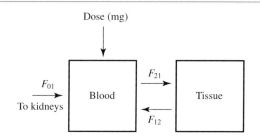

TABLE 8.5 Corn Data Reordered

n	1	2	3	4	5	6	7	8	9
Price	19.5	19.0	18.9	18.5	18.3	18.0	17.9	17.8	16.5
Quantity	998	1056	1042	1034	1053	1116	1098	1107	1243

containing D mg of a drug is injected into the bloodstream of a patient at time $t = 0$. If the flow rates are as indicated in Figure 8.13, find a system of differential equations giving the amount of drug present in the blood and in the tissue at any time. Let V_b and V_t denote the volumes of the blood and tissue, respectively.

8. A drug is taken orally at once. The mass of the drug at time t in hours after ingestion in the gastrointestinal tract and bloodstream is denoted by $m_g(t)$ and $m_b(t)$ respectively. The changes of these masses can be modeled as a compartment model with rate of change constant of k_{go} l/hr going from the gastrointestinal tract to the bloodstream, k_{bg} l/hr going from the bloodstream to the gastrointestinal tract, and k_{go} l/hr going from the gastrointestinal tract out of the system. Suppose $k_{gb} = 0.2$, $k_{bg} = 0.03$, and $k_{go} = 0.05$ and $m_g(0) = 1$mg, while $m_b(0) = 0$mg. Assume the volume of the gastrointestinal tract is 10 liters and remains constant. Assume the volume of the bloodstream is 5 liters an remains constant.

   a. Draw a sketch of the compartments and flows.
   b. From the description, describe what you think will happen to the drug in the person's gastrointestinal tract and bloodstream.
   c. Construct a model of a system of differential equations for the amount of drug in each of the two compartments at time t in hours.
   d. Now formulate a model in which there is a constant infusion of 0.02 mg/hr into the bloodstream and the same physiology as described above.
   e. If we were to model the basic problem (a-c) as a system of difference equations, what might have to change? Why?

9. A simple three-compartment model that describes nutrients in a food chain has been studied by M. R. Cullen.[†] The constant between components means that at any given time, nutrients pass from one compartment to the other at that rate. For example, the nutrient leaves from phytoplankton and moves to water at the rate of $0.06p$ per hour, where $p(t) =$ amount of nutrient in the phytoplankton. Suppose 50 kg of a harmful nutrient is dumped into the water at time $t = 0$ in hours (see Figure 8.14).

   a. Model the flow of the nutrient as a system of difference equations.
   b. Model the flow of the nutrient as a system of differential equations.
   c. Contrast the systems in (a) and (b).

[†]Cullen, M. R. 1983. Mathematics for the biosciences. Boston: PWS Publishing Company.

FIGURE 8.14 Phytoplankton Nutrient Model

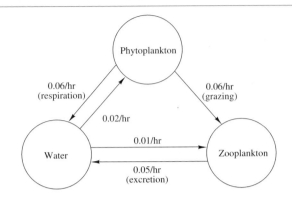

10. A chemical reaction, "A going to B (some back to A) and on to C," is depicted with reaction constants in l/sec. Originally we have 100 g of A and no B or C (see Figure 8.15).
    a. Model this reaction as a system of difference equations.
    b. Model this reaction as a system of differential equations.
    c. Why are (a) and (b) similar but the systems in Example 1 had different rate constants?

11. In 1868 the accidental introduction into the United States of the cottony cushion insect (*Icerya purchasi*) as a species of scale insect from Australia threatened to destroy the U.S. citrus industry. To counteract this situation, one of its natural Australian predators, a species of ladybird beetle, was imported. The beetles kept the scale insects down to a relatively low level. When DDT was discovered to kill scale insects, farmers applied it in the hopes of further reducing the scale insect population. However, DDT turned out to be fatal to the beetles as well, and the overall effect of using the insecticide was to *increase* the number of scale insects. Modify the competitive hunter model given in Example 3 to reflect a predator-prey system of two insect species where farmers apply an insecticide on a continuing basis that destroys both the insect predator and the insect prey at a common rate proportional to the numbers present.

12. Suppose most of the water flowing from Lake Shasta is from Bull Pond and that a small chemical company is dumping a pollutant into Bull Pond. Assume that 100% of the water in Lake Shasta comes from Bull Pond and that Shasta's water flows into the sea. Each year the percentage of water

FIGURE 8.15 Chemical Reaction

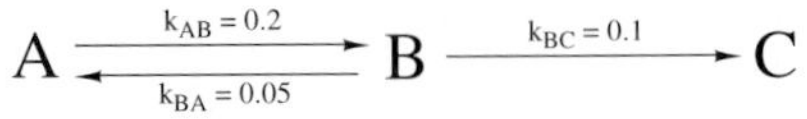

$$A \underset{k_{BA} = 0.05}{\overset{k_{AB} = 0.2}{\rightleftarrows}} B \xrightarrow{k_{BC} = 0.1} C$$

replaced in Bull Pond and Lake Shasta is approximately 25% and 5%, respectively. The chemical company dumps only 3 units of pollutant a year into Bull Pond. Initially there are 350 and 1200 units of pollutant in Bull Pond and Lake Shasta, respectively.

a. Construct a system of difference equations that models the units of pollutant in Bull Pond and Lake Shasta in the nth year.

b. What assumption(s) is (are) being made about mixing in your model above?

c. Construct a system of differential equations that models units of pollutant at time t in years.

d. Compare the two models in (a) and (c). Which seems most appropriate? Why?

13. In your house you have one air conditioner that must cool three rooms (see Figure 8.16). The air conditioner is in the dining room and is set at 75°F. Use Newton's law of cooling to monitor the cooling of the three rooms. The dining room has a cooling constant of 0.5 per minute when the air conditioner is on. Therefore, the change in temperature per minute for the dining room would be $\Delta T_d = 0.5(75 - T_d)\Delta t$. The cooling constants for the kitchen and living room are 0.25 per minute and 0.15 per minute, respectively.

a. Assume Δt is a minute and construct a model of the temperature in each of the rooms as a system of difference equations after the nth time period.

b. Construct a model of the cooling in each room as a system of differential equations at time t in minutes.

c. Make a list of assumptions that must be made to arrive at each of the models (a) and (b).

d. If Δt was changed to seconds in our models, what would change and how?

FIGURE 8.16 House Plan

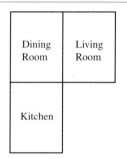

8.3

Numerical Solutions: Iteration and Euler's Method

In Chapter 7, we numerically approximated a single difference equation with an initial value. Similarly, if we are given starting values for the systems of difference equations we formulated in the previous section, we can build a table of values that **is** the solution for those particular starting values. That is, a system of difference equations tells us precisely how to calculate the next terms in the sequences under study, knowing the previous terms. For example, consider the model developed for competing species:

$$x_n = x_{n-1} + ax_{n-1} - bx_{n-1}y_{n-1}$$
$$y_n = y_{n-1} + cy_{n-1} - dx_{n-1}y_{n-1}.$$

That is, x_n and y_n depend only on the immediately preceding terms in the sequences $\{x_{n-1}\}$ and $\{y_{n-1}\}$. Let's illustrate by iterating numerical solutions to the examples of the previous section for particular starting values.

Iteration

E X A M P L E   1   **More Chemicals**

From Example 1 in Section 8.2, we are trying to model what is happening to chemicals A, B, and C over time, so we can determine the optimum time to stop the reaction. Our variables for the system of difference equations are

$$A_n \; = \; \text{the number of moles of chemical } A \text{ at the } n\text{th time period}$$
$$B_n \; = \; \text{the number of moles of chemical } B \text{ at the } n\text{th time period}$$
$$C_n \; = \; \text{the number of moles of chemical } C \text{ at the } n\text{th time period}$$

and our model is

$$A_n = A_{n-1} - 0.0484 A_{n-1}$$
$$B_n = B_{n-1} + 0.0484 A_{n-1} - 0.0221 B_{n-1}$$
$$C_n = C_{n-1} + 0.0221 B_{n-1},$$

where $A_1 = 1$, $B_1 = 0$, $C_1 = 0$.

As was stated in the previous section, this chemical reaction problem is better modeled continuously. As a matter of fact, for the given $k_A = 0.0484$ and $k_B = 0.0221$, iteration results in answers far removed from the data. Even reducing the k's to some form of the average slope from the data produces a result far removed from the data. The cause of this massive error may be the relatively large and unequal time intervals. Of course, the problem could be the model, but we will verify the model later by assuming a continuous solution and using Euler's method. For an exploration of this iteration, see Problem 14 of Exercises 8.3. In the meantime, let's look at iteration on a discrete model that works, such as the Republicans and Democrats.

E X A M P L E   2   **Two-Party Politics**

For the Republicans versus the Democrats modeled in Section 8.2, we want to know what the effect of this model is on the long-term ratio of Democrats to Republicans. Let's investigate. For this example, the variables are

$$D_n = \text{milllions of Democrats for the } n\text{th election}$$
$$R_n = \text{millions of Republicans for the } n\text{th election,}$$

and the equations are

$$D_n = 0.70 D_{n-1} + 0.25 R_{n-1}$$
$$R_n = 0.30 D_{n-1} + 0.75 R_{n-1}.$$

Let's assume the two parties are initially equal, say $D_1 = 100 = R_1$. Go to MODE and put Graph in SEQUENCE mode. Go to the Y= menu. Check that you are set to TIME mode in the AXES dialogue box. Then, for u1, type in

$$u1=0.70*u1(n-1)+0.25*u2(n-1).$$

This is the equation for D_n. Alternately, you can press F3, or Edit, and change the previous equation. Next, change the system so

$$ui1 = 100$$
$$u2 = .30*u1(n-1)+.75*u2(n-1)$$
$$ui2 = 100.$$

Now, let's go to TblSet and set

```
tblStart: 1,
Δ tbl: 1,
Graph <-> Table: ON, and
Independent: AUTO.
```

TABLE 8.6 Two-Party Percentages

n	D_n	R_n
1	100	100
2	95	105
3	92.75	107.25
4	91.738	108.26
5	91.282	108.72
6	91.077	108.92
7	90.985	109.02
8	90.943	109.06
9	90.924	109.08
10	90.916	109.08
11	90.912	109.09
12	90.91	109.09
13	90.91	109.09
14	90.91	109.09
15	90.909	109.09
16	90.909	109.09
17	90.909	109.09
18	90.909	109.09
19	90.909	109.09
20	90.909	109.09

Next, go to the TABLE editor. Your table should resemble Table 8.6. It looks as if after the fifteenth election there is no longer any change. This is called an equilibrium (rest) situation. The fact that the values near 90.909 and 109.09 drove the system back to these values indicates this rest point may be stable. So, our model suggests in the long term the Republicans will win. Before we analyze this situation further, let's draw a couple of graphs.

Go to the Y= menu and select Axes (F7). For now, let's choose 1: TIME, which means n will be on the x-axis and D_n and R_n will be on the y-axis. Next, go to the WINDOW menu. Set

```
nmin=1,nmax=20,
plotstrt=1,plotstep=1,
xmin=0,xmax=20,xscl=2,
ymin=80,ymax=120, and yscl=2.
```

Next, go to the GRAPH window and you should see the graph in Figure 8.17.

Next, select Axes from the Y= menu. Choose CUSTOM. For the X Axis choose u1, or Democrats, and for the Y Axis choose u2, or Republicans. Next, go to WINDOW and set

```
nmin=1,nmax=20,
plotstrt=1,plotstep=1,
xmin=80,xmax=120,xscl=2,
ymin=80,ymax=120, and xscl=2.
```

Next, press GRAPH and see the graph in Figure 8.18.

FIGURE 8.17 Time Axes: Democrats and Republicans

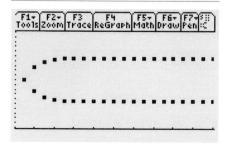

FIGURE 8.18   Equilibrium for Democrats vs. Republicans

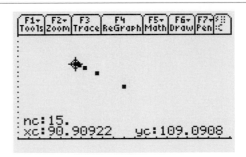

Think and Share

Break into groups. Notice how quickly the system moved to the equilibrium values. Investigate how this systems works when our starting values are [80, 120], [120, 80], [80, 80], or [120,120]. What does this occurrence mean as far as our model is concerned? Are the results realistic? Are our parameters accurate? How does an Independent Party affect this situation?

E X A M P L E   3   **Species in Competition**

The example of the competitive hunters is illustrated here. Fixing the time interval for the family of difference equations as one season (year) and measuring the populations of trout and bass in pounds gives us:

$$X_n = \text{thousands of trout in a pond after } n \text{ seasons}$$
$$Y_n = \text{thousands of bass in a pond after } n \text{ seasons,}$$

and

$$X_n = X_{n-1} + aX_{n-1} - bX_{n-1}Y_{n-1}$$
$$Y_n = Y_{n-1} + cY_{n-1} - dX_{n-1}Y_{n-1},$$

where a is the growth rate of the trout in isolation, c is the growth rate of the bass in isolation, b is the decrease in trout due to competition with the bass, and d is the decrease in bass due to competition with the trout. Let's try some numbers. Let $a = 0.15, b = 0.015, c = 0.1,$ and $d = 0.011$. The initial quantities of trout and bass are 15 and 15, respectively. Therefore, our equations are

$$X_n = 1.15X_{n-1} - 0.015X_{n-1}Y_{n-1}$$
$$Y_n = 1.1Y_{n-1} - 0.011X_{n-1}Y_{n-1} \tag{8.3}$$

and $X_1 = 15$ and $Y_1 = 15$.

Now, let's iterate. Again we are looking for the long-term behavior of our system, so we will follow the techniques and graphing we did in Example 2. In the Y= menu in SEQUENCE mode, the settings are as follows:

u1 is the trout (X_n), ui1=15, u2 is the bass (Y_n), and ui2=15

Next is `TblSet`. In the `TblSet` editor, set

> `tblStart: 1`,
> `Δtbl: 1`, and
> `Independent: AUTO`.

Now display the table. Table 8.7 displays the first 20 iterations. What's going on here? Both the number of trout and the number of bass decrease steadily until season 17, when the trout start to steadily increase. In fact, this trend continues until season 56 (not shown), when the bass become extinct. By this time the trout are up to 123. Let's see what this looks like graphically. Again, with the `CUSTOM Axes` set to `u1` on the `x`-axis and `u2` on the `y`-axis, enter the following WINDOW values:

> `nmin= 1, nmax= 42`,
> `plotstrt= 1, plotstep=1`,
> `xmin=5, xmax=25, xscl=5`,
> `ymin=0, ymax=15, yscl=1`

(see Figure 8.19).

TABLE 8.7 Pounds of Trout and Bass (in Thousands)

n	x_n	y_n	n	x_n	y_n
1	15	15	11	10.385	10.702
2	13.875	14.025	12	10.276	10.549
3	13.037	13.287	13	10.191	10.412
4	12.395	12.71	14	10.128	10.286
5	11.891	12.248	15	10.085	10.168
6	11.49	11.871	16	10.059	10.057
7	11.167	11.558	17	10.051	9.9501
8	10.906	11.294	18	10.058	9.845
9	10.695	11.068	19	10.082	9.7403
10	10.523	10.873	20	10.121	9.6341

Think and Share

Break into groups. This system is nonlinear because of the hunter interaction. Is there a rest point? The system seems headed towards a rest point and then suddenly shoots away. Try some different initial values like [5, 5], [15, 5], and [5, 15]. What happens in these cases? Can you find any nonzero rest values? How? Now, try some different values for the parameters a, b, c, and d. How do these affect the rest point(s) you found?

FIGURE 8.19 Equilibrium for Trout and Bass

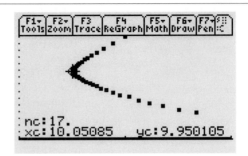

Euler's Method

Suppose we had a system of differential equations and could not find the symbolic form of the solution. Could we approximate a solution for particular starting values iterating as we did with the system of difference equations in Equation (8.3)? The answer is yes: we iterate a system of difference equations that *approximates* the system of differential equations just as we did for single differential equations in Chapter 6 (see 8.20) using Euler's method. For example, consider the system of differential equations representing competing species:

$$\frac{dx}{dt} = ax - bxy$$
$$\frac{dy}{dt} = cy - dxy. \tag{8.4}$$

Using the **approximations** $\frac{dx}{dt} \approx \frac{\Delta x}{\Delta t}$ and $\frac{dy}{dt} \approx \frac{\Delta y}{\Delta t}$, we have

$$\frac{\Delta x}{\Delta t} \approx ax - bxy$$
$$\frac{\Delta y}{\Delta t} \approx cy - dxy. \tag{8.5}$$

Multiplying by Δt and substituting $x = x_n$, $y = y_n$, $\Delta x_{n-1} = x_n - x_{n-1}$ and $\Delta y_{n-1} = y_n - y_{n-1}$ gives

$$\Delta x_{n-1} = x_n - x_{n-1}$$
$$= \Delta t (ax_{n-1} - bx_{n-1}y_{n-1})$$
$$\Delta y_{n-1} = y_n - y_{n-1}$$
$$= \Delta t (cy_{n-1} - dx_{n-1}y_{n-1}). \tag{8.6}$$

Or, in a form ready for iteration, we have

$$x_n = x_{n-1} + \Delta t (ax_{n-1} - bx_{n-1}y_{n-1})$$
$$y_n = y_{n-1} + \Delta t (cy_{n-1} - dx_{n-1}y_{n-1}). \tag{8.7}$$

FIGURE 8.20 Euler's Method at (t, y)

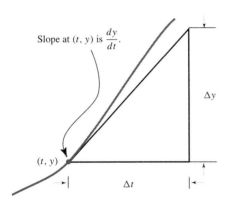

Slope at (t, y) is $\frac{dy}{dt}$.

(t, y)

Δy

Δt

Now, given starting values, we can iterate the system of difference equations as we did previously. Before generalizing, let's note the form of the equations in Equations (8.7). For example, the equation

$$x_n = x_{n-1} + \Delta t \, (a x_{n-1} - b x_{n-1} y_{n-1})$$

is in the form

$$x_n = x_{n-1} + \Delta t \cdot f(x_{n-1}, y_{n-1})$$

or using our standard modeling equation,

$$future = present + change,$$

where $f(x, y)$ is a function which represents the change per time period Δt. Let's generalize the procedure and solve several examples.

Euler's Method for Systems of Differential Equations

To solve the system of differential equations

$$\frac{dx}{dt} = g(t, x, y, z)$$

$$\frac{dy}{dt} = h(t, x, y, z)$$

$$\frac{dz}{dt} = l(t, x, y, z) \tag{8.8}$$

over an interval $t_0 \leq t \leq b$, given $x(t_0) = x_0$, $y(t_0) = y_0$ and $z(t_0) = z_0$, we proceed as follows.

Step 1 Divide the interval $t_0 \leq t \leq b$ into n equally spaced subintervals:

$$t_1 = t_0 + \Delta t, \ t_2 = t_1 + \Delta t, \ ..., \ t_n = t_{n-1} + \Delta t = b.$$

Step 2 Obtain the following sequences of approximations:

$$x_1 = x_0 + \Delta t \cdot g(t_0, x_0, y_0, z_0)$$
$$y_1 = y_0 + \Delta t \cdot h(t_0, x_0, y_0, z_0)$$
$$z_1 = z_0 + \Delta t \cdot l(t_0, x_0, y_0, z_0)$$

$$x_2 = x_1 + \Delta t \cdot g(t_1, x_1, y_1, z_1)$$
$$y_2 = y_1 + \Delta t \cdot h(t_1, x_1, y_1, z_1)$$
$$z_2 = z_1 + \Delta t \cdot l(t_1, x_1, y_1, z_1)$$

$$\vdots$$

$$x_n = x_{n-1} + \Delta t \cdot g(t_{n-1}, x_{n-1}, y_{n-1}, z_{n-1})$$
$$y_n = y_{n-1} + \Delta t \cdot h(t_{n-1}, x_{n-1}, y_{n-1}, z_{n-1})$$
$$z_n = z_{n-1} + \Delta t \cdot l(t_{n-1}, x_{n-1}, y_{n-1}, z_{n-1}).$$

EXAMPLE 4 **More Chemicals, Again**
_____

The system of equations that models the chemical reaction data is

$$\frac{da}{dt} = -0.0639a$$

$$\frac{db}{dt} = 0.0639a - 0.0214b$$

$$\frac{dc}{dt} = 0.0214b,$$

where $a(0) = 1$, $b(0) = 0$, and $c(0) = 0$. Remember, we wish to maximize profits when each mole of chemical A costs \$0.50, chemical B profits \$3.50 per mole, and chemical C profits \$1.00 per mole. We've assumed a continuously changing model, and we would like to get an estimate of what the situation will be for the full 200 seconds. We'll divide the period into 100 two-second intervals, or $\Delta t = 2$. A table helps us to apply Euler's method. But first, we will do a few iterations to see how Euler's method works.

We have completed Step 1 by dividing the 200 seconds into 100 subintervals. So, $t_0 = 0, t_1 = 0 + 2 = 2, t_2 = 4, t_3 = 6$, etc., out to $t_{100} = 200$. We will denote a_n as the approximate value of $a(t_n)$. Thus,

$$a_0 = a(t_0) = a(0) = 1.0$$
$$b_0 = b(t_0) = b(0) = 0.0$$
$$\text{and}$$
$$c_0 = c(t_0) = c(0) = 0.0.$$

Now, let's apply Step 2.

$$a_1 = a_0 + \Delta t(a'(0))$$
$$= 1.0 + 2(-0.0639 \cdot a_0) = 1.0 + 2(-0.0639 \cdot 1.0) = 0.8722$$
$$b_1 = b_0 + \Delta t(0.0639a_0 - 0.0214b_0) = 0 + 2(0.0639 \cdot 1.0 - 0.0214 \cdot 0) = 0.1278$$
$$\text{and}$$
$$c_1 = c_0 + \Delta t(0.0214b_0) = 0 + 2(0.0214 \cdot 0) = 0.$$

So, these values approximate $a(2)$, $b(2)$, and $c(2)$, respectively. Continuing the process

$$a_2 = a_1 + 2(-0.0639 \cdot a_1) = 0.8722 + 2(-0.0639 \cdot 0.8722)$$
$$= 0.76073$$
$$b_2 = b_1 + 2(0.0639a_1 - 0.0214b_1) = 0.1278 + 2(0.0639 \cdot 0.8722 - 0.0214 \cdot 0.1278$$
$$= 0.2338$$
$$\text{and}$$
$$c_2 = c_1 + 2(0.0214b_1) = 0 + 2(0.0214 \cdot 0.1278)$$
$$= 0.00547.$$

Next using these values, we compute a_3, b_3 and c_3 and so forth. The values computed for the first 20 seconds are listed in Table 8.8.

TABLE 8.8   Chemical Productions from Euler's Method for the First 20 Seconds

n	t_n	a_n	b_n	c_n
0	0	1.00	0.00	0.00
1	2	0.8722	0.1278	0.00
2	4	0.76073	0.2338	0.00547
3	6	0.66351	0.32101	0.01548
4	8	0.57871	0.39207	0.02922
5	10	0.50475	0.44925	0.04600
6	12	0.44025	0.49453	0.06522
7	14	0.38398	0.52963	0.08639
8	16	0.33491	0.55603	0.10906
9	18	0.29211	0.57503	0.13286
10	20	0.25478	0.58776	0.15747

After computing many of these values by hand, we realize Euler's method is just an iterative process that in effect approximates a system of differential equations with a system of difference equations. To set up our iteration, in the MODE menu set the Graph mode to SEQUENCE. In to the Y= menu enter

```
u1=u1(n-1)+2
ui1=0
u2=u2(n-1)+2*(-0.0639*u2(n-1)).
```

These are the equations for t_n, t_0, and a_n. Likewise, set

```
ui2=1.0
 u3=u3(n-1)+2*(0.0639*u2(n-1)-0.0214u3(n-1))
ui3=0
 u4=u4(n-1)+2*(0.0214*u3(n-1)), and
ui4=0
```

for a_0, b_n, b_0, c_n, and c_0, respectively. In the WINDOW editor set nmin to 0. Next, go to the TblSet editor and set

```
tblStart: 1
Δtbl: 1
Graph <-> Table: ON and
Independent: AUTO.
```

Next, go to the TABLE application. It should have the same values as Table 8.8. To see the full displayed accuracy for a number in a cell, simply highlight it and it will be displayed at the bottom of the table.

Think and Share

Break into groups. Let's investigate the graphs. Recall the data table Reaction. Select Plot SetUp and enable the graphs of c2, c3, and c4. Define each to be a Scatter plot versus c1, and c2 should be Pluses, c3 should be Boxes, and c4 should be Crosses. Select ZoomData and the graph of the data. Now, go to the Y= menu. Enable u2, u3, and u4. Select style Line for each of these. Select Axes Custom. Set xAxis to u1 and yAxis to u. Go to Window and let nmax=100. Go back to the Y= menu and select ZoomData. You should get a plot of the actual a, b, and c with appropriate symbols and the line graphs of their Euler approximations. How well did we do in approximating our data? What can we do to improve? What would the implications be for our chemist? When should we stop our reaction to maximize profits?

We can use Euler's method to explore our two other models from Section 8.2.

EXAMPLE 5　**More Elections!**

We look first at our continuous model of Democrats versus Republicans. The equations are

$$\frac{dd}{dt} = -0.30d + 0.25r$$
$$\frac{dr}{dt} = 0.30d - 0.25r, \tag{8.9}$$

where

$$d(t) = \text{the number of Democrats at any time } t$$
$$r(t) = \text{the number of Republicans at any time } t.$$

As in our discrete example, we take our units for $d(t)$ and $r(t)$ to be millions of persons. Therefore, $d(0) = 100$ and $r(0) = 100$. We are interested in numbers of voters four years from now, and decide to track the numbers of voters at three month intervals, so in Euler's method we set $\Delta t = 0.25$ to estimate $d(4)$ and $r(4)$. In the Y= editor we enter

```
u1=u1(n-1)+0.25*(-.3*u1(n-1)+.25*u2(n-1))
ui1=100
u2=u2(n-1)+0.25*(.3*u1(n-1)-.25*u2(n-1))
ui2=100
```

Here u1(n) represents Democrats, u2(n) represents Republicans, and n represents the number of 0.25 year steps from $t = 0$. The resulting table is shown in Table 8.9.

TABLE 8.9　Democrats and Republicans (in Millions)

t_n	D_n	R_n
0	100	100
0.25	98.75	101.25
0.5	97.672	102.33
0.75	96.742	103.26
1.0	95.94	104.06
1.25	95.248	104.75
1.50	94.652	105.35
1.75	94.137	105.86
2.0	93.693	106.31
2.25	93.31	106.69
2.5	92.98	107.02
2.75	92.695	107.3
3.0	92.45	107.55
3.25	92.238	107.76
3.5	92.055	107.94
3.75	91.898	108.1
4.0	91.762	108.24

With your experience in Example 5, where do you think these iterations are headed? Indeed, in Example 5, $D(4) = 91.282$ and $R(4) = 108.72$. This approximation is pretty close to our approximations here. Remember, this example had the stable rest point around [90, 109], but we didn't approach this value until $t = 15$. As a result we would have to calculate 61 iterations on our calculator. Actually, you can see these iterations rather quickly by going to TblSet and changing TblStart to 59.

Think and Share

Break into groups. In Example 3, we looked at the discrete competitive hunter model. Now, we want to look at the continuous model as trout and bass vie for the same food to see who survives. Our model continuous is

$$\frac{dx}{dt} = ax - bxy$$
$$\frac{dy}{dt} = cy - dxy, \qquad \textbf{(8.10)}$$

where $x(t)$ and $y(t)$ represent the thousands of trout and bass in a pond at any time t. As in Example 3, we will let $a = 0.15$, $b = 0.015$, $c = 0.1$, and $d = 0.011$. Our initial conditions are $x(0) = 15$ and $y(0) = 15$. In this example, we will let $\Delta t = 0.1$ and use Euler's method to approximate $x(2)$ and $y(2)$. Further, we will use the SEQUENCE mode. Thus, our equations for the Y= editor are

```
u1(n)=u1(n-1)+0.1*(0.15*u1(n-1)-0.015*u2(n-1)*u1(n-1))
   ui1=15
u2(n)=u2(n-1)+0.1*(0.1*u2(n-1)-0.011*u2(n-1)*u1(n-1))
   ui2=15,
```

where u1(n) would be x_n, or trout, u2(n) represents y_n, or bass, and n represents the number of 0.1 steps in time in years from $t = 0$. Again, tblStart should be 1 and Δtbl is 1. From this approximation $x(2) = 13.22$ and $y(2) = 13.44$. Comparing these values to the discrete model in Example 3, where $x_3 = x(2) = 13.037$ and $y_3 = y(2) = 13.287$, we are a few fish off, but in the same pond. On further investigation of Example 3, the trout population was decreasing until $t = 17$ when it started increasing. In the case of Euler's method, the turnaround point for the trout is $t = 16$. In both cases the turnaround values appear to be about [10, 10].

Putting it All Together

To this point we have learned to model with systems, both discrete and continuous. We have learned how to analyze these systems both graphically and numerically. Through systems of difference equations, we have either found the answer to our problem through a graphical picture or a table of values, or we have gained insight of an underlying continuous system. Further, we have seen how we can take a system of differential equations and get an approximation of the solution using Euler's method. In the next section, we will examine continuous solutions to systems of differential equations. These solutions will provide exact functions for something we tried earlier to approximate. However, these exact solutions are not always attainable, but looking at a few solutions we can solve will give us insight into the characteristics of the solutions we know we can at least approximate.

EXERCISE SET 8.3

In Exercises 1–3, use iteration and $\Delta t = 1$ to numerically solve the system of difference equations subject to the given initial conditions for the domain $t \in [0, 2]$.

1.

$$X_n = X_{n-1} + 2X_{n-1} - Y_{n-1} \quad X_1 = 1$$
$$Y_n = Y_{n-1} + 2Y_{n-1} \quad Y_1 = 1$$

2.

$$X_n = X_{n-1} + X_{n-1} + 3Y_{n-1} \quad X_1 = 0$$
$$Y_n = Y_{n-1} + X_{n-1} - Y_{n-1} + 2e^{t_n} \quad Y_1 = 2$$

3.

$$X_n = X_{n-1} + X_{n-1} - 2X_{n-1}Y_{n-1} \quad X_1 = 2$$
$$Y_n = Y_{n-1} + 2Y_{n-1} - X_{n-1}Y_{n-1} \quad Y_1 = 1$$

In Exercises 4–6 use Euler's method with $\Delta t = 0.5$, $\Delta t = 0.1$ and then $\Delta t = 0.01$ to solve the system of differential equations subject to the given initial conditions for the domain $t \in [0, 5]$. Compare your answer to the given symbolic solution, where available.

4.

$$\frac{dx}{dt} = 2x - y \quad x(0) = 1$$
$$\frac{dy}{dt} = 2y \quad y(0) = 1$$

$$x(t) = e^{2t} - te^{2t}$$
$$y(t) = e^{2t}$$

5.

$$\frac{dx}{dt} = x + 3y \quad x(0) = 0$$
$$\frac{dy}{dt} = x - y + 2e^t \quad y(0) = 2$$

$$x(t) = -2e^{-2t} + 3e^{2t} - 2e^t$$
$$y(t) = e^{-2t} + e^{2t}$$

6.

$$\frac{dx}{dt} = x - 2xy \quad x(0) = 2$$
$$\frac{dy}{dt} = 2y - xy \quad y(0) = 1$$

7. Compare your results from Exercise 1, Exercise 4, and the symbolic solution for $x(1)$, $y(1)$, $x(2)$, and $y(2)$. Can you describe which are good approximations and which are not?

8. Compare your results from Exercise 2, Exercise 5, and the symbolic solution for $x(2)$ and $y(2)$. Why is it only necessary to look at the end of the domain to determine if the approximation is good or not?

9. Compare the results from Exercise 3 and Exercise 6 for $x(1)$, $y(1)$, $x(2)$, and $y(2)$. Why is it necessary to look at all these values to determine if the approximation is good or not? Why is there no symbolic solution for these models?

10. Refer to Exercise 1 of Exercises 8.2. Let $P_n =$ the price in dollars per 1000 bushels of corn in the nth season, and let $Q_n =$ the quantity of 1000-bushel units supplied for the nth season. $k_P = -0.376$ and $k_Q = 0.00157$, and $P_1 = 20,000$ and $Q_1 = 1000$. Thus, your system of difference equations is

$$P_n = P_{n-1} - .376Q_{n-1} \quad P_1 = 20,000$$
$$Q_n = Q_{n-1} + .00157P_{n-1} \quad Q_1 = 1000.$$

Iterate this system of difference equations. Graph P_n and Q_n on the time axes for `nmax=40`, `xmin=0`, `xmax=40`, `ymin=0`, `ymax=20`. Will Q_n exceed 2000 and P_n go to 0? Is this realistic? Is the model flawed? Should we adjust our assumptions? What about changing the proportionality constants or initial conditions? You could check the model against the data in a graph by using `Custom` axes and `ZoomData`.

11. Refer to Exercise 6 of Exercises 8.2. Iterate the system of difference equations and ascertain the time at which the tanks contain the same amount of salt. Now use Euler's method and $\Delta t = 0.5$ min to numerically solve the system of differential equations. When does this system say the tanks will contain the same amount of salt?

12. Refer to Exercise 6 of Exercises 8.2 and to Exercise 11 above. Reformulate the system of differential equations for the system in Exercise 11 by adding a 0.025 lb/gal salt mixture into tank A at the constant rate of 4 gal/min. Now, numerically solve this system with $\Delta t = 0.5$ min. Again, when will the two tanks contain the same amount of salt.

13. Refer to Exercise 8 of Exercises 8.2. Numerically solve the system of differential equations from 8(c) using Euler's method and $\Delta t = 0.25$ hrs. How long before there is only 1/2 of the initial dosage in the gastrointestinal tract (try at

least 30 hours)? At the same time, what is the level of the drug in the bloodstream. Now, numerically solve the system of differential equations from 8(d) using Euler's method and $\Delta t = 0.25$hrs. Again, how long before there is only 1/2 of the initial dosage in the gastrointestinal tract? What is the level of the drug in the bloodstream at that time?

14. Refer to Example 1 of Section 8.2 and Example 1 of this section. Iterate the system of difference equations using $\Delta t = 10$. Compare this numerical solution with the real data, either graphically or numerically. What is wrong? Now try the iteration using $\Delta t = 2$ as an equal step size and compare with the actual data. Which iteration is better? Why?

8.4

Symbolic Solutions of Systems of Differential Equations

Solutions to Systems: Parametric Equations

In this section, we study solutions to linear systems. First, we need to know more precisely what we mean by a *symbolic solution* to a system of differential equations.

A SOLUTION TO A SYSTEM

A **solution** to the system

$$\frac{dx}{dt} = f(t, x, y)$$
$$\frac{dy}{dt} = g(t, x, y) \qquad\qquad (8.11)$$

is a pair of parametric equations

$$x = x(t) \quad \text{and} \quad y = y(t)$$

that when substituted in simplifies the System (8.11) to an identity.

Observe that, since $x = x(t)$ and $y = y(t)$ must be differentiable functions of t if they satisfy a system of differential equations, the solution curves to a system are necessarily continuous functions of t. Let us consider several examples illustrating the function of a solution concept.

EXAMPLE 1 **A Ferris Wheel**

If we wanted to describe the motion of a light on a Ferris wheel of radius 30 ft relative to the central axis, we would model this motion by the system of differential equations

$$\frac{dx}{dt} = -y$$

$$\frac{dy}{dt} = x,$$

(8.12)

where $x(t)$ is motion left (-) and right (+), $y(t)$ is motion up (-) and down (+), and t is time in seconds. The solution to this system $x(t) = 30\cos t$ and $y(t) = 30\sin t$ also models this motion. To verify this, find $\frac{dx}{dt}$ and $\frac{dy}{dt}$ from the solution and then substitute it into the system of differential equations. The result would be

$$\frac{dx}{dt} = -30\sin t = -(30\sin t) = -y$$

$$\frac{dy}{dt} = 30\cos t = 30\cos t = x,$$

which is an identity. By our definition, we have a solution to the system of differential equations.

▶ **FIRST REFLECTION**

If a is any constant, would $x = a\cos t$ and $y = a\sin t$ be a solution to the system of differential equations in Equations (8.12)? Verify this.   ◀

▶ **SECOND REFLECTION**

From MODE select graph PARAMETRIC. In the Y= menu input xt1=30cos t and yt1=30sin t. Input the following WINDOW parameters: tmin=0, tmax=7, tstep=.05, xmin=-120, xmax=120, xscl=10, ymin=-50, ymax=50, yscl=10. Graph the solution curve. Does this curve model the motion of a light on a Ferris wheel? Save a copy (F1) of this graph as a Picture in main named Ferris. If the equation of a circle of radius r is $x^2 + y^2 = r^2$, how does our solution $x(t) = a\cos t$ and $y(t) = a\sin t$ fit the graph of a circle?   ◀

▶ **THIRD REFLECTION**

Solve the system of differential equations in Equations (8.12) using Euler's method. Let $y(0) = 30$, $x(0) = 0$, and $\Delta t = .05$. You need to change the graph mode to Sequence. Input your sequences in Y= menu. Set the Style on both sequences to square and set the Axes to Custom and the axes to whatever you named your sequences. Go to Window and set nmin=1, nmax=140, plotstrt=1, plotstep=1, xmin=-120, xmax=120, xscl=10, ymin=-50, ymax=50, yscl=10. Now graph the Euler's solution. How did you do?

$\Delta t = .05$ is a pretty small step size. Now overlay your exact solution by opening (F1), `Picture`, `main`, `Ferris`. Is there a reason you might ever want an exact solution? ◀

E X A M P L E 2 **Verifying Some Symbolic Solutions**

A. Verify that the pair of functions

$$x(t) = -e^{-3t} + e^{4t}$$

$$y(t) = 8e^{-3t} - e^{4t}$$

(8.13)

is a solution to the system

$$\frac{dx}{dt} = 5x + y$$

$$\frac{dy}{dt} = -8x - 4y.$$

(8.14)

Using the formulas for $x(t)$ and $y(t)$ from Equation (8.13)

$$\frac{dx}{dt} = 3e^{-3t} + 4e^{4t}$$

$$\frac{dy}{dt} = -24e^{-3t} - 4e^{4t}$$

and

$$5x + y = 5(-e^{-3t} + e^{4t}) + (8e^{-3t} - e^{4t})$$
$$= 3e^{-3t} + 4e^{4t}$$
$$-8x - 4y = -8(-e^{-3t} + e^{4t}) - 4(8e^{-3t} - e^{4t})$$
$$= -24e^{-3t} - 4e^{4t}.$$

we have an identity. Therefore

$$x(t) = -e^{-3t} + e^{4t}$$
$$y(t) = 8e^{-3t} - e^{4t}$$

is a solution to the system of differential equations in Equations (8.14).

B. Verify that the pair of functions

$$x = (1 + t)e^t$$
$$y = -te^t$$

forms a solution to the system

$$\frac{dx}{dt} = 2x + y$$
$$\frac{dy}{dt} = -x.$$

for all real values of t. The derivative of

$$x = e^t + te^t$$

is

$$\frac{dx}{dt} = e^t + e^t + te^t = 2e^t + te^t.$$

At the same time,

$$2x + y = 2(1 + t)e^t - te^t = 2e^t + 2te^t - te^t = 2e^t + te^t.$$

Indeed, the left and right sides of the first equation are the same.
   Likewise, the derivative of

$$y = -te^t$$

is

$$\frac{dy}{dt} = -e^t - te^t$$

while

$$-x = -(1 + t)e^t = -e^t - te^t.$$

Again, we have an identity. Therefore, since polynomials and e^t are defined everywhere, this system of differential equations is satisfied for all t.

C. Verify that for any nonzero constant a, the pair of functions

$$x = a \sec(at)$$
$$y = a \tan(at)$$

forms a solution to the system

$$\frac{dx}{dt} = xy$$
$$\frac{dy}{dt} = x^2.$$

on the interval $-\pi/2a < t < \pi/2a$. The derivative of $a \sec(at)$ is $a^2 \sec(at) \tan(at)$ or xy. Likewise, the derivative of $a \tan(at)$ is $a^2 \sec^2(at)$, or x^2. Therefore, the system is satisfied. But what is the significance of the interval around t? The answer is that the interval around t is where both the $\sec(at)$ and $\tan(at)$ are defined.

Finding Symbolic Solutions

EXAMPLE 3 **At Last! When to Stop the Chemical Reaction**

The continuous model for the chemical reaction problem was the system of differential equations

$$\frac{da}{dt} = -0.0639a$$

$$\frac{db}{dt} = 0.0639a - .0214b \qquad \textbf{(8.15)}$$

$$\frac{dc}{dt} = 0.0214c,$$

where $a(0) = 1.00$, $b(0) = 0.00$, and $c(0) = 0.00$. Our variables were defined as

$$a(t) = \text{the number of moles of chemical } A \text{ at time } t \text{ in seconds}$$
$$b(t) = \text{the number of moles of chemical } B \text{ at time } t \text{ in seconds}$$
$$c(t) = \text{the number of moles of chemical } C \text{ at time } t \text{ in seconds.}$$

Our chemist friend wants to know the best time to stop the reaction if chemical A costs \$0.50 per mole, chemical B profits \$3.50 per mole, and chemical C profits \$1.00 per mole. We have analyzed this problem from discrete, numeric, and graphic points of view. Now, we will look at a symbolic solution, check our model and its solution against the data, and finally answer the question of when to stop the reaction.

We have already noted that $a(t) = e^{-0.0639t}$ is a solution to $\frac{da}{dt} = -0.0639a$.

▶ **FOURTH REFLECTION**
Verify that $a(t) = e^{-0.0639t}$ is the solution to the differential equation for chemical A. This solution must also satisfy the initial condition $a(0) = 1.00$. ◀

Since we have a solution for $a(t)$, we can substitute $a(t)$ into the equation for chemical B.

$$\frac{db}{dt} = 0.0639(e^{-0.0639t}) - 0.0214b. \qquad \textbf{(8.16)}$$

But this is just a linear differential equation. In Section 7.2 we developed the Method of Integrating Factors specifically to solve linear differential equations. We can use it now.

▶ **FIFTH REFLECTION**
Solve Equation (8.16). Remember $b(0) = 0$. Verify $a(t) = e^{-0.0639t}$ and $b(t) = -1.5035e^{-0.0639t} + 1.5035e^{-0.0214t}$ are solutions to the system

$$\frac{da}{dt} = -0.0639a$$

$$\frac{db}{dt} = 0.0639a - 0.0214b$$

where $a(0) = 1$ and $b(0) = 0$. ◀

Now that we have an explicit formula for $b(t)$, we should be able to find $c(t)$ by integrating

$$\frac{dc}{dt} = 0.0214b$$

with respect to t. The result of this operation should be $c(t) = 0.5035e^{-0.0639t} - 1.5035e^{-0.0214t}$. So our solution to the system of differential equations is

$$a(t) = e^{-0.0639t} \qquad\qquad\qquad (8.17)$$

$$b(t) = -1.5035e^{-0.0639t} + 1.5035e^{-0.0214t} \qquad\qquad (8.18)$$

$$c(t) = 0.5035e^{-0.0639t} - 1.5035e^{-0.0214t}. \qquad\qquad (8.19)$$

▶ **SIXTH REFLECTION**

Verify Equations (8.17)-(8.19) is the solution to the system of differential equation in Equations (8.16). Now compare the solution to this model with the actual data. How did we do?   ◀

Your assistant recommends you tell the chemist to stop the reaction at $t = 31.29$ seconds. Find your own model for the profit function and determine when the chemist will maximize profits. Compare your answer with your assistant's and the answer you got from Euler's method (Think and Share in Section 8.3).

▶ **SEVENTH REFLECTION**

What will you tell the chemist? Why?   ◀

Finding Symbolic Solutions: Separation of Variables

Suppose we are given the system

$$\frac{dx}{dt} = -2y$$

$$\frac{dy}{dt} = -8x.$$

Invoking the Chain Rule, we could obtain the first-order differential equation

$$\frac{dy}{dx} = \frac{\dfrac{dy}{dt}}{\dfrac{dx}{dt}} = \frac{4x}{y}.$$

Separating the variables yields

$$y\,dy = 4x\,dx,$$

and integrating this last equation leads to the family of solutions

$$y^2 - 4x^2 = C,$$

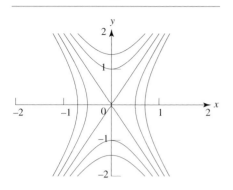

where C is an arbitrary constant that would be determined from the initial conditions. If $C = 0$, the solutions represent the two lines $y = \pm 2x$. If $C > 0$, the solutions are hyperbolas with intercepts on the y-axis. If $C < 0$, the hyperbolas intersect the x-axis (see Figure 8.21). Note the form of the solution tells us directly how y varies with x, but does not tell us directly how x and y vary with the independent variable t. Let's consider more examples.

E X A M P L E   4   **Example 1 Revisited**

Consider the linear system of Example 1.

$$\frac{dx}{dt} = -y$$

$$\frac{dy}{dt} = x.$$

Solving for dy/dx we have

$$\frac{dy}{dx} = \frac{\dfrac{dy}{dt}}{\dfrac{dx}{dt}} = \frac{-x}{y}.$$

Separating the variables yields

$$y\,dy = -x\,dx,$$

and integrating this last equation leads to the family of solutions

$$x^2 + y^2 = C$$

that are circles in the xy-plane. As we saw in Example 1, the solutions can be represented by the parametric equations

$$x = a\cos(t)$$
$$y = a\sin(t),$$

that tell us how x and y vary with t.

E X A M P L E   5   **Example 2 Revisited**

Consider the system

$$\frac{dx}{dt} = xy$$

$$\frac{dy}{dt} = x^2.$$

Solving for dy/dx we have

$$\frac{dy}{dx} = \frac{\frac{dy}{dt}}{\frac{dx}{dt}} = \frac{x^2}{xy} = \frac{x}{y}.$$

Separating the variables yields

$$y\,dy = x\,dx,$$

and integrating this last equation leads to the family of solutions

$$y^2 - x^2 = C,$$

that are hyperbolas in the xy-phase plane similar to Figure 8.21. As we saw in Example 2, the solutions can be represented by the parametric equations

$$x = a\sec(at)$$
$$y = a\tan(at).$$

The other parts of Example 2 (A and B) are not separable, and therefore we cannot find the symbolic solutions to these equations by this technique. Although we did provide solutions for parts A and B of Example 2, not being able to find a symbolic solution to a system of differential equations is fairly typical.

Putting it All Together

Although symbolic solutions to systems of differential equations are hard to come by, solving some and verifying they exist give us insight into their character. In general, these solutions provide a family of curves or surfaces (for three or more equations) that, when given a specific starting value, lead us to a specific curve, or solution. We are as far as we are going to go in our techniques of exploration of systems of difference and differential equations. In the next section, we will use the techniques we have learned to explore some models of systems generated from physical laws.

EXERCISE SET 8.4

In Exercises 1–5, verify that the given function pair is a solution to the system of differential equations given. If an initial condition is specified, verify that it too is satisfied.

1. $x = -e^t$, $y = e^t$, $\quad \dfrac{dx}{dt} = -y, \dfrac{dy}{dt} = -x$

2. $x = e^t\cos t$, $y = -e^t\sin t$, $\quad \dfrac{dx}{dt} = x+y, \dfrac{dy}{dt} = -x+y$

3. $x = -\dfrac{1}{2} + \dfrac{e^{2t}}{2}$, $y = -\dfrac{3}{4} + \dfrac{3e^{2t}}{8} + \dfrac{3e^{-2t}}{8}$, $\quad \dfrac{dx}{dt} = 2x+1, \dfrac{dy}{dt} = 3x-2y$

4. $x = 1 + \dfrac{e^{-t}}{3} - e^{-2t} + \dfrac{2e^{2t}}{3}$, $y = -\dfrac{2e^{-t}}{3} + e^{-2t} + \dfrac{2e^{2t}}{3}$, $\quad \dfrac{dx}{dt} = 2y+e^{-t}, \dfrac{dy}{dt} = 2x-2$

5. $x = (26t - 1)e^{4t}$, $y = (13t + 6)e^{4t}$, $\dfrac{dx}{dt} = 2x + 4y$, $\dfrac{dy}{dt} = -x + 6y$, $x(0) = -1$, $y(0) = 6$

6. Solve the system of differential equations symbolically for all possible solutions. What shape is this family of curves?

$$\frac{dx}{dt} = 3y$$
$$\frac{dy}{dt} = -6x$$

7. Solve the system of differential equations. What shape is this particular curve in the xy-plane?

$$\frac{dx}{dt} = 3y \quad \text{with } x(0) = 1$$
$$\frac{dy}{dt} = 9x \quad \text{with } y(0) = 2$$

8. Solve the system of differential equations. Trace this solution curve in the xy-plane by hand.

$$\frac{dx}{dt} = \frac{y}{x} \quad \text{with } x(0) = 1$$
$$\frac{dy}{dt} = x^2 \quad \text{with } y(0) = 2$$

9. Refer to Problem 12 of Exercises 8.2. If $b(t) =$ number of units of pollutant in Bull Pond at time t in years and $s(t) =$ number of units of pollutant in Lake Shasta at time t in years, then a model of the pollutant in the two-system might be

$$\frac{db}{dt} = -.25b + 3$$
$$\frac{ds}{dt} = .25b - .055$$

where $b(0) = 350$ and $s(0) = 1200$. Solve this system of differential equations to determine the amount of pollutant in each body of water at any time t. Assuming the water coming into Bull Pond is free of pollutant, determine when and if both bodies of water will be less than 20% of the current level of pollutant.

10. Refer to Problem 13 of Exercises 8.2. If $T_d(t)$ is the temperature ($^\circ F$) of the dining room at time t in minutes, $T_k(t)$ is the temperature ($^\circ F$) of the kitchen at time t in minutes, and $T_l(t)$ is the temperature ($^\circ F$) of the living room at time t in minutes, then a system of differential equations that models the temperature change in each room is

$$\frac{dT_d}{dt} = .5(75 - T_d)$$
$$\frac{dT_k}{dt} = .25(T_d - T_k)$$
$$\frac{dT_l}{dt} = .15(T_d - T_l)$$

where $T_d(0) = T_k(0) = T_l(0) = .85$. Determine the temperature in each room as a function of time. Determine the time when each room is cooled to at least $78^\circ F$.

8.5

Bungee Jumping

FIGURE 8.22 The Bungee Jumper

In this section, we will extend what we learned from modeling a skydiver in Chapter 7 and add another physical law called Hooke's Law to model the motion of a bungee jumper.

A Physical Model

Situation. You are about to jump off the Great Gorge Bridge over the Snake River, 1035 feet below. Of course, you are attached to a bungee cord. So that you have full confidence that you will not run into something during your trip, you need to solve a system of differential equations that model this situation.

Problem.   When you first start thinking about making this big leap, you have several questions. Having done some skydiving, you have some intuition about how the jump will go. Initially, you know there would be a free-fall phase until you get to the end of the cord. Then you expect the cord to work like a spring, with big oscillations at first, then smaller and smaller oscillations. You know you can model the free fall but need to learn how to model springs. Further, the following questions come to mind:

1.   Will you hit the water?
2.   Will you hit the bridge after that first big bounce?
3.   Since this model will have two phases, free fall and bouncing due to the cord, how fast will you be traveling when you get to the end of the cord?
4.   When will the bouncing stop?
5.   Finally, you have heard about these bungee jumping contests where the winner sees how close she can get to the water. How long does your cord need to be for you to just touch the water?

Our model should allow us to answer those questions. But first, we must learn something about springs.

The Theory of Springs: Hooke's Law

During your jump, not only will your height above the river be changing, your velocity will also be changing; that is, you will be subject to acceleration. According to Newton's second law of motion, your acceleration is proportional to the net external forces. There are three forces that must be considered: the force of gravity, the force of air resistance, and the restoring force of the cord. We know that the force of gravity can be considered to be constant and that one useful model for the force due to air resistance is proportional to the velocity. The force applied by the cord is not constant. Again, your intuition tells you that the farther you stretch a spring (cord), the harder the spring pulls back. We need to model how hard the spring will pull back.

Hooke's Law provides a model for this restoring force of a spring. In 1658, Robert Hooke found that the spring exerts a restoring force opposite to the direction of elongation and proportional to the amount of the elongation. Thus, the restoring force of the spring (cord) is $F_{sp} = -ky$ where y is the distance stretched or compressed beyond the natural equilibrium.

When the motion of a spring (cord) is operating up and down like you, our bungee jumper, the model is a little complicated, because the cord is stretched a little by your weight (the force due to gravity). If a spring (cord) and mass are set in motion by pulling the mass and stretching the spring, the mass will eventually return to what we call the equilibrium position, where no motion is taking place (see Figure 8.23). To simplify this spring-mass problem, we let $y =$ displacement from equilibrium. Therefore, for the second phase of your bungee jump, y will be different. Further, we will expect you to end up at this equilibrium position.

FIGURE 8.23 Cord With and Without Mass

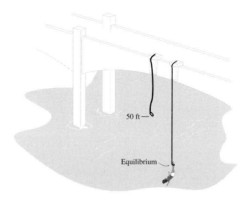

Assumptions. The assumptions for your model are the following.

1. The cord operates as a spring according to Hooke's Law and has negligible mass compared to you, the jumper.
2. The horizontal forces such as wind have no impact on the motion.
3. The air resistance is proportional to the velocity.
4. Up is the positive direction.
5. The mass is constant.

The Model. For phase I, the free fall can be modeled as a system of differential equations. Using the substitution $\frac{dy}{dt} = v$, the model we derived in Chapter 7 can be extended to the system of differential equations:

$$\frac{dy}{dt} = v$$

$$\frac{dv}{dt} = -g - \frac{b}{m}v \quad \text{with} \quad y(0) = y_0, \quad v(0) = v_0,$$

where y is the height above ground, v is the velocity, g is the acceleration due to gravity, b is the proportionality constant due to air resistance, and m is the mass of the jumper.

Situation Again. Let's say that you weigh 105 pounds. Your cord is 50 feet long with a spring constant (determined in the lab) of $k = 0.105$. Your proportionality constant for air resistance in the head-down position divided by the mass $\frac{b}{m} = 0.03$.

▶ **FIRST REFLECTION**

What is the time of travel to get to the end of the cord the first time? What is your velocity when you arrive at the end of the cord (50 feet down)? ◄

For phase II, once you've reached the end of the cord, the model becomes

$$\frac{dy}{dt} = v$$

$$\frac{dv}{dt} = -\frac{b}{m}v - \frac{k}{m}y \qquad y(0) = y_0, \quad v(0) = v_0,$$

where y is the displacement from the equilibrium position, v is the velocity, b is the proportionality constant due to air resistance, k is the spring constant for the cord, and m is the mass of the jumper.

▶ SECOND REFLECTION

If up is still positive, what is y_0 for phase II? Hint: what is the distance below the bridge of the equilibrium position?   ◄

▶ THIRD REFLECTION

What is v_0 for phase II? Is this a terminal velocity?   ◄

Think and Share

Break into groups of two. Where did $-g$ go in our model for phase II? Hint: at equilibrium

$mg = k \times$ (displacement due to placing the mass on the spring (cord)).

Using the information above, our model for phase II is

$$y' = v$$

$$v' = -0.03v - 0.032y$$

$$y(0) = 1000 \quad v(0) = -56.$$

Remember y is the displacement from the equilibrium position. Now we can answer the questions from the problem statement on page 537.

▶ FOURTH REFLECTION

Will you hit the water? If so, will you live? If not, what is your lowest point?◄

▶ FIFTH REFLECTION

Will you hit the bridge after the first big bounce? If not, what is your highest point?   ◄

You have already answered the third question on page 537 in the Third Reflection.

▶ SIXTH REFLECTION

When will the bouncing stop? Assume that if you are moving less than 10 feet, from a peak to a valley (bottom), you have stopped. Is this total time length realistic? Would this cause you to change your model?   ◄

Think and Share

Break into groups. Assuming the cord retains the same spring constant $k = 0.105$ and $k/m = 0.032$, how long must your cord be to just touch the water? Assume you reach 6 1/2 feet from the end of the cord to the tip of your hand. Outline a procedure for obtaining the answer first. What is the key parameter that determines your height in phase II? Is it really the length of the cord?

Think and Share

Break into groups. Now that you have explored these models for the bungee jumper, what assumptions would you change, add or delete, to make the results more realistic?

Putting it All Together

The skydiver and the bungee jumper are examples of models that come from physical laws. In science and engineering disciplines, these types of models are very common. However, more and more disciplines such as the social sciences, behavioral sciences, and economics use these models. In the meantime, all these problems can be validated by looking at differences and ratios of differences to conjecture models. You now have all the tools to tackle the major problems that are generated from changing phenomena. Go for it!

EXERCISE SET 8.5

1. Is the model of the bungee jumper realistic? How would you change the model to make it more realistic? What assumptions would you change and how?

2. Assuming an infinite-length cord, what is the terminal velocity of the bungee jumper in free fall? About when (within one second) would you achieve terminal velocity? How far from the water in the Snake River would you be? With your 50-foot cord, how could you achieve terminal velocity by the time you reach the end of the cord? Is this realistic? Why or why not?

† **3.** A weight of 10 grams is hung on a spring. The weight is pulled downward 5cm and released. Observing the mass, we notice that it moves up and down in a repeating pattern. By filming the experiment with a camera that takes 10 frames per second, we can determine the height of the mass every tenth of a second past the point of release. The data for the first second appears in Table 8.10. Note, the -5 at time $t = 0$ indicates the mass is 5cm below its equilibrium position.

A system of differential equations that model this motion is

$$\frac{dh}{dt} = v$$

$$\frac{dv}{dt} = \frac{-kh}{m}$$

TABLE 8.10   Spring Data

t	h(t)
0.0	−5
0.1	2.07033
0.2	3.23575
0.3	−4.72935
0.4	0.712973
0.5	4.09186
0.6	−4.09447
0.7	−0.066038
0.8	4.60061
0.9	−3.1561
1.0	−1.94117

where $h(t)$ is the displacement up from the equilibrium position, $v(t)$ is the velocity the mass is traveling, k is the spring constant, and m is the mass of the object on the end of the spring.

Here's a way to estimate a value for $\frac{-k}{m}$. The average velocity of the mass for the first tenth of a second can be computed directly from the data: $\frac{2.07033-(-5)}{0.1} = 70.07033$.

Since the initial velocity is 0, an estimate for the change in velocity or acceleration over the first tenth of a second is $\frac{70.07033-0}{0.1} = 700.07033$. The height after the first tenth of a second is 2.07033 cm. Thus, $700.07033 \approx \frac{-k}{m}2.07033$ so that $\frac{-k}{m} \approx \frac{700.07033}{2.07033} = 338.144$. Why is this an unreasonable estimate? What is wrong with the analysis?

4. A mass sitting on a horizontal surface is attached by two springs to opposite vertical surfaces (see Figure 8.24). If the

FIGURE 8.24   A Mass Attached to Two Springs

mass is pulled x_0 units to the right and released, determine a system of differential equations to model its motion. Explain what your variables represent. Give theoretical justification for your equation and be sure to point out any simplifying assumptions that you make.

5. A simple pendulum that consists of a light rigid rod of length h with a mass m attached at the end can be modeled by the system of differential equations

$$\frac{dx}{dt} = v$$
$$\frac{dv}{dt} = -\frac{g}{h}\sin(x),$$

where x is the angle the rod makes with the negative y-axis, v is the angular velocity, and g is the standard acceleration due to gravity. Assume that the rod is 32 inches long and that it is released at an angle of $\frac{\pi}{6}$. Use Euler's method with step sizes of 0.01 to graph the trajectory for these initial conditions. Describe the behavior of the trajectory. In what ways does it correspond to the actual pendulum? What aspects of the trajectory are unrealistic? Justify your responses.

6. A better model for the spring/mass system is given by the system of differential equations

$$\frac{dh}{dt} = v$$
$$\frac{dv}{dt} = -\frac{k}{m} - \frac{k_d}{m}\frac{dh}{dt}.$$

   a. Why is this a better model than the one in Problem 3? Justify the extra term.
   b. Substitute the estimate for $\frac{-k}{m}$ that you found in Problem 3. Pick various values for $\frac{-k_d}{m}$ and graph approximate solutions generated by Euler's method until you find one that is reasonably close to the actual data for the first second. What is your estimate for $\frac{-k_d}{m}$?

7. The acceleration of the skydiver in Section 7.5 due to air resistance varies more like the square of the velocity than like the velocity itself. Thus, a better model for the skydiver is given by the system of differential equations

$$\frac{dy}{dt} = v$$
$$\frac{dv}{dt} = -g - bv^2.$$

   a. Use the first 15 seconds of data in Section 7.5 for the skydiver to estimate the value of b in this equation. What is the sign of your estimate? Explain why this makes sense.
   b. Use Euler's method to generate an approximate graph of $y(t)$ in the same window as the actual data points. How does the graph compare to the data? Try adjusting the value of b to improve the match. What value of b seems to work best?

8. According to the Ideal Gas Law, the pressure P, volume V, temperature T, and the number of moles n of a gas in a closed container are related by the equation $PV = nRT$, where R is a constant. A cylinder containing an ideal gas has a piston closing from the top. Assume that the only forces acting on the piston are gravity and gas pressure. Find a system of differential equations that models the position of the piston as measured from the bottom of the cylinder. [†]

[†](Adapted from Borrelli, R. and Coleman, C. 1997. *Differential Equations — A Modeling Perspective*. PWS, Boston. page 222.

Summary

You have seen examples from biology, chemistry, population dynamics, and engineering mechanics. In addition to the above disciplines, you can now investigate problems in economics, life sciences, technology, heat transfer, and physics using the tools of calculus. You have modeled events in these disciplines with systems of difference and differential equations. The difference equations are not only helpful in giving insight to an analogous system of differential equations, but often provide a more intuitive and applicable model. Sometimes systems of differential equations can be solved symbolically, but usually numerical methods are required.

With Euler's method, a system of differential equations is transformed into a system of difference equations. The resulting table of values approximates the solution for the system of differential equations. This process is similar to the iteration of systems of difference equations that solves the system of difference equations. Often the graphs of these tables of values, either versus time or as an xy-plot, give a qualitative insight of the long-term behavior of the system and how two variables might interact.

Several of the systems of difference and differential equations we have looked at have involved three dependent variables. With the techniques we have learned in this chapter, except separation of variables, the number of equations in the model is only limited by the vision of the modeler.

Because of the insights from computer-generated graphics and the improvements in more advanced numerical solution techniques, the study of systems of difference and differential equations is a very active area of mathematical research. Change, in general, can be modeled with difference and differential equations. Systems allow us to look at more than one dependent variable and how these variables interact. Systems of difference and differential equations allow us to investigate many problems from many disciplines.

SUPPLEMENTARY EXERCISES

1. Two hamburger chains are in competition. If a person goes to A for a hamburger, the probability of going back to A for the next hamburger is 0.8, while the probability of going to B for the next hamburger is 0.2. But if a person goes to B for a hamburger, the probability of going back to B for the next hamburger is 0.7, while the probability of going to A for the next burger is 0.3.

   a. As the owner of A, I want to know the percentage of the market. How do I find that? Where do I start?

   b. Model this problem with a system of difference equations.

   c. What is the market percentage for owner A in the long term? How did you find this?

2. There are two age groups for a particular species of organism. Group I consists of all organisms aged under one year, while Group II consists of all organisms aged from one to two years. No organism survives more than two years. The average number of offspring per year born to each member of Group I is one, while the average number of organisms per year born to each member of Group II is two. Nine-tenths of Group I survive to enter Group II each year. If there are initially 450,000 organisms in Group I and 360,000 organisms in Group II, calculate the number of organisms in each group after one year, after two years, and after ten years. Are these numbers realistic? If not, why not?

3. In the scenario above in Exercise 2, suppose that at a certain time there were 810,000 organisms in Group I and 630,000 organisms in Group II. Determine the population of each group one year earlier.

4. A rental car agency divides its national market into three regions and has regional headquarters in Los Angeles,

Chicago, and Miami. The company inventories and records the location of its cars monthly. Historical records show that of the cars rented in the Los Angeles region each month, 60% remain in the Los Angeles region, and 30% go to the Chicago region, and 10% go to the Miami region. Of the cars rented in the Chicago region each month, 50% remain in the Chicago region, 10% go to the Los Angeles region, and 40% go to the Miami region. Of the cars rented in the Miami region each month, 50% remain in the Miami region, 20% go to the Los Angeles region, and 30% go to the Chicago region.

   a. Model this problem as a system of difference equations. It may help to sketch a diagram.

   b. What is the long-term distribution of cars between the three regions?

   c. Suppose there are originally 100 cars in the Los Angeles region, 4000 cars in the Chicago region, and 3000 cars in the Miami region. How many cars will be in each region after three months?

5. Two supermarket chains, called A and B, are in competition. Eighty% of the people who shop at Market A in one month return to Market A the next month, while the other 20% go to Market B the next month. Seventy% of the people who shop at Market B in one month return to Market B the next month, while the other 30% go to market A the next month. Assuming that these are the only two supermarkets available, what is chain A's percentage of the total customers in the long term?

6. An ecologist studies the population of rabbits and foxes in a contained valley. Initially, the rabbit population numbered 500 and the fox population numbered 20. Data indicate that the rabbit population is increased by a growth rate of 10% of the rabbits present and decreased by the rabbit-eating rate (0.15) of the foxes. The fox population has a constant growth rate (proportionality constant 1.0) and is bolstered by the population of rabbits (food) at a 0.85 rate. Model the rabbit and fox populations as a system of differential equations. Then determine the long-term population of foxes and rabbits. A graph in the xy-plane may provide the answer to this problem.

7. Suppose two neighboring countries are engaged in a competitive buildup of tanks. During each year Country A builds 0.25 tanks for every tank Country B had the previous year, while Country B, a poor country, builds only 0.1 new tanks for every tank Country A had the previous year. Country A's tank crews are poorly trained and lose 10% of their tanks each year to training accidents and maintenance problems. Country B trains their crews better and on average loses 5% of their tanks each year to training accidents and maintenance problems. Write the system of difference

equations modeling the buildup of tanks by countries A and B. If Country A has 1000 tanks and Country B has 750 tanks initially, determine the number of tanks each country will have after five years. Where is this system going and explain what this means in terms of the impact of wealth and training on a country's relative fighting power (in terms of tanks). Does this make practical sense? Why or why not?

8. You have inherited a small computer company and are trying to forecast sales of your top three systems. Every year, you know that your customers will buy a new system, with the choices as follows:

   - *System A:* The "Byte-Master 2000" File Server/Mini-Computer
   - *System B:* The "Data-Dude" Personal Desktop Computer
   - *System C:* The "Lap-Blaster" Portable Computer

You've determined that clients who just bought system A only buy another Byte-Master in the next year 10% of the time. Sixty % of them will instead buy a Data-Dude as a work station, with the rest buying Laptops for business trips. New owners of the Data-Dudes appear to be evenly split on their next purchase between another Data-Dude or a Lap-Blaster. Lap-Blaster owners will never buy the Byte-Master 2000, concentrating on the desktop model for 80% of their next buys and the other 20% purchasing another Laptop system. Last year's sales figures show 100 System A, 500 System B, and 300 System C were sold. How many of each system will we sell in the next 10 years? Twenty years? Describe in words the long-term behavior of our sales. Is there any one system we can stop selling? (We break even on at least five systems sold.)

9. Develop a model for the growth of trout and bass assuming that in isolation, trout demonstrate exponential decay (so $a < 0$ in the system of Equations (8.4)) and that the bass population grows logistically with a population limit M. Use the numbers in Example 3 from Section 8.3 and M=20 to give you some insight. Analyze graphically the motion in the vicinity of the rest points in your model. Is coexistence possible?

10. Consider the following economic model. Let P be the price of a single item on the market. Let Q be the quantity of the item available. Both P and Q are functions of time. If one considers price and quantity as two interacting species, the following model might be proposed:

$$\frac{dP}{dt} = aP\left(\frac{b}{Q} - P\right)$$

$$\frac{dQ}{dt} = cQ(fP - Q)$$

where a, b, c, and f are positive constants. Justify and discuss the adequacy of the model.

a. If $a = 1$, $b = 20,000$, $c = 1$, and $f = 30$, find the rest points of this system. If possible, classify each rest point with respect to its stability. If a point cannot be readily classified, give some explanation.

b. Perform a graphical analysis to determine what will happen to the levels of P and Q as time increases.

c. Give an economic interpretation of the curves that determine the rest points.

11. The horizontal motion of the skydiver in Section 7.4 is not affected by gravity. It is, however, affected by the force of air resistance. Suppose that the skydiver's initial horizontal velocity is 50 mph, and that air resistance varies with the square of velocity. Furthermore, suppose her horizontal velocity after one second is 49 mph.

a. Determine a system of differential equations that models her horizontal motion.

b. The skydiver plans to pull her rip cord at 2000 feet. Estimate her horizontal distance from the drop zone if she is to be directly over it when she deploys her chute.

12. An object having a mass m is suspended from the end of an elastic shaft as in Figure 8.25. The object is then twisted some initial angle q_0. We wish to find a system of differential equations which represents q, the amount of twist, as a function of time. Two forces act on the mass: the restoring

FIGURE 8.25   A Twisted Shaft

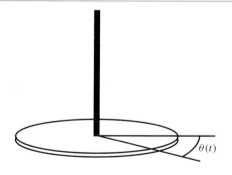

force that depends on the elasticity of the shaft and the size of q (the force is larger for larger values of q) and the damping force that always acts opposite to the direction of motion.

a. Develop a system of differential equations to model this situation. Explain how your equations follow from the described situation. Note any simplifying assumptions (factors that you are ignoring).

b. Choose values for your constants and initial values and use Euler's method to numerically solve your system until the reaction stops (be sure to use a scale appropriate for your choice of constants). Graph your results. How does your graph reflect the situation being modeled? What happens as t increases without bound? What other features of your graph seem relevant to the situation? Are there any features of your graph that do not seem to reflect reality?

c. Now symbolically solve the system you found in part b. Discuss how your solution corresponds to the graph you found in part b. Discuss any discrepancies between your symbolic and numerical solutions.

13. A rocket is shot vertically upward from the ground. We wish to obtain a system of differential equations that represent the height of the rocket as a function of time (see Figure 8.26).

FIGURE 8.26   A Rocket Launch

a. Develop a system of differential equations to model this situation. Explain how your system follows from the described situation. Note any simplifying assumptions (factors which you are ignoring).

b. Choose reasonable values for your constants and initial conditions and use Euler's method to numerically solve your system for a reasonable length of take-off (be sure to use a scale appropriate for your choice of constants). Graph your results. How does your graph reflect the situation being modeled? What happens as t grows larger and larger? What other features of your graph seem relevant to the situation? Are there any features of your graph that do not seem to reflect reality?

c. Now symbolically solve the system of differential equations you found in part b. Discuss how your solution corresponds to the field you found in part b. Discuss any discrepancies between your numerical and symbolic solutions.

d. Find the escape velocity of the rocket if it is launched from Earth. Find the escape velocity of the rocket if it is launched from Mars. Will any body, no matter how massive, always have an escape velocity? Explain why or why not.

14. Refer to Problem 12 of Exercises 8.2. We will solve this model numerically and compare to the symbolic solution found in Problem 9 of Exercises 8.4. Iterate the system until the reaction seems complete. Use Euler's method and a step size of $\Delta t = 0.5$ months until there is no change. Plot all three solutions (numeric and symbolic) on the same axes. How did the numerical methods do? If they are both way off the symbolic solution, then you should verify the symbolic solution.

15. Refer to Problem 13 of Exercises 8.2. We will solve this model numerically and compare to the symbolic solution found in Problem 10 of Exercises 8.4. Iterate the system until the reaction seems complete. Use Euler's method and a step size of $\Delta t = 0.5$ second until there is no change. Plot all three solutions (numeric and symbolic) on the same axes. How did the numerical methods do? If they are both way off the symbolic solution, then you should verify the symbolic solution.

16. Many models of physical phenomena are classically studied as higher order differential equations (equations with second and third derivatives in them). However, with the advent of computers and the difficulty faced in finding many symbolic solutions, many of these higher order equations are converted to a system of first order differential equations (equations with just a first derivative in them, like the ones we studied in Chapter 8) and solved symbolically. For example, in a physics class you may see the model of the skydiver with air-resistance as

$$\frac{md^2 y}{dt^2} = -mg - b\frac{dy}{dt}.$$

By dividing by m and letting $\frac{dy}{dt} = v$ we get the first order system of differential equations:

$$\frac{dy}{dt} = v$$

$$\frac{dv}{dt} = -g - \frac{b}{m}v$$

This also works for higher order (> 2) differential equations. Convert the following differential equations to a system of first-order differential equations.

a. $y'' + 2y' + 5y = 0$
b. $3y'' - 4y' + 6y = 0$
c. $y'' - y' + 2y = 0$
d. $2y'' - 3y' + 4y = 0$
e. $y''' - 3y' + y = 0$
f. $y''' - 5y'' + y = 0$
g. $y''' + 3y' - 2y = 0$
h. $y''' + y'' - 2y' = 0$

17. Let f be a polynomial function of degree n.
a. Write down a general formula for f.
b. Suppose that $(x_1, y_1), (x_2, y_2), \ldots (x_{n+1}, y_{n+1})$ are $n + 1$ points that lie on the graph of f. Write a system of $n + 1$ equations in $n + 1$ unknowns that has the coefficients of f as a solution.
c. Suppose that the coefficient of x^n for f is found to be zero. What does that imply about the points (x_1, y_1), $(x_2, y_2), \ldots (x_{n+1}, y_{n+1})$?

18. Follow the conclusions of Exercise 17.
a. Let $f(x) = e^x$. Find five points that lie on the graph of f and use those points to find the formula for a fifth degree polynomial function that approximates f.
b. Repeat part a for $f(x) = \cos(x)$.
c. Explain why it might be useful to approximate an exponential or trigonometric function with a polynomial function. (In the next chapter you will learn another way to find polynomial approximations of nonpolynomial functions.)

PROJECTS

Project One: Pollution in the Great Lakes

Problem Description.   Most of the water flowing into Lake Erie is from Lake Huron, and most of the water flowing into Lake Ontario is from Lake Erie (see Figure 8.27).

Suppose that pollution of the lakes ceased, except for pollution introduced by aluminum factories on Lake Huron and Lake Ontario. How long would it take for the pollution level in each lake to be reduced to 10% of its present level?

FIGURE 8.27   Lake Huron, Lake Erie, and Lake Ontario

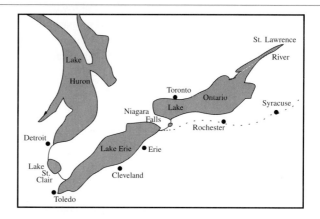

First, to simplify matters, let's assume that 100% of the water in Lake Erie comes from Lake Huron and 100% of the water in Lake Ontario comes from Lake Erie. Let $a(n)$, $b(n)$, and $c(n)$ be the total amount of pollution in Lake Huron, Lake Erie, and Lake Ontario, respectively, after n years. Since pollution has stopped, the concentration of pollution in the water coming into Lake Huron is zero. It has also been determined that, each year, the percentage of water replaced in Lakes Huron, Erie and Ontario is approximately 11, 36 and 12%, respectively. Additionally, suppose that aluminum factories on Lake Huron and Lake Ontario dump 30 units of pollutant directly into each lake each year. Initially, there are 3500, 1800, and 2400 units of pollutant in Lakes Huron, Erie and Ontario, respectively.

1.   Write a system of difference equations that models this process.
2.   Determine how long it would take for the pollution level in each lake to be reduced to 10% of its present level.
3.   Describe the long-term behavior of this system.
4.   If $a(t)$, $b(t)$, and $c(t)$ are the amount of pollution in Lake Huron, Lake Erie, and Lake Ontario, respectively, at t years, model this process as a system of differential equations.
5.   Use Euler's method to determine how long it would take for the pollution level in each lake to be reduced to 10% of its present level.
6.   Compare the two models. Which one seems most appropriate, the discrete model or the continuous model?

Project Report.   Write a report on your investigations. Address your report to your classmates, assuming the material is new to them. Describe your reasoning and include all necessary background information. A minimal project report must include:

1.   The construction and analysis of a difference equation model for pollution in the three lower Great Lakes.

2. The construction and analysis of a differential equation model for pollution in the lower Great Lakes.
3. A comparison of the results of the two models.
4. Justification and explanations for your calculations.
5. Suggestions for possible extensions of your observations and further questions for exploration.

Project Two: Chemical Reactions

Problem Description. Consider a chemical system involving three substances whose concentrations at any time t are $x(t)$, $y(t)$, and $z(t)$, respectively. We assume that each substance is converted to either of the other two by simple reactions involving only the substance being converted. Also, we assume that the rate of change per unit time from one substance a to substance b is proportional to the amount present of substance a with proportionality constant k_{ab}. Figure 8.28 below shows the situation for the three substances. To model the

FIGURE 8.28 A Three Substance Chemical Reaction Diagram

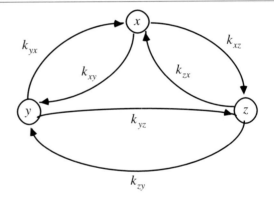

amount of substance x present in any time n, we must account for the amount of substance x lost to y and z, and the amount of substance x gained from y and z in the given time period. Thus, discretely the change in substance x can be modeled by the equation

$$x_n - x_{n-1} = k_{yx}y_{n-1} + k_{zx}z_{n-1} - k_{xy}x_{n-1} - k_{xz}x_{n-1}$$

or

$$\frac{dx}{dt} = k_{yx}y + k_{zx}z - k_{xy}x - k_{xz}x.$$

The difference equation can be written as

$$x_n = \left(1 - k_{xy} - k_{xz}\right)x_{n-1} + k_{yx}y_{n-1} + k_{zx}z_{n-1},$$

and the differential equation can be written as

$$\frac{dx}{dt} = \left(-k_{xy} - k_{xz}\right)x + k_{yz}y + k_{zx}z.$$

1. **a.** Write the equations for the other two substances (y and z) and give the first-order system of it difference equations that describes the reactions between the three chemicals for any time period n.

    b. Write the system of equations.

    c. Given the following proportionality constants,

    $$k_{xy} = 0.2 \ k_{xz} = 0.3 \ k_{yx} = 0.2 \ k_{yz} = 0.3 \ k_{zx} = 0.2 \ k_{zy} = 0.3,$$

    and that you initially have three units of substance x, two units of substance y, and four units of substance z, describe the long-term behavior of the chemical system. Does your answer make sense?

    d. Verify your conclusions by iterating the system on your calculator.

2. **a.** Write the transfer equations for the other two substances (y and z) and give the first-order system of *differential equations* which describes the reactions between the three chemicals for any time period n.

    b. Write the system of equations.

    c. Given the following proportionality constants

    $$k_{xy} = 0.2, \quad k_{xz} = 0.3, \quad k_{yx} = 0.2, \quad k_{yz} = 0.3, \quad k_{zx} = 0.2, \quad k_{zy} = 0.3,$$

    and assuming that you initially have three units of substance x, two units of substance y, and four units of substance z, describe the long-term behavior of the chemical system. Does your answer make sense?

    d. Verify your conclusions by Euler's Method.

3. Each time period an additional 0.8 units of substance x and substance z are introduced into the chemical reaction.

    a. Write transfer the equations for the three substances and give the system of difference equations or differential equations that describes the reactions between the three chemicals for any time period n or at any time t.

    b. Iterate the new system or use Euler's method on your calculator and describe the long-term behavior of the chemical system.

Project Report.   Write a report on your investigations. Address your report to your classmates, assuming the material is new to them. Describe your reasoning and include all necessary background information. A minimal project report must include:

1. The construction and analysis of a difference equation model for chemical reactions.

2. The construction and analysis of a differential equation model or chemical reactions.

3. A comparison of the results of the two models.

4. A discussion of your model of the second scenario.
5. Justification and explanations for your calculations.
6. Suggestions for possible extensions of your observations and further questions for exploration.

Project Three: Extending the Predator-Prey Model – TI-92

In this project, you are going to refine the predator-prey model of Section 8.1. You will be looking at two enhancements of the model and analyzing the results. Your project group will write a report on your findings and present the report, as a group, to your instructor and, possibly, to your class. In order to complete this project you will need the program, `Project3`. You can find a listing of the program at the end of this project

The Original Predator-Prey Model.   The predator-prey model of Section 8.1 can be extended. Given a predator, $\{x_n\}$, and a prey, $\{y_n\}$, a model based on mutual interaction was developed. Whether the interaction is beneficial depends on the viewpoint taken, *eater* or *eaten*. The model is

$$x_n = (1-a) \cdot x_{n-1} + b \cdot x_{n-1} \cdot y_{n-1} \tag{8.20}$$
$$y_n = (1+c) \cdot y_{n-1} - d \cdot x_{n-1} \cdot y_{n-1}$$
$$x_1 = Q$$
$$y_1 = R,$$

where $0 < a, b < 1$ and $0 < c, d, Q, R$.

Our model is based on three assumptions, two of which are that in each time period

1. the prey, *in the absence of predators*, will increase by a fixed proportion,
2. the predators, *in the absence of prey*, will decrease by a fixed proportion.

We'll be using the `Project3` to plot vector fields. The program works by using the recursion formulas as functions. We generate the next points x_n and y_n by using the current x and y values as x_{n-1} and y_{n-1} in the difference equations. Define the functions f for the predator difference equation and g for the prey difference equation as follows:

```
Define f(q,r) = (1-0.2)*q +0.002*q*r
```

```
Define g(q,r) = (1+0.5)*q +0.025*q*r.
```

These are the values used in Section 8.1.

To plot the vector field of the original model:

1. Obtain the program `Project3` from your instructor.
2. Set the graph mode to `SEQUENCE`.
3. Go to the Y= editor and enter `u1(n)=f(u1(n-1),u2(n-1)),ui1=25, u2(n)=g(u1(n-1),u2(n-1)),` and `ui2=120`.
4. In the Y= editor, from the `Axes...` dialogue box, choose `CUSTOM` with `u1` on the X Axis and `u2` on the Y Axis.

5. Use the WINDOW settings: `nmin=1,nmax=40,poltstrt=1,plotstep=1,`
   `xmin=0,xmax=40,xscl=10,ymin=-1,ymax=250,yscl=10.`

6. Return to the HOME screen and enter `Project3()`.

When the grid of (x, y) points has been drawn, the program *pauses*. At the bottom right corner of the screen, notice that the `BUSY` indicator has changed to `PAUSE`. To continue, press ENTER. See Figure 8.29.

When the program has finished, choose `Save Copy As...` from the Toolbox menu. Change the `Type` to `Picture` and enter the name `pic1` in the `Variable` edit box. (If you have access to a TI-GRAPHLINK, print your vector field for reference.)

FIGURE 8.29   Vector Field

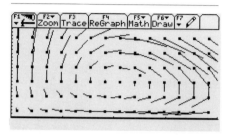

Limited Prey Growth.   In the absence of predators, the recursion formula for prey becomes

$$y_n = (1 + c) \cdot y_{n-1}.$$

This equation gives rise to exponential growth. As we know, exponential growth is not sustainable. Resources like food and space impose limits and the growth must slow or, if the population becomes too large, go negative. The Discrete Logistic model incorporates this limit with the *self-limiting* term $-m \cdot y_{n-1}^2$. Add the self-limiting term to the prey equation in the model.

$$
\begin{aligned}
x_n &= (1 - a) \cdot x_{n-1} + b \cdot x_{n-1} \cdot y_{n-1} \\
y_n &= (1 + c) \cdot y_{n-1} - d \cdot x_{n-1} \cdot y_{n-1} - m \cdot y_{n-1}^2 \\
x_1 &= Q \\
y_1 &= R,
\end{aligned}
\tag{8.21}
$$

where $0 < a, b < 1$ and $0 < c, d, m, Q, R$.

We use the same values for the parameters as before with and set $m = 0.001$. Redefine g as

`Define g(q,r) = (1+0.5)*q +0.025*q*r -0.001*r^2.`

Clear the GRAPH screen. Using the same WINDOW settings as above, draw a new vector field by returning to the HOME screen and re-executing `Project3()`. Save this image as `pic2`. (If you have access to a TI-GRAPHLINK, print your direction field for reference.)

1. How does the extra term affect the vector field?

2. Examine different starting values and describe the behavior of the sequences.

3. Determine the rest points of the altered model using the methods of Section 8.1.

Limited Prey and Limited Predator Growth.   Limited predator growth It is also reasonable to add a limiting term to the predator equation.

1. Add $-l \cdot x_{n-1}^2$ to the predator equation to represent the effect of internal competition. Set $l = 0.0005$.

   a. How does the extra term affect the vector field?
   b. Examine different starting values and describe the behavior you observe.
   c. Determine the rest points of the doubly limited model.

2. Redefine f to change the coefficient b to 0.0002. Change ymax to 1000 in the WINDOW editor.

   a. How does the new coefficient affect the vector field?
   b. Examine different starting values and describe the behavior of the sequences.
   c. Determine the stable points of this new model.

Program Listing.   The following is a listing of the program Project3.

PROJECT3

```
Project3()
Prgm
Local r,c,x,y
For c,7,238,17
 For r,10,102,17
  PxlCrcl r,c,1
 EndFor
EndFor
Pause
For c,7,238,17
 xmin+(xmax-xmin)/238*c → x
 For r,10,102,17
  ymax-(ymax-ymin)/102*r → y
  Line x,y,f(x,y),g(x,y)
 EndFor
EndFor
EndPrgm
```

Project Report.   Write a report on your investigations. Address your report to your classmates, assuming the material is new to them. Describe your reasoning and include all necessary background information. A minimal project report must include:

1. Description of the long-term behavior of your models.
2. Results (tabular and graphical) of experiments you ran with each model.
3. Illustrative and annotated vector fields showing trajectories from several of your experiments.
4. A discussion of any obvious shortcomings in your model.
5. Suggestions for possible refinements and extensions of your models.

6. A discussion as to how one might verify the validity of your model in either a natural or laboratory setting.

7. Justification and explanations for your calculations.

8. suggestions for possible extensions of your observations and further questions for exploration.

Extending the Predator-Prey Model – TI-89 and 92 Plus

In this project, you are going to refine the predator-prey model of Section 8.1. You will be looking at two enhancements of the model and analyzing the results. Your project group will write a report on your findings and present the report, as a group, to your instructor and, possibly, to your class.

The Original Predator-Prey Model. The predator-prey model of Section 8.1 can be extended. Given a predator, $\{x_n\}$, and a prey, $\{y_n\}$, a model based on mutual interaction was developed. Whether the interaction is beneficial depends on the viewpoint taken, *eater* or *eaten*. The model is

$$x_n = (1 - a) \cdot x_{n-1} + b \cdot x_{n-1} \cdot y_{n-1} \qquad \textbf{(8.22)}$$
$$y_n = (1 + c) \cdot y_{n-1} - d \cdot x_{n-1} \cdot y_{n-1}$$
$$x_1 = Q$$
$$y_1 = R,$$

where $0 < a, b < 1$ and $0 < c, d, Q, R$.

As usual this family of difference equation models give rise to the family of differential equations

$$\frac{dx}{dt} = -a \cdot x + b \cdot x \cdot y \qquad \textbf{(8.23)}$$
$$\frac{dy}{dt} = c \cdot y - d \cdot x \cdot y$$
$$x(0) = Q$$
$$y(0) = R,$$

where $0 < a, b < 1$ and $0 < c, d, Q, R$.

Our model is based on three assumptions, two of which are that in each time period

1. the prey, *in the absence of predators*, will increase exponentially,

2. the predators, *in the absence of prey*, will decrease exponentially.

For this investigation you will be using the values $a = 0.2$, $b = 0.002$, $c = 0.5$, and $d = 0.025$.

To plot the direction field of the continuous model:

1. Set the graph mode to DIFF EQUATIONS.

2. Go to the Y= editor and enter y1'=-0.2*y1+0.002*y1*y2, yi1=25, y2'=0.5*y2-0.025*y1*y2, and yi2=120.

FIGURE 8.30   Direction Field

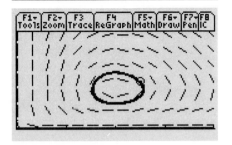

3. From the Tools menu open the Format... dialogue box and set Fields to DIRFLD.
4. From the Axes... dialogue box, choose CUSTOM with y1 on the X Axis and y2 on the Y Axis.
5. Use the WINDOW settings: t0=0, tmax=40, tstep=1, tplot=0, xmin=0, xmax=40, xscl=10, ymin=-1, ymax=250, yscl=10.
6. Go to the GRAPH screen. See Figure 8.30.

Choose Save Copy As... from the Toolbox menu. Change the Type to Picture and enter the name pic1 in the Variable edit box. (If you have access to a TI-GRAPHLINK, print your direction field for reference.)

Limited Prey Growth.   In the absence of predators, the recursion formula for prey becomes

$$\frac{dy}{dt} = c \cdot y.$$

This equation gives rise to exponential growth. As we know, exponential growth is not sustainable. Resources like food and space impose limits and the growth must slow or, if the population becomes too large, go negative. The Continuous Logistic model incorporates this limit with the *self-limiting* term $-m \cdot y^2$. Add the self-limiting term to the prey equation in the model.

$$\frac{dx}{dt} = -a \cdot x + b \cdot x \cdot y \qquad\qquad (8.24)$$
$$\frac{dy}{dt} = c \cdot y - d \cdot x \cdot y - m \cdot y^2$$
$$x(0) = Q$$
$$y(0) = R,$$

where $0 < a, b < 1$ and $0 < c, d, m, Q, R$.

We use the same values for the parameters as before with and set $m = 0.001$. Redefine y2 as

y2'=0.5*y2-0.025*y1*y2 - 0.001*y2^2.

Graph the new direction field using the same WINDOW settings as above. Save this image as pic2. (If you have access to a TI-GRAPHLINK, print your direction field for reference.)

1. How does the extra term affect the direction field?
2. Use the IC menu to examine different initial values and describe the behavior of the population trajectory.
3. Determine the rest points of the altered model.

Limited Prey and Limited Predator Growth.   It is also reasonable to add a limiting term to the predator equation.

1. Add $-l \cdot x^2$ to the predator equation to represent the effect of internal competition. Set $l = 0.0005$.

   a. How does the extra term affect the direction field?
   b. Examine different starting values and describe the behavior you observe.
   c. Determine the rest points of the doubly limited model.

2. Change the coefficient b to 0.0002. Change ymax to 1000 in the WINDOW editor.

   a. How does the new coefficient affect the direction field?
   b. Examine different starting values and describe the behavior of the population trajectory.
   c. Determine the stable points of this new model.

Project Report. Write a report on your investigations. Address your report to your classmates, assuming the material is new to them. Describe your reasoning and include all necessary background information. A minimal project report must include:

1. Description of the long-term behavior of your models.
2. Results (tabular and graphical) of experiments you ran with each model.
3. Illustrative and annotated direction fields showing trajectories from several of your experiments.
4. A discussion of any obvious shortcomings in your model.
5. Suggestions for possible refinements and extensions of your models.
6. A discussion as to how one might verify the validity of your model in either a natural or laboratory setting.
7. Justification and explanations for your calculations.
8. suggestions for possible extensions of your observations and further questions for exploration.

9

POWER SERIES: APPROXIMATING FUNCTIONS WITH FUNCTIONS

9.1
Polynomial Approximation of Functions

9.2
Using Polynomial Approximations

9.3
How Good Is a Good Polynomial Approximation?

9.4
Convergence of Series

9.5
Power Series Solutions of Differential Equations

Summary

9.1

Polynomial Approximation of Functions

In unconstrained population growth, the rate of change of a population is proportional to the size of the population. Suppose there are two neighboring islands with the same size bird population, $p = 100$. If 10 birds hatch the first year, then the population on each island grows by 10 in that year. If the islands become joined by tectonic activity, then the population of the new larger island is 200, and the growth in population is 20. Then, the *rate of change* is *grow by one tenth per year*, or, in the language of derivatives,

$$\frac{dp}{dt} = (0.1)\, p.$$

When the population is 100, then $dp/dt = 0.1p = 0.1(100) = 10$. When the population is 200, then $dp/dt = 0.1p = 0.1(200) = 20$. The change in population for a population of 100 and a population of 200 is different (10 and 20, respectively), but the rate of change in population is the same (0.1) for both.

This idea leads to the general differential equation

$$\frac{dp}{dt} = kp$$

for unconstrained or exponential population growth. We have studied this model in previous chapters, and solved the differential equation using separation of variables.

Approximating with Polynomials

Let's try to solve the initial value problem

$$\frac{dp}{dt} = kp \tag{9.1}$$
$$p(0) = p_0$$

by using another technique. We ask ourselves, "What is the best approximation to this function by a constant?"

Given we know one value of the solution, namely the initial value p_0, then a constant approximation that is true for time $t = 0$ is $f_0(t) = p_0$.

Since we know the population is growing, a constant won't be a good approximation for long. So we ask, "What is the best linear approximation for this function?" The solution will be of the form

We're using a subscript to indicate the degree of f.

$$f_1(t) = a + bt.$$

Once again, we would like this function to be true for $t = 0$.

$$f_1(0) = a + b \cdot 0 = p_0$$

So, $a = p_0$.

But what is the value of b? Let's assume we want the slope of the tangent line at $t = 0$ to be the same as it is for the actual function. This seems natural since the tangent line is a good approximation to the function. We know that

This is called "order of contact 1."

$$\frac{dp}{dt} = kp(t) \quad \text{and} \quad f_1'(t) = b.$$

Since, at t_0 we know $dp/dt = kp_0$, we choose $b = kp_0$. Thus, our linear approximation is

$$f_1(t) = p_0 + (kp_0)\, t.$$

Continuing in this way, we add the requirement that the second derivative at $t = 0$ be the same for the approximating second degree or quadratic function where the value of the function and its first derivative are also the same at $t = 0$. Thus,

This is called "order of contact 2."
What order of contact do you think f_0 has?

$$f_2(t) = a + bt + ct^2$$
$$f_2'(t) = b + 2ct$$
$$f_2''(t) = 2c. \tag{9.2}$$

Following what we have already done, $a = p_0$ and $b = kp_0$. How do we find the value of c?

The derivative of p is given by Equation (9.1). Differentiate both sides (using implicit differentiation since p is a function of t) to obtain

$$\frac{d^2 p}{dt^2} = k\,\frac{dp}{dt}.$$

But we know that $dp/dt = kp$, so we obtain

$$\frac{d^2 p}{dt^2} = k\,\frac{dp}{dt} = k^2 p, \tag{9.3}$$

and, at t_0, the value is $k^2 p_0$. To have the second derivatives match at t_0, choose $c = k^2 p_0 / 2$ (see Equation (9.2)). The quadratic best approximation is

$$f_2(t) = p_0 + (kp_0)\,t + \frac{k^2 p_0}{2}\,t^2.$$

There seems to be a pattern developing in the computation for these approximating functions.

The next approximation, f_3, will be a third-degree polynomial with the same constant value and first three derivatives as $p(t)$:

We don't use "d" for the t^3 coefficient, because it looks like a differential.

$$f_3(t) = a + bt + ct^2 + et^3$$
$$f_3'(t) = b + 2ct + 3et^2$$
$$f_3''(t) = 2c + 3 \cdot 2et$$
$$f_3'''(t) = 3 \cdot 2e.$$

We claim

$$f_3(t) = p_0 + (kp_0)\,t + \frac{k^2 p_0}{2}\,t^2 + \frac{k^3 p_0}{3 \cdot 2}\,t^3.$$

▶ **FIRST REFLECTION**

Can you verify this claim by implicitly differentiating Equation (9.3)?   ◀

The computation follows a pattern. Do the polynomials follow a pattern as well? We claim that the polynomials look like

$$p(t) \approx f_n(t) = p_0 + (kp_0)\,t + \frac{k^2 p_0}{2}\,t^2 + \frac{k^3 p_0}{3 \cdot 2}\,t^3 + \cdots$$
$$+ \frac{k^{n-1} p_0}{(n-1)(n-2)\cdots(3)(2)(1)}\,t^{n-1} + \frac{k^n p_0}{n(n-1)(n-2)\cdots(3)(2)(1)}\,t^n.$$

▶ **SECOND REFLECTION**

What would the approximating polynomial of degree four, $f_4(x)$ look like? ◀

Let's see what the first four of these functions look like against the slope field generated by our differential equation, $dp/dt = kp$, with $p_0 = 100$ and $k = 0.1$.

FIGURE 9.1   Constant Approximation   **FIGURE 9.2**   Linear Approximation

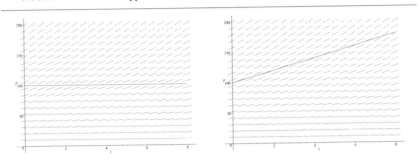

FIGURE 9.3   Quadratic Approximation   **FIGURE 9.4**   Cubic Approximation

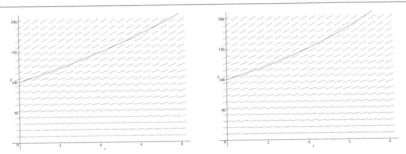

Summary

We constructed approximating polynomials for a function described by a differential equation. To get a better fit, we chose the polynomial's coefficents so that the polynomial matched the function's higher derivatives. By using implicit differentiation, we were able to obtain the higher derivatives of the original function that we wanted to approximate. There are two items to note. We could approximate a function that may have no symbolic formula with polynomials. We could get a better fit from the next higher degree polynomial, f_{n+1}, by adding one new term to f_n.

There are important questions that we'll need to investigate. Mostly these questions revolve around the basic need to know the answer to "How good is the approximation?"

EXERCISE SET 9.1

1. Use the techniques of this section to find a third-degree polynomial approximating the function described by

$$\frac{dp}{dt} = -0.2 \cdot p$$
$$p(0) = 500.$$

2. Use the techniques of this section to find a third-degree polynomial approximating the function described by the logistic equation

$$\frac{dp}{dt} = 0.15 \cdot p - 0.005 \cdot p^2$$
$$p(0) = 200.$$

3. Use the techniques of this section to find a third-degree polynomial approximating the function described by

$$\frac{dp}{dt} = p$$
$$p(0) = 1.$$

4. Use the techniques of this section to find a formula for the nth degree polynomials approximating the function described by

$$\frac{dp}{dt} = p$$
$$p(0) = 1.$$

5. Use the techniques of this section to find a formula for the nth degree polynomials approximating the function described by

$$\frac{dp}{dt} = -p$$
$$p(0) = 1.$$

6. Use the techniques of this section to find a third-degree polynomial approximating the function described by

$$\frac{dp}{dt} = \frac{1}{1-t}$$
$$p(0) = 1.$$

7. Use the techniques of this section to find a formula for the nth degree polynomials approximating the function described by

$$\frac{dp}{dt} = \frac{-1}{(1-t)^2}$$
$$p(0) = 1.$$

8. *Group Exercise* Adapt the techniques of this section to find approximating polynomials of degree 0, 1, 2, and 3 for the functions determined by the system

$$\frac{dp}{dt} = q$$
$$\frac{dq}{dt} = p$$
$$p(0) = 1$$
$$q(0) = 0.$$

9. *Group Exercise* Adapt the techniques of this section to find a formulas for the nth degree polynomials approximating the functions determined by the system

$$\frac{dp}{dt} = q$$
$$\frac{dq}{dt} = p$$
$$p(0) = 1$$
$$q(0) = 0.$$

10. *Group Exercise* Adapt the techniques of this section to find a formulas for the nth degree polynomials approximating the functions determined by the system

$$\frac{dp}{dt} = q$$
$$\frac{dq}{dt} = -p$$
$$p(0) = 1$$
$$q(0) = 0.$$

9.2

Using Polynomial Approximations

All numerical computation is at root made up of repeated additions and multiplications. In primary school we learned addition and multiplication tables, no others. Later we learned how procedures for subtraction, division, and, maybe, extraction of roots that used only the addition and multiplication facts we memorized in second and third grade. Polynomials too use only addition and multiplication (integral powers are simply repeated addition) for their evaluation. Therefore, polynomials have always played an important role in the theory and practice of computation. In the last section, we saw how a family of polynomials arises naturally as we approximate solutions to simple differential equations; in this section, we will see how the idea of "matching derivatives" allows us to build a sequence of polynomials to approximate functions that are not defined by an easily computable formula. These *Maclaurin polynomials* provide an introduction to the field of *numerical analysis*.

Trigonometric functions, for example, are defined in terms of arc length along the unit circle.

How Does One Compute e^2?

The number e^2, unlike 3^2, is very difficult to calculate by hand. We attack the problem by reasoning (hoping, really) that there most be *some* polynomial that approximates $f(x) = e^x$ well enough to supply us with a workable value for e^2.

We will address the sticky issues involved in the innocuous-looking phrases "approximated . . . well enough" and "workable" in the next section. For now, we'll proceed with optimism.

We can write our approximating polynomial, $P_n(x)$, for $f(x) = e^x$ as

$$P_n(x) = \sum_{k=0}^{n} a_k \cdot x^k.$$

Now all we need are numerical values for the degree and for the coefficients. Let's hold off on typing to find out about the degree for the moment and work on finding the coefficients. Our strategy will be to construct $P_n(x)$ so as to have the highest possible order of contact with e^x. That is, we will choose values for the coefficients a_k so that

$$\frac{d^m f}{dx^m}(0) = \frac{d^m P_n}{dx^m}(0), m = 0, 1, \ldots n. \tag{9.4}$$

The constant term is easy. Taking $m = 0$ in Equation (9.4), the left side evaluates to 1, and the right side to a_0. So now we may assert confidently that $a_0 = 1$. We only have n more terms to go. Letting $m = 1$ in Equation (9.4), we match the values of e^x and

$$\frac{dP_n}{dx} = a_1 + 2 \cdot a_2 \cdot x + 3 \cdot a_3 \cdot x^2 + 4 \cdot a_4 \cdot x^3 + 5 \cdot a_5 \cdot x^4 + \cdots + n \cdot a_n \cdot x^{n-1}$$

when $x = 0$. This time we find that $a_1 = 1$. To get the next coefficients, simply differentiate a second time to compare the values of e^x and

$$\frac{d^2 P_n}{dx^2} = 1 \cdot 2 \cdot a_2 + 2 \cdot 3 \cdot a_3 \cdot x + 3 \cdot 4 \cdot a_4 \cdot x^2$$
$$\dotplus 4 \cdot 5 \cdot a_5 \cdot x^3 + 5 \cdot 6 \cdot a_6 \cdot x^4 + \cdots + n \cdot (n-1) \cdot a_n \cdot x^{n-2}$$

Then we differentiate a third time,

$$\frac{d^3 P_n}{dx^3} = 1 \cdot 2 \cdot 3 \cdot a_3 + 2 \cdot 3 \cdot 4 \cdot a_4 \cdot x + 3 \cdot 4 \cdot 5 \cdot a_5 \cdot x^2$$
$$+ 4 \cdot 5 \cdot 6 \cdot a_6 \cdot x^3 + 5 \cdot 6 \cdot 7 \cdot a_7 \cdot x^4 + \cdots$$
$$+ n \cdot (n-1) \cdot (n-2) \cdot a_n \cdot x^{n-3}$$

and a fourth,

$$\frac{d^3 P_n}{dx^3} = 1 \cdot 2 \cdot 3 \cdot 4 \cdot a_4 + 2 \cdot 3 \cdot 4 \cdot 5 \cdot a_5 \cdot x + 3 \cdot 4 \cdot 5 \cdot 6 \cdot a_6 \cdot x^2$$
$$+ 4 \cdot 5 \cdot 6 \cdot 7 \cdot a_7 \cdot x^3 + 5 \cdot 6 \cdot 7 \cdot 8 \cdot a_8 \cdot x^4 + \cdots$$
$$\dotplus n \cdot (n-1) \cdot (n-2) \cdot (n-3) \cdot a_n \cdot x^{n-4}$$

▶ **FIRST REFLECTION**

Why doesn't $\dfrac{d^m f}{dx^m}(0)$ ever change? ◀

Setting $x = 0$ gets us $a_2 = \frac{1}{2}, a_3 = \frac{1}{6}$, and $a_4 = \frac{1}{24}$. But even better than getting the first five of n terms is getting a *pattern* for a_n. $a_n = \dfrac{1}{1 \cdot 2 \cdot 3 \cdots (k-1) \cdot k}$. Using this pattern, we write:

You may or may not recall the symbol for the extended product $1 \cdot 2 \cdot 3 \cdots (k-1) \cdot k$. It is k!. The symbol 0! is defined to be 1. That makes some formulas, like the one at the right, easier to write. On the TI calculator the exclamation point "!" is the second function of the "w" key.

$$P_n(x) = \sum_{k=0}^{n} \frac{x^k}{k!} \tag{9.5}$$

▶ **SECOND REFLECTION**

Calculate $P_1 1(2)$ by entering `Σ((2.0)^k/k!,k,0,11)`. How does you value compare with the calculator value of e^2? ◀

Table 9.1 shows the TI calculator values for e^x and the corresponding values of $P_5(x)$, $P_{10}(x)$, and $P_{15}(x)$, its approximating polynomials of degree 5, 10, and 15.

Maclaurin Polynomials

This process provides us with a family of polynomial approximations for $f(x)$ that often can be used to approximate a function $f(x)$ quickly and accurately,

TABLE 9.1 e^x and Approximating Polynomials

x	-3	-2	-1	0	1	2	3	4	5	6
e^x	0.049787	0.135335	0.367879	1	2.71828	7.38906	20.0855	54.5982	148.413	403.429
$P_5(x)$	-0.65	0.066667	0.366667	1	2.71667	7.2667	18.4	42.8667	91.4167	179.8
$P_{10}(x)$	0.053326	0.135379	0.367879	1	2.71828	7.38899	20.0797	54.4431	146.381	386.234
$P_{15}(x)$	0.049785	0.135335	0.367879	1	2.71828	7.38906	20.0855	54.5979	148.403	403.223

provided $f(x)$ has (potentially) infinitely many derivatives. Let's review our reasoning.

1. We optimistically assume that there really *is* a family of good polynomial approximations $P_n(x)$ for $f(x)$ and write

$$P_n(x) = a_0 + a_1 \cdot x + a_2 \cdot x^2 + a_3 \cdot x^3 + \cdots + a_n \cdot x^n. \qquad \textbf{(9.6)}$$

2. Our process for determining the coefficients a_k is to match values of $\dfrac{d^m f}{dx^m}(x)$ and $\dfrac{d^m P_n}{dx^m}(x)$ at $x = 0$.

3. We begin by setting $f(0) = P_n(0)$ gives us $a_0 = f(0)$.

4. The first derivative of $P_n(x)$ is

$$a_1 + 2 \cdot a_2 \cdot x + 3 \cdot a_3 \cdot x^2 + 4 \cdot a_4 \cdot x^3 \qquad \textbf{(9.7)}$$
$$+ 5 \cdot a_4 \cdot x^4 + \cdots + n \cdot a_n \cdot x^{n-1}. \qquad \textbf{(9.8)}$$

Substituting $x = 0$ into Equation (9.4) gives us $a_1 = f'(0)$.

5. Continuing the process, we differentiate Equation (9.7) to get

$$\frac{d^2 P_n}{dx^2}(x) = 2a_2 + 2 \cdot 3a_3 \cdot x + 3 \cdot 4a_4 \cdot x^2 + 4 \cdot 5a_5 \cdot x^3 \qquad \textbf{(9.9)}$$
$$+ 5 \cdot 4a_6 \cdot x^4 + \cdots + n \cdot (n - 1) \cdot a_n \cdot x^{n-2}. \qquad \textbf{(9.10)}$$

Again, substituting $x = 0$ into Equation (9.4) gives us

$$a_2 = \frac{f''(0)}{2}.$$

Recall that $f^{(k)}$ is an alternative notation for the kth derivative of f.

6. Let's think for a minute before we do any more computation. Differentiating Equation (9.9) again and letting $x = 0$ will yield $f^{(3)}(0) = 3! \cdot a_3$. The next time we differentiate and substitute zero we will get $f^{(4)}(0) = 4! \cdot a_4$, and then $f^{(5)}(0) = 5! \cdot a_5$; in general,

$$a_k = \frac{f^{(k)}(0)}{k!}.$$

We can summarize this in a definition.

MACLAURIN POLYNOMIALS

If a function $f(x)$ can be differentiated n times then the polynomial

$$f(0) + \sum_{k=1}^{n} \frac{f^{(k)}(0)}{k!} x^k \qquad (9.11)$$

is called the **Maclaurin polynomial** of degree n for $f(x)$.

On your TI calculator the syntax for the Maclaurin polynomial (in x, of degree n for a known function f(x)) is `taylor(f(x),x,n)`. *The function* `taylor(` *may be typed, or can be found in the* `Calc` *menu or in the* CATALOG. *The TI calculator's ability to generate Maclaurin polynomials is good for examples, but it can only do so for functions it already knows, functions we don't need to approximate.*

Whenever we can find the first n derivatives of a function (and for integral functions of initial value problems this can be a chore), we can write down its Maclaurin polynomial of degree n. Sometimes it is an excellent approximation of $f(x)$. Figure 9.5 shows e^x and its degree five Maclaurin polynomial.

FIGURE 9.5   The Exponential e^x Approximated by $P_5(x)$

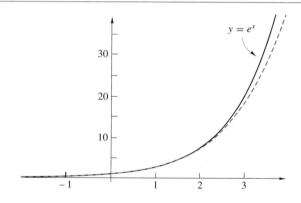

Limitations to Polynomial Approximations

Examining Table 9.1, we see that all three of the polynomials give fairly good approximations to e^x for values of x close to 0. But farther away from 0, reasonable approximations require polynomials of higher and higher degree. For the exponential we can get uniformly good approximations over large intervals, but for other functions there are intrinsic limits to the Maclaurin polynomials as approximations. Most obvious are limits imposed by continuity. The geometric series

$$\sum_{k=0}^{n} x^k = 1 + x + x^2 + x^3 + \cdots + x^n$$

gives us the Maclaurin series for the function $f(x) = 1/(1 - x)$, a function which has a discontinuity at $x = 1$. The Maclaurin polynomials, being polynomials, are continuous at $x = 1$. As a consequence of this mismatch in continuity, they are incapable of modeling the behavior of $f(x) = 1/(1 - x)$ in any open interval containing $x = 1$. Unfortunately, there are less apparent reasons for the Maclaurin polynomials for a function $g(x)$ to fail to give reasonable approximations to $g(x)$. The Maclaurin series for the arctangent function is

$$x - \frac{x^3}{3} + \frac{x^5}{5} - \frac{x^7}{7} + \cdots = \sum_{k=0}^{\infty} \frac{(-1)^n x^{2n+1}}{2n+1}. \qquad (9.12)$$

The Maclaurin polynomials can provide good approximations for $\tan^{-1}(x)$ but only for $-1 < x < 1$. All of the polynomials deviate wildly from $\tan^{-1}(x)$ as x approaches -1 or 1.

▶ **THIRD REFLECTION**

Can you verify this statement? Graph $P_5(x)$, $P_{15}(x)$, and $\tan^{-1}(x)$ on a single set of axes, with window size $-\pi \le x \le \pi$, $-\pi \le y \le \pi$.   ◀

For any given function $f(x)$, Maclaurin polynomials give useful approximations only within a fixed interval called the *interval of convergence*. When we are lucky the interval is infinite, but usually it is finite in extent. The interval of convergence is inherent in the function itself and cannot be finessed or avoided. In the next section, we begin to answer the question, "How good are Maclaurin polynomials as approximations?" We investigate the interval of convergence in Section 9.4.

EXERCISE SET 9.2

1. Determine the approximating Maclaurin polynomials for the function determined by the initial value problem

$$y' = 2y$$
$$y(0) = 1.$$

2. Determine the approximating Maclaurin polynomials for the function determined by the initial value problem

$$y'' = y$$
$$y(0) = 1$$
$$y'(0) = 0.$$

3. Determine the approximating Maclaurin polynomials for the function determined by the initial value problem

$$y'' = y$$
$$y(0) = 0$$
$$y'(0) = 1.$$

4. Determine the approximating Maclaurin polynomials for the function determined by the initial value problem

$$y'' = x \cdot y$$
$$y(0) = 1$$
$$y'(0) = 1.$$

5. Use the methods of this section to show that the Maclaurin polynomials for $\sin(x)$ are given by

$$\sum_{k=0}^{n} \frac{(-1)^k x^{2k+1}}{(2k+1)!}.$$

6. Use the graph of $y = \sin(x)$ to explain why only odd powers of x appear in the formula above.

7. Differentiate the formula for the Maclaurin polynomials approximating $\sin(x)$ to obtain a formula for polynomials approximating $\cos(x)$.

8. Use the methods of this section to show that the Maclaurin polynomials for $\cos(x)$ are given by

$$\sum_{k=0}^{n} \frac{(-1)^k x^{2k}}{(2k)!}.$$

9. Use the graph of $y = \cos(x)$ to explain why only even powers of x appear in the formula above.

10. Let $S_{2n+1}(x)$ and $C_{2n}(x)$ be the Maclaurin polynomials of degree $2n + 1$ and $2n$ for $\sin(x)$ and $\cos(x)$, respectively. Investigate the expressions $\left(S_{2n+1}(x)\right)^2 + \left(C_{2n}(x)\right)^2$.

11. Use the Maclaurin series for $\sin(x)$ and $\cos(x)$ to explain the limit

$$\lim_{x \to 0} \frac{\sin(x)}{x} = 1.$$

12. Use the Maclaurin series for $\sin(x)$ and $\cos(x)$ to explain the limit

$$\lim_{x \to 0} \frac{1 - \cos(x)}{x} = 0.$$

13. Use the Maclaurin series for $\sin(x)$ and $\cos(x)$ to explain the limit

$$\lim_{x \to 0} \frac{1 - \cos(x)}{x^2} = \frac{1}{2}.$$

†14. Substitute $x = t^2$ into the Maclaurin polynomials for e^x to obtain approximating polynomials for e^{-t^2}. Use your formula to obtain approximating polynomials for $f(x) = \int_0^x e^{-t^2} dt$. Prepare a table of values for $0 \le x \le 1$, comparing the values of the degree 7 and degree 11 approximating polynomials for $\int_0^x e^{-t^2} dt$ to the corresponding values of the integral.

15. Obtain a formula for polynomial approximations $\sin(t)/t$. Use your polynomials to obtain a Maclaurin series for $si(x) = \int_0^x \frac{\sin(t)}{t} dt$. Prepare a table of values for $0 \le x \le \pi$, comparing the values of the degree 7 and degree 11 polynomials for $\int_0^x \frac{\sin(t)}{t} dt$ to the corresponding values of the integral.

9.3

How Good Is a Good Polynomial Approximation?

The Setting

Suppose that we want or need to approximate the function $f(x)$ on an interval that contains $x = 0$, and that all of the infinitely many derivatives of $f(x)$, $f'(x)$, $f''(x)$, $f^{(3)}(x)$, ... exist and can be evaluated at $x = 0$. Under these circumstances, we can use any of the sequences of Maclaurin polynomials,

$$P_0(x) = f(0)$$

$$P_1(x) = f(0) + f'(0) \cdot x$$

$$P_2(x) = f(0) + f'(0) \cdot x + \frac{f''(0)}{2} \cdot x^2$$

$$P_3(x) = f(0) + f'(0) \cdot x + \frac{f''(0)}{2} \cdot x^2 + \frac{f^{(3)}(0)}{6} \cdot x^3$$

$$P_4(x) = f(0) + f'(0) \cdot x + \frac{f''(0)}{2} \cdot x^2 + \frac{f^{(3)}(0)}{6} \cdot x^3 + \frac{f^{(4)}(0)}{24} \cdot x^4$$

$$\vdots$$

We have seen that the interval of convergence of the function f imposes an intrinsic limit on the usefulness of the $P_n(x)$'s, so we require that x lie on an interval of the form $[-L, L]$ contained in the interval of convergence for f.

to approximate f. But which one should we use?

The Greek letter ϵ is the traditional symbol for such a fixed error.

In order to measure how well $f(x)$ is approximated by a given $P_k(x)$, we look at all of the values of $|f(x) - P_k(x)|$ for $-L \leq x \leq L$. If all of these values are less than a fixed error, ϵ, we say that "$f(x)$ is *uniformly* approximated by $P_k(x)$ on $[-L, L]$ with an accuracy of ϵ." Our task is to find an expression for $|f(x) - P_n(x)|$. Integration by parts is the key to that task.

The Strategy

This uniform upper bound is not the only way we can measure the accuracy of approximations. Both of $\int_{-L}^{L} |f(x) - P_k(x)| dx$ and $\sqrt{\int_{-L}^{L} \left(f(x) - P_k(x) \right)^2 dx}$ are sometimes used.

We need to rewrite $f(x)$ in a form that connects f to the Maclaurin polynomials. The first Maclaurin polynomial is simply $P_0(x) = f(0)$; so, using the Fundamental Theorem of Calculus, we start with the familiar and uncontroversial

$$f(x) = f(0) + \int_0^x f'(t)dt. \tag{9.13}$$

We have seen this representation of f before. It does get both the number $f(0)$ and the function f' into a formula for $f(x)$. Our overall strategy is going to be to rewrite the integral on the right side of Equation (9.13) so that the new form uses $f'(0)$ and f''. That done, we will rewrite the new formula so as to involve $f''(0)$ and $f^{(3)}$. The process we will use can be repeated as often as we want. The technique we use for the transformations is integration by parts. It has been a long time since we looked at integration by parts, so let's review the technique.

Integration by Parts: A Review

If u and v are functions of t, the Product Rule tells us that

$$\frac{d}{dt}(u \cdot v) = \frac{du}{dt} \cdot v + u \cdot \frac{dv}{dt}.$$

Taking antiderivatives with respect to t of both sides gives us

$$\int \frac{d}{dt}(u \cdot v)dt = \int \frac{du}{dt} \cdot vdt + \int u \cdot \frac{dv}{dt}dt.$$

Clearly, $u \cdot v$ is an antiderivative of $\int \frac{d}{dt}(u \cdot v)dt$, which gives

$$u \cdot v = \int \frac{du}{dt} \cdot vdt + \int u \cdot \frac{dv}{dt}dt.$$

Writing the differentials as

$$dv = \frac{dv}{dt}dt \text{ and } du = \frac{du}{dt}dt$$

for brevity and rearranging the term, we have the familiar

$$\int udv = u \cdot v - \int vdu. \tag{9.14}$$

We will also need the explicit definite integral form of this formula. First, we will need to recall from the Fundamental Theorem of Calculus that

$$\int_a^b f(t)\,dt = F(b) - F(a) \quad \text{where } F(t) \text{ is an antiderivative of } f(t).$$

Then, from Equation (9.14),

$$\int u\,dv = \int u(t)\frac{dv}{dt}\,dt = F(t) + C = u(t) \cdot v(t) - \int v(t)\frac{du}{dt}\,dt + C.$$

So,

$$\int_0^x u(t)\frac{dv}{dt}\,dt = F(x) - F(0) = \left[u(t) \cdot v(t) - \int v(t)\frac{du}{dt}\,dt\right]\Bigg|_0^x$$

$$= u(t) \cdot v(t)\Bigg|_0^x - \int_0^x v(t)\frac{du}{dt}\,dt.$$

▶ **FIRST REFLECTION**

What is $F(x)$ in this discussion? ◀

Recall that given du ($du = u'(t)\,dt$), we know u only up to a constant.

Four other Oysters followed them,
And yet another four;
And thick and fast they came at last,
And more, and more, and more. . .
The Walrus and the Carpenter
by Lewis Carroll.

The Attack

Integration by Parts: First came one . . . . We continue with Equation (9.13)

$$f(x) = f(0) + \int_0^x f'(t)\,dt.$$

We apply integration by parts to

$$\int_0^x f'(t)\,dt,$$

Patterns are easier to see when our
formulas have less clutter.

first as an indefinite integral. We take $u = f'(t)$ which gives us $du = f''(t)\,dt$. The choice of $dv = 1 \cdot dt$ is automatic but our choice of v is not. Instead of $v = t$, we take $v = t - x$ where x is a parameter independent of t, exploiting the flexibility inherent in the Fundamental Theorem of Calculus.[†]

▶ **SECOND REFLECTION**

Take the derivative, with respect to t, of $v(t) = t$ and $v(t) = (t - x)$. Are the results the same?

◀

[†] This use of the Fundamental Theorem of Calculus is due to Brook Taylor who we will encounter again later in this chapter.

With this choice of u and v we have

$$\int f'(t)dt = f'(t) \cdot (t-x) - \int (t-x) \cdot f''(t)dt.$$

Converting to definite integrals we get

$$\int_0^x f'(t)dt = f'(t) \cdot (t-x)\Big|_0^x - \int_0^x (t-x) \cdot f''(t)dt \qquad \textbf{(9.15)}$$

$$= f'(0) \cdot x - \int_0^x (t-x) \cdot f''(t)dt. \qquad \textbf{(9.16)}$$

▶ **THIRD REFLECTION**

When you explicitly make and carry out the substitution in $f'(t) \cdot (t-x)\big|_0^x$, is Equation (9.15) verified? ◀

Substituting this new expression for $\displaystyle\int_0^x f'(t)dt$ into Equation (9.13), we get

$$f(x) = f(0) + f'(0) \cdot x - \int_0^x (t-x) \cdot f''(t)dt. \qquad \textbf{(9.17)}$$

Now we are getting somewhere: the first two terms on the right are exactly $P_1(x)$ and that trailing integral involves f''.

Integration by Parts: ... another .... Let's see if integration by parts works again. We start with the integral

$$\int (t-x) \cdot f''(t)dt.$$

Following the same pattern as last time, we let $u = f''(t)$, which leaves $dv = (t-x)\,dt$. We get $du = f^{(3)}(t)\,dt$, and we take Taylor's antiderivative in choosing $v = \dfrac{(t-x)^2}{2}$ (here the obvious antiderivative works).

▶ **FOURTH REFLECTION**

Can you verify that the derivative, with respect to t, of $\dfrac{(t-x)^2}{2}$ is $(t-x)$? ◀

This brings us to:

$$\int (t-x) \cdot f''(t)dt = f''(t) \cdot \frac{(t-x)^2}{2} - \int \frac{(t-x)^2}{2} \cdot f^{(3)}(t)dt.$$

Interpreting this as a definite integral nets us

$$\int_0^x (t-x) \cdot f''(t)dt \;=\; f''(t) \cdot \frac{(t-x)^2}{2}\Big|_0^x - \int_0^x \frac{(t-x)^2}{2} \cdot f^{(3)}(t)dt$$

$$=\; f''(x) \cdot \frac{(x-x)^2}{2} - f''(0) \cdot \frac{(0-x)^2}{2}$$

$$-\int_0^x \frac{(t-x)^2}{2} \cdot f^{(3)}(t)dt$$

$$= -\frac{f''(0)}{2} \cdot x^2 - \int_0^x \frac{(t-x)^2}{2} \cdot f^{(3)}(t)dt.$$

Substituting our expression for $\int_0^x (t-x) \cdot f''(t)dt$ into Equation (9.17), we have

$$f(x) = f(0) + f'(0) \cdot x - \left(-\frac{f''(0)}{2} \cdot x^2 - \int_0^x \frac{(t-x)^2}{2} \cdot f^{(3)}(t)dt \right)$$

$$= f(0) + f'(0) \cdot x + \frac{f''(0)}{2} \cdot x^2 + \int_0^x \frac{(t-x)^2}{2} \cdot f^{(3)}(t)dt. \quad \textbf{(9.18)}$$

Integration by Parts: "... And yet another...".   We can proceed as follows. First, we integrate $\int \frac{(t-x)^2}{2} \cdot f^{(3)}(t)dt$ by parts.

Think and Share

In teams of two, carry out and write up the computations to show that

$$\int \frac{(t-x)^2}{2} \cdot f^{(3)}(t)dt = f^{(3)}(t) \cdot \frac{(t-x)^3}{6} - \int \frac{(t-x)^3}{6} \cdot f^{(4)}(t)dt.$$

Then we convert our computation to a definite integral and simplify.

$$\int_0^x \frac{(t-x)^2}{2} \cdot f^{(3)}(t)dt = f^{(3)}(t) \cdot \frac{(t-x)^3}{6} \Big|_0^x - \int_0^x \frac{(t-x)^3}{6} \cdot f^{(4)}(t)dt$$

$$= \frac{f^{(3)}(0)}{6} \cdot x^3 - \int_0^x \frac{(t-x)^3}{6} \cdot f^{(4)}(t)dt.$$

Substituting this computation into Equation (9.18), we get

$$f(x) = f(0) + f'(0) \cdot x + \frac{f''(0)}{2} \cdot x^2 + \left(\frac{f^{(3)}(0)}{6} \cdot x^3 - \int_0^x \frac{(t-x)^3}{6} \cdot f^{(4)}(t)dt \right)$$

$$= f(0) + f'(0) \cdot x + \frac{f''(0)}{2} \cdot x^2 + \frac{f^{(3)}(0)}{6} \cdot x^3 - \int_0^x \frac{(t-x)^3}{6} \cdot f^{(4)}(t)dt.$$

Or we can state this more elegantly as

$$f(x) = P_3(x) - \int_0^x \frac{(t-x)^3}{6} \cdot f^{(4)}(t)dt \quad \textbf{(9.19)}$$

"And thick and fast they came at last, and more, and more, and more.".   A pattern is emerging. We can continue to integrate by parts and build up a bigger polynomial in x. Each time we integrate, we gain another factor of $(t-x)$ in the integral and replace $f^{(n)}$ by the next higher derivative. The general formula is as follows.

TAYLOR'S FORMULA

$$f(x) = P_n(x) + (-1)^n \cdot \int_0^x \frac{(t-x)^n}{n!} \cdot f^{(n+1)}(t)\,dt \qquad \textbf{(9.20)}$$

▶ **FIFTH REFLECTION**

What is that $(-1)^n$ doing in Equation (9.20)? ◀

Though it may look funny, this is a useful formula (and the reason why there is a TI calculator command `taylor(` instead of a command `maclaurin(`). In fact, it is the formula that is going to let us get the estimate for $|f(x) - P_n(x)|$, we needed earlier. From Taylor's Formula, we have

$$|f(x) - P_n(x)| = \left| (-1)^n \cdot \int_0^x \frac{(t-x)^n}{n!} \cdot f^{(n+1)}(t)\,dt \right|$$

$$= \left| \int_0^x \frac{(t-x)^n}{n!} \cdot f^{(n+1)}(t)\,dt \right|.$$

If we replace the integrand with its absolute value, the value of the resulting integral can only go up. So,

$$|f(x) - P_n(x)| = \left| \int_0^x \frac{(t-x)^n}{n!} \cdot f^{(n+1)}(t)\,dt \right| \qquad \textbf{(9.21)}$$

$$\leq \int_0^x \left| \frac{(t-x)^n}{n!} \cdot f^{(n+1)}(t) \right| dt, \quad x \geq 0. \qquad \textbf{(9.22)}$$

We implicitly assumed continuity in "The Setting."

Since $f^{(n+1)}(t)$ is continuous on the interval $[-L, L]$, its values are bounded. We now take K_{n+1} big enough that $\left| f^{(n+1)}(t) \right| \leq K_{n+1}$. Incorporating this estimate and completing the calculation of the final integral, we can write

$$|f(x) - P_n(x)| \leq \int_0^x \left| \frac{(t-x)^n}{n!} \cdot f^{(n+1)}(t) \right| dt$$

We found similar bounds for $f''(x)$ to control the error in the Trapezoid Rule, and for $f^{(3)}(x)$ to bound the error in Simpson's Rule, and the Runge–Kutta method.

$$\leq \frac{K_{n+1}}{n!} \cdot \int_0^x \left| (t-x)^n \right| dt \qquad \textbf{(9.23)}$$

$$= \frac{K_{n+1}}{n!} \cdot \left| \frac{x^{n+1}}{n+1} \right|.$$

▶ **SIXTH REFLECTION**

Finish the computation on that last integral in Equation (9.23). Is that definite integral always positive? What changes in Equations (9.21 and 9.23) when $x < 0$? ◀

We tie this discussion up both at a point and on an interval with two observations.

1. For any particular x between $-L$ and L, the error in using $T_n(x)$ as an approximation to $f(x)$ is no more than $\dfrac{K_{n+1}}{(n+1)!} \cdot \left| x^{n+1} \right|$.

2. For $-L \leq x \leq L$,

$$|f(x) - P_n(x)| \leq \frac{K_{n+1}}{(n+1)!} \cdot L^{n+1}. \tag{9.24}$$

▶ SEVENTH REFLECTION

In the inequality above, the expected $|x^{n+1}|$ is replaced by L^{n+1}. Is this valid? Could there be a practical reason to do so? ◀

E X A M P L E 1 **The Approximation of** $\tan^{-1}(x)$

The Maclaurin polynomial of degree 9 for $\tan^{-1}(x)$,

$$T_9(x) = x - \frac{x^3}{3} + \frac{x^5}{5} - \frac{x^7}{7} + \frac{x^9}{9}$$

returns $T_9(0.5) \approx 0.463684$. Can we trust this as a value for $\tan^{-1}(0.5)$? To use the bound on the error given in our first observation, we graph

$$\left| \frac{d^{10}}{dx^{10}} \tan^{-1}(x) \right| \quad \text{for } -0.5 \leq x \leq 0.5$$

as shown in Figure 9.6.

From the image and from direct computation it is reasonable to take K_{10}

FIGURE 9.6 $y = \left| \frac{d^{10}}{dx^{10}} \tan^{-1}(x) \right|$

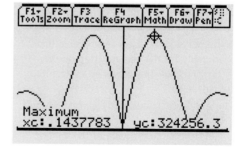

as 325,000 on $[-0.5, 0.5]$. Using this, the error in accepting 0.463684 for $\tan^{-1}(0.5)$ is no worse than $\dfrac{350,000 \cdot (0.5)^{10}}{10!} \approx 0.000094$. We can rest assured that $\tan^{-1}(0.5)$ lies between 0.46359 and 0.46378.

EXAMPLE 2   **The Approximation of** e^x

How good is the approximation of e^x given by $T_{15}(x)$, its Maclaurin polynomial of degree n for x on the interval $-2 \leq x \leq 2$?

To use the inequality in Equation (9.24), we need to find a value for K_{16}. Since $\frac{d^{16}}{dx^{16}}e^x = e^x$, and $e^2 \approx 7.38906$, we can safely adopt $K_{16} = 8$. The resulting bound puts $|f(x) - P_n(x)| \leq \dfrac{8 \cdot 2^{16}}{16!} \approx 0.0000000250582$. This error is sufficiently small to let us use values from $T_{15}(x)$ with a high degree of confidence.

EXAMPLE 3   **The Approximation of** $\sin(x)$

Verify this by entering the statements
3.2^(n+1)/(n+1)!|n=16 and
3.2^(n+1)/(n+1)!|n=17.

Find a polynomial that will approximate the sine function to six-digit accuracy for a full period.

What is the smallest Maclaurin polynomial that will do the trick? The sine function goes through a full period as x ranges from $-\pi$ to π. Thus, in Equation (9.24) we may take L = 3.2. Since the derivatives of $\sin(x)$ are all $\pm\cos(x)$ or $\pm\sin(x)$, we can take $K_{n+1} = 1$. The first value of n for which

$$\frac{3.2^{n+1}}{(n+1)!} < 0.0000005 \text{ is } n = 17. \text{ Use } P_{17}(x).$$

EXERCISE SET 9.3

1. What is the largest interval on which the Maclaurin polynomial of degree 9 for $\tan^{-1}(x)$ can be used to obtain a estimate for $\tan^{-1}(x)$ that is accurate to six decimal places?

2. What is the smallest Maclaurin polynomial for $\tan^{-1}(x)$ that can be used to obtain estimates accurate to six places on the interval $-0.5 \leq x \leq 0.5$?

3. What is the Maclaurin polynomial of smallest degree that will give six-digit accuracy for the sine function on the interval $0 \leq x \leq \frac{\pi}{2}$? Explain how this would be sufficient to calculate the sine of any angle.

4. Find a formula for the Maclaurin polynomials approximating the hyperbolic cosine, $\cosh(x)$.

5. What faith should you have in an estimate of the value of $\cosh(1)$ obtained from the Maclaurin polynomial for $\cosh(x)$ of degree 12?

6. On what interval does the Maclaurin polynomial of degree 13 for $\sinh(x)$, the hyperbolic sine function, supply six-place accuracy?

7. Here is an approach to computing the natural logarithm.
   a. Determine the series of Maclaurin polynomials for $\ln(1 + x)$.

   b. Use the Maclaurin polynomial of degree 9 for $\ln(1 + x)$ to determine an estimate for $\ln(1/2)$.
   c. How accurate is your estimate for $\ln(1/2)$?

8. Generate the Maclaurin polynomial of degree 9 for the tangent function, $\tan(x)$. Use inequality (9.24) to determine the largest interval on which this polynomial can be used to estimate the tangent with four-place accuracy.

9. The function $F(x) = \displaystyle\int_0^x e^{-t^2} dt$ is defined by an integral (which does not have a closed form).
   a. Develop the sequence of Maclaurin polynomials for $F(x)$.
   b. Find a bound for the error associated with using the Maclaurin polynomial of degree 9 to approximate $F(x)$ on the interval $[-1, 1]$.
   c. What is the maximum error associated with accepting $P_9(0.5)$ as $F(0.5)$?
   d. Determine the number of subdivisions necessary to obtain an approximation to $F(0.5)$ of equal accuracy by using Simpson's method.

10. The so-called Fresnel sine function is defined by
$$FS(x) = \int_0^x \sin(t^2)dt.$$
   a. Develop the sequence of Maclaurin polynomials for $FS(x)$.
   b. Use an appropriate Maclaurin polynomial to estimate $FS(2)$ to five-place accuracy.
   c. Fully explain your reasoning and justify your claim that your estimate is within the required accuracy.

11. The sine integral is defined by
$$si(x) = \int_0^x \frac{\sin(t)}{t}dt.$$

   a. Develop the sequence of Maclaurin polynomials for $si(x)$.
   b. Use an appropriate Maclaurin polynomial to estimate $si(2)$ to five-place accuracy.
   c. Fully explain your reasoning and justify your claim that your estimate is within the required accuracy.

12. Could we use $v(t) = t + x$ instead of $v(t) = t$ in the first round of using integration by parts?

13. Why can we use x as the defining variable in $f(x)$ but not require dx/dt in the differentiation of $v = t - x$. That is, why $d(t - x)/dt = 1$, not $dv/dt = 1 - dx/dt$?

14. Explain why $-L < x < L$ implies that $\left|x^{n+1}\right| < L^{n+1}$.

9.4

Convergence of Series

We have seen how the sequence of Maclaurin polynomials, $P_n(x)$ for a function $f(x)$, can sometimes be used to approximate values of f. Indeed, whenever a is in the interval of convergence of f, then $f(a)$ is the limit of the sequence $P_1(a)$, $P_2(a)$, $P_3(a)$, ... That is,

$$\lim_{n \to \infty} \sum_{k=0}^{n} \frac{f^{(k)}(0)}{k!} = f(a).$$

However, if a is *not* in the interval of convergence, then the values $P_1(a)$, $P_2(a)$, $P_3(a)$, ... *cannot* be used to approximate $f(a)$. (Well you can always use them but they won't be any good.) If we are going to use on Maclaurin polynomials to approximate values, $f(a)$, we had better have a way of determining when they can safely be used. That is, we need to have some way of determining which values a lie in the interval of convergence. In this section, we will develop a method that does that for us.

Notation

We have already seen that certain formulas make sense only for restricted values of x. In Chapter 1 we found the formula $\sum_{k=0}^{n} x^k = \frac{1 - x^{n+1}}{1 - x}$ for the finite geometric series. We noticed that whenever $-1 < x < 1$, $\lim_{n \to \infty} \frac{1 - x^{n+1}}{1 - x} = \frac{1}{1 - x}$. In light of this, it was reasonable to talk about the "infinite geometric series" $\sum_{k=0}^{\infty} x^k$ and declare $\sum_{k=0}^{\infty} x^k = \frac{1}{1 - x}$. With the Maclaurin polynomials,

$P_n(x) = \sum_{k=0}^{n} \dfrac{f^{(k)}(0)}{k!}$, we are faced with a similar situation. For some values of

x (at least for $x = 0$), the sequence of values $\left\{ \sum_{k=0}^{n} \dfrac{f^{(k)}(0)}{k!} \right\}_{n=0}^{\infty}$ has the limit

$f(x)$, so by analogy to the geometric series we talk about the infinite Maclaurin

series $\sum_{k=0}^{\infty} \dfrac{f^{(k)}(0)}{k!}$ being equal to $f(x)$.

CONVERGENCE OF A SERIES

If $\lim\limits_{n \to \infty} \sum_{k=0}^{n} a_k = L$, then we say the series $\sum_{k=0}^{\infty} a_k$ **converges** to L. If the sequence

of **partial sums** $\sum_{k=0}^{n} a_k$ does not have a limit, we say the series $\sum_{k=0}^{\infty} a_k$ **diverges**.

EXAMPLE 1

Show that

$$\sum_{k=1}^{\infty} \frac{1}{k^2 + k} = 1.$$

Examining the sequence of partial sums

$$\left\{ \sum_{k=1}^{n} \frac{1}{k^2 + k} \right\}_{n=1}^{\infty},$$

we get $\frac{1}{2}, \frac{2}{3}, \frac{3}{4}, \ldots 1 - \frac{1}{n+1}$. Indeed, using the method of partial fractions we see

$$\frac{1}{k^2 + k} = \frac{1}{k(k+1)} = \frac{1}{k} - \frac{1}{k+1}.$$

So that

$$\sum_{k=1}^{n} \frac{1}{k^2 + k} = \sum_{k=1}^{n} \left(\frac{1}{k+1} - \frac{1}{k} \right)$$

$$= \sum_{k=1}^{n} \frac{1}{k} - \sum_{k=1}^{n} \frac{1}{k+1}$$

$$= \frac{1}{1} + \frac{1}{2} + \ldots \frac{1}{3} + \frac{1}{n} - \frac{1}{2} - \frac{1}{3} - \ldots - \frac{1}{n} - \frac{1}{n+1}$$

$$= 1 - \frac{1}{n+1},$$

the observation that $\displaystyle\lim_{n\to\infty} 1 - \frac{1}{n+1} = 1$ completes the argument.

▶ **FIRST REFLECTION**

Does it matter that we have started the index at $k = 1$ in the above example rather than at $k = 0$ as in the definition?   ◀

In the above example, all the work was done in finding a closed form for the partial sums. Indeed, we were very lucky in that the form was both simple and easily justified and its limit easily found. Most sequences of partial sums do not have a simple closed form, just as most antiderivatives do not have a closed form. Usually we need to reason by analogy and pattern. In the next example we show that a closed form for the partial sums is neither possible nor necessary.

E X A M P L E 2 **The Harmonic Series**
_____

Let's show that the *harmonic series*, $\displaystyle\sum_{k=1}^{\infty} \frac{1}{k}$ diverges. The sequence of partial

sums $\displaystyle\left\{ \sum_{k=1}^{n} \frac{1}{k} \right\}_{n=1}^{\infty} = \left\{ 1, \frac{3}{2}, \frac{11}{6}, \frac{25}{12}, \frac{137}{60} \ldots \right\}$ has no easily found closed form.

However, we notice that the sequence of partial sums is increasing. An increasing sequence is either bounded above or not. If the sequence of partial sums is not bounded above, then the series must diverge. Let's look at some of the partial sums, for $n = 2, 4, 8, \ldots$

$$\sum_{k=1}^{2} \frac{1}{k} = 1 + \frac{1}{2}$$

$$\sum_{k=1}^{4} \frac{1}{k} = 1 + \frac{1}{2} + \frac{1}{3} + \frac{1}{4}$$

$$> 1 + \frac{1}{2} + \frac{1}{4} + \frac{1}{4} = 2$$

$$\sum_{k=1}^{8} \frac{1}{k} = 1 + \frac{1}{2} + \frac{1}{3} + \frac{1}{4} + \frac{1}{5} + \frac{1}{6} + \frac{1}{7} + \frac{1}{8}$$

$$> 1 + \frac{1}{2} + \frac{1}{4} + \frac{1}{4} + \frac{1}{8} + \frac{1}{8} + \frac{1}{8} + \frac{1}{8} = 2 + \frac{1}{2}$$

That the partial sum $\displaystyle\sum_{k=1}^{16} \frac{1}{k}$, is greater than 3 can be seen as follows: the sum involves 16 terms. The first eight add up to a number that exceeds two and a half. The last eight terms are $1/9+1/10+ \ldots + 1/16$. The smallest of these terms is $1/16$; therefore, the sum of the last eight terms is greater that $8/16 = 1/2$,

In fact, $\displaystyle\sum_{k=1}^{64} \frac{1}{k} =$
$4 + \frac{9771973120407246096664 3739}{1313629871225358075012 62400}$

and $\displaystyle\sum_{k=1}^{16} \frac{1}{k} > 3$. Similarly, $\displaystyle\sum_{k=1}^{32} \frac{1}{k} > 3 + \frac{1}{2}$, and $\displaystyle\sum_{k=1}^{64} \frac{1}{k} > 4$. In this fashion, we see that the partial sums for the harmonic sequence exceed any real number.

► **SECOND REFLECTION**
Use your calculator to find the exact partial sums for $n = 4, 8, 16, 32, 64$, and 128. Are these computations or the proceeding argument more persuasive? ◄

Existence of Limits

In the proceeding example, we alluded to a fundamental principle of the real numbers. This principle, The Completeness Property, follows.

THE COMPLETENESS PROPERTY

For any increasing sequence $u_1, u_2, u_3, \ldots$, if there is a real number M which is larger than every u_n, then there is a number L for which $\displaystyle\lim_{n\to\infty} u_n = L$.

► **THIRD REFLECTION**
Why is $L \le M$? ◄

EXAMPLE 3

Recall that $\displaystyle\sum_{n=0}^{\infty} a^n = \frac{1}{1-a}$ *whenever* $|a| < 1.$

Let us show that $\displaystyle\sum_{n=0}^{\infty} \frac{2}{3^n + 1}$ converges. The denominator of the *general term* $\frac{2}{3^n+1}$, containing a power of three, certainly suggests a geometric series. In fact, since both 3^n and $3^n + 1$ are positive and $3^n + 1 > 3^n$, we know that $\frac{1}{3^n+1} < \frac{1}{3^n}$.

But now, since the terms of the partial sums $\displaystyle\sum_{n=0}^{k} \frac{2}{3^n + 1}$ are, term by term, less than the terms of the sums $\displaystyle\sum_{n=0}^{k} \frac{2}{3^n}$, we see

$$\sum_{n=0}^{k} \frac{2}{3^n + 1} < \sum_{n=0}^{k} \frac{2}{3^n}$$

$$= 2 \cdot \sum_{n=0}^{k} \frac{1}{3^n}$$

$$\le 2 \cdot \sum_{n=0}^{\infty} \frac{1}{3^n}$$

$$= 3.$$

But now since each term $\frac{2}{3^n+1}$ is positive, the sequence of partial sums of the series $\sum_{n=0}^{\infty} \frac{2}{3^n + 1}$ is increasing and, as we have seen, bounded above by three. It follows from the Completeness Property that the series $\sum_{n=0}^{\infty} \frac{2}{3^n + 1}$ converges.

The series converges, but to what? Obviously to something less than three. We can approximate the infinite sum with some fairly large partial sum, but which one? $\sum_{n=0}^{10} \frac{2}{3^n + 1}, \sum_{n=0}^{100} \frac{2}{3^n + 1}$, or $\sum_{n=0}^{1000} \frac{2}{3^n + 1}$? And how well can we know just how accurate the approximation is? The comparison we made above between the geometric series $\sum_{n=0}^{\infty} \frac{2}{3^n}$ and $\sum_{n=0}^{\infty} \frac{2}{3^n + 1}$ can be exploited further.

EXAMPLE 4
_____

Let's approximate $\sum_{n=0}^{\infty} \frac{2}{3^n + 1}$ to five-decimal-place accuracy. We consider the series as comprised of two segments, one with finitely many (K) terms, which we simply add up and take as our approximation, and an infinite "tail" which we discard.

$$\sum_{n=0}^{\infty} \frac{2}{3^n + 1} = \sum_{n=0}^{K-1} \frac{2}{3^n + 1} + \sum_{n=K}^{\infty} \frac{2}{3^n + 1}$$

Our problem is determining how far to take K. Since we need a five-decimal-place accuracy, we need the discarded tail, $\sum_{n=K}^{\infty} \frac{2}{3^n + 1}$, to be less that 0.000005. But as we showed in the last example, each of the terms $\frac{2}{3^n+1}$ is less $\frac{2}{3^n}$ so any series $\sum_{n=K}^{\infty} \frac{2}{3^n + 1}$ is less than the corresponding series $\sum_{n=K}^{\infty} \frac{2}{3^n}$, whose value we

Simply input `Sigma(2/3 ^n,n,k,` `infinity)` *on your calculator.*

can calculate (in terms of K) as 3^{1-K}. Solving the equation $3^{1-K} < 0.000005$ to find that any value of K exceeding 12 will supply an approximation to the required accuracy. Finally, taking $K = 13$, we calculate $\sum_{n=0}^{12} \frac{2}{3^n + 1} = 1.80812$ as our approximation.

▶**Fourth Reflection**

Compare the results of calculating $\sum_{n=0}^{10} \frac{2}{3^n + 1}$, $\sum_{n=0}^{100} \frac{2}{3^n + 1}$, and $\sum_{n=0}^{1000} \frac{2}{3^n + 1}$ with our value of 1.80812. Are these computations or the proceeding argument more persuasive?   ◀

The Completeness Principle may seem to be overly specialized in that it deals only with increasing sequences. If a series has both positive and negative terms, its sequence of partial sums is not an increasing sequence. However, the following theorem (stated without proof) allows us to handle many sequences that contain both positive and negative terms.

ABSOLUTELY CONVERGENT SERIES

If the series $\sum_{n=0}^{\infty} |u_n|$, then so does $\sum_{n=0}^{\infty} u_n$. Further,

$$\sum_{n=0}^{\infty} |u_n| \geq \left| \sum_{n=0}^{\infty} u_n \right|,$$

A sequence for which $\sum_{n=0}^{\infty} |u_n|$ converges is said to be **absolutely convergent**.

Think and Share

Break into groups and adapt the computations of the last two examples to show that $\sum_{n=0}^{\infty} \frac{(-2)^n}{3^n + 1}$ converges absolute and approximate its sum to four-decimal-place accuracy.

Comparing a Series to Itself

E X A M P L E   5

Let us show that $\sum_{n=1}^{\infty} \frac{n^2}{3^n}$ converges. The series looks like it is related to the convergent geometric series with ratio 1/3, but just how is somewhat problematical. The n^2 in the numerator presents a difficulty. Rather than try to force our series to look like $\sum_{n=1}^{\infty} \frac{1}{3^n}$ we are going to rethink the comparison from scratch. Recall that a series is *geometric* if the *ratio* between consecutive terms is *constant*. Consider the ratio between consecutive terms:

$$\frac{u_{n+1}}{u_n} = \frac{\frac{(n+1)^2}{3^{(n+1)}}}{\frac{n^2}{3^n}}$$

$$= \frac{(n+1)^2}{3^{(n+1)}} \cdot \frac{3^n}{n^2}$$

$$= \frac{(n+1)^2}{n^2} \cdot \frac{3^n}{3^{(n+1)}}$$

$$= \left(\frac{n+1}{n}\right)^2 \cdot \frac{1}{3}.$$

The ratio may not be constant, but for large values of n (certainly for $n = 10$) the ratio is less than $1/2$. What does that tell us?

▶ **FIFTH REFLECTION**

What is the first value of n for which $\dfrac{u_{n+1}}{u_n} \leq \dfrac{1}{2}$? ◀

The ratio tells us that except possibly for the first few terms, the sequence of general terms, the u_n's, shrink fast. Specifically,

$$u_{11} < \frac{1}{2}u_{10}$$

$$u_{12} < \frac{1}{2}u_{11} < \left(\frac{1}{2}\right)^2 u_{10}$$

$$u_{13} < \frac{1}{2}u_{12} < \left(\frac{1}{2}\right)^2 u_{11} < \left(\frac{1}{2}\right)^3 u_{10}$$

$$u_{14} < \frac{1}{2}u_{13} < \left(\frac{1}{2}\right)^2 u_{12} < \left(\frac{1}{2}\right)^3 u_{11} < \left(\frac{1}{2}\right)^4 u_{10}$$

$$\vdots$$

In short, $u_{10+m} < \left(\frac{1}{2}\right)^m u_{10}$, or $\frac{(10+m)^2}{3^{10+m}} < \left(\frac{1}{2}\right)^m \frac{10^2}{3^{10}}$. This observation lets us use a convergent geometric series with ratio $1/2$, to bound the sequence of partial sums. Since the sequence of partial sums is increasing, that is enough to guarantee that the series converges. Here are the details.

$$\sum_{n=1}^{K} \frac{n^2}{3^n} = \sum_{n=1}^{9} \frac{n^2}{3^n} + \sum_{n=10}^{K} \frac{n^2}{3^n}$$

$$= \frac{9823}{6561} + \sum_{n=10}^{K} \frac{n^2}{3^n}$$

$$< \frac{9823}{6561} + \frac{10^2}{3^{10}} \sum_{m=0}^{K-10} \left(\frac{1}{2}\right)^m$$

$$< \frac{9823}{6561} + \frac{10^2}{3^{10}} \sum_{m=0}^{\infty} \left(\frac{1}{2}\right)^m$$

$$= \frac{9823}{6561} + 2 \cdot \frac{10^2}{3^{10}}$$

Let's review what we just did. First, in order to determine if a series of positive terms $\sum_{n=0}^{\infty} u_n$ was convergent, we looked at the ratios of consecutive terms, $\frac{u_{n+1}}{u_n}$. We found that eventually (for n big enough) this ratio was less than 1/2 and so were able to find an upper bound for the (increasing) sequence of partial sums. What was essential in the argument? We compared the tail of the series to a geometric series with ratio 1/2, a geometric series that converges. This forces the tail of the series to be bounded and, thus, the series itself to be convergent. Any convergent geometric series would have worked in the comparison as long as we could show that the ratio of consecutive terms was always less than one constant r, for which the geometric series with ratio r converged. The tail of the series would still be bounded above.

Second, if the terms in the series were not all positive, we use the same attack on the related series $\sum_{n=0}^{\infty} |u_n|$. If the series of absolute values converged, then the original series would too. Finally, we note again that if the ratio is not easy to analyze, we can look at $\lim_{n\to\infty} \left| \frac{u_{n+1}}{u_n} \right| = L$. If L is less than one, then we can bound the tail by comparison to a geometric series in $\frac{L+1}{2}$, the midpoint between L and 1. We formalize our discussion below.

THE RATIO TEST

If $\lim_{n\to\infty} \frac{|u_{n+1}|}{|u_n|} = L$, and $L < 1$, $\sum_{n=0}^{\infty} u_n$ converges absolutely. On the other hand, if $L > 1$, then $\sum_{n=0}^{\infty} u_n$ diverges.

EXAMPLE 6

We can use the Ratio Test to determine the interval of convergence of the Maclaurin series for $\tan^{-1}(x) = \sum_{k=1}^{\infty} \frac{x^{2k-1}}{2k-1}$. Computing the limit called for in the Ratio Test, we get

$$\lim_{k\to\infty} \frac{|a_{2k+1} x^{2k+1}|}{|a_{2k-1} x^{2k-1}|} = \lim_{k\to\infty} \frac{\left(\dfrac{|x^{2k+1}|}{2k+1} \right)}{\left(\dfrac{|x^{2k-1}|}{2k-1} \right)}$$

$$= \lim_{k\to\infty} \frac{|x^{2k+1}|}{|x^{2k-1}|} \frac{2k-1}{2k+1}$$

$$= \lim_{k \to \infty} x^2 \frac{2k-1}{2k+1}$$

$$= x^2 \cdot \lim_{k \to \infty} \frac{2k-1}{2k+1}$$

$$= x^2.$$

Since $x^2 < 1$ for $-1 < x < 1$, we find that the interval of convergence of the Maclaurin series for the $-1 < x < 1$.

E X A M P L E   7   **Radius of Convergence for the Exponential Function**

To determine the radius of convergence of the Maclaurin series for e^x, we evaluate

$$\lim_{k \to \infty} \frac{|a_{k-1} x^{k+1}|}{|a_k x^k|} = \lim_{k \to \infty} \frac{\left(\frac{|x^{k+1}|}{(k+1)!} \right)}{\left(\frac{|x^k|}{k!} \right)}$$

$$= \lim_{k \to \infty} \frac{|x^{k+1}|}{|x^k|} \frac{k!}{(k+1)!}$$

$$= \lim_{k \to \infty} \frac{|x|}{k+1}$$

$$= |x| \cdot \lim_{k \to \infty} \frac{1}{k+1}$$

$$= 0$$

Thus, the Maclaurin series for e^x converges everywhere.

E X E R C I S E   S E T   9 . 4

1. Show that the Maclaurin series for $\cos(x)$ is

$$\sum_{k=0}^{\infty} \frac{(-1)^k x^{2k}}{(2k)!} = 1 - \frac{x^2}{2!} + \frac{x^4}{4!} - \frac{x^5}{5!} + \cdots .$$

Determine the interval of convergence of the Maclaurin series for $\cos(x)$.

2. Determine the Maclaurin series and its radius of convergence for $\sin(x)$.

3. Euler used the geometric series as justification for his claim,

$$1 - 1 + 1 - 1 + 1 - 1 + \cdots = \frac{1}{2}.$$

Explain his probable reasoning. Decide if his claim is reasonable.

4. The Ratio Test is mute when $L = 1$. Apply the Ratio Test to both $\sum_{n=1}^{\infty} \frac{1}{n}$ and $\sum_{n=1}^{\infty} \frac{1}{n^2}$ and discuss what can be concluded when the limit of the ratio test is one.

5. In this exercise, we will create new series from the Geometric series.
   a. Substitute $x = -t$ in the Geometric series to obtain a series for $f(t) = 1/(1+t)$.
   b. Substitute for x in the Geometric series to obtain a series for $f(t) = 1/(1+t^2)$

c. Integrate both $1/(1+t^2)$ and your series in part b from 0 to x, to obtain the power series for $\tan^{-1}(x)$.

6. In this exercise, we will create new series from the Maclaurin series of e^x.
   a. Use a substitution to obtain a Maclaurin series for e^{-t^2}.
   b. Use your series to obtain a Maclaurin series for
   $$f(x) = \int_0^x e^{-t^2} dt.$$

7. In this exercise, we will obtain a Maclaurin series for a function defined in terms of an integral.
   a. Obtain a Maclaurin series for $\sin(t)/t$.
   b. Use your series to obtain a Maclaurin series for
   $$si(x) = \int_0^x \frac{\sin(t)}{t} dt.$$

8. Show that the series $\displaystyle\sum_{k=0}^{\infty} \frac{k^2}{k!}$ converges.

9. Show that the series $\displaystyle\sum_{k=0}^{\infty} \frac{k^3}{3^k}$ converges.

10. Show that the series $\displaystyle\sum_{k=0}^{\infty} \frac{k!}{(2k)!}$ converges.

11. Approximate the series $\displaystyle\sum_{k=0}^{\infty} \frac{k^2}{k!}$ to an accuracy of five decimal places.

12. Approximate the series $\displaystyle\sum_{k=0}^{\infty} \frac{k^3}{3^k}$ to an accuracy of five decimal places.

13. Approximate the series $\displaystyle\sum_{k=0}^{\infty} \frac{k!}{(2k)!}$ to an accuracy of five decimal places.

9.5

Power Series Solutions of Differential Equations

We know that $y' = cy$ has a solution that can be represented as a power series

$$y = f(x) = \sum_{k=0}^{\infty} \frac{c^k}{k!} x^k$$

and also as

$$y = f(x) = e^{cx}.$$

We are able to determine the number of terms of this infinite power series we need to compute to achieve a given bound on the error of the finite polynomial approximation to the solution function. In this section, we will create a power series representation for the solution to a more complicated differential equation which has no representation as a combination of elementary functions.

Finding a Power Series Solution

We begin with the differential equation $y' - 3xy = 0$. We assume that the solution to this differential equation is the power series

$$y = f(x) = \sum_{k=0}^{\infty} a_k x^k = a_0 + a_1 x + a_2 x^2 + a_3 x^3 + a_4 x^4 \cdots$$

We differentiate this formal solution as follows:

$$y' = a_1 + 2a_2 x + 3a_3 x^2 + 4a + 4x^3 + \cdots$$

$$= \sum_{k=0}^{\infty} a_k x^{k-1}$$

$$= \sum_{k=1}^{\infty} k a_k x^{k-1}.$$

In the last step, we have left off the first term in the series because it was zero by changing the starting index value from zero to one.

Thus, by substituting

$$y = \sum_{k=0}^{\infty} a_k x^k \text{ and also } \sum_{k=1}^{\infty} k a_k x^{k-1}$$

into the differential equation

$$y' - 3xy = 0$$

we get

$$\sum_{k=1}^{\infty} k a_k x^{k-1} - 3x \sum_{k=0}^{\infty} a_k x^k = 0$$

(multiplying through by $3x$)

$$\sum_{k=1}^{\infty} k a_k x^{k-1} - \sum_{k=0}^{\infty} 3 a_k x^{k+1} = 0$$

(using the distributive law on the second term)

$$\sum_{k=0}^{\infty} (k+1) a_{k+1} x^k - \sum_{k=1}^{\infty} 3 a_{k-1} x^k = 0 \qquad \textbf{(9.25)}$$

where we have changed the indexing of the power series representation so that the powers (x^k) match. You can use various starting values of the index variable to represent a power series in several ways, using the summation notation. A $\sum$-notation representation of a power series can be changed by increasing the index starting value and decreasing each occurrence of the index variable in the summation expression. The new $\sum$-notation still represents the original power series. For example, when a $\sum$-notation is expanded as a sum, we can see how to change the starting index value and the summation expression.

$$\sum_{k=0}^{\infty} a_k x^k = a_0 + a_1 x + a_2 x^2 + a_3 x^3 + a_4 x^4 \cdots = \sum_{k=1}^{\infty} a_{k-1} x^{k-1}$$

$$= \sum_{k=2}^{\infty} a_{k-2} x^{k-2}$$

▶ **FIRST REFLECTION**

Why does decreasing the $\sum$-notation index starting value by two for a given sum require that each occurrence of the index must be *increased* so that the new $\sum$-notation represents the same sum?   ◀

We write the two sums of Equation (9.25) with different starting values as a single term and two sums with the same starting index value and the same powers of x.

$$\sum_{k=0}^{\infty}(k+1)a_{k+1}x^k - \sum_{k=1}^{\infty}3a_{k-1}x^k = 0$$

$$(0+1)a_1x^1 + \sum_{k=1}^{\infty}(k+1)a_{k+1}x^k - \sum_{k=1}^{\infty}3a_{k-1}x^k = 0$$

(where the first term of the first sum is written out)

$$(0+1)a_1x^1 + \sum_{k=1}^{\infty}\left((k+1)a_{k+1} - 3a_{k-1}\right)x^k = 0 \qquad \textbf{(9.26)}$$

The zero function on the right side of Equation (9.26) is represented as a power series by

$$0 = 0 + 0 \cdot x^1 + 0 \cdot x^2 + 0 \cdot x^3 + 0 \cdot x^4 + \cdots$$

Comparing coefficients for the power series for the zero function and those on the left side of (9.26), we get

$$(0+1)a_1x^1 = a_1x^1 = 0 \qquad \textbf{(9.27)}$$

and

$$\left[(k+1)a_{k+1} - 3a_{k-1}\right] = 0 \quad \text{for } k = 1, 2, 3, \cdots \qquad \textbf{(9.28)}$$

From Equation (9.27), we have that

$$a_1 = 0. \qquad \textbf{(9.29)}$$

We have no initial constraints on a_0. We can use Equation (9.28) to create a recursion relationship between

$$a_{k+1} \text{ and } a_{k-1}$$

as follows:

$$(k+1)a_{k+1} - 3a_{k-1} = 0$$

$$a_{k+1} = \frac{3a_{k-1}}{k+1}. \qquad \textbf{(9.30)}$$

In fact, we have a difference equation for the coefficients a_n.

We know a great deal about the coeffients a_k of the power series $y = \sum_{k=0}^{\infty}a_kx^k$ that is a solution to the differential equation $y' - 3xy = 0$. Now we need to create the power series by computing values of the a_k's for small values of k. Our goal is to find a pattern in these values of a_k so that we can obtain a general expression for the terms of the power series. We restate Equations (9.29) and (9.30) for convenience.

$$a_1 = 0$$

$$a_{k+1} = \frac{3a_{k-1}}{k+1}$$

We computer values as follows.

$$a_0 = a_0 \qquad\qquad \text{(There is no initial constraint on } a_0.)$$

$$a_1 = 0$$

$$a_2 = \frac{3 \cdot a_0}{2} = a_0$$

$$a_3 = \frac{3 \cdot a_1}{3} = a_1 = 0$$

$$a_4 = \frac{3 \cdot a_2}{4} = \frac{3}{4}a_0$$

$$a_5 = \frac{3 \cdot a_3}{5} = \frac{3}{5} \cdot 0 = 0$$

$$a_6 = \frac{3 \cdot a_4}{6} = \frac{3}{6} \cdot \frac{3}{4} = \frac{3^2}{6 \cdot 4}a_0$$

$$a_7 = = 0$$

$$a_8 = \frac{3 \cdot a_6}{8} = \frac{3}{8} \cdot \frac{3^2}{6 \cdot 4}a_0 = \frac{3^3}{8 \cdot 6 \cdot 4}a_0$$

$$a_9 = 0$$

$$a_{10} = \frac{3^4}{10 \cdot 8 \cdot 6 \cdot 4}$$

and in general

$$a_k = \frac{3^{k/2-1}}{k \cdot (k-2) \cdot (k-4) \cdots 4} \qquad \text{for even values of } k.$$

Thus,

$$y = \sum_{k=0}^{\infty} a_k x^k$$

where $a_k = 0$ if k is odd and

$$a_k = \frac{3^{k/2-1}}{k \cdot (k-2) \cdot (k-4) \cdots 4}$$

if k is even.

Finding the Radius of Convergence

Does this power series converge and what is its radius of convergence? Computing the ratios of consecutive nonzero terms and comparing them to the corresponding ratios of a geometric series, we have the following:

$$\frac{a_{k+2}x^{k+2}}{a_k x^k} = \frac{\dfrac{3^{(k+2)/2-1}}{(k+2)\cdot k\cdot(k-2)\cdots 4}x^{k+2}}{\dfrac{3^{k/2-1}}{k\cdot(k-2)\cdot(k-4)\cdots 4}x^k}$$

$$= \frac{3^{(k+2)/2-1}}{(k+2)\cdot k\cdot(k-2)\cdots 4}\cdot\frac{k\cdot(k-2)\cdot(k-4)\cdots 4}{3^{k/2-1}}\cdot x^2$$

$$= \frac{3}{k+2}x^2.$$

We seek all those values of x for which the limit of $3/(k+3)$ is less than one (geometric series with ratios less than one are convergent).

$$\lim_{k\to\infty}\frac{3}{k+2} = 0$$

Since the limit is zero, we have that our power series's radius of convergence is infinite and that it is convergent for all real values.

► **SECOND REFLECTION**

Do the first five even terms of our power series solution fit the direction field for the differential equation on the interval $(-5, 5)$ with $a_0 = 2$? Will more terms give a better fit?   ◄

► **THIRD REFLECTION**

How are a_0 and the constant C in $\int f(x)\,dx = F(x) + C$ related?   ◄

E X E R C I S E   S E T   9 . 5

In Exercises 1–6, find the power series solution to the differential equation.

1. $y' - 2xy = 0$
2. $y' - x^2 y = 0$
3. $y' = 3y$
4. $y' - 4xy + 3 = 0$
5. $3x^2 y - y' = 0$
6. $y'' = y$
7. Why might the series method fail for the differential equation $y' - \frac{3y}{x}y = 0$.
8. Reindex each of the following series so that power of x is the index of summation:

   a. $\displaystyle\sum_{n=1}^{\infty} n\cdot a_n\cdot x^{n-1}$

   b. $\displaystyle\sum_{n=1}^{\infty} n\cdot(n-1)\cdot a_n\cdot x^{n-2}$

   c. $\displaystyle\sum_{n=2}^{\infty} n\cdot(n-1)\cdot a_n\cdot(n-2)\cdot x^{n-3}$

   d. $\displaystyle\sum_{n=0}^{\infty} a_n x^{n+3}$

9. Solve the initial value problem $(1+4x^2)\cdot y'' - 8y = 0$, $y(0) = 1$, and $y'(0) = -1$. Determine the interval of convergence of your series.

10. $(4+x^2)\cdot y'' + 3\cdot x\cdot y' - 3y = 0$, $y(0) = 1$, and $y'(0) = 0$. Determine the interval of convergence of your series.

11. Use the fact that we can expand e^x as a power series to find a series solution to the equation $y'' + (x-1)\cdot y = e^x$. Determine the interval of convergence of your series.

12. Use the fact that we can expand e^x as a power series to find a series solution to the equation $y'' + x\cdot y' = e^{-x}$. Determine the interval of convergence of your series.

Summary

In this chapter we have seen how one function may be used to approximate another and stand as its surrogate over an entire interval. Approximating functions are relatively easy to invent. More challenging is determining whether or not the approximating function is sufficiently close over a sufficiently large interval to be of use. In order to determine this, we must find some intrinsic connection between the approximating function and the function of interest. As we saw in the development of Taylor's Formula, building this connection can be a formidable though necessary task, both in theory and in practice.

Polynomials expressions, whose evaluation involves only addition and multiplication, play a special role in any computational scheme. Whenever anyone has seriously needed to evaluate lots of polynomials, they have replaced the explicit computation of powers with multiplications by using *Horner's method.* In Horner's method we simply rewrite $p(x) = a_0 + a_1 \cdot x + a_2 \cdot x^2 + \cdots + a_n \cdot$ as $p(x) = a_0 + x \cdot (a_1 + x \cdot (a_2 + x(\cdot a_3 + x(\cdots (a_{n-1} + xa_n)))))$. This simple device also reduces the number of multiplications used in any one evaluation. Evaluating a cubic polynomial in standard expanded form requires six separate multiplications; Horner's method requires three. All of our hand computations, and most machine computations, are based only on addition and multiplication. The addition and multiplication tables were the only numerical facts we learned in school. The procedures for subtraction and division we practiced in elementary school are later elaborations that used our knowledge of addition and multiplication.

Maclaurin (or Taylor) polynomials have been invented (or discovered) numerous times. We have been developed them by looking at initial value problems. Today, Taylor polynomials are only rarely the computational scheme used in mature numerical and symbolic computational technologies; details of hardware, software, and intended field of application usually allow clever devices to speed computation. But such polynomials still provide the conceptual frameworks used to investigate and understand new problems.

SUPPLEMENTARY EXERCISES

1. Find the Maclaurin Polynomial of degree six for the function determined by the initial value problem

$$y' = y \cdot \sin(2x)$$
$$y(0) = 2.$$

Determine how well this polynomial approximates $y(x)$ on the interval $[-1, 1]$.

2. Find the Maclaurin Polynomial of degree six for the function determined by the initial value problem

$$y' = 2t \cdot (1 + y)$$
$$y(0) = 0.$$

Determine an interval on which this polynomial approximates $y(t)$ to an accuracy of 0.00005.

3. Find the sequence of Maclaurin Polynomials for the function defined by $f(x) = \int_0^x \cos(t)dt$. Find the Maclaurin Polynomial of least degree that approximates $f(x)$ on the interval $[-1, 1]$ to an accuracy of 0.0005.

4. Use the Maclaurin series for functions like $\sin(x)$, $\tan(x)$, and e^x to show each of the following limits:

$$\lim_{x \to 0} \frac{\tan(x) - \sin(x)}{x^3} = \frac{1}{2}$$

$$\lim_{x \to 0} \frac{e^x - e^{-x}}{x} = 2$$

$$\lim_{x \to 0} \left(\frac{1}{x} - \frac{1}{\sin(x)} \right) = 0$$

5. Use the Maclaurin series for $\tan^{-1}(x)$ to argue that
$$\sum_{n=1}^{\infty} \frac{(-1)^{n+1}}{n} = \frac{\pi}{4}$$ *Provided the series converges.*

6. Use the Completeness Property to show the following. For any increasing sequence $v_1, v_2, v_3, \dots$, if there is a real number P which is smaller than every v_n then there is a number Q for which $\lim_{n \to \infty} v_n = Q$. (Hint: Look at the sequence $-(v_n)$.)

7. Show that the sequence $u_k = \sum_{n=1}^{2k-1} \frac{(-1)^{n+1}}{n}$ is a decreasing sequence, and that the sequence $v_k = \sum_{n=1}^{2k} \frac{(-1)^{n+1}}{n}$ is an increasing sequence.

8. With u_k and v_k defined as in the proceeding exercise, show that $v_k > u_k$. (Hint: Show that in fact $v_k - u_k = \frac{1}{2k \cdot (2k-1)}$). Use this to show that $v_k > 1 - \frac{1}{2}$ and that $u_k < 1$.

9. Use the results of the proceeding exercises to argue that the sequence $v_1, v_2, v_3, \dots$ converges, the sequence

$u_1, u_2, u_3, \dots$ and that they converge to the same limit. Explain how this shows that $\sum_{n=1}^{\infty} \frac{(-1)^{n+1}}{n}$ converges.

10. Find the radius of convergence for the power series listed below.

$$\sum_{n=1}^{\infty} \frac{(-1)^{n+1} x^n}{n}$$

$$\sum_{n=1}^{\infty} (\ln(n)) x^n$$

$$\sum_{n=0}^{\infty} (2x)^n$$

11. Euler decided, after examining the Maclaurin series expansions for $\cos(x)$, $\sin(x)$ and e^x, that, even though the square root of -1 was an imaginary fiction of the algebraists, $\cos(x) + i \cdot \sin(x) = e^{i \cdot x}$. Explain why he may have thought that. If we were to accept Euler's identity, what value would we assign to $1 + e^{i \cdot \pi}$?

PROJECTS

Project One: Taylor Series

In Sections 9.1, 9.2, and 9.3 we investigated power series arising from initial value problems. In particular we saw that initial value problems of the form

$$y'(x) = F(x, y(x))$$
$$y(0) = a_0$$

can give rise to the Maclaurin series for $y(x)$, $\sum_{k=0}^{\infty} a_k x^k$. Revisit our discussion and develop the formulas corresponding to Equations (9.6), (9.11), (9.20), and others as appropriate for a series of the form $\sum_{k=0}^{\infty} b_k (x - a)^k$ arising from initial value problems of the form

$$y'(x) = F(x, y(x))$$
$$y(a) = b_0.$$

Series of the form $\sum_{k=0}^{\infty} b_k (x - a)^k$ are known as *Taylor series*.

Project Report. Write a report on your investigations. Address your report to your classmates, assuming the material is new to them. Describe your reasoning and include all necessary background information. A minimal project report must include the following:

1. A discussion leading to a formula corresponding to Equation (9.6)
2. A discussion leading to a formula corresponding to Equation (9.11)
3. A discussion leading to a formula corresponding to Equation (9.20)
4. Justification and explanations for your calculations
5. Suggestions for possible extensions of your observations and further questions for exploration

Project Two: Computing Natural Logs

The computation of logarithms by Maclaurin series poses a problem. In this project we outline one classic response to this challenge.

1. Integrate the power series for $\dfrac{1}{1-x}$ and $\dfrac{1}{1+x}$ to obtain power series for $-\ln(1-x)$ and $\ln(1+x)$.
2. Find a power series and determine the radius of convergence for

$$F(x) = \frac{1+x}{1-x}$$

.
3. Find an rational expression for the nth derivative of $F(x)$ and use it to bound the error associated with accepting values of the $(n-1)$st Maclaurin polynomial for the values of $F(x)$ on the interval $-0.5 \le x \le 0.5$.
4. Determine the smallest Maclaurin polynomial for F that will give six-place accuracy on the interval $-0.5 \le x \le 0.5$.
5. Use your Maclaurin polynomial to approximate $\ln(2)$ to six-place accuracy.
6. Determine the range of values of y for which $y = \dfrac{1+x}{1-x}$ has solutions in the interval $-0.5 \le x \le 0.5$. In particular, show that e lies in this interval.
7. Use the identity $\ln(e^n \cdot t) = n + \ln(t)$ to develop a scheme that would allow you to approximate the natural logarithm of any positive number using your Maclaurin polynomial.

Project Report. Write a report on your investigations. Address your report to your classmates, assuming the material is new to them. Describe your reasoning and include all necessary background information. A minimal project report must include the following:

1. Written responses to items one through seven above
2. Justification and explanations for your calculations and derivations
3. Suggestions for possible extensions of your observations and further questions for exploration

Project Three: Picard Iterates

In the Interlude, "Existence and Uniqueness Theorems," we used the technique of Picard Iterates to prove that the autonomous initial value problem

$$\frac{d}{dt} f(t) = G(f(t))$$
$$f(t_0) = a$$

determines a function f on an interval containing t_0. This technique constructs a *sequence of functions*, $f_1(t)$, $f_2(t)$, $f_3(t)$, $\cdots$ of increasingly good approximations to $f(t)$. The Picard Iterates, $f_n(t)$, are generated by repetition of a two-step process. We first use G to compute a new function $G(f_{n-1}(t))$, then compute the antiderivative of $G(f_{n-1}(t))$ that assumes the value a at t_0, specifically:

$$f_n(t) = \int_{t_0}^{t} G(f_{n-1}(t)) \, dx + a$$
$$f_1(t) = a.$$

Whenever the function G is is simple enough to allow us to explicitly calculate the required antiderivatives, this process can provide useful approximations for $f(t)$.

1. Calculate the first six Picard iterates for

$$\frac{d}{dt} f(t) = 5 - f(t)$$
$$f(0) = 1.$$

2. Graphically compare $f_n(t)$ to the exact solution to the initial value problem in (1) above given by $f(t) = 5 - 4 \cdot e^{-t}$.

3. Calculate the first six Picard iterates for the logistic equation

$$\frac{d}{dt} f(t) = 0.1 \cdot f(t) \cdot (1 - 0.25 \cdot f(t))$$
$$f(0) = 0.1.$$

4. Do these functions give reasonable approximation to the sigmoid you expect?

5. Calculate the first four Picard iterates for

$$x \cdot y' + 0.2 y^2 = 3x^2$$
$$y(10) = 1.$$

6. How do the Picard iterates compare to the sequential solutions we get from our Euler and Runge–Kutte Methods?

Project Report.   Write a report on your investigations. Address your report to your classmates, assuming the material is new to them. Describe your reasoning and include all necessary background information. A minimal project report must include the following:

1. Written responses to items one through six above
2. Justification and explanations for your calculations and derivations
3. Suggestions for possible extensions of your observations and further questions for exploration

10

OPTIMIZATION OF FUNCTIONS OF TWO VARIABLES

10.1
Optimization with Two Variables

10.2
Vectors, Lines and Planes

10.3
Tangent Vectors and Tangent Lines in Three Dimensions

10.4
Tangent Planes

10.5
The Gradient Search

Summary

10.1

Optimization with Two Variables

We have become accustomed to seeing more than one symbol in the expressions we have used to model different situations. Up to now, the models we have constructed have used a single independent variable, usually time or a single linear measurement. Sometimes, however, as when we need to consider position in the plane or in three-space, it is not possible to construct an adequate model based on a single independent variable. In that case, we need to extend the notions of calculus to functions of more than one variable.

We approach the calculus of two variables in this chapter by extending what we have already learned about calculus of one variable in the first part of this text. We will apply the technique of partial differentiation (first presented in Section 3.9) to solve problems in two variables.

We begin with an optimization problem in two variables that requires us to find a minimum distance. In Section 3.6, whenever a problem involved two variables we found a way to represent one value in terms of the other, reducing the problem to a single variable. We will not do that here; rather, we will begin to develop the concepts of the calculus of two variables. As usual, we will analyze the problem graphically, symbolically, and numerically, extending the concepts we introduced in Section 3.9.

E X A M P L E   1   **The Distribution Center—Graphics Approach**

WGA Pharmaceuticals operates five drugstores in the Waycross metropolitan area. The pharmacies are marked by small black rectangles in Figure 10.1. A new distribution center/warehouse is being built to update and enhance the service to the stores. Where should WGA build the center?

FIGURE 10.1   WGA Pharmaceuticals in Waycross[†]

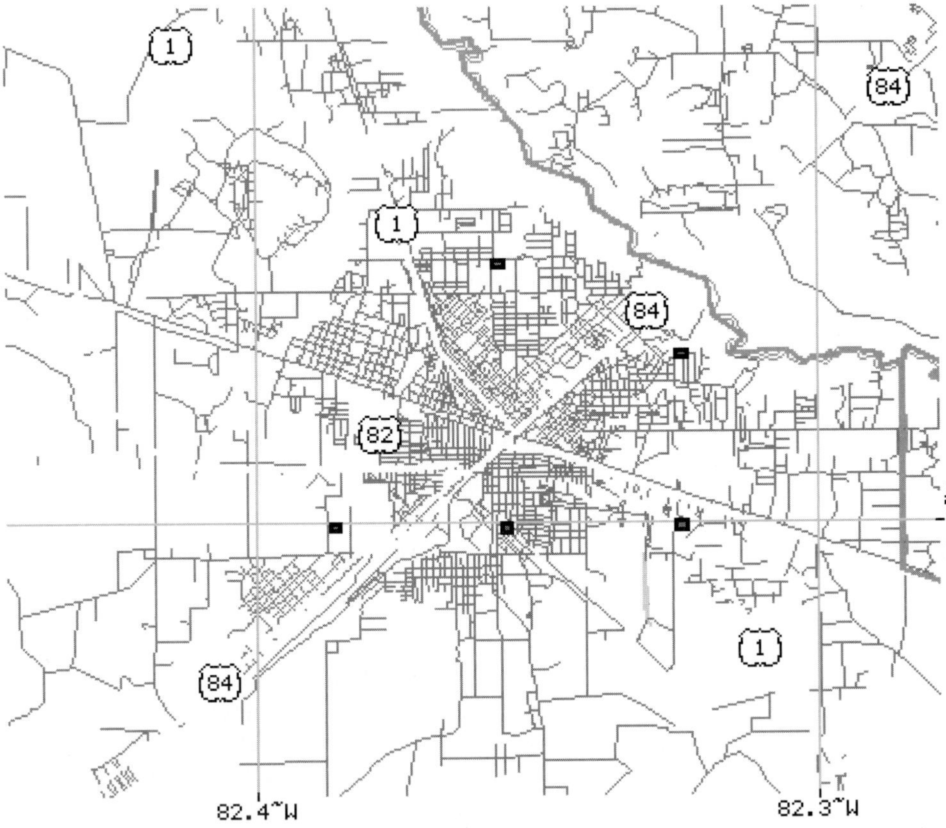

WGA Pharmaceuticals wants to locate a distribution center/warehouse in a location that will, among other things, minimize their ongoing costs of

---

[†]This map was produced by the Tiger Map Server of the Office of Statistics of the U. S. Census Bureau.

delivering inventory to the five stores. These costs include transportation and delivery (time and personnel) expenses. There are a number of factors that affect these costs: the size of each store, the volume of sales at each store, each store's storage capacity, the quality of the road system leading to each store, to name a few. In order to identify a cost-effective location for the new distribution center, we need a model for the ongoing delivery costs associated with any particular location. On the other hand, it is reasonable to assume that the cost of delivery to each store will depend on the distance between it and the new distribution center. This suggests that we can initially model the ongoing delivery costs associated with a particular location by summing the distances from the proposed location to each of the five existing stores. We can refine the model later to take other factors into account.

► **FIRST REFLECTION**

Make a list of additional factors that might affect the choice of a site for the proposed distribution center. ◄

   Distance is a quantity that we can easily compute once we choose coordinate axes. We arbitrarily choose our axes so that stores lie in the first quadrant, giving all five stores positive coordinates.

FIGURE 10.2   Coordinates of the Drugstores

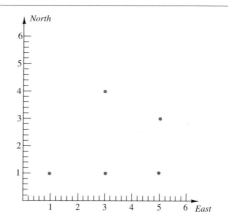

Reading coordinates (rounded to the nearest integer) from the graph gives

$$(1, 1), (3, 1), (3, 4), (5, 1), (5, 3)$$

as the coordinates of the five stores. We use (x, y) to represent the point at which the warehouse is located. We can now use the distance formula to calculate the distance from (x, y) to each of the drugstores and add the results to get the function representing the total distance from the proposed warehouse to the five stores. Algebraically stated, our task becomes

$$\text{minimize: } d(x, y) = \sum_{i=1}^{5} \sqrt{(x - x_i)^2 + (y - y_i)^2}.$$

FIGURE 10.3 Distance Function Surface

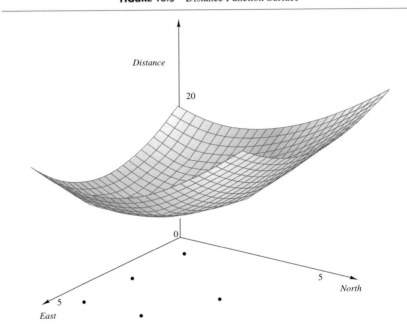

Geometrically, the problem is simple to understand: find the lowest point on the surface

$$z = f(x, y) = \sum_{i=1}^{5} \sqrt{(x - x_i)^2 + (y - y_i)^2}.$$

▶ **SECOND REFLECTION**

How do we know there is a lowest point? Here's a "thought experiment": Describe what happens when a marble is put inside the surface. ◀

The Trace *feature automatically displays the x-, y-, and z-coordinates when the* Format *option* Coordinates *is set to* RECT.

We can quickly find a graphical solution to the problem the model poses by using the graph to estimate the x- and y-coordinates corresponding to the minimum z-value.

Break into groups. On your TI calculator, define the function

$$f(x, y) = \sqrt{(x-1)^2 + (y-1)^2} + \sqrt{(x-3)^2 + (y-1)^2}$$
$$+ \sqrt{(x-3)^2 + (y-4)^2} + \sqrt{(x-5)^2 + (y-1)^2}$$
$$+ \sqrt{(x-5)^2 + (y-3)^2}.$$

Think and Share

Use the 3D graphing mode to obtain a plot of the surface `z1 = f(x,y)` over the square in the (x, y) plane with corners at $(0, 0)$, $(6, 0)$, $(6, 6)$, and $(0, 6)$. Using the `Trace` feature, estimate the coordinates at which `z1` achieves its smallest value. You can obtain finer grids to examine by increasing the `xgrid` and `ygrid` settings or choosing a smaller domain over which to graph. Compare your group's result with that of other groups. How sensitive is the value $f(x, y)$ to small changes in x and y?

We may congratulate ourselves at having obtained a graphical solution to the problem posed by the model. In the remainder of the chapter we use the ideas of tangent lines and their two-dimensional analogues, *tangent planes*, to develop additional algebraic tools with which to analyze and refine our solution. In Sections 10.2 and 10.3, we use *vectors* to assist us in describing lines and planes in three-dimensional space.

EXERCISE SET 10.1

1. Use your TI calculator to plot the surface $z = x + y$ from several viewing angles. Produce a hand-drawn graph of the surface that faithfully represents its shape.

2. Use your TI calculator to plot the surface $z = 3$ from several viewing angles. Produce a hand-drawn graph of the surface that faithfully represents its shape. Describe, and produce hand drawn graphs of the surfaces $x = -2$ and $y = 5$.

3. Describe and produce a hand-drawn graph of the surface $x + y = -2$.

4. Use your TI calculator to plot the surface $z = x^2$ from several viewing angles. Produce a hand-drawn graph of the surface that faithfully represents its shape. Determine the point(s) (x, y) at which z achieves its minimum and maximum values. Describe how a graph of the surface $z = y^2$ will differ from your graph.

5. Use your TI calculator to plot the surface $z = (x - 2)^2 + (y + 1)^2$ from several viewing angles. Produce a hand-drawn graph of the surface that faithfully represents its shape. Determine the point(s) (x, y) at which z achieves its minimum values.

6. Use your TI calculator to plot the surface $z = (x + 1)^2 - (y + 1)^2$ from several viewing angles. Produce a hand-drawn graph of the surface that faithfully represents its shape. Determine the point(s) (x, y) at which z achieves its minimum values.

7. Use your TI calculator to plot the surface $z^2 = x^2 - y^2$ from several viewing angles. (You will need to plot the surface in two stages, $z = \sqrt{x^2 - y^2}$ and $z = -\sqrt{x^2 - y^2}$. You may want to `Save` one image as a `Picture` and superimpose it on the other graph.) Produce a hand-drawn graph of the surface that faithfully represents its shape.

8. Use your TI calculator to plot the surface $z^2 = (x - 3)^2 + (y + 4)^2$ from several viewing angles. (You will need to plot the surface in two stages, $z = \sqrt{x^2 + y^2}$ and $z = -\sqrt{x^2 + y^2}$. You may want to `Save` one image as a `Picture` and superimpose it on the other graph.) Produce a hand-drawn graph of the surface that faithfully represents its shape.

9. Solve the problem posed by our model, minimizing the sum of the distances between a point (x, y) and store locations, if there are only two stores located at $(1, 1)$ and $(5, 3)$.

†**10.** Estimate the point (x, y) at which the sum of the distances between (x, y) and the points $(-2, -4)$, $(-1, -3)$, $(-1, -4)$, $(5, 6)$, and $(7, 8)$ is minimized.

11. *Group Exercise* Develop a critique of the model we have proposed to locate the new distribution center. Discuss the strengths and weaknesses of the model as a practical guide to locating new facilities or evaluating proposed locations. You may wish to contact local business managers about their practices or your local police and/or fire departments about their procedures for locating new facilities.

†**12.** *Group Exercise* The accounting office has supplied data in Table 10.1 to assist you in refining the model of this section. Clearly, some of the stores require more frequent deliveries to maintain their stock. Add a "weight" to the distance model developed in this section to account for this difference in demand and estimate the optimal position for the distribution center under your new model.

TABLE 10.1   Delivery Statistics for 1996

Store Coordinates	Number Deliveries 1996
$(1, 1)$	137
$(3, 1)$	89
$(3, 4)$	222
$(5, 1)$	103
$(5, 3)$	187

10.2

Vectors, Lines and Planes

The *vector* concept was developed by nineteenth century physicists to describe motion from one point to another in space or in the plane. In this section we introduce vectors in two and three dimensions and use them to facilitate finding algebraic descriptions of lines in two- and three-dimensional space and planes in three-dimensional space.

Directed Motion in Two and Three Dimensions

We use an arrow as the *geometric representation* of a vector in order to convey the idea of motion from one point to another. In Figure 10.4, two arrows representing motion are shown. The arrow on the left represents motion from the *initial point* $(-2, -1)$ to the *terminal point* $(2, 4)$ while the other represents motion from initial point $(2, -3)$ to the terminal point $(6, 2)$. Both motions can be described as a move of four units to the right and five units up. This common motion is the vector $\mathbf{A} = \langle 4, 5 \rangle$. A **vector** in the plane is formally defined as an ordered pair. A vector in three-dimensional space is defined as an ordered triple. In order to distinguish vectors from points, we use the angle brackets $\langle \ldots \rangle$ and print the variables representing vectors in upper case **bold** type.

Many other notations for vectors are common. You might see square brackets $[\ldots]$ or even parentheses $(\ldots)$. Or, you might find variables with bars or arrows over them such as $\overline{A}$ or $\vec{A}$.

▶ **FIRST REFLECTION**

What is the vector that represents the motion in three dimensions from the initial point $(2, -1, 5)$ to the terminal point $(5, 8, -3)$?   ◀

FIGURE 10.4 Two Representations of the Vector $\langle 4, 5 \rangle$

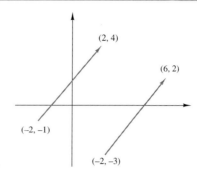

The opposite of the motion described by the vector **A** is denoted as the negative of **A**. We write $-\mathbf{A} = \langle -4, -5 \rangle$. The vector $2 \cdot \mathbf{A} = \langle 8, 10 \rangle$, represents a motion in the plane of 8 units to the right and 10 units up, a motion in the same direction as **A** but twice as far. The operation of *scalar multiplication* $t\mathbf{A}$ (multiplying a vector **A** by a scalar t) is illustrated in Figures 10.5 and 10.6 below.

FIGURE 10.5 The vectors **A** and $2 \cdot \mathbf{A}$ **FIGURE 10.6** The vectors **B** and $-1/2 \cdot \mathbf{B}$

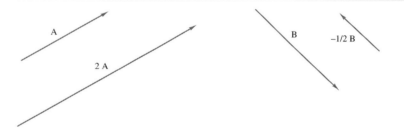

The *vector sum* of two vectors $\mathbf{A} = \langle a_1, a_2 \rangle$ and $\mathbf{B} = \langle b_1, b_2 \rangle$ is the vector $\mathbf{A} + \mathbf{B} = \langle a_1 + b_2, a_2 + b_2 \rangle$. This vector represents the result of executing the directed motion described by **A** followed by the directed motion described by **B**.

▶ SECOND REFLECTION

How would you describe the result of executing the directed motion described by **B** followed by the directed motion described by **A**? How does this differ from $\mathbf{A} + \mathbf{B}$? ◀

Figure 10.7 illustrates the vector sum of two vectors. Note that in this diagram, the representation of $\mathbf{A} + \mathbf{B}$ forms the diagonal of the parallelogram formed by two representations of the vector **A** and two representations of the vector **B**. Because of this geometric interpretation, vector addition is said to obey the *parallelogram rule*. We collect some of our observations on vectors as a set of definitions.

*For two three-dimensional vectors **A** and **B**, the geometric representation of their vector sum, **A** + **B**, lies in the plane determined by geometric representations of **A** and **B**.*

FIGURE 10.7   The Vector Sum **A** + **B**

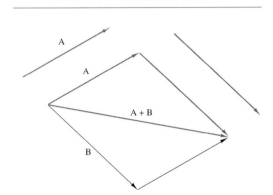

VECTOR ADDITION AND SCALAR MULTIPLICATION

A vector **A** in two dimensions is an ordered pair $\langle a_1, a_2 \rangle$. For any real number t,

$$t \cdot \mathbf{A} = \langle t \cdot a_1, t \cdot a_2 \rangle.$$

For $\mathbf{B} = \langle b_1, b_2 \rangle$ the vector sum of **A** and **B** is given by

$$\mathbf{A} + \mathbf{B} = \langle a_1 + b_1, a_2 + b_2 \rangle.$$

Similarly, a vector **A** in three-dimensional space is an ordered triple $\langle a_1, a_2, a_3 \rangle$. For any real number t,

$$t \cdot \mathbf{A} = \langle t \cdot a_1, t \cdot a_2, t \cdot a_3 \rangle.$$

For $\mathbf{B} = \langle b_1, b_2, b_3 \rangle$ the vector sum of **A** and **B** is given by

$$\mathbf{A} + \mathbf{B} = \langle a_1 + b_1, a_2 + b_2, a_3 + b_3 \rangle.$$

On the TI calculator, there are two ways to enter a vector $\langle x, y \rangle$. One is to use the square brackets [x,y]; the other is to use the curly braces {x,y}. We find, on balance, the curly braces easier to use and adopt them in this chapter.

Any vector in two-space has infinitely many geometric representations, one for every possible initial point. The representation with initial point at the origin is known as a *position vector*. Figure 10.8 illustrates the position vector $\langle -5, -3 \rangle$. We frequently describe how to get somewhere by giving a direction and a distance: "To get to Cleveland from Columbus, Ohio, you need to go northeast about 140 miles." Vectors are sometimes described using this approach.

EXAMPLE 1 **A Vector Has Length and Direction**

A position vector **A** can be described in terms of its length and the angle θ it makes with the positive x axis in a counterclockwise direction to the terminal side of **A** (see Figure 10.8). The length of the vector $\langle -5, -3 \rangle$ is the distance from $(0, 0)$ and $(-5, -3)$. Calculating the distance is a simple application of

FIGURE 10.8 The Position Vector $\langle -5, -3 \rangle$

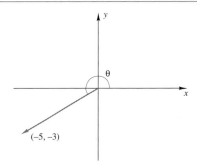

the Pythagorean Theorem (or distance formula) $\sqrt{(-5)^2 + (-3)^2} = \sqrt{34} \approx$ 5.93095. In order to obtain a value for the angle θ, we recognize that θ is a third-quadrant angle in standard position whose tangent is $-3/-5$. Thus, we can compute θ as $\arctan\left(\frac{3}{5}\right) + \pi \approx 3.6801$.

MAGNITUDE OF A VECTOR

The magnitude of a vector **A**, denoted as $\|\mathbf{A}\|$, is the length of its representation as a position vector. For a two-dimensional vector, $\mathbf{A} = \langle a_1, a_2 \rangle$, $\|\mathbf{A}\| = \sqrt{a_1^2 + a_2^2}$. For a three-dimensional vector, $\mathbf{A} = \langle a_1, a_2, a_3 \rangle$, $\|\mathbf{A}\| = \sqrt{a_1^2 + a_2^2 + a_3^2}$.

▶ THIRD REFLECTION

Is the magnitude of a vector **A** the length of the other geometric representations of **A** or only the position vector representation? ◀

Parametric Representations of Lines and Planes

There is a simple relationship between the initial and terminal points of a vector representation and the vector itself. The vector with initial point (x_1, y_1) and terminal point (x_2, y_2) is $\langle (x_2 - x_1), (y_2 - y_1) \rangle$. The vector **V** that is represented by the arrow from point A to point B may be thought of as being formed by a difference of the points B and A. Conversely, it is convenient to think of adding a vector to a point resulting in a new point. This relationship is illustrated in Figure 10.9 where $\mathbf{V} = \langle 11 - 3, 5 - 1 \rangle = \langle 8, 4 \rangle$. Let's consider

FIGURE 10.9   A + V = B

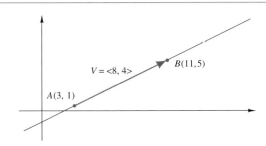

the entire line through the points $A(3, 1)$ and $B(11, 5)$. Other points on the line can be represented as A plus some multiple of the vector $\mathbf{V}$. For example, the point midway between A and B is $A + \frac{1}{2} \cdot \mathbf{V}$, while the point opposite B (obtained by reflecting the point B through A) is $A - \mathbf{V} = (-5, -3)$ as shown in Figure 10.10

▶ **FOURTH REFLECTION**
What multiple of $\mathbf{V}$ represents A? What multiple represents B?   ◀

This observation allows us to represent the line through A and B in *vector parametric* form $A + t \cdot \mathbf{V} = (3 + 8 \cdot t, 1 + 4 \cdot t)$. This point $(x(t), y(t))$ leads to the familiar parametric equations

$$x(t) = 3 + 8 \cdot t$$
$$y(t) = 1 + 4 \cdot t.$$

▶ **FIFTH REFLECTION**
How can you recover the slope of the line through A and B from the vector $\mathbf{V}$?   ◀

E X A M P L E   2   **A Line in Three-Space**

Find the parametric equations for the line in three-space that passes through the points $A(-3, 4, -5)$ and $B(0, 6, -8)$.

We take A as the initial point for the line and think of the line being generated by multiples of $\mathbf{V} = B - A = \langle 3, 2, -3 \rangle$. Taking t as our parameter, the vector parametric representation of the line is

$$A + t \cdot \mathbf{V} = (-3 + 3 \cdot t, 4 + 2 \cdot t, -5 - 3 \cdot t).$$

FIGURE 10.10 The Line through A and B

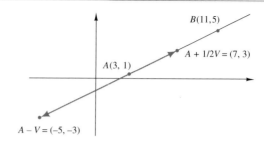

From this we can read off the parametric equations of the line as

$$x(t) = -3 + 3 \cdot t$$
$$y(t) = 4 + 2 \cdot t$$
$$z(t) = -5 - 3 \cdot t.$$

Equations of Planes in Three-Space

Vector representations are helpful in describing planes in three-dimensional space as well as lines. Suppose we want to find an algebraic description for the plane containing the point $A(a_1, a_2, a_3)$ and any two vectors $\mathbf{V} = \langle v_1, v_2, v_3 \rangle$ and $\mathbf{W} = \langle w_1, w_2, w_3 \rangle$. (This plane certainly contains the lines through A generated by the vectors $\mathbf{V}$ and $\mathbf{W}$.) In fact, for any point $P(x, y, z)$ in three-dimensional space, we can form a parallelogram with A and P at opposite corners as in Figure 10.11. Two sides of the parallelogram are multiples of $\mathbf{V}$ and can be written as $s \cdot \mathbf{V}$, for some real number s, while the other two sides of the parallelogram will be vectors of the form $t \cdot \mathbf{W}$ for some real number t. The parallelogram rule for vector addition tells us that

$$P = A + s \cdot \mathbf{V} + t \cdot \mathbf{W}.$$

From this observation, we can read off the parametric equations for the plane.

Notice that the parametric equations for a (one-dimensional) line use one parameter, while the parametric equations for a (two-dimensional) plane need two parameters.

$$x = a_1 + s \cdot v_1 + t \cdot w_1 \qquad \textbf{(10.1)}$$
$$y = a_2 + s \cdot v_2 + t \cdot w_2 \qquad \textbf{(10.2)}$$
$$z = a_3 + s \cdot v_3 + t \cdot w_3 \qquad \textbf{(10.3)}$$

EXAMPLE 3 **A Plane in Three-Space**

Find parametric equations for the plane that contains the points (with coordinates) $A(3, 0, 8)$, $B(1, 3, 0)$, and $C(-2, 6, 10)$.

FIGURE 10.11   The Plane Containing A, $\mathbf{V}$, and $\mathbf{W}$

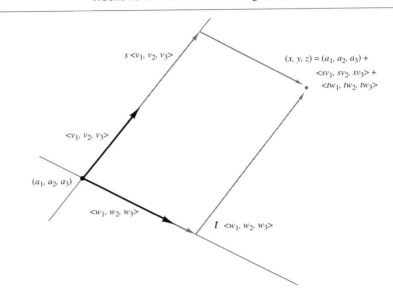

We take the point $A(3, 0, 8)$ as the point at which to begin our computations, as illustrated in Figure 10.12. The required plane contains the vectors $B - A = \langle -2, 3, -8 \rangle$ and $C - A = \langle -5, 6, 2 \rangle$. Any other point in the plane $P(x, y, z)$ can be represented as the sum of A and appropriate multiples of these two vectors; that is,

$$P(x, y, z) = A(3, 0, 8) + s \cdot \langle -2, 3, -8 \rangle + t \cdot \langle -5, 6, 2 \rangle.$$

From this equation, we can obtain the parametric equations

$$x = 3 - 2 \cdot s - 5 \cdot t$$
$$y = 3 \cdot s + 6 \cdot t$$
$$z = 8 - 8 \cdot s + 2 \cdot t.$$

▶ SIXTH REFLECTION

Repeat the calculations of the previous example using $B(1, 3, 0)$ as your initial point to obtain a different set of parametric equations for the plane containing $A(3, 0, 8)$, $B(1, 3, 0)$, and $C(-2, 6, 10)$. Record your equations in the margin for future reference. Compare the two parametric representations.   ◀

Cartesian Representations of Lines and Planes

It is very difficult to determine if planes described by different parametric representations are the same or different. Converting the parametric representations to a standard Cartesian form is a powerful way to determine whether the two planes are the same or different.

FIGURE 10.12   The Plane Containing $A(3, 0, 8)$, $B(1, 3, 0)$ and $C(-2, 6, 10)$

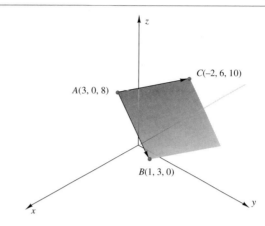

EXAMPLE 4 **From Parametric to Cartesian Representation**

In Section 3.8 we learned how to convert parametric equations for a line to Cartesian representation by eliminating the single parameter t. To convert the parametric equations of a plane to Cartesian representation we need to eliminate two parameters.

We begin with the parametric equations

$$x = 3 - 2 \cdot s - 5 \cdot t \qquad \textbf{(10.4)}$$

$$y = 3 \cdot s + 6 \cdot t \qquad \textbf{(10.5)}$$

$$z = 8 - 8 \cdot s + 2 \cdot t \qquad \textbf{(10.6)}$$

of the plane of Example 3.

Equations (10.4)–(10.6)

relate x, y, and z to s and t. We need only eliminate s and t from the equations to get a single equation involving x, y, and z. Solving Equation (10.4) for s we get

$$s = \frac{3}{2} - \frac{5}{2}t - \frac{1}{2}x. \qquad \textbf{(10.7)}$$

Substituting this value for s into Equation (10.5) gives us

$$y = \frac{9}{2} - \frac{3}{2}t - \frac{3}{2}x. \qquad \textbf{(10.8)}$$

Solving Equation (10.8) for t, we get

$$t = 3 - x - \frac{2}{3}y. \qquad \textbf{(10.9)}$$

Substituting this value for t into Equation (10.7) gives us

$$s = -6 + 2x + \frac{5}{3}y. \tag{10.10}$$

Finally, substituting Equation (10.9) and Equation (10.10) into Equation (10.6) and bringing all the variables to the left results in

$$18x + \frac{44}{3}y + z = 62. \tag{10.11}$$

▶ **SEVENTH REFLECTION**

Take your parametric equations from the previous reflection and eliminate the parameters s and t. Compare the resulting equation with Equation (10.11). What does your comparison tell you?   ◀

Comparing Equations

The slope-intercept equations for lines in the plane permit us to see by inspection whether two lines are the same, parallel, or perpendicular. Similarly, the Cartesian equations for planes in three-space allow us to see immediately whether two planes are the same or parallel.

▶ **EIGHTH REFLECTION**

How would the constant terms be related in two cartesian equations with the same coefficient for x, y, and z if (1) the planes were to be the same or (2) the planes were to be parallel?   ◀

Unfortunately, the parametric equations of a line in three-space do not allow us to see what the slope-intercept form of a line in two-space shows us by inspection. We can, however, rewrite the parametric equations of a line in three-space by eliminating the single parameter t.

Recall that the parametric equations (with parameter t) for the line through $P(a_1, a_2, a_3)$ generated by the vector $\mathbf{V} = \langle v_1, v_2, v_3 \rangle$ are

$$x = a_1 + v_1 \cdot t$$
$$y = a_2 + v_2 \cdot t$$
$$z = a_3 + v_3 \cdot t.$$

If none of the v_i's is zero, we can solve each of these equations for t obtaining three different values for the single parameter t. Equating these different expressions gives us the *symmetric equations* for a line in three-space.

$$\frac{x - a_1}{v_1} = \frac{y - a_2}{v_2} = \frac{z - a_2}{v_3}$$

▶ **NINTH REFLECTION**

Why do we need the restriction that $v_i \neq 0$? What would it mean if, say, $v_2 = 0$? Can all three of v_1, v_2, and v_3 equal zero? How could the symmetric equations of two lines in three-space reveal that the lines were parallel?   ◀

EXERCISE SET 10.2

In Exercises 1–6, take $\mathbf{A} = \langle 2, -3, 4 \rangle$, $\mathbf{B} = \langle 0, 7, -4 \rangle$, and $\mathbf{C} = \langle -3, 13, 5 \rangle$.

1. Calculate $\mathbf{A} - 3 \cdot \mathbf{B}$.
2. Calculate $2 \cdot \mathbf{A} + \mathbf{B} - 3 \cdot \mathbf{C}$.
3. Calculate $\|\mathbf{A}\|$.
4. Calculate $\|\mathbf{A} + \mathbf{B}\|$.
5. Calculate $\|\mathbf{A}\| + \|\mathbf{B}\|$.
6. Find values of s, t, and u for which $s \cdot \mathbf{A} + t \cdot \mathbf{B} - u \cdot \mathbf{C} = \langle 1, 0, 4 \rangle$.
7. Give parametric equations for the line containing the point $(-1, 3)$ generated by the vector $\langle -2, 5 \rangle$.
8. Give parametric equations for the line containing the points $(1, 2, -3)$ and $(1, -2, 3)$.
9. Give equations for the line that is parallel to

$$\frac{x-2}{3} = \frac{y+1}{-2} = \frac{z-3}{\frac{2}{3}}$$

and contains the point $(5, 3, -2)$.
10. Give the Cartesian equation for the plane that contains the point $(0, 2, 1)$ and the two vectors $\langle -1, 3, 0 \rangle$ and $\langle 3, 2, 1 \rangle$.

11. Give the Cartesian equation for the plane that contains the points $(0, 0, 0)$, $(2, 1, 3)$ and $(-2, -3, 6)$.
12. Find three points that lie on the plane whose Cartesian equation is $x + 2y - z = 5$.
13. Find the Cartesian equation of the plane that is parallel to the plane $x + 2y - z = 5$ and contains the point $(0, 3, 0)$.
14. Find the Cartesian equation of the plane that contains the parallel lines

$$\frac{x-1}{2} = \frac{y+1}{3} = \frac{z-3}{3}$$

and

$$\frac{x-3}{2} = \frac{y-4}{3} = \frac{z+5}{3}.$$

15. Show, either by computing the Cartesian equations or by substitution, that the Cartesian equation for the plane of Equations (10.1)–(10.3) is

$$(v_2 \cdot w_3 - v_3 \cdot w_2)x - (v_1 \cdot w_3 - v_3 \cdot w_1)y$$
$$+ (v_1 \cdot w_2 - v_2 \cdot w_1)z = 0.$$

10.3

Tangent Vectors and Tangent Lines in Three Dimensions

In this section we will revisit the topics of vector parametric curves, vector parametric differentiation, and partial derivatives from Part I. We apply our new tool of vectors to extend the idea of tangent lines to curves and surfaces in three-dimensional space.

Vector Parametric Functions in Two and Three Dimensions

You are familiar with two ways to describe and plot curves in two dimensions. Simple graphs that pass the vertical line test can often be described in rectangular or Cartesian coordinates as the graphs of the form $y = f(x)$. You have also used vector parametric functions as a convenience to describe graphs such as the cycloid, that are awkward to describe in Cartesian coordinates, or those, like ellipses, that fail the vertical line test. In three dimensions, vector parametric equations are indispensable. These equations are by far the easiest way to describe curves in three dimensions.

VECTOR PARAMETRIC FUNCTIONS

A function $\mathbf{F}(t)$ that accepts as input a single variable t and gives as output a vector in the plane $\langle x(t), y(t) \rangle$ is a **vector parametric function** in two dimensions. A function $\mathbf{F}(t)$ that accepts as input a single variable t and gives as output a vector in three space $\langle x(t), y(t), z(t) \rangle$ is a **vector parametric function** in three dimensions.

Unfortunately, your TI calculator has no three-dimensional curve plotting capabilities. Other, computer based, graphing and algebra utilities such as Maple, Mathematica, and MathCad do.

We graph a vector parametric function for $a \le t \le b$ by interpreting the vectors $\mathbf{F}(t)$ as position vectors and tracing out the path of their terminal points as t ranges from a to b.

When we first start graphing curves in three dimensions, we use the same technique that we used when we started graphing in two dimensions. We calculate a lot of points by substituting in values of t, plot them in three dimensions and then connect them in order to approximate a smooth curve. However, with time and experience we can learn to recognize some graphs.

TABLE 10.2

t	−2	−1	0	1	2
x(t)	0	1	2	3	4
y(t)	−9	−5	−1	3	7
z(t)	1	3	5	7	9

FIGURE 10.13   The Line
$\mathbf{F}(t) = \langle 2+t, -1+4t, 5+2t \rangle$

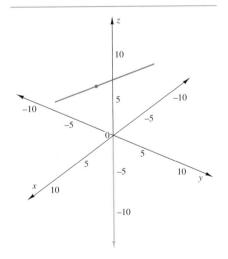

EXAMPLE 1   A Linear Vector Parametric Function

To graph the vector parametric function $\mathbf{F}(t) = \langle 2 + t, -1 + 4t, 5 + 2t \rangle$, we begin by making a table of values for $x(t) = 2 + t$, $y(t) = -1 + 4t$, and $z(t) = 5 + 2t$ (see Table 10.2). We let t range from -2 to 2 in steps of one. We note that Δx, Δy, and Δz from point to point are constant, $\Delta x = 1$, $\Delta y = 4$, and $\Delta z = 2$. This suggests a straight line. On the other hand, from our work with lines we know that the graph of $\mathbf{F}(t) = \langle 2+t, -1+4t, 5+2t \rangle$ should be the line through $(2, -1, 5)$ generated by the vector $\langle 1, 4, 2 \rangle$ (see Figure 10.13).

Our experience with vectors in two dimensions can also help us analyze some graphs. We look at pairs of variables to see if we can recognize circles or ellipses (or anything at all). As we saw in Example 1, it is a lot easier to recognize the vector parametric equation for a line than to try to recognize a shape by plotting points; we would plot points as a last resort.

EXAMPLE 2   An Ellipse

To graph the function $\mathbf{G}(t) = \langle 2, 2\sin(t) + 4, 5\cos(t) + 4 \rangle$, we begin by realizing the graph must lie in the plane $x = 2$ because $x(t) = 2$ is constant. The terms $2\sin(t)$ and $5\cos(t)$ in $y(t) = 2\sin(t) + 4$ and $z(t) = 5\cos(t) + 4$ suggest that the curve ought to be an ellipse in the yz-plane with major and

FIGURE 10.14 An Ellipse in Three-Space

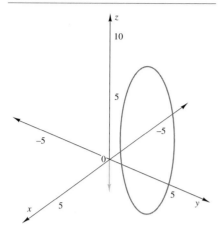

minor semi-axes of lengths 5 and 2, respectively. The center of the ellipse is located at the point $(2, +4, +4)$. The graph is shown in Figure 10.14.

E X A M P L E   3   **A Single Helix**

We employ similar reasoning to graph the function $\mathbf{H}(t) = \langle t, 3\sin(t), 3\cos(t)\rangle$. Looking at the second two coordinates

$$y^2 + z^2 = 9$$

we see a circular motion, while the uniform change in the x-coordinate suggests a spiral. $\mathbf{H}(t)$ traces a path along the surface of a right circular cylinder of radius three lying along the x-axis, as we see in Figure 10.15. We call this figure a **helix** with the x-axis as the axis of symmetry.

E X A M P L E   4   **An Increasing Spiral**

For the graph of $\mathbf{K}(t) = \langle t\cos(t), t\sin(t), t\rangle$ we detect the same sort of spiraling motion that we saw in the graph of $\mathbf{H}(t)$ in the previous example. Since $x(t) = t\cos(t)$, $y(t) = t\sin(t)$, and $z(t) = t$, we know the graph will spiral about the z-axis. The use of t as a multiplier on both $\cos(t)$ and $\sin(t)$ in the first two coordinates tells us that the radius of the spiral will increase as t grows larger (see Figure 10.16).

FIGURE 10.15 A Single Helix   **FIGURE 10.16** An Increasing Spiral

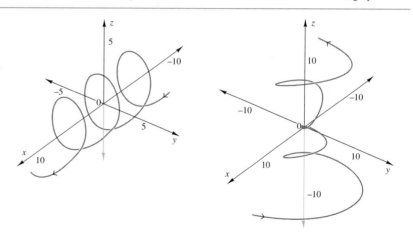

Tangent Vectors

The original use of vector parametric functions was to describe the motion of an object through three-space. We have looked carefully at the paths of objects in three-space at least twice in this text. In Chapter 1 we developed the Galilean model for the trajectory of a cannonball. In Chapter 8 we refined this model to analyze the motion of a skydiver. In both instances, velocity (the rate of change in position) and acceleration (the rate of change of velocity) are the fundamental tools of the physical analysis. The derivatives of vector parametric functions supply a natural language for velocity and acceleration.

Suppose the function $\mathbf{F}(t) = \langle x(t), y(t), z(t) \rangle$ is used to model the position of a particle of smoke in three-space for time t on the interval $0 \le t \le b$. Consider the change in the particle's position over a small change in time, Δt. The vector difference $\mathbf{F}(t + \Delta t) - \mathbf{F}(t)$ represents this change. To describe the average rate of change of $\mathbf{F}$ over the interval from t to $t + \Delta t$, we need to divide by Δt. This leads us to define the *average rate of change of* $\mathbf{F}$ as

$$\frac{1}{\Delta t}(\mathbf{F}(t+\Delta t) - \mathbf{F}(t)) =$$
$$\left\langle \frac{x(t + \Delta t) - x(t)}{\Delta t}, \frac{y(t + \Delta t) - y(t)}{\Delta t}, \frac{z(t + \Delta t) - z(t)}{\Delta t} \right\rangle.$$

FIGURE 10.17   Vector Change in Position

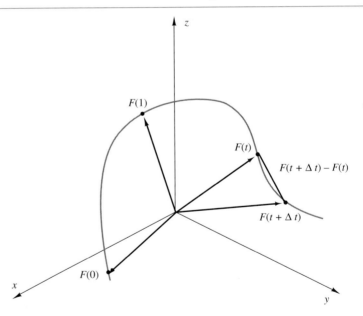

Taking the limit as Δt approaches 0, we make the following definition of the derivative of $\mathbf{F}(t)$.

VECTOR PARAMETRIC DERIVATIVES

The derivative of the function $\mathbf{F}(t) = \langle x(t), y(t) \rangle$ is given by

$$\frac{d}{dt}\mathbf{F}(t) = \lim_{\Delta t \to 0} \frac{1}{\Delta t}(\mathbf{F}(t + \Delta t) - \mathbf{F}(t))$$

$$= \left\langle \lim_{\Delta t \to 0} \frac{x(t + \Delta t) - x(t)}{\Delta t}, \lim_{\Delta t \to 0} \frac{y(t + \Delta t) - y(t)}{\Delta t} \right\rangle$$

$$= \langle x'(t), y'(t) \rangle.$$

The derivative of the function $\mathbf{F}(t) = \langle x(t), y(t), z(t) \rangle$ is given by

$$\frac{d}{dt}\mathbf{F}(t) = \lim_{\Delta t \to 0} \frac{1}{\Delta t}(\mathbf{F}(t + \Delta t) - \mathbf{F}(t))$$

$$= \left\langle \lim_{\Delta t \to 0} \frac{x(t + \Delta t) - x(t)}{\Delta t}, \lim_{\Delta t \to 0} \frac{y(t + \Delta t) - y(t)}{\Delta t}, \right.$$

$$\left. \lim_{\Delta t \to 0} \frac{z(t + \Delta t) - z(t)}{\Delta t} \right\rangle$$

$$= \langle x'(t), y'(t), z'(t) \rangle.$$

THE TANGENT VECTOR

At $t = a$, the vector $\frac{d}{dt}\mathbf{F}(a)$ is tangent to the curve $\mathbf{F}(t)$. Any nonzero multiple of the derivative vector is a *tangent vector* for the curve $\mathbf{F}(t)$ at $t = a$.

E X A M P L E 5   **Graphing the Tangent and Acceleration Vectors**

If we graph the path of the particle whose motion is given by

$$\mathbf{P}(t) = \langle t \cdot \cos(t), t \cdot \sin(t), t \rangle$$

for $0 < t < 4\pi$, we have an increasing spiral. We now find vector parametric functions representing velocity and acceleration of the particle and sketch the position of the particle along with its velocity and acceleration vectors at $t = 3\pi/2$.

Since velocity is the rate of change of position, the vector parametric function representing velocity $\mathbf{V}(t)$ is

$$\mathbf{P}'(t) = \langle \cos(t) - t \cdot \sin(t), \sin(t) + t \cdot \cos(t), 1 \rangle.$$

Acceleration, $\mathbf{A}(t)$, is the rate of change of *velocity*, and thus is computed as the second derivative of position.

$$\mathbf{P}''(t) = \langle -2 \cdot \sin(t) - t \cdot \cos(t), 2 \cdot \cos(t) - t \cdot \sin(t), 0 \rangle.$$

FIGURE 10.18   Position, Velocity and Acceleration

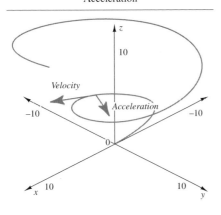

If you have forgotten the idea of partial derivative, you should review Section 3.7.

FIGURE 10.19   The Graph of $z = f(x, y)$

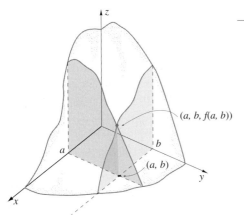

The position, velocity, and acceleration vectors of the curve at $t = 3\pi/2$ are

$$\mathbf{P}(3\pi/2) = \langle 0, -3/2, 3\pi/2 \rangle$$
$$\mathbf{V}(3\pi/2) = \langle 3\pi/2, -1, 1 \rangle$$
$$\mathbf{A}(3\pi/2) = \langle 2, 3\pi/2, 0 \rangle.$$

The curve, point, velocity, and acceleration vectors are graphed in Figure 10.18.

▶ **FIRST REFLECTION**

Is it reasonable that a rising particle should have a zero acceleration in the z-direction?   ◀

Tangents to Surfaces

In Chapter 3 we saw how functions of two variables $f(x, y)$ could be graphed as surfaces in three dimensions, as $z = f(x, y)$ as in Figure 10.19. We also saw how to calculate the partial derivatives $\frac{\partial f}{\partial x}(x, y)$ and $\frac{\partial f}{\partial y}(x, y)$. We interpreted $\frac{\partial f}{\partial x}(a, b)$ as the slope of the surface $z = f(x, y)$ at the point $Q(a, b, f(a, b))$ in the direction of x, and $\frac{\partial f}{\partial y}(a, b)$ as the slope of the surface $z = f(x, y)$ at the point $Q(a, b, f(a, b))$ in the direction of y as illustrated in Figures 10.20 and 10.21.

FIGURE 10.20   Slope in the x-"Direction",   **FIGURE 10.21**   Slope in the y-"Direction", $y = b$ $x = a$

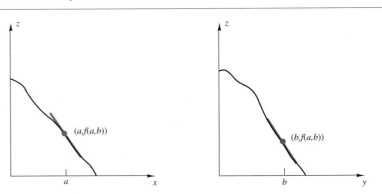

▶ **SECOND REFLECTION**

Why are the coordinates of the point $Q(a, b, f(a, b))$ given as $(b, f(a, b))$ in the $x = a$ plane in Figure 10.21 and $(a, f(a, b))$ in the $y = b$ plane in Figure 10.20?   ◀

We may be more interested in going "uphill" to get to the top of a mountain or "downhill" to ski to the bottom than in going in the x- or y-directions. In

Figure 10.22 we use the angle θ to describe just such a new direction. We can use our new understanding of tangent vectors to calculate the slope of $f(x, y)$ in the direction of θ. Our strategy is to find a vector parametric representation of the curve on $z = f(x, y)$ that lies over the line in the plane through (a, b) in the direction of θ.

FIGURE 10.22   Traveling in the Direction of θ

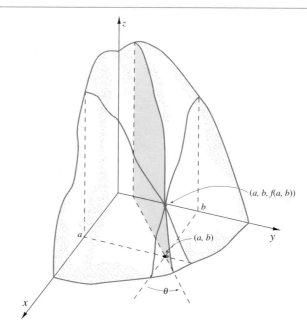

FIGURE 10.23   The Slope in the Direction of θ

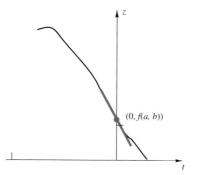

$(0, f(a, b))$

The choice of $\mathbf{U} = \langle \cos\theta, \sin\theta \rangle$ *as our generating vector is not arbitrary. Since* $\|\mathbf{U}\| = 1$, *a unit change in our parameter t will result in a unit change in the horizontal* (x, y)-*plane.*

We reason as follows: one vector in the plane in the direction of θ is $\langle \cos(\theta), \sin(\theta) \rangle$. The line in the plane that passes through (a, b) in the direction of θ has the vector parametric representation $\mathbf{G}(t) = \langle a + t \cdot \cos(\theta), b + t \cdot \sin(\theta) \rangle$. The points on the surface $z = f(x, y)$ that lie directly over the line $\mathbf{G}(t) = \langle a + t \cdot \cos(\theta), b + t \cdot \sin(\theta) \rangle$ are exactly the points on the graph of the vector parametric function

$$\mathbf{H}(t) = \langle a + t \cdot \cos(\theta), b + t \cdot \sin(\theta), f(a + t \cdot \cos(\theta), b + t \cdot \sin(\theta)) \rangle.$$

▶ **THIRD REFLECTION**

Why are the coordinates of the point $Q(a, b, f(a, b))$ given as $(0, f(a, b))$ when viewed in the tz-plane in Figure 10.23? ◀

The third coordinate of $\mathbf{H}'(t)$ gives us rate of change of $f(a + t \cdot \cos(\theta), b + t \cdot \sin(\theta))$ with respect to t. Since, as we have noted, a unit change in t represents a unit change in the horizontal (x, y)-plane, $\frac{d}{dt} f(a + t \cdot \cos(\theta), b + t \cdot \sin(\theta))$ is the slope of $f(x, y)$ in the direction of θ. We summarize this work with a definition.

THE DIRECTIONAL DERIVATIVE

$$D_{[\theta]}f(a, b) = \frac{d}{dt}f(a + t \cdot \cos(\theta), b + t \cdot \sin(\theta)) \Big| \text{ evaluated at } t = 0$$

is known as the **directional derivative** of $f(x, y)$ at (a, b) in the direction of θ.

E X A M P L E   6   **Computing a Directional Derivative**

Let's compute the directional derivative of $f(x, y) = 9 - x^2 - y^2$ at the point $(2, 1)$ in the direction of $\pi/3$.

The unit vector $\langle 1/2, \sqrt{3}/2 \rangle$ gives us the required direction. Substituting $x = 2 + (1/2)t$ and $y = 1 + (\sqrt{3}/2)t$ into $9 - x^2 - y^2$ results in $4 - (2 + \sqrt{3}) \cdot t - t^2$. Taking the derivative of this expression with respect to t gives $-(2 + \sqrt{3}) - t$. Setting $t = 0$, we find $D_{[\pi/3]}f(2, 1) = -(2 + \sqrt{3})$.

E X E R C I S E   S E T   1 0 . 3

1. Describe and graph the curve determined by

$$\mathbf{F}(t) = \langle \cos(t), -1, 4\sin(t) \rangle.$$

2. Describe and graph the curve determined by

$$\mathbf{F}(t) = \langle 2 + 3t, -1 + 6t, 5t \rangle.$$

3. Describe and graph the curve determined by

$$\mathbf{F}(t) = \langle \sin(2t), t\cos(2t) \rangle.$$

4. Describe and graph the curve determined by

$$\mathbf{F}(t) = \langle \sin(t), t/2, \cos(t) \rangle.$$

5. Describe and graph the curve determined by

$$\mathbf{F}(t) = \langle t\sin(t), t, 2\cos(t) \rangle.$$

6. Find the vector parametric equations for the line tangent to

$$\mathbf{P}(t) = \langle t \cdot \cos(t), t \cdot \sin(t), t \rangle \text{ at } t = \pi/3.$$

7. Determine the velocity and acceleration for a particle traveling along the path determined by

$$\mathbf{P}(t) = \langle t^2 - 1, t, t^2 + 3 \rangle.$$

† 8. For $f(x, y) = x^2 - 4x + y^2 - 2y + 9$:

   a. Calculate the directional derivative $D_\theta f(1, 2)$.
   b. Plot $D_{[\theta]}f(1, 2)$ for $0 \le \theta \le 2\pi$.
   c. Which direction is uphill from $(1, 2, 6)$? Which direction is downhill? Explain your answer.

† 9. For $f(x, y) = x^2 - 4x + y^2 - 2y + 9$:

   a. Calculate the directional derivative $D_\theta f(2, 1)$.
   b. Plot $D_{[\theta]}f(1, 2)$ for $0 \le \theta \le 2\pi$.
   c. Which direction is uphill from $(1, 2, 6)$? Which direction is downhill? Explain your answer.

10. Explain how to represent the partial derivatives $\frac{\partial f}{\partial x}(x, y)$ and $\frac{\partial f}{\partial y}(x, y)$ as directional derivatives.

10.4

Tangent Planes

In Chapter 2 we looked at the rate of change of a function with respect to its independent variable and the slope of a tangent line as our motivating

FIGURE 10.24 $z = x^2 - 4x + y^2 - 2x + 9$
with `zmin=-10`, and `zmax=200`

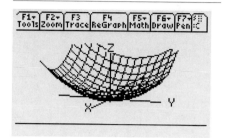

FIGURE 10.25 `ZoomIn` Once on $(2, 2, 5)$

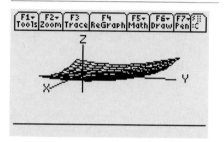

FIGURE 10.26 ttsmall `ZoomIn` Twice on $(2, 2, 5)$

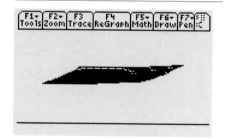

examples for the derivative. We have generalized these notions to functions of two variables by defining the partial derivative and directional derivatives. These are sufficient to define lines tangent to a surface $z = f(x, y)$. Throughout the second part of this text, we have used the tangent line as a "best local linear approximation" to the function. However, no *line*, tangent or otherwise, can be an adequate approximation to a *surface*. Lines are only one-dimensional, while surfaces have two dimensions. We need two-dimensional objects to approximate surfaces. Planes, as simple two-dimensional objects, suggest themselves. Consequently, our next objective is to understand how to find that locally approximate surfaces.

Local Linearity

Is it at all reasonable to locally approximate a relatively simple surface with a plane? We used lines as local approximations for curves in Chapter 2. There we found that for differentiable curves, at most points, repeatedly zooming in on the curve resulted in a graph that looked like a straight line. We referred to this property as *local linearity*.

▶ **First Reflection**
What are some examples of curves and points at which local linearity fails? ◀

In order to discover if there is a similar phenomenon of local linearity in three dimensions, we again try zooming in on a simple surface. Figures 10.24,

10.25, 10.26, and 10.27 show the result of zooming in on the surface

$$z = x^2 - 4x + y^2 - 2x + 9$$

starting with the domain

$$-10 \leq x \leq 10 \text{ and } -10 \leq y \leq 10.$$

Apparently, viewed in sufficient detail, the surface around $(2, 2, 5)$ does appear to flatten into a plane.

FIGURE 10.27 `ZoomIn` Three Times on $(2, 2, 5)$

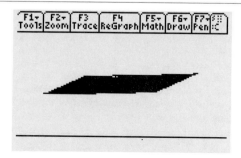

Think and Share

Break into groups of four with teams of two. One team should zoom in three times on (2, 1, 4) for the surface

$$z = x^2 - 4x + y^2 - 2x + 9$$

and the other team should zoom in three times on (0, 0, 0) for the surface

$$z = \sqrt{x^2 + y^2}.$$

Then compare the results of these two experiments.

Tangent Planes From Tangent Lines

Take a pear, apple, or orange and roll it across a hard flat table. Balance the fruit at one point and look closely at the relation between the skin of the fruit and the plane of the tabletop where the two surfaces meet. The tabletop is flush to the side of

the fruit. The plane of the tabletop is the plane tangent to the surface of the fruit. Look carefully at the outline of the fruit with your eye right at the edge of the table. What you see is something like the silhouette in Figure 10.28. The line you see as the surface of the table is the line tangent to the outline of the fruit. Move around the table 90 degrees to your left or right and sight along the top of the table again. You should see a silhouette like that in Figure 10.29. Again, the line that you see as the top of table is tangent to the outline of the fruit.

Note that the plane tangent to the surface at our balance point on the tabletop contains two lines. Each of these lines is tangent to mutually perpendicular cross sections of the surface. You may snack on the fruit now while we complete the following example.

Don't do this with a pineapple. A pineapple has too many prickles and cusps, points where the surface of the pineapple is not "smooth."

FIGURE 10.28   The First Cross Section of Our Pear

FIGURE 10.29   A Perpendicular Cross Section of Our Pear

E X A M P L E   1   **Finding Equations of Tangent Planes**

Let's find parametric and Cartesian equations for the plane tangent to

$$z = f(x, y) = 4 - x^2 + 4x - y^2 + 2y$$

at the point (1, 2, 7).

We begin with the following observations. The plane $y = 2$ intersects the surface in a curve as shown in Figure 10.30. This curve passes through the point $(1, 2, 7)$. The line tangent to the curve at that point must lie in the required tangent plane. A parametric function for this curve is $\mathbf{P}(x) = \langle x, 2, f(x, 2) \rangle$. The tangent vector to $\mathbf{P}(x)$ is $\langle 1, 0, \frac{\partial f}{\partial x}(x, 2) \rangle$. Setting $x = 1$, we get one vector in the tangent plane as $\langle 1, 0, 2 \rangle$.

▶ **SECOND REFLECTION**
How did we get this vector $\langle 1, 0, 2 \rangle$?   ◀

Similarly, the tangent plane contains the line tangent to the curve $\mathbf{Q}(y) = \langle 1, y, f(1, y) \rangle$ at $y = 2$. The vector that generates this line is $\langle 0, 1, -2 \rangle$. Recalling the method we developed in Section 11.2 to write the equation of a plane, we let $A(1, 2, 7)$ serve as an initial point and we use the two tangent vectors $\mathbf{v} = \langle 1, 0, 2 \rangle$ and $\mathbf{w} = \langle 0, 1, -2 \rangle$ as two vectors in the tangent plane. Thus, taking s and t as parameters, we may write the equations for any other point (x, y, z) in the tangent plane as

$$x = 1 + s \qquad\qquad\qquad\qquad \textbf{(10.12)}$$

$$y = 2 + t \qquad\qquad\qquad\qquad \textbf{(10.13)}$$

$$z = 7 + 2s - 2t. \qquad\qquad\qquad \textbf{(10.14)}$$

To obtain the Cartesian equation for the tangent plane, we need only to substitute $s = x - 1$ and $t = y - 2$ from Equations (10.12) and (10.13) into Equation (10.14), yielding

$$z = 7 + 2 \cdot (x - 1) - 2 \cdot (y - 2).$$

In fact, the computations of the preceding example can be made completely general. To find the plane tangent to the surface $z = f(x, y)$ at the point $(a, b, f(a, b))$, we need only determine two vectors from that tangent plane. We use the tangent vectors from two curves on the surface, one along the x-direction, one along the y-direction:

$$\mathbf{P}_x(x) = \langle x, b, f(x, b) \rangle$$
$$\mathbf{P}_y(y) = \langle a, y, f(a, y) \rangle.$$

The tangent vectors to $P_x(x)$ and $P_y(y)$ at $x = a$ and $y = b$, respectively, are

$$\mathbf{V}_x = \langle 1, 0, \frac{\partial f}{\partial x}(a, b) \rangle$$

$$\mathbf{V}_y = \langle 0, 1, \frac{\partial f}{\partial y}(a, b) \rangle.$$

The tangent plane is the plane that contains the point $(a, b, f(a, b))$ and the vectors $\mathbf{V}_x$ and $\mathbf{V}_y$. The parametric equations for this plane are

$$x = a + s$$

FIGURE 10.30   Vectors Tangent to $f(x, y)$

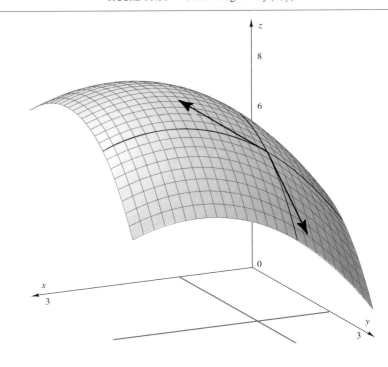

$$y = b + t$$

$$z = f(a, b) + \frac{\partial f}{\partial x}(a, b) \cdot s + \frac{\partial f}{\partial y}(a, b) \cdot t.$$

By eliminating s and t, we get the Cartesian equation of the plane tangent to the surface $z = f(x, y)$ at the point $(a, b, f(a, b))$:

$$z = f(a, b) + \frac{\partial f}{\partial x}(a, b) \cdot (x - a) + \frac{\partial f}{\partial y}(a, b) \cdot (y - b). \tag{10.15}$$

Tops and Bottoms

Equation (10.15) lets us give a second answer to the question we began this chapter with: "How can we identify the point at which $f(x, y)$ achieves its minimum value?" Assuming the surface has a tangent plane at its minimum, that plane will be perfectly horizontal. A horizontal plane, one perpendicular to the z-axis, has a Cartesian equation of the form $z = C$ for some constant C. Comparing this to Equation (10.15) we state the following.

A maximum point that has a tangent plane will also have a horizontal tangent plane.

THE FIRST PARTIAL DERIVATIVE TEST

If $z = f(x, y)$ has a tangent plane everywhere, $f(x, y)$ achieves its minimum or maximum values at simultaneous solutions to

$$\frac{\partial f}{\partial x}(x, y) = 0$$

$$\frac{\partial f}{\partial y}(x, y) = 0.$$

A point that satisfies the first partial derivative test is not necessarily a maximum or minimum. Points with a horizontal tangent planes may be a maximum, as in Figure 10.31, or a minimum, as in Figure 10.32, or neither, as depicted in Figure 10.33. Points such as those in Figure 10.33 that have horizontal tangent planes but are neither a maximum nor a minimum are called *saddle points*.

Think and Share

Break into groups. Describe the intersection of a surface and its tangent plane at a minimum or a maximum. Describe the intersection of a surface and its tangent plane at a saddle point. Will a point whose tangent plane is horizontal but which is neither a maximum nor a minimum always look like a saddle?

FIGURE 10.31 Maximum Point   **FIGURE 10.32** Minimum Point

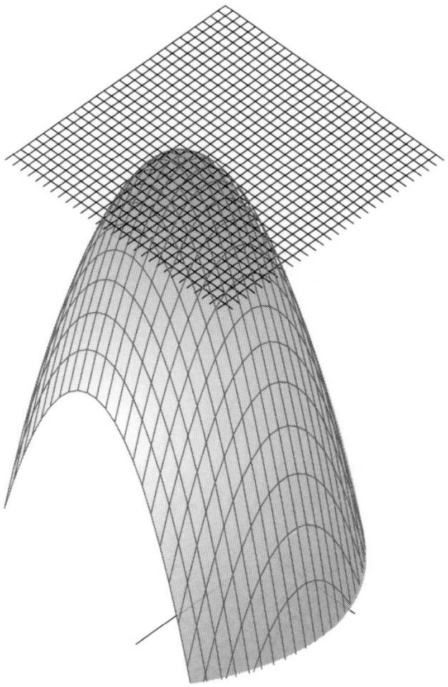

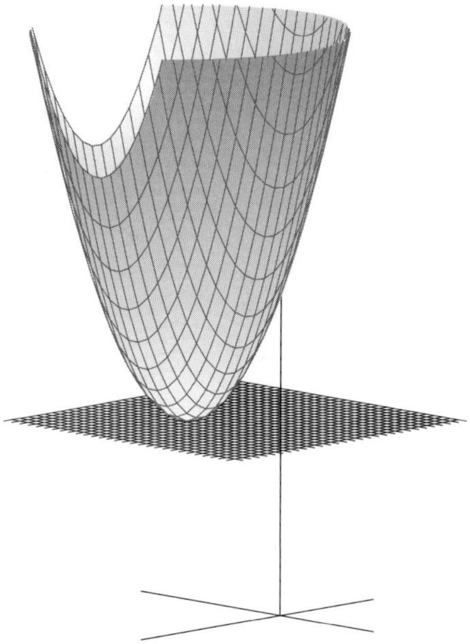

FIGURE 10.33   Saddle Point

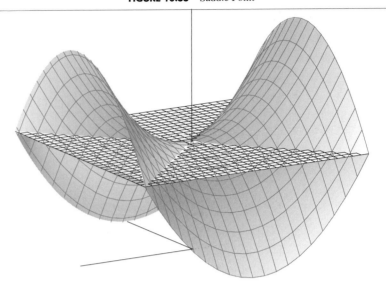

E X A M P L E   2   **Applying the First Partial Derivative Test**

Let's find the maximum and minimum values on the surface $z = x \cdot y \cdot e^{-x^2-y^2}$.

To identify candidates for maximum and minimum values, we find those points at which $\partial z / \partial x = 0$ and $\partial z / \partial x = 0$. That is, we solve the system of equations

$$y \cdot (1 - 2x^2) \cdot e^{-x^2-y^2} = 0$$
$$x \cdot (1 - 2y^2) \cdot e^{-x^2-y^2} = 0.$$

Since $e^{-x^2-y^2}$ is never zero, these equations reduce to the system

$$y \cdot (1 - 2x^2) = 0 \qquad \textbf{(10.16)}$$
$$x \cdot (1 - 2y^2) = 0. \qquad \textbf{(10.17)}$$

FIGURE 10.34 $z = x \cdot y \cdot e^{-x^2-y^2}$

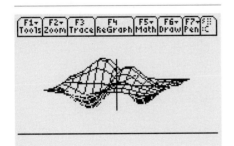

Equation (10.16) requires that either $y = 0$ or $x^2 = 1/2$. Considering first $y = 0$ in Equation (10.17), we find that $x = 0$. So we have $(0, 0, 0)$ as our first point at which the surface has a horizontal tangent plane. Considering the second possibility, we find $x = \pm \frac{1}{\sqrt{2}}$, which in conjunction with Equation (10.17) requires that $y^2 = 1/2$. So we have four additional points at which there is a horizontal line tangent to the surface. They are $(\frac{1}{\sqrt{2}}, \frac{1}{\sqrt{2}}, \frac{1}{2e})$, $(\frac{1}{\sqrt{2}}, \frac{-1}{\sqrt{2}}, \frac{-1}{2e})$, $(\frac{-1}{\sqrt{2}}, \frac{1}{\sqrt{2}}, \frac{-1}{2e})$, and $(\frac{-1}{\sqrt{2}}, \frac{-1}{\sqrt{2}}, \frac{1}{2e})$. It is not clear whether each point is a minimum, a maximum, or a saddle point until we look at the graph in Figure 10.34. By inspecting the graph we conclude that $(0, 0, 0)$ is a saddle point, that $(\frac{1}{\sqrt{2}}, \frac{1}{\sqrt{2}}, \frac{1}{2e})$ and $(\frac{-1}{\sqrt{2}}, \frac{-1}{\sqrt{2}}, \frac{1}{2e})$ are maxima, while $(\frac{-1}{\sqrt{2}}, \frac{1}{\sqrt{2}}, \frac{-1}{2e})$ and $(\frac{1}{\sqrt{2}}, \frac{-1}{\sqrt{2}}, \frac{-1}{2e})$ are minima.

E X A M P L E 3 **The Distribution Center: A Symbolic Approach**

The first partial derivative test supplies a theoretical technique for isolating and identifying extreme points of functions of two variables. Unfortunately, when we apply the test to our function from Section 10.1, we are led to seek the simultaneous solutions to the system of equations.

$$\frac{(2x-2)}{2(x^2-2x+2+y^2-2y)^{1/2}} + \frac{(2x-6)}{2(x^2-6x+10+y^2-2y)^{1/2}} +$$
$$\frac{(2x-6)}{2(x^2-6x+25+y^2-8y)^{1/2}} + \frac{(2x-10)}{2(x^2-10x+26+y^2-2y)^{1/2}} +$$
$$\frac{(2x-10)}{2(x^2-10x+34+y^2-6y)^{1/2}} = 0$$

$$\frac{(2y-2)}{2(x^2-2x+2+y^2-2y)^{1/2}} + \frac{(2y-2)}{2(x^2-6x+10+y^2-2y)^{1/2}} +$$
$$\frac{(2y-8)}{2(x^2-6x+25+y^2-8y)^{1/2}} + \frac{(2y-2)}{2(x^2-10x+26+y^2-2y)^{1/2}} +$$
$$\frac{(2y-6)}{2(x^2-10x+34+y^2-6y^{1/2}} = 0$$

Algebraic solutions to systems of this complexity are out of the question and the presence of so many square roots in the denominators presents serious challenges to even the most powerful and sophisticated numerical algorithms. The algebraic solution to the warehouse distribution problem is theoretically simple and beautiful but in practice cannot be implemented. In the last section we will look again to curves to give us a numerical procedure to reliably approximate the minimum value of functions.

E X E R C I S E S E T 1 0 . 4

1. Find the equation of the plane tangent to $z = x^2 - y^2$ at $(0, 0, 0)$.

2. Find the equation of the plane tangent to $z = x \cdot y \cdot e^{-x^2-y^2}$ at $(1, 1, e^{-2})$.

3. Find the equation of the plane tangent to $z = \sqrt{r^2 - x^2 - y^2}$ at $(a, b, \sqrt{r^2 - a^2 - b^2})$.

† 4. *Group Exercise* The surface $z = \sqrt{x^2 + y^2}$ is a cone. It comes to a sharp point at $(0, 0, 0)$. Consequently, it fails to have a tangent plane at the origin. Using $f(x, y) = \sqrt{x^2 + y^2}$, repeat the computations that lead to Equation (10.15). What is the result of these computations?

5. Find all points on the surface $z = 6x^2 - 8x^3y + y^4$ at which the tangent plane is horizontal.

6. Find the smallest value obtained by the function $f(x, y) = x^2 + 3y^3 - 8x + 7y$.

† 7. Define the term "local extreme point" for a function of two variables.

8. Find all local extreme points of the function

$$f(x, y) = x^3 - y^3 + 3xy.$$

9. Find the largest and smallest values of

$$f(x, y) = \frac{100y}{1 + x^2 + y^2}.$$

10. Find the largest and smallest values of

$$f(x, y) = \frac{100y}{1 + x^2 + y^2},$$

if the point (x, y) is required to lie on the circle $x^2 + y^2 = 4$.

11. Find the largest and smallest values of

$$f(x, y) = \frac{100y}{1 + x^2 + y^2},$$

if the point (x, y) is required to lie on the ellipse $x^2 + 4y^2 = 4$. Suggestion: Parameterize the ellipse.

12. As we saw in Chapter 1, the points $\{(x_j, y_j)\} = \{(1, 210),$ $(2, 538), (3, 1052), (4, 1765), (5, 2466), (6, 3221),$

$(7, 3589), (8, 4309), (9, 4690), (10, 5902), (11, 7095)\}$ lie near a line. A linear model for these data is a function of the form $L(x) = m \cdot x + b$. The residuals of the model are the values $\{L(x_j) - y_j\}$. The model is considered a "good" fit if the sum of the squared residuals, the SSR, is relatively small. The formula for SSR, $\sum_{j=1}^{11} \left(L(x_j) - y_j\right)^2$ is a function of m and b. Find the values of m and b that minimize the value of SSR.

10.5

The Gradient Search

We close our discussion of surfaces with a simple idea. To reach the bottom of a valley, adopt the strategy that water does: just go downhill from wherever you are. Once you start using this approach, you will move toward the bottom of *some* valley, possibly not the valley you had hoped for. So care must be taken with this "method of steepest descent."

The Directional Derivative Revisited

If we are to travel "downhill," we need to be able to answer the question "which way is downhill?" Specifically, given a function $f(x, y)$ and a point (a, b) in its domain, how do we determine the direction in which $D_{[\theta]} f(a, b)$ is least? One entirely reasonable approach to finding θ is to graph $D_{[\theta]} f(a, b)$ for $0 \leq \theta \leq 2\pi$ and then use the techniques of Part I of the book to determine the value of θ at the minimum.

▶ **FIRST REFLECTION**
Why restrict the domain of $D_{[\theta]} f(a, b)$ to $0 \leq \theta \leq 2\pi$?   ◀

EXAMPLE 1

Calculate the directional derivative of

$$f(x, y) = \frac{-10x}{3 + x^2 + y^2}$$

at the point $(-2, 2)$ as a function of θ and determine the value of θ that minimizes $D_{[\theta]} f(-2, 2)$.

   We recall from Section 11.3 that the directional derivative to a function $f(x, y)$ at the point $(-2, 2)$ is given by

$$D_{[\theta]} f(-2, 2) = \frac{d}{dt}(f(-2 + t \cos\theta, \ 2 + t \sin\theta)).$$

FIGURE 10.35 The Directional Derivative for $0 \le \theta \le 2\pi$

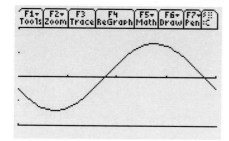

Letting the TI calculator do our symbolic operations for us and setting $t = 0$ in the resulting formula, we get the directional derivative of $f(x, y)$ at $(-2, 2)$,

$$\frac{-30}{121} \cos(\theta) - \frac{80}{121} \sin(\theta).$$

The directional derivative reduces to the sum of two sinusoidal functions, which is itself a sinusoidal function as shown in 10.35. The minimum of $D_{[\theta]} f(-2, 2)$ will be the smaller solution to

$$\frac{d D_{[\theta]} f(-2, 2)}{d\theta}(\theta) = 0$$

on the interval $0 \le \theta \le 2\pi$. Solving $\frac{30}{121} \cdot \sin(\theta) - \frac{80}{121} \cdot \cos(\theta) = 0$ for θ gives $\theta = \arctan\left(\frac{8}{3}\right) \approx 1.21203$. The maximum value of the directional derivative is achieved at the other root to the equation, $\theta = \arctan\left(\frac{8}{3}\right) + \pi$.

▶ **SECOND REFLECTION**

Does it make physical sense for the directions of steepest descent (the minimum of the directional derivative) and the direction of steepest ascent (the maximum of the directional derivative) to be 180 degrees apart? ◀

Think and Share

In groups of four with two teams of two, repeat the computation for $D_{[\theta]} f(a, b)$ at the points $(-1, 1)$, $(2, 2)$, $(-3, 0)$, and $(-1, -1)$. In each case, describe the form of the function and its graph. Why is the graph reasonable? For one point chosen by your instructor, determine the angle θ that minimizes the directional derivative.

The Gradient

In the example above and in the Think and Share, all of the formulas for the directional derivative were of the form $A \cdot \cos(\theta) + B \cdot \sin(\theta)$. This is no accident. In fact, the values for A and B have significance, because they are $\frac{\partial f}{\partial x}(a, b)$ and $\frac{\partial f}{\partial y}(a, b)$, respectively. Recall from Section 10.3 that we computed the directional derivative of $f(x, y)$ at (a, b) from the tangent vector to

$$\mathbf{H}(t) = \langle a + t\cos(\theta), b + t\sin(\theta), f(a + t\cos(\theta), b + t\sin(\theta))\rangle.$$

The derivative of $\mathbf{H}$ at $t = 0$, $\langle \cos(\theta), \sin(\theta), D_{[\theta]} f(a, b)\rangle$, generates a line tangent to the surface $z = f(x, y)$ at the point $(a, b, f(a, b))$. Since this tangent line lies in the tangent plane, its coordinates must satisfy the equation for the tangent plane,

$$z = f(a, b) + \frac{\partial f}{\partial x}(a, b) \cdot (x - a) + \frac{\partial f}{\partial y}(a, b) \cdot (y - b).$$

Check this by computing $\dfrac{\partial f}{\partial x}(-2, 2)$ and $\dfrac{\partial f}{\partial x}(-2, 2)$ for the function $f(x, y) = \dfrac{-10x}{3 + x^2 + y^2}$ of our example.

The parametric equations for the tangent line are
$$x = a + t \cdot \cos(\theta),$$
$$y = b + t \cdot \sin(\theta),$$
$$z = f(a, b) + t \cdot D_{[\theta]} f(a, b).$$

Substituting the parametric values for x, y, and z, shown in the margin note, we get the equation

$$f(a, b) + t \cdot D_{[\theta]}f(a, b) = f(a, b) + \frac{\partial f}{\partial x}(a, b) \cdot ((a + t \cdot \cos(\theta)) - a)$$
$$+ \frac{\partial f}{\partial y}(a, b) \cdot ((b + t \cdot \sin(\theta)) - b).$$

But this reduces to the equation

$$D_{[\theta]}f(a, b) = \frac{\partial f}{\partial x}(a, b) \cdot \cos(\theta) + \frac{\partial f}{\partial y}(a, b) \cdot \sin(\theta). \qquad \textbf{(10.18)}$$

We can now use Equation (10.18) to simplify our search for the direction of steepest change of a function. Specifically, we seek the values of θ, for $0 \le \theta \le 2\pi$, that maximize and minimize the function

$$\frac{\partial f}{\partial x}(a, b) \cdot \cos(\theta) + \frac{\partial f}{\partial y}(a, b) \cdot \sin(\theta).$$

The function has period 2π. Thus, it has a single maximum and a single minimum (determined using the first dervative test) in the interval. The extreme values of the function are π radians apart and are the roots of the equation

$$-\frac{\partial f}{\partial x}(a, b) \cdot \sin(\theta) + \frac{\partial f}{\partial y}(a, b) \cdot \cos(\theta) = 0.$$

The values of θ that maximize and minimize our directional derivative are those that satisfy

$$\tan(\theta) = \frac{\dfrac{\partial f}{\partial y}(a, b)}{\dfrac{\partial f}{\partial x}(a, b)}.$$

Recall that for a vector $\mathbf{v} = \langle a, b \rangle$, $\tan \theta = b/a$.

Therefore, the vector $\left\langle \dfrac{\partial f}{\partial x}, \dfrac{\partial f}{\partial y} \right\rangle$, called the **gradient vector**, is a vector in the direction of greatest change.

THE GRADIENT

For a function of two variables $f(x, y)$, the gradient of f, denoted by $\nabla f(x, y)$, is defined to be $\nabla f(x, y) = \left\langle \dfrac{\partial f}{\partial x}(x, y), \dfrac{\partial f}{\partial y}(x, y) \right\rangle$

EXAMPLE 2   **Distribution Center: The Gradient Approach**

It is time to give a third solution to the problem of minimizing a function of two variables. To estimate where the minimum value of

$$f(x, y) = \sqrt{(x-1)^2 + (y-1)^2} + \sqrt{(x-3)^2 + (y-1)^2}$$
$$+ \sqrt{(x-3)^2 + (y-4)^2} + \sqrt{(x-5)^2 + (y-1)^2}$$
$$+ \sqrt{(x-5)^2 + (y-3)^2}$$

occurs, we begin somewhere (why not at the origin $(0, 0)$ where $f(0, 0) \approx$ 20.50646263?) and travel downhill.

Stage I. The direction of greatest descent at $(0, 0)$ is given by

$$\nabla f(0, 0) \approx \langle -4.09386, -2.53395 \rangle.$$

The path along the surface in the direction opposite the gradient is given by

$$\langle 0 - 4.09386 \cdot t, 0 - 2.53395 \cdot t, f(-4.09386 \cdot t, -2.53395 \cdot t) \rangle.$$

We can see the shape of this curve by plotting

$$f(-4.09386 \cdot t, -2.53395 \cdot t),$$

the height of f above the xy-plane in the direction $\langle -4.09386, -2.53395 \rangle$, as displayed in Figure 10.36 as t varies from -2 to 2.

In the function graphing mode, define
y1(x)=f(-4.09386*x,-2.53395*x)
on the Home *screen. Graph* y1(x).
The window is $[-2, 2] \times [0, 30]$.

FIGURE 10.36 Following the First Gradient in $[-2, 2] \times [0, 30]$

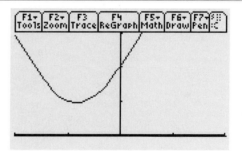

▶ **THIRD REFLECTION**
As t increases, in which direction is the path in Figure 10.36 drawn.   ◀

The graph shows a clear minimum in this direction. At that point we should abandon our first path and pick a new direction. Determining that the coordinates of the minimum point in the graph is a problem from the calculus of one variable, we speed the process by using the minimum command on our TI calculator, as shown in Figure 10.37. The coordinates of the minimum tell us that when $t \approx -0.824132$, and consequently $(x, y) \approx (3.37388, 2.08831)$, the value of $f(3.37388, 2.08831)$ is approximately 9.53107.

Stage II.   Now we repeat the process to refine our estimate for the minimum. We determine the direction of greatest descent as

$$-\nabla f(3.37388, 2.08831) \approx \langle -0.277447, 0.448243 \rangle.$$

FIGURE 10.37   Stopping along the First Gradient

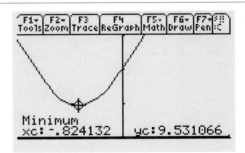

► **FOURTH REFLECTION**

Compare the size of the coordinates in the first and second gradients we have computed. Do you expect as large an improvement in Stage II as in Stage I? ◄

The path along the surface in the direction of the gradient is given by

$$\mathbf{F}(t) \approx \langle 3.37388 - 0.277447 \cdot t, 2.08831 + 0.448243 \cdot t,$$
$$f(3.37388 - 0.277447 \cdot t, 2.08831 + 0.448243 \cdot t) \rangle.$$

To determine the minimum value of f along this path, we graph

$$f(3.37388 - 0.277447 \cdot t, 2.08831 + 0.448243 \cdot t)$$

as in Figure 10.38. The minimum value of the graph tells us that when $t \approx -0.577851$, that is when

$$(x, y) \approx 3.37374 - 0.277441 \cdot (-0.577851),$$
$$2.08831 + 0.448237 \cdot (-0.577851))$$
$$\approx (3.53420, 1.82929)$$

and the value of f is 9.44843. This process of iteration in which we produce a sequence of points $\{x_j, y_j\}$ by following along the direction of the gradient at (x_j, y_j) to a new point (x_{j+1}, y_{j+1}) at which f has a smaller value, is a variation of the *method of steepest descent* for optimizing a function of two variables. The method is widely used and easily adapted to situations in which the goal is to minimize (or maximize) a function of three or more variables. As with Newton's Method, the convergence of the method of steepest descent is sensitive to the initial values (x_1, y_1).

The question of when to stop (how many iterations should we take?) is a knotty one. Generally, one stops when the last iteration or two has produced no

FIGURE 10.38 Following the Second Gradient on $[-2, 2] \times [8, 10]$

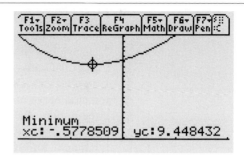

"significant" change in the value of f. What constitutes a significant change depends on context. In our example, the change in the value of f produced by the second iteration was roughly 0.083. The significance of this change is impossible to determine without knowledge of the units associated with f. If f is measured in \$1000/year, the change represents a yearly expense of only \$83. On the other hand, if f is measured in \$1,000,000/year, the change represents a yearly expense of \$83,000 and further iterations may be justified.

EXERCISE SET 10.5

1. Describe two scenarios in which a bad starting point and an awkward function could cause the method of steepest descent to fail to find the desired minimum.

2. Carry out Step III for our distribution problem.

† **3. Group Exercise** The accounting office has supplied data in Table 10.3 of data to assist you in refining the model of this section. Clearly some of the stores require more frequent

TABLE 10.3 Delivery Statistics for 1996

Store Coordinates	Number Deliveries 1996
(1, 1)	137
(3, 1)	89
(3, 4)	222
(5, 1)	103
(5, 3)	187

deliveries to maintain their stock. Add a "weight" to the distance model developed in this section to account for the difference in demand and use the method of steepest descent to estimate the optimal position for the distribution center under your new model.

† **4.** Use the method of steepest descent to estimate the point (x, y) at which the sum of the distances between (x, y) and the points $(-2, -4)$, $(-1, -3)$, $(-1, -4)$, $(5, 6)$, and $(7, 8)$ is minimized.

5. Use the function 1+rand() to assign weights for the five distances in the preceding exercise and determine the point (x, y) in the plane that minimizes the sum of the weighted distances.

6. Use the method of steepest ascent (descent) to estimate the maximum values for the function $f(x, y) = \dfrac{10xy}{4 + x^4 + y^4}$.

Summary

In this chapter we have seen the ideas and tools of the differential calculus of one variable (local linearity and tangent relationships) used to investigate the behavior of functions of two variables. Functions of two (or more) variables arise naturally in many contexts. The graph of a function of two variables, $z = f(x, y)$ is a surface in three-space, a much more difficult object to analyze than a curve in two-space. In order to carry out that analysis, we introduced the notion of a parametric curve in three-space. The notion of vectors prove extremely useful in linking our geometric insight to the algebraic precision required for effective computation. In order to carry forward the graphical, analytical, and numerical analysis of just one model, we have had to develop considerable machinery and in the process reached areas in which all three approaches *must* be used in concert. No one of these complementary modes of thinking was powerful enough to satisfactorily resolve the problem of finding extrema for functions of two variables.

SUPPLEMENTARY EXERCISES

1. Give equations for the plane that contains the line
$$\frac{x - 2}{2} = \frac{y + 3}{5} = \frac{z - 1}{7}$$
and the point $(0, 0, 0)$.

2. Find the Cartesian equation for the plane containing the lines
$$\frac{x - 2}{2} = \frac{y + 1}{3} = \frac{z - 1}{-5} \text{ and } \frac{x - 2}{-3} = \frac{y + 1}{2} = \frac{z - 1}{3}.$$

3. Find the point of intersection of the lines
$$\frac{x}{-2} = \frac{y + 3}{-4} = \frac{z - 11}{6} \text{ and } x - 4 = \frac{y - 11}{5} = \frac{z - 1}{-2}.$$

4. Find the velocity and acceleration vectors at $t = \pi/3$ for the curve $\mathbf{H}(t) = \langle 3 \cdot \cos(t), 4 \cdot \sin(t), t/2 \rangle$.

5. Find an equation for the plane tangent to the surface $z = 3x^2 - 4y^3 + 7$ at the point $(1, 2)$.

6. Find an equation for the plane tangent to the surface $z = 3x - 4y + 7$ at the point $(1, 2)$. Explain your result in terms of local linearity.

7. Find the minimum and maximum values of the function
$$f(x, y) = (4y^2 - x^2)e^{-2x^2 - y^2}.$$

8. Show that the curve $\mathbf{F}(t) = \langle t \cdot \cos(t), t \cdot \sin(t), t \rangle$ lies on the cone $z = \sqrt{x^2 + y^2}$.

9. Find a parametric representation for the curve on the surface $z = \sqrt{5 - x^2 + y^2}$ that lies over the circle $x^2 + y^2 = 1$.

10. Find the point in two-space whose average distance from all of the points $(1, 2)$, $(-3, 4)$, $(-1, 5)$, $(3, -8)$, $(3, 3)$, $(5, 6)$ is least.

11. Describe how one could use the minimization techniques of this chapter to find the "best least squares fit" for a model of the form $G(t) = a \cdot b^t$ for a set of data (t_j, y_j), $j = 1 \ldots n$ whose graph suggested exponential decay.

12. Discuss how the magnitude of the gradient vector might be used to determine a stopping criterion for the method of steepest descent presented in Section 10.5.

13. By analogy, to the results of Section 10.4, present algebraic conditions to identify and find the extreme points of a function of three variables, $f(x, y, z)$.

14. Use the criteria you developed in the preceding exercise to determine the minimum value and the values of x, y, and z for $f(x, y, z) = x^2 - 4x + y^2 - 2y + z^2 + 10z + 50$.

15. Use the method of completing the square to verify the result you obtained in the proceeding exercise.

PROJECTS

Project One: The Geometry of Vectors in Three-Space

Consult a text on three-dimensional geometry, or on the calculus of several variables and prepare a report on the *dot product* and *cross product* of vectors in three-space.

The TI calculator commands for these are dotP(*and* crossP(, *respectively.*

Project Report.   Write a report on your investigations. Address your report to your classmates, assuming the material is new to them. Describe your reasoning and include all necessary background information. A minimal project report must include:

1.   A discussion of the relation of the cross and dot products to the trigonometric functions and to *orthogonality*
2.   A discussion of the relation of these "products" to areas and volumes
3.   Definitions or descriptions for all new terms
4.   A discussion as to how the major results can be proved
5.   Illustrative examples; these will form a crucial part of your report
6.   Suggestions for possible extensions of your observations and further questions for exploration

Project Two: Level Curves and Gradients

Consult a text on the calculus of several variables (or a calculus text that includes a full development of the topic) and prepare a report on *level curves* and the gradient.

Project Report.   Write a report on your investigations. Address your report to your classmates, assuming the material is new to them. Describe your reasoning and include all necessary background information. A minimal project report must include:

1.   A discussion of the relation between the gradient and tangents to level curves
2.   The use of level curves and gradients in Lagrange Multipliers
3.   Definitions or descriptions for all new terms
4.   A discussion as to how the major results can be proved
5.   Illustrative examples; these will form a crucial part of your report (You will find the use of a more powerful computer algebra system such as Maple useful in preparing graphical illustrations.)
6.   Suggestions for possible extensions of your observations and further questions for exploration

Project Three: Volumes by Monte Carlo Techniques

In Chapter 5 you calculated the volumes of various solids of revolution. Functions of two variables are used to describe other surfaces and solids, like the ellipsoid bounded by $z = \pm\sqrt{36 - 4x^2 - 9y^2}$. We can use the functions `when(` and `rand(` in conjunction with `seq(` and `sum(` to estimate the volumes of such solids. That portion of the ellipsoid that lies in the first octant, where all three variables are positive, represents one-eighth of the total ellipsoid. We can estimate the volume of this solid and multiply by eight to obtain an estimate for the volume of the entire ellipsoid.

Our process for estimating the volume of the solid is as follows. We generate 50 points (x_j, y_j, z_j) at random in some box that contains the ellipsoid. We test each point to see if it lies inside or outside to the solid. We then count the number of points that lie inside the ellipse. The ratio of the number of points interior to the solid to the total number of points times the volume of the bounding box is one estimate of the volume of the solid.

On the TI calculator, first define $f(x, y) = \sqrt{(36 - 4x^2 - 9y^2)}$. The solid in the first octant lies in the box $0 \le x \le 3$, $0 \le y \le 2$, and $0 \le z \le 6$. A random point within this box is generated as `[3 rand(), 2 rand(), 6 rand()]`. We can create the function `test() = when(f(3 rand(), 2 rand()) >= 6 rand(), 1, 0, 0)` to count when the random point (x_j, y_j, z_j) lies below the surface $z = f(x, y)$. To see how many points out of a sequence of 50 random points lie interior to the solid, we use `sum(seq(test(), 1, 0, 0)`. Using random numbers to estimate volume is called a *Monte Carlo* technique in honor of the gaming houses in the Principality of Monaco.

Use a Monte Carlo technique to estimate the volume of the solids listed below:

- The ellipse bounded by $z = \pm\sqrt{36 - 4x^2 - 9y^2}$
- The region above the (x, y)-plane and below the surface $z = 36 - 4x^2 - 9y^2$
- The region above the cone $z = \sqrt{x^2 + y^2}$ and below the sphere

$$x^2 + y^2 + z^2 = 1$$

Project Report.   Write a report on your investigations. Address your report to your classmates, assuming the material is new to them. Describe your reasoning and include all necessary background information. A minimal project report must include:

1. A full explanation of all of the TI calculator commands and functions you use and a description of how they apply to each of you solids; this should include the initial value of `RandSeed` you used
2. A statistical summary and analysis of the experiments you have run
3. Justification and explanations for your calculations
4. Suggestions for possible extensions of your observations and further questions for exploration

A

TI-89 COMPUTER ALGEBRA SYSTEM TUTORIAL

A.1

The TI Computer Algebra System Tutorial

Until recently, Computer Algebra Systems (CASs) capable of manipulating symbols as well as numbers were available only on high-powered computers costing thousands of dollars. With the introduction of the TI-92, and more recently the TI-89, a CAS is now available at a fraction of the cost in a conveniently portable package. Although in the past you may have expected your calculator to perform tedious numeric calculations (such as multiplying two five-digit numbers together), with these TI calculators you can now perform operations involving complex symbol manipulation (such as multiplying two fifth degree polynomials together). Despite its computational power, a CAS cannot do your thinking. To make proper use of the technology made available by the TI-92 and TI-89, it is essential that you understand the basic ideas of the mathematics you are using. Now more than ever, doing mathematics requires thinking, not just computing.

The TI-89 keypad

The TI-89 keypad looks much like a standard graphing calculator keypad even though the TI-89's CAS greatly extends the range of mathematical computations available to you. The additional functionality is made available through a flexible menuing system rather than an extended keypad. The keypad and menu system of the TI-89 lets you quickly and easily perform numeric, graphic, and symbolic mathematical computations. A great way to learn to use the TI-89 is to play with it. You will not break it by trying it out, and you are likely to learn as much when it doesn't do what you expect as when it does.

Study the image of the keypad and display window that appears in Figure A.1.

FIGURE A.1   TI-89 Keypad

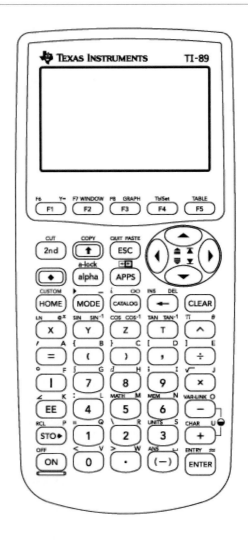

Let's look carefully at the keypad.

Screen Intensity.   Locate the ON button and use it to turn on your calculator. Adjust the contrast by holding down the green ◇ key in the upper left corner just below the yellow 2nd button while pressing the + or − key on the lower right of the calculator. The green $\boxed{\diamond}$ key is used to activate any feature written in green on the keypad. We will sometimes refer to it as $\boxed{\diamond}$ or as the Green Diamond key. We will use $\boxed{\diamond}$ often with Y=, WINDOW and GRAPH to access the Y= and WINDOW editors, and the GRAPH screen.

The 2nd Key.   The yellow $\boxed{\text{2nd}}$ key is found in the upper left corner of the kcypad. The $\boxed{\text{2nd}}$ key will activate any feature on the keypad written in orange.

Function Keys.   Across the top of the keypad, you will see five blue function keys marked $\boxed{\text{F1}}$ through $\boxed{\text{F5}}$. The first three of these keys also have a $\boxed{\text{2nd}}$ function $\boxed{\text{F6}}$, $\boxed{\text{F7}}$, and $\boxed{\text{F8}}$. The function keys are used in conjunction with the various menus that appear at the top of the display window. Look at the display screen. Notice that the items at the top of the screen have words indicating the kind of operations that are available within each menu. You will use these function keys often to access the menu system. For example, $\boxed{\text{F1}}$ is associated with the label `Tools` and contains a toolbox of both editing commands (like `Copy` or `Paste`) and system commands (such as `Save Copy As`). You will use $\boxed{\text{F1}}$ to access this set of `Tools` when you want to clear the screen.

The Cursor Arrows.   The large blue keys at the top of the keypad are called the Cursor arrows. They are used to control cursor movement. Pressing the button with the arrowhead that points left moves the cursor left, an action which will be indicated by the phrase "Cursor Left." Pressing the button with the arrowhead that points down moves the cursor down, indicated by "Cursor Down." The other two buttons (right and top) are similar.

The Escape and Applications Keys.   Just to the left of the Cursor arrows are blue keys marked $\boxed{\text{ESC}}$ and $\boxed{\text{APPS}}$. The Escape key let's you "escape" from almost any screen or menu you may be "stuck" in. The Applications key, lets you access the `Data/Matrix`, `Program`, and `Text` editors. You will use these editors to create *data tables* and edit sets of commands called *scripts*.

The Alpha and Editing Keys.   To the left of the Escape and Applications keys locate the Shift key, $\boxed{\Uparrow}$, and the purple Alpha key, $\boxed{\text{alpha}}$. Pressing the Alpha key once allows you to type any one of the purple letters indicated above the keys in the bottom five rows. Pressing the Alpha key twice will "lock" these keys into their alphabetical mode until you press the Alpha key a third time. The keys marked X, Y, Z, and T always give the (lower case) variables x, y, z, and t, respectively. Used in conjunction with the Alpha key, the Shift key gives uppercase letters. Additionally, the Shift key allows you to select text in order to copy it and paste it into a later expression. Depressing the Shift key while cursoring left or right will highlight (select) the desired text. The editing commands CUT, COPY and PASTE are the green $\boxed{\diamond}$ functions of the Second, Shift and Escape keys, respectively.

The Clear Key and Backspace Keys.   The $\boxed{\text{CLEAR}}$ key is used to clear lines of type. The Backspace key ($\boxed{\longleftarrow}$) is used to delete single characters or highlighted expressions. Both are found directly beneath the Cursor arrows.

The Mode Key.   The $\boxed{\text{MODE}}$ key, directly below the Alpha key, determines the current calculator settings such as whether angles are read in degrees or radians. The screen will inform you that you can cancel the $\boxed{\text{MODE}}$ menu with the $\boxed{\text{ESC}}$ key.

Arithmetical Calculations.   The bottom five rows of the keypad are similar to a standard scientific calculator with numbers on gray keys and with arithmetic operations and punctuation on black keys. The $\boxed{\wedge}$ exponentiation key appears at the top of the column of arithmetic operation keys, and the blue $\boxed{\text{ENTER}}$ key appears at the bottom. Using the "with" key $\boxed{\mathbf{|}}$ you can make temporary substitutions or add conditions to expressions.

Going HOME**.**   The $\boxed{\text{HOME}}$ key is just below the Green Diamond key. The Home screen displayed in Figure A.2 is where you will do most of your arithmetic and algebraic computations. It houses the Computer Algebra System on your TI-89. Pressing $\boxed{\text{HOME}}$ will take you to the Home screen from any other application.

The Home Screen

Figure A.2 indicates the various regions of the Home screen (which you access by pressing $\boxed{\text{HOME}}$). Enter calculations on the Entry line; the results appear in the History area. The Toolbar menu lets you access various features and operations contained in Tools ($\boxed{\text{F1}}$) and Algebra ($\boxed{\text{F2}}$) and so forth. The Status line at the bottom of the screen displays information about the state of the calculator. Look at the screen often for helpful information displayed in pop-down menus, dialogue boxes, and in the History area.

FIGURE A.2

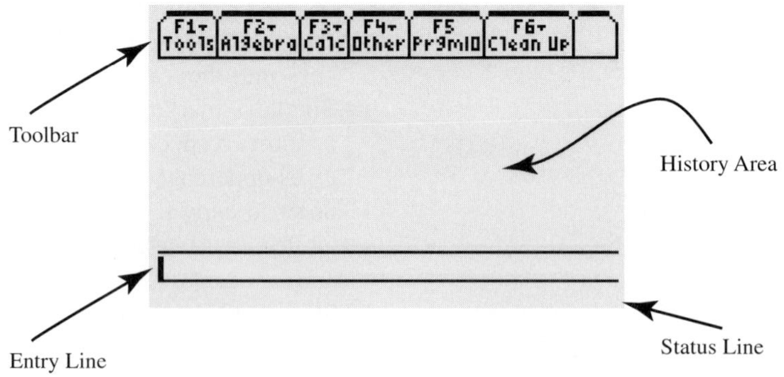

Toolbar

History Area

Entry Line

Status Line

The Menu System

Turn on your calculator if you have not done so already.

After you turn on your calculator, press the $\boxed{\text{F1}}$ key. Your display screen should look something like Figure A.3. How would you characterize the options or

FIGURE A.3

items available on this Tools menu? Do you know what item 2 might be used for? Press $\boxed{\text{ESC}}$ to make the menu disappear. Now press $\boxed{\text{F2}}$. What is the most interesting option on this list? Press $\boxed{\text{ESC}}$ to make this menu disappear.

Now press the Applications key ($\boxed{\text{APPS}}$). Which number on the list refers to the Numeric Solver? Do the $\boxed{\text{APPS}}$ menu items have anything in common? Escape from this menu.

Although there are many more menus, we will now make some simple calculations and then investigate an algebra problem to see how the keypad and menu items work together to assist you in doing mathematics.

Calculating with the TI-89

Clearing the History Area and Entry Line.   You can clear the History area or window (if necessary) by pressing $\boxed{\text{F1}}$ and selecting the Clear Home menu option. One way to do this is to Cursor Down until the menu option is highlighted and then press $\boxed{\text{ENTER}}$. Another way is to press $\boxed{8}$. To clear any expressions on the Entry line (if necessary), press $\boxed{\text{CLEAR}}$. We will describe these procedures as "Clear the Home screen and Entry line."

Arithmetic Calculations.   Type 2+3 on the Entry line and press $\boxed{\text{ENTER}}$. This process will be called "Enter 2+3." When you are asked to enter any expression, you will be expected to type in the expression and press $\boxed{\text{ENTER}}$. Your display should look something like Figure A.4.

FIGURE A.4

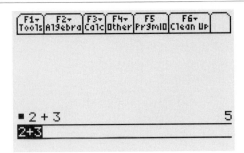

Notice that there is now a small box just above the Entry line followed by 2+3. On the far right of the screen on the same line as 2+3, the number 5 appears as the result of the computation indicated on the left. These expressions are referred to as the "last entry" and the "last answer," respectively. Notice also that 2+3 is highlighted on the Entry line.

Last Entry and Last Answer

Enter the expression 2+3*4. Did you remember to press ENTER? The answer, 14, results from an application of the order of operations—multiplication and division first, followed by addition and subtraction, both left to right.

Enter 10!. The exclamation point (!) can be found as the top entry of the Punctuation *submenu* of the Character menu. The Character menu is opened as the 2nd function of the + key. To open the Punctuation submenu, Cursor Down to the third line of the Character menu and then Cursor Right as suggested by the right pointing arrowhead. Your display should look something like Figure A.5. Pressing ENTER will paste the exclamation point (!) into the

Let's try a more complicated computation.

10! = 3628800

FIGURE A.5

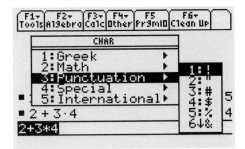

entry line of your Home screen. Press ENTER again.

Editing on the Entry Line.   We would like to compute 100!. You can do this using the editing features of your calculator. Press Cursor Right to place the cursor just after the exclamation point on the Entry line. Press Cursor Left to move the cursor to the left so that it is between 0 and the exclamation point. Now type zero. Compute 100! by pressing ENTER. A highlighted BUSY box appears on the right of the Status line briefly. Is the number as displayed a correct answer? The number is only 38 characters long, but it should be at least 90 characters long. The number 100! is the product of 100 numbers, 90 of which are larger than 10. You can see all the digits of the number by pressing Cursor Up to highlight the answer and then pressing Cursor Right to display more digits to the right. How do you know when there are no more digits to the right?

Entering Fractions. Let's try several ways to enter $\dfrac{1}{x+1}$. Type 1⁄x+1. Typed in on one line, this expression looks like it might work. Press ENTER. Look on the left side or "last entry" of the History area. The expression $\dfrac{1}{x} + 1$ is not correct. Now type 1⁄(x+1) and press ENTER. The History area now shows the correct value.

Storing a Value to a Variable. Clear the Home screen and Entry line. Find the STO key in the left of the keypad, just above the ON key.

Press 2, STO, X, and ENTER. Your display should look something like Figure A.6.

FIGURE A.6

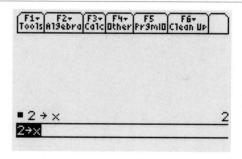

You can approximate the value of an expression.

Enter $\dfrac{1}{x+1}$ correctly as you did above. Find the green $\approx$ above the ENTER key. To approximate this expression, press green ◇ ≈. Notice the approximation that is displayed for one third is $\dfrac{1}{3} \approx .333333$.

Let's change the way decimal numbers are displayed.

Changing the Number of Digits Displayed. Press the MODE key. Your display should look something like Figure A.7.

FIGURE A.7

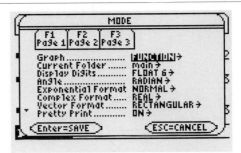

The $\boxed{\text{MODE}}$ dialogue box has three pages. The first is displayed when you press the $\boxed{\text{MODE}}$ key. The second (and third) can be displayed by pressing the $\boxed{\text{F2}}$ ($\boxed{\text{F3}}$) key, as shown at the top left of the menu. On the first $\boxed{\text{MODE}}$ screen, use Cursor Down to the Display Digits ... item. Now Cursor Right to open a submenu. (*This* submenu is indicated by the presence of an arrow rather than an arrowhead.) Cursor Down the submenu to highlight P:FLOAT 11. Press $\boxed{\text{ENTER}}$ twice to change the number of floating point digits displayed to 11—once to accept FLOAT 11, once to save the dialogue box information.

Return to the Home screen. Press $\boxed{\diamond}$ $\boxed{\approx}$ again. The statement that is highlighted on the Entry line is performed again. An approximation for one-third is displayed as the last answer. How does the new approximation for 1/3 differ from the previous one?

When you complete these operations, your display should look like Figure A.8.

FIGURE A.8

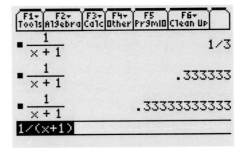

Go back to the $\boxed{\text{Mode}}$ dialogue box and change the floating point digits displayed back to 6. Check that you have successfully done this, by returning to the Home screen and pressing $\boxed{\diamond}$ $\boxed{\approx}$ to see that only six digits are now displayed. Currently x has the value 2. To turn x back into a "variable", press $\boxed{\text{F4}}$, highlight DelVar, press $\boxed{\text{ENTER}}$ and type x. (From now on, we will say "select the DelVar command from the F4 menu".) Notice that there is a space between DelVar and x. Press ENTER and the calculator responds with "Done".

Default Settings.   We will now set the calculator to its default settings—the way your calculator was set at the factory. The factory settings assure us that modes such as floating point digits displayed are set in a standard way. We use the scientific calculator key with the orange MEM above it to do this. Press 2nd 6 to open the MEMORY dialogue box. Your display should look something like Figure A.9.

FIGURE A.9

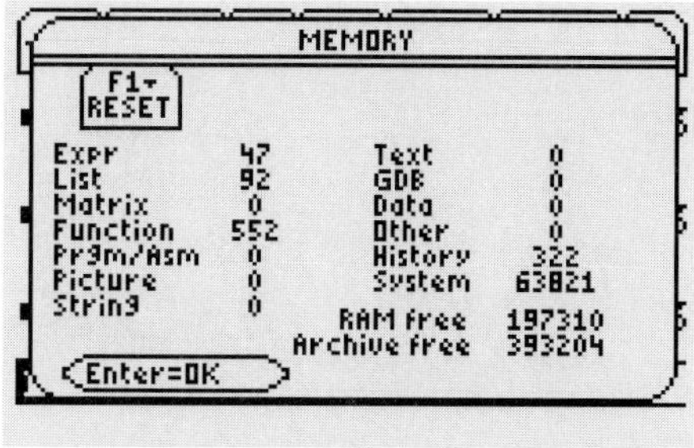

Now press F1 to open the RESET. Press 3 to choose the Default option. A dialogue box will appear. To reset all system variables and modes to their original factory settings without affecting any user-defined variables, functions, or folders, press ENTER twice. The first time to reset to the factory defaults; the second time to confirm that you know the modes are now set to the default or factory settings. You may want to reset your calculator to the default settings when you begin a new activity.

Solving a Quadratic Inequality

The following problem from precalculus or advanced algebra demonstrates the flexibility you will enjoy by using the features of the TI-89 in concert to explore mathematical ideas graphically, numerically, and symbolically. Before you start an activity, such as this one, that involves variables you should consider using the NewProb command from the Home screen Clean Up menu.

New Problem

Bring up the Home screen. Clear the Home screen, the Entry line and any assigned variables by first pressing 2nd F6, and then selecting and entering the NewProb command. Your display should look like Figure A.10.

FIGURE A.10

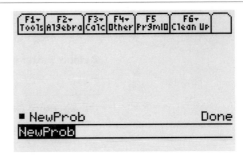

Graphing an Inequality.   We will investigate the quadratic inequality

$$x^2 - 3x - 10 < 0$$

from several vantage points. First, we will look at the inequality as a question about when the polynomial is positive or negative. Then we will look at it from the point of view of the factors of the polynomial. Finally, we will look at the inequality from a symbolic point of view.

Let's graph this function.

Press the ◇ key and then the Y= key (in green above F1) to bring up the Y= editor as shown in Figure A.11.

FIGURE A.11

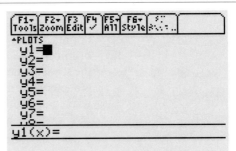

Press ENTER to move the cursor to the Entry line of the Y= editor. Enter the polynomial $x^2 - 3x - 10$.

The polynomial moves up to the line y1= when you press ENTER . Now press F2 to display the Zoom menu as shown in Figure A.12.

FIGURE A.12

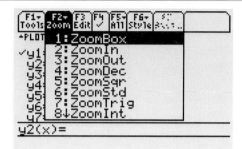

Cursor Down to highlight `ZoomStd`. Press ENTER to select it. (Pressing 6 gives the same result.) We will refer to this activity in the future as "Select `ZoomStd` from the `Zoom` menu." Your graph should look like the one in Figure A.13.

FIGURE A.13

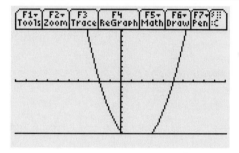

From the graph of this function, we can see that the inequality appears to be satisfied, that is, $x^2 - 3x - 10$ is negative, when x is between -2 and 5 because then the graph lies below the x-axis. To see this more clearly, look at the menus at the top of the screen. The F3 key is associated with the `Trace` feature. Press F3 and notice that a round crosshair cursor called the **Tracing cursor** appears on the graph of the function. Notice also that numbers appear at the bottom of the screen next to `xc:` and `yc:` representing the x-coordinate and y-coordinate of the cursor location.

Cursor Left several times. Notice that the cursor traces along the graph to the left and that the numbers at the bottom of the screen associated with `xc` and `yc` change. Now type `4.3`. What happened as you typed the number? What happens when you press ENTER? Is the inequality satisfied when x is 4.3? Is the inequality satisfied when x is 5?

Return to the Home screen by pressing HOME. Select `NewProb` from the `Clean Up` menu unless you have already done this.

Type in x^2−3x−10<0 on the Entry line and press ENTER. In this book, we will refer to this activity by saying "Enter the expression $x^2 - 3x - 10 < 0$." You will be expected to enter the expressions correctly using the caret (^) for exponentiation, the < sign (above the 0), and parentheses if needed. When you press ENTER, your display should look like Figure A.14.

FIGURE A.14

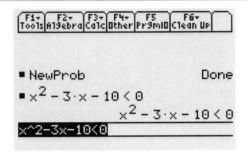

You can also use the CATALOG key to access factor(.

Select factor(from the Algebra menu (See Figure A.15). Cursor Up to highlight the inequality in the History area. Press ENTER to paste the inequality into the Entry line. Complete the command by typing).

Pasting

FIGURE A.15

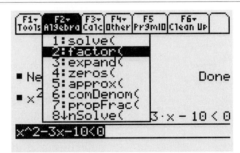

Press ENTER to factor the polynomial. Your display should look something like Figure A.16. Notice that the command you entered appears in the "last entry" position (as a check on what you typed) and its factored form appears in the "last answer" position.

FIGURE A.16

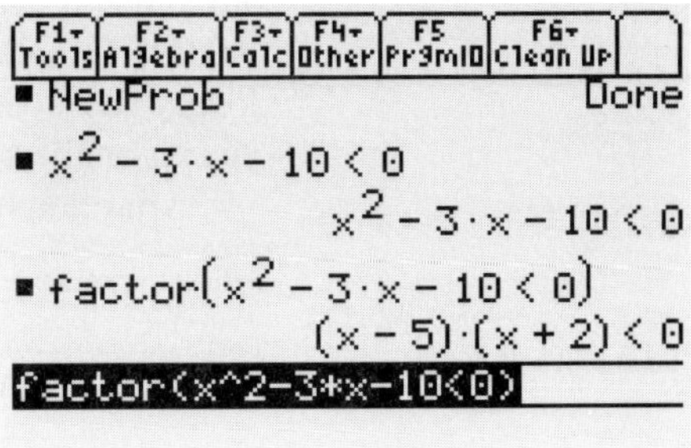

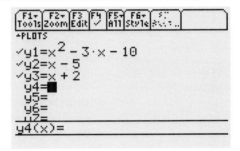

Let's enter these two polynomial factors as y2(x) and y3(x) in the Y= editor.

Go to the $\boxed{Y=}$ editor. Enter $x - 5$ for y2(x) and $x + 2$ for y3(x). Did you remember to press $\boxed{\text{ENTER}}$ each time? Your display should look something like Figure A.17.

FIGURE A.17

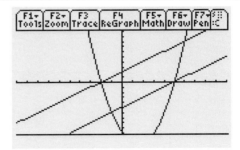

Select ZoomStd from the Zoom menu. Notice that the graph of the first function is a parabola and the graph of the next two functions are straight lines with slope 1. Your graph should look something like Figure A.18.

FIGURE A.18

Activating a Function for Graphs and
Tables

Return to the [Y=] editor. Use the Cursor button to highlight y1(x). Look at the menu items at the top of the screen. What do you think the check mark on the F4 Toolbar item means? Press [F4] to deselect y1(x). Notice that y1 no longer has a check by it. Graph the functions ([◇] [GRAPH]). Which functions appear on the screen?

Your graph should look something like Figure A.19.

FIGURE A.19

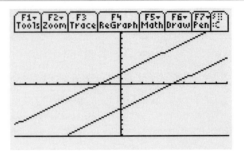

Go back to the [Y=] editor and use [F4] to "check" the y1(x) function so that it will be graphed.

The WINDOW editor

Press [◇] [WINDOW] to display the Window editor. In this editor, you can set the values for various aspects of the graphing window. For example, entering −4 for xmin makes −4 the minimum x-coordinate that will be graphed. Remember, entering includes pressing the [ENTER] key. Enter the following values for the Window editor items:

xmin=-4

Be sure to use the gray [(−)] negative key, not the black [−] subtraction key.

xmax=6

Press [ENTER] or Cursor Down to ignore the "xscl" item.

ymin=-14

ymax=20

Your display should look like Figure A.20.

FIGURE A.20

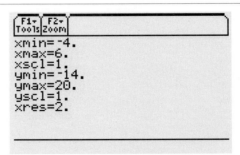

The xscl and yscl items are the distances between tick marks on the x- and y-axes. The xres item determines whether function values will be plotted for every pixel along the x-axis (xres=1) or every second pixel (xres=2), etc.

Press ◇ GRAPH to graph the functions in the new graphing window. Notice that the vertex of the parabola is now visible in this graphing window.

Activate the Trace feature. A 1 (one) should be in the upper right corner of the graphing area to indicate that the Tracing cursor is on the first function in the Y= menu. Cursor Left on the first function until the cursor is to the left of -2. Cursor Down to move to the second function $x - 5$. Is the function value positive or negative? Look just above the Status line for the value of yc. Cursor down to the third function. Is the function value positive or negative? Since the function $x^2 - 3x - 10$ is the product of $x - 5$ and $x + 2$, should this function be positive or negative? Check your result by moving the cursor to the first function using Cursor down.

Let's look at tables of function values.

Bring up the Y= editor and verify that all three functions we have entered are checked. Now press ◇ TABLE . You should see a table of values appear on the screen. From left to right, the columns are x-values, and function values for y1, y2, and y3. You can move around in this table using the Cursor button. Use the Cursor button to scroll up until both y2 and y3 are negative. Notice that y1 is positive.

Changing the Increment in a Table.

Change the increment of the x-values in the far left column of the table by pressing F2 for Setup. This displays the Table Setup dialogue box. You can also do this by pressing ◇ TblSet keys. In the Table Setup dialogue box, set the Δtbl to 0.5. Return to the TABLE window by pressing ENTER (perhaps twice). Notice the increment from one x-value to the next is now 0.5. Go back to the TblSet menu and change the value for Δtbl by entering 1 for Δtbl. Press ENTER . Return to the TABLE screen to see how this affects the increment in the x-values.

Calculating Function Values.   Go to the Y= editor. Check that y1, y2 and y3 are the expressions displayed in Figure A.17. Go to the Home screen by pressing HOME . Type y1(-3)/3 on the Entry line and press ENTER . What is the function value? If you got an error message, you probably entered the -3 with the subtraction key − rather than the gray (−) negative sign. You should always be careful to press the appropriate key when dealing with subtraction and with negatives of numbers.

Let's compute function values for y1, y2, and y3.

Now enter y1(-3.)/3. How does the displayed value on the right differ from the value that you got for y1(-3)/3 without the decimal point?

Rational or Floating Point Arithmetic?

Symbolic Computations.   What is the value of y3(y2(-2))? If you do not remember the expressions for y2 and y3, display them by pressing ◇ Y= . You can find the value of this expression by entering it. As always, the expression that you enter on the Entry line is displayed on the left of the History area while its value is displayed on the right.

Clear the Home screen and Entry line.

Type `y2(x)*y3(x)` and press ENTER. Now press F2 to display the `Algebra` menu, and select the `expand(` option. Use the Cursor button to select the expression at the right of the History area. Press ENTER to bring that result to the cursor position on the Entry line. Now type in the closing `)` and press ENTER to multiply the expression $(x + 2)(x - 5)$.

If you press F2 again, you can see displayed some of the many algebraic operations that are possible. You will use these capabilities of the TI-89 to help you in investigating mathematical problems. You can investigate and solve more involved mathematical problems using these features.

The Data/Matrix Editor

Creating a Data Table.   Table A.1 gives the weight of a colony of yeast as a function of time. We would like to enter this data in a table and plot it using the `Data/Matrix` editor.

TABLE A.1   Time vs. Weight

Time	10	20	30	40	50	60	70	80	90	100
Weight	9.6	29.0	71.1	175	351	513	594	640	656	662

To construct the table, go to the APPS menu and from the `Data/Matrix` editor select `New`. In the `New` dialogue box, submenu selections are made by displaying the submenu items with the Cursor Right button and then selecting a submenu item by highlighting the item and pressing ENTER. Make the following submenu selections and fill in the edit box:

Select `Data` for `Type:`

Select `main` for `Folder:`

Fill in `yeast` in the `Variable:` edit box.

The `Variable` item in this and other editor dialogue boxes is the name of the table (or file) you are creating. It is quite confusing to have this edit box labeled `Variable:` even though a table name varies. Your display should look like Figure A.21.

Press alpha *twice for* a-lock.

FIGURE A.21

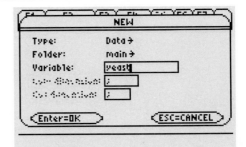

When these settings are correct, press ENTER twice—once to make sure the last submenu item has been chosen, once to save the entries in the dialogue box. Your screen should look something like Figure A.22.

FIGURE A.22

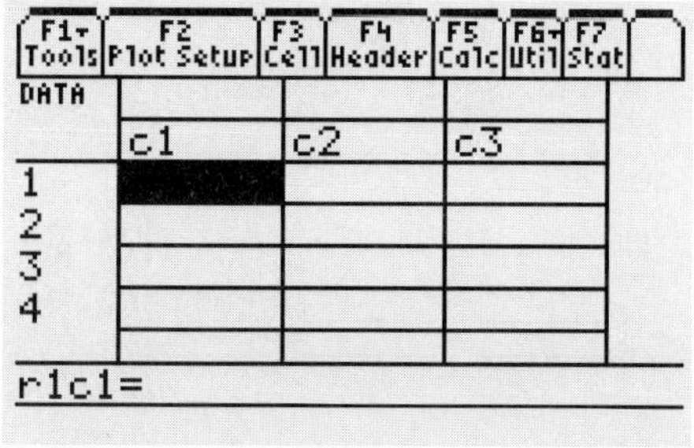

Successively enter 10, 20, 30, ..., 100 in the column *below* the heading c1. Your screen should look something like Figure A.23.

FIGURE A.23

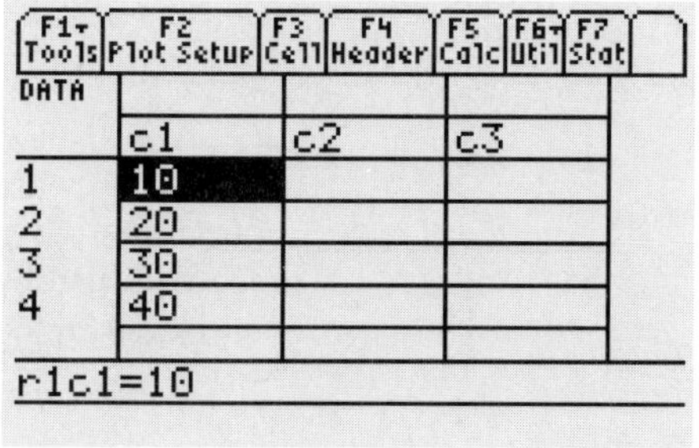

Cursor up to the blank cell *above* the heading c1, press enter and type the title time and enter, as shown in Figure A.24. Title the second column wght.

FIGURE A.24

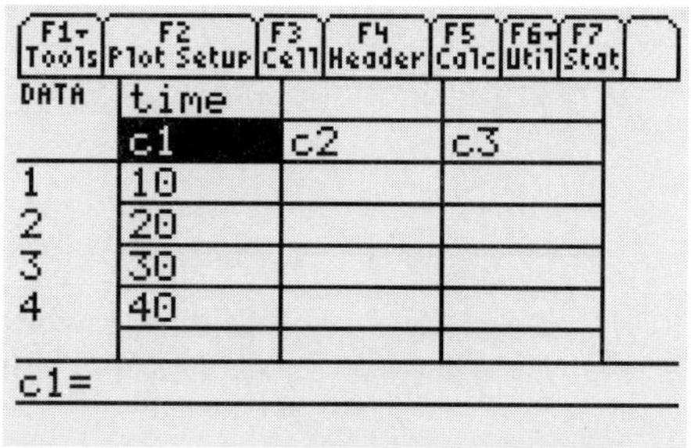

and enter the weights from Table A.1 in the cells below c2. This creates the table of values with the name yeast if you cursor up to the top of your screen. Your screen should look something like Figure A.25. Notice that not all the values are visible.

Go to the Home screen by pressing HOME. The information in an editor is automatically saved as you enter it. Thus, the information in the data matrix named yeast has been saved.

Plotting Data from a Data Table.   The Plot Setup menu in the Data/Matrix editor lets you create a definition for a plot. This process is similar to, but more complicated than, setting up the definition of a function in the Y= editor. The GRAPH command will display this data plot in much the same way as it graphs a function.

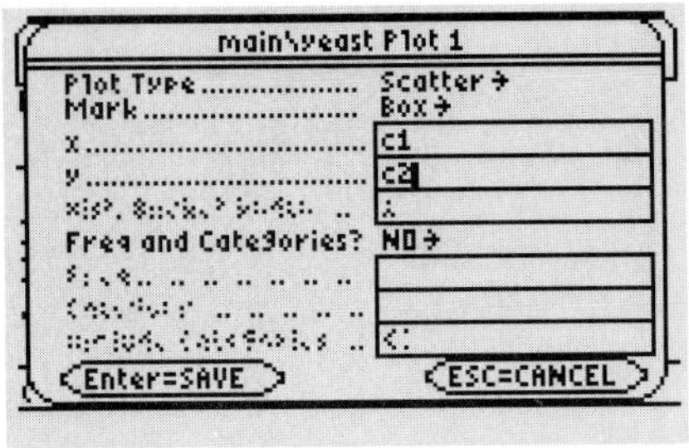

Begin by opening the yeast data table in the Data/Matrix editor. Directions for opening the editor and entering the table are given in the subsection entitled "Creating a Data Table" on page 646. Next, open the Plot Setup menu and select Define. In the resulting dialogue box select

    Plot Type ...Scatter
    Mark ...Box
    x ... c1
    y ...c2

Your display should look something like Figure A.26. Pressing ENTER three times—once to complete the assignment of c2 to y, once to save the entries in the Define dialogue box, and once to save the Plot 1 definition in the Plot Setup dialogue box—returns us to the data table yeast.

Go to the Y= editor, deselect or clear all the functions. Cursor Up to Plot 1 (found immediately above y1) to make sure it is selected, that is, has a check mark by it. Finally, from the Zoom menu select ZoomData which plots the data using the highest and lowest table values for ymax and ymin and similarly for xmax and xmin. The data is displayed in Figure A.27 and the plot in Figure A.28.

FIGURE A.26

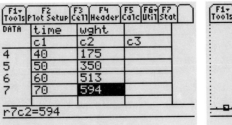

FIGURE A.27   Table of Data for yeast

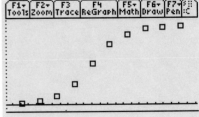

FIGURE A.28   Scatter Plot for yeast

Summary

In this very brief tutorial, we described the keypad, saw how to clear the Home screen and Entry line, and learned how to enter expressions. We have also

FIGURE A.25

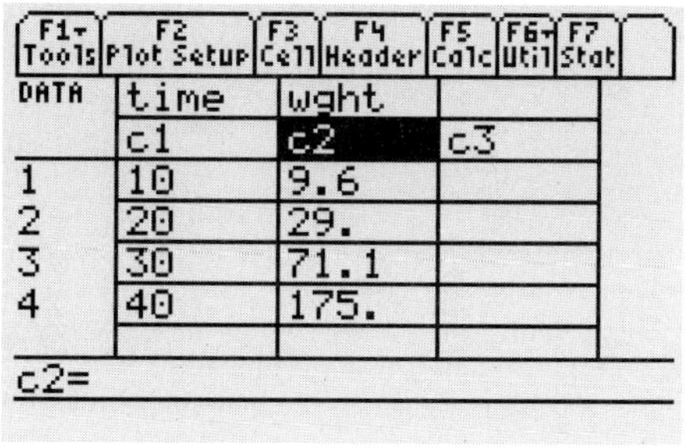

introduced the symbolic capabilities of a CAS through the algebraic operations of factoring and expanding.[†] We saw how the TI-89 handles graphing and we introduced the Data/Matrix editor. From this experience, you discovered how to look carefully at the keypad, the menus, the dialogue boxes, and the editor windows for help in performing whatever mathematical operations you might want.

The TI-89 is a natural, user-friendly extension of graphing calculators like the TI-83. A CAS enables us to investigate a problem from graphical, numerical, or symbolic points of view without changing computing environments. We believe this calculator will help you to concentrate on the concepts of this course while it performs much of the computation for you.

A.2

Troubleshooting: Things that Go Bump in the Night

In this section, we will discuss problems frequently encountered in using a TI-89 calculator. Many of these difficulties occur because of very slight differences between what you intended to enter and what you actually entered. Select NewProb from the Clean Up menu, as usual.

Subtraction and the Negative of a Number

There are two keys that have what appear to be dashes on them: $-$ and $(-)$. The first of these is for subtracting one number from another while the second is to indicate the negative of a number.

[†] There are many, many more symbolic operations available which we will introduce in the text as needed.

To see what these symbols look like on the screen, enter $-2 - 3$. The first of these two minus signs is entered with $\boxed{(-)}$ and the second with the black $\boxed{-}$ key. Your Entry line should look something like this:

$$^{-}2-3$$

The result is $^{-}5$.

Now replace $^{-}$ with $-$ using the subtraction operator key $\boxed{-}$ so that the Entry line looks like $-2 - 3$. Press $\boxed{\text{ENTER}}$. Notice that the machine complains. (If you try to type $-2 - 3$ directly, `ans(1)-` will appear since $-$ is the subtraction operator which requires two numbers or arguments, one to its left and one to its right.)

Exiting a Menu or Dialogue Box, Interrupting a Command

You will sometimes find that you have erroneously brought up a menu or dialogue box. To exit such boxes, you can often press $\boxed{\text{ESC}}$ or $\boxed{\text{ENTER}}$. To exit a dialogue box you may need to press $\boxed{\text{ENTER}}$ two or more times to get out of the activity.

There is no harm in pressing $\boxed{\text{ENTER}}$ many times.

To interrupt the drawing of a graph, you can press $\boxed{\text{ENTER}}$.

This simply interrupts the drawing of the graph (and everything else). To continue the drawing of the graph, press $\boxed{\text{ENTER}}$ again. You can also press $\boxed{\text{ON}}$ to stop the graphing process (or any ongoing computation) although you cannot continue graphing when you interrupt graphing in this way.

Locked Columns in the Data/Matrix Editor

You may not be able to change entries in a column in the `Data/Matrix` editor if that column has been created based on some other column (or list). For example, in the `yeast` data table, you can make the `c3` column equal to `2*c1`. Do so by highlighting `c3` and pressing $\boxed{\text{ENTER}}$ to place the cursor on the Entry line. Enter the expression `2*c1` as shown in Figure A.29.

FIGURE A.29

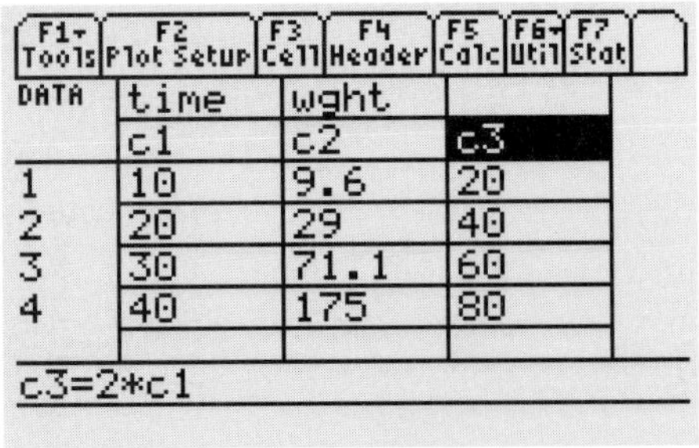

F1▾ Tools	F2 Plot Setup	F3 Cell	F4 Header	F5 Calc	F6▾ Util	F7 Stat
DATA	time	wght				
	c1	c2	c3			
1	10	9.6	20			
2	20	29	40			
3	30	71.1	60			
4	40	175	80			

`c3=2*c1`

Now try to change the entry in the third row of column c3. You cannot do so because these entries are derived from entries in column c1. You can change the entries in c1 and, when you do, the corresponding entry in column c3 will change. But you cannot change an entry in c3 by going to that entry and editing it on the Entry line. You may unlock the column by selecting the cell c3 and deleting the expression 2*c1.

Deleting the Last Entry in a Column in the Data/Matrix Editor

You delete the last entry in a column using Delete from the Util menu, 2nd F6 (see Figure A.29), in the Data/Matrix editor. To delete the last item in a column, highlight that item, bring down the Util menu, select Delete, then select cell, and press ENTER.

Numerical Calculations Involving the Euler Number e

Clear the Home screen and Entry line. Find the green e^x above the X key. Press ◇ e^x and type 3. Complete the expression by typing). Press ENTER. Notice that the italicized e with a raised 3 appears on the screen, as both the last entry and the last answer.

To approximate e^3, press ◇ ≈. Now use the alpha E to type e^3. Press ◇ ≈. What happened? The mathematical Euler number e is accessed using the ◇ e^x key. We will sometimes refer to this combination as the e^x key.

The Sigma-Notation

You may not see this notational problem until Chapter 4. In using the Σ-Notation, you must enter a lower limit and an upper limit. Although it makes sense mathematically to have the upper limit smaller than the lower limit, with the TI-89 the upper limit must be larger than the lower limit. For example, clear the Home screen and Entry line and enter $\sum_{k=1}^{5} k^2$ by selecting ∑(sum from the Calc menu and typing k^2, k, 1, 5). Notice that the sum is 55. Now enter $\sum_{k=5}^{1} k^2$. The result is -29. This sum should be positive since we are adding all positive numbers. You should be aware of this difficulty as you use the Σ-Notation.

Using the Complex Number i

You display the complex number i by typing the 2nd CATALOG key combination. We may call this i since there is an orange i above CATALOG. The letter "i" alone is not the complex root of unity. To show this, type 2nd CATALOG ^ 2 and press ENTER. The last answer is -1. Now type alpha I ^ 2. The result is not -1.

Graphing with the ZoomFit Feature

You may have discovered `ZoomFit` option in the `Zoom` menu. This option sometimes takes a long time to plot the graph. It is often better to think about using the WINDOW editor to create the graphing window that would best display the information you want from the graph.

B

TI-92 COMPUTER ALGEBRA SYSTEM TUTORIAL

B.1

The TI Computer Algebra System Tutorial

Until recently, Computer Algebra Systems (CASs) capable of manipulating symbols as well as numbers were only available on high-powered computers costing thousands of dollars. With the introduction of the TI-92, and more recently the TI-89, a CAS is now available at a fraction of the cost in a conveniently portable package. Although in the past you may have expected your calculator to perform tedious numeric calculations (such as multiplying two five-digit numbers together), with these TI calculators you can now perform operations involving complex symbol manipulation (such as multiplying two fifth degree polynomials together). Despite its power, a CAS cannot do your thinking. To make proper use of the technology made available by the TI-92 and TI-89, it is essential that you understand the basic ideas of the mathematics you are using. Now more than ever, doing mathematics requires thinking, not just computing.

The TI-92 Keyboard

We will base this tutorial on the TI-92. The keyboard and menu system of the TI-92 lets you quickly and easily perform numeric, graphic, and symbolic mathematical computations. A great way to learn to use the TI-92 is to play with it. You will not break it by trying it out, and you are liable to learn as much when it doesn't do what you expect as when it does.

Study the image of the keyboard and display window that appears in Figure B.1.

FIGURE B.1

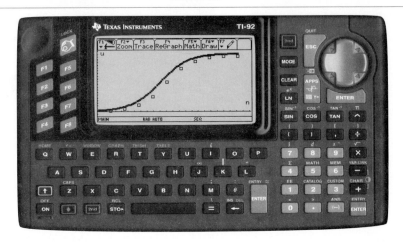

Let's look carefully at the keyboard.

Screen Intensity.   Locate the ON button and use it to turn on your calculator. Adjust the contrast by holding down the green ◇ key next to the ON button while pressing the + or − key on the far right of the calculator. The green ◇ key is used to activate any feature written in green on the keyboard. We will use ◇ often with HOME, Y=, WINDOW, and GRAPH to access the HOME screen, the Y= and WINDOW editors, and the GRAPH screen, respectively.

Function Keys.   In the upper left of the keyboard, you will see eight blue function keys marked F1 through F8. These keys are used in conjunction with the various menus that appear at the top of the display window. Look at the display screen. Notice that the items at the top of the screen have words or icons (pictures) indicating the kind of operations that are available with each menu. You will use these function keys often to access the menu system. For example, F1 is associated with the saw and hammer icon (meant to suggest a Toolbox) in the upper left corner of the screen picture. You will use F1 to access this Toolbox when you want to clear the screen.

QWERTY Keypad.   Below the function keys and the display window is a QWERTY typewriter keypad—so named for the letters at the beginning of the first row of keys. You may use these keys to type in any command or function, even $\boxed{\text{sin}}$ and $\boxed{\text{cos}}$. Locate the Shift key, $\boxed{\Uparrow}$, on the left of the third row, the large blue rectangular $\boxed{\text{ENTER}}$ key at the right of the typewriter keyboard, the $\boxed{\text{2nd}}$ key and the Backspace key ($\boxed{\leftarrow}$) along the bottom row of the typewriter keyboard.

Scientific Calculator.   To the right of the QWERTY keyboard is a standard scientific calculator with numbers on gray keys and with arithmetic operations, trigonometry, exponential, and logarithmic functions, and punctuation on black keys. The $\boxed{\wedge}$ exponentiation key appears at the top of the column of arithmetic operation keys, and a small blue rectangular $\boxed{\text{ENTER}}$ key appears at the bottom. Can you find the third $\boxed{\text{ENTER}}$ key on the keyboard?

The Cursor Button.   The large round blue key at the top of the keyboard is called the Cursor button. The raised cross on this button is used to control cursor movement. By pressing the left arm of the cross, you can move the cursor left, an action which will be indicated by the phrase "Cursor Left". By pressing on the bottom arm of the cross you can move the cursor down, indicated by "Cursor Down". The other two arms (right and top) are similar.

The Escape Key.   Just to the left of the Cursor button is a large blue key marked $\boxed{\text{ESC}}$. This Escape key let's you "escape" from almost any screen or menu you may be "stuck" in.

The Mode Key.   The $\boxed{\text{MODE}}$ key, just to the left of the Cursor button, determines the current calculator settings, such as whether angles are read in degrees or radians. The screen will inform you that you can cancel the $\boxed{\text{MODE}}$ menu with the $\boxed{\text{ESC}}$ key.

The Clear and Applications Keys.   The $\boxed{\text{CLEAR}}$ key, just below the $\boxed{\text{MODE}}$ key, is used to clear lines of type. The large blue $\boxed{\text{APPS}}$ key, the Applications key, let's you access the `Text`, `Program`, and `Data/Matrix` editors. Use these editors to create *data tables* and edit sets of commands called *scripts*.

The 2nd Key.   Can you find the two yellow (orange) $\boxed{\text{2nd}}$ keys on the keyboard? Either $\boxed{\text{2nd}}$ key will activate any feature on the keyboard written in yellow (orange).

The Home Screen

Figure B.2 indicates the various regions of the Home screen. Enter calculations on the Entry line; the results appear in the History area. The Toolbox menu lets you access various features and operations. The Status line at the bottom of the screen displays information about the state of the calculator. Look at the screen often for helpful information displayed in pop-menus, dialogue boxes, and in the History area.

FIGURE B.2

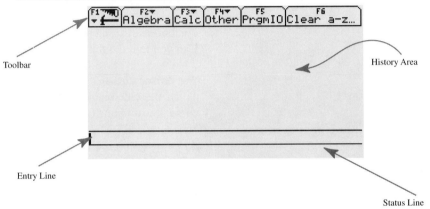

The Menu System

Turn on your calculator.

After you turn on your calculator, press the F1 key. Your display screen should look something like Figure B.3. How would you characterize the options or

FIGURE B.3

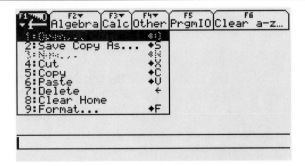

items available on this Toolkit menu? Do you know what item 2 might be used for? Press ESC to make the menu disappear. Now press F2. What is the most interesting option on this list? Press ESC to make this menu disappear.

Now press the Applications key (APPS). Which number on the list refers to the Geometry package? Do the APPS menu items have anything in common? Escape from this menu.

Although there are many more menus, we will now make some simple calculations and then investigate an algebra problem to see how the keyboard and menu items work together to assist you in doing mathematics.

Calculating with the TI-92

Clearing the History Area and Entry Line. You can clear the History window (if necessary) by pressing F1 and selecting the Clear Home menu option. One way to do this is to Cursor Down until the menu option is highlighted and then press ENTER. Another way is to press 8. To clear any expressions on the Entry line (if necessary), press CLEAR. We will describe these procedures as "Clear the Home screen and Entry line."

Arithmetic Calculations. Type 2+3 on the Entry line and press ENTER. This process will be called "Enter 2+3." When you are asked to enter any expression, you will be expected to type in the expression and press ENTER. Your display should look something like Figure B.4.

FIGURE B.4

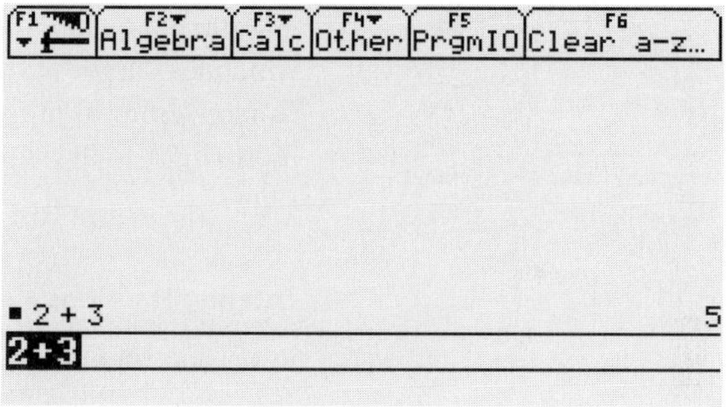

Notice that there is now a small box just above the Entry line followed by 2+3. On the far right of the screen on the same line as 2+3, the number 5 appears as the result of the computation indicated on the left. These expressions are referred to as the "last entry" and the "last answer," respectively. Notice also that 2+3 is highlighted on the Entry line.

Last Entry and Last Answer

Enter the expression 2+3*4. Did you remember to press ENTER? The answer, 14, results from an application of the order of operations—multiplication and division first, followed by addition and subtraction, both left to right.

Let's try a more complicated computation.

Enter 10!. The exclamation point (!) is displayed by typing 2nd W. Press ◇ K to see some other symbols that are available on the TI-92 using 2nd in combination with a letter key. Your display should look something like Figure B.5.

FIGURE B.5

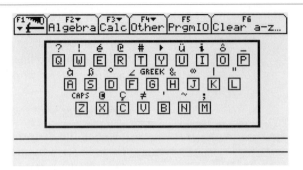

How do you think you can get the QWERTY keypad image to disappear? You want to escape from this display. Press the ESC key.

Editing on the Entry Line.   We would like to compute 100!. You can do this using the editing features of your calculator. Press Cursor Right to place the cursor just after the exclamation point on the Entry line. Press Cursor Left to move the cursor to the left so that it is between 0 and the exclamation point. Now type zero. Compute 100! by pressing ENTER. A highlighted BUSY box appears on the right of the Status line briefly. Is the number as displayed a correct answer? The number is only 38 characters long, but it should be at least 90 characters long. The number 100! is the product of 100 numbers, 90 of which are larger than 10. You can see all the digits of the number by pressing Cursor Up to highlight the answer and then pressing Cursor Right to display more digits to the right. How do you know when there are no more digits to the right?

Entering Fractions.   Let's try several ways to enter $\dfrac{1}{x+1}$. Type 1/x+1. Typed in on one line, this expression looks like it might work. Press ENTER. Look on the left side or "last entry" of the History area. The expression $\dfrac{1}{x} + 1$ is not correct. Now type 1/(x+1) and press ENTER. The History area now shows the correct value.

Storing a Value to a Variable.   Clear the Home screen and Entry line. Find the STO key just to the left of the Spacebar on the QWERTY keypad.

Press 2, STO, X and ENTER. Your display should look something like Figure B.6.

FIGURE B.6

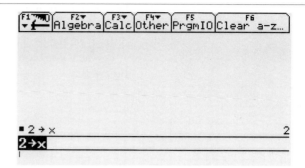

You can approximate the value of an expression.

Enter $\dfrac{1}{x+1}$ correctly as you did above. Find a green $\approx$ above an $\boxed{\text{ENTER}}$ key. To approximate this expression, press green $\boxed{\diamond}$ $\boxed{\approx}$. Notice the approximation that is displayed for one third is $\dfrac{1}{3} \approx .333333$.

Changing the Number of Digits Displayed.   Press the $\boxed{\text{MODE}}$ key. Your display should look something like Figure B.7.

Let's change the way decimal numbers are displayed.

FIGURE B.7

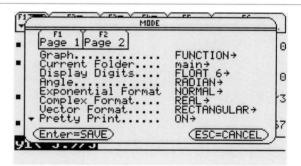

The $\boxed{\text{MODE}}$ dialogue box has two pages. The first is displayed when you press the $\boxed{\text{MODE}}$ key. The second can be displayed by pressing the $\boxed{\text{F2}}$ key, as shown at the top left of the menu. Use Cursor Down to scroll down to the Display Digits ... item. On the first $\boxed{\text{MODE}}$ screen, Cursor Right to open a submenu. (A *submenu* is indicated by the presence of an arrow.) Scroll down the submenu to highlight P:FLOAT 11. Press $\boxed{\text{ENTER}}$ twice to change the number of floating point digits displayed to 11—once to accept FLOAT 11, once to save the dialogue box information.

Return to the Home screen. Press ◇ ≈ again. The statement that is highlighted on the Entry line is performed again. An approximation for one-third is displayed as the last answer. How does the new approximation for 1/3 differ from the previous one?

When you complete these operations, your display should look like Figure B.8.

FIGURE B.8

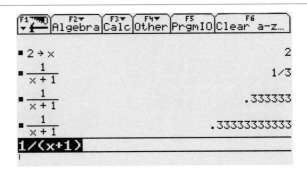

Go back to the Mode dialogue box and change the floating point digits displayed back to 6. Check that you have successfully done this, by returning to the Home screen and pressing ◇ ≈ to see that only six digits are now displayed. Currently x has the value 2. We can turn x back into a "variable" by typing DelVar x now. Notice that there is a space between DelVar and x.

Default Settings.   We will now set the calculator to its default settings—the way your calculator was set at the factory. The factory settings assure us that modes such as floating point digits displayed are set in a standard way. We use the scientific calculator key with the yellow (orange) MEM above it to do this. Press 2nd 6 to open the MEMORY dialogue box. Your display should look something like Figure B.9.

FIGURE B.9

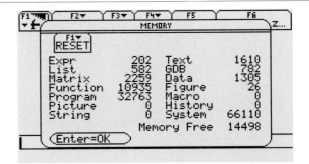

Now press $\boxed{\text{F1}}$ to open the RESET. Press $\boxed{3}$ to choose the Default option. A dialogue box will appear. To reset all system variables and modes to their original factory settings without affecting any user-defined variables, functions, or folders, press $\boxed{\text{ENTER}}$ twice. The first time to reset to the factory defaults; the second time to confirm that you know the modes are now set to the default or factory settings. You may want to reset your calculator to the default settings when you begin a new activity.

Solving a Quadratic Inequality

Resetting Variables

The following problem from precalculus or advanced algebra demonstrates the flexibility you will enjoy by using the features of the TI-92 in concert to explore mathematical ideas graphically, numerically and symbolically. Before you start an activity, such as this one, that involves variables you should consider using the Clear a-z... (Clear Up on the TI-92 Plus) Home screen menu item to reset all single letter variables. You can reset single variable, like x, by using the command DelVar x.

Bring up the Home screen. Clear the Home screen and Entry line using the Toolkit ($\boxed{\text{F1}}$) and $\boxed{\text{CLEAR}}$. Your display should look like Figure B.10.

FIGURE B.10

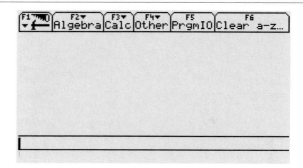

Graphing an Inequality.   We will investigate the quadratic inequality

$$x^2 - 3x - 10 < 0$$

from several vantage points. First, we will look at the inequality as a question about when the polynomial is positive or negative. Then we will look at it from the point of view of the factors of the polynomial. Finally, we will look at the inequality from a symbolic point of view.

Let's graph this function.

Press the $\boxed{\diamond}$ key and then the $\boxed{\text{Y=}}$ key to bring up the Y= editor as shown in Figure B.11.

FIGURE B.11

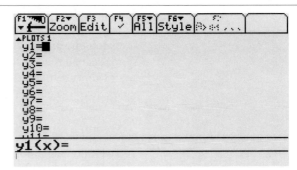

Press $\boxed{\text{ENTER}}$ to move the cursor to the Entry line of the $\boxed{\text{Y=}}$ editor. Enter the polynomial $x^2 - 3x - 10$.

The polynomial moves up to the line y1= when you press $\boxed{\text{ENTER}}$. Now press $\boxed{\text{F2}}$ to display the Zoom menu as shown in Figure B.12.

FIGURE B.12

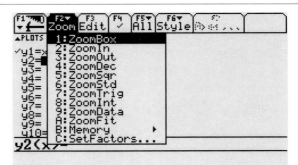

Cursor Down to highlight ZoomStd. Press $\boxed{\text{ENTER}}$ to select it. (Pressing $\boxed{6}$ gives the same result.) We will refer to this activity in the future as "Select ZoomStd from the Zoom menu." Your graph should look like the one in Figure B.13.

FIGURE B.13

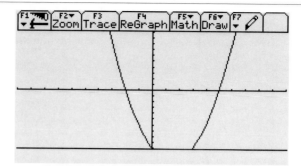

From the graph of this function, we can see that the inequality appears to be satisfied, that is, $x^2 - 3x - 10$ is negative, when x is between -2 and 5 because then the graph lies below the x-axis. To see this more clearly, look at the menus at the top of the screen. The [F3] key is associated with the Trace feature. Press [F3] and notice that a round crosshair cursor called the **Tracing Cursor** appears on the graph of the function. Notice also that numbers appear at the bottom of the screen next to xc: and yc: representing the x-coordinate and y-coordinate of the cursor location.

Cursor Left several times. Notice that the cursor traces along the graph to the left and that the numbers at the bottom of the screen associated with xc and yc change. Now type 4.3. What happened as you typed the number? What happens when you press [ENTER]? Is the inequality satisfied when x is 4.3? Is the inequality satisfied when x is 5?

Return to the Home screen by pressing [◇] [HOME]. Now clear the Home screen and Entry line unless you have already done this.

Type in x^2-3x-10<0 on the Entry line and press [ENTER]. In this book, we will refer to this activity by saying "Enter the expression $x^2 - 3x - 10 < 0$." You will be expected to enter the expressions correctly using the caret (^) for exponentiation, the < sign (above the 0), and parentheses where needed. When you press [ENTER], your display should look like Figure B.14.

FIGURE B.14

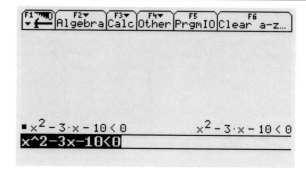

Use the QWERTY keyboard to type factor(. Cursor Up to highlight the inequality in the History area. You can also use the Algebra menu or the CATALOG (above the [2] key) feature to display factor(. Press [ENTER] to paste the inequality into the Entry line. Complete the command by typing).

Pasting

FIGURE B.15

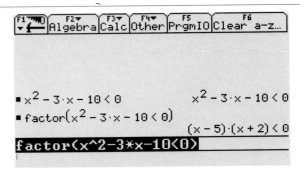

Press [ENTER] to factor the polynomial. Your display should look something like Figure B.15. Notice that the command you entered appears in the "last entry" position (as a check on what you typed) and its factored form appears in the "last answer" position.

Go to the [Y=] editor. Enter $x - 5$ for y2(x) and $x + 2$ for y3(x). Did you remember to press [ENTER]? Your display should look something like Figure B.16.

Let's enter these two polynomial factors as y2(x) and y3(x) in the Y= editor.

FIGURE B.16

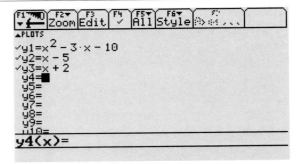

Select ZoomStd from the Zoom menu. Notice that the graph of the first function is a parabola and the graph of the next two functions are straight lines with slope 1. Your graph should look something like Figure B.17.

FIGURE B.17

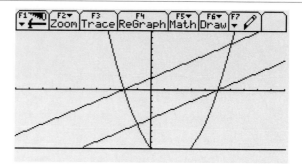

Activating a Function for Graphs and Tables

Return to the $\boxed{\text{Y=}}$ editor. Use the Cursor button to highlight y1(x). Look at the menu items at the top of the screen. What do you think the check mark on the **F4** Toolbar item means? Press $\boxed{\text{F4}}$ to deselect y1(x). Notice that y1 no longer has a check by it. Graph the functions ($\boxed{\diamond}$ $\boxed{\text{GRAPH}}$). Which functions appear on the screen?

Your graph should look something like Figure B.18.

FIGURE B.18

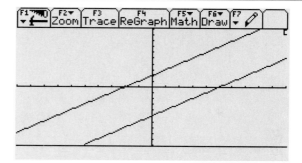

Go back to the $\boxed{\text{Y=}}$ editor and use $\boxed{\text{F4}}$ to "check" the y1(x) function so that it will be graphed.

The WINDOW editor

Press $\boxed{\diamond}$ $\boxed{\text{WINDOW}}$ to display the Window editor. In this editor, you can set the values for various aspects of the graphing window. For example, entering -4 for xmin makes -4 the minimum x-coordinate that will be graphed. Remember, entering includes pressing the $\boxed{\text{ENTER}}$ key. Enter the following values for the Window editor items:

    xmin=-4

Be sure to use the gray $\boxed{(-)}$ negative key, not the black $\boxed{-}$ subtraction key.

    xmax=6

Press $\boxed{\text{ENTER}}$ or Cursor Down to ignore the "xscl" item.

    ymin=-14

ymax=20

Your display should look like Figure B.19.

FIGURE B.19

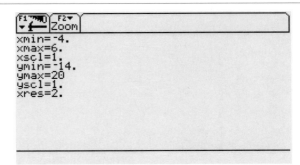

The xscl and yscl items are the distances between tick marks on the *x*- and *y*-axes. The xres item determines whether function values will be plotted for every pixel along the *x*-axis (xres=1) or every second pixel (xres=2), etc.

Press ◇ GRAPH to graph the functions in the new graphing window. Notice that the vertex of the parabola is now visible in this graphing window.

Activate the Trace feature. A **1** (one) should be in the upper right corner of the graphing area to indicate that the Tracing cursor is on the first function in the Y= menu. Cursor Left on the first function until the cursor is to the left of −2 on the *x*-axis. Cursor Down, to move to the second function $x - 5$. Is the function value positive or negative? Look just above the Status line for the value of yc. Cursor down to the third function. Is the function value positive or negative? Since the function $x^2 - 3x - 10$ is the product of $x - 5$ and $x + 2$, should this function be positive or negative? Check your result by moving the cursor to the first function using Cursor down.

Let's look at tables of function values.

Bring up the Y= editor and verify that all three functions we have entered are checked. Now press ◇ TABLE . You should see a table of values appear on the screen. From left to right, the columns are x-values, and function values for y1, y2, and y3. You can move around in this table using the Cursor button. Use the Cursor button to scroll up until both y2 and y3 are negative. Notice that y1 is positive.

Changing the Increment in a Table.

Change the increment of the x-values in the far left column of the table by pressing F2 for Setup. This displays the Table Setup dialogue box. You can also do this by pressing ◇ TblSet keys. In the Table Setup dialogue box, set the Δtbl to 0.5. Return to the TABLE window by pressing ENTER (perhaps twice). Notice the increment from one x-value to the next is now 0.5. Go back to the TblSet menu and change the value for Δtbl by entering 1 for Δtbl. Press ENTER . Return to the TABLE screen to see how this affects the increment in the x-values.

Let's compute function values for y1, y2, *and* y3.

Calculating Function Values.   Go to the [Y=] editor. Check that y1, y2 and y3 are the expressions displayed in Figure B.16. Go to the Home screen by pressing [◇] [HOME]. Type y1(-3)/3 on the Entry line and press [ENTER]. What is the function value? If you got an error message, you probably entered the −3 with the subtraction key [−] rather than the gray [(-)] negative sign. You should always be careful to strike the appropriate key when dealing with subtraction and with negatives of numbers.

Now enter y1(-3.)/3. How does the displayed value on the right differ from the value that you got for y1(-3)/3 without the decimal point?

Rational or Floating Point Arithmetic?

Symbolic Computations.   What is the value of y3(y2(-2))? If you do not remember the expressions for y2 and y3, display them by pressing [◇] [Y=]. You can find the value of this expression by entering it. As always, the expression that you enter on the Entry line is displayed on the left of the History area while its value is displayed on the right.

Clear the Home screen and Entry line.

Type y2(x)*y3(x) and press [ENTER]. Now press [F2] to display the Algebra menu, and select the expand(option. Use the Cursor button to select the expression at the right of the History area. Press [ENTER] to bring that result to the cursor position on the Entry line. Now type in the closing) and press [ENTER] to multiply the expression $(x + 2)(x - 5)$.

If you press [F2] again, you can see displayed some of the many algebraic operations that are possible. You will use these capabilities of the TI-92 to help you in investigating mathematical problems. You can investigate and solve more involved mathematical problems using these features.

The Data/Matrix Editor

Creating a Data Table.   Table B.1 gives the weight of a colony of yeast as a function of time. We would like to enter this data in a table and plot it using the

TABLE B.1   Time vs. Weight

Time	10	20	30	40	50	60	70	80	90	100
Weight	9.6	29.0	71.1	175	351	513	594	640	656	662

Data/Matrix Editor.

To construct the table, go to the [APPS] menu and from the Data/Matrix editor select New. In the New dialogue box, you make submenu selections by display the submenu items with the Cursor Right button and then selecting a submenu item by highlighting the item and pressing [ENTER]. Make the following submenu selections and fill in the edit box:

Select Data for Type:
Select main for Folder:

Fill in yeast in the Variable: edit box.

The Variable item in this and other editor dialogue boxes is the name of the table (or file) you are creating. It is quite confusing to have this edit box labeled Variable: even though a table name varies. Your display should look like Figure B.20. When these settings are correct, press ENTER twice—once

FIGURE B.20

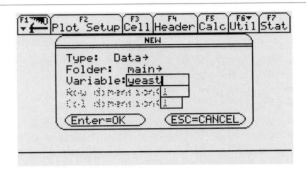

to make sure the last submenu item has been chosen, once to save the entries in the dialogue box. Your screen should look something like Figure B.21.

FIGURE B.21

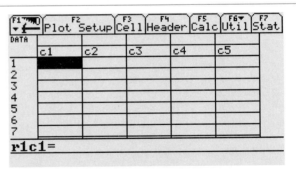

Successively enter 10, 20, 30, . . . , 100 in the column *below* the heading c1. Your screen should look something like Figure B.22.

FIGURE B.22

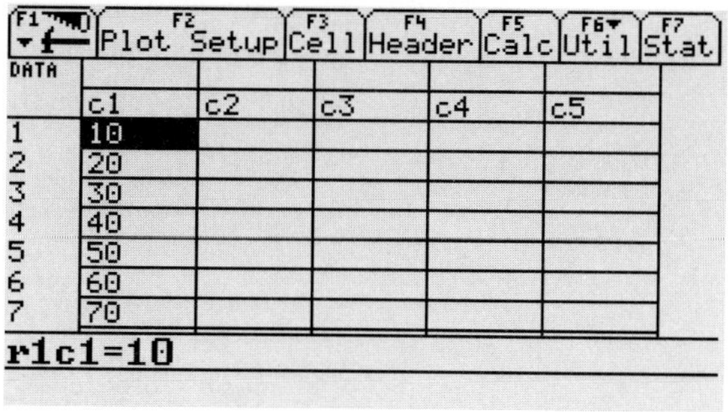

Cursor up to the blank cell *above* the heading $c1$, and enter the title $time$ as in Figure B.23.

FIGURE B.23

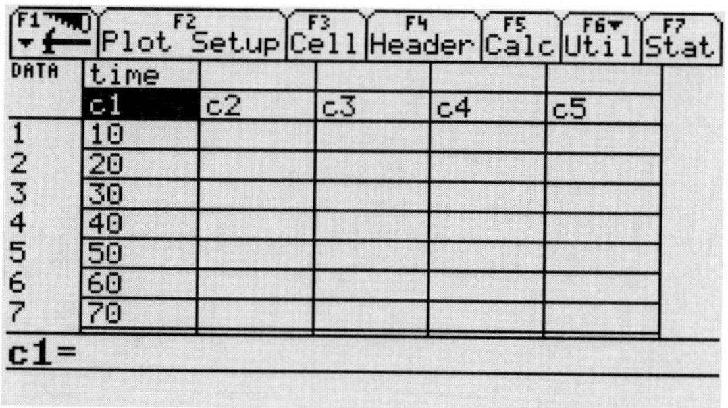

Title the second column $wght$ and enter the weights from Table B.1 in the cells below $c2$. This creates the table of values with the name $yeast$ if you cursor up to the top of your screen. Your screen should look something like Figure B.24. Notice that not all the values are visible.

FIGURE B.24

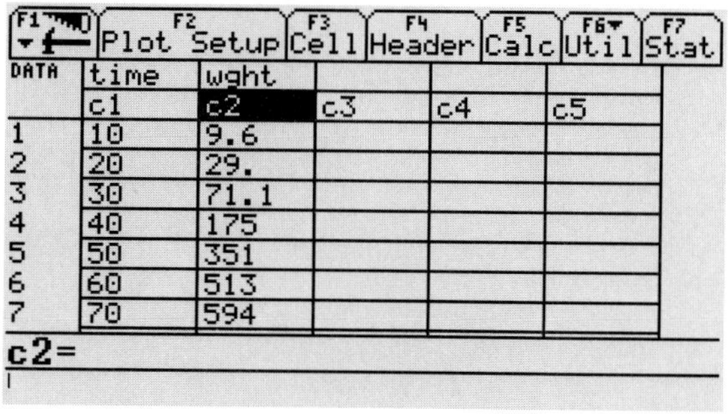

Go to the Home screen by pressing ◇ HOME . The information in an editor is automatically saved as you enter it. Thus, the information in the data matrix named yeast has been saved.

Plotting Data from a Data Table.   The Plot Setup menu in the Data/Matrix editor lets you create a definition for a plot. This process is similar to, but more complicated than, setting up the definition of a function in the Y= editor. The GRAPH command will display this data plot in much the same way as it graphs a function.

Begin by opening the yeast data table in the Data/Matrix editor. Directions for opening the editor and entering the table are given in "Creating a Data Table." Next, open the Plot Setup menu and select Define. In the resulting dialogue box select

Plot Type ... Scatter
Mark ... Box
x ... c1
y ... c2

and press ENTER .

Your display should look something like Figure B.25. Pressing ENTER

FIGURE B.25

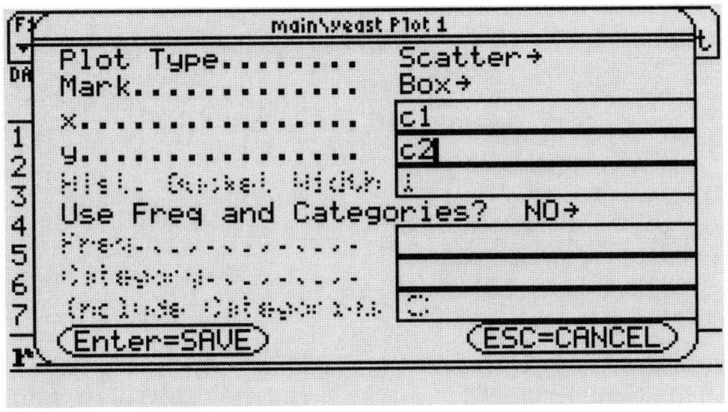

three times—once to complete the assignment of c2 to y, once to save the entries in the Define dialogue box, and once to save the Plot 1 definition in the Plot Setup dialogue box—returns us to the data table yeast.

Go to the Y = editor, deselect or clear all the functions. Cursor Up to Plot 1 (found immediately above y1) to make sure it is selected, that is, has a check mark by it. Finally, from the Zoom menu select ZoomData which plots the data using the highest and lowest table values for ymax and ymin and similarly for xmax and xmin. The data is displayed in Figure B.26 and the plot in Figure B.27.

FIGURE B.26   Table of Data for 𝚢𝚎𝚊𝚜𝚝   **FIGURE B.27**   Scatter Plot for 𝚢𝚎𝚊𝚜𝚝

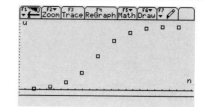

Summary

In this very brief tutorial, we have described the keyboard, seen how to clear the Home screen and Entry line, and how to enter expressions. We have also introduced the symbolic capabilities of a CAS through the algebraic operations of factoring and expanding. We saw how the TI-92 handles graphing and we introduced the 𝙳𝚊𝚝𝚊⁄𝙼𝚊𝚝𝚛𝚒𝚡 editor.[†] From this experience, you discovered how to look carefully at the keyboard, the menus, the dialogue boxes, and the editor windows for help in performing whatever mathematical operations you might want.

The TI-92 is a natural extension of graphing calculators like the TI-82 and is at least as user friendly. A CAS enables us to investigate a problem from graphical, numerical, or symbolic points of view without changing computing environments. We believe this calculator will help you to concentrate on the concepts of this course while it performs much of the computation for you.

[†]There are many, many more symbolic operations available, which we will introduce in the text as needed.

B.2

Troubleshooting: Things that Go Bump in the Night

In this section, we will discuss problems frequently encountered in using a TI-92 calculator. Many of these difficulties occur because of very slight differences between what you intended to enter and what you actually entered.

Subtraction and the Negative of a Number

You should be aware that there are two keys that have what appear to be dashes on them: $\boxed{-}$ and $\boxed{(-)}$. You should know that the first of these is for subtracting one number from another while the second is to indicate the negative of a number.

To see what these symbols look like on the screen, enter $-2 - 3$. The first of these two minus signs is entered with $\boxed{(-)}$ and the second with the black $\boxed{-}$ key. Your Entry line should look something like this:

$$^-2 - 3$$

The result is $^-5$.

Now replace $^-$ with $-$ using the subtraction operator key $\boxed{-}$ so that the Entry line looks like $-2 - 3$. Press $\boxed{\text{ENTER}}$. Notice that the machine complains. (If you try to type $-2 - 3$ directly, ans(1)- will appear since $-$ is the subtraction operator and requires a left and right argument.)

Defining a Function

You can define functions using the Define command (type define or use 2nd CATALOG). The variable you use in defining a function can then no longer be used as an input or argument for that function. For example, enter:

$$\text{define}\quad f(x)=2x+3$$

Now enter $f(x)$. Notice that this action results in a syntax error.[†]

To avoid this problem, we will use q as the variable for defining functions of one variable. Enter:

$$\text{define } f(q)=2q+3$$

We can now enter f(x) without obtaining a syntax error. A syntax error does occur, however, if we enter f(q) where q is the variable we used in defining f.

[†]No error message is displayed on machines built before 1996.

Exiting a Menu or Dialogue Box, Interrupting a Command

You will sometimes find that you have erroneously brought up a menu or dialogue box. To exit such boxes you can often press ESC or ENTER. To exit a dialogue box you may need to press ENTER two or more times to get out of the activity. There is no harm in pressing ENTER many times.

To interrupt the drawing of a graph, you can press ENTER. This simply interrupts the drawing of the graph (and everything else). To continue the drawing of the graph, press ENTER again. You can also press ON to stop the graphing process (or any ongoing computation) although you cannot continue graphing when you interrupt graphing in this way.

Locked Columns in the Data/Matrix Editor

You may not be able to change entries in a column in the Data/Matrix editor if that column has been created based on some other column (or list). For example, in the yeast data table, you can make the c3 column equal to $2*c1$. Do so by highlighting c3 and pressing ENTER to place the cursor on the Entry line. Enter the expression $2*c1$ as shown in Figure B.28.

FIGURE B.28

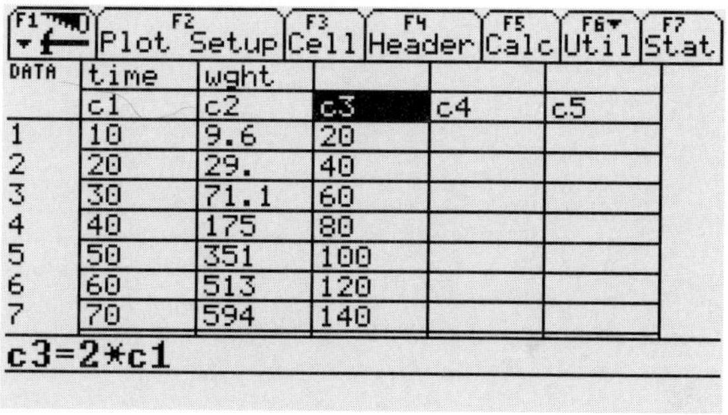

Now try to change the entry in the third row of column c3. You cannot do so because these entries are derived from entries in column c1. You can change the entries in c1 and, when you do, the corresponding entry in column c3 will change. But you cannot change an entry in c3 by going to that entry and editing it on the Entry line. You may unlock the column by selecting the cell c3 and deleting the expression $2*c1$.

Deleting the Last Entry in a Column in the Data/Matrix Editor

You delete the last entry in a column using Delete item from the Util menu, F6 (see Figure B.28), in the Data/Matrix editor. To delete the last item in a column, highlight that item, bring down the Util menu, select Delete, then select cell, and press ENTER. Using the Backspace key (←) or the CLEAR key will *appear* to delete the last entry in a column from the Entry line, but when you press ENTER the entry remains in place. You can delete an entry in the data table by highlighting the entry and pressing the Backspace key (($\boxed{\leftarrow}$)) without going to the Entry line.

Numerical Calculations Involving the Euler Number e

Clear the Home screen and Entry line. Find the yellow (orange) e^x near the $\boxed{\text{APPS}}$ key. Press $\boxed{\text{2nd}}$ $\boxed{e^x}$ and type 3. Complete the expression by typing). Press $\boxed{\text{ENTER}}$. Notice that the italicized *e* with a raised 3 appears on the screen, as both the last entry and the last answer.

To approximate e^3, press $\boxed{\diamond}$ $\boxed{\approx}$. Now use the $\boxed{\text{E}}$ key from the QWERTY keypad to type e^3. Press $\boxed{\diamond}$ $\boxed{\approx}$. What happened? The mathematical Euler number *e* is accessed using the $\boxed{\text{2nd}}$ $\boxed{e^x}$ key. We will sometimes call this the $\boxed{e^x}$ key.

The Sigma-Notation

You may not see this notational problem until Chapter 4. In using the Σ-Notation, you must enter a lower limit and an upper limit. Although it makes sense mathematically to have the upper limit smaller than the lower limit, with the TI-92 the upper limit must be larger than the lower limit. For example, Clear the Home screen and Entry line and enter $\Sigma_{k=1}^{5}k^2$. Notice that the sum is 55. Now enter $\Sigma_{k=5}^{1}k^2$. The result is -29. This sum should be positive since we are adding all positive numbers. You should be aware of this difficulty as you use the Σ-Notation.

Using the Complex Number i

You display the complex number *i* by typing the $\boxed{\text{2nd}}$ $\boxed{\text{I}}$ key combination. We may call this $\boxed{i}$ since there is an yellow (orange) *i* above I. The letter "i" alone is not the complex root of unity. To show this, type $\boxed{\text{2nd}}$ $\boxed{i}$ $\boxed{\wedge}$ 2 and press $\boxed{\text{ENTER}}$. The last answer is -1. Now type $\boxed{\text{I}}$ $\boxed{\wedge}$ 2. The result is not -1.

Graphing with the ZoomFit Feature

You may have discovered ZoomFit option in the Zoom menu. This option sometimes takes a long time to plot the graph. It is often better to think about using the WINDOW editor to create the graphing window that would best display the information you want from the graph.

The *TI-92 Guidebook*

Please look in your *TI-92 Guidebook* for further information about how to use your calculator. In particular, you can sometimes unlock a "locked up" calculator by following the instructions on page 498.

C

SOLVING EQUATIONS WITH THE TI CALCULATOR

In previous courses, you have learned to solve a variety of equations using algebraic techniques and your graphing calculator. In this Interlude, we will investigate how the TI calculator can assist us in solving such equations.

Polynomial Equations

First we need an equation to solve. Let's take the linear equation

$$a(x + b) = c - dx$$

as our first example. We can give the equation a name for future reference. Enter the line:

`Define` *is available from the* `Other` *menu, the* CATALOG *or, of course, may simply be typed.*

$$\texttt{Define eq1 = a*(x + b) = c - d*x}$$

The command to solve `eq1` for *x* is pretty much what you might expect:

$$\texttt{solve(eq1,x)}$$

`solve(` *is available from the* `Algebra` *menu, the* CATALOG *or may be typed.*

as are the commands to solve `eq1` for *a*, *b*, *c*, or *d*: `solve(eq1,a)`, etc.

To see the solution process rather than merely the final value for *x*, enter the following sequence of commands yourself.

```
expand(eq1)
ans(1)+d*x
ans(1)-a*b
ans(1)/(a+d)
```

expand(*is available from the* Algebra *menu, the* CATALOG *or may be typed.*

The results are shown in Figure C.1.

FIGURE C.1   Solving Equations the Old-fashioned Way

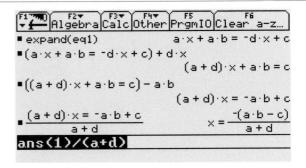

The TI calculator will symbolically solve any linear or quadratic equation in x. It will even solve our old favorite $a \cdot x^2 + b \cdot x + c = 0$, via

$$\text{solve(a*x^2+b*x+c=0, x)}$$

This symbolic capability allows the TI calculator to give exact solutions to *any* quadratic equation. Entering the commands

```
Define eq2=(x^2+7*x=18)
solve(eq2, x)
```

shows that x must be either 2 or -9.

Verify that $x = -9$ satisfies eq2 and that $x = -2$ does not, by entering the commands:

The with symbol "|" is the second function of the k-key.

```
eq2 | x=-9
eq2 | x=-2
```

What does your machine respond to these commands?

Of course, most quadratic equations do not have integer, or even rational solutions. The solutions to $x^2 + 7x = 17$ surely ought to be close to 2 and -9. (Why?) But the responses to

$$\text{solve(x^2+7*x=17, x)}$$

If you got decimal rather than radical answers, please go to the MODE *dialogue box and on* Page 2 *change your* Exact/Approx... *setting to* AUTO.

are not recognizably close to either 2 or -9. Frequently, we really do want the decimal approximation, which we get by pressing the green ◇ before pressing the ENTER key.

The TI calculator uses the built-in assumption that we are primarily interested in solutions that are Real numbers. What is your machine's response to the instruction

$$\texttt{solve(x\textasciicircum 2+7x=-18, x)?}$$

In order to get our machine to report complex solutions, we must ask specifically for them. Compare the last response to the result of entering

$$\texttt{cSolve(x\textasciicircum 2+7x=-18, x).}$$

The command cSolve(*may be typed, selected from the* CATALOG, *or can be found in the* Complex *submenu of the* Algebra *menu.*

The Theory of Equations.   Formulas are known for polynomials of degree 3, 4, and 5, but the TI calculator does not implement them symbolically. The formulas for polynomials of degree 3 and 4 were discovered amid the intrigue of the Italian Renaissance[†] and use complex cube and square roots of complex numbers. The formula for polynomials of degree 5 was discovered near the end of the nineteenth century and uses "Jacobi functions." While these symbolic solutions are of great theoretical interest, their practical significance has declined with the advent of cheap portable calculation devices. The lack of formulas however is not a fatal deficit. There is a considerable body of knowledge, known collectively as the Theory of Equations, about the solutions of polynomial functions. In order to understand what we can expect of any automatic solver, we need to have some understanding of the limitations that the theory imposes. Whenever

$$p(x) = a_n x^n + a_{n-1} x^{n-1} + \cdots + a_1 x + a_0$$

is a polynomial function with all the a_n's real numbers, the following facts are known:

Many of these ideas you may recall from previous courses.

- If r is a zero of $p(x)$, that is, if $p(r) = 0$, then $(x - r)$ is a factor of $p(x)$. If $(x - r)^k$ is a factor of $p(x)$ then r is called a zero of multiplicity at least k.

- If n is odd, there is at least one real zero of $p(x)$.

- The complex roots of $p(x)$ come in pairs. Whenever $a + bi$ is a zero of $p(x)$, so is $a - bi$.

- There are, counting multiplicity, exactly n real and/or complex zeros of $p(x)$.

- Given sufficient time, energy, diligence, or computer memory, the real roots of $p(x)$ can be identified and approximated to any required accuracy. These procedures use what are called *Sturm functions*.

For polynomials encountered in well-mannered textbook exercises, the preferred method of solution is surely that of factoring. The command

---

[†]A lively retelling of the tale can be found in *Journey Through Genius* by Bill Dunham.

$$\mathtt{factor(2x^3 + 13x^2 - 26x - 16)}$$

shows that

$$2x^3 + 13x^2 - 26x - 16 = (x - 1)(x + 8)(2x + 1)$$

Since the only way a product of two or more terms can be zero is for one of the terms itself to be zero, we can simply read the values of x that make $2x^3 + 13x^2 - 26x - 16$ equal to zero. The roots are 1, -8, and $-1/2$. The factors of a polynomial may sometimes give us partial information, as with $2x^3 + 13x^2 - 25x - 8$. Enter the command:

$$\mathtt{factor(2x^3 + 13x^2 - 25x - 8)}$$

Knowing that

$$2x^3 + 13x^2 - 25x - 8 = (x + 8)(2x^2 - 3x - 1)$$

tells us that if x is a zero of $2x^3 + 13x^2 - 25x - 8$, then either $x = -8$, or x is a zero of the quadratic $2x^2 - 3x - 1$. The zeros of the quadratic are available from the command

$$\mathtt{zeros(2x^2-3x-1,\ x)}$$

The command `zeros(` *is available under the* `Algebra` *menu, from the* CATALOG, *or may be typed.*

On the TI calculator, the commands `zeros(` and `solve(` are closely related but have different syntaxes. The command `solve(` must be given an equation as its first argument and will return solutions as equations joined by the logical `or`. The command `zeros(` requires an expression as its first argument and returns a list of values. With practice, you will develop your own sense of when to use one or the other. Everyone seems to have a personal preference.

Solving Equations Using Polynomial Methods.   Computer Algebra Systems, including the TI calculator, solve some equations by transforming them to polynomials and applying the methods we've seen. As an example, one can solve a rational equation by "clearing the denominator" and solving the resulting polynomial. Likewise, one can solve some radical equations by "squaring" to clear radicals. Both of these processes can introduce extraneous roots. The calculator, as part of its solution routine, checks the calculated solutions against the original problem.

Use the `solve(` command to solve each of the equations

$$x - 2 = \sqrt{19 + 2x}$$
$$\frac{x^2}{x^2 - 4} = \frac{x + 6}{x + 2}.$$

Check that you have entered each expression correctly by examining the output on the left above the Entry line to make sure it matches the equation given in the text.

Then transform each equation to a polynomial before solving. Compare the results of the two procedures. Explain how the TI calculator eliminates the extraneous roots.

We have paid a lot of attention to zeros of polynomials, not necessarily because polynomials are the most interesting class of functions, but because they are the class that is most tractable. There are more "handles" on polynomials than on other classes of functions. We have already noted that polynomials (which are made from positive powers of the variable) are used to find zeros for rational functions (negative powers) and for radical functions (fractional powers). However, new techniques are needed to search for zeros of exponential, logarithmic, and trigonometric functions.

Exponential and Logarithmic Functions

Logarithms were invented, in part, to turn exponential problems into simple polynomial problems. The zeros of exponential functions are often expressed in terms of ln, the natural logarithm. When we solve the equation

$$10^x = 5,$$

the TI calculator returns the potentially unsatisfying result

$$x = \frac{\ln(5)}{\ln(10)}.$$

This rarely happens in the context of an applied problem because we usually have one or more decimal numbers in the equation. These decimals (formally called *floating point numbers*) are "infective" and force the TI calculator to return answers that are themselves decimals. Compare the result above with what happens if we solve

$$(1.087)^t = 2$$

to find the doubling time for an investment at 8.7% interest.

Using logarithms doesn't help if we mix exponential and polynomial functions. There is no closed form solution to the equation $2^x = x^2$; but your TI calculator will try to solve it for you. Enter the command

$$\mathtt{solve(2\char94 x = x\char94 2, x)}.$$

More precisely, by "closed form" we mean an expression that can be constructed from finitely many powers, logarithms, exponential, trigonometric, and inverse trigonometric functions.

When you request solutions to an equation that does not have a closed form solution, the machine simply goes and looks for solutions and reports back with floating point approximations that either exactly or nearly satisfy the equation it was given. Well and good. In this case, two solutions have been reported, one positive, and one negative. We can check to see if these constitute reasonable solutions with

$$\mathtt{2\char94 x = x\char94 2 \mid x = 4}$$

and

$$\mathtt{2\char94 x = x\char94 2 \mid x = -.766665}$$

as shown in Figure C.2

FIGURE C.2   Floating Point Solutions are not Exact

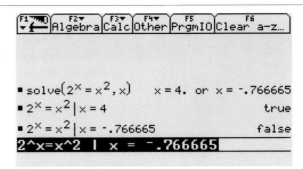

The results are what we might expect. Since $x = 4$ is an exact solution, the machine reports `true`. The machine is fussy, however, and reports that -0.766665 is not an exact solution. We might have expected that. We would like to see whether 0.766665 is really an approximate solution. The TABLE editor is a useful place to find out. First go to the Y = editor and enter `y1(x)=2^x` and `y2(x)=x^2`. Next go to the TblSet dialogue box and set `tblStart:` to -0.76667 and Δ`tbl:` to 0.000001, as shown in Figure C.3.

FIGURE C.3   Setting up a TABLE

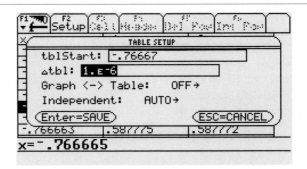

The TABLE SETUP *dialogue box is also available as* Setup *in the* TABLE *editor.*

Finally, go to the TABLE window (see Figure C.4).

We see that 2^x and x^2 agree to six places for $x = -0.766665$. We also see that, for $x = -0.766666$, 2^x is the smaller function, while for $x = -0.766664$, x^2 gives the smaller value. This is sufficient to guarantee that the graphs of $y = 2^x$ and $y = x^2$ do have a point of intersection between $x = -0.766666$ and $x = -0.766664$. We can confidently accept $x = -0.766665$ as an approximate root of the equation $2^x = x^2$.

There is, however, one further question that may, or may not, be of concern to us. We have found two solutions to the equation $2^x = x^2$. Are there others, or have we identified them all? The process the TI calculator uses to find solutions

FIGURE C.4 The TABLE showing values of x, 2^x and x^2.

x	y1	y2
⁻.76667	.587773	.587783
⁻.766669	.587773	.587781
⁻.766668	.587773	.58778
⁻.766667	.587774	.587778
⁻.766666	.587774	.587777
⁻.766665	.587775	.587775
⁻.766664	.587775	.587774
⁻.766663	.587775	.587772

x=⁻.766665

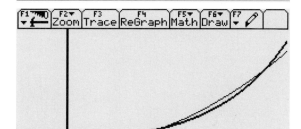

to equations that are not polynomial is simply to go and search. We cannot guarantee that a machine will search everywhere. If we want (or need) to know that the search was complete and thorough, we will have to do a follow-up search ourselves. Figure C.5 clearly indicates that there are at least three points of intersection, one of which (the one between 0 and 4) the TI calculator did not find.

`y1(x)=2^x` *is graphed in* `Style Bold` *with* WINDOW *settings that include* `xmin=-1`, `xmax=4.5`, `ymin=0`, *and* `ymax=25`.

FIGURE C.5 The Graphs of $y = 2^x$, and $y = x^2$

Enter the values 1 and 3 as the upper and lower bounds when prompted.

We can find the third point of intersection while in the GRAPH screen by using the `Intersection` command from the `Math` menu. The resulting dialogue asks us to confirm `y1` and `y2` as the functions whose intersection we seek, and prompts us for lower (left endpoint) and upper (right endpoint) bound for the *search interval*. The resulting screen, Figure C.6, reveals $x = 2$ as the third solution to the equation $2^x = x^2$.

The decimal point following the numbers "2" and "4" indicate that the TI calculator has computed the point of intersection as a floating point approximation.

The value "2." is stored automatically as `xc`. We verify the solution on the HOME screen by entering `y1(xc)` and `y2(xc)`. What happens when you enter `y1(xc)=y2(xc)`?

FIGURE C.6 $x = 2$ is a solution to $2^x = x^2$

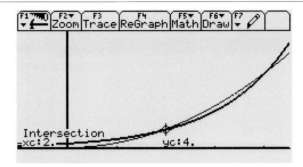

Solving Equation Involving Trigonometric Functions

Periodic functions pose a special problem for machines. If a periodic equation has one solution, it may well have *infinitely* many. Consider the equation $\sin(x) = \cos(x)$. This is equivalent (divide both sides by $\cos(x)$) to the equation $\tan(x) = 1$. This transformed equation is satisfied when $x = \pi/4$ and, since $\tan(x)$ has period π, for all x of the form $\pi/4 + n \cdot \pi$. The TI calculator is able to communicate this, but does so in a somewhat novel fashion. The TI calculator uses the special names @n1, @n2, @n3, etc., to indicate variables that can take on any integer value.

FIGURE C.7   The Equation $\sin(x) = \cos(x)$ has Infinitely Many Solutions

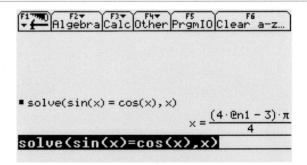

The situation is subtly different for the equation $\tan(x) = x$. Like the equation $\sin(x) = \cos(x)$, the equation $\tan(x) = x$ has infinitely many solutions, one in every period of $\tan(x)$. But, unlike the solutions to $\sin(x) = \cos(x)$, the solutions to $\tan(x) = x$ are not distributed at regular intervals. If you simply ask the TI calculator to solve $\tan(x) = x$, it will return several of the infinitely many soltions, but not necessarily any of the ones you are interested in. We can, however, ask for numerical solutions and give restrictions to select which solution will be computed. The third positive solution is computed by entering the command.

```
nSolve(tan(x)=x, x) | 5π/2<x and x<7π/2
```

Summary

You can use the `solve(`, `cSolve(`, `Zeros(`, and `cZeros(` commands to find roots or solutions of equations and the zeros of functions. These commands sometimes give exact solutions for polynomial functions and equations. The commands can also be used to solve radical, rational, exponential, or logarithmic equations. The Theory of Equations provides a theoretical foundation for your search for solutions and zeros for polynomials. Graphing provides a check in the search for solutions and zeros for polynomials and other types of equations and functions. The `Intersection` graphics command finds the point of intersection of two functions. This process is useful when the functions graphed are the left and right sides of an equation. You can also check approximate roots or zeros using the `Table` editor.

You may also find the zeros of the function that is the difference of the two sides of an equation. For example, the commands

```
x^2-5x+3=3-2x → eqn
left(eqn)-right(eqn) → eqn1
Define y1(x)=eqn1
```

The symbol → indicates the use of the store key [STO].

when entered as shown in Figure C.8.

FIGURE C.8   Using [STO] and `Define`

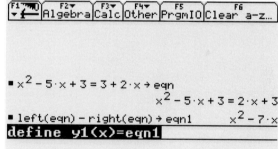

FIGURE C.9   The graph of $y = x^2 - 3x$

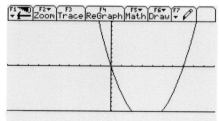

The solution to the equation `eqn` can be determined from the graph of `y1(x)=eqn1` given in Figure C.9.

E X E R C I S E   S E T   C . 0

1. Solve the equation $x^2 - 3x = 5x - 2$.

2. Solve the equation $x^2 - \ln(x) = 5$.

3. Solve the equation $\sin(x) = \sin^2(x)$ and interpret the calculator output.

4. Find the zeros of the rational function

$$f(x) = \frac{x^2 - 6x - 5}{x - 5}$$

. Be sure to check your result by graphing.

5. Find the complex roots of $x^3 - 3x^2 + 5x - 15 = 0$.

6. Factor $x^4 - 4x^3 - 6x^2 + 71x - 132$.

7. Calculate all the real roots of the equation $x^3 = 3^x$.

D

THE SLOPEFLD PROGRAM: WHAT TO DO 'TILL THE PLUS MODULE COMES

Follow the directions in the *Entering The SlopeFld Program* Technology Note, to create your own slope field program. Be careful as you enter the program. Syntax matters! One missing parenthesis can throw off the whole program. Think about the purpose of each statement as you enter it.

The program takes two arguments: a, the number of horizontal centers for the grid, and b, the number of vertical centers for the grid. The horizontal length of the graph is calculated by computing xmax minus xmin. This number is divided by 3a and stored in h to set the maximum horizontal span of a line segment to be two-thirds the distance between centers. Both xmin and xmax are *system variables* that you have been setting with the WINDOW editor. The same is done for the vertical dimension using ymin and ymax and stored in k. The two For statements set up a pair of nested loops. The outer loop moves horizontally while the inner loop moves vertically. Each time x is incremented by the outer loop, the inner loop runs through all the y-values from bottom to top. You can actually observe this process taking place as the program runs.

Inside the inner loop, the slope at the grid point is determined. The program assumes that the slope function m(x,y) has been defined as f(x,y) from the differential equation where *x* is the independent variable and *y* is the dependent

ENTERING THE SLOPEFLD PROGRAM

Choose `Program Editor` from the `APPS` menu. Select `New` and press ENTER. In the `NEW` dialog box, type the name `SlopeFld` in the `Variable` edit box. Press ENTER twice. The `Program` editor is ready for your program. Use the Cursor button to insert the arguments (line 1) and then to move to line 3. In the `Program` editor, press ENTER to finish a line and go to the next line. Here's the program:

`:SlopeFld(a,b)` *Insert the arguments a and b. The number a specifies the number of columns in the grid and b the number of rows.*

`:Prgm` *It's a program, not a function.*

`:Local x, y, h, k, s` *These variables can be used inside the program without affecting their values outside of the program.*

`:(xmax-xmin)/(3a) → h` *Maximum horizontal segment length for one of the short line segments.*

`:(ymax-ymin)/(3b) → k` *Maximum vertical length for one of the short line segments.*

`:For x, xmin, xmax, (xmax-xmin)/a`

Begin a loop. The variable x will start with the value xmin and go up through xmax by adding (xmax-xmin)/a each time. Each choice of x corresponds to a grid column

`:For y, ymin, ymax, (ymax-ymin)/b`

Begin an inner loop. The variable y will start with the value ymin and go up through ymax by adding (ymax-ymin)/b each time. Each choice of y will correspond to a grid row

`:m(x,y) → s` *STOre the slope at (x,y) in s.*

`:If abs(s*h)<k then` *If abs(s*h) is less than k, then*

`:Line x-h,y-s*h, x+h,y+s*h:Else`

*draw a segment from (x-h,y-s*h) to (x+h,y+s*h). Otherwise,*

`:Line x-k/s,y-k, x+k/s,y+k`

draw a line from (x-k/s,y-k) to (x+k/s,y+k).

`:EndIf: EndFor: EndFor: EndPrgm`

Mark the end of the If, the end of the inner For loop that changes y, the end of the outer For loop that changes x. End the program.

When you're finished typing the program, return to the HOME screen.

variable. The `If` statement's condition, `abs(s*h)<k`, tests to see whether the line segment extends too far, greater than one-third the way to the next center. If the condition is true, then we use the first `Line` statement, `Line x-h,y-s*h, x+h,y+s*h`, which plots a segment centered on the grid point from the tangent line. The colon separates the two statements, the `Line` and the `Else`. (We put two statements on the same line so the program would fit on a single `Program Editor` screen.) If the condition is false, then we use the second statement, `Line x-k/s,y-k, x+k/s,y+k`. Now we close the `If` statement, the two `For` loops, and end the program. (Figure D.1 shows the entered program.)

Show that the line segment from the second `Line` command also lies on the tangent line.

FIGURE D.1   The `SlopeFld` Program

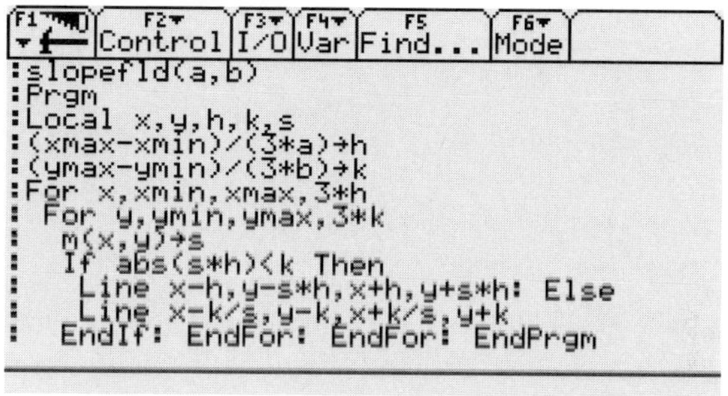

Let's see the program in work on the last example from Section 6.4.

$$\frac{dy}{dt} = \frac{t^2 - y^2}{2}.$$

Before we start examining slope fields, set the MODE to `Sequence`. Now, go to the GRAPH screen, select `ClrDraw` from the `Draw` menu, and deselect all sequences/functions and plots with the Y= editor. Define `m(q,r) = (q^2-r^2)/2` and in the WINDOW editor set up a $[-3.1, 3.1] \times [-3, 3]$ viewing window. To draw the slope field, enter `SlopeFld(20,10)` in the HOME screen.

An more informative picture, Figure D.2, results from using Euler's method to add several representative solution curves to the image. We can add the first solution curve by entering

u1	=	u1(n-1) + h
ui1	=	0
u2	=	u2(n-1) + h * m(u1(n-1),u2(n-1))
ui2	=	-3

in the Y= editor. The simplest method of adding additional solution curves is to make a new sequence, say u3, for the y value using `u3= u3(n-1) + h*m(u1(n-1),u3(n-1))`. Set the WINDOW parameters `nmin=1` and `nmax=60`. Add four solutions by defining the sequences u1 and u2 as shown, then copy u2 into u3, u4, and u5, changing the subscript 2's to 3, 4, and 5, respectively.

Use the initial values $ui1=-3$, $ui2=7.5$, $ui3=-1$, $ui4=-2.5$, and $ui5=-2.6$.
Enter SlopeFld(20,10) in the HOME screen.

In order to get all the sequences to plot, set the Axes to CUSTOM. Keep the
x-axis set to u1, but change the y-axis to u. A side effect is that u1 will also be
plotted on the y-axis, so deselect u1 to prevent this. Set h to 0.2.

FIGURE D.2   A Slope Field with Four Solution Curves

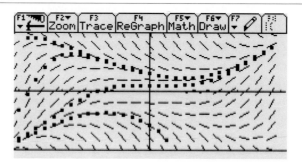

INDEX

$\boxed{\leftarrow}$, 634, 655
(Σ), 17
$\Sigma \langle$ command, 264
$\sum \langle$ command, 243–244
$\int \langle$ **integrate** feature, 291
$\boxed{(-)}$, 644, 665
$\boxed{\Uparrow}$, (shift) , 633, 655
$\boxed{\char`^}$, exponentiation, 634, 655
$\boxed{-}$, 644, 665
Σ-Notation, 651, 674
2nd key, 633, 655

a-lock, 633
Absolute
  convergence, 449
  extrema, 182
  maximum, 182
  minimum, 182
  value function, 116, 138
Absolutely convergent series, 578
Acceleration vector, 612
Accumulated quantity, 311
Accumulating distance, 314
Accumulation, 15
  average rate of, 239
  exact, 257
  function, 312, 339, 363
    derivative, 324
  of a quantity, 243
  of barrels of oil, 239
  of distance, 237
  of drugs, 15
  of the rate function, 313
  of the sine function, 288
  patterns of, 15
  rate of, 227

Accuracy
  improving, 233, 250
  of estimates, 243
  of limit values, 39
  of the model, 436
Active window, 136
Actual
  average value, 249
  volume of a cone, 272
Adequate model, 30, 593
Adjusting the contrast, 632, 654
Adventure, 365
Aggressor, 30
Air resistance, 430
Algebraic
  approach, 55
  insights, 305
  representation, 26
  simplification of a limit, 259
Algebra menu, 646, 667
Alpha key, 633
Ambient temperature, 402, 465
Ambiguity, 17
Amount function, 102
Amoxicillin, 9
Analysis
  symbolic, 162
  of a continuous function, 159
  of data, 273
Analytic method, 11
Analyze systems
  graphically, 526
  numerically, 526
Analyzing
  a function, 103
  continuous functions, 118
  discrete functions, 118
  function behavior, 117

the kth strip, 280, 334, 338
  the model, 503
Anesthetic, 6
Angle of entry, 162
Annual percentage rate, 62
Antibiotic level, 9
Antiderivative, 143, 288, 290, 314, 319, 356, 364
  corresponding, 321
  elementary, 293
  general, 321, 363
  graph of, 323
  graphical interpretation of the, 321
  guessing an, 292
  original, 351
  particular, 364
Antiderivatives
  families of, 319
  finding, 324, 348, 352
  guessing and checking, 349
  properties of, 328
  rules of, 328
  tables of, 291
Antidifferentiation, 143
  extending, 455
Applications, 3, 270
  key, 633, 655
Applied force, 284
Appropriate
  approach, 388
  domain, 184
Approximate area, 283
Approximating
  an expression, 637, 659
  area, 334
  numerically, 303
  the value of an integral, 304

Approximation
  decimal, 676
  discrete, 344
  good, 11
  of a definite integral, 288
  of limits, 304
APPS menu, 135, 283, 635, 646, 657, 667
Area, 281
  approximating, 283, 334
  exact, 335
  between two curves, 334
  of a plane region, 303, 335
  of a trapezoid, 263
  under a curve, 250, 281, 334
Art of modeling, 391
Assumptions for a model, 538
Asymptote
  horizontal, 27
Autonomous initial value problem, 444
Average
  actual value, 249
  daily cost, 185, 186
  rate, 228, 312
    of accumulation, 239
    of change, 91–92, 239, 246, 399, 610
    of speed, 233. 234
  velocity, 255
Averaging
  method, 230, 263
  rates, 230
Axis of symmetry, 201

Backspace key, 137, 634, 655
Bacterial infection, 9
Barometer, 489
Barrels of oil
  accumulation of, 239
Barrels per hour, 238
Basic
  formulas for integrals, 331
  pattern, 29
Behavior of a function, 160
Better fit, 558
Billing data, 241
Biological
  growth, 32
  processes, 425
Birth rate, 340
Blood, 5
Blood system, 10
Borda's constant, 490
Borda's law, 490
Bounded sequence, 33
Breaks, 47

Building the model, 26, 29, 496
BUSY box, 637, 658

caffeine, 5
Calc menu, 18, 50, 291
Calculating
  definite integrals, 257
  derivatives, 141, 305
  function values, 645, 667
  integrals, 287
  limits, 304
  Riemann sums, 243
Calculation errors, 488
Calculus, 3, 45, 280
  integral, 243
  of two variables, 593
Carle Runge, 481
Carrying capacity, 32, 360, 410
Cartesian representation
  of a line, 604
  of a plane, 604
CAS, 631, 653
CATALOG, 642, 664
Census
  bureau, 273
  figures, 274
Center of mass, 276, 278–279
  two-dimensional, 341
Central angle, 209
Certificate of deposit, 176
Chain rule, 152, 196, 348
  in Leibniz notation, 156
Change, 1
  average rate of, 92, 239, 399, 610
  describing, 219
  exact rate of, 106
  incremental, 15
  instantaneous rate of, 96, 111, 399
  measuring, 3, 219
  measuring rate of, 67, 98, 117, 165, 213,
Changes
  in slopes, 73
  in temperature, 171
  of variable, 352, 381
Characteristics of the data, 56
Circular functions, 208
Class of functions, 321
Clear a-z..., 661
Clear command, 135
Clear Home menu, 635, 657
CLEAR key, 634, 655
Clearing
  the Entry line, 635, 657
  the History window, 635, 657
  variables, 18

Clever observations, 305
Closed form, 8, 39, 56
  expression, 18
  formula, 15
  solution, 679
comDenom command, 18, 145
Command menu, 135
Commands
  $\sum$ (, 264
  seq(, 243
  $\sum$ (, 243–244
  Clear, 135
  comDenom, 18, 145
  Define, 672
  DelVar x, 661
  executing, 136
  interrupting, 650, 673
  Intersection, 681
  limit(, 50
  NewProb, 639
  nDeriv(, 113
  nSolve(, 683
  Save Copy As..., 282
  solve(, 171, 195, 678
  taylor(, 570
  when, 49
  zeros(, 678
Compact notation for $F(b) - F(a)$, 291
Competitive interaction, 30
Completeness Property, 576
Completing the square, 371, 383
Complex
  numbers, 651, 674
  roots, 677
  solutions, 677
Complicated limits, 305
Components of velocity, 207
Composite functions, 155, 348
Compound interest, 177
Computations
  symbolic, 645, 667
Computer-generated graphics, 542
Computer algebra system, 134
  tutorial, 631, 653
Concave
  down, 27, 73, 164, 267, 323
  up, 27, 73, 165, 267, 323
Concavity, 73, 164, 265
Concentration of molecules, 6
Concept of limit, 45, 127
Conic sections, 200
Conjecture, 42, 66
  a model, 8

Constant
    function, 101, 251
    multiplier rule, 350
    of proportionality, 26, 30, 206
    rate, 101
Constructing
    models, 78, 393, 399, 506
    sequences, 39
Construction materials, 162
Consumer demand, 185
Continuity at a point, 54
Continuity property, 54
Continuous
    function, 47, 54, 92, 162
    model, 4, 46, 56, 280, 397, 525
        logistic, 408
    process, 4
    quantities, 3
Contrast
    adjusting the, 632, 654
Convenient pattern, 377
Convergence
    absolute, 449
    interval of, 564, 573
    of a series, 574
    of Riemann sums, 303
    radius of, 585
    uniform, 449
Convergent geometric series, 578
Corresponding antiderivative, 321
Cosine function, 43
Cost-effective location, 595
Cost of an inventory cycle, 186
Creating
    a model, 25, 205
    a script, 135
Critical
    point, 167, 173
    questions, 46
Cursor
    arrows, 633
    button, 136, 655
    tracing, 641, 663
Curvature, 162
    of the earth, 204
Curve fitting, 428
CUSTOM axes menu, 470

$d\langle$ operator, 142
$\frac{d^2 y}{dx^2}$, 164
Data
    analysis, 273
    billing, 241
    characteristics of, 56

collection errors, 488
    discrete, 3
    tables, 633, 646, 655, 667
Data/Matrix editor, 29, 633, 646, 655, 667
Data plot
    population, 274
$\frac{d}{dx}\left(\frac{d}{dx}f(x)\right) = \frac{d^2}{dx^2}f(x)$, 164
Death rate, 340
Decay
    exponential, 21
Decimal approximation, 676
Decomposing functions, 155
Decreasing
    derivative, 119
    functions, 162
    the error, 262
Default option, 639, 661
Default settings, 639, 660
Define command, 672
Define dialogue box, 648, 670
Defining a function, 672
Definite integral, 254, 257, 270, 281, 312, 335, 363
    approximation of, 288
    computing, 257
    properties, 294
    transforming, 295
Definition
    of a limit, 135
    of the derivative, 134
Delete item, 651, 673
Delivery costs, 186
DelVar x command, 638, 660, 661
Demand
    function, 162
    rate, 187
Density, 280
    population, 273
    uniform, 280, 343
Dental procedure, 6
Dependent variable, 497
Derivative, 67, 111, 155, 196, 314
    accumulation function, 324
    as a tool, 219
    at a point, 128
    calculating, 141, 305
    decreasing, 119
    definition of, 134
    directional, 614, 622
    first, 119, 251
    for specific functions, 326
    function, 113, 117, 141
    increasing, 119
    notation, 101

of a function, 135
    partial, 213, 217, 612
    second, 164, 265
    sign of, 160
    tests, 170
    the constant multiplier rule, 143
    the product rule, 144
    the quotient rule, 144
    the sum rule, 143
    vector parametric, 611
Derived function, 111, 113
Describing change, 219
Deselect, 644, 665
Developing
    a model, 399
    pattern, 15
DIFF EQUATIONS mode, 431
Difference equations, 5, 6, 21, 33, 56, 398
    discrete systems of, 496
    iterating a system of, 521
    model, 26, 507
    systems of, 495, 506
Difference
    model, 403
    quotient, 131
    tests, 74
Differentiable, 117, 133, 138
    function, 164, 528
Differential, 352, 358, 380
    calculus, 87
Differential equations
    closed form solutions, 467
    Euler's method for systems of, 522
    first order, 435
    general solution, 435
    linear, 462, 467
    model, 404, 508, 512
    solution of, 414, 435, 453
    particular solution, 435
    power series solutions of, 582
    second order, 435
    symbolic
        solutions, 467
        solutions of systems of, 528
    systems of, 495, 506, 531, 538
Differentiation, 138
    implicit, 193, 202
    operator, 141, 173
    parametric, 209
    product rule, 356
    rules of, 141, 146, 325
Directed motion, 598
Directional derivative, 614, 622
    maximum value of, 623

Discontinuity
  jump, 49–50
  removable, 51
Discontinuous functions, 54, 132
Discount, 62
Discovery activities, vii
Discrete
  approximation, 344
  data, 3
    sample, 274
  functions, 67
  logistic
    equation, 63
    model, 32, 407
  models, 4, 33, 397
  phenomenon, 56
  point masses, 279
  quantities, 3
  systems of difference equations, 496
`Display Digits ...`, 638, 659
Display
  screen, 635, 656
  window, 632, 654
Distance
  accumulation of, 237, 314
  estimate, 229
  formula, 601
  traveled, 230, 234, 237, 313
    estimating, 231
    horizontal, 205
    vertical, 205
Distribution center, 595
Divergence of a series, 574
Divided differences, 89–90, 394, 430
Domain
  appropriate, 184
  of a function, 215
  of interest, 181
Dominating term, 361
Double angle identity, 378
Drug
  dosages, 5
  levels, 15
    stabilizing, 10
  overdose, 11
  regimen, 10
Drugs
  accumulation of, 15
$\frac{dy}{dx}$ notation, 138

Ecological system, 510
Editing a script, 135
Editing on the Entry line, 636, 658

Editors
  `Data/Matrix`, 29, 633, 646, 655, 667
  `Program`, 633, 655
  `Text`, 135, 633, 655
  WINDOW, 7, 69
  Y =, 6
Educated guess, 349
Effect of gravity, 205, 206
Elapsed time, 228, 401
Electricity flow, 241
Electric power plant, 241
Elementary antiderivative, 293
Eliminating drugs, 9
Ellipse, 202
Ellipsis, 16
Endpoints of a subinterval, 243
Energy use
  predicting rate of, 241
Entering fractions, 637, 658
ENTER key, 655
Entry line, 646, 667
  clearing, 635, 657
  editing on, 636, 658
Environment, 32
Equation of value, 62
Equations
  difference, 5, 21, 33, 56
  logistic, 40, 63
  of a plane, 603
  parametric, 204–205, 207–208
  periodic, 682
  rectangular, 208
  recursion, 11, 40
  solving trigonometric, 682
Error
  bounds, 246, 248, 260, 582
  calculation, 488
  data collection, 488
  Euler's method, 469, 472
  in model construction, 427
  left endpoint method, 251
  midpoint method, 266
  modified Euler's method, 479
  round-off, 453, 472, 488
  truncation, 488
  Runge–Kutta, 484
  Simpson's method, 269
  total, 249
  trapezoid method, 265
  upper bound, 250
ESC key, 633, 655
Estimate, 26
  accuracy of, 243
  area under a curve, 250
  distance, 229

inaccurate, 274
midpoint, 268
population, 273
the slope, 105, 111
total
  distance, 255
  error, 249
  population, 274
  trapezoid, 268
Estimated
  lower bound, 240
  upper bound, 242
Estimating
  average rate, 230
  distance traveled, 231
  number of interactions, 30
  parameters, 30, 408
  rectangles
    number of, 250
  total accumulation, 227, 246
  velocity, 231
Euler's method, 413–414, 417–418, 521
  error in, 469, 472
  error in modified, 479
  for systems of differential equations, 522
  improving, 474
  iteration
    numerical solutions, 515
  modifing, 475, 477
  theoretical error in, 473
Evaluating integrals, 287
Evangelista Torricelli, 489
e^x, 651, 674
Exact
  accumulation, 257, 289
  area, 335
  local extrema, 172
  moment of mass, 345
  rate of change, 106
  total accumulation, 227, 312
  value of a quantity, 270
Examples
  A Bouncing Golf Ball: Graphics Approach, 82
  A Bug on the Sphere, 218
  A Chaotic Population, 32
  A Discontinuous Function, 52
  A Ferris Wheel, 529
  A Hamburger Box for Dave, 182
  A Linear Vector Parametric Function, 608
  A Line in Three-Space, 602
  A Missing Constant Multiplier, 330
  A Mix Requiring Integration by Parts, 374
  A More Complicated Integrand, 259
  An Ellipse, 608

An Extended Standard Form, 373
An Increasing Spiral, 609
An Integral That Can Use Trigonometric Substitution, 380
An Inventory Problem: A Single Warehouse, 185
An Inventory Problem: The General Case, 187
Another Extended Standard Power Form, 376
Antibiotics — The Amoxicillin Example, 9
A Plane in Three-Space, 603
A Power Form, 373
Applying the First Partial Derivative Test, 620
Applying the Fundamental Theorem, 292
Applying the Fundamental Theorem, Part I, 291
A Question of Balance, 341
Area Between Curves, 334
A Simple Slope Field, 421
A Single Helix, 609
A Symbolic Solution, 457
At Last! When to Stop the Chemical Reaction, 531
A Trout Farm, 458
A Two-Party System, 510
A Vector Has Length and Direction, 601
Average Rates of Change for a Linear Function, 94
A Volume of Revolution, 338
Center of Mass: General Case of n Point Masses, 278
Coffee at Sunrise, 465
Completing and Analyzing a Difference Table, 75
Computing a Directional Derivative, 614
Daring Turnpike Drive, 91
Distance a Car Travels, 231
Distance Traveled by a Car, 246
Distribution Center: The Gradient Approach, 624
Dripping Coffee via Simpson's Method, 268
Dripping Coffee via the Trapezoid Method, 263
Energy Use, 241
Estimating a City's Population, 273
Estimating Extreme Values Graphically, 172
Example 1 Revisited, 534
Example 2 Revisited, 534
Existence of the Derivative, 115
Finding a Limit of a Function at a Point, 309
Finding a Limit of a Function at Infinity, 309
Finding A Limit Symbolically, 55
Finding an Antiderivative, 322

Finding Equations of Tangent Planes, 616
Finding Extrema and Inflection Points Exactly, 172
Finding Limits from a Graph, 53
Finding the Limit of a Sequence, 308
Find the Area of a Region: A Graphing Approach, 281
From Parametric to Cartesian Representation, 605
Graphing a Function and Its Derivative, 160
Graphing the Tangent and Acceleration Vectors, 611
Heat Shield for a Space Shuttle, 161
Inflating a Balloon, 199
Integrating a Polynomial, 329
Integrating Term by Term, 330
Integration by Parts Method, 357
Interaction of Individuals — The Rumor Example, 25
Interest Paid by a Bank, 176
Intersecting Graphs, 336
Logistic Solution Curve, 423
Mercury Pollution, 397
Mercury Pollution, Again, 422
Modeling a Chemical Reaction, 393
More Chemicals, 506, 516
More Chemicals, Again, 523
More Elections, 525
Painless Dentistry—The Novocaine Example, 6
Ponderosa Pines, 89
Population Prediction, 340
Pumping Water, 284
Purification Tank, 416
Radius of Convergence for the Exponential Function, 581
Recognizing the Chain Rule, 331
Recognizing the Integrand, 332
Richer or Broker, 67
Slope: the Implicit Way, 195
Slope: the Old-Fashioned Way, 193
Slope: the Parametric Way, 209
Species in Competition, 510, 518
Sun Microsystems's Revenues, 78
Sun Microsystems's Revenues Revisited, 81
Sun Microsystems's Yearly Revenues, 78
The Area Under a Curve, 252
The Brine Model Solved, 463
The Cannonball Question, 205
The Center of Mass of a Thin Plate, 343
The Chemical Reaction's Slope Field, 422
The Coalrus Project, 255
The Cone Example, 270
The Derived Function for the Bombay Tides, 113

The Derived Function of the Sine, 111
The Distribution Center—Graphics Approach, 594
The Distribution Center: A Symbolic Approach, 621
The Double Angle Pattern, 378
The Guess and Check Method, 349
The Harmonic Series, 575
The Heat Shield Revisited, 171
The Lion Model, 47
The Logistics of Bread and Wine — the Yeast Example, 29
The Non-Leaking Water Tank, 101
The Path of an Arrow in Flight — The Arrow Example, 207
The Pipeline Example, 238
The Power Form for Secant, 379
The Power Form for Tangent, 379
The Power Form Pattern, 377
The Power Rule for Functions, 329
The Same Derivative?, 114
The Signum Function, 138
The Simple Substitution Method, 353
The Sphere Example, 214
The Student and the Kite, 197
The Tapered Beam, 280
Three on a Seesaw, 277
Two-Party Politics, 516
Verifying Some Symbolic Solutionsl, 530
Yeast Data, 423
Executing commands, 136
Exercises
    group, x
    required, vii
    supplementary, viii
Existence
    theorem, 446, 447, 450
    of limits, 576
Exiting a menu, 650, 673
expand(option, 646, 667
Explicitly defined, 194
Explicit relationship, 194
Exploring the model, 27, 497
Exponential
    decay, 21
    decline, 274
    fit, 274, 340
    function, 404, 679
    function model, 274
    growth pattern, 21, 30
    model, 340
Exponents
    negative, 144

Extended standard
form integral, 371, 388
logarithm form, 376, 382, 386
power form, 377, 386
Extending
antidifferentiation, 455
integration by parts, 374
Extraneous roots, 678
Extrema
absolute, 182
global, 182
Extreme value property, 119, 178, 183
Extreme values, 179, 624

f'', 164
$f''(x)$, 161
$f'(x)$, 113, 141
`factor(`, 664
Factorial, 636, 657
Factors of polynomial functions, 677–678
False patterns, 17
Family
of antiderivatives, 319
of continuous functions, 56
of functions, 363
of solutions, 535
Features of a continuous function, 159
Field of slopes, 201
Financial calculations, 61
Finding
antiderivatives, 324, 348, 352
patterns, 16
symbolic solutions, 531, 533
First derivative, 119, 160, 251
test, 166, 173
First differences, 69, 78, 85, 118
formal definitions, 76
First order differential equation, 435
Fixed precision, 472
Flight
of a projectile, 204
path, 206
Floating Point Arithmetic, 645, 667
Flow
of electricity, 241
rate, 239
Force
air resistance, 537
applied, 284
net external, 537
of gravity, 537
restoring, 537
`For...EndFor statement`, 283

Formal definitions
first differences, 76
second differences, 76
sequences, 75
Forward differences, 70
Fractions
entering, 637, 658
Free fall
principle of, 206
Fulcrum, 276, 279
Function
absolute value, 116, 138
amount, 102
analyzing a, 103
analyzing behavior of, 117
constant, 101, 251
continuous, 47, 54, 92, 162
analysis of a, 118, 159
cosine, 43
demand, 162
derivative, 113, 117, 141
derivative of, 135
derived, 111, 113
differentiable, 164, 528
discontinuous, 54, 132
discrete, 67
domain of, 215
exponential, 404, 679
extreme values of, 624
ill-behaved, 117
limit of, 50–51, 303
linear, 102, 251
model, 8
nonlinear, 103
optimizing, 626
periodic, 112
rate, 102, 111
`sign(x)`, 139
slope, 106, 111, 113
steepest change of, 624
strictly decreasing, 251
strictly increasing, 251
vector parametric, 613
well-behaved, 111, 117
Function keys, 633, 654
Functions
accumulation, 312, 339
analyzing discrete, 118
behavior of, 160
circular, 208
class of, 321
composite, 155, 348
decomposing, 155
decreasing, 162
defining, 672

factors of polynomial, 677–678
families of continuous, 56
features of a continuous, 159
graphing, 644, 665
increasing, 162, 248
integrals of rational, 385
logarithmic, 679
maximizing, 182
minimizing, 182
of one variable, 213
of two variables, 214
piecewise continuous, 54
polynomial, 56
approximation of, 555, 561
properties
of continuous, 170, 176
of limits at a point, 308
of limits at infinity, 307
quadratic, 85, 231
regression, 231
`rand(`, 258
rate, 268, 288, 311, 324
undefined, 52
uniformly approximated, 566
vector parametric, 608
velocity, 233, 255
zeros of, 178
of a polynomial, 677
Function values
calculating, 645, 667
Fundamental relations, 87
Fundamental Theorem
of Algebra, 384
of Calculus, 363
Part I, 287, 290
Part II, 318, 324
Future
revenues, 78
term, 6
value, 61

Galileo, 64, 206
General
antiderivative, 321
error bound, 251
solution, 187
Generate tables, 101
Geometric
series, 18–19, 303
sum, 61
Global
extrema, 182
maximum, 171
picture, 420
Good approximation, 11

Gottfried Leibniz, 138
Gradient, 623
  vector, 624
GRAPH , 645, 666
Graph
  increasing, 27
  sigmoid, 28
Graphical
  interpretation, 72
    of the antiderivative, 321
  solution to a problem, 596
Graphing
  an inequality, 640, 661
  functions, 644, 665
  the antiderivative, 323
GRAPH key, 633, 654
Graph of a sequence
  properties of, 72
Graph of the slopes, 202
Graphs, 39
  mportant features of, 162
  parametric, 205
GRAPH screen, 633, 654
Gravitational constant, 489
Gravity, 204
  effects of, 205, 206
Greatest value, 181
green ≈, 637, 659
green ◇ key, 632, 654
Group exercises, x
Growth
  biological, 32
  exponential, 21
  inhibit, 30, 32
  in isolation, 511
  limited, 21
    predator, 550
    prey, 550
  rate, 360
    in isolation, 518
  unlimited, 511
Guarantee, 43
Guess, 42
  a model, 8
  and check process, 359
Guessing
  an antiderivative, 292
  and checking antiderivatives, 349
  limits of sequences, 37

History
  area, 195, 634, 646, 656, 667
  window
    clearing, 635, 657
Holes, 47, 51

HOME key, 634, 654
Home screen, 6, 634, 656
Hooke's law, 536–537
Horizontal
  asymptote, 27
  component, 205, 206
  distance traveled, 205
  tangent line, 160
Horner's method, 587
Hyperbola, 200
  upper branch of, 201
Hypercritical points, 171, 173

Identity
  double angle, 378
Ill-behaved function, 117
Image
  of the keyboard, 654
  of the keypad, 632
Implicit
  differentiation, 193, 202
  relationship, 194
Implicitly defined, 194
Important features of a graph, 162
Improper integral, 293
Improving
  accuracy, 233, 250
  Euler's method, 474
  the fit, 46
  the model, 430
Inaccurate estimate, 274
Increasing
  derivative, 119
  function, 162, 248
  graph, 27
  without bound, 21
Incremental changes, 15
Indefinite integral, 321, 332, 348, 356, 363,
    371
Independent variable, 142, 196, 243, 497
Index of summation variable, 17
Infinite
  geometric series, 573
  Maclaurin series, 574
Inflection point, 27, 73, 169, 173
Inhibit growth, 30, 32
Initial
  condition, 6
  constraints, 584
  position, 205, 207
  value problem, 417, 454
  velocity, 204
Input variable, 213
Installment loans, 62

Instantaneous
  rate, 229
    of change, 96, 111, 312, 399
    of change at a point, 131
  velocity, 303
Integral
  approximating the value of, 304
  definite, 254, 257, 270, 281, 312, 335
  extended standard form, 371, 388
  improper, 293
  indefinite, 321, 332, 348, 356, 371
  standard exponential form, 373
  standard form, 371, 388
  standard trigonometric form, 377
Integral calculus, 243
Integrals
  basic formulas, 331
  evaluating, 287
  of rational functions, 360, 385
  transforming, 348, 353, 356, 360, 378
Integrand, 257, 349, 358, 380
  original, 350
  splitting, 360
Integrating factors, 467
Integration, 287
  by parts, 356, 364, 371
    extending, 374
  lower limit of, 257, 292
  process, 385
  strategy, 372
  techniques of, 356
  upper limit of, 257, 292
  using partial fractions, 360, 364, 371
  variable of, 313
Intense harvesting, 492
Intensity of the screen, 632, 654
Interactions
  competitive, 30
Interest rate, 177
Interludes, viii, 675
Intermediate value property, 119, 177
Interpretation, 48
Interrupting a command, 650, 673
Intersection command, 681
Interval
  time, 243
  length, 69
  of convergence, 564, 573
Inventory
  cycles, 186
  strategy, 189
Investigating limits, 36
Irregular data, 89
Iteration process, 6, 524, 626

Jean Borda, 490
Jump discontinuity, 49–50

Keyboard
    image, 654
    TI-92, 653
    Map, 658
    system, 653
Keypad
    image, 632
    system, 631
    TI-89, 631
Keys
    ◇ , 632, 654
    2nd, 633, 655
    alpha , 633
    APPS , 633, 655
    Backspace 137
    CLEAR, 634, 655
    ENTER, 655
    ESC, 633, 655
    escape, 633, 655
    function, 633, 654
    GRAPH, 633, 654
    HOME, 634, 654
    MODE, 634, 655
    shift, 633, 655
    WINDOW, 633, 654
    Y=, 633, 654
Kidney function, 5, 9
Known limits, 305
kth strip analysis, 280, 334, 338

Language of limits, 20, 45
Learning
    by writing, vi
    process, x
Least squares line, 79
Left endpoint method, 229
    error, 251
Left hand limit, 50
Length
    of arc, 209
    of a stage, 228
    of interval, 69
Leonhard Euler, 414
Level of drugs, 15
    Novocaine, 6
Limit
    at infinity, 20
    concept, 5, 45, 127
    definition, 135
    existence of, 576
    left hand, 50
    of a function, 50, 51, 303

of an expression, 290
of a Riemann sum, 281, 287, 335, 339, 363
of a sequence, 40, 45, 303
of a sum, 256, 272, 304, 363
of a summation
    lower, 17
    upper, 17
point, 11
right hand, 50
two-sided, 132
value, 11, 20, 40
    estimate, 11
    for a sequence, 22
    accuracy of, 39
limit(command, 50
Limitations
    of the limit, 258
    to polynomial approximations, 563
Limited growth, 21
    prey growth, 550
Limiting behavior
    predict, 30
Limiting values, 303
Limits
    algebraic simplification of, 259
    approximating, 304
    calculating, 304
    complicated, 305
    investigating, 36
    known, 305
    language of, 20, 45
    limitations of, 258
    new from old, 305
Line, 686
Linear
    differential equations, 462, 467
    equations
        solving, 676
    factors, 384
    function, 102, 251
    model, 79
    polynomials
        denominator, 361
    regression, 79
Linearity
    local, 103, 106, 615
Lines
    cartesian representation of, 604
    parametric representation of, 601
    symmetric equations for, 606
    tangent, 607
LinReg, 90
Lipschitz condition, 443
Liver, 10
Local, 686

Local
    extreme values, 166
        exact, 172
    linearity, 103, 106, 615
    maximum, 10, 73
    minimum, 73
Locally linear, 414
Locating
    leaks, 238
    relative extrema, 168
Locked columns, 650, 673
Logarithm form
    extended standard, 376, 382, 386
Logarithmic functions, 679
Logistic
    equation, 40
    model, elaborations, 491
Long-term behavior, 495
Look ahead approach, 229
Look back approach, 229
Low-level harvesting, 492
Lower
    bound
        estimated, 240
    limit of a summation, 17
    limit of integration, 257, 292

Maclaurin polynomials, 560, 587
Magnitude
    of a vector, 601
    of velocity, 205
Marie Curie, 439
Martin Kutta, 481
Mass
    moment of, 342
    total, 345
    total moment of, 342
Mathematical
    description, 4
    induction, 66
        principle of, 449
    model, 46, 59, 67, 189, 391, 420
Math menu, 681
Maximizing
    a function, 182
    a volume, 182
Maximum, 73, 168
    absolute, 182
    global, 171
    point, 166
    relative, 323
    stability, 280
    value, 178
        relative, 166
        on a surface, 620

Mean, 81
Mean Value Theorem, 96, 119, 289–290, 445, 471
Measuring change, 3, 219
    rate of, 117
Medication, 10
MEM , 639, 660
MEMORY dialogue box, 639, 660
Menus, 635, 656
    APPS, 135, 635, 657
    Calc, 18, 50
    Clear Home, 635, 657
    Command, 135
    CUSTOM axes, 470
    exiting, 650, 673
    Math, 681
    RESET, 639, 661
    system, 631, 633, 653–654
    Toolbar, 634
    Toolbox, 135, 656
    Toolkit, 656
    Tools, 635
    View, 136
    Zoom, 69, 640, 662
Mercury pollution, 414
Method
    of integrating factors, 462, 467
    of steepest descent, 626
Methods of integration, 364
Midpoint
    estimate, 268
    method, 266
        error, 266
Minimizing a function, 182
Minimum, 73, 168
    absolute, 182
    relative, 323
    point, 166
    value, 160, 178
        relative, 166
        on a surface, 620
MODE dialogue box, 6
MODE key, 634, 655
Model
    accuracy of, 436
    adequate, 30, 593
    analyzing the, 503
    assumptions for, 538
    behavioral sciences, 540
    building the, 496
    conjecture, 540
    construct, 506
    constructing a, 25, 26, 29, 78, 205, 393, 399

continuous, 4, 46, 56, 280, 397, 525
    logistic, 408
    piecewise, 46
developing a, 399
difference, 403
    equation, 26, 507
differential equation, 404, 420, 508, 512
discrete, 4, 33, 397
    logistic, 32, 407
economics, 540
errors in construction, 427
exploring the, 27, 497
exponential, 340
exponential function, 274
function, 8
reasonable, 48
how well does it fit, 31
improving the, 430
insights into, 405
linear, 79
mathematical, 46, 59, 67, 189, 391, 420
new from old, 403
physical, 536
population growth, 360
predictions, 436
quadratic regression, 84
refining the, 391, 436
simple, 427
social sciences, 540
test and refine the, 488
to predict the future, 495
useless, 57
visualizing the, 499
with systems, 526
Model-building process, 7, 392
Modeling, 428
    art of, 391
    change, 1
    data, 231
    motion, 203
    process, 4, 303
Modifing Euler's method, 475, 477
Moment of mass, 277–279, 342
    exact, 345
Monthly interest rate, 62
Motion
    modeling, 203

Natural rate of increase, 32
nDeriv command, 113
Negative
    exponents, 144
    key, 644, 665
    slope, 163
Net external force, 537

New dialogue box, 646, 667
New limits from old, 305
New models from old, 403
NewProb command, 639
Newton's law of cooling, 405, 465
Newton's second law of motion, 430, 537
nicotine, 5
Nonlinear function, 103
Notation
    sigma, 17
    summation, 17
Novocaine, 5
    level of , 6
nSolve(command, 683
Number
    of digits displayed, 637, 659
    of interactions
        estimating, 30
Numerical
    analysis, 414, 473, 560
    methods, 475
        stable, 467
    solutions, 515, 542
Numerically approximate, 303

Ocean tides, 98
Octants, 214
Odd integer pattern, 17
Oil
    flow rate, 238
    viscosity of, 238
Opening a script, 135
Optimal inventory cycle, 186
Optimization, 67, 181
    with two variables, 593
Optimizing a function, 626
Original
    antiderivative, 351
    integrand, 350
    rate function, 316
Osmotic pressure, 5
Output variable, 213
Overestimate, 249

P.F. Verhulst, 360
Parallelogram rule, 600, 603
Parameters, 27, 162, 187
    estimating, 30, 408
Parametric
    approach, 207
    differentiation, 209
        vector, 611
    equations, 204–205, 207–208
    graphs, 205

representation, 211
  of a lines, 601
  of a plane, 601
  system, 206
Partial
  derivative, 213, 217, 612
  fractions, 387
    decomposition, 385
  sums
    sequence of, 574–575, 577
Particular solution, 187
Partition, 290, 344
  of a interval, 312
  regular, 257
Passive filtration, 5
Path of a projectile, 204
Pattern
  basic, 29
  exponential growth, 30
  false, 17
  finding a, 16
  from graphs, 78
  odd integer, 17
  of accumulation, 15
  recognition, 87, 355
Payment size, 62
Pearl's Yeast Experiment, 407
Pendulum clock, 206
Penicillin, 5
Periodic
  equation, 682
  function, 112
Physical
  model, 536
  processes, 425
Picard's successive approximations, 447
Piecewise continuous
  functions, 54
  models, 46
Pierre-Francois Verhulst, 411
Pixels, 47, 645, 666
Planes
  Cartesian representation of, 604
  equations of, 603
  parametric representation of, 601
  tangent, 614
Planetary motion, 287
Plot of the population data, 274
Plot Setup menu, 648, 670
Plot Setup window, 69
Plotting
  data from a data table, 648, 670
  the model, 31

Point
  of diminishing returns, 27
  of inflection, 73
Poison, 9
Polynomial
  approximation of functions, 555, 561, 582
    limitations to, 563
  equations
    solving, 675
  functions, 56
  Maclaurin, 560, 587
  Taylor, 587
Population
  density, 273
  estimates, 273
  growth rate
    estimated, 339
  of interest, 26
  outbreaks, 32
  projections, 339
Position, 204
  initial, 207
  vector, 600–601, 612
Positive slope, 163
Power
  form
    extended standard, 377, 386
  rule, 322
  series, 582
    solutions of differential equations, 582
Precise value, 55
Predator/prey interaction, 496–497
Predator population, 497
Predicted levels, 7
Predicting rate of energy use, 241
Prediction and correction, 476
Predict limiting behavior, 30
Predictor-corrector, 482
Present
  term, 6
  value, 61
Prey population, 497
Principle
  of free fall, 206
  of Mathemacical Induction, 449
Problem-solving skills, 371, 387
Product rule for differentiation, 356
Products
  sum of, 243
Program editor, 633, 655
Projectile
  flight of, 204
  path of, 204

Projects, vii
  An Odd Sum and Mathematical Induction, 64
  Chemical Reactions, 547
  Computing Natural Logs, 589
  Detecting Leaks in a Heating System, 440
  Differences and Polynomials, 124
  Extending the Logistic Model, 491
  Extending the Predator-Prey Model – TI-92, 549
  Flying off on a Tangent—Again, 222
  Geometric Series and Financial Calculations, 61
  Level Curves and Gradients, 629
  Picard Iterates, 590
  Pollution in the Great Lakes, 545
  Predicting Populations, 438
  Radioactive Pathways, 439
  Refraction, 222
  Surfing the Tangent Lines — Finding the Roots, 123
  Surveying, 224
  Taylor Series, 588
  The Geometry of Vectors in Three-Space, 629
  The Rotating Skydiver Model, 493
  Torricelli's Law, 489
  Variety in Behavior, 63
  Volumes by Monte Carlo Techniques, 629
Proof, 146
Properties
  limits
    of functions at a point, 308
    of functions at infinity, 307
    of sequences, 306
  of antiderivatives, 328
  of continuous functions, 170, 176
  of the definite integral, 294
  of the graph of a sequence, 72
Proportionality constant, 26, 30
Pythagorean Theorem, 601

Quadratic
  equations
    solving, 676
  factors, 386
  function, 85, 231
  Quadratic regression
    function, 231
  model, 84
QuadReg, 90
QWERTY keypad, 655

R. Carlson, 29
Radius of convergence, 585
rand(function, 258

Random number generator, 258
Rate
  average, 228, 230, 312
  barrels per hour, 238
  constant, 101
  demand, 187
  estimating the average, 230
  instantaneous, 229
  monthly interest, 62
  of accumulation, 227
  of change, 67, 98, 165, 213
    average, 246
    instantaneous, 131, 312
  of flow, 239
  of growth, 360
    in isolation, 518
  of increase
    natural, 32
  of interest, 177
  of the temperature change, 161
  population growth, 339
  reproductive, 32
Rate function, 102, 111, 268, 288, 311, 324,
    363
  accumulation of the, 313
  original, 316
Rational functions
  integrals of, 360
Ratio test, 580
Raw data, 78
Real
  roots, 677
  world phenomena, 4
Reasonable
  model, 48
  predictions, 60
Recognizing
  patterns, 87, 355
  trends, 407
Rectangular equation, 208
Recursion
  equation, 11, 40
  formula, 6, 19
Reentry, 161
Refining the model, 391, 436
Reflections, vi, 8
Regression
  linear, 79
  quadratic, 90
Regular
  partition, 257
  subdivisions, 260
Relations
  fundamental, 87

Relative
  extrema, 323
    locating, 168
  maximum, 323
    value, 166
  minimum, 323
    value, 166
Removable discontinuity, 51
Repeated doses, 12
Reproductive rate, 32
Required
  accuracy, 677
  exercises, vii
Reset all system variables, 639, 661
RESET menu, 639, 661
Resolution of the screen, 47
Resource allocations, 339
Rest
  point, 11, 19, 22, 40, 409
  value, 410
Revenues
  future, 78
Riemann sum, 243, 246, 260, 272, 281
  limit, 335
Right
  endpoint method, 229, 231, 239, 247–248,
    272
  hand limit, 50
Roots
  complex, 677
  extraneous, 678
  real, 677
Rotation of the earth, 204
Round-off error, 453, 472, 488
Rules
  of antiderivatives, 328
  of differentiation, 141
Runge–Kutta
  error, 484
  method, 481, 486

Saddle points, 619
Save Copy As... command, 282
Scalar multiplication, 599
Scatter plot, 9, 69, 99, 507, 648, 670
Scientific calculator, 634, 655
Screen
  display, 635, 656
  intensity, 632, 654
  resolution, 47
  splitting the, 136
Script, 172, 633, 655
  commands, 136
  creating a, 135

  editing a, 135
  opening a, 135
Script View option, 136
Search interval, 681
Secant line, 263
Second derivative, 160–161, 265
  function, 164
  test for concavity, 165
Second differences, 70, 78, 85, 119
  formal definitions, 76
Second order differential equation, 435
Separation of variables, 360, 456, 533
seq(command, 232
Sequence
  bounded, 33
  limit of, 22, 40, 45, 303
  of partial sums, 574–575, 577
  of values, 6
  sum of, 231–232
  terms, 8
  unbounded, 33
Sequences
  constructing, 39
  formal definitions, 75
  guessing limits of, 37
  properties of limits, 306
Series
  absolutely convergent, 578
  convergence of, 574
  convergent geometric, 18–19, 303, 578
  divergence of, 574
  infinite geometric, 573
  infinite Maclaurin, 574
  power, 582
Shares of stock, 68
Sharp corners, 117
Shift key, 633, 655
Shortcut techniques, 141
Sigma-notation, 17, 651, 674
Sigmoid
  graphs, 28
  shape, 56, 63
sign(x) function, 139
Sign of the derivative, 160
Simple
  models, 427
  substitution, 352
Simplifying assumptions, 427
Simpson's method, 366
Simpson's rule, 481
Simulations, 27
Sine function
  accumulation of, 288
Skydiver, 485, 537

Slope
change, 73
estimating, 105, 111
fields, 201, 420, 423
function, 111, 113
lines, 201
negative, 163
of a function, 106
of a surface, 612
of a tangent line, 131, 165, 303, 315
positive, 163
tangent line, 193
`SlopeFld`, 686
Solution
closed form, 679
complex, 677
curve, 417, 423, 435, 528
family of, 535
general, 187
graphical, 596
particular, 187
process, 675
unique, 40
`solve(` command, 171, 195, 678
Solving
differential equations, 414, 453
linear equations, 676
polynomial equations, 675
quadratic equations, 676
quadratic inequalities, 639, 661
trigonometric equations, 682
Special trigonometric combinations, 377, 381
Speed
average, 234
rate of, 233
Sphere, 214
Split
screen toggle, 136
the screen, 136
Splitting the integrand, 360
Stabilizing drug levels, 10
Stable numerical methods, 467
Stage
length of, 228
Standard
exponential form integral, 373
form integral, 371, 388
trigonometric form integral, 377
Starting conditions, 495
Status line, 634, 656
Steepest change of a function, 624
`STO` , 637, 658
Stock market, 67
Storage costs, 185
Storing a value to a variable, 637, 658

Straight line, 104
Strategy
workable, 387
Strictly
decreasing function, 251
increasing function, 251
Student
outcomes, v
participation, x
Subdividing an interval, 255
Subdivisions
regular, 260
Subintervals, 312
endpoints of, 243
number of, 250
standard width, 249
Submenu, 636, 638, 659
Substitution
trigonometric, 380, 383, 386
Subtraction key, 644, 665
Sum
geometric, 61
limit of, 256, 272, 304
limit of a Riemann, 287
of a sequence, 231–232
of the masses, 279
of the products, 237, 243, 266, 303, 312
Riemann, 243, 246, 272, 281, 303
telescoping, 289
`sum(` command, 232
Summary remarks, viii
Summation notation, 17, 264
Supplementary exercises, viii
Surface, 215
in three-space, 214
maximum value on, 620
minimum value on, 620
Surgical procedure, 6
Survey results, 241
Symbolic
analysis, 162
approach, 55
calculations, 219, 645, 667
form, 117
representation, 145
solutions of systems of differential equations, 528
Symbol substitution, 358
Symmetric equations for a line, 606
Systems
of difference equations, 495, 506
of differential equations, 495, 506, 531, 538

`TABLE` , 645, 666
Table feature, 6

Table generatation, 101
Tables, 39
data, 633, 655
of antiderivatives, 291
`TABLE SETUP` dialogue box, 645, 666, 680
`TABLE` window, 645, 666 ,680
Tangent/secant combinations, 379
Tangent line, 163, 607
horizontal, 160
planes, 614
slope of, 165, 193, 303, 315
vectors, 607, 609, 617
Taylor's formula, 587
`taylor(` command, 570
Taylor polynomials, 587
Δ`tbl`, 645, 666
`TblSet` , 645, 666
TblSet dialogue box, 680
Techniques of integration, 356, 364
Technology
use of, 631, 653
Telescoping sum, 289
Temperature
ambient, 402, 465
change, 171
rate of, 161
distribution, 161
variation in, 238
Terms in a sequence, 8
Test and refine, 488
Tests for increasing and decreasing, 163
`Text` editor, 135, 283, 633, 655
Theoretical error in Euler's method, 473
Theory
of equations, 677
of springs, 537
The power rule, 141
The trapezoid method, 262
Think and share, vii
activities, 12
Three-space
surfaces in, 214
Threshold term, 492
TI-89, 631, 653
calculator, 6
keypad, 631
TI-92, 631, 653
calculator, 6
guidebook, 674
keyboard, 653
TI differentiation operator, 141
Time
interval, 243
value of money, 61
`Toolbox` menu, 135, 283, 634, 656

`Toolkit` menu, 656
`Tools` menu, 635
Torricelli–Borda law, 490
Total
    accumulation, 248, 288
        estimating, 227, 246
        exact, 227, 312
    distance estimate, 255
    error estimate, 249
    mass, 279–280, 345
    moment, 344
        of mass, 342
    population, 275
        estimate, 274
    volume of an object, 339
Tour de France, 228
Toxic substance, 11
`Trace` feature, 9, 104, 316, 641, 645, 663, 666
Tracing cursor, 641, 663
Trajectory, 193
Transformations, 364
    techniques, 348
Transforming
    definite integrals, 295
    integrals, 348, 353, 356, 360, 366, 378
Trapezoid estimate, 265, 268
Trial and error, 41
Trigonometric
    combinations, 371, 375
        special, 377, 381
    form
        standard, 377
    identities, 375
    substitution, 371, 380, 383, 386
Troubleshooting, xii, 631, 649, 653, 672
Truncation errors, 488
Tutorial
    computer algebra system, 631, 653
Two-dimensional center of mass, 341
Two-sided limit, 132

u-substitution, 354
Unambiguous, 17
Unbounded
    increase, 21
    sequence, 33
Unboundedness of the integers, 44
Undefined function, 52
Underestimate, 249
Uniform
    convergence, 449
    density, 280, 343
Uniformly approximated functions, 566
Unique solution, 40
Uniqueness Theorem, 444, 450

Unit circle, 208
Unlimited growth, 511
Upper
    bound, 249
        estimated, 242
    branch of a hyperbola, 201
    limit
        of a summation, 17
        of integration, 257, 292
Useless model, 57
Use of technology, 631, 653
Using secant lines to approximate the slope of
        the tangent line, 128

Value
    extreme, 179
    maximum, 178
    minimum, 178
    of a limit, 20
`Variable`, 646, 668
Variable, 26
    change of, 352, 381
    dependent, 497
    independent, 142, 196, 243, 497
    index of summation, 17
    input, 213
    of integration, 313
    output, 213
    storing a value to, 637, 658
`Variable` edit box, 135
Variables
    calculus of two, 593
    clearing, 18
    optimization with two, 593
    reset all system, 639, 661
    separable, 360
Variation in temperature, 238
Vector
    acceleration, 612
    addition, 603
    geometric representation of, 598
    gradient, 624
    initial point, 598
    magnitude of, 601
    parametric
        derivatives, 611
        function, 608, 613
    position, 600–601, 612
    sum, 599
    tangent, 607, 609, 617
    terminal point, 598
    velocity, 204, 612
Velocity
    average, 255
    components of, 207

estimating, 231
function, 233, 255
initial, 204
instantaneous, 303
magnitude of, 205
vector, 612
Verifiable predictions, 60
Vertical
    component, 205–206
    distance traveled, 205
    shift, 321
    translations, 115
Victim, 30
`View` menu, 136
Viscosity of oil, 238
Visual
    impressions, 87
    representation, 239
Visualizing the model, 499
Volume of a cone, 271
Volumes of revolution, 337

Weighted average, 268
Well-behaved function, 111, 117
Wesley C. Salmon, 36
`when` command, 49
Wildlife, 511
$\boxed{\text{WINDOW}}$, 644, 665
WINDOW editor, 7, 69
WINDOW key, 633, 654
Windows
    active, 136
    display, 632, 654
    `Plot Setup`, 69
    TABLE, 680
Wind resistance, 204
Work, 284
Workable strategy, 387

`xmax`, 644, 648, 665, 670
`xmin`, 644, 648, 665, 670
`xres`, 645, 666
`xscl`, 645, 666
xy-plane, 213
xz-plane, 214

y'', 164
$\boxed{\text{Y=}}$, 640, 661
Y = editor, 6
Y= key, 633, 654
Yeast colonies, 29
`ymax`, 644, 648, 666, 670
`ymin`, 644, 648, 665, 670
`yscl`, 645, 666
yz-plane, 214

`zeros(` command, 678
Zeno's paradoxes, 36
Zeros
  of a function, 178
  of polynomial functions, 677
`Zoom` menu, 69, 640, 662
`ZoomData`, 69
`ZoomFit` feature, 652, 674
`ZoomFit` option, 652, 674
Zooming in, 104